Published in co-operation with the
United Nations
Economic and Social Commission
for Asia and the Pacific

Mineral Concentrations and Hydrocarbon
Accumulations in the ESCAP Region Volume 3

C. S. Hutchison (Ed.)

# Geology of Tin Deposits

## in Asia and the Pacific

Selected Papers from the International Symposium on the Geology of Tin Deposits held in Nanning, China, October 26–30, 1984, jointly sponsored by ESCAP/RMRDC and the Ministry of Geology, People's Republic of China

With 404 Figures

Springer-Verlag
Berlin Heidelberg New York
London Paris Tokyo

Professor Dr. Charles S. Hutchison
Committee for Co-ordination of Joint Prospecting for Mineral Resources in Asian Offshore Areas (CCOP)
CCOP Technical Secretariat
Offshore Mining Organization Building
110/2 Sathorn Nua Road, Bangkok 10500
Thailand



This volume has been edited by Charles S. Hutchison, Ph.D. and has been issued by ESCAP without formal editing.

St/ESCAP/547

ISBN-13: 978-3-642-72767-2    e-ISBN-13: 978-3-642-72765-8
DOI: 10.1007/978-3-642-72765-8

Library of Congress Cataloging-in-Publication Data. Tin mineralization, with emphasis on Asian deposits / C.S. Hutchison (ed.). p. cm. Selected papers presented at the International Symposium on the Geology of Tin Deposits held in Oct. 1984 in Nanning City, China. "United Nations Economics and Social Commission for Asia and the Pacific"–P. [ii] 1. Tin ores–China–Congresses. 2. Tin ores–Congresses. 3. Tin ores–Asia–Congresses. I. Hutchison, Charles S. (Charles Strachan), 1933-. II. International Symposium on the Geology of Tin Deposits (1984 : Nan-ning shih, China) III. United Nations. Economic and Social Commission for Asia and the Pacific. IV. Title: Tin mineralization. TN406.C6T56 1988 553.4'53'0951–dc 19
88-12340



Softcover reprint of the hardcover 1st edition 1988



2132/3130-543210 – Printed on acid-free paper

# Preface

This volume represents an edited selection of papers presented at the International symposium on the geology of tin deposits held in Nanning City in October 1984. It documents a great advance in our knowledge of tin deposits, particularly of the People's Republic of China. Details are presented in English for the first time on the major tin-polymetallic sulphide deposits of Dachang and Gejiu, which bear similarities to the deposits of Tasmania, but are little known to the geological community outside of China.

The publication of this volume was sponsored by the United Nations ESCAP Regional Mineral Resources Development Centre (RMRDC), now a Regional Mineral Resources Development Project (RMRDP) within ESCAP. The Centre had previously published a report on the Symposium in Nanning City and the following field trip to the Dachang tin-polymetallic sulphide deposit of Guangxi, entitled "Report on the International Symposium on the Geology of Tin Deposits: Nanning and Dachang, China, 27 October – 8 November 1984".

It is my privilege to acknowledge the help provided by Dr. J.F. McDivitt and Dr. H.W. Gebert, co-ordinator of ESCAP-RMRDC.

Publication of this volume would not have been possible without the dedicated work of Dr. Zhang Zhongmin of the editorial department of *Mineral Deposits*, Institute of Mineral Deposits, Chinese Academy of Geological Sciences, Beijing, People's Republic of China. I am extremely grateful to him for providing excellent translations of the manuscripts which were originally written in Chinese. He will be responsible for editing a Chinese language version of this publication.

The modern English transliteration of Chinese place names may present some confusion to readers already familiar with the older English literature on specific deposits. For example the very famous Gejiu deposit of Yunnan may be familiar to some readers by the older spelling "Kochiu".

During the process of editing, it became clear that a properly documented reference list could not be achieved for most of the Chinese papers. Dr. Zhang experienced great difficulty in individually contacting the authors, many of whom are currently working in widely dispersed areas. It was therefore decided to use the term *"relevant literature"* rather than *"references"* for such papers.

This volume is strongly complemented by *"Tungsten Geology Jiangxi, China"*, Proceedings of a symposium also jointly sponsored by the Ministry of Geology, People's Republic of China and ESCAP-RMRDC. Copies are still available from ESCAP, Natural Resources Division, United Nations Building, Rajadamnern Avenue, Bangkok 10200, Thailand.

Kuala Lumpur, January 1988 C. S. HUTCHISON

# Contents

# 1 Worldwide

# 1.1 The World's Major Types of Tin Deposit

K.F.G. Hosking[1]


## Abstract

This paper is primarily for those concerned with the search for and exploitation of tin deposits: facts are presented rather than hypotheses.

A classification of tin deposits is necessary before the major types can be discussed in a paper of reasonable length. The difficulties of constructing such a classification are noted and a modification of my earlier classification (Hosking, 1979, pp. 25–26) is presented. The major groups, which are sub-divided in the paper, are:

(1) Disseminations other than those in placers and that are not included in the other major groups.
(2) Pegmatites/aplites.
(3) Skarns (pyrometasomatic deposits).
(4) Hydrothermal breccias.
(5) Deposits associated with greisenised and/or albitised country rock.
(6) Stanniferous veins other than those of Group 5.
(7) Lodes of the Cornish type.
(8) Replacement (metasomatic) deposits that cannot be satisfactorily placed in any of the other groups.
(9) Telescoped, mineralogically complex deposits (largely xenothermal or sub-volcanic).
(10) Deposits of the Mexican type (epithermal or fumarole).
(11) Stanniferous massive sulphide and massive iron oxide deposits.
(12) 'Ancient', variously modified, stanniferous sedimentary deposits.
(13) 'Modern' placers.

The characteristics of the economically important types of deposit within each of these groups are provided, and the observations are supported by reference to specific deposits which have been selected, as far as possible, from those with which the writer has personal experience.


[1] Camborne School of Mines, Pool, Redruth, Cornwall, United Kingdom

**Background Generalisations**

The following facts are relevant when surveying the world's major types of tin deposit:

(1) Primary tin deposits have developed from the Precambrian to the Tertiary.
(2) As the geological time-scale is ascended tin deposits become more plentiful and more economically important.
(3) In several regions tin deposits of a number of widely different ages occur, perhaps due to recycling.
(4) The style of primary tin mineralisation presently encountered tends to vary with its age. Schuiling (1967, pp. 539—540) notes that "whereas in the Tertiary the association of tin deposits or groups of tin deposits with volcanic rocks is predominant, there is a shift towards greisen and other contact deposits in the Mesozoic, towards deposits in the granites in the Palaeozoic and towards pegmatite deposits in the pre-Cambrian". However, this pattern may be the result of progressive erosion. I am of the opinion that all types of primary tin deposit have been repeatedly generated since very early times. This view accords with that of Hunter (1973, p. 53). He states that "despite the great antiquity of granite rocks in South Africa, the localisation of tin mineralisation is related to processes of concentration similar to those that are believed to have been operative in younger, and economically more important, tin-bearing provinces".
(5) The world's tin deposits have a spotty distribution but Schuiling (1967) shows that they are restricted to elongate zones (tin belts).
(6) Most primary tin deposits are spatially and probably genetically related to granitoids or their volcanic equivalents.
(6i) About 75% of the world's tin reserves are associated with granitoids in mountain belts.
(6ii) Deposits associated with anorogenic (ring-structured) granites occur, for example, in Africa and Brazil. The primary deposits (greisen types and disseminations) are of very limited economic importance but are the parents of important placers.
(7) Some of the massive stanniferous base-metal sulphide and iron oxide deposits such as the Sullivan deposit (British Columbia) and others in Canada (Mulligan, 1975, pp. 50—51) may be synsedimentary, formed by chemical precipitation from metalliferous hydrothermal brines near fumarolic vents on ancient seafloors as suggested by Hutchinson (1981). Others, such as the somewhat similar stratiform/stratabound stanniferous deposits of Cleveland, Renison Bell and Mount Bischoff, in Tasmania, and certain deposits in Southeast Asia, notably the bedding plane lodes of Kelapa Kampit and the Selumar deposit, both in Belitung, *may* have been generated in the way suggested by Hutchinson (1981), but I adopt the view that they are replacement deposits.
(8) The overwhelming majority of tin deposits are associated with S-type granites, or types closely akin to them, and not to I-type granites. These granites are more-or-less equivalent to Ishihara's (1977) ilmenite-series and magnetite-series granites respectively.
(9) The majority of the major types of primary tin deposits are associated with granitoid cusps and ridges. In polyphase granitoids the major stanniferous

deposits are commonly, perhaps invariably, related to the latest phase, but this may be difficult to establish.

(10) Within a stanniferous province or field the centres of mineralisation may occur at the intersections of two sets of planes of preferred mineralisation as in the Southeast of England, Portugal, Queensland and Rondônia.

(11) Within a stanniferous area of limited extent a number of quite distinct types of primary tin deposits may occur. For example, in Cornwall, greisen-bordered vein-swarms and complex lodes are often spatially quite closely associated, whilst, in Bolivia, the same relationship exists between porphyry tin deposits and xenothermal lodes.

(12) Faults have played major roles in determining the distribution, nature and size of primary tin deposit and so have impounding bodies.

(13) Zoning is common amongst most of the primary types of deposit but telescoping is more widespread than was once believed.

(14) Cassiterite is the only tin species of major economic importance but stannite, teallite and even a little malayaite, the last as a minor component in an essentially cassiterite concentrate, have been utilised by the smelter.

(15) Placers are by far the most important sources of cassiterite. Next in importance are lodes and 'bedded' deposits. All other types of primary deposit discussed below are being, or have been, exploited.

(16) Presently, in addition to placers, low-grade (c. 0.1% Sn), high tonnage (c. 40 Mt) greisen, vein-swarm deposits and skarns are amongst the most important targets.

## The Classification of Tin Deposits

As far as the *applied* geologist is concerned the major object of classifying tin deposits is to facilitate the search for deposits which will be worth exploiting. Their presence may also suggest that the area might contain other, perhaps economically more important types. Furthermore a good classification should assist the trained searcher for tin deposits, after examination of a limited exposure and/or a modest quantity of drill core, to forecast, with some confidence, something of its overall mineralogical character, and to make intelligent guesses of the grade and tonnage of ore that it might contain.

A number of classifications of tin deposits occurring within a given stanniferous province have been published that have proved reasonably satisfactory. I think my classifications of the tin deposits of Malaysia (Hosking, 1974, pp. 42—43) and my pictorial one of the Southwest of England (Hosking, 1970a) (Fig. 1), fall into this category, although it has been necessary to modify the latter to accomodate the hydrothermal breccia type of deposit whose true character has only recently been appreciated. However, a classification of all the known types of tin deposit of the world which would wholly satisfy the majority of those concerned with the worldwide search for viable deposits and/or with problems of oregenesis, still remains to be developed. It can be stated dogmatically that the data are not available to erect a satisfactory genetic classification of the *primary* tin deposits nor are the data likely to be available in the foreseeable future. Indeed, any genetic classification adopted now is likely to retard

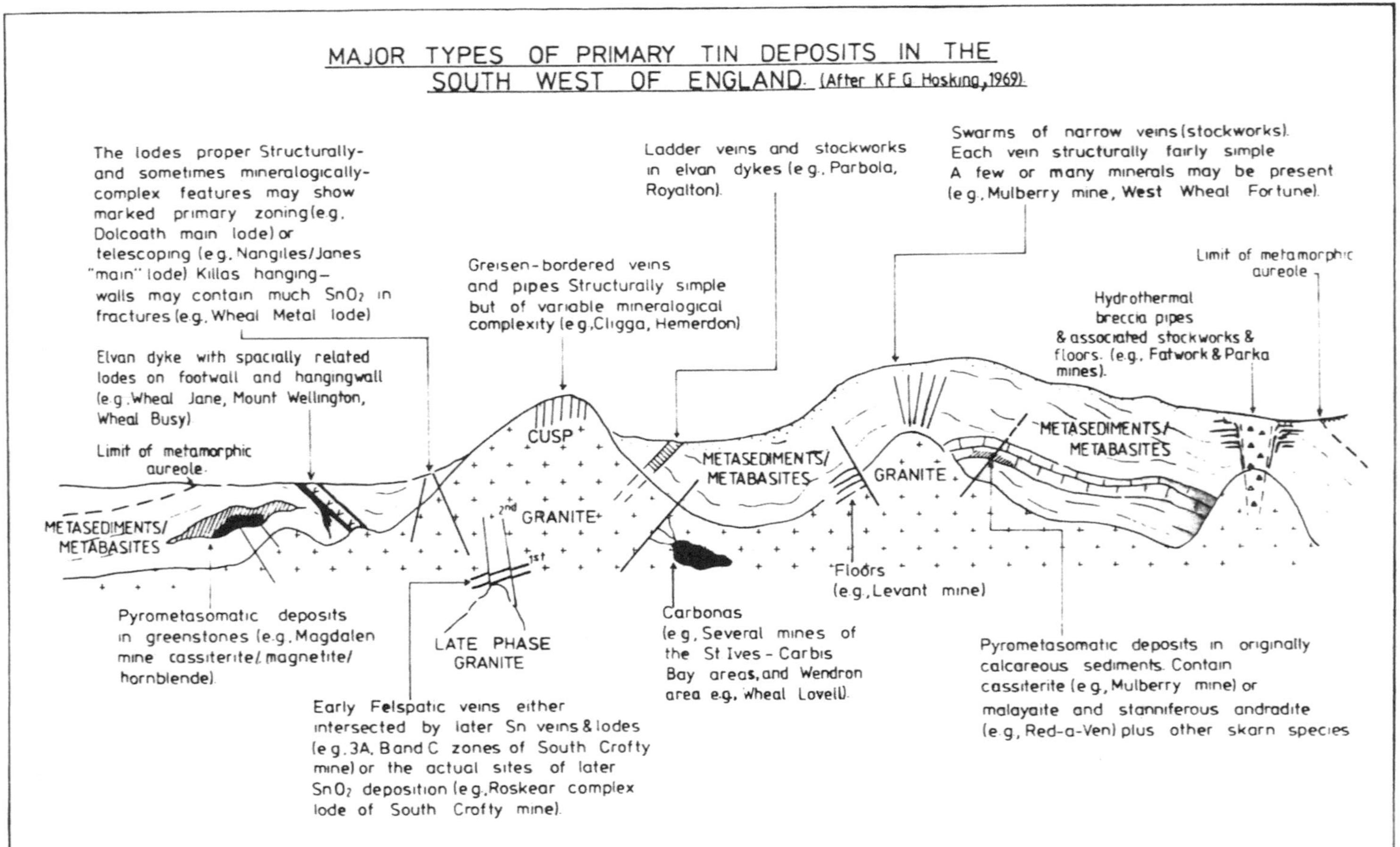


**Fig. 1.** Major types of primary tin deposits in the South West of England (after K.F.G. Hosking, 1969)

the advancement of understanding how the primary tin deposits were generated and how best to search for them. This is precisely what Lindgren's (1933) genetic classification of mineral deposit did. We should do well to remember that the following remark of Beverage (1955) about the 'hypothesis' applies well to any classification: "It is generally agreed that unverified assumptions should be kept down to the bare minimum and the *hypothesis* with the fewest assumptions is to be preferred". Those who favour a genetic classification of the primary tin deposits whould also take note of Bernard's (1865) view that "Men who have excessive faith in their theories or ideas are not only ill prepared for making discoveries; they also make bad observations".

Gilmour (1962), when proposing a non-genetic classification of copper deposits, provided the following comments with which I am largely in agreement and which, I think, are equally relevant to the question of the classification of tin deposits: "In English-speaking countries, at least, the systems of classification most widely used for copper and other mineral deposits are based on the supposed mode of origin of the deposits (the present writer, K.H., is sure that as far as tin deposits are concerned this is no longer true). Instead of recognizing the genetic basis as a defect some authors consider it an asset – yet in our present state of knowledge the application of a classification based on mode of origin must be subjective and arbitrary. A genetic classification is comparable, say, to a zoological classification derived from some worker's notion of phylogeny rather than from observed anatomical features. Obviously, a classification of ore deposits should be made as objective and descriptive as possible and with this object descriptive criteria, such as the form and composition of the deposits and the type and setting of the host rocks, should be employed". Later, Gilmour opines that "one of the principal reasons for making any classification as objective as possible is that classification may stand no matter what happens to theories of origin".

Taylor (1979, p. 12) pointed out that Russian and other classifications of primary tin deposits, which are based essentially on their mineralogical characteristics, are not of great value, neither to those concerned with the genesis of the deposits nor to those searching for them. He concluded that a much more useful classification might be developed by approaching the problem from an environmental point of view such as had been attempted by Itsikson (1960) and pictorially by me (Hosking, 1965) (Fig. 2). As a result of this view Taylor (1979, pp. 27–29) provided a classification of 'Environments containing significant primary concentrations of tin'. This forward looking approach, which should be given wide approval, will require international co-operation in order to convert it into a major exploration aid: I am quite sure that in the course of time that is what it will become. In the meantime I have found the imperfect, essentially non-genetic classification (Table 1) of some practical value and the remainder of this paper is based on it. My earlier pictorial classification (Fig. 2) might be readily modified to accord with the classification I am presently using, but it is still worthwhile referring to. In addition to broadly relating the deposits to their geologic environment, it has the merit of being remembered without much difficulty.

The major characteristics of the deposits, particularly those of considerable economic importance, will be presented and supported by reference to known examples, especially to those with which I have personal experience. I have deliberately refrained from referring to Chinese deposits although I have seen, albeit briefly, a few of them, believing that they will be best dealt with by those who have spent much time and effort in investigating them.

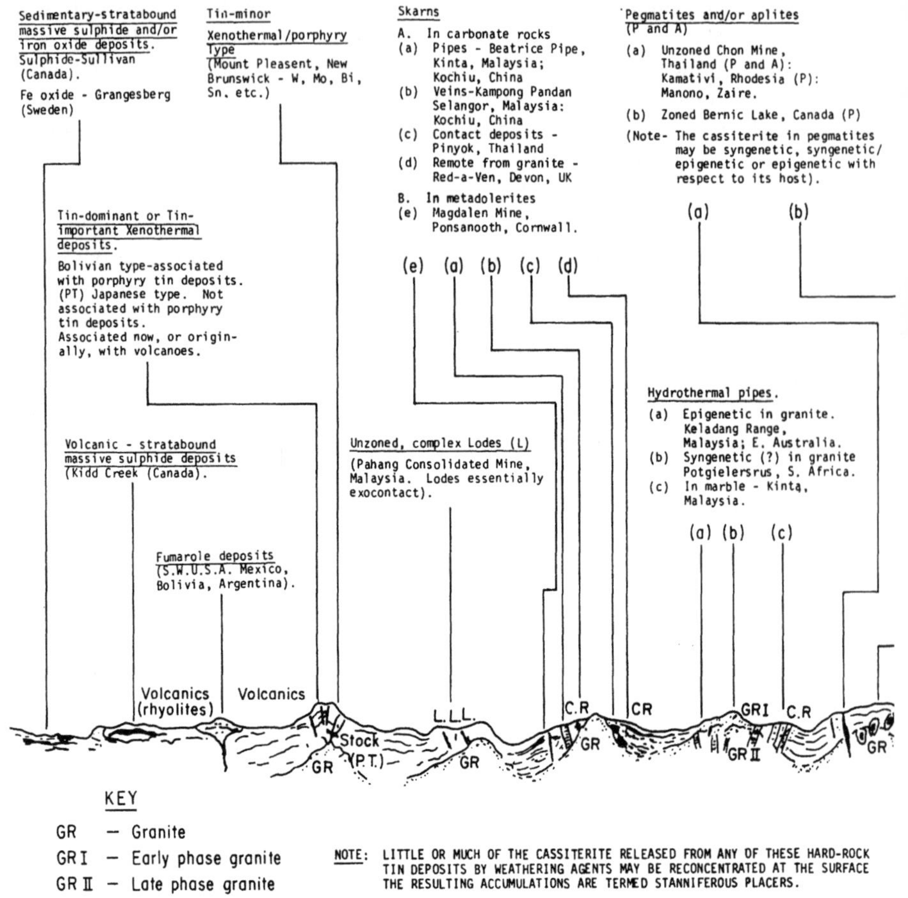


**Table 1.** A classification of tin deposits into major groups

| | |
|---|---|
| Group 1. | Disseminations, other than those in placers, that are not included in the other major groups. |
| Group 2. | Pegmatites/aplites. |
| Group 3. | Skarns (pyrometasomatic deposits). |
| Group 4. | Hydrothermal breccias. |
| Group 5. | Deposits associated with greisenised and/or albitised country rock. |
| Group 6. | Stanniferous veins other than those of Group 5. |
| Group 7. | Lodes of Cornish type. |
| Group 8. | Replacement (metasomatic) deposits, that cannot be satisfactorily placed in any of the other groups. |
| Group 9. | Telescoped, mineralogically complex deposits (largely xenothermal or sub-volcanic). |
| Group 10. | Deposits of the Mexican type (epithermal or fumarole). |
| Group 11. | Stanniferous massive sulphide and massive iron oxide deposits. |
| Group 12. | 'Ancient', variously modified, stanniferous sedimentary deposits. |
| Group 13. | 'Modern' placers. |

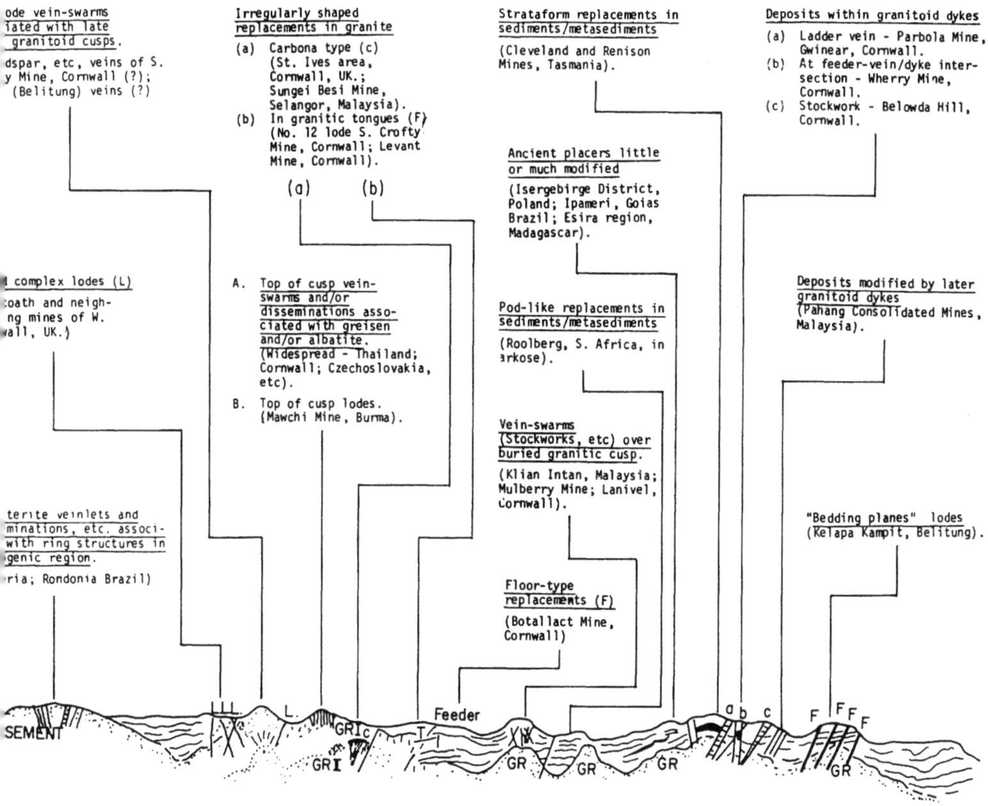


**Fig. 2.** Pictorial classification of the major types of tin deposits of the world (Hosking, 1965)

Although I have avoided genetic concepts as far as possible in my classification, I cannot completely avoid the question of genesis when discussing some of the deposits. In marked contrast to the classification of the primary deposits, it would be out of the question to ignore the genetic aspect when classifying and discussing 'modern' placers, as the processes leading to the development of these bodies are available for all to see.

## 1. Disseminations, Other Than Those in Placers, That Are Not Included in the Other Major Groups

The members of this group are of very minor importance as direct sources of tin although they *may* have been locally important contributors of cassiterite to certain placer deposits, for example, in Nigeria.

In this group the cassiterite is syngenetic with respect to the host rock, and in particular granitoids.

If Emmons' view (1940, pp. 185—194) is generally valid, then during the consolidation of a granitoid magma, and due to the lack of fracture systems developing, much of its tin content would eventually report as cassiterite crystals disseminated in a granitoid matrix. In reality it is very difficult to establish whether disseminated cassiterite in any given granitoid is truly syngenetic or epigenetic with respect to the host. It is thought by some that syngenetic disseminated cassiterite in economically interesting concentrations occurs in some of the 'younger' (Jurassic) granites of Nigeria, for example, at Odegi. There, cassiterite, in part intergrown with columbite, is confined to a granitoid porphyry characterised by the presence of bipyramidal quartz crystals and thorite. As these minerals do not appear to be related to any vein system they are thought to be syngenetic.

Bradford (1961) noted that a limited number of analyses from ordinary granite suggested "that tin-ore is finely disseminated through many Malayan granites" and he also thought that some of the placer deposits in the vicinity of Bukit Yong might be "derived from cassiterite disseminated throughout the local granite" since there is a paucity of known stanniferous lodes and stringers in the area (op. cit., p. 388). However, in recent years further field and laboratory work have provided little support for Bradford's views, as, except on the close vicinity of epigenetic deposits, the tin content of the granite averages 10, or less, ppm. It can be concluded that in Malaysia, although accessory syngenetic cassiterite may, and probably does occur in the granitoids, it never reaches a concentration of any economic interest.

Aubert (1969) records that at Montebras, France, there is a granite which has been exploited, in part, for its lithium and tin content. It consists of quartz, microcline, albite and muscovite with minor topaz, apatite, cassiterite and niobotantalite. If, as Aubert believes, the body is a magmatic intrusion containing syngenetic cassiterite it constitutes an example of a deposit falling into Group 1 of my classification.

In the Potgeitersrus tin-fields of South Africa a band of early-developed, possibly syngenetic disseminated cassiterite occurs in the Bobbejaankop granite (Fig. 3). Strauss (1954, p. 135) remarks that clearly the zone of low-grade ore in the Bobbejaankop granite – must have contained disseminated cassiterite during the late magmatic stages (when the lower portions of the Lease granite were still in process of formation) and before the hydrothermal stage (that is, when Strauss thinks the stanniferous pipes of the field (Fig. 4) were developed). However, Sohnge (1944), according to Taylor (1979, pp. 172—173) claims that "the branching worm-like system of pipes – appear to have developed prior to the formation of fractures in the granite host". Taylor further observes that like certain Mo-Bi pipes in East Australia they "may represent elongate bubble trails where an immiscible low density aqueous phase was in the process of separating from a silicate melt, but became entrapped by crystallisation before reaching any overlying plumbing system".

This last example is introduced to demonstrate that if genetic concepts are introduced, the pipes must be classed as members of Group 1. I will ignore such genetic considerations and place them in Group 8.

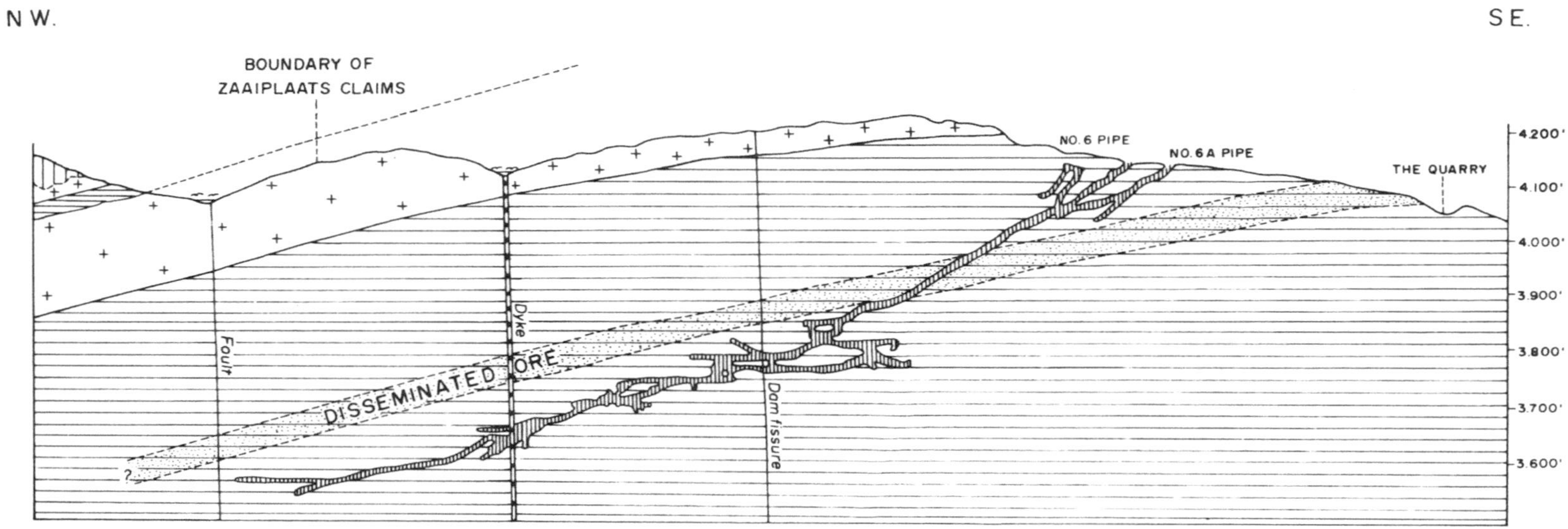


**Fig. 3.** Geological section across Zaaiplaats tin mine (after Strauss, 1954)

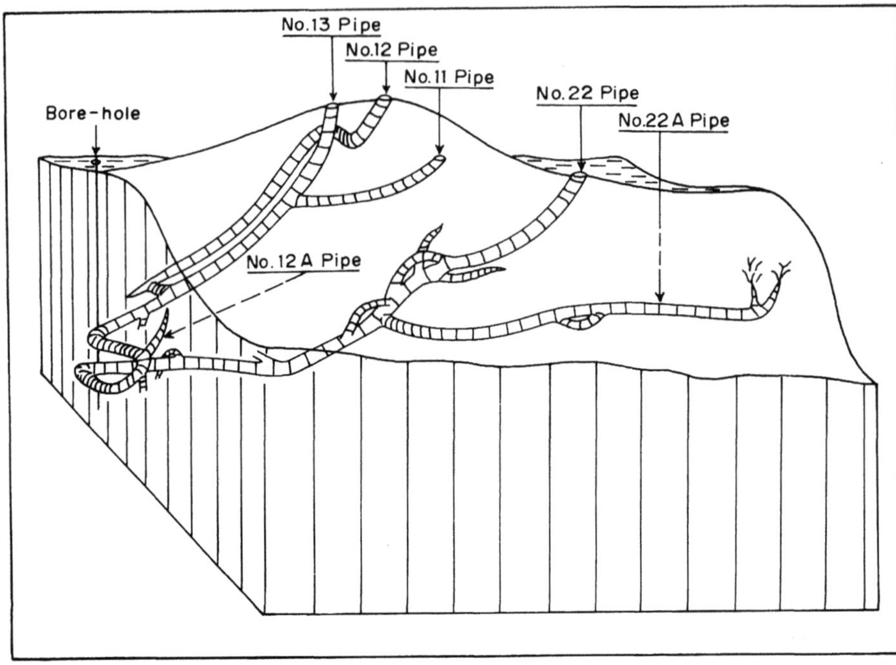


**Fig. 4.** Block diagram of the No. 22 pipe-system, Groenfontein Mine. Note the loops formed by some branches (after Strauss, 1954)

## 2. Pegmatites/Aplites

From the exploration point of view, only granitoid pegmatites and their fine-grained common partners, the aplites, are of economic importance, although the tin mineral nordenskiolite occurs, albeit only in academically interesting amounts, in a nepheline-syenite pegmatite in Norway (Mulligan, 1975, p. 9). In this paper the terms 'pegmatite' and 'aplite' will refer only to the granitoid varieties. The term 'pegmatite' has often been misused: I shall use it as Holmes (1965, p. 84) does, for a rock whose bulk composition is granitoid and in which "the constituent minerals have crystallised to an unusually large size –". Many so-called pegmatite bodies are, in fact, pegmatite/aplite units as is, for example, the stanniferous Chon 'pegmatite' of Peninsular Thailand. In the West Australian Moolyella tinfield "the only rocks in the area of the field in which primary cassiterite has been identified are the layered aplite-pegmatites" (Blockley, 1980, p. 20).

Stanniferous pegmatites varying in age from the Precambrian to the Tertiary are known, but the largest and economically the most important, are restricted to the Precambrian. These have been exploited, for example, in Central Africa (Zaire, Zimbabwe, etc.) and West Australia (Greenbushes). In some instances the economic importance of the bodies has depended on the presence of species other than cassiterite, particularly tantalite and columbite, and in some instances, as at Bikita (Zimbabwe) the pegmatite is worked primarily for its lithium species. The Bikita body did contain a little cassiterite but weathering and erosion effected its transference to the superficial deposits. Younger stanniferous pegmatites are usually rather small units whose tin-grades are rather low. They have been exploited by opencast means in Southeast Asia and in small underground mines, for example, in the Southwest of England and Portugal, where their economic potential was often enhanced by the presence of wolframite.

Some of the large Precambrian pegmatites have been the parents of significant Sn/Nb/Ta placers, as at Greenbushes, and often the superficial parts of such pegmatites have been so argillitised that they also are easily exploited. On occasion such bodies have been mined by underground methods, as at Kamativi, and recently, a decline was sunk in the Greenbushes pegmatite which has revealed not only the presence of cassiterite but also what may prove to be a significant concentration of spodumene.

The fact that the large stanniferous pegmatites are restricted to the Precambrian suggests that they were generally emplaced at greater depths and so were the last to be subjected to the destructive influence of weathering and erosion. However the reasons for their survival are probably considerably more complex.

In a given pegmatite field the composition and/or the importance of individual pegmatites may vary considerably as in the examples at Phuket (Thailand) and in Northern Nigeria. In Northern Nigeria, where the pegmatites are associated with the Older Granites, one finds, according to Jacobson and Webb (1946, p. 12), Interior Pegmatites (core pegmatites), Marginal Pegmatites (hood pegmatites), and Exterior Pegmatites (roof pegmatites). Of these, the first two are mineralogically and structurally rather simple and virtually or completely devoid of heavy minerals. The External Pegmatites, on the other hand, are mineralogically quite variable, but the members of major economic importance are complex albitised dykes which locally contain cassiterite and columbite/tantalite. Areas of intense albitisation often correspond to centres of rich mineralisation. Such a situation is encountered in certain pegmatites elsewhere, and is reminiscent of the cassiterite/albitite association, discussed later, which occurs in the cusps of certain granitoids. These Nigerian pegmatites show little or no zoning and their paragenesis is presented in Figure 5. It is of interest to note that the great pegmatites of Manono and Kitotolo, each of which has a strike-length of c. 5 km, and which are important direct and indirect sources of tin and tantalum/niobium minerals, probably lie over, not in, a granitic ridge and, at least at Manono, close to a greenstone/schist contact.

Stanniferous pegmatites may be zoned or unzoned and the former, unlike the latter, probably developed largely by fractional crystallisation within a closed system. In the extreme case, the zoning may be quite complex as in the Bernic Lake body of Canada and as illustrated by the now well-publicised sections of the Bikita pegmatite (Zimbabwe). Usually, some or all of the cassiterite of the great zoned pegmatites displays the characteristic bipyramidal habit, and in thin section it is characterised by an intense deep red to pale pleochroism, due to the presence of tantalum (and niobium?) in its structure.

The suite of lithium-rich pegmatites, which are locally stanniferous and occupy portions of a NE-SW fault system in Peninsular Thailand, constitute an excellent example of the unzoned type. The cassiterite does not provide the intense red to pale pleochroism which one might expect if it were a product of the pegmatite magmatic fraction, and particularly as the pegmatites occur in a region noted for the abundance of Ta/Nb-rich cassiterite in the placers. It is therefore possible that the cassiterite was introduced later, the pegmatite providing a favoured host for its deposition. Indeed,

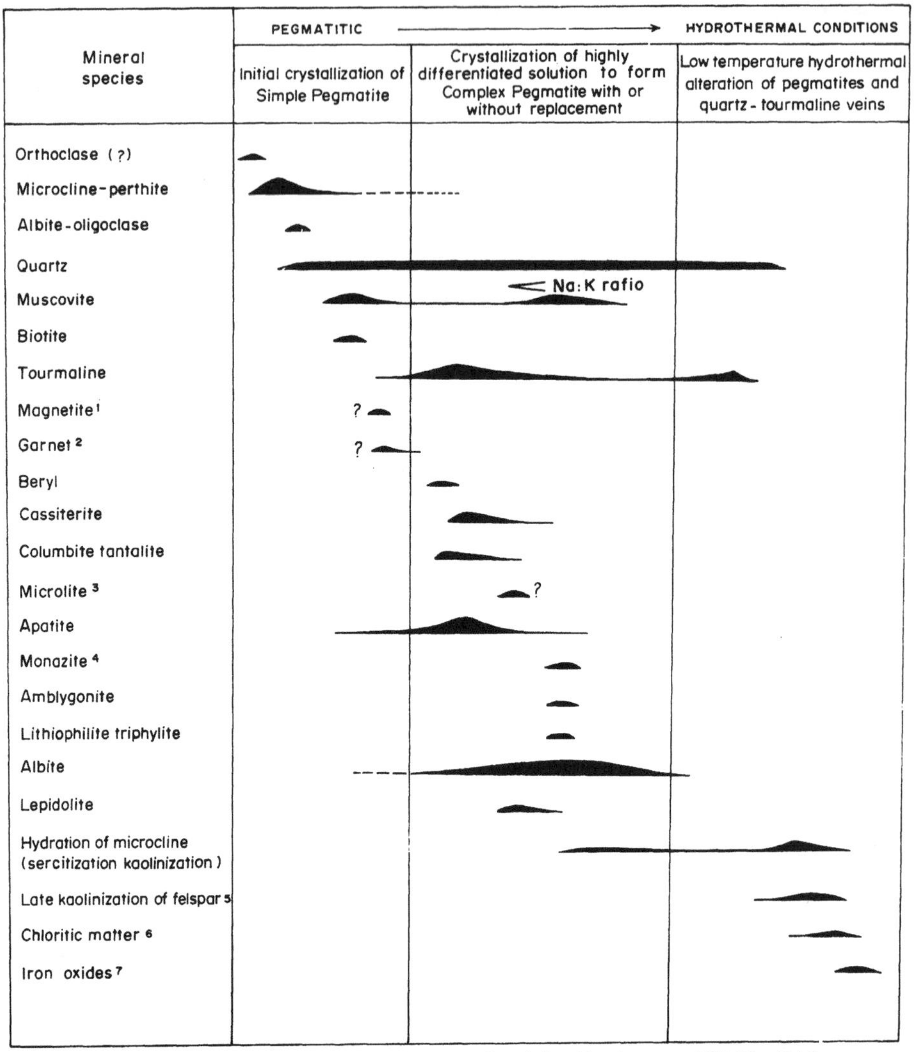


**Fig. 5.** Mineral paragenesis in the feldspathic pegmatites (after Jacobson and Webb, 1946)

there is little doubt that pegmatites, which appear to be unusually rich in cassiterite and sometimes base metal sulphides, frequently owe their richness to the introduction of some or all of these species after the pegmatite has developed. A number of stanniferous pegmatites in Malaysia, have, I think, been so developed. One such occurs at Gunong Baku, where, in addition to much tourmaline and cassiterite, it also contains pyrite, arsenopyrite, chalcopyrite, and galena (Roe, 1951).

Although it is generally true that large pegmatites and large important hydrothermal lodes and vein systems tend to be mutually exclusive, stanniferous pegmatites and veins can be closely associated. Thus many of the tin-fields of Western Australia consist of stanniferous pegmatites associated with later tin-bearing veins whose source may have been independent of that of the pegmatites. It is also of interest to note that Haapala (1966), commenting on the genesis of the zoned Haapalaoma pegmatite (Finland), concluded that after the body had developed in a closed system in which crystallisation proceeded from the walls to the core, fluids worked their way into the pegmatite via fractures and caused the generation, largely by replacement, of a wide variety of species including cassiterite, columbite and loellingite. Establishment of the paragenesis of a pegmatite body, in common with all other primary mineral deposits, is rarely easy, and most that have been published are doubtless in some measure defective, although they often serve a useful purpose. One paragenesis of a stanniferous pegmatite has already been presented; another is provided here (Fig. 6) without comment for comparison and in the hope that it will generate further study of the subject.

The search for stanniferous pegmatites is fraught with difficulty. This is due to the unpredictable distribution of cassiterite in these bodies, that no cassiterite may be found in the outcrops of stanniferous bodies, and that non-stanniferous pegmatites may be associated with those that are stanniferous. On occasion, the presence of indicator minerals and/or the trace tin content of the feldspar *may* facilitate the selection of stanniferous pegmatites but no certain means has yet been devised. For further discussion of this problem, see Hosking (1965).

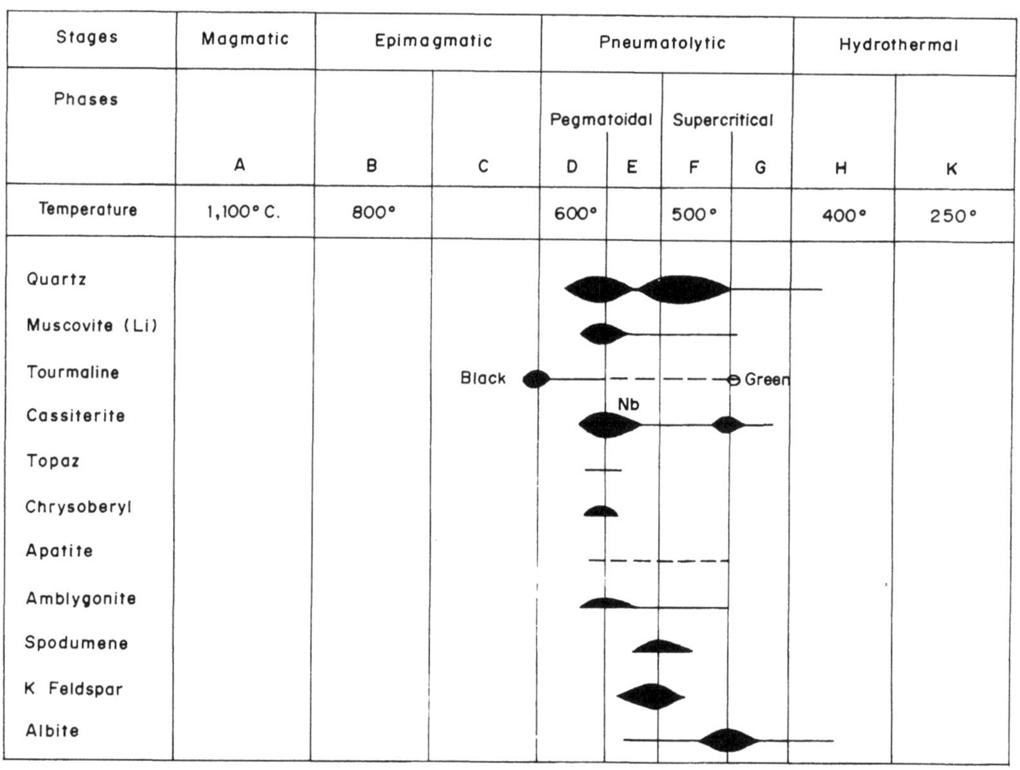


**Fig. 6.** Paragenesis of certain stanniferous pegmatites of Mumba-Numbi, Kivu, Congo (after Agassiz, 1954)

Finally, although the exploitation of pegmatites may be locally important, they have not been collectively major direct sources of tin but they have, doubtless, made considerable contributions to placers, some of greater value than the parents.

### 3. Skarns (Pyrometasomatic Deposits)

In common with many geologic terms, skarn' has come to mean different things to different people. I think that a true skarn *must* contain calcium or magnesium silicates and that even when a replacement mineral deposit is hosted by marble and is close to a granitoid contact, it must be termed a metasomatic deposit if it lacks these silicates. Kwak (1983, p. 1440) is one who is prepared to extend the meaning of skarn, for when discussing the zoned Sn/W/F/Be skarns' of Mt Lindsay, Tasmania, he states that "the term skarn is used here in a broad sense meaning essentially replacement of carbonate or a replacement of a previous replacement carbonate, whether or not Ca or Mg silicates are extensive – – –". However, when Kwak and Askins (1981, p. 443) describe the skarn (wrigglite) of Moina, Tasmania, they remark that 'skarn here refers to all calc-silicate-bearing rocks derived from Ca-rich sedimentary rocks". Burt (1977), noted that 'skarn' was generally used in Northern Europe for the silicates associated with ore replacing limestone and dolomite near igneous contacts. He declared that 'skarn deposit' should only be used for ore deposits possessing 'skarn' as a gangue and emphasised that "the only truly distinctive feature of these deposits is not the metals mined, nor the presence of an intrusive, but rather the gangue called skarn".

Stanniferous skarn deposits have generally developed from carbonate rocks (usually, but not invariably impure ones) and occasionally from meta-basites (as in West Cornwall) as a result of them being subjected to heat and later hydrothermal mineralising agents from an invading granitoid.

These deposits occur near the high parts of granitoids in roof pendants and enclaves, as border deposits to the granitoid (as at Pinyok, Thailand), as bedded deposits bordering a granitoid (as, locally in the Kuala Lumpur, Malaysia, tin-field) or overlying but separated from the granitoid by non-calcareous deposits (as at Moina, Tasmania), as pipes (for example, the Beatrice Pipe, Perak), as veins (as locally in the Kuala Lumpur tin-field) and as deposits bordering granitoid dykes and sills (as locally in the Kuala Lumpur field).

Stanniferous skarns range from Precambrian to Tertiary, but because they developed at fairly high levels and so were often rapidly attacked and destroyed by weathering and erosion, most of the economically important ones are relatively young.

Skarn deposits are often zoned, and the Moina deposit ofTasmania is an excellent example of such a body.

Several classifications of skarns have been designed based on their mineralogical character. However, the mineralogy of the skarn deposits is complex, and as yet far too few tin-bearing skarns have been subjected to the detailed mineralogical/chemical examination that they merit, that is, of the type that has been employed by Burt (1977), Kwak and Askins (1981) and Kwak (1983).

It is outside the scope of this paper to discuss in any but the briefest detail the mineralogy and genesis of the stanniferous skarns. Readers seeking an in depth knowledge of the subject should refer to the papers quoted immediately above.

From the point of view of the tin exploration, the most important aspects of tin skarn mineralogy are how tin occurs in the deposits and what fraction of the tin can be expected to be recovered economically.

Stanniferous skarns may contain species in which tin is an essential component and/or others in which it is found as a non-essential element, although sometimes in surprisingly high concentrations. Of the former, only cassiterite is commonly encountered, and this is followed by malayaite ($CaO.SnO_2.SiO_2$). Stannite and a number of tin silicates and borates have been recorded from skarns, but only rarely, and when they occur they are usually only present in very modest concentrations.

Skarns are known in which malayaite is the only species containing essential tin (as in the Langkawi Islands, Malaysia). Others occur in which cassiterite is the only tin species (as in the Beatrice Pipe, Perak [Fig. 7]), whilst still others contain both cas-

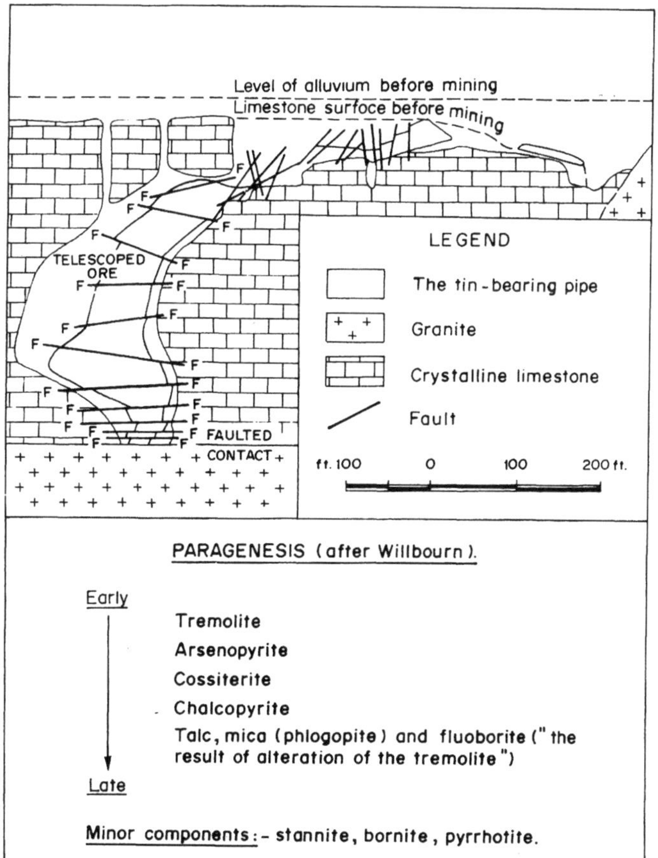


**Fig. 7.** Section of the Beatrice Mine, Selibin, Perak, and the paragenesis of the ore (after Willbourn, 1931)

siterite and malayaite (as at Pinyok, Thailand, and Chendriang, Perak, Malaysia). Malayaite may convert to an aggregate of varlamoffite, quartz and calcite as in the skarns of Sungai Goh (Perak) and at Pinyok. It is worth noting as an aid to exploration that malayaite usually, but not invariably, fluoresces a greenish-yellow under short-wave ultraviolet light.
Some skarn minerals which may contain appreciable amount of non-essential tin are sphene, epidote, ferroactinolite, ludwigite-type minerals, magnetite (which locally in northeast U.S.S.R. contains up to 0.45% Sn), axinite and andradite (which may contain up to 2%, or more Sn) (see Mulligan, 1975, p. 29, and Hosking, 1975, pp. 15–16).

The cassiterite in skarns may be recovered only with difficulty or not at all by conventional means. For example, at Pinyok in Thailand, the skarn not only contains readily recoverable cassiterite but also acicular crystals, only a micron or so in width, that report in the mill tailings. The skarn at Otero, Spain, has not been exploited, in part because its appreciable cassiterite, which is associated with considerable chalcopyrite, sphalerite, and scheelite, would be extremely difficult to recover by conventional means as all of it is extremely fine.

It is evident that during the early stages of examination of a stanniferous skarn, an endeavour must be made to determine how and in what species the tin is partitioned. Chemical analyses, without mineralogical investigations can be quite misleading as they have been in Pinyok and elsewhere in recent years. In addition, it is most important to look for other species, notably scheelite, which might increase the value of the ore. It is also important to note that the superficial parts of these skarns may be so modified by weathering processes that their true nature may be obscured. Finally, the analysis of residual soils over stanniferous skarns during geochemical prospecting may provide a distinct tin anomaly as they did at Red-a-ven, Devon, England, even though no cassiterite is present in the rock. At Red-a-ven the tin in the skarn was restricted to malayaite and garnets, particularly andradite.

Variations in mineralogy have been used as a basis for the classification of skarn deposits. Thus Shcherba (1970), according to Taylor (1979, pp. 204–205) notes that cassiterite may occur in phenacite-mica-fluorite and beryl-mica-fluorite skarns, but he classifies the skarns in which cassiterite is a component of major economic importance as cassiterite-skarn, cassiterite-skarn-sulphide and cassiterite-scheelite-skarn. He also notes that whilst the first two are only of minor importance, the third is of moderate economic importance. All three types of skarn possess a complex mineralogy due to no small degree to their genesis which involves replacements which are effected by the introduction of a number of pulses of mineralising fluids.

In spite of Shcherba's views concerning the relative importance of his three types of tin skarn, both Taylor (1979, pp. 207–209) and I consider the sulphide-rich and the magnetite (iron-oxide)-rich stanniferous skarns are deserving of special attention.

Possibly the sulphide-rich skarns are more common than the magnetic-rich ones, and in the Tin Province of Southeast Asia the former are confined to what I term the Western Tin-Belt. Unfortunately only a few have been sufficiently large to be of economic importance. The most important of these is at Pinyok, where, in addition to cassiterite the skarn contains malayaite, stanniferous garnet and sulphides including galena, which has, in the superficial parts, been largely converted to pyromorphite. Second, in order of importance, is the Beatrice Pipe, Perak (Fig. 7). The following will serve as examples of the many small sulphide-rich skarns in Malaysia: At Teh Wan

Seng No. 4 mine (Sungai Besi, Selangor) a skarn borders a granitoid dyke. The skarn contains, in addition to cassiterite and malayaite, a variety of sulphides, scheelite, fluorite, diopside, vesuvianite, grossularite, calcite and quartz. At the Melor Syndicate mine (near Kuala Lumpur) there are a number of different types of skarn, but one in contact with the granitoid lacks cassiterite but contains malayaite associated with scheelite, sulphides, fluorite, quartz, diopside, epidote, chlorite, vesuvianite and grossularite.

Examples of sulphide-rich skarns might be described from other countries but limitations of space will not permit it.

Tin-bearing magnetite-skarns are not uncommon in the Southeast Asia Tin Province but are virtually confined to the Eastern Tin Belt. At Batu Besi, Trengganu, Malaysia, where iron ore has been mined from such a skarn that borders the granite, the ore contains an average grade of 0.7% tin, but it is locally enriched. Some of the tin is present as cassiterite in and bordering fractures, but most is probably present as exsolved bodies in the magnetite and in the structures of one or more of the associated species. In addition to the species noted above, locally garnet, pyroxene, amphibole, pyrite, pyrrhotite and secondary iron oxides are present in the ore whose generation commenced by the skarnification of limestones and calcareous shales. (For further details of this and other magnetite-skarns occurring in Malaysia see Bean [1969] and Hosking [1970b]).

Considerable interest has been shown in a number of stanniferous magnetite-rich skarns in Australia. Taylor (1979, p. 208) for example, remarks that in North Queensland two such small bodies (in the 2–5 Mt range) are known. The major minerals in these bodies are "fluorite, idocrase, magnatite/hematite, spinel and biotite with minor feldspar, garnet, chlorite, muscovite, scheelite, cassiterite, tourmaline, ilvaite, epidote, pyrrhotite, pyrite, chalcopyrite, arsenopyrite and galena". Both locally display "a peculiar rhythmically finely layered contorted texture where the bands consist of fluorite, magnetite and vesuvianite. This material is 'wrigglite' which, apart from occurring in Australia, is also locally present in the stanniferous magnetite-rich skarn of Pelepah Kanan (Johore) (Santokh Singh. Personal communication).

At Moina, Tasmania, a major magnetite/sulphide (wrigglite) skarn, similar but much larger than the North Queensland ones, has been subjected to a most detailed investigation by Kwak (1981). The skarn, together with associated Sn/W/F veins and greisen, all of which seem to be inter-related, margins the Dolcoath granite and occurs as a sizeable horizontal plate. The plate is separated from the underlying granite by about 200 m of Moina sandstone and is overlain by the Gordon Marble. The plate itself consists of stanniferous wrigglite skarn sandwiched between calc-silicate skarn. Taylor (1979, p. 208) remarks that the deposit is on the scale of the somewhat similar one at Lost River, Alaska, which contains 33 Mt of ore grading 0.27% Sn, 0.0037% $WO_3$ and 15.6% fluorite. Taylor, however, is of the view that the Australian deposits will not be exploited in the near future because not all their tin occurs as cassiterite, and the fluorite is difficult to extract.

Finally, the bedding plane lodes of Kelapa Kampit (Belitung) might reasonably be termed stanniferous magnetite/sulphide skarns. Adam (1960) was well aware of this. These 'lodes' will be referred to later, but the question still needs to be answered 'are they carbonate replacement bodies?'.

## 4. Hydrothermal Breccias

Hydrothermal breccias may be sub-divided into explosive hydrothermal breccias and collapse breccias.

Although explosive hydrothermal breccias are probably present in many tinfields, in some cases without their true identity being recognized, the best documented ones are in southern and central Bolivia. There, high-level Tertiary stocks of intermediate composition, which were probably overlain by stratavolcanoes now largely removed by erosion, locally consist of a stanniferous quartz-tourmaline breccia which, together with any coeval stockwork and disseminated tin mineralisation, have been collectively designated as porphyry tin deposits by Sillitoe, Halls and Grant (1975). These deposits have many of the characteristics of porphyry copper deposits in that, for example, associated with them are shells of sericite and propylite alteration (Fig. 8). However, they differ from the porphyry coppers in lacking potassium silicate alteration and in occurring in inverted conical stocks rather than in upright cylinders. They predate the xenothermal-type lodes of major economic importance which are associated with them. In course of time these low grade/high tonnage deposits may be mined. It is recorded by Sillitoe et al. (1975, p. 916) that in the Llallagua porphyry tin deposit, in which cassiterite occurs largely in breccia and a stockwork, the overall grade is c. 0.3% Sn.

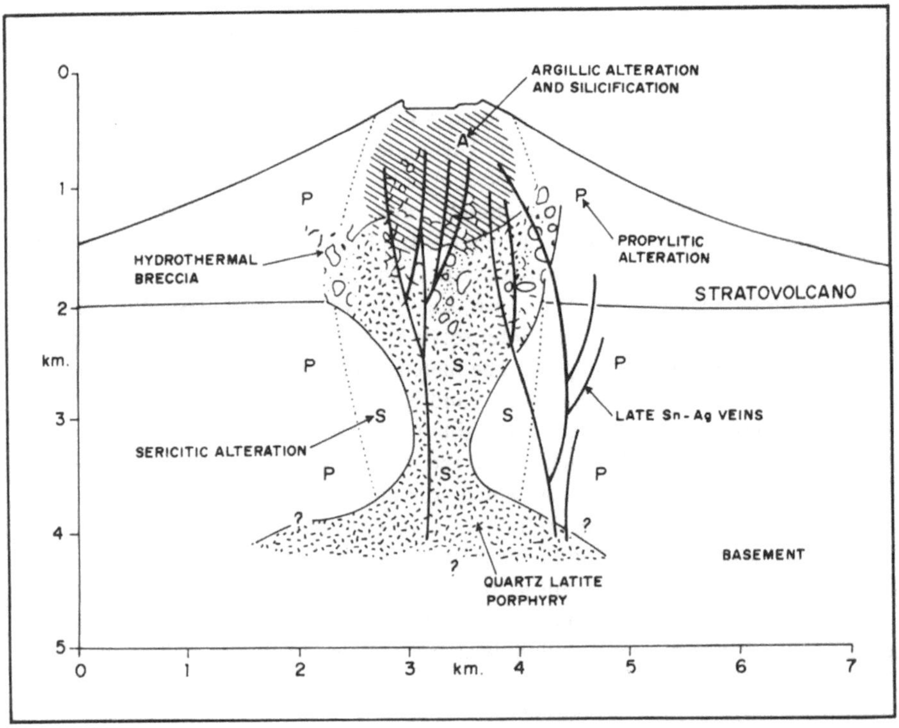


**Fig. 8.** Composite reconstruction of typical Bolivian tin deposit system (from Silitoe et al., 1975)

Sillitoe et al. (1975, p. 935) mention the occurrence of tin in the porphyry type ore deposits of Climax, Colorado, and at Mount Pleasant, New Brunswick, and note certain areas in which porphyry tin deposits, judging by published descriptions, may occur. To these may be added the similar hydrothermal tourmaline breccia deposits of Carbo-Permian age of Central Cornwall, some of which are stanniferous. Of these breccia deposits the one that has received most attention is at the Wheal Remfry china-clay pit. It is well exposed in a wall of the pit and is, clearly, an inverted cone consisting essentially of an outer zone of tourmalinised granitoid masses cemented by quartz and tourmaline. The outer zone encloses a core of similarly altered granitoid fragments and tourmalinised metaargillites, and these are also cemented by quartz and tourmaline. These breccias are virtually identical to those from Bolivia that have been described by Sillitoe et al. (1975). The Wheal Remfry example contains very little cassiterite. There is the possibility that the tin in such deposits largely moved beyond the zone of intense brecciation and locally, at any rate, was emplaced in fractures collectively forming closely sheeted vein systems and stockworks. These are chiefly characterised by marked tourmalinisation of the metasedimentary wall rocks, and these vein systems have been exploited at the small Parka mine and other opencast and underground mines of the district. Material I collected from one of the sites consisted of a vein of quartz and cassiterite in a host consisting of highly tourmalinised and somewhat brecciated metasediment in which small crystals of rutile were much in evidence. The cassiterite in the vein occurred as aggregates of finely zoned cassiterite, of orange-vermilion hues not displayed by any other recorded Cornish cassiterites, but reminiscent, in colour, of certain cassiterites I have seen from the Mexican volcanic deposits! It is also relevant that ore from the same geologic environment, and from the same locality, also contains wood-tin. Thus, one is forced to the conclusion that in central Cornwall one has an example of what may well be sub-volcanic mineralisation similar to the Bolivian type. No evidence of volcanic deposits of about the same age as the tin deposits under review occur in the area, but the porphyry dykes there, which are about the same age as the breccias, may well have originally been channels along which magmatic material travelled and reported eventually as volcanic deposits, which were subsequently removed completely by denudation.

In Rondônia, Brazil, the Morro Potosi hill is composed of an exogreisen containing cassiterite, stannite and other sulphides. Its apex consists of an inverted breccia cone, the gneissic fragments of which are cemented essentially by topaz, quartz, cassiterite and pre-cassiterite wolframite. This very rich stanniferous cone is surely a Precambrian hydrothermal breccia, but of a type differing significantly from those described above.

Collapse stanniferous breccias appear to be rather rare. Taylor (1979, p. 172) mentions that the White Crystal orebody at Ardlethan, Australia, "has many features suggestive of a collapse breccia, that is, large vugs and brecciated marginal zones where replacement via fractures is clearly seen – – –".

Yeap (1978) has described two stanniferous collapse breccia ore bodies in granite which occurred at the Yap Peng Mines, Sungai Besi, Selangor, Malaysia, and had been exploited. Yeap visualises the breccias as having been formed as follows: Marble was invaded by a granite magma which crystallised centripetally forming a carapace that was locally microclinised, tourmalinised and silicified. Ascending hydrothermal fluid collected at two sites under the roof. Fracture occurred, the entrapped fluid escaped and the roof collapsed. Later the fallen fragments were silicified and sericitised and then quartz, beryl, cassiterite, arsenopyrite, pyrite, sphalerite, and chalcopyrite were deposited as a cement. Then kaolinisation and further sericitisation occurred and this was followed by the replacement of some feldspars by rhodochrosite and rhodonite, the deposition of chalcedony and opal, and a little fluorite and pyrite.

Yeap (1978, p. 375) claims that these breccias developed in a mezozonal environment. This I do not believe as the bodies show all the features of a high-level genesis.

## 5. Deposits Associated with Greisenised and/or Albitised Country Rocks

Greisen deposits are now divided into endogreisens (occurring within what is generally regarded as the genetically related granitoid) and exogreisens (occurring in the granitoid-invaded host).

Greisen consists essentially of quartz and sericite and within the granitoids it is the product of alteration of the feldspars and biotite.

Greisenisation and/or albitisation takes place within and in the close vicinity of granitoid cusps and also within minor granitoid intrusives (porphyry dykes, pegmatite and aplite bodies). The greisen-quartz-association found in these geologic environments provides a wide variety of distribution patterns which have been collected by Rundquist et al. (1971) and which have been reproduced in a simplified form by Taylor (1979, pp. 58—60). In some fields, for example in the Southwest of England and Rondônia, greisen deposits are distributed in a pattern which appears to stem from their development at the intersections of two sets of planes of preferred mineralisation (Fig. 9).

In some fields, as for example in Czechoslovakia and North Queensland, the uppermost parts of cusps have been subjected to massive greisenisation with attendant albitisation and microlinisation, and over such cusps greisen-bordered vein-swarms may occur. Within the greisen of the cusps, which may locally contain disseminated cassiterite, closely-sheeted cassiterite-bearing veins may also be present. Figure 10, after Shcherba (1970) provides a general picture of the scene.

In addition to the rather complex situation described above, in some tin-fields such as those of the Southwest of England and Southeast Asia, greisen is almost solely confined to the borders of veins which are commonly impounded within the apices of granitoid cusps, for example at St. Michael's Mount and Cligga in Cornwall and at Mawchi Mine (Burma), or to the fringes of veins haloing a granitoid body as at Redmoor, Cornwall. These veins are commonly rather narrow and may contain, in additon to cassiterite, such minerals as wolframite, topaz, apatite, stannite and other sulphides. Their strike length is often of modest dimensions and their extension in depth is generally 200 m or less. In addition to the cassiterite in the vein proper, some may be disseminated in the greisen which may be further modified by the addition of such minerals as chlorite, fluorite, topaz and zinnwaldite. Occasionally the status of the cassiterite in such vein-swarm deposits may be relegated to that of a by-product with wolframite being of major importance. This is so at Hemerdon (Devon, England) where the greisen-bordered veins plus their host rock constitute a deposit of 45 Mt grading 0.17% $WO_3$ and 0.025% (Mining Journal, 1979, p. 485). In Wicklow, Ireland,

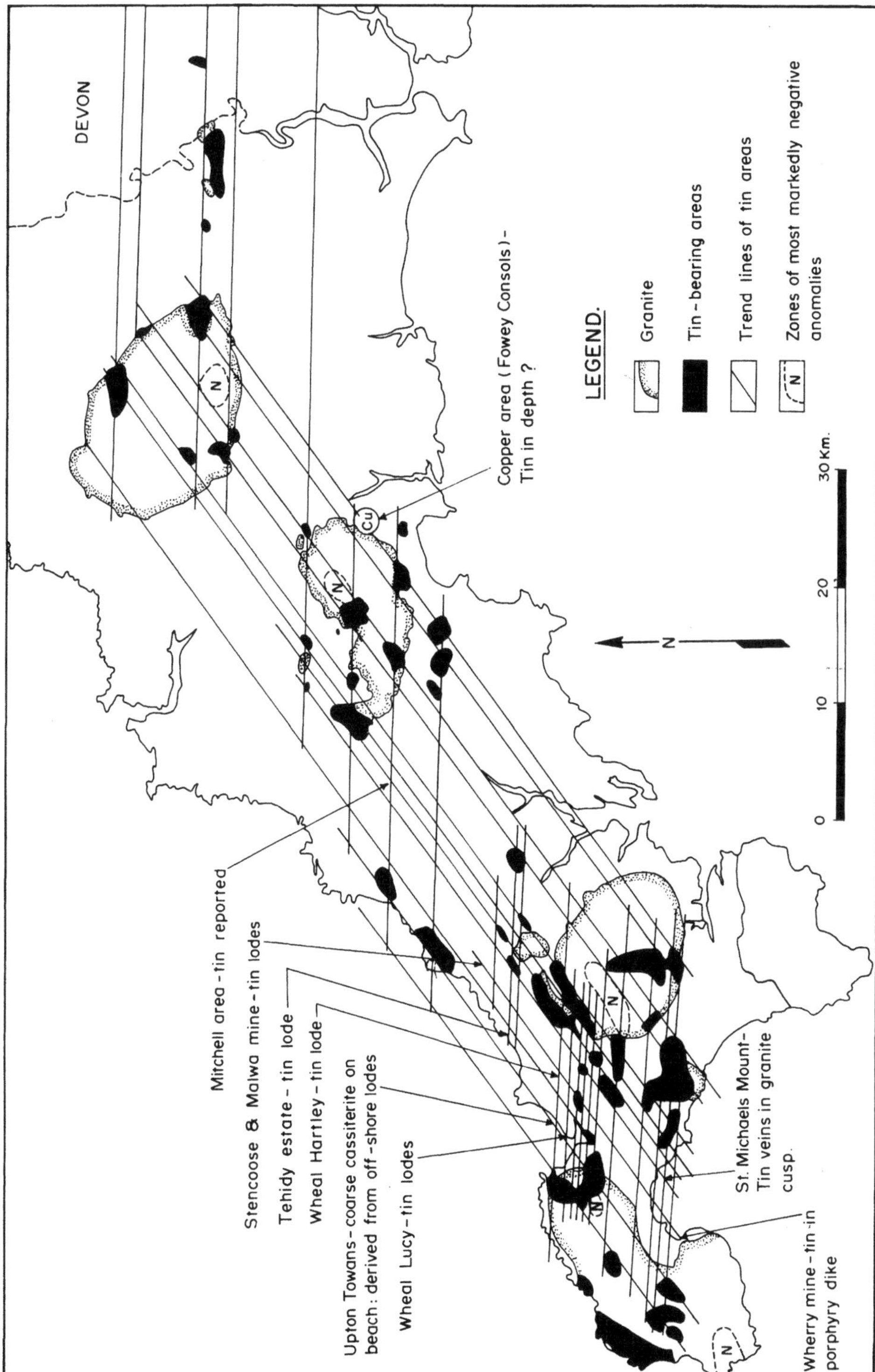


**Fig. 9.** Distribution pattern of the tin bearing areas and of the zones of biggest negative anomalies (after Hosking, 1964)

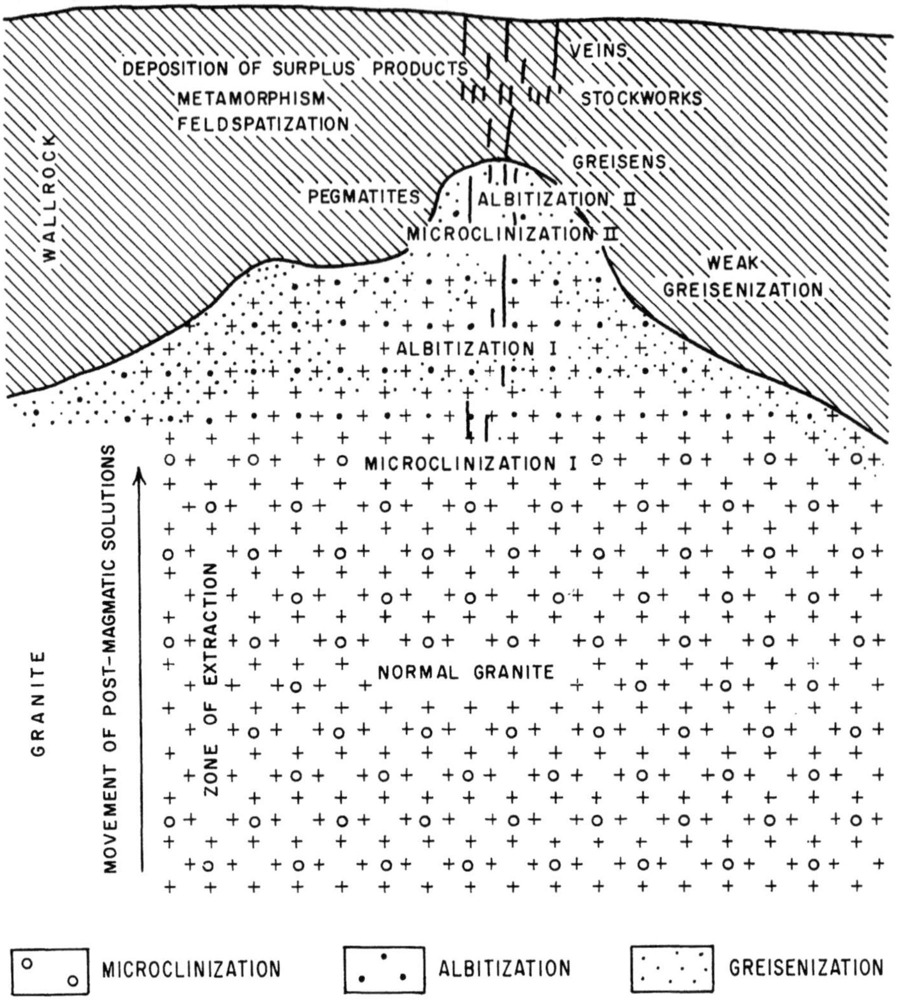


**Fig. 10.** The greisen model of Shcherba (1970)

greisen-bordered vein swarms contain scheelite in economically interesting concentrations, accompanied by sulphides which include stannite. But the tin content is of no value.

Such greisen-bordered vein swarms may display a degree of primary zoning. At Cligga, Cornwall, the portions of the veins nearest the source consist solely of tourmaline and quartz. Further from the source they contain cassiterite, wolframite, stannite and a number of other sulphides. In the deeper parts of the granitoid cusps these veins give way to others containing chalcopyrite, sphalerite, pyrite, chlorite, etc., indicating reverse zoning.

Greisen veins may also occur in marble, and excellent examples have been described by Yew (1971) from the eastern Kuala Lumpur area (Malaysia). They are bordered by sericite, generally banded and multi-mineralic (Fig. 11). They contain cassiterite, wolframite, scheelite, beryl, helvite, tourmaline fluorite, chlorite and a number of sulphides. Their texture and mineralogy vary markedly over short strike lengths. They are markedly telescoped and so might reasonably also be placed in Group 9 of my classification.

Stanniferous bodies containing albite but no mica are known, but they are of minimal economic importance and rare. Such a body occurs at Levant Mine (Cornwall).

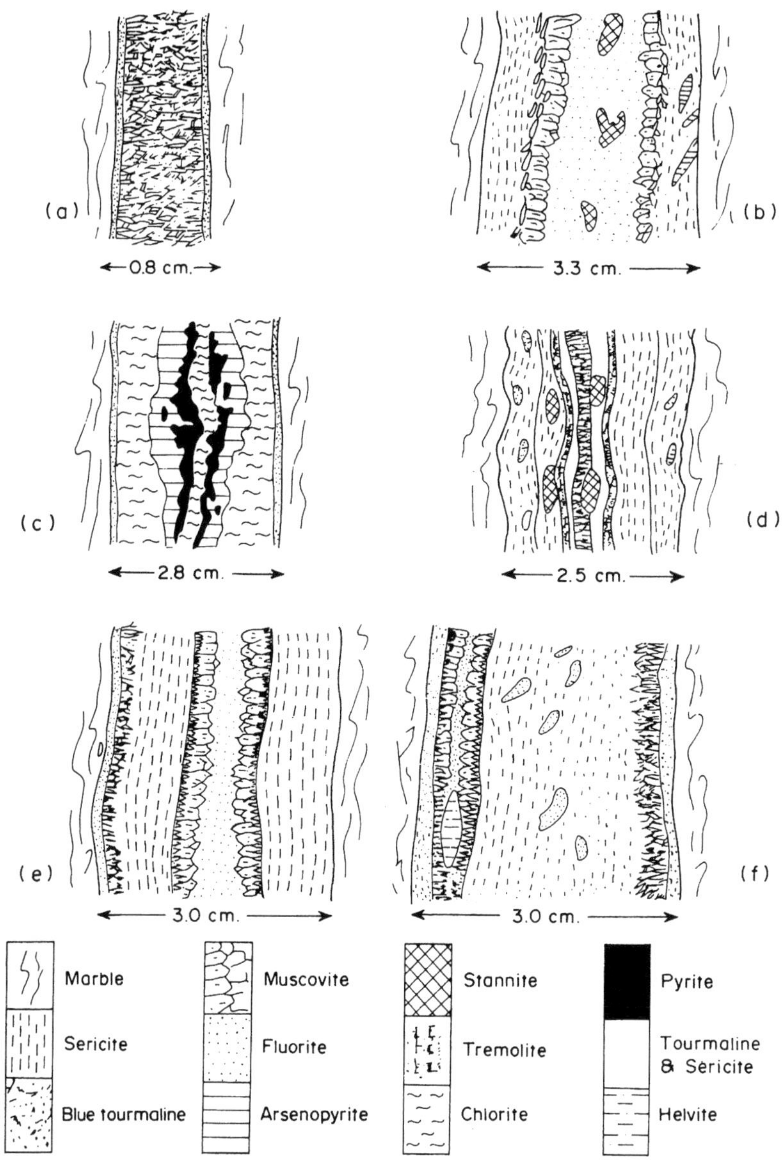


**Fig. 11 a–f.** Illustrating symmetrically crustified veins in marble (**a–c**) and asymetrically crustified veins in marble (**d–f**). *Locality:* **a, c** Galian Berjuntai, Kg. Pandan, **b, d, e, f** Hock Ann Mine No. 1, Kg. Pandan, Malaysia (after Yew, 1971)

It is an irregularly shaped replacement body, occurring within a granite tongue where mineralising agents from a feeder channel were impounded. Jackson (1979, p. 222) notes that the "mineralisation comprises disseminated cassiterite, sulphides (pyrite, chalcopyrite, arsenopyrite) and massive cassiterite and sulphides associated with intense alteration, *albite*, quartz, chlorite, tourmaline and fluorite.

A variation of the theme is to be found in the Greenbushes tin-field, Australia. Cassiterite, apart from occurring in pegmatites, "also occurs in foliated greisenous rock – – – which, in depth, passes into compact, banded albite-rich rock with quartz, tourmaline and cassiterite" (Liddy, 1977, p. 53).

Taylor (1979) should be consulted for a comprehensive account of the complex stanniferous massive greisen-albite assemblages. Here it must suffice to mention the East Kemptville tin deposit of Nova Scotia, discovered in 1978, which falls into this category. It contains estimated reserves of 56 Mt grading 0.15% Sn (Mining Journal, 1984, p. 242). Whilst cassiterite occurs in the Davies Lake pluton in argillized, sericitized, silicified granites and albitites, economic concentrations are restricted to greisens and greisenised granitoid rocks. Associated with the cassiterite are pyrite, pyrrhotite, chalcopyrite, sphalerite, tetrahedrite, chalcostibite (?), arsenopyrite, molybdenite, stannite and bismuthinite (Chattergee and Strong, 1984).

Finally, while small greisen type deposits will continue to be mined in Southeast Asia and elsewhere by underground methods, but more particularly by opencast means, especially where the host rock is soft, they are of slight interest to large mining companies. The latter would only be interested in a deposit, as Taylor (1979) so rightly suggests, exceeding 40 Mt, grading about 0.2% Sn (60% recovery) or one, such as Hemerdon, that contains recoverable cassiterite and wolframite that might yield about the same profit. Such large deposits are attractive exploration targets, and the recent success at East Kemptville suggests that there may be similar deposits elsewhere awaiting discovery.

## 6. Stanniferous Veins Other Than Those of Group 5

Stanniferous veins other than greisen-related ones are known, and when they form a closely sheeted vein system or a stockwork, they collectively constitute a deposit that might be worth exploiting in either a selective or non-selective manner. The stanniferous quartz-tourmaline veins, which may halo tourmaline-breccias fall into this category in Cornwall. So also do simple quartz/cassiterite veins that occur in sandstone which have been exploited in several localities in Belitung, for example at Teba. As might be expected, the only wallrock alteration of this inert host is a slight colour change near the veins.

Veins of this type often occur close to or in contact with major lodes. Thus, in the underground mine of the Menglembu Lode Mining Company Limited, in the Kledang granitoid range (Malaysia), two types of ore occurrence have been recognized, the first consisted of parallel quartz/cassiterite veins with which were associated cassiterite-rich wallrocks, while the second was an earlier closely sheeted system, about 4 m wide, composed of exceedingly thin veins. According to Ingham and Bradford (1960, p. 123) this latter type "was known locally as 'streaky bacon' ore and carried cassiterite, brown biotite, tourmaline, pyrite, magnetite, chlorite and muscovite. The veinlets cut straight through crystals of feldspar and biotite, without apparently displacing the portions of the penetrated crystals and without causing any recognizable alteration of them. There was no sign of a zone of mineralized granite on either side of each veinlet".

At Wheal Coates, St. Agnes, Cornwall, the Towanrath Lode, which is from c. 0.6 to 3.7 m in width, consists of quartz with locally rich pockets of cassiterite, and is hosted by non-calcareous metasediments. In its hanging wall, there is a very rich stockwork of veinlets filled with white clay and cassiterite (Jones, 1925, p. 72).

Numerous other vein systems have been exploited for their tin content. For instance, I have cited a number of Malaysian examples (Hosking, 1973, pp. 371–373).

Finally, during the underground search for lodes, veinlets occurring in exploratory crosscuts or drill core may, by their mineralogical character, trace metal content, and their tendency to increase in concentration as a lode is approached, provide an important aid to exploration.

## 7. Lodes of the Cornish Type

Lodes in the type region, Cornwall, are spatially, and one supposes genetically related to a Carbo-Permian polyphase batholith whose original surfaces were characterised by a marked topography consisting of distinct ridges with undulating crest lines. Porphyry dykes are abundant and they and the tin-bearing lodes tend to strike approximately parallel to the ridges. The lodes and dykes provide some interesting spatial relations, shown in Fig. 12a.

These early lodes post-date the dykes, greisen-bordered vein swarms, tourmaline-breccias, and their related vein systems. Associated with them are major wrench faults, which strike at about right angles to the tin lodes. The wrench faults are sometimes barren, and sometimes contain a late mesothermal assemblage of galena, sphalerite, siderite, etc.. Whilst some of these faults post-date and disrupt the tin lodes, others clearly pre-date them. I think the lodes originated by the generation of fractures between pairs of wrench faults, and the mineralogical and textural complexity of the lodes is due, to no small degree, to them being repeatedly reopened as a result of continuing activity along the wrench faults. However, the lodes which are generally steeply dipping, have also been subject to normal and occasionally other forms of faulting. As a result of repeated movement, the texture of the lodes is often complex, but generally each displays locally either a banded and/or brecciated texture (Fig. 12b). The wallrocks may be ornamented by vein systems which may be so dense and mineralised that they can be mined with the lode proper.

Individual lodes may have strike lengths of several kilometres (for example, the Great Flat Lode is about 5 km long), and they may be mineralised for dip lengths of 1,000 m or more (Dolcoath Main Lode, for example, was worked for copper and tin to a vertical depth of the order of 1,000 m). Some of the lodes are 15 m wide, and in the presently operating Geevor Mine the average width is from 0.45 to 0.6 m.

The mineralogical development of the lodes has resulted from the ascent into the open faults of a succession of mineralising agents whose character varied with time. Where within a fault a given mineral was deposited doubtless depended on a number

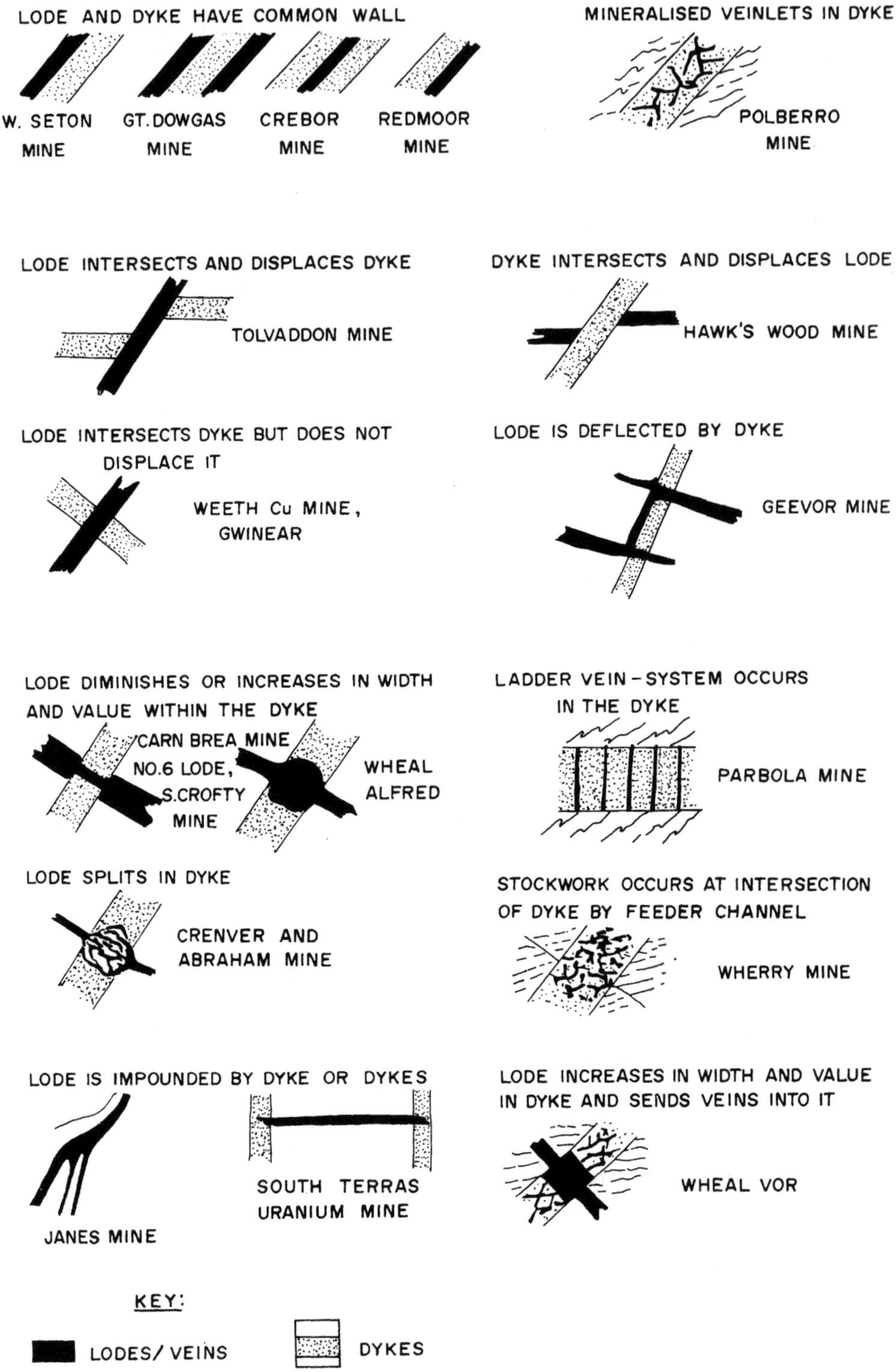


**Fig. 12a.** Diagrams illustrating Cornish relationships between lodes/veins and elvans

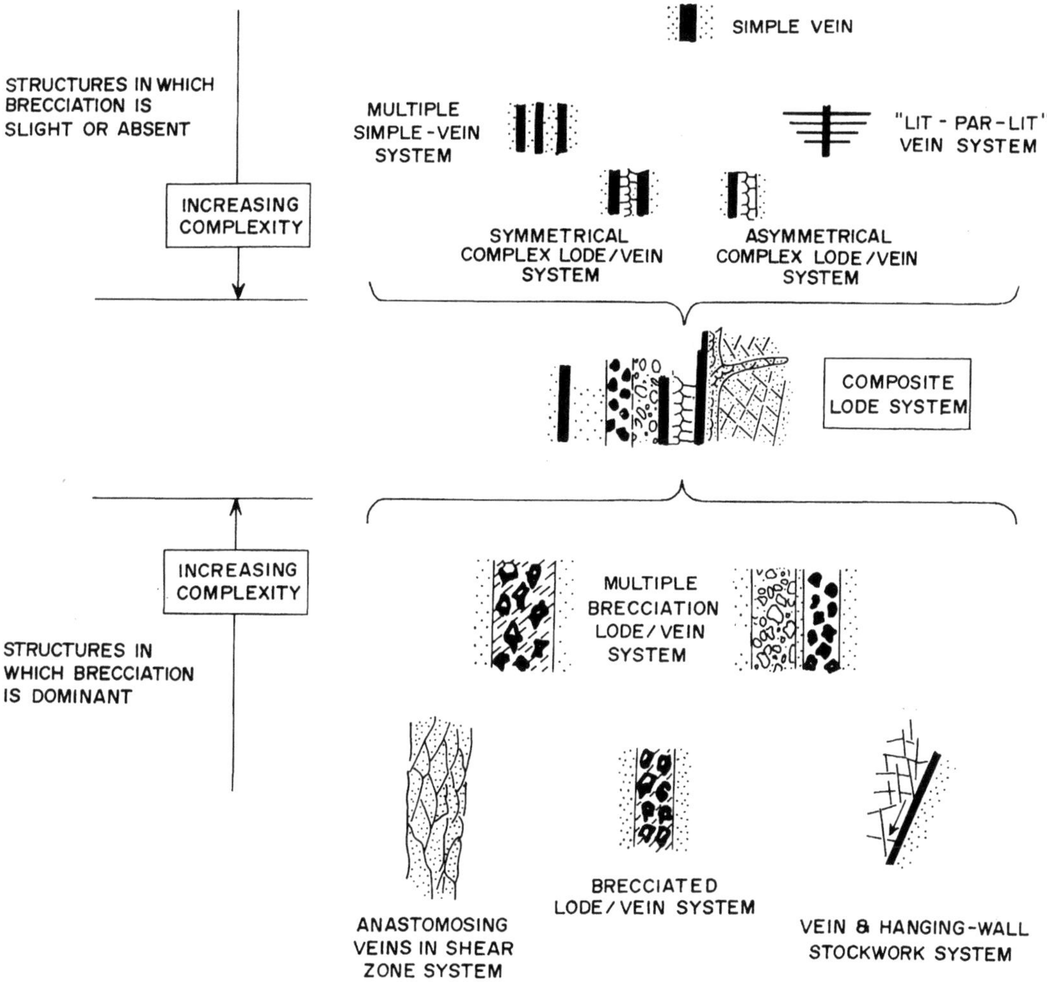


**Fig. 12b.** Some fundamental lode/vein structures (after Hosking, 1969)

of factors, but temperature was probably the most important. Cassiterite was deposited in the deeper horizons and this was followed by chalcopyrite and iron-rich sphalerite in a higher zone which considerably overlapped the tin zone. Later, in a still higher zone, which overlapped the copper one, galena and sphalerite were deposited. Each zone tended to spread over a greater strike length than the one beneath it (Fig. 13a). Furthermore, the zones in lodes which crossed the granite/country rock contact were usually much flatter than the neighbouring contact. The Cornish primary zoning pattern is only one of a number of types which have been reported. Some of these I have described elsewhere (Hosking, 1979, pp. 34–41). Due particularly to Pliocene marine planation, many of the Cornish lodes have lost some of their higher primary zones, and the zonal pattern has been modified in a way that deceived the miners of the last century by the development of secondary zones caused by deep oxi-

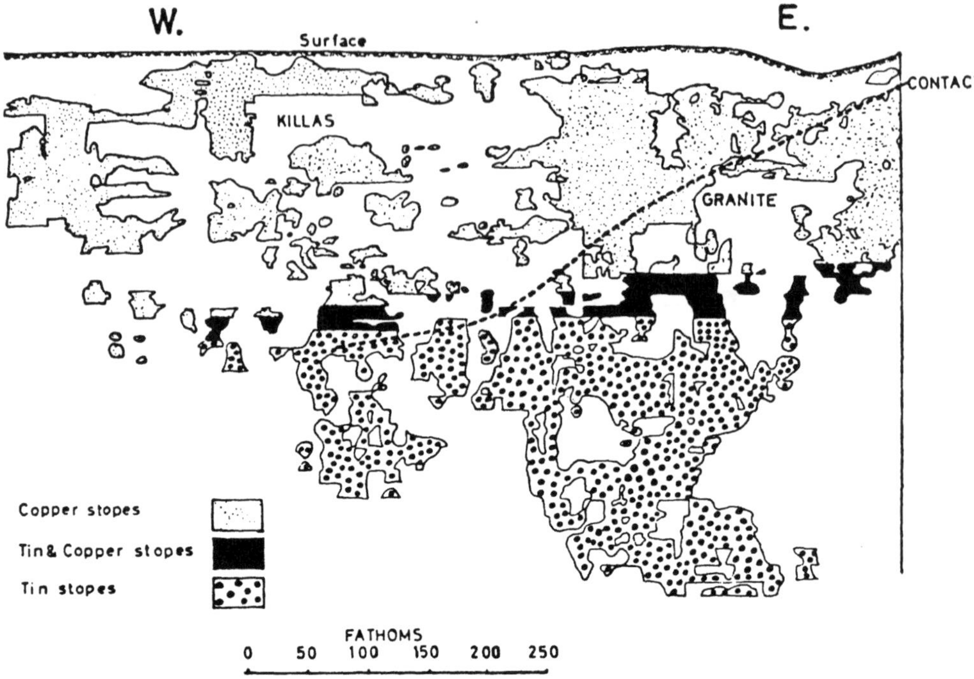


**Fig. 13a.** Longitudinal section of Dolcoath Mine Lode, Cornwall, England

dation which commenced during the Mesozoic when arid conditions prevailed (Fig. 13b). In some instances tin is found solely in lodes in the country rock (essentially noncalcareous metasediments and intercalcated metabasites), while in others it is confined to lodes in the granite. In still other lodes, tin may occur both within granite and within country rock.

It is also relevant to note that certain lodes show distinctly telescoped characteristics. The presence of woodtin in some lodes and the nature of the tourmaline-breccias, collectively serve to suggest that the mineralisation was of a fairly high-level type.

Of course many minerals other than those noted above were deposited in the tin lodes and Figure 14 provides a general paragenesis of the Cornish deposits. This table is a compromise, and many more species, albeit often in small amount, occur in the lodes. Electron probe investigations have recently located species in them which had hitherto been overlooked. The table also provides something about the nature of the wallrock alteration associated with the lodes. Within the lodes, concentrations usually occur in well defined shoots. However, the pattern of distribution of tin in the shoots is complex and is not really understood. A number of factors apart from temperature and chemical control the distribution, size, and shape of the shoots. These include changes in dip and strike of the lode, and impounding features such as faults, dykes and granite/metasediment contacts. Many of the shoots exploited during the last hundred years provided ore grading c. 1% Sn. Of course, both considerably richer and considerably poorer ones have been exploited.

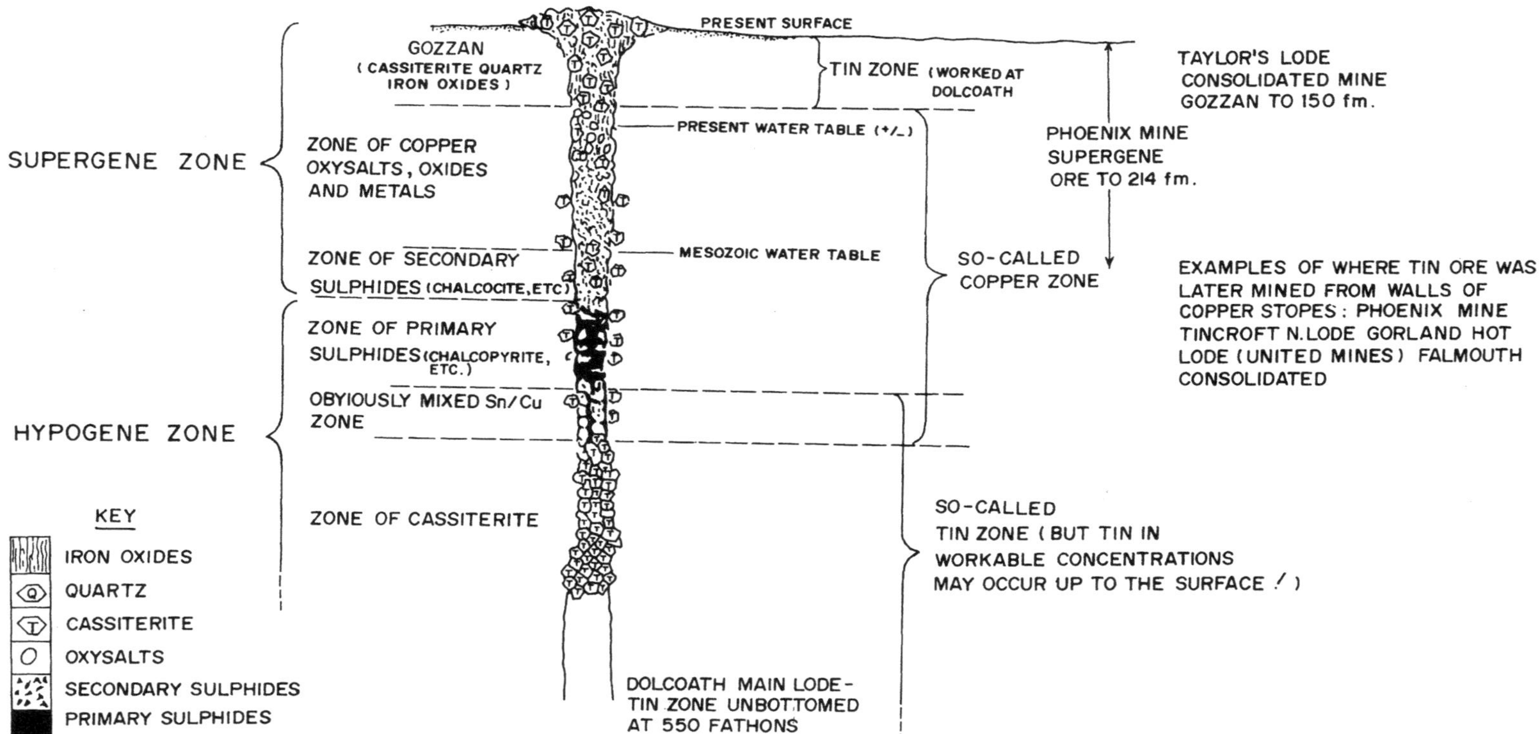


**Fig. 13b.** Primary and secondary zoning as found in some of the lodes of the Southwest of England (Hosking, 1981)

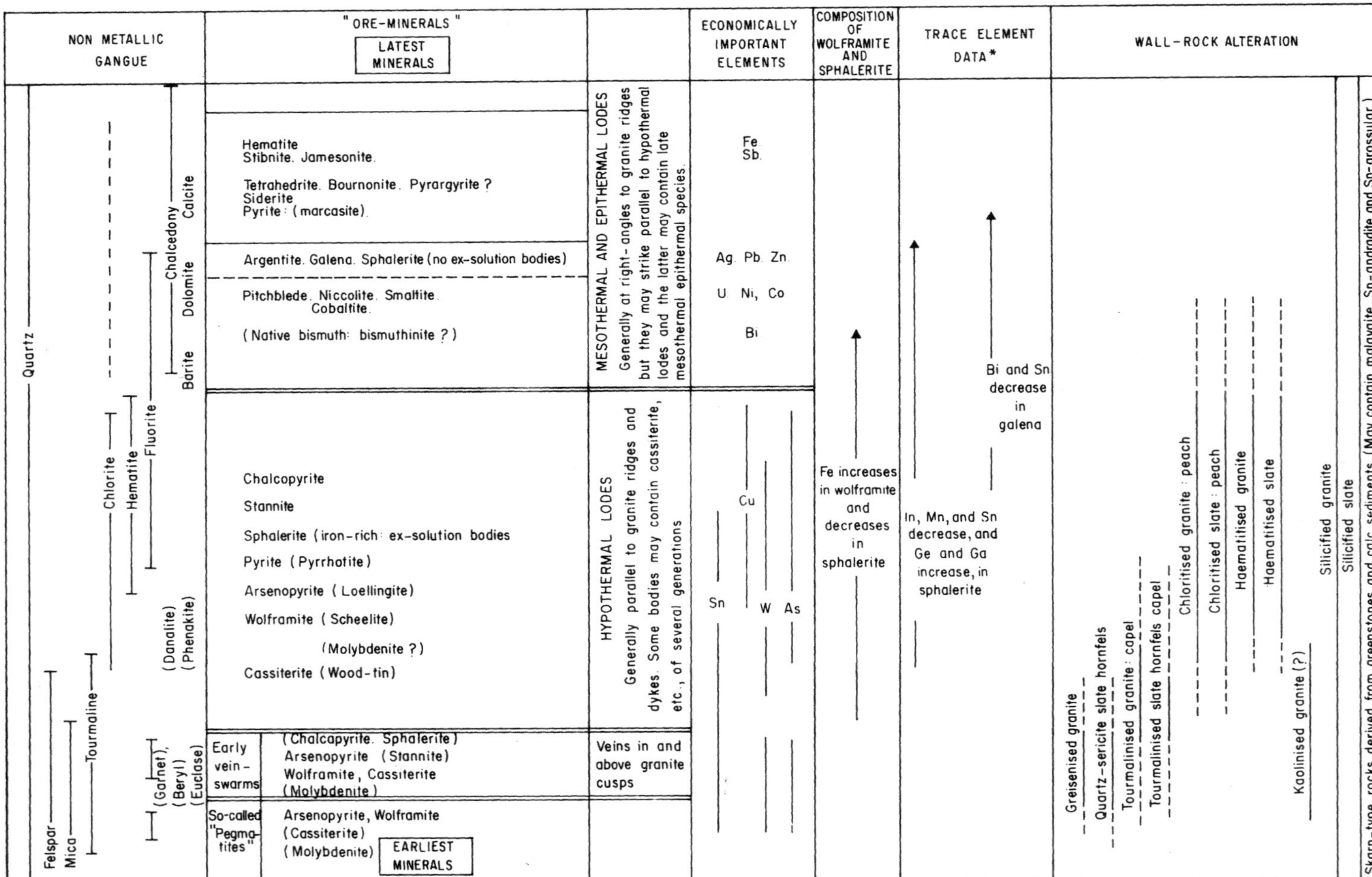


*See El Shazly, E. C., Webb, J. S., and Williams, D. Trace Elements in Sphalerite, Galena from the British Isles, (Trans. Instn. Min. Metall., 66, pp. 241–271)

**Fig. 14.** Generalized mineral paragenesis of the mineral deposits of Southwest England

Lodes similar to the Cornish ones are known elsewhere, for example at P.C.C.L. mine in Pahang, Malaysia. There, the lodes, which are in metasediments overlying granite, are only zoned in that the upper horizons are somewhat richer in copper.

## 8. Replacement (Metasomatic) Deposits That Cannot Be Satisfactorily Placed in Any of the Other Groups

In that part of my classification that is concerned with primary deposits I have endeavoured to keep genetic considerations to a minimum. However, I have placed in this group those skarn-free deposits that I consider to have been formed by replacements effected by hydrothermal mineralising agents derived from granites. They are epigenetic with respect to their host rock. Stanniferous skarns have also been formed in this way but I have placed them in an earlier group although the division between some of them and some of the metasomatic deposits noted below is somewhat diffuse. Others, especially Hutchinson (1981, pp. 81–97), hold that a number of deposits I include in this group are "synsedimentary and were formed by chemical precipitation from exhaled metalliferous hydrothermal brines near fumarolic vents on ancient sea floors". I think that particularly in Canada stanniferous bodies are to be found that may well have been generated in the way Hutchinson suggests, and these I have placed in Group 11 of this classification.

Deposits falling into this present group may be divided into two sub-groups, namely those of small or moderate size and those of large size. Examples falling into each sub-group are briefly mentioned.

### Small or Moderate Size Replacement Bodies

Carbonas is a Cornish word, which has been accepted for irregularly shaped bodies of stanniferous, highly-altered granite. A number of them have been recorded from mines in the Lands End peninsula of Cornwall. They are connected to normal lodes by one or more feeder channels. Figure 15 provides an example of a carbona. It lacked clearly defined walls, consisted of tourmaline, cassiterite, fluorite, pyrite and chalcopyrite, and occurred at the junction of the north-south striking barren fault and east-west veins. I think that the orebody developed as a result of ore-forming agents being impounded beneath a granite that was an earlier phase than the host. At Sungai Besi (Malaysia) a number of stanniferous carbonas developed in the granite just below a marble cover, and one was an extremely rich deposit although its extent down dip was only about 30 m.

Small replacements which have been exploited occur in the metamorphic rocks near to the granites in Cornwall. Thus, at Grills Bunny Mine, Botallack, a flat-dipping, metabasalt-pelite horizon locally contains four exposed flat, lenticular bodies, varying in thickness from 1 to 4 m. Each one "consists of an upper 0.3–1 m thick bed of tourmaline-quartz-cassiterite (thin sections of which indicated its origin by replacement) and a lower less than 1 m thick zone of chloritised amphibolite containing disseminated cassiterite (Jackson, 1979, p. 222).

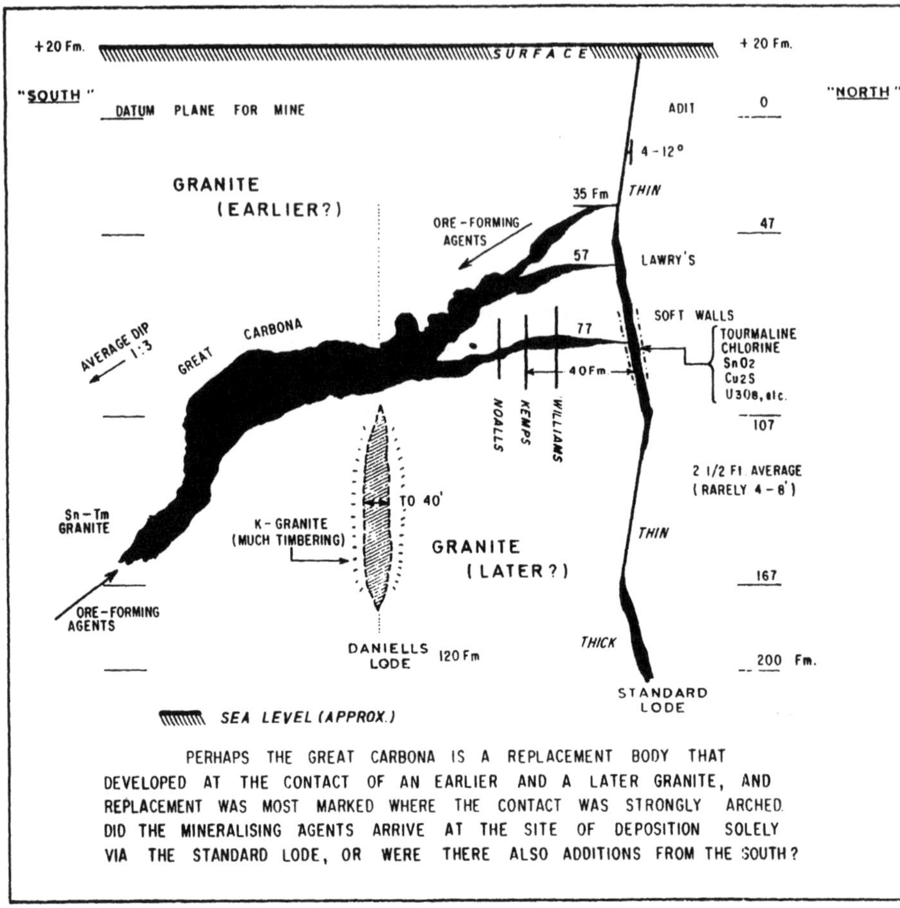


**Fig. 15.** The Great Carbona of St. Ives Consols Mine

In mid-Cornwall, associated with a plexus of quartz-tourmaline-cassiterite veins in the Parka Mines, are lenticular replacements of cassiterite and quartz which are 2—3 m thick x 15 m x 20 m. These made the small mine a viable proposition (Collins, 1912).

In the Rooiberg field (South Africa) tin mineralisation occurs as "fracture-ore-bodies", and replacement ore-bodies. The latter, occurring in quartzite, can be subdivided into pocket ore-bodies and bedding ore-bodies. Leube and Stumpfl (1963, pp. 398—401) note that the pockets occur either on the steep fractures or in the quartzites up to 6 m on either side. They are connected to the steep fractures (the feeder channels) by very thin, usually unmineralised fissures. They usually vary in diameter "from a few inches to about 20 feet", but on rare occasions they may reach 30 m. They are usually concentrated in groups. "In some places the pockets are lined up closely one behind the other forming pipe-shaped ore-bodies up to 200 feet in length and 15 feet in diameter". In addition, "impregnation and replacement bodies occur parallel to the stratification forming sheetlike ore-bodies of varying but generally limited extension of not more than 50 feet and a width of up to 3 feet".

Pockets in all stages of development are encountered. The most primitive consist solely of a ring of tourmaline and all the pockets display an annular structure. The well-developed pockets are often surrounded "by an outer ring of red quartzite followed by an inner ring of white, highly silicified quartzite. The centre of the pockets is commonly formed of tourmaline, sulphides (mainly pyrite) and, in places, also feldspar. The centre is surrounded by rings composed of cassiterite or carbonates". (Leube and Stumpfl, 1963, p. 400) (Fig. 16).

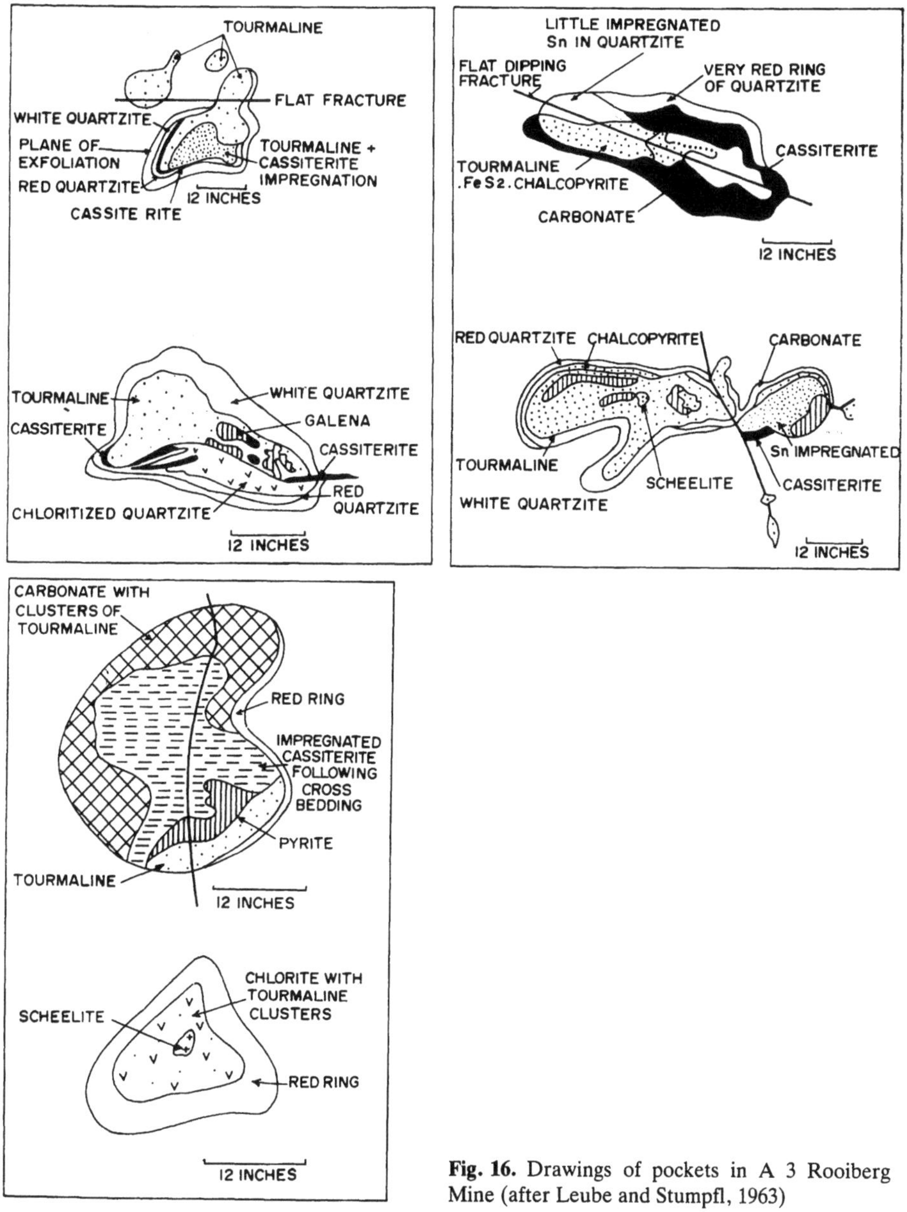


**Fig. 16.** Drawings of pockets in A 3 Rooiberg Mine (after Leube and Stumpfl, 1963)

## Large Replacement Bodies

Essentially there are two types of large replacement bodies, namely pipes and stratiform/stratabound bodies.

Amongst the known stanniferous pipes, only some fall, beyond doubt, into the classification group under discussion. Some, such as the much branched and zoned pipes of the Potgeitersrus field (South Africa) can only be placed here if they are regarded as epigenic. Others such as the breccia pipes of Bolivia and the Beatrice skarn pipe of Malaysia also fall into other groups. In Malaysia, however, some twenty or thirty other stanniferous pipes have been examined on the west side of the Kinta Valley and only the Beatrice Pipe qualifies on mineralogical grounds to be classified as a skarn deposit. The remainder contain cassiterite with a variety of sulphides. In all of them calcite and dolomite form the chief non-metallic gangue (Willbourn, 1931/32). I think that the pipes are replacement bodies which probably developed in the vicinity of intersecting feeder channels that possibly overlie high-spots of the underlying granitoid. They, unlike the Beatrice Pipe, have proved only of slight economic importance.

In eastern Australia there are many well-documented pipe-shaped tin deposits (Taylor, 1979, p. 166) and some of these may be rightfully placed in the present classification group. However, the reconstruction of the development of a given pipe is often difficult so that it is not always easy to decide how to classify it. Taylor remarks that some so-called pipes are not, in fact, pipes at all, the ore deposit having "acquired a pipe-like shape due to mining restrictions, grade restrictions or post-mineralisation faulting". He cites the so-called pipes of the Wild Cherry Mine in the Albury-Ardlethan province as examples of pseudo-pipes (Taylor, 1979, p. 167).

From an economic point of view by far the most important ore-bodies falling into this group are the stratabound/stratiform deposits occurring in Tasmania, and in par-

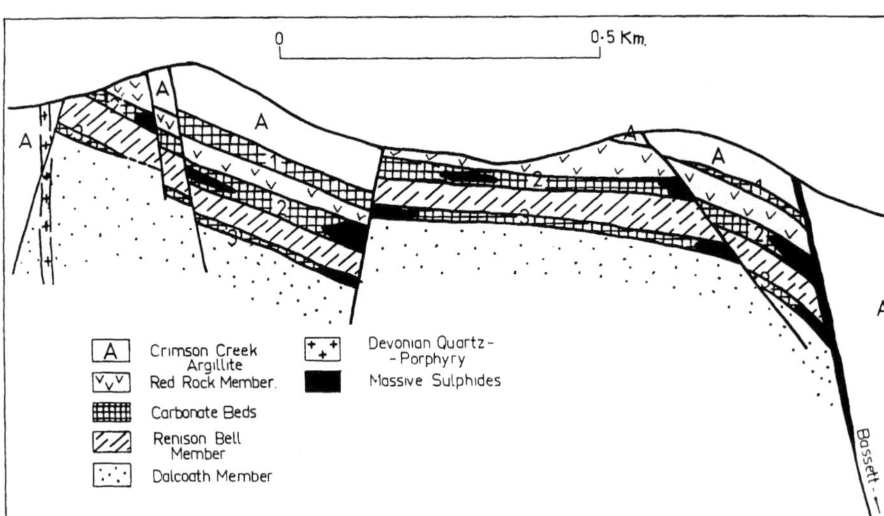


**Fig. 17.** Cross section of the Renison Mine area, Tasmania (after Newnham, 1973)

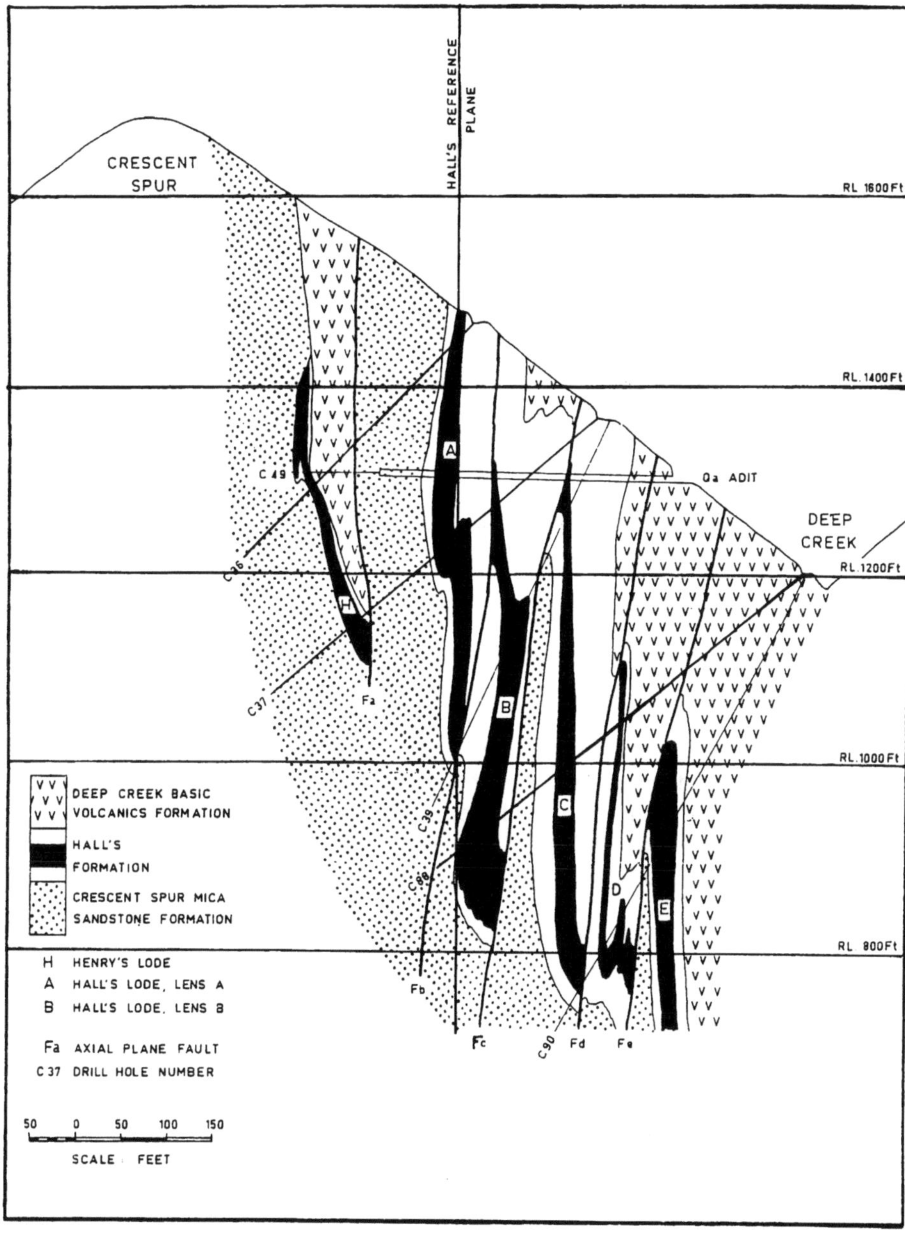


**Fig. 18.** Cleveland Mine, Tasmania, Cross section "Qa", looking North-east (after Cox and Glasson, 1967)

ticular those of Renison Bell, Cleveland and Mt. Bischoff. Figure 17, 18, and 19, provide broad details of the geology of the three mines. In each case ascending mineralising agents have probably migrated from buried granites. At Mount Bischoff these agents caused marked greisenisation of the porphyries "with the formation of topaz, tourmaline, muscovite and cassiterite pseudomorphing primary feldspar". They also

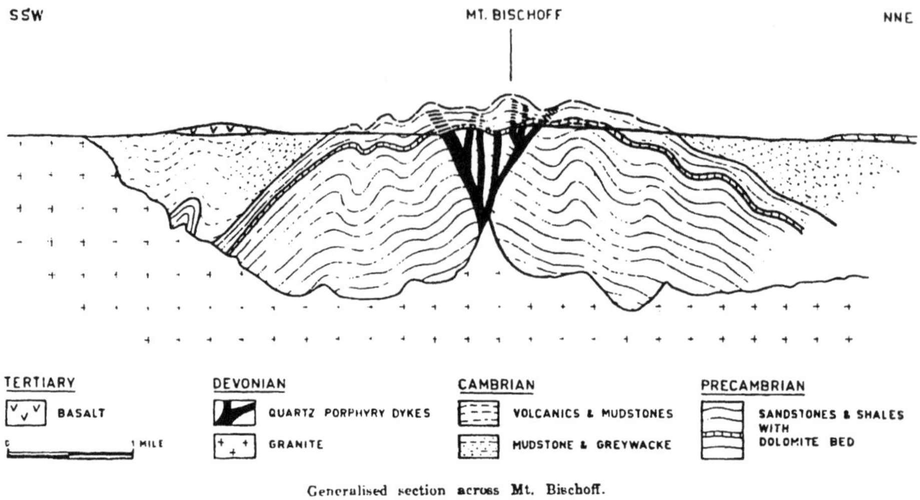


**Fig. 19.** Generalized section across Mt. Bischoff (after Groves and Solomon, 1964)

caused "selective replacement of the dolomite bed – – – by iron sulphides, talc, quartz, carbonate and cassiterite, and the filling of tension fractures throughout the area" (Groves and Solomon, 1964).

At Cleveland the Lode Bed was originally a finely layered calcareous shale which was later folded and faulted. "Cassiterite, wolframite, fluorite, tourmaline, sulphides, chlorite and further dolomite and quartz were introduced into the Lode Bed which acted as a permeable channel-way and chemically reactive host for mineralising solutions" (Cox and Glasson, 1967, p. 11). In 1966, before production commenced at Cleveland, indicated and inferred ore reserves totalling approximately 3 Mt of average grade 1.02% Sn and 0.43% Cu were established.

At Renison Bell both fault-confined and stratabound ore bodies occur. Of the faults, only the Basset-Federal structure contains economic mineralisation. It is a lode of the Cornish type containing, in addition to sulphides and a variety of non-metallic gangue minerals, fine-grained cassiterite which occurs both in the latter and the sulphides.

The stratabound deposits have developed locally within three major carbonate beds in the succession. In these three beds "the primary carbonate is a manganiferous dolomite or pistomesite, usually well bedded and becoming coarsely crystalline and sideritic adjacent to the sulphide ore bodies" (Newnham, 1973, p. 2). The "ore bodies invariably lie adjacent to mineralized fault zones which are regarded as the feeder channels". "Sulphides typically form 60–75% of the rock and consist of pyrrhotite with minor inclusions of pyrite, chalcopyrite and arsenopyrite. Gangue is composed of secondary quartz, carbonate, fluorite, tourmaline, talc and chlorite". "Cassiterite is included in both sulphides and gangue minerals, and has an average 50–100 microns grain size. Typical No. 2 and No. 3 horizon ores average 1.3 and 1.8 per cent tin respectively" (Newnham, 1973, p. 5).

All these three Tasmanian deposits have supported or are supporting major tin-mining operations. Unfortunately stanniferous deposits of the type described above and of comparable tonnages and grade do not appear to have been discovered outside of Tasmania, except at Dachang, China.

## 9. Telescoped, Mineralogically Complex Deposits (Largely Xenothermal or Sub-Volcanic)

Of the sub-volcanic deposits falling into this group the best documented are those of Central and South Bolivia (see, for example, Kelly and Turneaure [1970] and Sillitoe et al. [1975]) where they are of major economic importance. They occur as lode systems in and around high-level late Tertiary stocks of intermediate composition (Fig. 8). Locally they are associated with somewhat earlier porphyry tin deposits, but seem to have developed by mineralising agents which had a deeper source than those responsible for the porphyry tin deposits.

The lodes are often considerably branched and often display quite a complex pattern both in plan and section. The lodes usually have a complex character with banding, brecciation, colloform, and replacement textures locally much in evidence. Their mineralogy is complex characterised by the not uncommon presence of wood-tin, tin and silver sulphosalts (stannite, teallite, cylindrite, ruby silver, etc.). This complexity is due to telescoping as a result of deposition near the surface. An indication of the mineralogical complexity is provided by Table 2 in which the sequence of vein minerals in certain xenothermal deposits of Central Bolivia is presented.

In view of the conditions under which these deposits were formed, it might be reasonable to expect that if zoning occurs at all it would be the lateral type that was domi-

**Table 2.** Sequence of vein minerals in the xenothermal deposits of central Bolivia (After Turneaure, 1960, 577)

| Stages | 1 | | 2 | 3 | 4 | 5 |
|---|---|---|---|---|---|---|
| Llallagua | q, (tm). | bm, cs, (wf), (ap). | frk, po, asp. | sl, (st), (cp). | mc, (asp), py, sd. | sd, sl py, wl (?) |
| Huanuni | q, tm, (ap), (fl). | cs. | (tl), po, asp. | sl, st, (cp). | py, mc, sd. | sd. |
| Morococala | q, tm, (tz). | cs. | tl, pa, (asp). | sl, (st). | py, (mc), sd. | sd, sl, (py). |
| Potosi | q, py, cs, (asp), (wf), (bm). | | st, (cp). | td, (sl), (mt), ad. | fo, rs, (gl). | aln. |
| Oruro | q, py, cs, (asp). | | (sl), (cp), st. | td, ad. | fo, gl, frk. | kn, dk, aln, (q), (mc). |

| | | |
|---|---|---|
| ad – andorite | frk – franckeite | sd – siderite |
| aln – alunite | gl – galena | sl – sphalerite |
| ap – apatite | kn – kaolinite | st – stannite |
| asp – arsenopyrite | mc – marcasite | td – tetrahedrite |
| bm – bismuthinite | mt – matildite | tl – teallite |
| cp – chalcopyrite | po – pyrrhotite | tm – tourmaline |
| cs – cassiterite | py – parite | tz – topaz |
| dk – dickite | q – quartz | wf – wolframite |
| fl – fluorite | rs – ruby silber | wv – wavellite |
| fo – lead sulphosalts | | |

nant. However, Turneaure (1960, p. 583) makes the following pertinent remarks: "Vertical zoning, marked by a consistent change in mineralogy with depth has not been clearly demonstrated in the (xenothermal) deposits of Central Bolivia, but some semblance of vertical zoning is noted in the tin-silver deposits of Potosi and Oruro. Lateral zoning, of the regional type, so well illustrated by the ore deposits of Northern Bolivia – – – is not evident in the mineralised belt of Central Bolivia, but lateral zoning on a local scale is suggested by mineral distribution at Llallagua and Huanuni". The vertical zoning to which he refers is indicated at Potosi, for example, by the occurrence of silver above tin and of cassiterite above stannite. Concerning lateral zoning Turneaure states (1960, p. 588) that at Llallagua the tin-bearing zone "is partly surrounded by one in which several low-grade sulphide veins have been explored", and at Huanuni, also, a tin zone is surrounded by one in which sulphides are dominant. Finally in order to emphasise the economic importance of these xenothermal Bolivian lodes it can be said that their ore shoots are of about the same size as those of the Cornish lodes. Turneaure (1960, pp. 583–584) notes that in Central Bolivia numerous ore shoots in major lodes have stope lengths of 200 to 300 m, and that the San Fermi shoot, with a maximum stope length of 700 m, was undoubtedly one of the greatest tin shoots ever to be discovered.

Somewhat similar deposits are known in volcanic environments elsewhere, and those in the Ashio mine provide a good example (Nakamuara, 1970, pp. 231–246).

Locally within the porphyry multi-metal deposit of Mount Pleasant, New Brunswick, stanniferous ore occurs which shows some lateral zoning, and consists essentially of cassiterite, stannite, and a considerable number of other sulphosalts and sulphides, and so is distinctly xenothermal in character (Hosking, 1963).

In Malaysia a number of telescoped deposits are known. One of the most interesting is the Manson Lode in Kelantan. It is associated with limestone, indurated shale, phyllite, meta-acid volcanic rock, tuff and quartz porphyry dykes, all of which *may* be of Permian age. The nearest outcropping granite is about 9 km away. The area is patterned by faults in a complex way and it is thought that the lode itself is a mineralised shear zone that has been repeatedly reopened. The lode contains, in addition to cassiterite, stannite and a wide variety of other sulphides which include cinnabar (Hosking et al., 1970 and Hosking, 1973). It is not surprising to find this xenothermal body in a geologic environment in which volcanics and porphyry dykes are much in evidence and it may be that the ore deposit and the igneous units are of about the same age.

In the western tin belt of Malaysia a number of complex and distinctly telescoped stanniferous veins occur, that are not in volcanic settings (Hosking, 1973, pp. 374–379). Doubtless they developed at a fairly high level, and I think they must be genetically related to a very late-phase of the wide-spread Triassic granite or even to a Cretaceous granitoid which does not happen to outcrop (see Hosking, 1974, pp. 38–39 and Fig. 9).

Telescoped stanniferous bodies are not uncommon throughout the world and further testify to the probability that most primary tin deposits have developed at rather modest depths; the large Precambrian pegmatites are possibly the only major exception.

## 10. Deposits of the Mexican Type (Epithermal or Fumarole)

Deposits of this group occur in Mexico, Bolivia, Argentinia, Nevada and New Mexico (U.S.A.), and Eastern U.S.S.R. (Maly Khingan). Probably the wood-tin in the Alaskan gold placers were derived from the same type of deposit (Mulligan, 1975, p. 48).

Although some of these deposits have been worked by opencast and underground methods in a small way they are of only slight economic importance.

The deposits of Durango, Mexico, indicate the major characteristics of the group. They occur in a belt of rhyolites and in the same belt silver deposits are found, but at a lower level than the tin ones, and since the rhyolites are underlain by basalts, andesites and dacites, it is in these that the silver deposits are generally situated.

According to Ypma and Simons (1969), the tin deposits are apparently confined to a rhyolite which possesses steeply-dipping flow banding and which seems to be an intrusive stock or dyke rather than a true extrusive. This tin host rock possesses a chilled margin consisting of a very dense breccia (Fig. 20). In it the mineralisation is confined to vein, disseminated and breccia deposits.
Individual veins are small and any one vein would be regarded as significant were it, say, 0.2 m wide, 30 m long and 15 m deep. The veins are steep-dipping and generally occur near the top of the intrusive. The grade of ore in these bodies varies from 2—10% $SnO_2$.

In the vicinity of the veins both syngenetic and epigenetic disseminations of cassiterite are found, the former being associated with fayalite and zircon and locally reaching concentrations of 0.1—0.15% Sn.

The breccia type of mineralisation may be quite extensive and stopes up to 25 m wide, 8—10 m high and 100 m long have been excavated in it. Ore grade is about 0.4% Sn and apparently "the tin values are confined to a fine stockwork of veinlets".

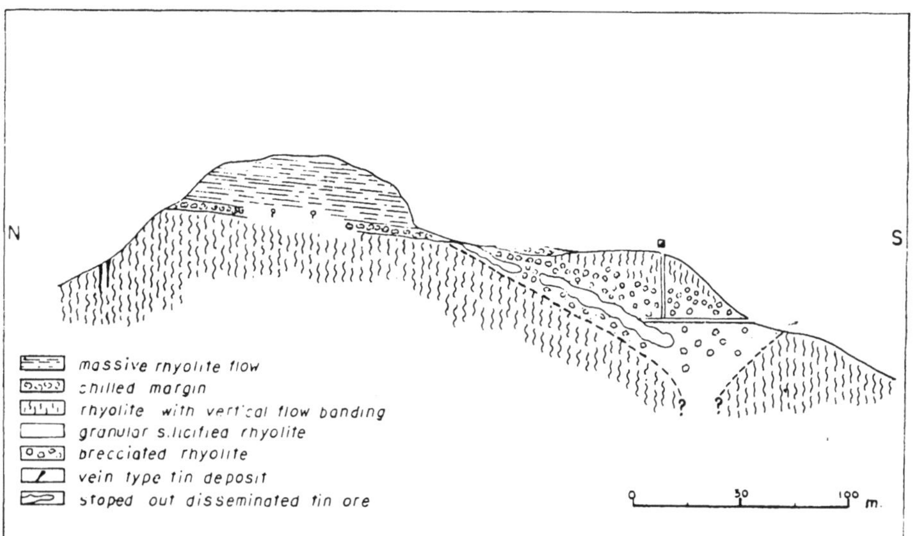


**Fig. 20.** N-S section through the Esperanza breccia type deposit (southern half of the figure) and a regular vein deposit (at northern end of section). Note the massive rhyolite cap, chilled margin, and possibly funnel shaped outline of the breccia pipe. Durango, Mexico (after Ypma and Simons, 1969)

These deposits appear to have developed just beneath a roof of rhyolite with horizontal flow banding.

Both the vein and breccia ore contain specularite and $SnO_2$ which occurs as 'normal' cassiterite and wood-tin. In addition to these major components, native bismuth, mimetite, and antimony (up to 1.0%, and perhaps as iron-antimony-oxide) are of common occurrence. Non-metallic associates are tridymite, quartz, chalcedony, opal, sanidine, white mica, zeolites, and fluorite, and these minerals are also present in the rhyolite vesicles.

The cassiterite in the breccias is much finer than in the veins and difficult to recover unless flotation is employed. However large tonnages of breccia ore may be present in Mexico and elsewhere.

## 11. Stanniferous Massive Sulphide and Massive Iron Oxide Deposits

This group has been introduced primarily to accomodate those large stanniferous deposits that are stratabound and consist largely of sulphides with interbands of detrital grains of gangue minerals. Although they are not confined to Canada, a number of excellent, much investigated examples occur there. In some of these Canadian deposits the tin concentration is only of nuisance value, but at Sullivan and Kidd Creek, where the deposits are mined primarily for their base metal content, tin is recovered as an important by-product. At present they are Canada's only demestic sources of tin, and together they produce about 360 t of metal annually.

As Mulligan (1975, p. 50) notes "many (of the deposits of this group) are underlain by or grade stratigraphically downwards into zones of lower grade pyritic or pyrrhotite-bearing 'disseminated' or 'vein' ore that is obviously not 'stratabound'". They are often considerably deformed and metamorphosed.

Mulligan (1975, p. 50) points out that this group may be sub-divided into volcanic-stratabound and sedimentary-stratabound. Deposits of the volcanic sub-group are generally "associated with rhyolite or other alkali-siliceous rocks interbedded with more basic volcanic rocks, and thought to mark the end-stage of basic-intermediate-acid volcanic cycles". In Canada many of these deposits, such as those of "Kidd Creek and South Bay are near margins of sedimentary depositional basins, and the sedimentary-stratabound deposits like Manitouwadge (but not Sullivan) are commonly in sedimentary depositional basins surrounded by volcanic rocks" (Mulligan, 1975, p. 50). He also notes that in the Canadian sedimentary-stratabound assemblages iron-formation is often represented, and is much in evidence at Manitouwadge. He also notes that some sulphide deposits grade laterally into iron-formation, and he suggests that the iron oxide deposits may be the shallower water equivalents of the massive sulphides. The iron-formation deposits, in addition to the massive sulphide ones, may be stanniferous. That of Grandesburg is a case in point.

I think that these massive Canadian deposits and others in a closely similar geological setting, are genetically different from those stratabound stanniferous deposits which have been described earlier and which I believe to be replacement deposits. I agree with Hutchinson (1981, p. 87) that there is compelling evidence for believing that the Canadian-type deposits are synsedimentary and that they were formed in much the same way as the Sullivan deposit was generated. Hutchinson states that the deposit in question "was formed by sea floor exhalations of metalliferous hydrothermal brine from a fumarolic vent, now marked by the discordant footwall breccia zone, followed by the precipitation of the stratiform massive sulphides on the surrounding sea floor. – – – –The presence of cassiterite as a minor, but locally enriched mineral in the Sullivan ores is compelling evidence that tin, like iron, zinc, lead and silver was mobilised from the footwall rocks by this fluid, transported upwards towards the seafloor, and precipitated, by cooling, Eh and pH changes, both in the feeder fractures just beneath the seafloor and on the seafloor itself above the vent".

In the Eastern Tin Belt of Southeast Asia (Hosking, 1977) particularly in Trengganu and Belitung, there are a number of concordant and distinctly stanniferous iron oxide deposits, and some of these pass laterally, or in depth, into bodies which are essentially sulphidic. A good example is the Selumar deposit of Belitung which has been locally exploited because of its tin content. While the mineralogical character of some of these deposits, together with their geometry and geological setting, provide some reason for classifying some as skarn deposits and others as metasomatic replacements, they may have been generated in much the same way as the Canadian deposits noted above. However, it is premature to make dogmatic statements concerning the genesis of the Asian deposits because they have not been adequately investigated. I can still conclude, as I did some years ago (Hosking, 1979, p. 30) "that the geneses of these iron/tin deposits of Southeast Asia are still matters of uncertainty, and it is at least probable that from the point of view of genesis, a number of different types exist. It is uncertain at this stage if any of them are strictly comparable to the Canadian deposits noted above".

## 12. 'Ancient', Variously Modified, Stanniferous, Sedimentary Deposits

I cannot do better than to record much of what I wrote about this group a few years ago (Hosking, 1979, pp. 30–31) and to add something of relevance that has come to light lately. "Because until comparatively recently, there was little evidence of 'ancient' stanniferous sedimentary deposits, of which ancient tin placers are members, such deposits were not usually provided for in classifications of tin deposits. I think it is likely that such deposits are considerably less rare than one used to believe to be the case, and that we have not recognised some of them because regional metamorphism and/or the heat and chemical components derived from neighbouring invading magma have so modified them by effecting remobilisation, recrystallisation, etc., that they have lost much of their original identity".

"The difficulty in recognising the true identity of such deposits is made particularly clear by considering the data relating to the 'schists stanniferes' of the Isergebirge district of Poland which have been provided by Mulligan (1975, pp. 51–52). According to Mulligan these stanniferous bodies occupy a stratigraphic zone several metres thick in a series of ancient crystalline chlorite schists intruded by Hercynian granites. The zone extends intermittently for some 14 kilometres. It consists of chlorite-garnet schists believed to have resulted from metamorphism of originally argillaceous and limy sedimentary rocks. Cassiterite, accompanied by sulphides and secondary iron-rich chlorite, which is thought to have developed from garnet and biotite by hydrothermal processes, occurs in quartz veins in schist".

"Because of the complex association, and the confinement of the zone to a specific stratigraphic horizon, it is concluded that the mineralisation could not have been introduced by normal hydrothermal processes. The deposit is believed to be a placer enriched in iron and other heavy elements".

The problem of establishing whether a given deposit is, or is not an ancient stanniferous placer, has recently been highlighted by Matthews (1982), who discusses a deposit in South Goias (Brazil). This deposit, which I have seen, is probably a single tin-bearing horizon that has been exploited on the northern and southern limits respectively of a westward plunging anticline, composed of closely-banded Upper Precambrian paragneisses. Matthews (1982, p. 467) observes that at one of the mines "Encruzilhada there is no marked hydrothermal alteration at surface. The cassiterite is fine-grained and disseminated – in a – 0.1—1 m thick – continuous schist band intercalated with lighter coloured more arenaceous muscovite-bearing schistose horizons". Matthews remarks that although the deposit has been considered to be a fossil placer it is, in his view, "not very logical to suppose that the cassiterite" which occurs as euhedral crystals "was deposited in what, before granitisation, must have been a thin argillaceous horizon in a well-bedded packet of arenaceous and argillaceous sediments". Matthews concludes that the tin deposits may be directly related to granitisation (which resulted in the generation of granite gneiss in the core of the dome) and which caused "remobilisation, transport and redistribution of tin – on a wide scale – and lithological and local structural features in the paragneisses determined the geometry of the existing deposits".

Matthews' view of the genesis of the deposit echoes similar views which were advanced long ago by the Swedes to account for some of their base metal deposits. If it is correct, then perhaps another group should be added to this classification to accomodate such deposits.

Matthew's work certainly supports my contention that it is probably more difficult to assign a given deposit to this group than to any of the others. Although it has been my aim to erect a classification that is as free as possible from genetical considerations, of necessity one would have to investigate the origin of a given deposit before placing it in this group.

## 13. 'Modern' Placers

'Modern' placers are by far the most important sources of tin.

Members of the group may be further classified according to whether they are residual, eluvial, colluvial, or alluvial. The latter may be sub-divided according to the nature of the site of accumulation (fluviatile, lacustrine, estuarine, marine). They may also be classified according to whether they are submarine or terrestrial. A further classification of the placers in general may be founded on mineralogical considerations.

For the purposes of this paper it must suffice to present a pictorial classification of the stanniferous placers (Fig. 21). Whilst it is out of the question to discuss the nature of the placers in any real detail in this paper, it is pertinent to make the following observations:

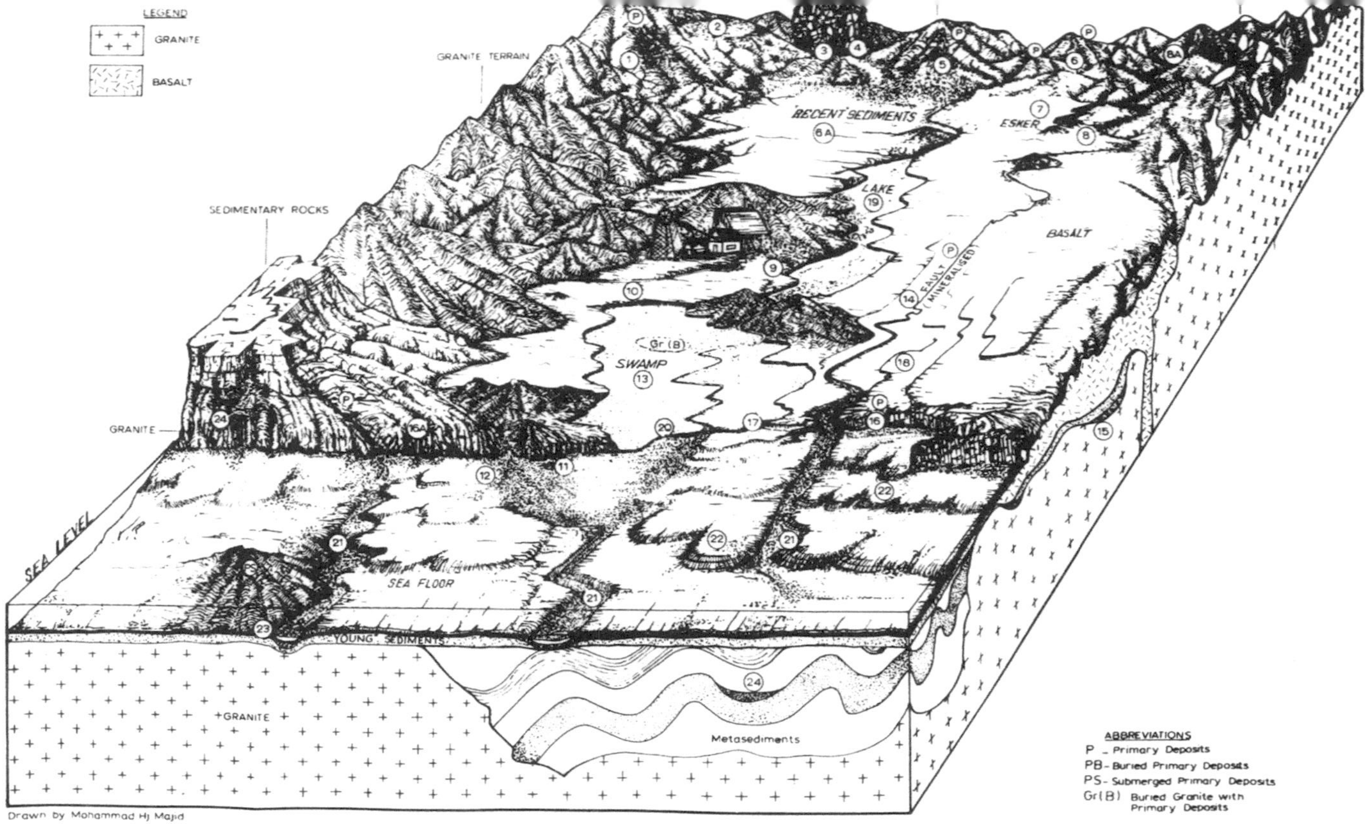


**Fig. 21.** Patterns of stanniferous placers (Hosking, 1973). *1* Eluvial (Kulit) deposits – Belitung; *2* High-level basin accumulations (eluvial/colluvial) – Goss Moor. Cornwall: North Portugal; *3* Accumulations (alluvial etc.) at granite/limestone contact and placers on buried karstic platforms – Malaysia; *4* Placers in limestone caves – Kaki Bukit, Malaysia; *5* Stanniferous outwash fans – Gambang area Malaysia; *6* Eluvial/colluvial placers – Nigeria; *6A* Alluvial deposits covered by 'recent' sediments – South East Asian Tin belt; *7* Stanniferous esker – Example (?); *8* Deep leads beneath glacial debris – New Brunswick (?); *8A* Sn-bearing moraine – Andes; *9* Mine tailings (man-made placer) – Cornwall; *10* Tailings – derived placer – Red River, Cornwall; *11* Tailings – derived beach placer – Gwithian, Cornwall; *12* Tailings – derived submarine placer – Gwithian, Cornwall; *13* Alluvial, etc., placers derived from primary sourse (PB) in buried granitic (or other) hill – Malaysia; *14* Placer along strike of tin-lode in vallcy floor – kaksa type deposit of Bangka; *15* 'Deep lead' beneath basalt – Nigeria; *16* Beach placers: cassiterite from primary deposits (P) in adjacent cliffs – Cligga, Cornwall; *16A* Beach placers: cassiterite transported by rivers from inland sources – Peninsula Burma; *17* Raised beach placer – Cornwall, Bangka/Belitung; *18* Terrace placers – Example (?); *19* Lacustrine placers – Loo Pool, Cornwall; *20* Estuarine placers – West coast Burma: Folmouth Estuary, Cornwall; *21* Submarine placers in drowned river systems – Bangka/Belitung; *22* Submarine placers on submerged beaches – S.E. Asia Tin Belt; *23* Submarine placers (eluvial/colluvial/alluvial) derived from submerged primary sources – Singkep/Bangka/Belitung; *24* Ancient placers (+/– metamorphosed) – Madagascar: Brazil

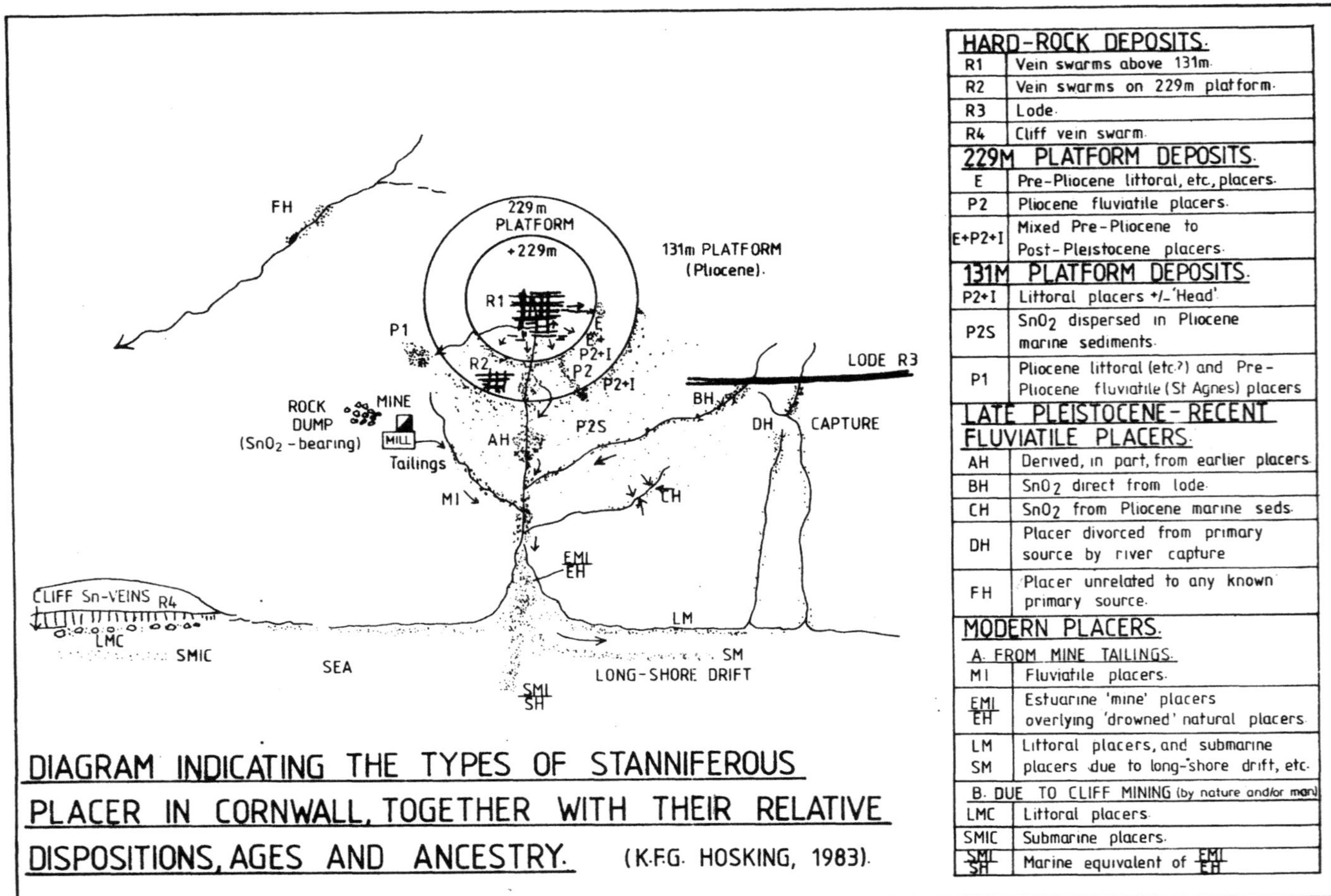


**Fig. 22.** Diagram indicating the types of stanniferous placer in Cornwall together with their relative dispositions, ages and ancestry (Hosking, 1983)

(1) Of the various types of placer the alluvial members (on-shore and submarine) are the most important sources of tin.
(2) Using conventional bucket dredges, submarine ore to about 50 m below sea level can be recovered.
(3) The grade of ore worth recovering by large dredges may be at least as low as 0.08 kg Sn $m^{-3}$. However, what constitutes an economic grade will depend on such considerations as the size of the cassiterite grains, the matrix in which they occur and the depth of overburden.
(4) In order to maximize the results of exploration, evaluation and exploitation, major efforts must be made to establish the genesis of any deposit. That means considering its development from the time when its cassiterite was first released from the primary parent. In many instances the developmental history of such a deposit is quite complex as Fig. 22 shows. Whenever possible detailed studies should be made in order to establish the distribution pattern of the cassiterite in the placer. I know of economically viable stanniferous horizons in certain placers that were ignored until this was done.
(5) Finally, all the mineral species in the placer must be established and an assessment must be made of the amount of each mineral of value (monazite, zircon, xenotime, ilmenite, etc.) likely to be present in the crude concentrate recovered. It is also important to realise that by making separate concentrates of the by-products the mining operation will be more profitable, albeit, largely by the increased recovery of cassiterite which ensues.

## References


Adam, J.W.H., 1960. On the geology of the primary tin deposits in the sedimentary formation of Billiton. *Geologie Mijnb., 39,* 405–426.

Agassiz, J., 1954. Géologie et pegmatites stannifères de la région Mumba-Numbi, Kivu (Congo Belge). Comité National du Kivu. Bruxelles.

Aubert, G., 1969. Les coupoles granitiques de Montebras et d'Echassières (Massif Central Français) et la genèse de Leurs Mineralisations en étain, tungstène, lithium et beryllium. Memoires du Bureau de Recherches Géologiques et Minières, No. 46 (345 pages).

Bean, J.H., 1969. The iron-ore deposits of West Malaysia. *Econ. Bulletin 2,* Ipoh, Geol. Surv., W. Malaysia.

Bernard, C., 1865. An introduction to the study of experimental medicine (English translation). MacMillan and Co. N. York. 1927.

Beveridge, W.I.B., 1955. The art of scientific investigation. The Scientific Book Club, London.

Blockley, J.G., 1980. The tin deposits of Western Australia. *Mineral Resources Bulletin 12,* Geol. Surv. Western Australia (184 pages).

Bradford, E.F., 1961. The occurrence of tin and tungsten in Malaya. *Proc. Ninth Pacific Sci. Congress,* 1957, 12, 378–398.

Burt, D.M., 1977. Mineralogy and petrology of skarn deposits. *Rendiconti Societa Italiana di Mineralogia e Petrologia,* 33, (2), 859–873.

Chatterjee, A.K. and Strong, D.F., 1984. Rare-earth and other element variations in greisens and granites associated with East Kemptville tin deposit, Nova Scotia, Canada. *Trans. Instn. Min. Metall.* (Sect. B: Appl. earth sci.), 93, B59–70.

Collins, J.H., 1912. Observations on the West of England mining region. *Trans. R. Geol. Soc. Cornwall, 14,* 1912.

Cox, R. and Glasson, K.R., 1967. The geology and mineralisation of Cleveland Mine. Reprint from symposium *'The geology of Western Tasmania',* held at the University of Tasmania, Nov. 1967 (11 pages).

Emmons, W.H., 1940. *The principles of economic geology.* McGraw-Hill, N. York.
Gilmour, P., 1962. Notes on a non-genetic classification of copper deposits. *Econ. Geol., 57,* 450–455.
Groves, D.I. and Solomon, M., 1964. The geology of the Mt. Bischoff District. Publication 142, *Dept. Geology, University of Tasmania,* Hobart.
Haapala, I., 1966. On the granite pegmatites of the Peräseinäjoki-Alavus area, South Pohjanmaa, Finland. *Bull. Comm. Geol. Finlande,* No. 224.
Holmes, A., 1965. *Principles of physical geology.* Nelson (1288 pages).
Hosking, K.F.G., 1963. Geology, mineralogy and paragenesis of the Mount Pleasant tin deposits. *Can. Mining J.,* 84, 95–102.
Hosking, K.F.G., 1964. Permo-Carboniferous and later primary mineralization of Cornwall and South-west Devon. Hosking, K.F.G. and Shrimpton, G.W. (eds.). *Present views of some aspects of the geology of Cornwall and Devon.* Blackford Ltd. Truro, Cornwall, 201–245.
Hosking, K.F.G., 1965. The search for tin. *Mining Mag.,* 113, 261–275, 368–383, 448–481.
Hosking, K.F.G., 1969. The nature of the primary tin ores of the south-west of England. *Pre-print prepared for the 2nd Technical Conf. on Tin. Intl. Tin Council, 1969, Bangkok.*
Hosking, K.F.G., 1970a. The nature of the primary tin ores of the Southwest of England. *A second technical conference on tin, Bangkok, 1969.* I.T.C., 1155–1243.
Hosking, K.F.G., 1970b. Aspects of the geology of the tin fields of Southeast Asia. *A second tech. conference on tin, Bangkok, 1969.* I.T.C., 39–79.
Hosking, K.F.G., 1973. Primary mineral deposits. In Gobbet, D.H. and Hutchison, C.S. (eds.), Chapt. 11 of *'Geology of the Malay Peninsula',* Wiley-Interscience, N. York, 335–390.
Hosking, K.F.G., 1974. The search for deposits from which tin can be profitably recovered now and in the foreseeable future. *Fourth world conference on tin, Kuala Lumpur,* 1974., I.T.C. 21–76.
Hosking, K.F.G., 1975. *The search for tin deposits.* Intl. Tin Council, London.
Hosking, K.F.G., 1979. Tin distribution patterns. *Geol. Soc. Malaysia, Bulletin* 11, 1–70.
Hosking, K.F.G., Leow, J.H.; Chin, L.S.; Wong, Y.F., 1970. The nature and significance of the Manson Orebody, Ulu Ketubong, Kelantan: an interim report. *Malayan Scientist,* 5, 32–41.
Hunter, D.R., 1973. The localisation of tin mineralization with reference to South Africa. *Minerals Sci. Engng.,* No. 1, 53–77.
Hutchinson, R.W., 1981. Lode tin deposits of exhalative origin. Pp. 81–106 of *'Complex tin ores and related problems'.* Tech. Publication No. 2 of SEATRAD Centre, Ipoh, Malaysia, 81–97.
Ingham, F.T. and Bradford, E.F., 1960. Geology and mineral resources of the Kinta Valley, Perak, Federation of Malaya. *Geol. Surv. District Memoir 9.*
Ishihara, S., 1977. The magnetite-series and ilmenite-series granitic rocks. *Mining Geol.,* 27, 293–305.
Jackson, N.J., 1979. Geology of the Cornubian tin fiels - 'a review'. *Geol. Soc. Malaysia, Bulletin* 11, 209–237.
Jacobson, R. and Webb, J.S., 1946. The pegmatites of Central Nigeria. *Geol. Surv. Nigeria Bulletin 17* (61 pages).
Jones, W.R., 1925. *Tin fields of the world.* Mining Publications Ltd. London (423 pages).
Kelly, W.C. and Turneaure, F.S., 1970. Mineralogy, paragenesis and geothermometry of the tin and tungsten deposits of the Eastern Andes, Bolivia. *Econ. Geol.,* 65, 609–680.
Kwak, T.A.P., 1983. The geology and geochemistry of the zoned, Sn-W-F-Be skarns at Mt. Lindsay, Tasmania, Australia. *Econ. Geol.,* 78, 1440–1465.
Kwak, T.A.P. and Askins, P.W., 1981. Geology and genesis of the F-Sn-W (Be-Zn) skarn (Wrigglite) at Moina, Tasmania. *Econ. Geol.* 76, 439–467.
Leube, A. and Stumpfl, E.F., 1963. The Rooiberg and Leeuwpoort tin mines, Transvaal, South Africa. *Econ. Geol.* 58, 391–418 and 527–557.
Liddy, J.C., 1977. The tin deposits and potential of Australia. *Australian Mining Journal,* 17–54.
Lindgren, W., 1933. *Mineral Deposits.* McGraw-Hill, New York.
Matthews, P.F.P., 1982. Tin mineralization in Central Goias, Brazil. *Mining Mag.,* 461–467.
Mining Journal, 1979. Hemerdom reserves doubled. *Mining Journal,* London, 292, 485–487.
Mining Journal, 1984. East Kemptville update. *Mining Journal,* London, 303, 242–243.
Mulligan, R., 1975. Geology of Canadian tin occurrences. *Geol. Surv., Canada Econ. Geol. Rept.,* 28 (155 pages).

Nakamura, T., 1970. Mineral zoning and characteristic minerals in the polymetallic veins of the Ashio copper mine. In Tatsumii, T. (ed.) *'Volcanism and Ore genesis'*, Univ. Tokio Press, Tokyo, 231–246.

Newnham, L.A., 1973. *Geology of the Renison Bell Tinfield.* Unpublished account (6 pages).

Reed, B.L., 1982. Tin greisen model. Pp. 55–61 of *'Characteristics of Mineral Deposit Occurrences'*, compiled by R.L. Erickson. Open file Report 82–795, U.S.G.S.

Roberts, J.S., 1983. Controls of tin mineralization with special reference to the Rooiberg Tin Mines. Re-print of paper presented at the *15th G.F.S.A. Geological Colloquium* (15 pages).

Roe, F.W., 1951. The geology and mineral resources of the Fraser's Hill area, Selangor, Perak and Pahang, Federation of Malaya, with an account of the mineral resources. *Mem. Geol. Survey Dept.* Fed. Malaya, 5 (138 pages).

Rundquist, D.V.; Denisenko, V.R. and Pavlova, I.G., 1971. *Greisen deposits.* (In Russian) NEDRA Publishing House, Moscow (328 pages).

Schuiling, R.D., 1967. Tin belts around the Atlantic Ocean: some aspects of the geochemistry of tin. Pp. 531–547 of *"A technical conference on tin:* London." I.T.C., London.

Shcherba, G.N., 1970. Greisens. *International Geological Review,* 12, 114–151, 239–254.

Sillitoe, R.H.; Halls, C. and Grant, J.N., 1975. Porphyry tin deposits of Bolivia. *Econ. Geol., 70,* 913–927.

Sohnge, P.G., 1963. Genetic problems of pipe deposits in South Africa. *Proc. Geol. Soc. S. Africa, 66,* 19–72.

Strauss, C.A., 1954. The geology and mineral deposits of the Potgietesrus tinfields. *Mem. Geol. Surv., South Africa, 46* (252 pages).

Taylor, R.G., 1979. Geology of tin deposits. Elsevier Scientific Publishing Co., Oxford (543 pages).

Turneaure, F.S., 1960. A comparative study of major ore deposits of Central Bolivia. *Econ. Geol.* 55, 217–254 and 574–606.

Willbourn, E.S., 1931. The Beatrice mine, Selibin, F.M.S., *Mining Mag.,* 45, 338–341 and 1932, *Mining Mag.,* 46, 20–24.

Yeap, E.B., 1978. Hydrothermal tin-bearing breccia of the Yap Peng Mine, Sungai Besi, Selangor, Peninsular Malaysia. Pp. 367–375 of *'Geology and Mineral Resources of South-east Asia'.* Pub. A.I.T., Bangkok.

Yew, C.C., 1971. The geology and mineralization of the eastern Kuala Lumpur area, West Malaysia. *Unpub. B.Sc. (Hons) Thesis, Dept. of Geol., Univ. Malaya* (111 pages).

Ypma, J.M. and Simons, J.H., 1969. Genetic aspects of the tin mineralization in Durango, Mexico. *A second technical conference on tin,* Bangkok. International Tin Council (Pre-print).

# 1.2 The Problems of Tin Metallogeny

Guo Wenkui[1]

**Abstract**

On the basis of the spatial and temporal distribution of tin deposits in the world and their relation to different kinds of igneous rocks, the sources of tin are discussed according to the available tin abundance in meteorites and ultramafic, mafic and felsic rocks in combination with the siderophile habit of tin. It is proposed that tin is derived from the mantle rather than from the crust.

It has been estimated that the gross production of tin metal of the world was roughly around 10 million t up to 1980. Of this, about 54% was produced from Southeast Asian countries such as Thailand, Malaysia and Indonesia and South America, mainly Bolivia, whereas the remaining 46% was from more than 40 separate regions located in different continents of Oceania, Africa, Europe, North America, Latin-America as well as the other parts of Asia (Taylor, 1979). Several larger tin deposits in East Asia are located east of longitude 105°, up to north Kamchatka (Yakutia) of the U.S.S.R. and down to South China (Fig. 1). China plays a rather important part in the world tin production. The works on particular tin deposits of China by Meng and Chang (1935) and Hsieh (1963) have been repeatedly cited by a few geologists abroad up to the eighties, e.g. Lehmann (1982). Since the founding of the People's Republic of China, especially in the last seven years, remarkable progress has been made in the field of tin prospecting, exploration, mining and study.

In China, there are both placer and hypogene tin deposits. In ancient time, tin was mined predominantly from placers; later, placer mining was replaced gradually by hypogene tin excavating. At present, the amount of tin produced from hypogene deposits is far more than that from placers. The hypogene tin deposits are of great variety and associated with varied metals. Temperature of major metallization ranges generally from 150° to 550°C. Four common types have been recognized, i.e. (1) pegmatite (360–620°C); (2) contact metasomatic (340–550°C); (3) porphyry, including xenothermal, (100–500°C) and (4) hydrothermal veins (250–500°C), usually 300–400°C. A characteristic feature is the occurrence of two or three types of deposits within a single tin district or two or three types of orebodies within a single deposit. Some districts and deposits show deposit zoning and orebody zoning respectively. In other districts, different types of deposits may be superimposed upon one another.

[1] Institute of Geology, Chinese Academy of Geological Sciences, Beijing

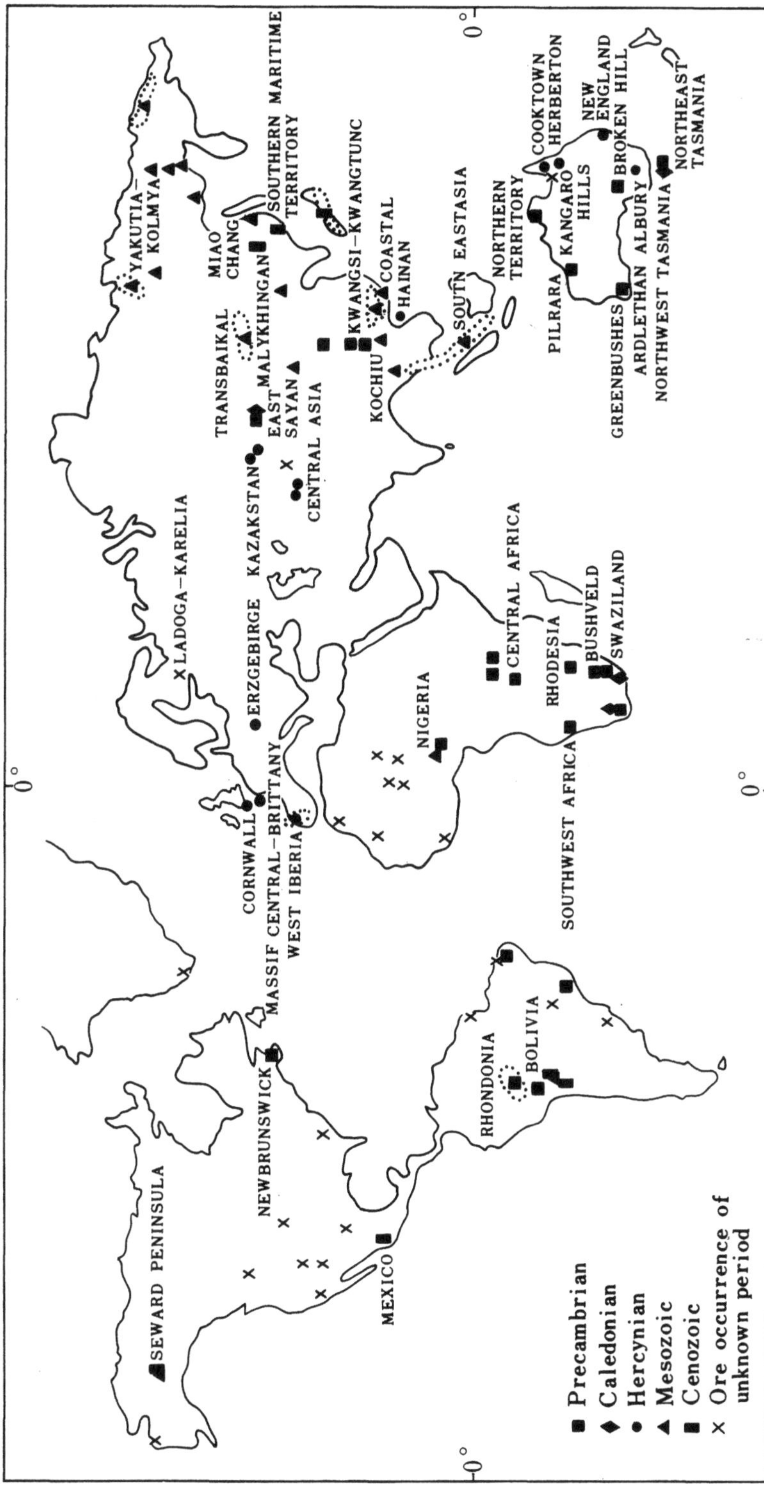


**Fig. 1.** Schematic map of the world showing distribution of tin deposits formed during different metallogenic epochs

Spatially, tin deposits may occur in different geotectonic units such as craton, shield, platform, fold-belt, continental margin and island arc (Jones, 1920). Tin deposits in different geotectonic units are characterized by different types and sizes. The deposits present in relatively stable units, such as the craton and shield of Africa and North America, are mostly of pegmatite type and lesser in size, whereas the larger deposits of types other than pegmatite are usually located in fold-belts and/or continental margins. Therefore, the latter geotectonic units are the main sites of productive tin deposits.

Temporally, there are five metallogenic epochs as follows:

(1) Proterozoic, including tin deposits in Africa (Pelletier, 1964), DGR, Brazil, North Territory and Western part of Australia, Eastern Sayan in the U.S.S.R., Kangdian axis of China and probably part of Bolivia;
(2) Caledonian, such as southern part of South Africa, northwest Tasmania of Australia and part of China;
(3) Hercynian, including most deposits in Europe, Queensland and New South Wales in Australia, Central Asia and Kazakstan in the U.S.S.R., Eastern Canada and Northeast China;
(4) Mesozoic, examplified by tin provinces of the Far East, Lesser Khingan (Ichekson et al., 1959) and Yakutia in the U.S.S.R., Bolivia and South China and the tin belt from western Yunnan of China through Burma, Thailand to Malaysia;
(5) Cenozoic, represented by some deposits in Peru, Bolivia, Mexico and Shihot-Alin in the U.S.S.R. and Japan.

The number of deposits formed in the Mesozoic and Cenozoic exceeds half the total number of deposits formed in all the five metallogenic epochs. Therefore, Mesozoic and Cenozoic are the major metallogenic epochs for tin. Of them, Mesozoic is most important in tin metallization (Ichekson et al., 1959). Such associated metallic elements as lead, zinc, silver and antimony seem to increase with time.

Most of the hypogene tin deposits appear to be related, both in space and in time, to granitic rocks including plutonic, hypabyssal, subvolcanic and even volcanic phases of such varieties as granodiorite, quartz monzonite, common granite and the granite associated with the Bushveld stratiform complex (Pelletier, 1964). Among them, the silica-oversaturated, high alumina and alkali-rich granite shows an intimate relation with tin deposits. Many large tin deposits occur inside, proximal or occasionally distal to satellitic bodies of or cupolas above big granitic batholiths in regional metamorphic regions, and stocks along fractures and anorogenic small intrusives. The tin content of granite is highest at the late stage of magmatic evolution. In addition, these tin-rich granites are also rich in "incompatible elements" such as fluorine, lithium, boron, and caesium. Granites with a higher content of fluorine and lithium are usually in association with metallization in Sn-W-Mo (Bailey, 1977). The metallization of tin often lags about twenty million years behind the consolidation of the intrusive body concerned. The lag time tends to vary with depth of intrusion. This probably implies the relationship between the evolution of magmatic fractionation and the metallogeny of tin.

The Gejiu ore district may be used to illustrate this time span between granite and tin metallization. In the Gejiu ore district, there are several granitic intrusive bodies of which the representative and well studied ones are Longchahe, Malage and Laochang. The Longchahe intrusive mass is of porphyritic biotite granite with some xenoliths of granodiorite near the margin. Its average Rb-Sr age (from eight samples) is 147 ± 3 Ma. It contains Ti 0.182%, F 0.15%, K 4.7% and Sn 4ppm. So far no economic tin deposit has been found in association with the Longchahe intrusive mass. Also made up of porphyritic biotite granite, the Malage intrusive body contains Ti 0.175%, F 0.26%, K 4.87% and Sn 10 ppm. The Rb-Sr isochron dating of nine samples gives an age of 90.4 ± 6.3 Ma. At the contact of the Malage granite with Triassic limestone there exist productive skarn tin deposits. The Laochang granite body is a highly albitized and K-feldspathized equigranular granite containing Ti 0.022%, F 0.377%, K 5.517% and Sn 20 ppm. The Rb-Sr isochron dating of seven samples gives an age of 81.0 ± 4.9 Ma. A large tin and polymetallic deposit occurs in association with the Laochang igneous body. A detailed study of Rb-Sr isotopes shows the initial $^{87}Sr/^{86}Sr$ values of the three granitic intrusives to be more or less consistent. It is therefore believed that they are the products of magmatic differentiation and evolution from a co-magmatic source at different periods and different stages, although the difference of the ages of petrogenesis is about 70 Ma (Wu Qinsheng et al., 1984). Potassium feldspathization of the granite usually marks the beginning of tin metallization.

Although tin deposits are distributed over every continent of the world (Taylor, 1979), they are usually confined to certain tin provinces or tin belts. In these, it is not uncommon that tin deposits of different metallogenic epochs occur next to one another or are superimposed at the same site, as exemplified by the deposits in Bolivia. This leads to the formulation of the concept of geochemical inhomogeneity of tin in the crust and metallogenic heritage.

Based on a great amount of information, Lehmann (1982) argues against the concept of geochemical heritage of tin, i.e., the assumption of a primary crustal tin anomaly in tin provinces (Pollard et al., 1983). On the contrary, he suggests that the controlling factor for the generation of tin granites is a special bulk distribution coefficient of tin, probably dependent on the oxygen fugacity of the melt resulting from crust remelting (Lehmann, 1982). This suggestion gives support to the argument that tin originates from the continental crust (Hutchison and Chakraborty, 1979), a rather popular concept in the circle of Chinese tin geologists.

On the other hand, the relationship between tin deposits and mafic or ultramafic rocks has also been repeatedly noticed by many geologists. The deposits of cassiterite-silicate type in the south coastal region of the Far East of the U.S.S.R. are not in direct association with granite, but are often related to basic dykes or intrusive diorite bodies (Radkevich et al., 1974). Other Russian information shows that the tin content in ultramafic rocks may be rather high. For example, the tin content of lherzolite reaches 3.8 ppm; harzburgite, 2.0 ppm; dunite, 0.6 ppm; eclogite, 1.4 ppm and fortunite, 1.9 ppm (Stemprok, 1977). In addition, there is also information about high to very high tin contents in magnesian and calcic skarn minerals (Nekrasov, 1971). In China, there are many skarn deposits of which the important ones are tungsten, molybdenum, bismuth and tin deposits in the Nanling region. Among various skarn minerals, hornblende has the highest tin content (6577 ppm), the next is garnet (1525 ppm), followed in order by muscovite (535 ppm), vesuvianite (318 ppm), epidote (305 ppm), magnetite (224 ppm), protolithionite (95 ppm), and pyroxene (92 ppm). The tin content of other skarn minerals such as feldspar, quartz and carbonates is all less than 5 ppm (Liu Yingjun et al., 1983). From the foregoing, it is obvious that tin metallization began at the stage of the crystallization of anhydrous silicates, attaining calmination at the stage of the crystallization of hydrous ferro-magnesian silicates and gradually decreased in intensity until the final deposition of carbonate minerals. Such a process of tin metallization is, to a great extent, consistent with that of porphyry and hydrothermal vein type of tin deposits.

The recent discovery of cassiterite-sulphide orebodies both in faults and in breccias within the ultramafic and mafic complex in northern Guangxi, though unexpected, is very interesting. The cassiterite there contains chromium, nickel, cobalt, titanium and copper instead of the niobium and tantalum commonly found in granite-associated cassiterite.

The tin deposit occurs here both at the contact zone or in the inner part of an elongated ultramafic intrusive body along a N-S trending anticline 1 km east of a major anticlinorium intruded by a granite body dated at 780—909 Ma by the zircon U-Pb method and 322—448 Ma by the K:Ar method on biotite. In the ultramafic body there are three petrological phases - hornblende peridotite, amphibolite and diabasegabbro. All the rocks are altered to variable degrees. Therefore, most of the rock-forming minerals have turned into serpentine, tremolite, actinolite, talc, chlorite, clinozoisite and epidote. In the less altered part, however, pseudocrysts of olivine and common hornblende can be found. Probably owing to the tectonic effect, rocks at the western margin of the ultramafic body are foliated, and locally there occur schistose tremolitite and chloritite. At its contact with country rocks, a layer of biotite schist is observed everywhere. The age of the ultramafic intrusion is supposed to be older than the emplacement of the granite.

According to the occurrence of orebody and the association of ore minerals, four types of tin ore may be identified, viz. (1) cassiterite-quartz-tourmaline-sulphide ore at the contact of the ultramafic igneous body; (2) cassiterite-sulphide ore at the contact; (3) cassiterite-sulphide ore within the ultramafic mass, and (4) vein ore of cassiterite-asbestos association within the ultramafic body.

Of significance are types 3 and 4, whose tenor of tin is much higher but the size is rather smaller than types 1 and 2. The orebodies occur in the northwest trending joints of tremolite serpentine formed from the alteration of hornblende peridotite. The boundary between the orebody and the country rocks is fairly sharp. The wall rocks surrounding the orebody are highly altered, accompanied by the generation of fibrous tremolite, fibrous serpentine, chlorite, carbonates and the introduction of sulphides. The nearer to the orebody, the more the fibrous minerals. In direct contact with the vein orebody are white fibrous serpentine and talc filling in the joint fissures.

The ore of the massive cassiterite-sulphide type consists of such minerals as cassiterite, pyrrhotite, pentlandite, chalcopyrite and sphalerite. Sometimes cassiterite is included in the sulphides, suggesting its deposition in a stage earlier than the sulphides. The common associated vein minerals are talc, serpentine, chlorite, actinolite and tremolite.

The ore of cassiterite-asbestos vein type is composed chiefly of cassiterite and locally of minor chalcopyrite and pyrrhotite. Sometimes the orebody occurs as a nest consisting of single cassiterite whose tenor of tin may attain 50—60%. There are two kinds of cassiterite. Under the condition of less asbestosized alteration, the cassiterite of the ore is grey in colour and very fine grained, occurring either as aggregates or as separate minute euhedral slender prisms with a grain diameter of 0.01 mm ±, while under the condition of highly asbestosized alteration, the cassiterite is light pink and coarse grained, occurring as either crystal stocks or tetragonal dipyramidal coarse grains of diameter 1–2 cm.

A preliminary study of the tin geochemistry of the country rocks suggests that the metallization of the deposit might have resulted from the leaching of tin out from the ultramafic rocks. Fifteen samples from both the hanging and the foot walls of the ore vein give tin contents of 10–200 ppm, averaging 60 ppm. Five samples from the asbestos vein, after sorting out coarse cassiterite, give tin contents of 20–60 ppm, averaging 38 ppm. Farther from the orebody, the rocks are richer in tin. Several samples 80–100 m away from the orebody contain tin values of 60–450 ppm, averaging 165 ppm.

In this tin district, the whole spectrum of rocks seems to be rather rich in tin, e. g., granite, 11–17 ppm; schist, 19.5 ppm (19 samples); and ultramafic rocks, 20.5 ppm (22 samples).

A strange fact is that this district is impoverished in elements generally associated with tin metallization in granite terrains such as Li, Be, Nb, Ta, W, Bi, Mo, etc. On the contrary, it shows a background value fairly high in Ni, Cr, Co, and V, which are commonly not in affinity with granite. During the tin metallization, lithophile elements remained almost unchanged whereas chalcophile elements increased a great deal. The tin metallization of this district is a special case which deserves due consideration.

All these facts, together with the tin deposit associated with the Bushveld stratiform complex (Pelletier, 1964) and the tin-mineralization present at the Atlantic mid-ocean ridge (Evans, 1980), lead us to consider the possibility of an oceanic crustal source of tin.

The known tin-bearing minerals amount to more than fifty species belonging, separately, to eight different classes, i. e., oxide, hydroxide, sulphide, sulphosalt, silicate, borate, alloy and native tin. Minerals of the first six classes are found in tin deposits or tin-bearing polymetallic deposits. Minerals of the last two classes, though only of mineralogical or byproduct significance, can indicate peculiar tin occurrence in nature. There is a group of alloy tin minerals including both named and unnamend ones (Fig. 2). The significant tin-copper alloy minerals and lead-bearing tin-copper alloy minerals have been found in trap rocks of the U.S.S.R. and in kakortokite of Greenland (Karup-Moller, 1978). The main named alloy tin minerals are atokite, stannopalladinite, niggliite, paolovite and rustenburgite, etc., all of which occur in ultramafic rocks. Metallic tin has been found in basalt, dolerite (Stemprok, 1977) and lunar rocks. Such occurrences of tin probably imply that in ultramafic and mafic rocks tin can combine with copper and/or precious metallic elements intrinsic in these rocks to form alloy minerals under anhydrous, reductive and volatile-free conditions. Native tin may occur under the strong reductive condition during the formation of lunar rocks.

In addition, mineralizations of tin, copper and zinc have been reported in Proterozoic (700 Ma) spilitic metavolcanics in Democratic Germany. Tin-bearing polymetallic deposits have also been found in a Lower Palaeozoic spilite-keratophyric volcano-sedimentary complex in China. These two examples shed light on the events of fossil metallization of tin at spreading axes of ancient oceanic crust.

The data of tin abundance of different rocks collected by different geologists from different territories in different periods fluctuate considerably (Table 1). This might

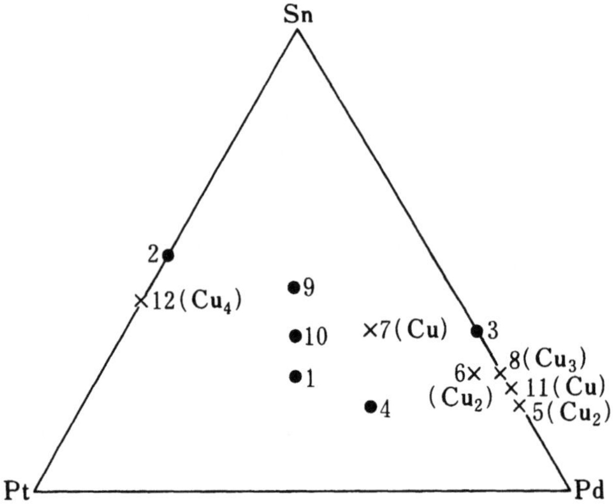


1. Atokite $(Pd,Pt)_3Sn$ 2. Niggliite PtSn
3. Paolovite $Pd_2Sn$ 4. Unnamed $Pd_3Pt_2Sn_2$
5. Unnamed $Pd_5Cu_2Sn_2$ 6. Unnamed $Pd_7PtCuSn_3$
7. Unnamed $Pd_2PtCuSn$ 8. Stannopalladinite $(PdCu)_3Sn$
9. Unnamed PtPdSn 10. Unnamed $Pt_3Pd_3Sn_2$
11. Unnamed $Pd_4CuSn$ 12. Unnamed $Pt_4Cu_4Sn_3$

**Fig. 2.** Triangular diagram showing occurrence of various alloy tin minerals

result either from unrepresentative sampling and analytical methods with inadequate accuracy or from inhomogeneity of tin content of the crust. As for the tin content of meteorites, it is certain that nickel-iron meteorites and iron meteorites have much higher tin contents than chondrite and silicate meteorites. A record shows that nickel-iron meteorites contain tin up to 100 ppm (Heide, 1957) while iron meteorites contain tin ranging from 0.20 ppm to 7.72 ppm (Wedepohl, 1969). There are two average values for the tin content of iron meteorite, i.e., 6.70 ppm (Winchester and Aten, 1975) and 2.00 ppm (Onishi and Sandell, 1957). All these values surpass the tin contents of ultramafic, mafic (Durasova, 1967) and intermediate crustal rocks (Wedepohl, 1969). Some even surpass the average value of tin content of granite. The variation in tin content of different meteorites indicates the siderophile habit of tin. The frequent existence of tin-magnetite orebodies in tin deposits of fold-belts is a manifestation of this conclusion. In the tin district of north Guangxi mentioned above, a tentative group analyses of geochemical data on minor elements in twelve samples from hornblende peridotite reveals the intimate relationship among the elements Sn, Ti, V, Cr and Co, implying that tin is intrinsic in the ultramafic rocks and has not been contaminated by other sources (Ma Hong, 1985). In other Precambrian tin deposits associated with granitoids in the tectonic unit of the Kangdian axis, the REE of the granitoids has been systematically analysed. The chondrite-normalized REE patterns are inclined to the right or towards the heavy REE. It is thus supposed that the source of ore might have originated from the mafic rocks in the deep crust (Jin Mingxia, 1985). In the

**Table 1.** Abundance of tin in different rocks

| | | | |
|---|---|---|---|
| I. | Iron Meteorites | 0.20–7.72 ppm | (Wedepohl, 1969) |
| | Average | 6.70 ppm | (Winchester and Aten Jr., 1957) |
| | | 2.00 ppm | (Onishi and Sandell, 1957) |
| II. | Nickel-Iron Meteorites | 100 ppm | (Heide, 1957) |
| III. | Troilite | 15 ppm | (Heide, 1957) |
| IV. | Chondrite | 0.07–2.40 ppm | (Wedepohl, 1969) |
| | | 1.00 ppm | (Vinogradov, 1962) |
| | Average | 1.00–0.50 ppm | (Onishi and Sandell, 1957) |
| V. | Silicate-Meteorites | 5.00 ppm | (Heide, 1957) |
| | | 20.40 ppm | (Heide, 1957) |
| | Total Abundance of Tin in Meteorites | 20.00 ppm | (Rankama and Sahama, 1950) |
| VI. | Granitic Rocks, Average | 3.50–3.60 ppm | (Wedepohl, 1969) |
| VII. | Intermediatee Igneous, Average | 1.30–1.50 ppm | (Wedepohl, 1969) |
| VIII. | Mafic Rocks | 0.90–1.20 ppm | (Durasova, 1967) |
| IX. | Ultramafic Rocks | 0.35–0.50 ppm | |
| | Continental Crust, Average | 2.30 ppm | (Durasova, 1967) |
| | Crust, Average | 2.10 ppm | |

well-known Gejiu Deposit, the lead isotopes of galena in association with the tin ore have been studied in greater detail. They suggest that the lead was derived from both the crust and from the mantle, but predominantly from the crust. The lead isotopic components relevant to magmatism, when plotted, are concentrated in a narrow area where lead isotopic components indicate island arc and marine sediments types (Wu Qinsheng et al., 1984). The results show that at least part of the ore materials originate from the mantle. Therefore, the inference that the tin source is the mantle rather than the crust seems to be quite reasonable.

Being variable in valence ($Sn^{2+}$, $Sn^{4+}$), with rather a high charge and activity in ionic state, tin is therefore unlikely to be held in the crystal structures of minerals crystallized during early magmatic crystallization stages. Tin seems to have remained in the melt till late stages of crystallization ranging from ultra-silicic to ultramafic varieties (Haapala and Kinnunen, 1982; U Khin Law et al., 1983). It could then combine with relevant elements and, under favourable reductive or oxidizing conditions, form the different above-mentioned tin-bearing minerals. In tin prospecting, apart from emphasizing the geological survey of felsic igneous areas, regions with oceanic extrusions and intrusions of ultramafic and mafic rocks are also potential targets deserving of attention.

## References


Bailey, J.C., 1977. Fluorine in Granitic Rocks and Melts: A Review. *Chemical Geol.,* 19, 1–24.

Durasova, R.R., 1967. Some problems of the geochemistry of tin. *Geochemistry International,* 4, 671–681.

Evans, A.M., 1980. *An introduction to ore geology.* Blackwell Publications, England, 197–204.

Haapala, E.T., Kinnunen, K., 1982. Fluid Inclusion on the Genesis of Tin Deposits, in: *Metallization associated with acid magmatism,* 101–109.

Heide, F., 1957. *Kleine Meteoritenkunde,* 2nd. Edn., Springer-Verlag, Berlin.

Hsieh, C.Y., 1963. A Study of Tin Deposits in China. *Sci. Sinica,* Vol. 12, 373–390.

Hutchison, C.S. and Chakraborty, K.R., 1979. Tin: a mantle or crustal source? *Geol. Soc. Malaysia Bull.* 11, 71–79.

Ichekson, G.B. et al., 1959. Tin Deposits of Lesser Khingan, *Bull. BSEGEI,* Vol. 27 (in Russian). New series, Vol. 27, 10–11 and 315–317.

Jin Mingxia, Shen Su, Huang Yonghe, and Yang Yanhua, 1985. Evolution of anatectic granites and their tin-tungsten minero-genetic series in the Kangdian Axis. *Geological Review,* Vol. 31, No. 3, 240–252 (in Chinese).

Jones, W.R., 1920. Tin and Tungsten Deposits: The Economic Significance of their Relative Temperature of Formation, *Trans. IMM,* 29, 320–376.

Karup-Moller, 1978. The Ore Mineral of the Ilimaussaq Intrusion: Their Mode of Occurrence and Their Condition of Formation. *Gronland Geol. Undersøgelso Bull.,* 127, 1–51.

Lehmann, Bernd, 1982. Metallogeny of Tin: Magmatic Differentiation versus Geochemical Heritage. *Econ. Geol.,* Vol. 77, No. 1, 50–59.

Liu Yingjun, Zhang Jingrong and Yang Zhi, 1984. An approach to metallogenic problems of Shizhuyuan polymetallic ore deposit of tungsten, molybdenum, bismuth, and tin (beryllium). *Geology and Exploration,* No. 5, 8–14.

Ma Hong, 1985. The ore forming condition of Jiumeo copper and tin deposit and analytic study of its metallogenic mechanism – the relationship between tin metallization and basic, ultrabasic rocks. *M.Sc. Thesis Kunming College of Technology* (in Chinese).

Meng, H.H. and Chang, K., 1935. Geology of Hsianghualing Tin Deposits, Linwu, Hunan. *Acad. Sinica, Natl. Research Inst. Geol.,* No. 15, 15–72.

Nekrasov, I.Y., 1971. Features of tin mineralization in carbonate deposits as in eastern Siberia. *International Geol. Rev.,* 13, 1532–1542.

Onishi, H. and Sandell, E.B., 1957. Meteoric and terrestrial abundances of tin. *Geochim. et Cosmochim.* Acta, 12, p. 262.

Pelletier, R.A., 1964. *Mineral Resources of South Central Africa.* Oxford Univ. Press, Oxford.

Pollard, P.J. et al., 1983. Metallogeny of Tin: Magmatic Differentiation versus Geochemical Heritage – A Discussion. *Econ. Geol.,* Vol. 78, 543–545.

Radkevich, E.A., Gonovchuk, V.G., Kokarin, A.M., and Korostelev, R.G., 1974. The age and space relation of the tin deposit of cassiterite-silicate formation to granites (Far East, U.S.S.R.). Stemprok, M. (ed.). *Symposium Metallization associated with acid magmatism (MAWAM).* Geol. Surv. Prague, 1, 348–350.

Rankama, K. and Sahama, T., 1950. *Geochemistry.* Univ. Chicago Press.

Stemprok, M., 1977. The source of tin, tungsten and molybdenum of primary ore deposits. Stemprok, M., Burnol, L., Tischendorf, G. (eds.) *Symposium Metallization associated with acid magmatism (MAWAM),* Geol. Surv., Prague, 2, 127–166.

Taylor, R.G., 1979. *Geology of Tin Deposits.* Elsevier Scientific Publishing Company, Amsterdam-Oxford-New York.

U Khin Law et al., 1983. A note on a fluid inclusion study of tin-tungsten mineralization at Mawchi Mine, Kayah State, Burma. *Econ. Geol.,* Vol. 78, 530–534.

Vinogradov, A.P., 1960. Atomic abundance of the chemical elements in the sun and in stony meteorites. *Geokhimiya,* 4, p. 291.

Wedepohl, K.H., 1969. *Handbook of geochemistry,* Vols. 1–4, Springer-Verlag, Berlin.

Winchester, J.W. and Aten, A.H.W., Jr., 1957. The content of tin in iron meteorites. *Geochim. et Cosmochim. Acta,* 12, p. 57.

Wu Qinsheng, Xu Junzhen, and Yang Zhi, 1984. Sr isotopic characteristics of Gejiu Sn-bearing granites and a study of ore search indicators. *Geochimica,* No. 4, 393–402 (in Chinese).

# 1.3 Genetic Modelling of Greisen-Style Tin Systems

P.J. POLLARD, R.G. TAYLOR, and C. CUFF[1]


## Abstract

The general features of the greisen systems include the occurrence of lenticular to massive alteration zones contained within cuspate protruberances from the apical zones of late stage geochemically specialised granitoids. The systems consist of an upper outer zone of minor barren pegmatite development, often associated with fine grained pegmatitic granite. The mineralisation occurs as irregular, to massive, or sheet-like bodies with the zone immediately below the contact extending for some 10–100 m. The bodies are essentially zones of fluorine rich sericitic-silicic alteration with associated cassiterite. The mineralised zone merges downwards into highly altered granite dominated by feldspathic types of alteration which in turn grade into mildly altered "fresh" granite.

The postmagmatic nature of the alteration has led to the development of a model of mineralisation resulting from the upward passage of post magmatic fluids which leach tin from the lower feldspathic horizon and precipitate cassiterite in the sericitic zone. This model currently holds dominance over concepts which invoke fluids derived from unspecified deeper sources ultimately concentrating in structural traps beneath the pegmatite, and models which favour processes of magmatic fractional crystallisation/melt diffusion.

Major questions that require solution before acceptance of any current modelling include:

1) What is the significance and origin of the pegmatitic marginal zones? These are early crystallising components and suggest consistent specialised melt-hydrothermal conditions.
2) Why are the early formed pegmatitic components usually devoid of cassiterite?
3) How are the postmagmatic mineralising fluids evolved, and why should they consistently occupy the apical region?
4) Why do postmagmatic fluids normally fail to cause any alteration within the overlying thin zone of pegmatite?
5) How do postmagmatic fluids form the frequently well developed curving lensoid zones which mirror the apical contact shape?
6) Why do the curving lensoid bodies frequently become smaller and more dispersed with depth?


[1] Department of Geology, P.O. James Cook University, Queensland, Australia, 4811

7) Why do the systems generally exhibit a total lack of structural control via major brittle fracture systems?
8) Why is the lower feldspathic zone ("Leaching Zone") consistently more anomalous in tin and related volatile elements than the underlying granite?

Consideration of these points suggests that a theory of an early magmatic preconcentration of the ore elements within apical melt zones, followed by an essentially in-situ development of mineralisation during late stage crystallisation, would resolve many of the problems.

## Introduction

Tin mineralisation associated with acid magmatism occurs within several different environments of granite emplacement (Mitchell and Garson, 1976; Sillitoe, 1974; Taylor, 1979a). In plutonic-dominated environments, tin mineralisation shows a pronounced tendency to occur within or adjacent to roof zones of intrusives (Hosking, 1969, 1970; Taylor, 1979a). Tin deposits are frequently associated with dome or cusp-like features on the batholith roof, and this feature is utilised extensively in prospecting (Hosking, 1962; Taylor and Steveson, 1972; Beer et al., 1975). Detailed mapping of granitic complexes frequently demonstrates that tin mineralisation is spatially related to specific minor phases of the granitoids, which themselves often occupy a high level in the igneous complex (Groves and McCarthy, 1978; Kozlov, 1978).

The ultimate characteristics of individual tin systems are strongly influenced by their structural environment, such that evolving fluids are dispersed along faults and fractures where brittle fracture predominates, or are "contained" within the granitoids where permeable release zones are not available (Fig. 1). Most systems lie between these extremes, but the "contained" systems offer the best opportunity to observe fluid evolution closest to the point of origin.

The contained systems are characterised by substantial hydrothermal alteration occurring within apical regions of granitoids. In most cases they are related to cusps

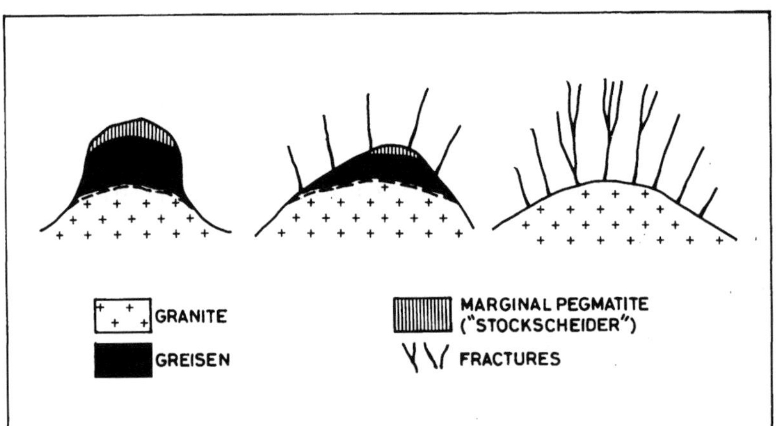


**Fig. 1.** Structural types of tin systems (after Baumann, 1970)

of specific minor phases, whilst in other examples they relate to portions of the upper endocontact zones of large intrusions.

## General Features of Greisen-Style Tin Systems

The general features of greisen-style tin systems include the occurrence of lenticular to massive alteration zones associated with cusps on the upper surface of late-stage, geochemically specialised granites. The system consists of an upper and outer zone of minor, barren pegmatite development, often associated with fine grained pegmatitic granite.

The mineralisation occurs as irregular, massive or sheet-like bodies immediately beneath the contact, and extending downwards for some 10 to 100 m. The mineralised bodies are essentially zones of fluorine-rich, sericitic-silicic alteration with associated cassiterite ± sulphides. The mineralisation grades downwards into a zone of feldspathic alteration (albitisation, microclinisation), which in turn grades into "fresh" granite.

It should be noted that there is no generally accepted definition of greisen systems, and similar configurations occur within feldspathic and boron-dominated tin systems (Zalashkova and Gerasimovskii, 1974; Taylor, 1979b).

## Morphology and Problems Associated with Massive-Disseminated Ore Zones

Ore zones in contained greisen systems are usually located beneath the upper contact of the intrusion, particularly within local high points (cusps, domes) on the granite surface (Fig. 2). In detail, the ore zones are often overlain by a series of pegmatites and/or pegmatitic rocks which are developed as a contact facies.

Despite the broad scale similarities, the general form of massive mineralisation varies widely. Systems are depicted (Fig. 2) ranging from massive ± root zones to lensoidal sheets. The latter are frequently curved, often irregularly stacked, with lenses becoming smaller with depth. The arcuate lenses invariably mirror portions of the overlying granite roof.

While there is little doubt that the above general patterns exist, it is crucial to any genetic modelling that the shape and distribution of ore zones be accurately defined. Several problems are encountered in achieving this objective.

### a) Problems with Massive Systems

Within the literature, there are no detailed accounts of massive greisen systems, and upon inspection it is often apparent that they are not massive *sensu stricto*. The amount and intensity of alteration is variable, and there are frequently substantial areas of relatively unaltered granite. The distribution of tin values is similarly uneven, and normally undocumented in other than a broad sense. There is no distinction made between alteration and infill mineralogy, and features such as veining or infill vug distribution are not recorded.

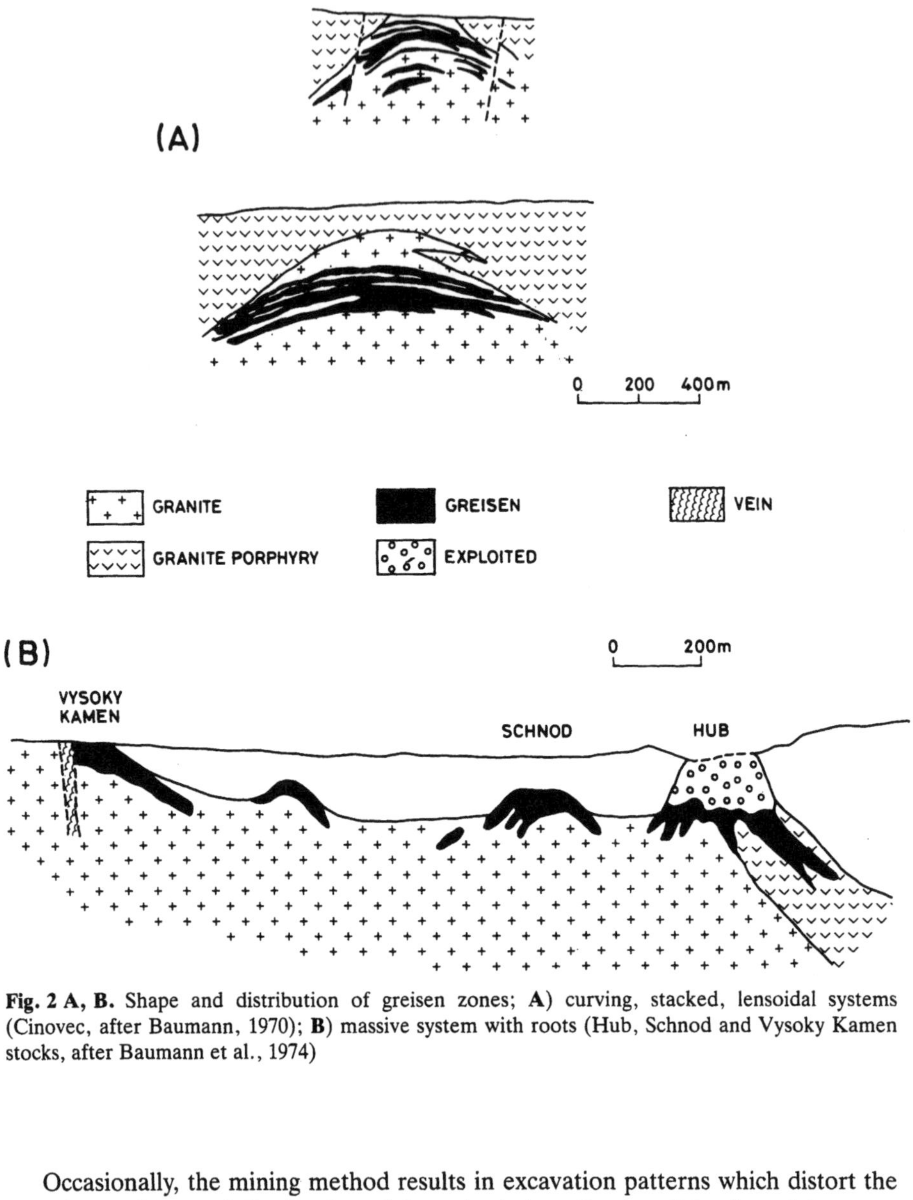


**Fig. 2 A, B.** Shape and distribution of greisen zones; **A**) curving, stacked, lensoidal systems (Cinovec, after Baumann, 1970); **B**) massive system with roots (Hub, Schnod and Vysoky Kamen stocks, after Baumann et al., 1974)

Occasionally, the mining method results in excavation patterns which distort the true shape of the total distribution of alteration/mineralisation. Selective mining operations only highlight areas of high grade distribution, while bulk mining obscures the presence of low grade and/or barren regions.

## b) Problems Associated with Sheeted Systems

Most of the above problems are equally true at smaller scales within sheeted systems, and this is well illustrated at the Zaaiplaats tin mine (South Africa). The mining pat-

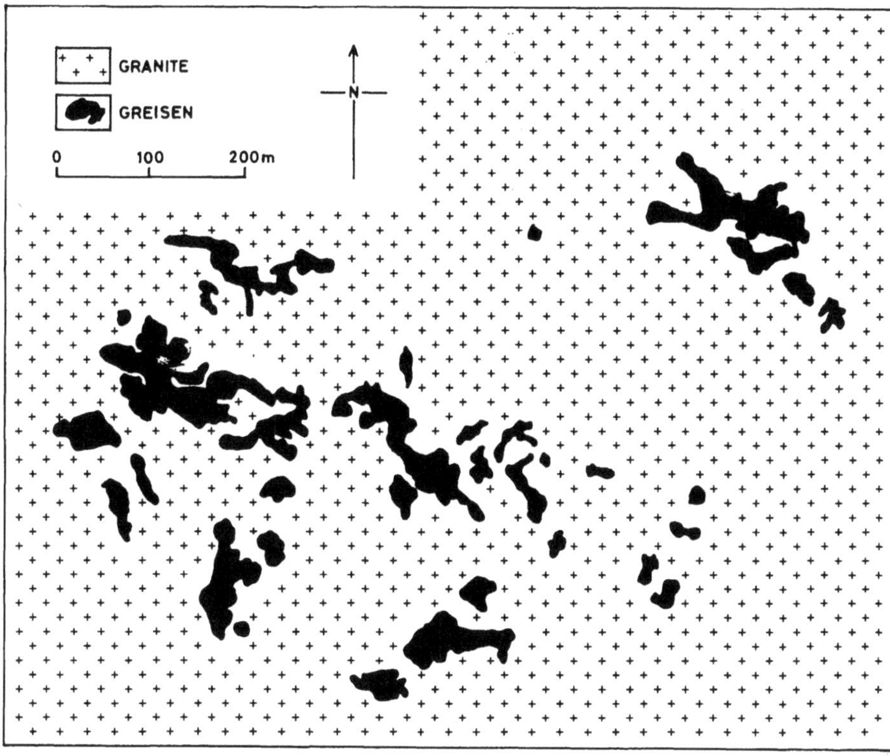


**Fig. 3.** Distribution of massive greisen zones within the Lease Granite, Zaaiplaats mine, South Africa. Black shading corresponds to mine excavations, however both feldspathisation and greisenisation extend beyond the ore limits, i.e. this does not represent the true pattern of alteration/mineralisation

tern (Fig. 3) depicts areas of flat, lensoidal greisen immediately below the granite contact which are regarded as massive ore.

However, inspection of the mining faces clearly demonstrates that the individual lenses are composed of irregularly distributed "fresh" granite, altered granite, vugs and tin values. It is also apparent from casual inspection that areas between and adjacent to the stopes are composed of similar material, although both alteration and mineralisation are less pervasive. A general impression is therefore gained that the entire region below the contact is irregularly altered/mineralised, and the existing pattern of mineralisation simply reflects economic criteria.

This problem is also encountered in interpretation of drilling results within these systems. The temptation to join high grade zones or intense alteration between holes frequently results in apparently well-defined, lensoidal patterns (Fig. 4). Although these reflect a general truth, they are usually not an accurate picture of the real distribution.

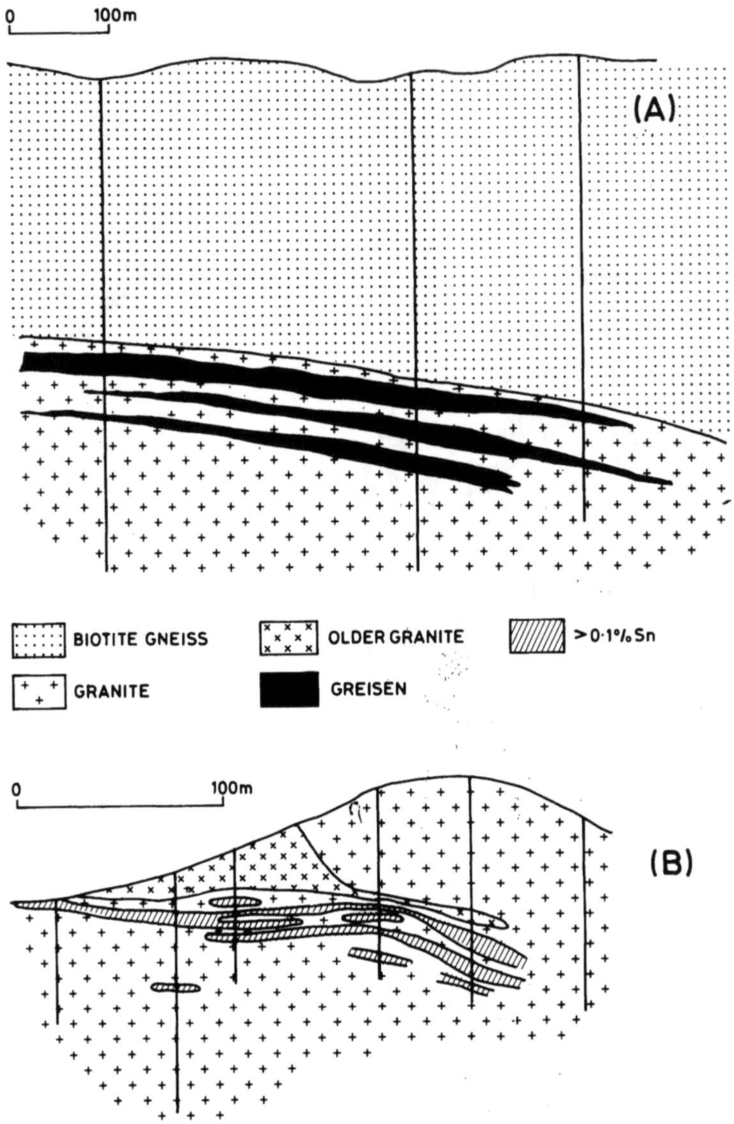


**Fig. 4 A,B.** Lensoidal patterns derived from **A**) intense alteration (Knotel, after Baumann et al., 1974) and **B**) grade (Anchor, after Groves and Taylor, 1973). Although true in a general sense, such patterns are not an accurate reflection of the actual distribution of alteration mineralisation

## Problems Associated with Fluid Access

Perhaps the most serious problem relating to genetic interpretation of contained greisen systems is the question of the source and access of hydrothermal fluids responsible for alteration/mineralisation. The problem of fluid access is rarely considered, and the collection of relevant data is currently inadequate. Close detail is required to

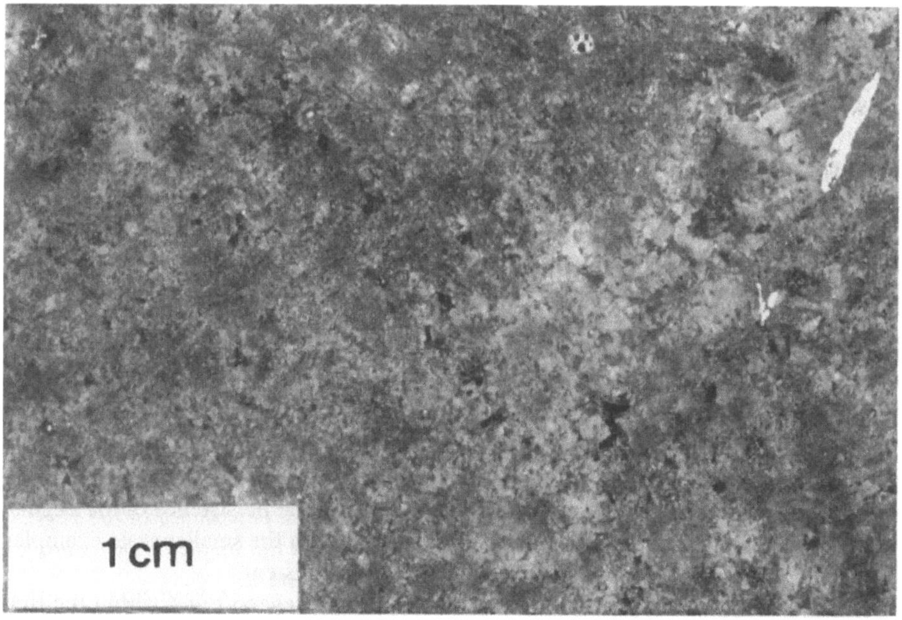


**Fig. 5.** Partially greisenised miarolitic, fine grained granite, Lease granite, Zaaiplaats mine. The general texture of a network of coarser miarolitic material within fine grained granite is ubiquitous within the upper region of the Lease granite. Greisenisation overprints earlier feldspathic (albitic) alteration

accurately define the distinction between alteration and infill. This is a problem which relates both to scale and mineralogy.

The mineralogical problem presents severe difficulties in that the infill mineralogy is frequently identical to the alteration mineralogy. Without careful observation, this obscures sites of fluid activity. This is especially true at the small scale where dispersed miarolitic cavities occur within totally altered granite. This is illustrated in a partially altered specimen in Fig. 5 where small infilled cavities are readily recognised, however with increasing alteration or smaller vug size, this recognition becomes increasingly difficult.

At a larger scale, massively brecciated rock presents similar problems, for example totally silicified rock fragments with a predominantly silica infill. Many seemingly homogeneous mineralised bodies are heavily disguised breccias. Similarly, although the more obvious features of fluid access zones (veins, stockworks, etc.) are present, their distribution and detailed relationships remain undocumented.

Prior to testing of genetic models concerning fluid access, a rigorous examination of textures of alteration/infill zones is required. There are clearly some massive greisen zones which are controlled by large scale brittle fracturing (brecciation, veining), for example the pipe- and vein-like ore-bodies in many such deposits (Plimer, 1974). Equally, there are some massive greisen zones which show extensive evidence for the presence of vug infill, but which lack evidence of brittle fracture fluid access. These appear to represent the *in situ* development of hydrothermal fluids during late stages of crystallisation.

It seems probable that many massive greisen systems may contain both brittle fracture and crystallisation components, and the proportions and distribution of these need to be clearly established.

## Problems Associated with the Development of Pegmatite/Marginal Facies Zones

Despite the very strong association of unusual marginal facies zones with massive greisen systems, their nature and origin has received very little attention. In fact there is confusion concerning almost every aspect of their petrogenesis.

Within the current data base it is difficult to accurately define their morphological characteristics. However several variations on a major theme are apparent on a broader scale. The majority of pegmatites occur as layers located immediately at the upper granite contact, particularly in localised apical high spots.

The pegmatite zone varies considerably in size and extent. Some large scale developments reach thicknesses of several metres (Strauss, 1954) whilst others are on the scale of centimetres (Groves and Taylor, 1973). Lateral persistence of the pegmatite layer is usually unrecorded, but it is noted that within the smaller scale examples the pegmatite development is often irregular and impersistent.

In some cases the pegmatite layer may be repeated several times within the first few metres. In other examples the pegmatite component forms a fine to coarse grained network within granophyric textured granite.

Some systems contain a mixture of the above components. At Zaaiplaats, an upper, layered facies is associated with pegmatitic granite which grades downwards into miarolitic granite hosting the mineralisation (Fig. 5).

Within large, well developed pegmatite facies, the entire zone is characterised by widespread and frequently abrupt mineralogical and textural variation. In detail, the pegmatite facies is often internally layered, with individual layers of quartz-feldspar, quartz, feldspar, and fine grained granite. Individual layers are sometimes vuggy. An example of this textural complexity is shown in Fig. 6.

Individual crystals within the pegmatite facies are often oriented perpendicular to the layering. Large, euhedral quartz crystals, and branching, plumose alkali feldspars are common, apparently nucleating on the upper surface of the layer. Similar features in porphyry molybdenum systems have been interpreted to result from inward crystallisation from the upper surface of the intrusion (Shannon et al., 1982), and plumose feldspars similar to those in the pegmatite facies are produced frequently in experimental studies on the effects of supercooling in magmas (Lofgren and Donaldson, 1975).

The petrogenesis of the pegmatite facies is seldom critically evaluated, and the contribution of direct crystallisation from the magma, or from emerging hydrothermal fluids has not been closely studied. The majority of textural features in the pegmatite facies are consistent with a direct magmatic crystallisation model, however the minor occurrences of vugs and open space filling texture suggests a contribution from emerging hydrothermal fluids. There is little direct evidence for the formation of the pegmatite facies via metasomatic replacement of pre-existing material.

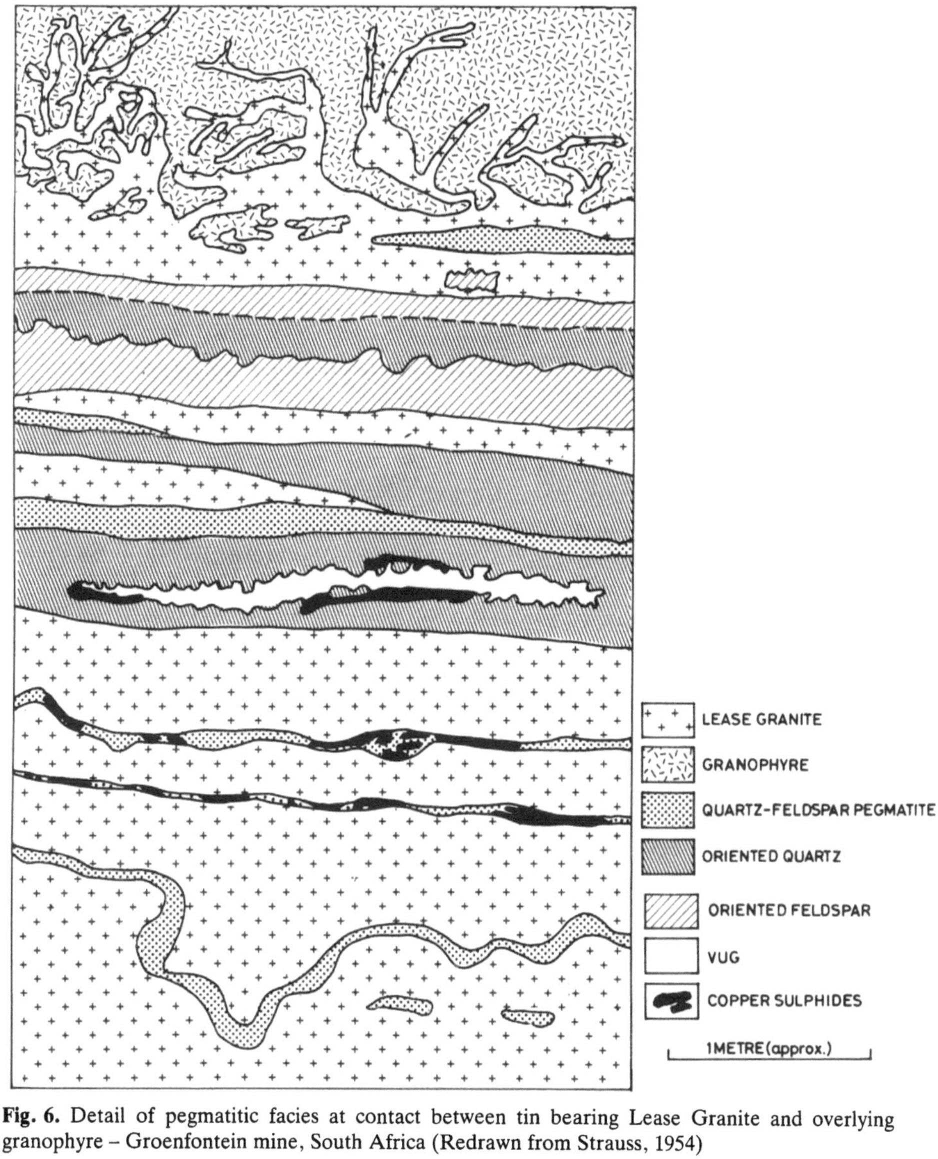


**Fig. 6.** Detail of pegmatitic facies at contact between tin bearing Lease Granite and overlying granophyre – Groenfontein mine, South Africa (Redrawn from Strauss, 1954)

## Relationships Between Pegmatite Facies and Mineralisation

The pegmatite facies appears to be the earliest crystallising component in greisen systems, and is noteable for the general absence of cassiterite mineralisation. Massive, greisen-style mineralisation occurs below the pegmatite zone, and is characterised by major metasomatic/hydrothermal alteration. Despite the proximity of these two components, the pegmatite facies is usually unaffected by greisen-style alteration. This critical timing relationship is rarely carefully examined, however several examples are known where the pegmatite facies is partially overprinted by the greisen-style alteration/mineralisation, for example at Zaaiplaats (Strauss, 1954), Sadisdorf (Baumann, 1970) and Penfeunteun (Charoy and Weisbrod, 1974).

The implication of this relationship poses some major problems, as it suggests that the earliest evolution of a volatile-rich magmatic/hydrothermal component is not associated with significant cassiterite mineralisation. This phase also contains only minor base metal and sulphur components. Further research is required to develop base data and hypotheses which may help to resolve this situation, since it could contain important genetic implications.

## Problems Concerning Broad Scale, Vertical Alteration Zoning in Greisen Systems

Vertical zonation within tin greisen systems is commonly reported, and is also apparent in related W, Mo and Ta-Nb systems (Syritso and Chernik, 1967; Zalashkova and Gerasimovskii, 1974; Smirnov, 1976; Kozlov, 1978; Ontoev, 1978; Xu Keqin et al., 1980). Within tin systems, the intensity of greisenisation diminishes downwards, however details concerning the lower portions of the system are less well understood. Zones of pervasive feldspathic alteration (albitisation, microclinisation) commonly occur beneath the main greisen zone, with a tendency for microclinisation to predominate in the deeper levels. The zone of feldspathic alteration is often underlain by relatively unaltered biotite granite (Beus and Zalashkova, 1964; Shcherba, 1979; Smirnov, 1976; Kozlov, 1978; Ontoev, 1978).

Details of the size, shape and extent of these deeper alteration zones are poorly documented. Ontoev (1978) reports a depth of 450 m to the biotite granite zone, but places the albitised zone beneath the microclinised granites. Stemprok and Sulcek (1969) however report abundant albitisation to a depth of 730 m at Cinovec. Conceptual models for southern China deposits (Xu Keqin et al., 1980) do not indicate biotite granite at depth.

Despite the general features outlined above, it seems there is considerable confusion concerning the size, shape, vertical and lateral extent of various feldspathic facies beneath massive greisen systems. Similarly, the definition of unaltered granite is confused. For instance, pervasive alteration has recently been reported occurring within all members of a major suite of tin-bearing granites on the Herberton tinfield, Australia (Pollard et al., 1983), and this alteration includes microclinisation, albitisation and greisenisation. The relationship between this type of regionally distributed alteration and the alteration zones occurring beneath massive greisen deposits is unknown, and it seems likely that this broad scale alteration is present in many other tin-bearing batholiths (e.g. Martin and Bowden, 1981; Charoy, 1979). Understanding of both the broad scale and localised alteration patterns is crucial to petrogenetic interpretation of fluid evolution, and there is a need for clearer international definition of terminology, together with well controlled three dimensional studies.

The timing relationships between alteration stages seem consistent (Pollard, 1983) with an early development of feldspathic alteration being overprinted by greisenisation within the massive greisen systems. A similar relationship has been reported on a pervasive, regional scale (Pollard et al., 1983).

The development of alteration zoning in massive greisen systems indicates widespread interaction with late- to post-magmatic fluids. The derivation and factors controlling fluid distribution constitute a major, unresolved problem. Within the current data base there seems to be little evidence to relate the strong vertical zoning with any major, underlying brittle fracture feeder system. This lack of fracture control suggests an essentially *in situ* development of fluids within the upper portion of the crystallising granite.

The understanding of broad scale vertical alteration zoning in greisen systems is confused. It seems that while there is general aggreement concerning the upper portions of the alteration system, the deeper levels are less well documented. The lack of spatial data, and of close consideration of factors concerning fluid evolution and permeability control are major impediments to genetic modelling.

## Problems Associated with Genetic Modelling

Current concepts concerning the derivation of the ore-bearing fluids responsible for the development of massive greisen systems are varied (Stemprok, 1979), and directly reflect many of the problems discussed above.

The leaching-reprecipitation model has received widespread support (Korzhinsky, 1964, Beus, 1962) and envisages that hydrothermal fluids evolve from a crystallising pluton, and ascend towards the apical regions. The zone of feldspathic alteration represents a region where tin is taken into solution, with precipitation occurring in the upper zone of massive greisen development. This model is attractive conceptually, but faces many unresolved problems.

This model conceives that fluids emerge from regions below the feldspathic zone by a loosely specified crystallisation process, and traverse essentially solid granite to consistently produce the characteristic geometry of the massive greisen systems. Derivation of fluids from deeper levels poses several problems. The lower, feldspathic zone is essentially late- to postmagmatic in origin with very little evidence of major brittle fracture controls to provide the essential permeability required for pervasive alteration. Notwithstanding the problems of fluid access to the upper, greisenised regions, it is also difficult to envisage why they should not cause widespread alteration in transit to the feldspathic zone, and why alteration/precipitation should occur with such consistent geometry only in the apical zones.

The derivation of the tin content of the greisen zone via leaching within the feldspar zone (Aktanov, 1971) also poses many problems, and is difficult to prove. The case is based largely upon the concept that feldspathic replacement and associated alteration of primary tin-bearing minerals must result in tin being released into solution. The concept is also based largely upon the observation that the tin content of the upper massive greisen zones and associated rocks is considerably higher than tin contents within the lower feldspathic zones. However, it should be noted that the tin content of the feldspathic zones is usually relatively high (Stemprok and Sulcek, 1969) and well in excess of the content of both the underlying biotite granites and the associated granitic suite. It could equally be argued that the feldspathic zone is an area of tin enrichment. Clearly, there are many problems requiring clarification before this model can be accepted.

The postmagmatic nature of alteration in massive greisen systems has also led to suggestions that the fluids are channelled into the apical configuration by major fracture systems, with fluids emenating from very deep sources (Stemprok and Sulcek, 1969). This model faces similar difficulties to those discussed above. The fluids are required to precipitate in a consistent position, and the existence of major fluid-bearing fault systems beneath massive greisen zones requires verification.

Given the current constraints, models which involve mechanisms to produce early concentration of appropriate volatiles within the apical zone have many advantages. One concept has been to envisage the evolution of an immiscible fluid phase during early crystallisation, which rises to collect within the apical zone (see Stemprok, 1979). This model faces several difficulties in that the fluids need to reach the apical position before substantial crystallisation has occurred, whereas the postmagmatic nature of the majority of the alteration (i.e. replacement of granite) suggests fluid evolution at later stages of granite crystallisation. Whilst there is textural evidence for fluid immiscibility on a small scale during late stages of crystallisation (Fig. 5), there are few data to suggest large scale fluid accumulation in structural traps beneath unbroken roof rocks (i.e. large sheet-like zones of hydrothermal infill/precipitation).

Another mechanism for producing the early concentration of volatile and ore elements within the apical zone is through concentration at the magmatic stage via essentially magmatic processes (Cuff et al., in prep.), i.e. the production of a stratified melt. This concept contains considerable advantages in that it becomes relatively easy to rationalise the major features of the system. Within this model, crystallisation of the apical portion results in localised fluid evolution at late stages of crystallisation, giving rise to intense autometasomatic replacement. The shape and distribution of the ore zones would reflect a balance between crystallisation and the process controlling melt stratification. Production of curved, lensoid alteration zones parallel to the roof could easily be envisaged by this process.

Processes which produce stratified magmas, including diffusive transfer and thermogravitational diffusion (e.g. Hildreth, 1979) are not well accepted, and examples of vertical stratification have only recently been documented in detail (Hildreth, 1979; Volkov and Garbacheva, 1980). Since rates of diffusion in volatile-rich magmas are poorly known, the validity of this model remains speculative, and further detailed study is required.

## Conclusions

The current status of genetic modelling of massive tin-greisen systems can only be regarded as unsatisfactory. The situation is hampered by lack of detailed observations upon greisen systems, especially with reference to the problems of mineralogical and textural evolution of the granites, and the mechanisms of fluid release and fluid transfer. Many of the problems require close examination of field evidence, especially detailed studies of the distribution of alteration and mineralisation within both massive ore zones and associated host rocks. The lower limits of the system also require sharper definition, and the processes of leaching versus volatile accumulation require clarification.

The melt stratification concept is consistent with both the broad scale and detailed textural features, and has the potential to produce systems of the appropriate geometry. However, considerably more research directed towards the major problems of greisen systems is required before current hypotheses can be critically evaluated.

*Acknowledgements.* The manuscript was kindly prepared by P. Bristow, and the diagrams by the cartographic section of the James Cook University of North Queensland.

## References


Aktanov, M.T., 1971. Behaviour of tin in postmagmatic alteration of granitoids of southern Tien Shan (southern Kirghiziya). *Geochem. Int.* 8, 367–370.

Baumann, L., 1970. Tin deposits of the Erzgebirge. *Trans. I.M.M.* Sect. B. (Appl. Earth Sci.) 79, 68–75.

Baumann, L., Stemprok, M., Tischendorf, G. and Zoubek, V., 1974. Metallogeny of tin and tungsten in the Krusne Hory-Erzgebirge. Pre Symp. *Excursion Guide IGCP-MAWAM Symp.* Karlovy Vary (Geol. Surv. Prague), 66 p.

Beer, K.E., Burley, A.J. and Tombs, J.M.C., 1975. The concealed granite roof in southeast Cornwall. *Trans. I.M.M.* Sect. B (Appl. Earth Sci.), 84, 24–25.

Beus, A.A., 1962. Wall rock alteration of hydrothermal-pneumatolytic deposits of rare elements. *Int. Geol. Rev.* 4, 1144–1153.

Beus, A.A. and Zalashkova, N.Ye, 1964. Postmagmatic high temperature metasomatic processes in granitic rocks. *Int. Geol. Rev.* 6, 668–681.

Charoy, B., 1979. Définition et importance des phénomènes deuteriques et les fluides associés dans les granites, conséquences métallogéniques. *Sci. de la Terre, Memoire* 37, 364 p.

Charoy, B. and Weisbrod, A., 1974. Interactions between rocks and solutions in mineralised greisens from St. Renan (Brittany, France). In M. Stemprok (ed.) *Metallization Associated with Acid Magmatism* Vol. 1 (Geol. Surv. Prague), 254–261.

Cuff, C., Taylor, R.G. and Pollard, P.J., in prep. Diffusion: a mechanism for element migration and enrichment with special reference to tin.

Groves, D.I. and McCarthy, T.S., 1978. Fractional crystallisation and the origin of tin deposits in granitoids. *Miner. Deposita* 13, 11–26.

Groves, D.I. and Taylor, R.G., 1973. Greisenization and mineralization at Anchor tin mine, northeast Tasmania. *Trans. I.M.M.* Sect. B (Appl. Earth Sci.), 82, 135–146.

Hildreth, W., 1979. The Bishop Tuff: evidence for the origin of compositional zonation in silicic magma chambers. *Spec. Paper Geol. Soc. Am.* 180, 43–75.

Hosking, K.F.G., 1962. The relationship between the primary mineralisation and the structure of the south-west of England. In K. Coe (ed.) *Some aspects of the geology of the Variscan Fold Belt.* (9th Intervarsity Geol. Cong., Manchester), 135–153.

Hosking, K.F.G., 1969. The nature of the primary tin ores of the South-west of England. In *A second technical conference on tin* (Int. Tin Council, London, 1969), Vol. 3, 1157–1243.

Hosking, K.F.G., 1970. The primary tin deposits of South East Asia. *Minerals Sci. Engin.* 1, 24–50.

Korzhinsky, D.S., 1964. Acidity conditions in post-magmatic processes. Rep. *22nd Int. Geol. Congress,* India, pt V. 16–27.

Kozlov, V.D., 1978. The sequence of phases and facies in the massifs of rare-metal granites in Transbaikalia and the problem of their ore-bearing capacity. In M. Stemprok, L. Burnol and G. Tischendorf (eds.) *Metallization Associated with Acid Magmatism* (Geol. Surv. Prague), Vol. 3, 249–255.

Lofgren, G.E. and Donaldson, C.H., 1975. Curved branching crystals and differentiation in comb-layered rocks. *Contrib. Mineral. Petrol.* 49, 309–319.

Martin, R.F. and Bowden, P., 1981. Peraluminous granites produced by fluid-rock interaction in the Ririwai nonorogenic ring complex, Nigeria: mineralogical evidence. *Canad. Mineral.* 19, 65–82.

Mitchell, A.H.G. and Garson, M.S., 1976. Mineralisation at plate boundaries. *Minerals Sci. Engn.* 8, 129–169.

Ontoev, D.O., 1978. Relation of multi-stage deposits of tungsten, molybdenum and tin to the history of granitoid formation. In M. Stemprok, L. Burnol and G. Tischendorf, *Metallization Associated with Acid Magmatism* (Geol. Surv. Prague) Vol. 3, 97—108.

Plimer, I.R., 1974. Pipe-like molybdenite-wolframite-bismuth deposits of Wolfram Camp, north Queensland, Australia. *Mineral. Deposita* 9, 95—104.

Pollard, P.J., 1983. Magmatic and postmagmatic processes in the development of rocks associated with rare element deposits. *Trans. I.M.M.* Sect. B (Appl. Earth Sci.), 92, 1—9.

Pollard, P.J., Milburn, D., Taylor, R.G. and Cuff, C., 1983. Mineralogical and textural modification in granites associated with tin mineralisation, Herberton-Mount Garnet tinfield, Queensland. *Proc. Symp. Permian Geology of Queensland,* Geol. Soc. Aust. Qld. Div. 413—429.

Shannon, J.R., Walker, B.M., Carten, R.B. and Geraghty, E.P., 1982. Unidirectional solidification textures and their significance in determining relative ages of intrusions at the Henderson Mine, Colorado. *Geology* 10, 293—297.

Shcherba, G.N., 1970. Greisens. *Int. Geol. Rev.* 12, 114—150, 239—255.

Sillitoe, R.H., 1974. Tin mineralisation above mantle hotspots. *Nature* 248, 497—499.

Smirnov, V.I., 1976. *Geology of mineral deposits.* (Moscow Min. Publishers) 520 p.

Stemprok, M., 1979. Mineralised granites and their origin. *Episodes,* 2, 20—24.

Stemprok, M. and Sulcek, Z., 1969. Geochemical profile through an ore-bearing lithium granite. *Econ. Geol.* 64, 392—404.

Strauss, C.A., 1954. The geology and mineral deposits of the Potgietersrus tinfields. *Mem. Geol. Surv. South Africa* 46, 241 p.

Syritso, L.F. and Chernik, L.N., 1967. Evolution in accessory mineral paragenesis during metasomatic alteration of granites in eastern Transbaykalia massifs. *Int. Geol. Rev.* 9, 814—827.

Taylor, R.G., 1979a. *Geology of tin deposits.* (Elsevier, Amsterdam), 543 p.

Taylor, R.G., 1979b. Some observations upon the tin deposits of Australia. *Geol. Soc. Malaysia Bull.* 11, 181—207.

Taylor, R.G. and Steveson, B.G., 1972. An analysis of metal distribution and zoning in the Herberton tinfield, north Queensland. *Econ. Geol.* 67, 1234—1240.

Volkov, V.N. and Gorbacheva, S.A., 1980. Composition of rockforming biotite and the variation in crystallisation conditions in a vertically zoned pluton. *Geochem. Int.* 17, 75—79.

Xu Keqin, Guo Lingzhi, Hu Shouxi, Ji Shouyuan, Shi Yangshen, Wang Zhengran, Sun Mingzhi, Mou Weixi, Lin Chengyi and Ye Jun, 1980. Investigation of the time and spatial distribution of the granitic rocks of southeastern China; their petrographic evolution, petrogenetic types and metallogenic relations. *Dept. Geol. Nanjing University,* 56 p.

Zalashokova, N.E. and Gerasimovskii, V.V., 1974. Petrographic and geochemical features of rare metal amazonite granites. In M. Stemprok (ed.) *Metallization Associated with Acid Magmatism* (Schweizerbart, Stuttgart) Vol. 1, 232—236.

# 1.4 Models of Grades and Tonnages of Some Lode Tin Deposits

W.D. Menzie[1], B.L. Reed[2], and D.A. Singer[1]

## Abstract

Descriptive and grade/tonnage models have recently been built for many types of deposits. Such models consist of descriptions of mineralogy, host rocks, ore textures, controls, alteration, geochemical signatures, age, and tectonic settings, together with statistical models of grades, tonnages, and contained metal of deposits of each type. The models are used to identify areas that may contain undiscovered deposits of given types, to convey to non-geologists an idea of the importance of such deposits, and to test and refine classifications of mineral deposits.

Descriptive and grade/tonnage models have recently been built for five types of primary tin deposits: rhyolite-hosted such as in Mexico; hydrothermal lodes such as in Cornwall, England, and the Herberton district, Queensland; replacement (or exhalative?) such as Renison Bell, Tasmania; skarn such as at Lost River, Alaska; and greisen such as in the Erzgebirge. Analyses of frequency distributions of tonnage, contained metal, tin grades and the relationships between these variables show that the deposits fall into four well-defined domains that have definite geological characteristics. Rhyolite-hosted , or Mexican, deposits contain a median of 4 t of tin and have a median grade of 0.4% Sn. Hydrothermal lode deposits have the highest grades. Half of such deposits have grades over 1.0% Sn, and the majority contain more than 1,000 t Sn. Large hydrothermal vein deposits contain more than 50,000 t Sn. Replacement (or exhalative?) deposits contain the largest amount of tin (median = 40,000 t). They are only of slightly lower grade (median = 0.80% Sn) than the hydrothermal lodes. Greisen or stockwork deposits have larger tonnages than replacement deposits, but contain less tin (median = 25,000 t). They are also of much lower grade (median = 0.3% Sn). Though grades and tonnages are available for only four skarn deposits, they appear to be more like greisen deposits than replacement deposits when compared using grades, tonnage and contained tin.

Although these individual models of primary tin deposits must be regarded as preliminary because of the relatively small number of deposits upon which they are built, they clearly demonstrate differences among types and provide basic information that can be useful in making decisions about exploration strategy, land classification, and tin supply.

[1] U.S. Geological Survey, 345 Middlefield Road, Menlo Park, California 94025, USA
[2] U.S. Geological Survey, 4200 University Drive, Anchorage, Alaska 99508, USA

## Introduction

Descriptive and grade/tonnage models have recently been built for many types of ore deposit (Cox, 1983, a, b, and Singer and Mosier, 1983 a, b). Such models consist of descriptions of the mineralogy, host rocks, ore textures, controls, alteration, geochemical signatures, ages and tectonic settings, together with associated statistical models of grades and tonnages. While some data on the grades and tonnages of tin deposits have been published (Taylor, 1979), descriptive and grade/tonnage models have not been built. This paper presents preliminary grade/tonnage models of five types of primary tin deposits: rhyolite-hosted, greisen, hydrothermal, skarn, and carbonate replacement (exhalative?). Descriptive and grade tonnage models are useful for evaluating exploration strategies, estimating future supplies of resources, assessing undiscovered resources and for testing and refining classifications of ore deposits.

## Building Grade/Tonnage Models

The first step in constructing a grade/tonnage model is to identify a set of well-explored deposits of the type one wishes to model. A correct classification of the deposits is important because differences in the grades and tonnages of different types usually will be apparent. For example, Figures 1 and 2 show tonnages and grades of fluorine-rich Climax type, and fluorine-deficient porphyry molybdenum deposits (Singer et al., 1983; Menzie and Theodore, 1983). These plots show that the fluorine rich granite-hosted porphyry molybdenum deposits have larger tonnages and higher average grades than the low-fluorine quartz monzonite-hosted deposits. The process of building grade/tonnage models is usually iterative and during the course of building

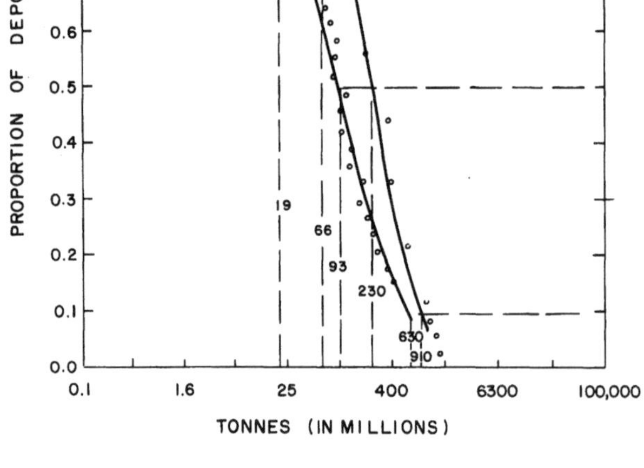


**Fig. 1.** Tonnages of ore of low fluorine and Climax types of porphyry molybdenum deposits

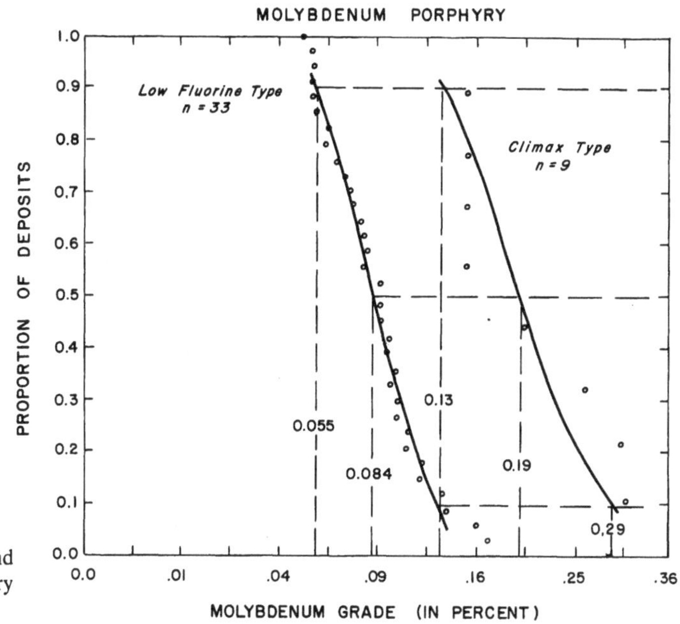


**Fig. 2.** Molybdenum grades of low fluorine and Climax types of porphyry molybdenum deposits

the model, deposits which are incorrectly classified may be identified as statistical out-liers.

Because grade/tonnage models are used to represent undiscovered deposits, it is desirable that the data in the models include past production, reserves and resources at a uniform cut-off grade. Although this ideal is rarely met completely for any model, it is important to understand the nature of the data used so that the causes of unusual results can be identified.

The second step in building the model is to analyze the data statistically. This involves fitting distributions to the observed grades and tonnages and testing the correlations of the grade and tonnage. In order to build a grade/tonnage model for a deposit type which is mined for two commodities, one must investigate the distributions of three variables: deposit tonnage and two commodity grades. Three sets of correlations must be investigated: the correlations of tonnage with each grade and the correlation between the grades. For most deposit types, the frequency distributions of grade and tonnage can be modelled by lognormal distributions. That is, if histograms are made of the logarithms of grades or tonnages of deposits, most of the plotted histograms will have the bell shape of the Gaussian, or normal distribution. When the logarithms of tonnage and average grade are plotted against each other, they frequently show a wide scatter that indicates the tonnages and grades are not correlated. For a few deposit types a significant correlation of tonnage and grade does exist, but for most deposit types investigated to date (Singer and Mosier, 1983 a, b) tonnage and grade are independent. Logarithms of the grades of two commodities are also likely to be statistically independent unless the commodities occur in the same or closely linked minerals; then, in most cases they will be significantly correlated. For example, lead and silver are correlated in sedimentary exhalative deposits (Menzie and Mosier, 1983).

Grade/tonnage models are usually presented as a series of plots, one for tonnage, with additional plots for each grade, and a listing of those correlation coefficients that are significant. To make the plots, data for each variable are sorted from smallest to largest and the proportion of the deposits that are as large as each deposit are calculated. The logarithms of tonnage or grade of deposits are plotted versus the calculated proportion. The mean and standard deviation of the logarithm of tonnage, or grade, and a table of areas of the normal curve (Arkin and Colton, 1963) are used to fit a curve to the observed data points. The curves usually have a backward "S" shape. If the size of the sample is small, the individual data points will plot above the fitted curve. This is an artifact of the method of plotting individual values and the effect diminishes with increasing sample size. The fitted curve provides the basis for estimating grades and tonnages of undiscovered deposits. The mean of the variable in logarithms can be determined from the curve by taking the logarithm of the median tonnage (50% value), and the standard deviation of the variable in logarithms can be determined by dividing the absolute value of the difference between the logarithms of the 50 and 90% values by 1.28, the number of standard deviations between the 50th and 90th percentiles of a normal distribution.

## Preliminary Grade/Tonnage Models of Some Primary Tin Deposits

### Rhyolite-Hosted Deposits

Rhyolite-hosted tin deposits generally occur as discontinuous veinlets of cassiterite and wood-tin in flow-dome complexes composed of alkali-feldspar rhyolite. Most

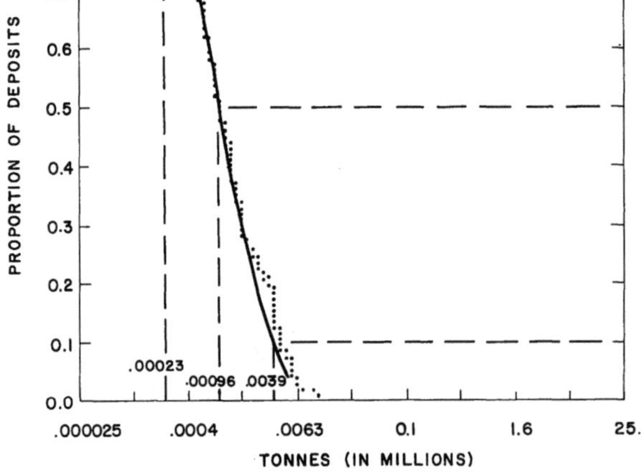


**Fig. 3.** Tonnages of ore of rhyolite-hosted tin deposits

veins contain cassiterite, specular hematite, and varying amounts of cristobalite, fluorite, tridymite, opal, chalcedony, beudantite/mimetite, adularia, durangite, and zeolite minerals. The veins are commonly 0.1 to 10 cm wide and rarely extend more than 75 m along strike or down dip. The tin is generally considered to be derived from the rhyolite in which it occurs, principally during degassing and attendant vapor phase alteration as the rhyolite solidified, cooled, and fractured. Many rhyolite-hosted deposits lie in the mid-Tertiary volcanic province of the Sierra Madre Occidental located in central and north central Mexico. A few occur in middle Tertiary rhyolite terranes in the western United States, such as those in New Mexico's Black Range.

The grade/tonnage model of rhyolite-hosted deposits was built with data on reserves of 131 deposits from Aquascalientes, Durango and Zacatecas, Mexico (Bracho Valle, 1960 and 1961). Though some of these deposits have produced, data on production was not available. Analysis of the distributions of the tonnages and tin grades of these deposits indicates they can be fitted by lognormal distributions, and examination of the correlation of grades and tonnages indicates they are positively correlated (correlation coefficient r = 0.36). Such a correlation is unusual and may be an artifact of the way the data were collected.

Figure 3 presents the plot for tonnage, and Figure 4 the tin grade. These figures show that 80% of rhyolite-hosted tin deposits have between 230 and 3,900 t of ore and that 80% of them have grades between 0.14 and 1.04% Sn. The curves, together with the correlation, may be used to calculate the probability that an undiscovered rhyolite-hosted tin deposit will possess at least a given grade and tonnage. Using standard statistical procedures for correlated variables (see Netter and Wasserman, 1974, 399—401) there is a 5% chance that an undiscovered rhyolite-hosted tin deposit will contain more than 2014 t of ore and also have a grade greater than 0.86% Sn.

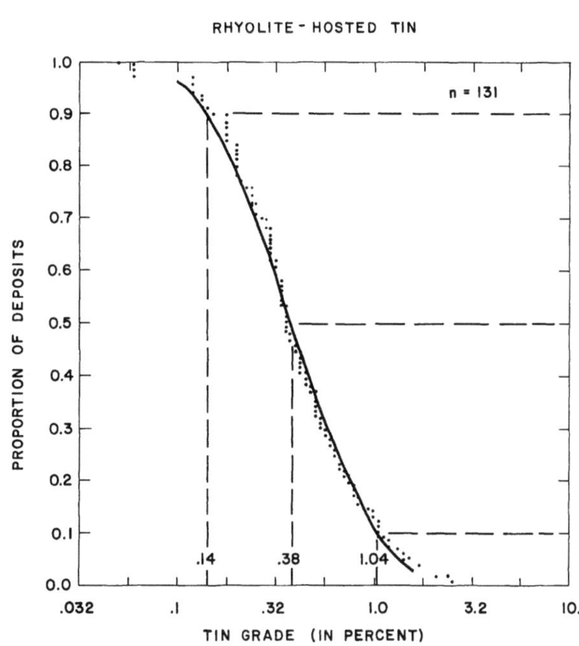


**Fig. 4.** Tin grades of rhyolite-hosted tin deposits

**Table 1.** Deposits used to construct the replacement, tin skarn, greisen tin and hydrothermal tin lodes grade/tonnage models. Location codes: AUNS – New South Wales, Australia, AUNT – Northern Territories, Australia, AUQL – Queensland, Australia, AUTS – Tasmania, Australia, BLVA – Bolivia, BRZL – Brazil, CNNS – Nova Scotia, Canada, CZCL – Czechoslovakia, GRBR – Great Britain, INDS – Indonesia, MLYS – Malaysia, USAK – Alaska, United States. Deposits are listed for largest to smaller tonnage

| Replacement Deposit | |
|---|---|
| Deposit | Location |
| Renison Bell | AUTS |
| Mount Bischoff | AUTS |
| Queen Hill | AUTS |
| St. Dizier | AUTS |
| Cleveland | AUTS |
| Razorback | AUTS |

| Skarn Deposits | |
|---|---|
| Deposit | Location |
| Lost River | USAK |
| Moina | AUTS |
| Pinnacles | AUQL |
| Gilliam | AUQL |

| Greisen Deposits | |
|---|---|
| Deposit | Location |
| Altenberg | GRME |
| East Kempville | CNNS |
| Hub | CZCL |
| Cinovec | CZCL |
| Ardlethan | AUNS |
| Potosi | BRZL |
| Coal Creek | USAK |
| Anchor | AUTS |
| Archer | AUTS |
| Cista | CZCL |
| Prebuz | CZCL |

| Hydrothermal Vein Deposits | |
|---|---|
| Deposit | Location |
| Pahang | MLYS |
| Wheal Jane | GRBR |
| Carocoles | BLVA |
| Mount Wellington | GRBR |
| Geevor | GRBR |
| Carn Brea-Tincroft | GRBR |
| Dolcoath | GRBR |
| Kelapa Kampit | INDS |
| Basset | GRBR |
| Krupka | CZCL |
| Grenville | GRBR |
| Aberfoyle | AUTS |
| Mawchi | BRMA |
| Levant | GRBR |
| Irvinebank | AUQL |
| Wheal Kitty-Penhalls | GRBR |
| South Crofty | GRBR |
| Herberton | AUQL |
| Royal George | AUTS |
| Stannary Hills | AUQL |
| Killifreth | GRBR |
| Watsonville | AUQL |
| Mount Wells | AUNT |
| Ottery Lode | AUNS |
| Mount Paynter | AUNS |
| Mowbray Creek | AUQL |
| Coolgarra District | AUQL |
| Mount Nolan District | AUQL |
| Emuford District | AUQL |
| Nymbool District | AUQL |
| Bakerville | AUQL |
| Adventure Creek | AUQL |
| Maranboy | AUNT |
| Dargo Range District | AUQL |
| Glenlindale District | AUQL |
| Brownsville | AUQL |
| Hales Siding | AUQL |
| Emu Creek | AUQL |
| Gurrumba District | AUQL |
| Silver Valley | AUQL |
| Conrad Lodes | AUNS |
| Bloodwood Creek | AUQL |
| Gundle | AUNS |

## Replacement (Exhalative [?]) Deposits

Although the genesis of replacement, or exhaltive (?), tin deposits has been debated in recent years, many characteristics of the deposits are not in dispute. The deposits are stratabound bodies of cassiterite and sulphide minerals, that replace carbonate rocks. The deposits are associated with faults that, depending upon interpretation, were either paths for movement of synsedimentary hydrothermal brines (Hutchison, 1979, 1981 and 1982; Plimer, 1980), or alternatively were paths for the migration of

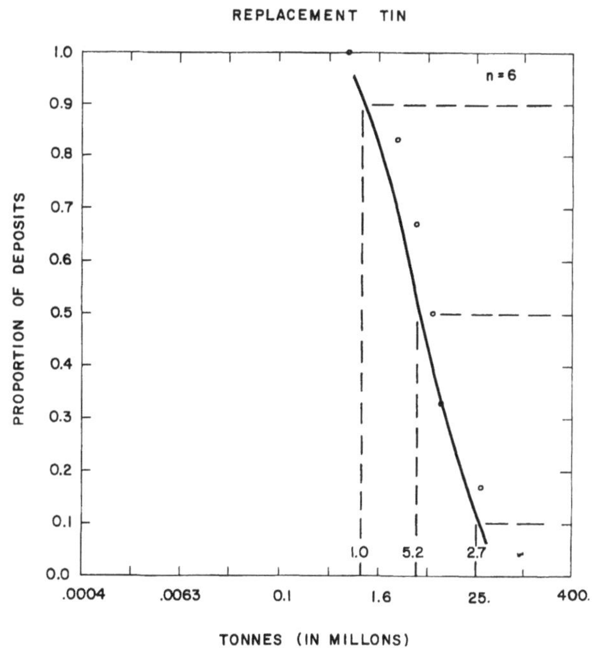


**Fig. 5.** Tonnages of ore of replacement, or exhalative (?), tin deposits

magmatic aqueous fluids related to the intrusion of underlying granites (Patterson et al., 1981; Patterson, 1982). The most common ore minerals in these deposits are pyrrhotite, arsenopyrite, cassiterite, chalcopyrite, ilmenite and fluorite. Minor minerals include pyrite, sphalerite, galena, stannite, tetrahedrite and magnetite. The best known examples of this type are the Renison Bell, Cleveland and Mt. Bischoff deposits, Tasmania. Replacement deposits also occur in the Dachang and Gejiu ore fields, People's Republic of China. Each of these fields contains a number of deposits, not all of which are replacement tin deposits. Although no official reserves or resources are available, the Changpo-Tongkeng deposit, a replacement deposit in the Dachang field, may contain 35 to 50 Mt of ore with a grade of 0.7 to 1.0%. This would exceed the announced reserves, resources and past production for Renison Bell.

The grade/tonnage model of replacement, or exhalative (?), tin deposits is built with data for six deposits from Tasmania (Table 1). The data for individual deposits include, where applicable, past production, reserves and resources. Sources of data include Newnham (1975), Govett and Robinson (1980), Roskill (1981), Ranson and Hunt (1975), Knight (1975), Ingram (1977) and the Mining Magazine. It cannot be determined if the resources were calculated at the same cut-off grade, but they were calculated within a few years of each other under presumably similar economic conditions. Though the number of deposits used to build the model is small, lognormal distributions fit moderately well observed tonnages and grades. Tonnage and grade are not significantly correlated (r = 0.26). Figure 5 presents the plot for tonnage which shows that 90% of the deposits contain at least one million tonnes of ore, half of the deposits contain 5.2 Mt or more and 10% contain at least 27 Mt. 90% of the deposits have grades of at least 0.55% Sn, half of the deposits have grades of 0.80% or more and 10% of the deposits have grades of at least 1.2% Sn (Figure 6).

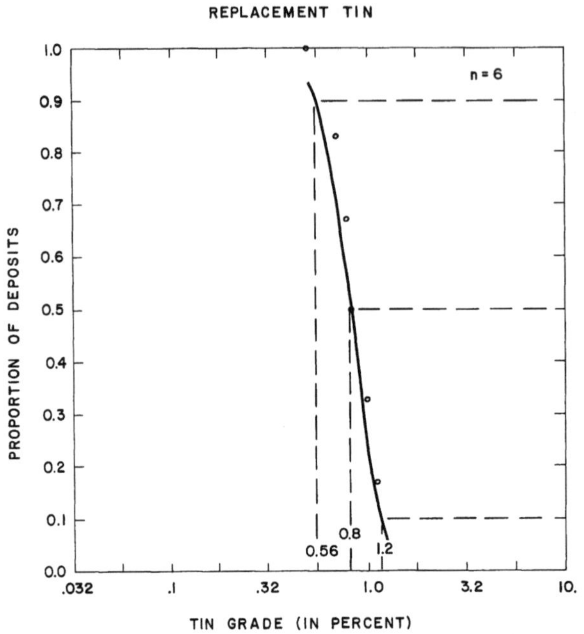


**Fig. 6.** Tin grades of replacement, or exhalative (?), tin deposits

## Tin Skarn Deposits

Tin skarn deposits occur at or near the contacts of carbonate rocks and, generally, leucogranites. Most major skarn is developed within 300 m of the pluton and may be controlled by open fractures and felsic dykes. Ore minerals include cassiterite, scheelite, sphalerite, pyrrhotite, magnetite, pyrite, chalcopyrite, arsenopyrite and fluorite. Skarn formation is multistage and the mineralogy is complicated by repetitive phases of replacement. Commonly, massive – granular skarn is composed of garnet, vesuvianite, magnetite, pyroxene and fluorite. A distinctive laminar ("wrigglite") magnetite-fluorite-vesuvianite skarn may be present – such as described at Moina, Tasmania (Kwak and Askins, 1981) and Lost River, Alaska (Dobson, 1982). Associated granites may be greisenized near their margins. In many cases the commercial viablility of tin skarns is presently limited because tin is not present as an economically recoverable metal. Much tin may be trapped within silicate structures (garnet, malayaite, amphibole, helvite, etc.) or occur as non-recoverable exsolution blebs of cassiterite in magnetite.

Four skarn deposits with reported tonnages and grades were used for the present study (Table 1). Sources of data include Taylor (1979), Govett and Robinson (1981), and Northern Miner. Normally a grade/tonnage model would not be built with so few data, but in the process of building the present models, it became apparent that the skarn deposits have grade/tonnage as well as mineralogical characteristics that are distinctly different from replacement (exhalative) deposits. While replacement and skarn tin deposits are approximately of the same size (Figures 5 and 7), skarn deposits are lower grade than replacement deposits (Figures 6 and 8).

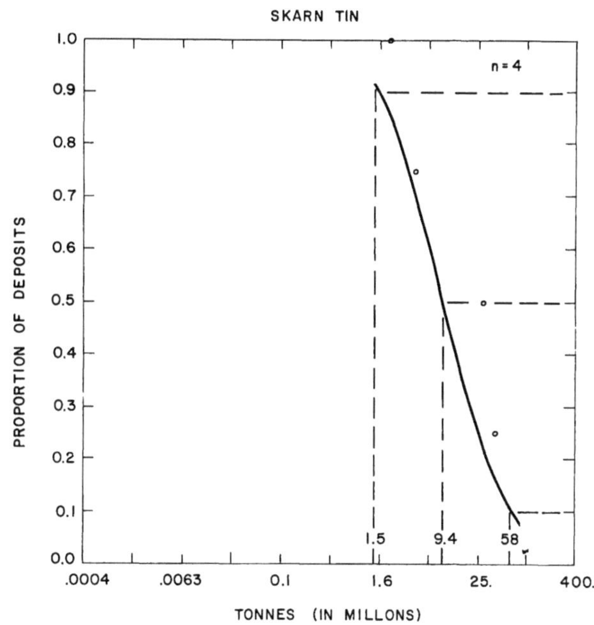


**Fig. 7.** Tonnages of ore of skarn tin deposits

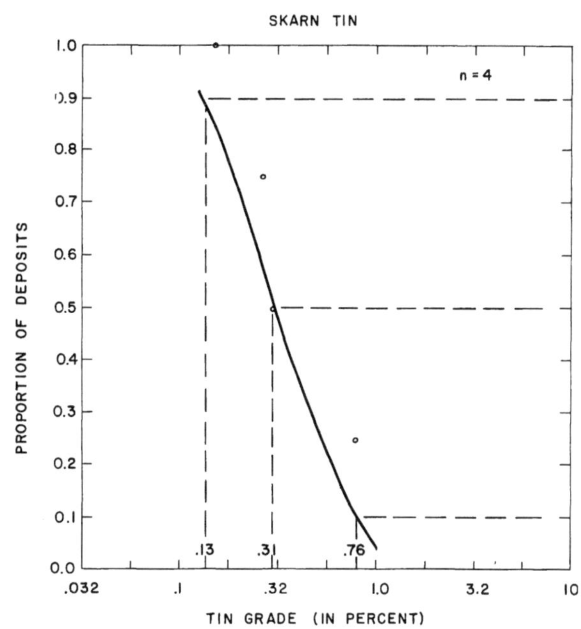


**Fig. 8.** Tin grades of skarn tin deposits

## Greisen Deposits

Tin greisen deposits consist of disseminated cassiterite and cassiterite-bearing veinlets, lenses, pipes and breccias in greisenized granite (Reed, 1982). The deposits are associated with late phases of specialized biotite and/or muscovite leucogranites. Lodes are preferentially located in or near apical portions or along the margins of granitoids. Fractures and faults may be important ore controls. Ore and associated minerals, which commonly are zonally arranged, include cassiterite, molybdenite, arsenopyrite, beryl, wolframite, bismuthinite, base-metal sulphides, sulphostannates, quartz, fluorite, topaz, tourmaline, calcite and pyrite. Alteration is varied and, in granites, ranges from incipient greisen (muscovite ± chlorite, tourmaline and fluorite) to massive greisen (quartz-muscovite-topaz ± fluorite and tourmaline) in which no original textures are preserved. Well known greisen deposits include the Anchor deposits, Australia, the classic massive greisen deposits in the German Democratic Republic and Czechoslovakia and the East Kempville deposit in Nova Scotia.

The grade/tonnage model of greisen deposits is based on data from eleven deposits in six different countries (Table 1). Only a few have been producers; for those deposits, the estimates of tonnage and grade are based upon past production, reserves and resources. Estimates for unmined deposits are based upon reported reserves and resources. Sources of data include Ingram (1977), Taylor (1979), Laznicka (1973), and Mining Magazine. Ninety percent of these deposits contain at least 0.87 Mt, half contain at least 7.0 Mt and 10% contain at least 57 Mt (Fig. 9). Figure 10 shows that 80% of these deposits contain between 0.17 and 0.54% Sn. Lognormal distributions appear to adequately fit the observed tonnages and grades; tonnage and grade are not significantly correlated (r = –0.32) in the greisen deposits.

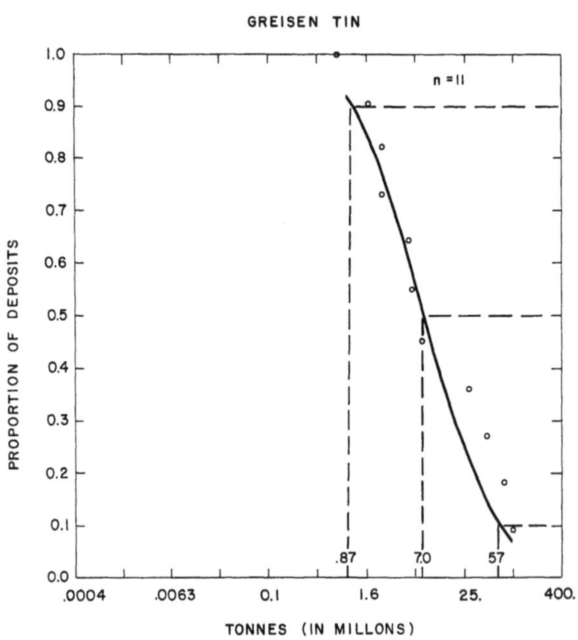


**Fig. 9.** Tonnages of ore of greisen tin deposits

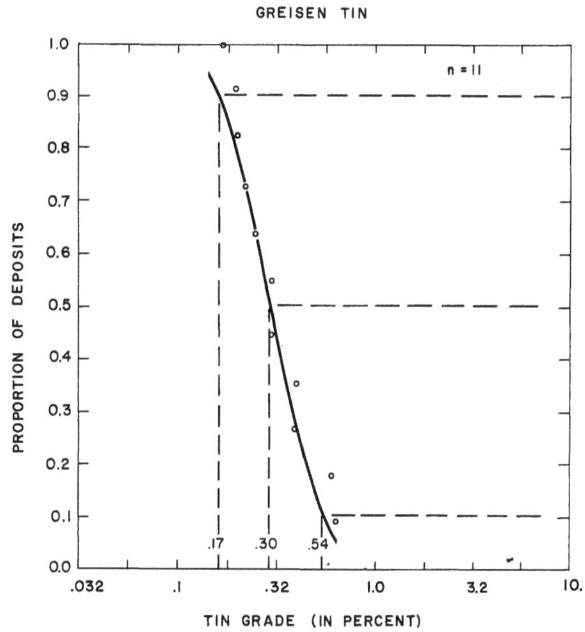


**Fig. 10.** Tin grades of greisen tin deposits

## Hydrothermal Lodes

Hydrothermal, or Cornish type, lodes are fissure fillings or replacement lodes that form in or near multiphase, specialized biotite and/or muscovite leucogranites. The lodes tend to occur within or above the apical portions of the granite cusps. Local controls include variations in vein structure, lithological and structural changes and intersections of veins with other veins, dykes or cross-faults. Mineralogy of the lodes is extremely varied; cassiterite may be accompanied by wolframite, arsenopyrite, molybdenite, hematite, scheelite, beryl, galena, chalcopyrite, sphalerite, stannite, and/or bismuthinite. Ore minerals may be zonaly arranged (see Hosking, 1969). Rocks adjacent to the lodes are usually hydrothermally altered; sericitization (greisen development), tourmalization, silicification, chloritization and hematitization are the commonly recognized types of alteration. The classic lodes of Cornwall and those at Herberton are examples.

The grade/tonnage model of hydrothermal lodes is based on data from forty-three deposits, or in some cases, districts (Table 1). Data include eight deposits from Cornwall, nineteen "districts" from the Herberton area, eight deposits from elsewhere in Australia, and one deposit each from Malaysia, Bolivia, Indonesia, Czechoslovakia, and Burma. The imprecise definition of what constitutes a deposit arises because many of the lodes were mined by a number of operators during the second half on the nineteenth century. Consequently, because properties changed hands and merged, records for individual deposits were rarely kept or are no longer available. Thus, data on the deposits and districts in the model consist mainly of recorded past production, though for some deposits, especially those still operating, reserves are also included. Sources of data include Bawden (1929), Blake (1972), Dines

(1956), Govett and Robinson (1980), Ingram (1977), Laznicka (1973), Norman and Trangcotchasan (1982), Taylor (1979), annual reports of the Queensland Department of Mines, and various copies of the Queensland Government Mining Journal, Mining Magazine and Mining Yearbook.

The fit of the observed grades and tonnages of the hydrothermal lode deposits to lognormal distributions is not good. Deposit grades are slightly peaked; too many deposits have grades between 1.25 and 1.5% Sn. This is probably the result of inadequate records for earlier mined deposits. The distribution of deposit tonnages appears to be slightly bimodal (Fig. 11). This is probably the result of mixing data from Australia, where both small and large deposits are reported, with data from other areas, where there appears to be a tendency to report grades and tonnages of only larger than average deposits. For example, if data were available for the many small deposits from Cornwall, they would plot in the portion of the curve presently dominated by Australian deposits. Tonnages and grades are not significantly correlated (r = 0.18) in this model. Ninety percent of the hydrothermal lodes have at least 12 kt, half of them have at least 240 kt and ten percent have at least 4.5 Mt of ore (Figure 11). Figure 12 shows that ninety percent of these deposits have grades of at least 0.7% Sn, half have grades of at least 1.3% and 10% have grades as great or greater than 2.3% Sn.

## Uses of Grade/Tonnage Models

During the last ten years grade/tonnage models have been used increasingly to convey to non-geologists an idea of the importance of undiscovered deposits in land classifica-

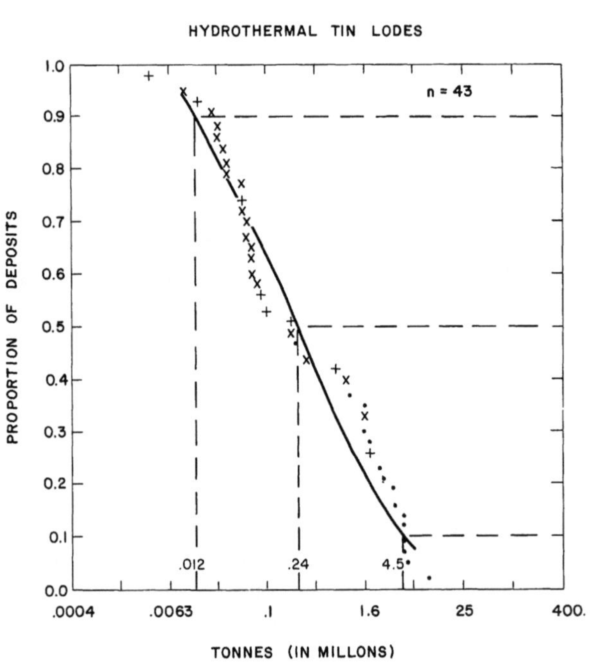


**Fig. 11.** Tonnages of ore of hydrothermal tin lodes (x deposit from Herberton district, + deposit from elsewhere in Australia, • other deposits)

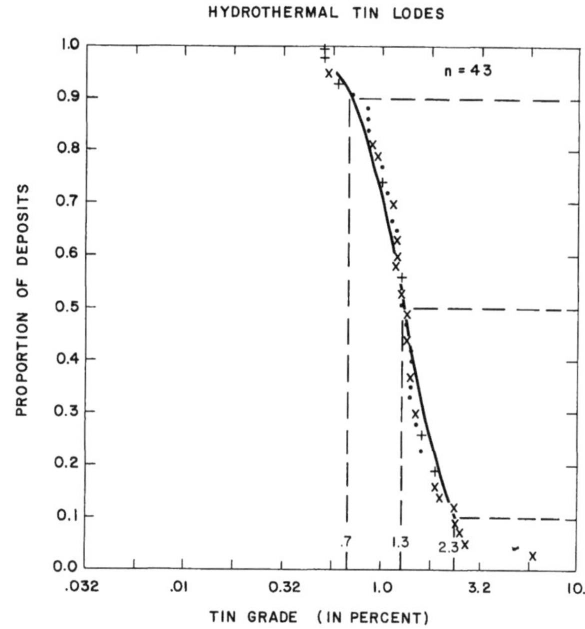


**Fig. 12.** Tin grades of hydrothermal tin lodes (x deposit from Herberton district, + deposit from elsewhere in Australia, • other deposits)

tion, mineral exploration strategy, and estimates of long term resources available. Grade/tonnage models can provide a basis for quantifying undiscovered resources which commonly are ignored unless they are estimated in a quantitative manner.

Classification of lands involves placing individual tracts into a category that permits certain uses and prohibits others. A decision-maker faced with a land use decision asks geologists to delineate areas that may contain undiscovered deposits, provide grade/tonnage models of the types of deposit present and estimate the number of undiscovered deposits of each type that may be present. A mineral economist can use this information, together with assumptions about commodity prices, the amount and efficienty of exploration and the cost of producing the minerals, to estimate the value of lands under different classifications (Singer, 1975).

Delineation of tracts of land that may contain undiscovered mineral deposits is accomplished by comparing the characteristics of descriptive deposit models with geological information from the area being assessed. Grade/tonnage models are built with data for well explored deposits. Estimates of the number of undiscovered deposits present in the tracts may be made by subjective probability or multivariate statistical models. Examples of these types of mineral assessment are given by Singer and Ovenshine (1979), and Menzie et al. (1983).

The use of grade/tonnage models in formulating exploration strategy can range from strategic decisions of what type of deposits will be the most profitable for exploration, to tactical decisions such as estimating the likely outcome of a particular exploration programme. Figure 13 shows, for each of the five primary tin deposit types, domains defined by the mean plus and minus one standard deviation of the logarithms of grade and contained metal. Each domain should contain approximately 45 percent of deposit of its type. Figure 13 indicates that, if other factors such as loca-

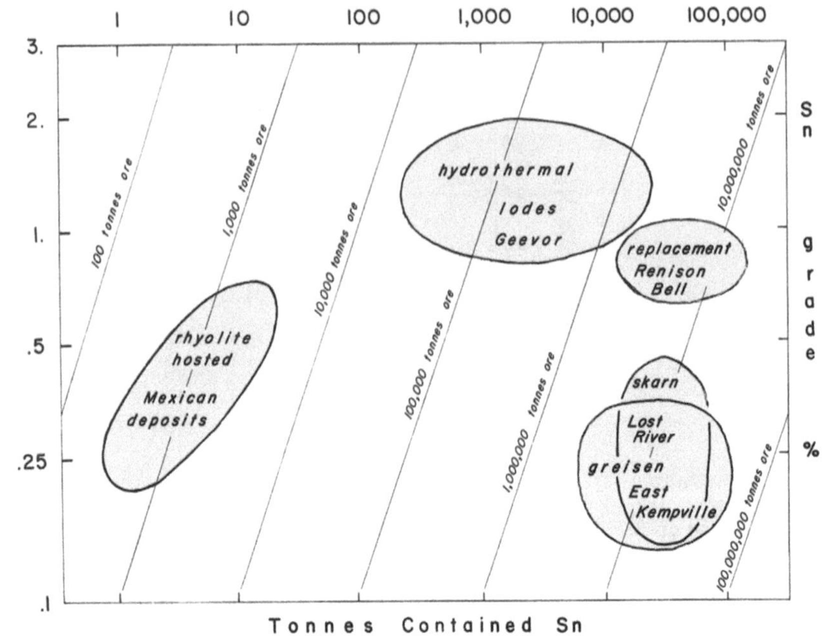


**Fig. 13.** Grade and contained tin domains for five types of primary tin deposits. Domains are defined by the mean plus and minus one standard deviation of the logarithms of grade and contained metal

tion, mining costs and probability of discovery are equal, replacement deposits are, by virtue of their higher grades and economically extractable tin content, more attractive exploration targets than skarn deposits.

Fig. 13 can also be used to gain perspective on questions of national mineral supply. For example, the United States, which has a few discovered rhyolite-hosted skarn and greisen deposits, is more likely to effect its national supply of tin by encouraging exploration for additional greisen, or replacement deposits rather than rhyolite-hosted deposits. The part of the Seward Peninsula (Alaska) Paleozoic carbonate terrane that is cut by the Late Cretaceous tin-granite belt could well host replacement tin deposits.

## References


Arkin, H. and Colton, R.R., 1963. *Tables for statisticians;* New York, Barnes and Noble, 168 p.

Bawden, E.R., 1929. The Killifreth Mine, Cornwall: *Mining Magazine* (May), 279—286.

Blake, D.H., 1972. Regional and economic geology of the Herberton tinfield, north Queensland. *Australia Bureau of Mineral Resources, Bull.* 124, 265 p.

Bracho Valle, Filipe, 1960. Yaciemientos de estano en La Sierra de Chapultepec, Zuc., La Ochoa, Dgo. y Cosio. *Ags.: Consejo de Recursos Naturales no renovables, Boletin* 48, 116 p.

Bracho Valle, Filipe, 1961. Yaciemientos de estano en La Ochoa, Dgo. y Juan Aldama, Zac.: *Consejo de recursos Naturales no Renovables, Boletin* 60, 87 p.

Cox, D.P., ed., 1983a. U.S. Geological Survey – INGEOMINAS mineral resource assessment of Columbia. Ore deposit models. *U.S. Geological Survey Open-File Report* 83—423, 64 p.

Cox, D.P., ed., 1983b. U.S. Geological Survey - INGEOMINAS mineral resource assessment of Columbia. Additional ore deposit models. *U.S. Geological Survey Open-File Report* 83—423, 31 p.

Dines, H.G., 1956. The Metalliferous Mining Region of South-West England. *Memoirs of the Geological Survey of Great Britain, England and Wales,* volumes 1 and 2, 795 p.

Govett, M.H. and Robinson, H.A., compilers, 1980. *The world tin industry – supply and demand.* Australian Mineral Economics Pty Ltd., 278 p.

Hosking, K.F.G., 1969. The nature of primary tin ores of the South-West of England; in Fox, W., ed., *A second technical conference on tin,* Bangkok, vol. II, London ITC, 1157—1243.

Hutchinson, R.W., 1979. Evidence of exhaltive origin for Tasmanian tin deposits. *Canadian Inst. Mining Metall. Bull.,* v. 72, 90—104.

Hutchinson, R.W., 1981. Lode tin deposits of exhalative origin; in Haji Hassan and van Wees (eds.) *Complex tin ores and related problems,* Technical Publication 2, SEATRAD CENTRE, Ipoh, Malaysia, 81—106.

Hutchinson, R.W., 1982. Geologic setting and genesis of cassiterite-sulfide mineralization at Renison Bell, Western Tasmania – a discussion. *Economic Geology,* v. 77, 199—206.

Ingram, J.A., 1977. Australian tin deposits. *Australia Bureau of Mineral Resources, Report* No. 7, 77 p.

Knight, C.L., 1975. Mount Bischoff Tin Orebody; in Knight, C.L. (ed.) *Economic Geology of Australia and Papua New Guinea, 1. Metals,* Monograph Series No. 5, Australian Institute of Mining and Metallurgy, 591—2.

Kwak, T.A.P. and Askins, P.W., 1981. Geology and genesis of the F-Sn-W (-Be-Zn) skarn (Wrigglite) at Moina, Tasmania. *Economic Geology,* v. 76, 439—467.

Laznicka, Peter, compiler, 1973. *Manifile; the University of Manitoba File of Non-ferrous Metal Deposits of the World.* Manitoba University, Dept. of Earth Sciences, Centre Precambrian Studies Pub. 2, 3, v.: Pt. 1., 533 p; Pt. 2., 698 p; Pt. 3, 767 p.

Menzie, W.D. and Mosier, D.L., 1983. Sediment-hosted exhalative zinc-lead; in Singer, D.A. and Mosier, D.L. (eds.) mineral deposit grade-tonnage models. *U.S. Geological Survey Open-File Report* 83—623, 69—73.

Menzie, W.D. and Theodore, T.G., 1983. Molybdenum porphyry (low F type): in Singer, D.A. and Mosier, D.L. (eds.) Mineral Deposit Grade-Tonnage Models, *U.S. Geological Survey Open File Report* 83—623, 31—33.

Menzie, W.D., Foster, H.L., Tripp, R.B., and Yeend, W.E., 1983. Mineral resource assessment of the Circle quadrangle, Alaska. *U.S. Geological Survey Open-File Report* 83-170-B, 57 p.

Netter, J. and Wasserman, W., 1974. *Applied linear statistical models.* Homewood, Ill., Irwin, 399—401.

Newnham, L.A., 1975. Renison Bell Tinfield: in Knight, C.L. (ed.) *Economic Geology of Australia and Papua New Guinea,* 1. Metals, Monograph Series No. 5, Australasian Institute of Mining and Metallurgy, 581—583.

Norman, D.I. and Trangcotchasan, Y., 1982. Mineralization and fluid inclusion studies of the Yod Nam Mine, Southern Thailand: in Evans, A.M., ed., *Metallization Associated with Acid Magmatism,* John Wiley and Sons, 261—272.

Patterson, D.J., 1982. Geologic setting and genesis of cassiterite-sulfide mineralization at Renison Bell, Tasmania – a reply. *Economic Geology,* v. 77, 203—206.

Patterson, D.J., Ohmoto, H., and Solomon, M., 1981. Geologic setting and genesis of cassiterite-sulfide mineralization at Renison Bell, Tasmania. *Economic Geology,* v. 76, 393—438.

Plimer, I.R., 1980. Exhalative Sn and W deposits associated with mafic volcanism as precursors to Sn and W deposits associated with granites. Mineralium Deposita, v. 15, n. 3, 275—289.

Ransom, D.M. and Hunt, F.L., 1975. Cleveland Tine Mine: in Knight, C.L. (ed.) *Economic Geology of Australia and Papua New Guinea,* 1. Metals Monograph Series No. 5, Australasian Institute of Mining and Metallurgy, 584—591.

Reed, B.L., 1982. Tin greisen model: in Erickson, R.L., compiler, Characteristics of Mineral Deposits and Occurrences. *U.S. Geological Survey Open-File Report* 82—795, 55—61.

Roskill Information Services, Ltd., 1981. *The Economics of Tin,* Third Edition, 354 p.

Singer, D.A., 1975. Mineral resource models and the Alaskan Mineral Resource Assessment Program: in Vogely, W.A. (ed.) *Mineral Materials Modeling, A state-of-the-art review, Resources for the future,* Washington, D.C., 370—382.

Singer, D.A. and Mosier, D.L. (eds)., 1983a. Mineral deposit grade-tonnage models. *U.S. Geological Survey Open-File Report* 83—623, 100 p.

Singer, D.A. and Mosier, D.L. (eds.), 1983b. Mineral deposit grade-tonnage models II. *U.S. Geological Survey Open-File Report* 83—902, 101 p.

Singer, D.A. and Ovenshine, A.T., 1979. Assessing metallic resources in Alaska. *American Scientist,* v. 67, 582—589.

Singer, D.A., Theodore, T.G. and Mosier, D.L., 1983. Molybdenum porphyry - Climax: in Singer, D.A. and Mosier, D.L. (eds.) Mineral Deposit Grade-Tonnage Models. *U.S. Geological Survey Open-File Report* 83—623, 28—30.

Taylor, R.G., 1979. *Geology of tin deposits.* Elsevier, Amsterdam, 543 p.

# 1.5 Exploration Strategies for Primary Tin Deposits

C. PREMOLI[1]


## Abstract

Primary tin deposits are becoming an increasingly attractive exploration target, particularly for well capitalized, and technologically advanced expoloration companies.

The exploration procedures for such tin deposits are rather different from the ones utilized for other mineral deposits; although they are still tentatively known they are developing fast.

The paper briefly reviews exploration targets, field techniques and costs in the search for hard-rock, possibly low-grade, tin resources.


## Purpose

With the inevitable depletion of alluvial tin deposits in SE Asia and the progressively lower grades of the Bolivian tin mines, the interest worldwide is increasingly focussing on the exploration of new tin resources amenable to modern, low-cost mining. Large, low grade, primary tin deposits can be a target for such exploration projects and under certain conditions these deposits can compare favourably with deep alluvial mining whether on-shore or off-shore (Taylor, 1979; Rich, 1980).

Unfortunately comparatively little is known of exploration and mining for primary tin, particularly in tropical rain forest conditions. For instance, although Indonesia, Malaysia and Thailand have supplied more than 60% of the western world's tin, hard rock mining has yielded barely 2% of the recorded production.

When a major company has decided to include tin in its exploration portfolio (and an encreasing number of them are presently doing so) the first problem consists in defining the minimum grade and tonnage required to support a large tin mining operation at the projected tin prices. Figure 1 gives the tonnages and grades of several major primary tin deposits of the western world. Various authors have recently proposed a target scale in the order of 40–60 million t of ore at a head grade of at least 0.20% Sn and a projected recovery of 60%. However the presently depressed world tin consumption suggests that only higher grades will be economic in this century, probably of the order of 30–40% Sn depending on available infrastructure, mining and benefication costs. In any case these targets would mean a recoverable resource of at least 60,000 t of tin. If we project a mine life of 20 years we obtain a projection

[1] International Mineral Research, 16/3-5 St. Neots Avenue, Potts Point, Sydney, Australia 2011

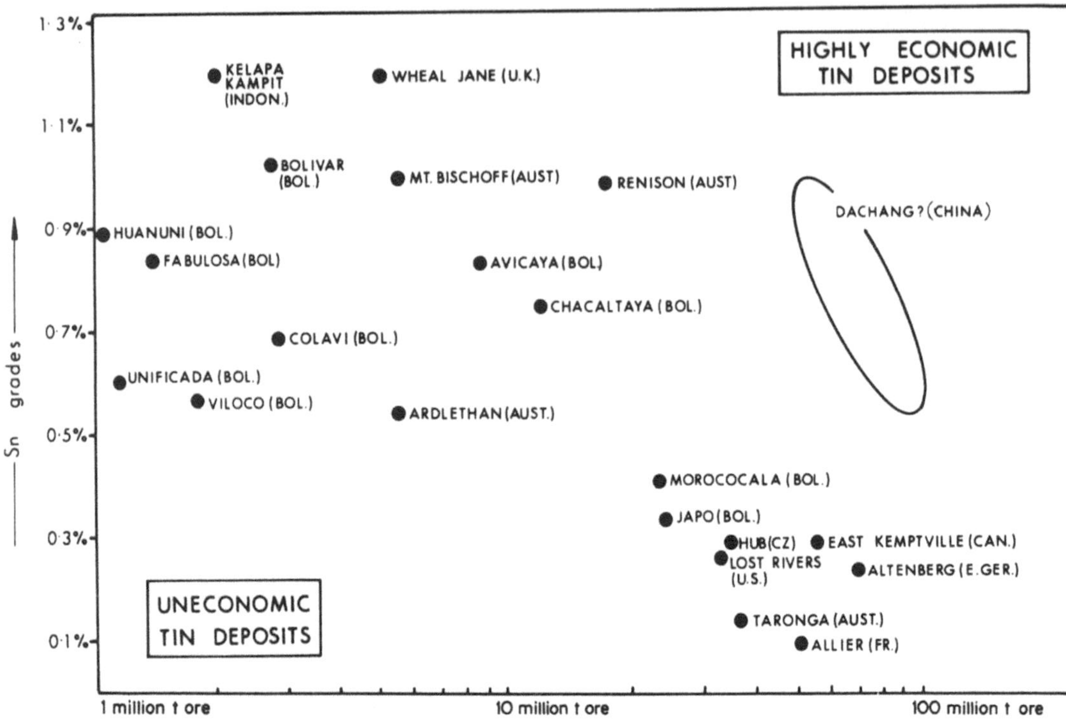


**Fig. 1.** Tonnage and grade of some major primary tin deposits of the Western World. The clustering of the Bolivian deposits to the left of the graph indicates a serious ore reserve problem in this country

of 3,000 $ty^{-1}$ of tin or about 2% of the present western world's tin production (see Fig. 2). This represents a comparatively new and challenging exploration target for which, at present, little case history exists.

The prime objective of a company which decides to explore for primary tin is to find such an orebody at a minimum cost and within a minimum time-frame.

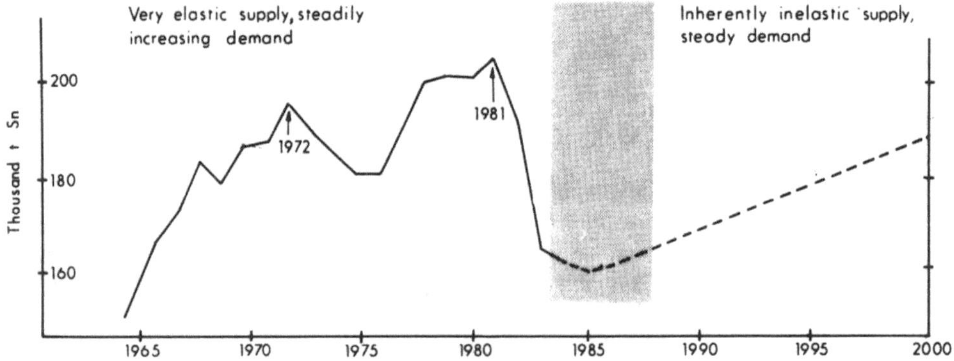


**Fig. 2.** Past and projected tin production in Market Economy Countries. The dashed areas represent a period of probable phasing-out of high-cost producers

**Procedures**

Several countries around the world (and particularly certain developing countries) present the rare combination of overall geological favourability and a propitious investment climate (Hutchison and Taylor, 1978; Bettencourt et al., 1981; Suensilpong et al., 1983). This combination makes a commitment to exploration for major primary tin deposits an acceptable risk venture. Obviously the exploration strategies appropriate to the different countries vary considerably and, with few exceptions, it is a prerequisite to start with carefully prepared negotiations with the various governments. The mining code, the fiscal regime, and above all the willingness of the various governments to encourage or even permit foreign investments in this field is all important.

Geologically the type of primary tin deposit that could come within the grade-tonnage parameters mentioned and, above all, be amenable to low-cost, bulk mining either by underground or opencast methods, are:

1. Subvolcanic hydrothermal deposits of stockwork type (also called 'porphyry tin' by some authors, e.g. Sillitoe et al., 1975).
2. Greisen-type deposit (in all its numerous subtypes) and limestone replacement deposits.
3. Pegmatite-type deposits.

These three broad types of deposits are essentially defined in function of the depth of emplacement of the tin-bearing granites (see Fig. 3).

Only the first two types of deposit are economic and realistic for mining, the third type still belongs to the sub-economic resource class. Each of these three types of tin deposit tend to have rather distinctive orebody geometry, mining byproducts and tonnage-grade curves. Obviously these factors must be carefully considered before deciding on any new major exploration development.

The tonnage-grade curve of major tin deposit types (see Fig. 4) is particularly significant. Obviously these curves largely depend on the mining methods selected. Traditionally the hydrothermal vein system' deposits have been the most attractive exploitation targets. This has sometimes generated much necessary development capital for more costly bulk exploration methods, but more often hopelessly exhausting the potential for very large, if low grade, tin resources. On the contrary large, pegmatitic tin deposits have been shown to be particularly difficult to be selectively mined (other than in their superficial, weathered part) and the raising of their development capital has proved, at least in the market economies, invariably a slow and difficult operation.

**Methods**

The exploration methodology for large, low grade, tin deposits is still in its infancy. The model orebody, although large by tin standards, is still one order of magnitude smaller than, for instance, porphyry copper deposits (see Fig. 5).

Its identification requires highly sophisticated methods, particularly in tropical rain forests conditions prevailing in the majority of the most potential tin provinces of the world (see Fig. 6).

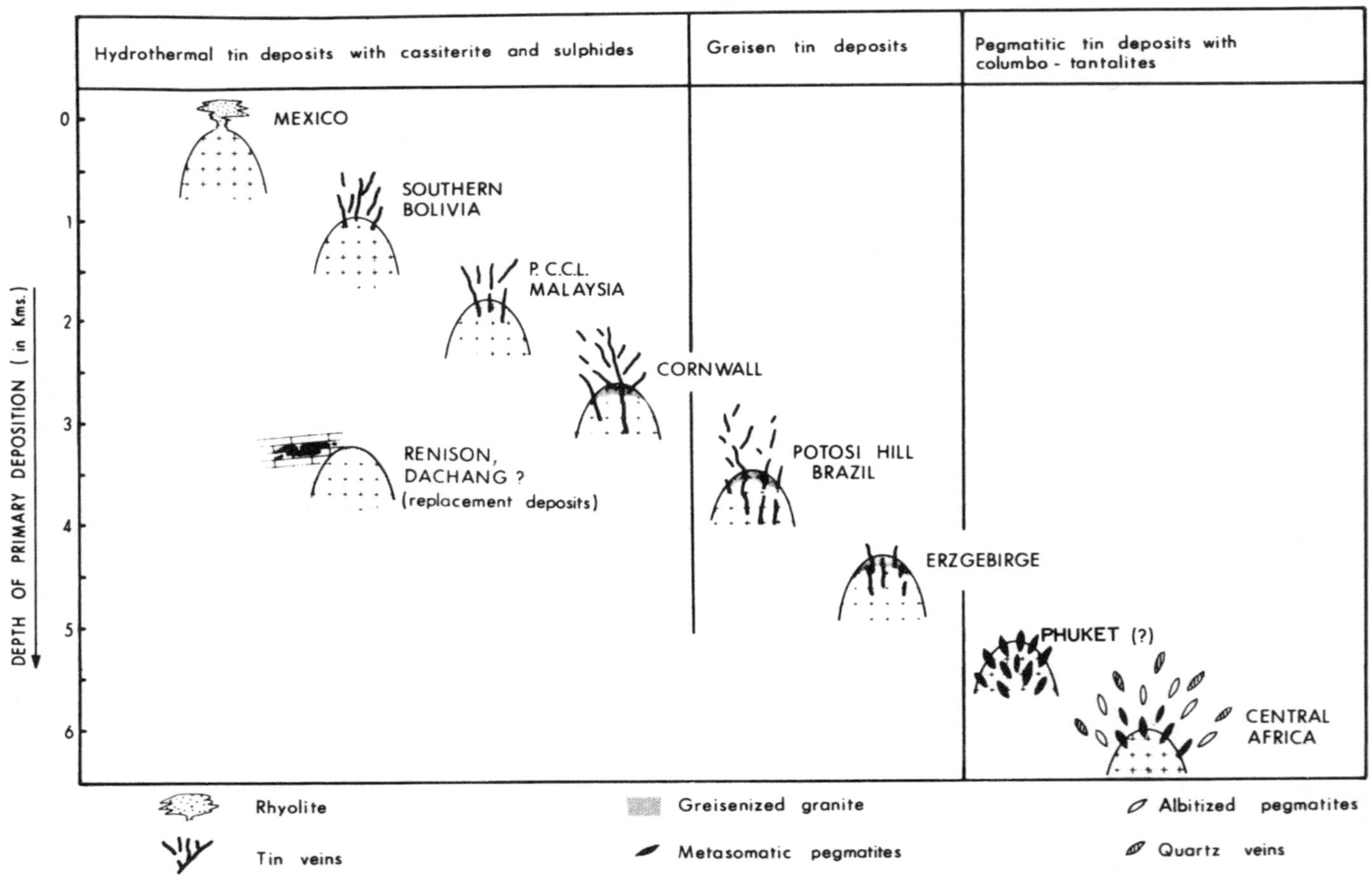


**Fig. 3.** Classification of tin deposits according to the depth of emplacement of the associated granitoids. At the present time only epizonal and mesozonal deposits are mining realities as large primary tin sources

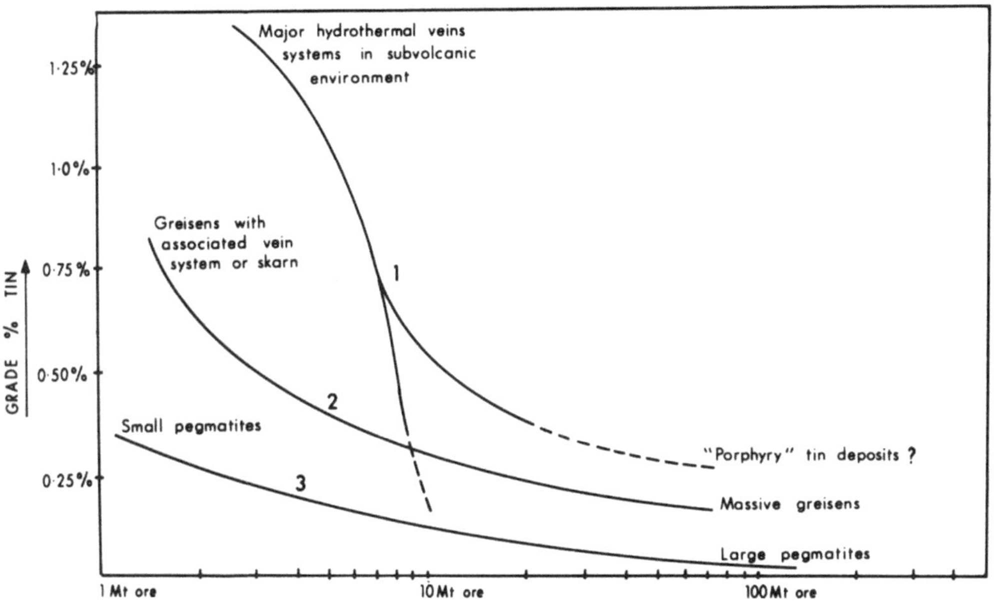


**Fig. 4.** Possible grade-tonnage curves for three main types of large, primary tin deposits

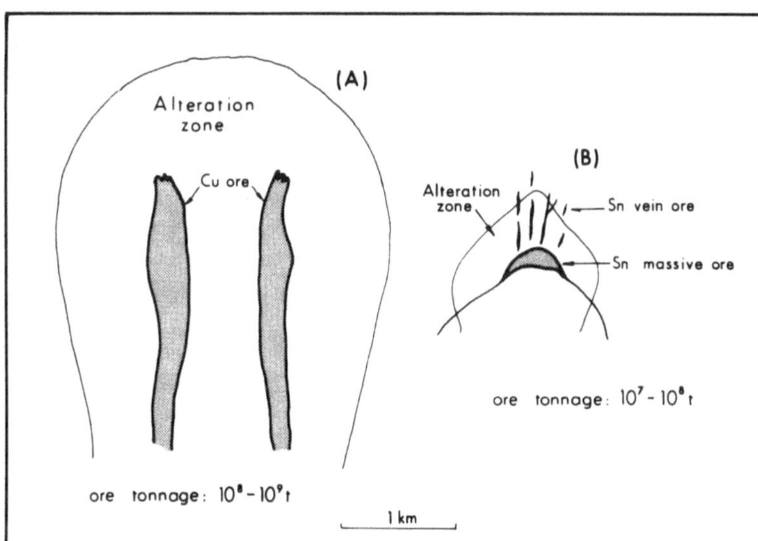


**Fig. 5.** Comparison between size and geometry of disseminated Cu and Sn mineralizations. (A): porphyry Cu system (From Lowell and Guilbert, 1980), (B): massive greisen Sn system (freely adapted from Shcherba, 1970, and other authors)

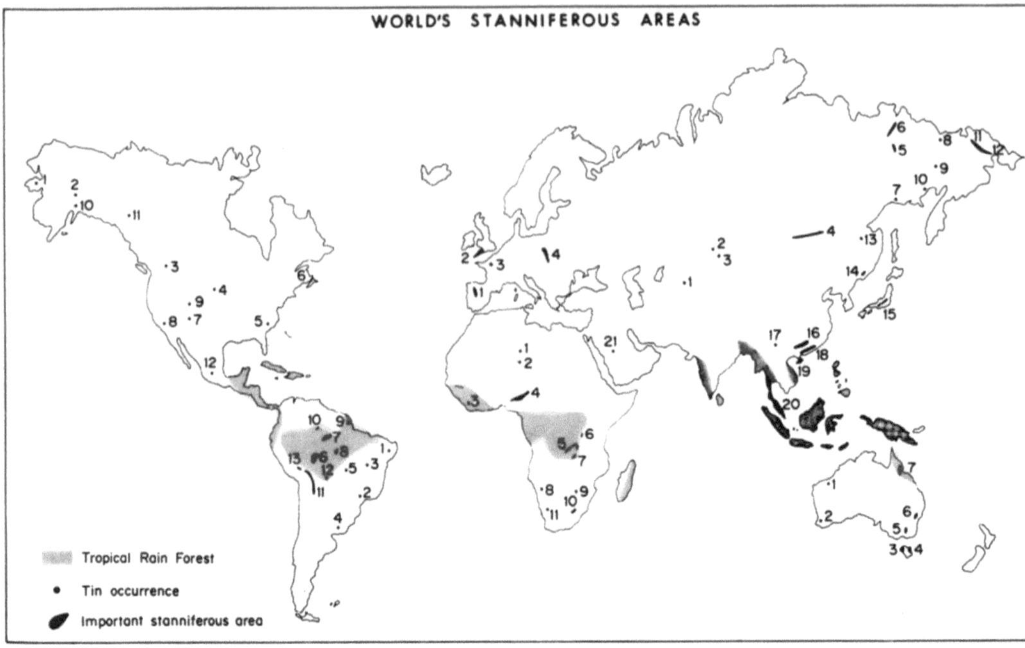


**Fig. 6.** World's stanniferous areas (adapted from Sainsbury, 1969)

The obvious need to develop new methods for discovering blind tin deposits that have no surface expression has resulted in the formulation of the modern 'regional' approach to exploration. Given the highly potential greisen model (Shcherba, 1970), 'cupola detection' plays a prominent role in the exploration for primary tin orebodies.

It is only very recently that the search for large, low grade tin deposits was started, by developing regional statistical methods which take into account what is known of the geometry of large tin deposits, their relationship to granite bodies, and other factors. Figure 7 summarizes the modern methods used in tin exploration from reconnaissance through evaluation.

A typical exploration sequence for primary tin (see Fig. 8) may involve several different techniques.

As a regional exploration tool, remote sensing has been widely used. LANDSAT interpretation has proved helpful. For instance, the recently discovered Potosi Hill greisen tin deposit in Rondonia has a faint LANDSAT expression. Various other remote sensing techniques are useful in the exploration for primary tin deposits, particularly when coupled with an airborne geophysical survey; these methods however, are more applicable for the definition of broadly favourable areas (tin granites) rather than direct target identification.

Geochemistry invariably plays a major role in any exploration programme for primary tin deposits. Its role is twofold; first, rock geochemistry for the identification of tin granites (Ishihara et al., 1979; Plimer, 1980), and secondly, soil or stream sediment sampling for direct target identification. In stream sediment geochemistry, a polymetallic approach is advisable, particularly for remote or little explored areas; in Brazil for instance, several alluvial tin and gold provinces are known to overlap. Sev-

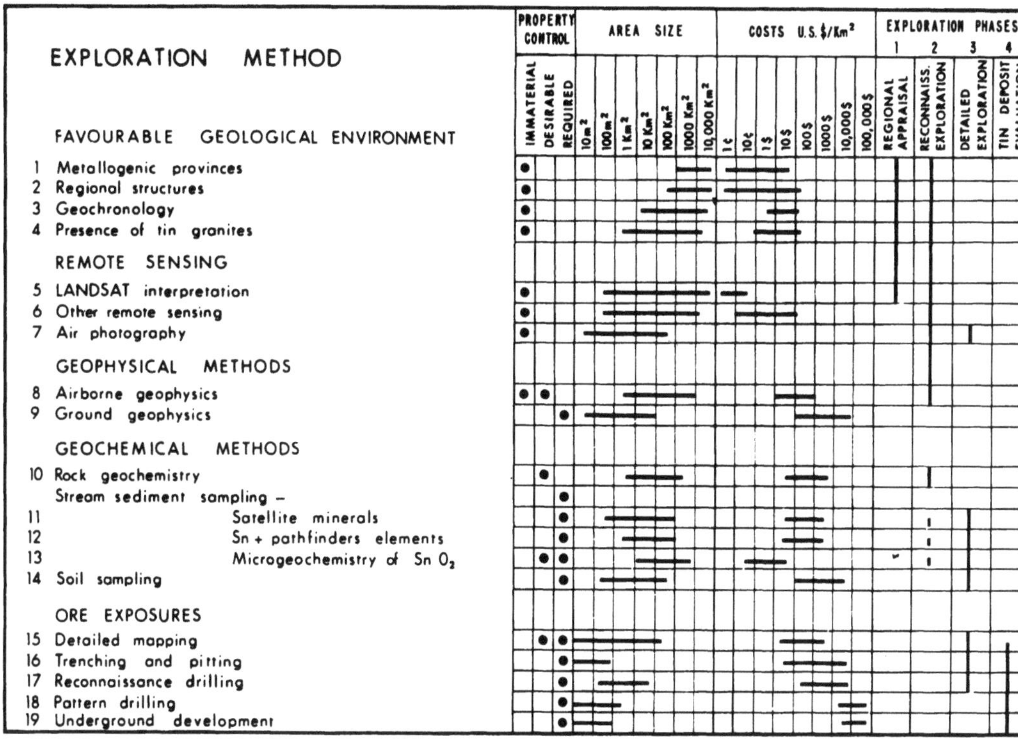


**Fig. 7.** Criteria in the selection of exploration methods for primary tin deposits

eral pathfinder elements for primary tin mineralizations have been recognized and are gaining widespread acceptance. Significantly there are different pathfinder elements depending on whether the tin deposit is sought in hydrothermal, greisen or pegmatitic environments.

Other and more sophisticated geochemical methods are increasingly used in exploration for primary tin deposits; identification of detrital indicator minerals (tourmaline, topaz, lithium mica, etc.), anomalous tin content in common resistates (ilmenite, magnetite, etc.), microchemistry of individual cassiterite grains, etc.

Technologically advanced laboratories, possibly equipped with X-Ray Fluorescence (XRF) and Inductively Coupled Plasma (ICP) facilities are becoming essential in modern tin exploration. Most systematic geochemical surveys must be preceded by thorough and imaginative orientation surveys. This is particularly true for new or little explored areas.

Geophysics has so far proved less useful for the identification of either tin granites or individual tin deposits. However, several methods based either on the magnetic susceptibility or the spectrometric signatures of the tin granites have been tested at regional level both in SE Asia and in Australia, with encouraging results (Webster, 1984; Yeates et al., 1982). This is obviously a field of great potential which has seen significant development in the last few years.

At the evaluation stage tri-dimensional sampling of a tin deposit requires special precautions. Several tin specialists have pointed out the unusual difficulties in grade

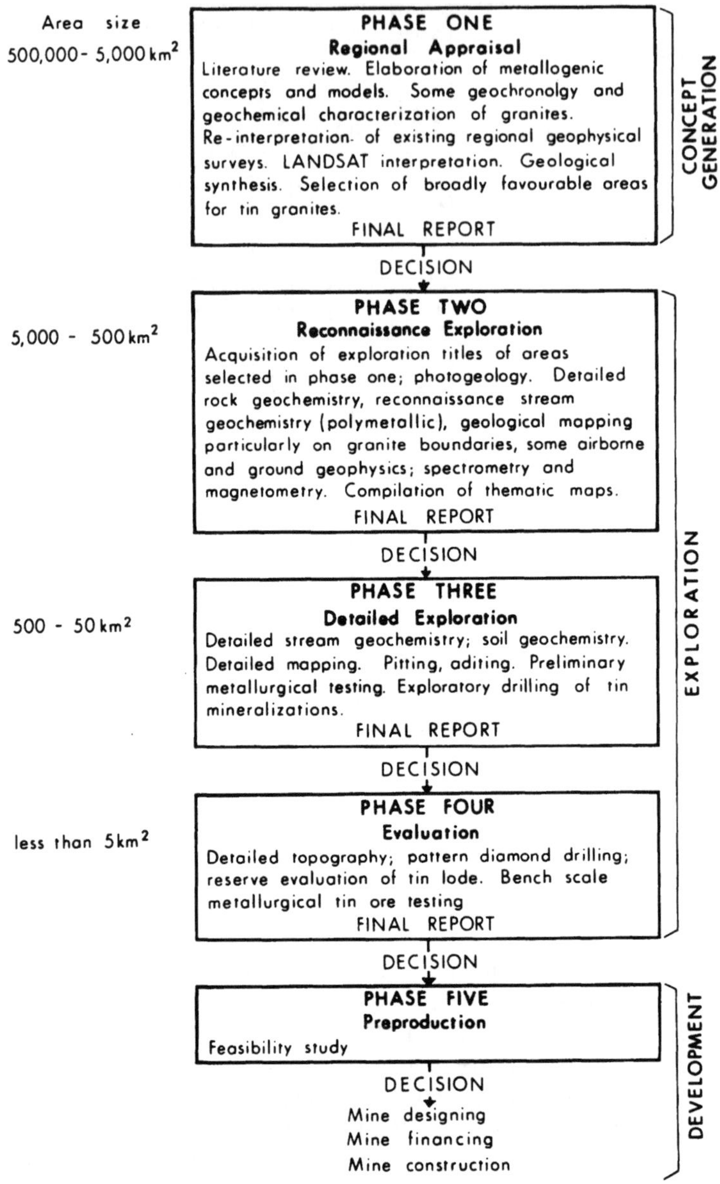


**Fig. 8.** Normal exploration and development sequence for large, primary tin deposits

assessment of tin deposits by drill hole intersections (Hosking, 1979). Here automatic data processing, although unquestionably a very valuable tool in ordinary circumstances, can at times be utterly misleading as it tends to reduce the all-important role of personal experience and shrewd guesswork as to whether the tin deposit being drilled can be economically developed or not. At this stage some emphasis is generally given to exploratory underground development in order to obtain large and truly representative samples for metallurgical test work: invariably a very important aspect in exploring and developing a primary tin orebody.

It has been noted that primary tin deposits, particularly large, low grade ones, are essentially geological objects which may have only an attenuated geophysical and geochemical expression and highly erratic assay data. Furthermore, their case history is very limited and they can be plagued with an uncommon number of development problems. Hence the person responsible for making decisions on their exploration must have an above-average knowledge of ore geology, resource management and mineral economics.

## Exploration Costs

Exploration costs for primary tin deposits can vary considerably. Each country has its own legal and logistic problems and a very different cost structure. Here analogies are not very applicable. In countries with abundant man-power a 'hybrid' exploration strategy can be developed. The regional exploration strategy and targets can be selected utilizing the most sophisticated state-of-the-art, whether plate tectonics, advanced metallogenetic criteria, airborne geophysics, computer modelling, etc. On the other hand the actual, and indispensable ground work on the selected areas can be successfully carried out by numerous but unsophisticated prospectors who have been trained in elementary tin prospecting techniques. After all, tin is the second metallic commodity, immediately after gold, which has traditionally been explored for, worldwide, by individual prospectors.

Estimating overall costs for regional exploration campaigns for primary tin deposits is difficult, and costs tend to escalate quite dramatically in tropical rain forest conditions. For the Market Economy Countries it has been calculated that to successfully reduce a general area of interest from 5,000 $km^2$ to a mineralized area of 1 $km^2$ or less may require from 3 to 5 years and costs between 2 M and 20 M U.S. dollars. These costs can be significantly reduced should the exploration start at the evaluation stage rather than 'concept generation' stage (Bailly, 1972; Rongfu et al., 1983).

Several possibilities to evaluate newly found prospects or old, and possibly misunderstood tin mines, do exist in several countries in Latin America or S.E. Asia.

## References

Bailly, P.A., 1972. Mineral exploration philosophy: *Mining Cong. Jour.*, v. 58, No. 4, 31–35.

Bettencourt, J.S., Damasceno, E.C., Fontanelli, W.S., Franco, J.R.M. and Pereira, N.M., 1981. Brazilian tin deposits and potential. *I.T.C. Fifth World Conference on Tin,* Kuala Lumpur, 1981 (pre-print, 69 p.).

Ishihara, S., Sawata, H., Arpornsuwan, S., Busaracome, P., Bungbrakearti, N., 1979. The magnetite-series and ilmenite-series granitoids and their bearing on tin mineralization, particularly of the Malay Peninsula region. *Geol. Soc. Malaysia Bull.* 11, 103–110.

Hosking, K.F.G., 1979. Tin distribution patterns. *Geol. Soc. Malaysia Bull.* Vol. 11, 1–70.

Hutchison, C.S. and Taylor, D., 1978. Metallogenesis in SE Asia, *J. Geol. Soc. London.* Vol. 135, 407–428.

Plimer, I.R., 1980. A review of the principal types of tin deposits in the world and models for their genesis in *"Tin in South Eastern Australia"* S.M.E.D.G. Seminar, 1982, 1–3.

Rich, P.J.H., 1980. Future of tin as a tonnage commodity. *Trans. Inst. Min. Metall.*, Sect. A, 88: A8—A16.
Rongfu, P., Zhizong and Mingke, Fu., 1983. A preliminary discussion on the reasonable sequence of solid mineral prospecting exploration and development. *Bulletin of* the Chinese Academy of Geol., Sciences, N. 5 Beijing, 13—16.
Sainsbury, C.L., 1969. Tin resources of the world, *Bulletin U.S. Geol. Survey,* 1301, 55 p.
Shcherba, G.N., 1970. Greisens. *International Geological Review,* 12; 114—151, 239—254.
Sillitoe, R.H., Halls, C. and Grant, J.N., 1975. Porphyry tin deposits in Bolivia. *Economic Geology,* 70: 913—927.
Suensilpong, S., Putthapiban, P., Mantajit, N., 1983. Some aspects of tin granite and its relationship to tectonic setting. *Geol. Soc. of America Memoir* 159, 77—85.
Taylor, R.G., 1979. *Geology of tin deposits.* Amsterdam, Elsevier, 1979, 543 p.
Webster, S.S., 1984. A Magnetic Signature for Tin Deposits in South-East Australia. *Expl. Geoph.* 15, 15—31.
Yeates, A.N., Whyatt, B.W. and Tucker, D.H., 1982. Application of gamma-ray spectrometry to prospecting for tin and tungsten granites, particularly within the Lachlan Fold Belt, N.S.W., *Econ. Geol.* v. 77, 1725—1738.

# 2 Australia

# 2.1 The Western Tasmanian Tin Province with Special Reference to the Renison Mine

L.A. Newnham[1]

## 1 Introduction

Tasmania is the dominant tin producing state in Australia, with current production coming from two major underground tin mines situated in the Western Tasmanian tin province.

Most of the tin deposits in this province occur either as replacement deposits in Lower Palaeozoic sediments adjacent to Upper Palaeozoic granites or to a lesser extent within these granites themselves. Small alluvial deposits are also developed from material derived from the granites.

Due to the complex nature of their geology, exploration for these types of deposit in Western Tasmania is difficult and because of their depth and mineralogically complex nature, expensive high technology mines and treatment plants are required to extract the tin.

This paper provides a brief description of the geology of the Western Tasmanian tin deposits, together with a more detailed description of the Renison deposits, followed by some generalised comments on exploration methods and problems encountered in searching for deposits of this kind.

## 2 Western Tasmanian Tin Province

### 2.1 Location and History

The State of Tasmania consists of a group of islands lying off the South-East corner of the main Australian continent. It lies at the Southern end of a 3,500 km long geosyncline which runs the full length of the East Coast of Australia (Fig. 1).

The majority of Australia's tin deposits lie within this geosynclinal trough.

Exploration for mineral deposits only commenced in Tasmania in the second half of the nineteenth century. During that period exploration teams, working in virtually uninhabited country, discovered the outcrops of several tin deposits all of which are shown on Fig. 2.

A major mine was established at Bischoff and smaller mines were initially developed on the other deposits during this early period. However most of these oper-

[1] Gold Fields Exploration, Burnie, Tasmania, Australia 7320

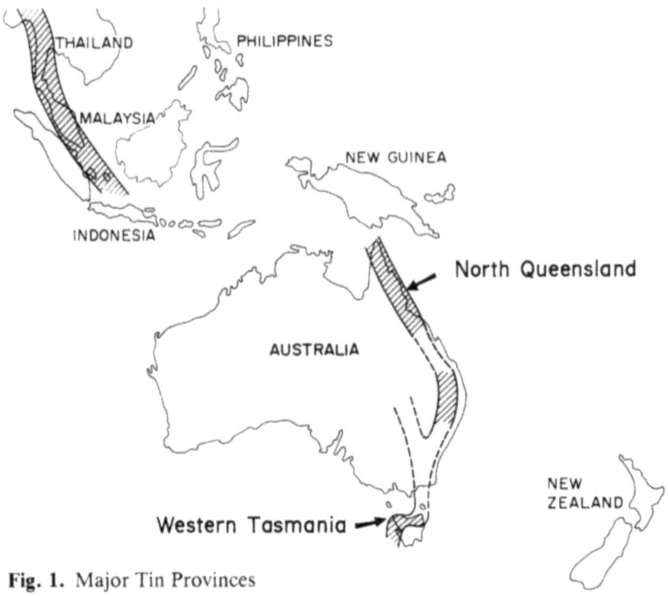


**Fig. 1.** Major Tin Provinces

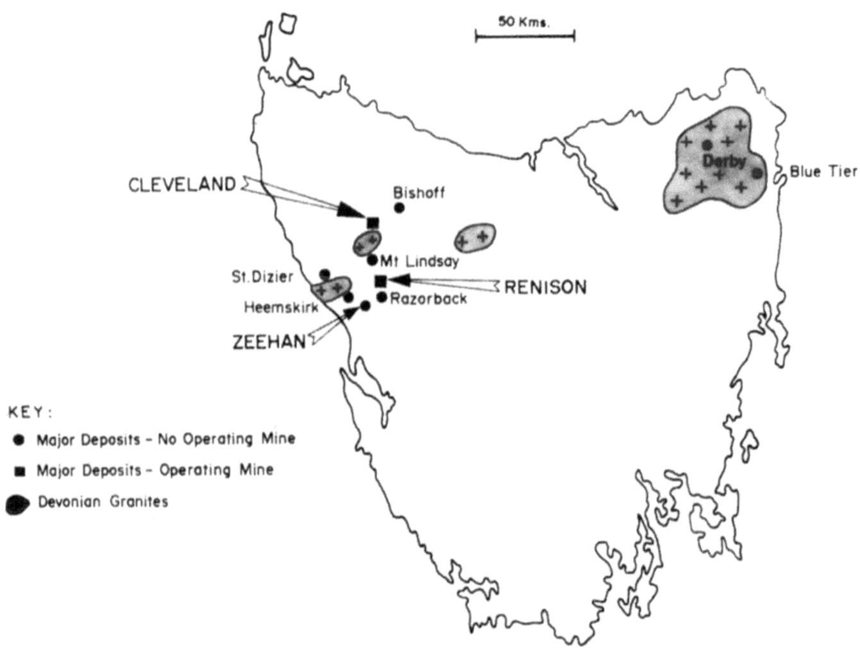


**Fig. 2.** Major Tasmanian Tin Deposits

ations were closed down by 1930, and it was not until the 1960's, that major new mines were developed on the Renison and Cleveland deposits.

## 2.2 Regional Geology

All of the tin deposits of Western Tasmania lie in a 20 km wide 200 km long trough of Palaeozoic sediments and volcaniclastic sediments, wedged between two blocks of Proterozoic rocks. This trough runs up the West Coast of Tasmania and curves along the North Coast where it is terminated by a major cross cutting faulted graben structure.

The sedimentary succession, which is about 5 km thick, is represented by a wide range of sediments including conglomerate, sandstone, chert, shale, greywacke and carbonates (dolomite and limestone). A minor component of lavas and tuffaceous material within these sediments was derived from a coeval belt of island-arc volcanism which occupied the eastern half of the trough.

Minor ultramafic bodies were emplaced into the sedimentary pile in the Upper Cambrian.

During the Upper Palaeozoic, the region underwent a period of strong tectonic deformation which resulted in broad folding and severe faulting of the sedimentary pile, accompanied by the intrusion of several significant granite bodies. Only minor sedimentation and igneous activity occurred in Western Tasmania after this major period of tectonism and plutonism.

The land form was extensively modified by Pleistocene glaciation.

## 2.3 The Tin Deposits

The major tin deposits are shown on Fig. 2. All the deposits are primary, located either in granite or sediments.

Major operating tin mines are located at Renison and Cleveland. A third major deposit is located at Bischoff but active mining ceased there in the 1920's.

A large deposit is located at Zeehan but is currently not mined. Smaller deposits at Mt. Lindsay, Razorback, St. Dizier and Heemskirk have all produced small amounts of tin but are not producing at present.

In addition to the above, there is a large number of smaller primary and alluvial deposits, too numerous to represent on this figure.

The total tin content of these deposits is estimated as 60 Mt of 0.9% tin. This figure is the sum of past production and present reserves and does not include an estimate of additional mineralisation which may be found in the future.

The individual deposit sizes are graphically represented in Fig. 3. Again these data indicate the original size of the deposits and is the sum of past production and present reserves.

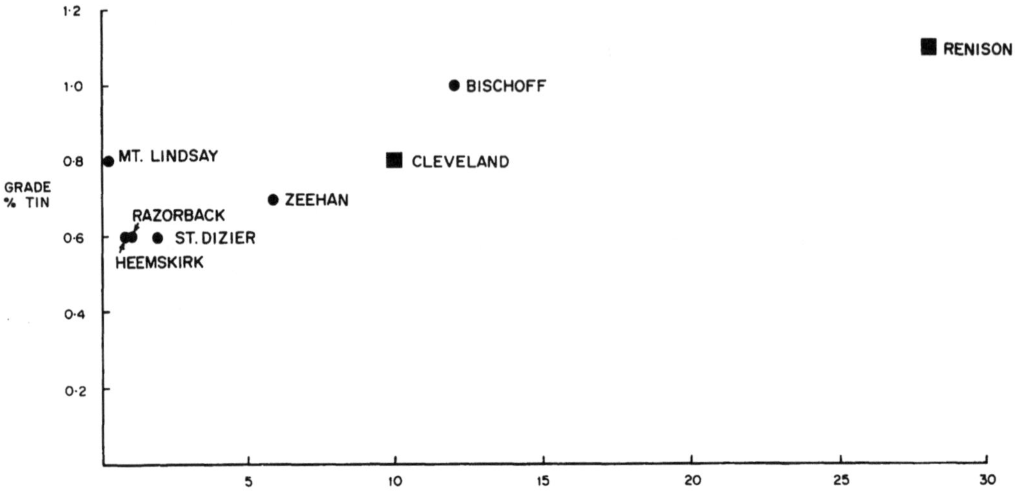


**Fig. 3.** Western Tasmanian Tin Mines. Deposit Sizes

A very brief description of each of the main deposits follows:

*Bischoff*

Tin was discovered at Bischoff in 1871 and in the period between 1873—1942 approximately 56,000 t of tin were produced (Groves et al., 1972).

Recent drilling in the vicinity of the former mine has established a possible resource of 6.1 Mt of ore of average grade 0.49% tin (Collins and Jennings, 1982).

The deposits occur within a folded and faulted sequence of Cambrian sediments which have been intruded by an Upper Palaeozoic swarm of granitic dykes (Groves et al., 1972).

Three main types of tin mineralisation are recognised: stratabound sulphide-cassiterite deposits which replace dolomite beds; fault or fissure filling quartz-sulphide-cassiterite bodies; and cassiterite-sulphide deposits within altered quartz-feldspar dykes.

*Cleveland*

The Cleveland primary tin deposits, which lie a short distance to the South West of Bischoff, were first discovered in 1893 but large scale mining was not commenced until 1968. Since then the deposits have produced 24,600 t of tin and 9,592 t copper from 5.5 Mt of ore (Cox and Dronsecka, this volume).

The orebodies are stratabound and were formed by the hydrothermal replacement of carbonate units within a steeply dipping Cambrian carbonate-chert sequence. Mineralisation is predominantly of pyrrhotite-pyrite-chalcopyrite-stannite and cassiterite (Cox and Dronsecka, this volume).

### *Mt. Lindsay*

The Mt. Lindsay primary tin deposits occur midway between the Cleveland and Renison deposits.

They were discovered about 1909, but have only produced several hundred tonnes of tin from a series of shallow underground and surface workings. Reserves are small, although a potentially large tonnage of refractory low grade mineralisation also exists in this region.

The mineralisation occurs in a series of parallel, steeply dipping skarn bodies. These skarns, which are typically 10—30 m thick, several hundreds of metres long and deep, were formed by the metasomatic effects of hydrothermal fluids rising from an underlying granite. The skarn bodies are zoned in a rather complex manner (Kwak, 1982), but most of the cassiterite occurs in localised assemblages of pyrrhotite-magnetite-amphibole mineralisation.

### *Renison*

South of Mt. Lindsay are the large Renison tin deposits. These are discussed in more detail later in this paper.

### *Razorback*

A few km south of Renison are the Razorback tin deposits. These deposits were mined on a small scale in the first instance by shallow tunnels and more recently by open-cut methods. The mine is currently closed and reserves are small. The mineralisation occurs predominantly as replacement bodies in an ultramafic sill, and to a lesser extent in fault zones and as minor veins and disseminations in conglomerate.

The mineralisation in the ultramafic replacement bodies is a pyrite-pyrrhotite-talc-cassiterite assemblage, deposited when hydrothermal fluids rose along fault structures from an underlying granite, extensively altering the ultramafic rock to talc and depositing sulphides and cassiterite.

### *Zeehan*

The recently discovered tin deposits at Zeehan occur within an older silver-lead-zinc mining field which was discovered in 1882 but exhausted by about 1913. Guided by extensive geophysical and geological surveys, recent deeper drilling beneath some of these old workings has succeeded in defining 7 Mt of ore of grade 0.7% Sn in several deposits of complex and variable mineralogy (Skey, 1983).

The deposits occur in a folded and strongly faulted sequence of Cambrian and Proterozoic (?) sediments and volcanics, and are generally considered to be hydrothermal replacement deposits, which may have originated from an underlying or adjacent granite.

Mineralisation is variable but is basically a complex mixture of sulphides (pyrite and phyrrhotite, sphalerite, galena and chalcopyrite), and gangue minerals (carbonates and quartz). Tin occurs dominantly as fine cassiterite with some minor stannite.

Mining operations have not yet commenced on these deposits.

*Heemskirk*

A variety of small tin deposits occurs within the southern margin of the multi-phase Upper Paleozoic Heemskirk Granite.

Approximately 300 t of tin were produced between 1880 and 1940 from many small workings. No mining is currently in progress and reserves are small, probably less than 1 Mt of ore of grade 0.6% Sn.

The mineralisation occurs both in quartz-tourmaline-topaz dykes and veins, and in sulphide rich pipe-like bodies (Roberts, 1984). The former type is generally of small, discontinuous bodies in which the tin occurs as coarse cassiterite.

The latter pipe-like bodies are more complex both in their shape and mineralogy, which typically consists of a mixture of pyrite, sphalerite and chalcopyrite with tin occurring both as fine cassiterite and stannite.

*St. Dizier*

The St. Dizier deposits lie 20 km north-west of Zeehan near the northern margin of the Heemskirk Granite.

They were discovered in 1877 and mined in a small way by shallow open-cut methods. No mining has taken place now on this property for many years.

The deposits which occur in a faulted, steeply dipping Cambrian-Precambrian carbonate bed adjacent to an Upper Palaeozoic granite, consist of several discontinuous lenses of magnetite and minor iron sulphides.

It is thought that the carbonate bed was extensively altered to serpentinite by the intrusion of the granite and that the tin was emplaced within this serpentinite during a later phase of metasomatism.

The tin occurs both as cassiterite and as a complex tin-iron-manganese hydroxide.

Reserves are estimated as 1 or 2 Mt of ore of grade 0.6% Sn.

## 3 The Renison Deposits

### 3.1 History

The Renison orebodies were discovered in 1890 both in alluvial and gossan forms.

Small scale mining of the gossanous or leached outcrops continued until 1920 when the only tin bearing material remaining was massive iron sulphides from which it was difficult to recover the tin. However, these problems were overcome in the 1930's and small scale open cut and underground mining of the cassiterite-massive pyrrhotite bodies continued.

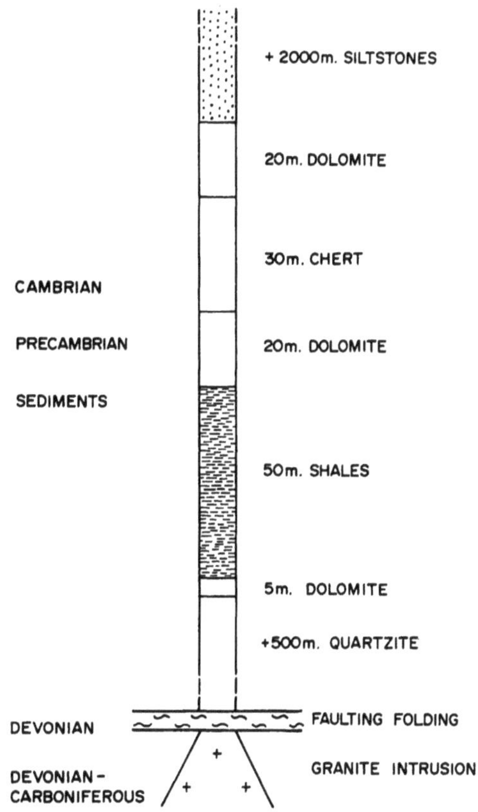


**Fig. 4.** Renison Mine Stratigraphy

Extensive exploration in the 1950's and 1960's successfully outlined substantial reserves and a large scale underground mining operation was commenced in 1967.

To date, approximately 10 Mt of ore have been mined for the production of 80,000 t of tin. Reserves are estimated as 17 Mt of ore of grade 1.1 % Sn.

## 3.2 Geology

The orebodies occur within a strongly faulted, gently folded sequence of Lower Palaeozoic sediments and minor volcanics, which was intruded during the Upper Palaeozoic by a granite (Patterson et al., 1981).

The basic stratigraphy of the Renison Mine is shown in Fig. 4. It consists of a thick sequence of quartzites and shales overlain by a metre thick dolomite which in turn is overlain by 50 m of shale, then 20 m of dolomite, then 30 m of hematitic chert and volcaniclastic sediments, then a further 20 m of dolomite, then finally a very thick sequence of silt-stone, mudstone and mafic volcaniclastic sediments. All of the above units are conformable and probably of Cambrian age.

In the Upper Palaeozoic, probably Devonian, the sediments were folded into a broad North-West trending anticline and extensively faulted. One major fault system

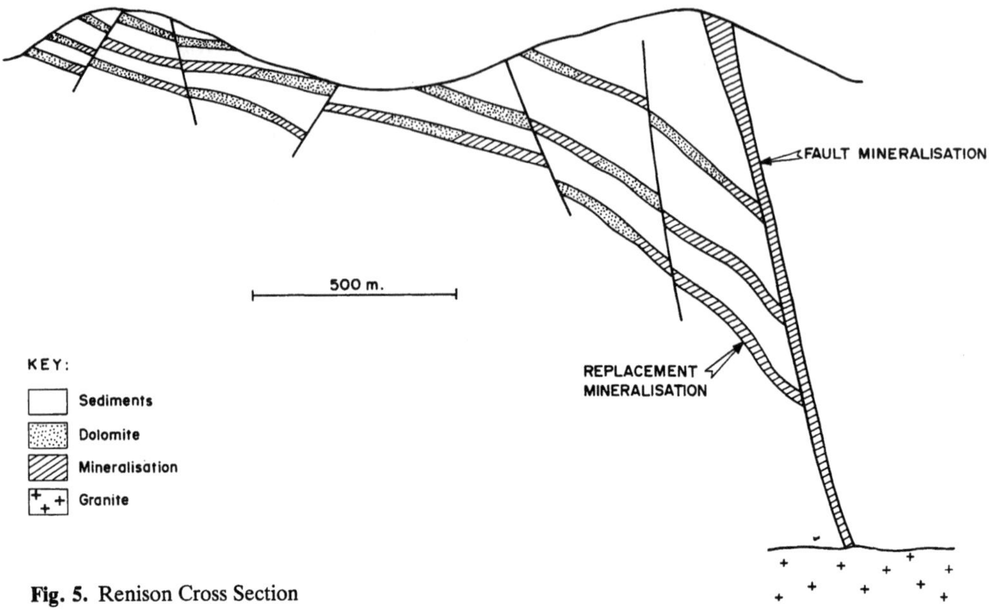


**Fig. 5.** Renison Cross Section

developed along the eastern flank of the anticline, whilst another important set of faults was developed across the anticlinal axis. Most, if no all, of this faulting is normal or strike-slip faulting.

Following this deformation phase, a granite was intruded into the area. It outcrops several km south of the mines, and lies 1,000 m beneath the surface in the vicinity of the major orebodies.

A typical cross section through the mine is shown in Fig. 5. A wide zone of thermal metamorphism, characterised by the presence of fine brown tourmaline, has been developed in the hornfelsed sediments adjacent to the granite.

### 3.3 Mineralisation

Three styles of mineralisation are of economic importance in the mining operation:

a) dolomite replacement bodies
b) fault infillings
c) breccia zones

The mineralisation for all three types is considered to have originated in the underlying granitic bodies and migrated in several phases into the sedimentary sequence along the abundant fault zones, at a temperature of between 300–350°C (Patterson et al., 1981).

When these hydrothermal fluids contacted the thin dolomite beds in the sequence, they replaced them by a process of microfracturing induced by gaseous overpressures developed during the reaction of hot fluids with dolomite.

Elsewhere the mineralisation was deposited in large quantities within fault zones, and within brecciated sediments.

**Table 1.** Renison Ore Mineralogy

| Dolomite Replacement | Fault Infilling | Breccia |
|---|---|---|
| Pyrrhotite | Pyrrhotite | Arsenopyrite |
| | Pyrite | Pyrite |
| | Arsenopyrite | |
| | Chalcopyrite | |
| Talc | Quartz | Quartz |
| Quartz | | Tourmaline |
| Cassiterite | Cassiterite | Cassiterite |

The mineralogy and hence the treatment characteristics of the three styles of mineralisation vary considerably. The main components of each type are shown in Table 1. The replacement bodies tend to be of very massive pyrrhotite with lesser amounts of quartz and talc. The tin in these bodies occurs as cassiterite and is typically in the size range of 0.05—0.20 mm. The fault infilling orebodies normally are more siliceous and accompanied by a more complex sulphide assemblage of pyrite, pyrrhotite, arsenopyrite and chalcopyrite. The tin in the faults is present as both cassiterite and minor stannite. The cassiterite grain size is typically in the 0.01—0.10 mm size range. The breccia zone mineralogy is more siliceous again, consisting of quartz and tourmaline with the sulphides arsenopyrite and pyrite only constituting 5—10% of the rock. Tin is present only as cassiterite which tends to be reasonably coarse in the 0.10—0.20 mm size range.

The sizes of the individual orebodies vary considerably. The stratabound replacement deposits can range from bodies of a few thousand tonnes up to massive sheets of pyrrhotite measuring 300 m down dip, 200 m along strike and from 5—20 m thick. The fault infilling bodies occur mainly scattered along the large fault system developed on the eastern limb of the anticline. This fault system is variably mineralised over a strike length of 2,000 m, a dip length of 1,200 m, and varies in width from 4—30 m.

## 3.4 Resource Extraction

The Renison mining method is a highly mechanised underground cut and fill system of ore extraction, accessed by a network of ramps or declines. All mining and ore haulage is undertaken by large scale rubber tyred diesel equipment (Goodman et al., 1982).

The ore is treated on site in a high technology mill by a sequence of heavy media separation upgrading followed by sulphide flotation, gravity recovery of cassiterite, flotation recovery of cassiterite and final upgrading in an acid leach (Goodsman et al., 1982).

Approximately 74% of the cassiterite is recovered into a 50% concentrate which is freighted by road and sea to smelters in Australia and Malaysia.

The current production from the mine is summarised in Table 2. Production is limited to 60% of capacity by the existing over supply of tin on world markets.

**Table 2.** Renison Mine 1984

| | | | |
|---|---|---|---|
| Tonnes | Mined | : | 500.000 |
| Grade | | : | 1–0% tin |
| Mill | Recovery | : | 74% |
| Tin | Produced | : | 3.800 tonnes |
| Reserves | | : | 17 Million tonnes at 1–1% tin |
| Workforce | | : | 500 |

## 4 Exploration Methods

The Western part of Tasmania is quite mountainous, heavily vegetated, cold, and has a high rainfall. These factors combine to make exploration for further tin deposits both difficult and expensive.

Exploration normally consists of a co-ordinated programme of geological mapping, geophysical and geochemical surveying followed by core drilling.

Because of the high rainfall, the water table is high and therefore weathering is not deep. This is a considerable benefit to mapping and geophysical surveys.

Most of the deposit styles sought respond to a wide range of geophysical techniques such as magnetics, electromagnetics and induced polarisation. Hence all these methods are widely used.

Geochemical methods, mainly by soil sampling, are also widely applied but with varying degrees of success.

Core drilling is perhaps the most widely used method of tin exploration in Tasmania. The combined amount of drilling undertaken by all tin exploring organisations in the area would be in the vicinity of 30–40,000 m per year.

Exploration in the region is expensive. Access is difficult and wages are high. Remote areas are today normally accessed either by helicopter or by road.

## 5. Conclusions

Western Tasmania is the site of one of the world's major primary tin provinces. The deposits are difficult to locate, expensive to mine and complex to treat but through the appication of modern exploration, mining and metallurgical methods, the area has become a substantial contributor to the market, further deposits will be located as a result of the substantial exploration effort currently taking place in the region.

**Acknowledgements.** The author wishes to acknowledge the co-operation of Rension Limited who gave permission for the preparation of this paper, and also to representatives of other companies operating in Western Tasmania for their continual assistance in exploring many of the deposits described.

Permission from the Chinese Academy of Geological Sciences to present the paper at the International Symposium on Geology of Tin Deposits is also gratefully acknowledged.

## References

Collins,.P.L.F. and Jennings, 1982. *A Review of Tasmania's Tin Resources and their Mineralogy.* Tasmanian Mines Department publication 1982/23.

Cox, R. and Dronsecka, E.V. (this volume ). Cleveland Stratabound Tin Deposits, Tasmania, Australia.

Goodman, R.H.; Hylton, J.; Selby, D.W., 1982. Current Expansion at the Renison Limited Tin Mine, Tasmania. Trans. Inst. Mining and Metallurgy, 91.

Groves, D.I., 1972. A Century of Tin Mining at Mount Bischoff,.1871—1971. *Tasmanian Department of Mines, Geological Survey Bulletin* No. 54.

Kwak, T.A.P., 1982. *The Geology and Geochemistry of the Zoned, Sn-W-F-Be Skarns at Mt. Lindsay, Tasmania, Australia.* LaTrobe University, Victoria, Australia.

Patterson, D.H.; Ohmoto, H.; Solomon, M., 1981. Geological Setting and Genesis of Cassiterite-Sulfide Mineralisation at Renison Bell, Western Tasmania. *Economic Geology,* 76.

Roberts, P.A., 1984. Geology and Exploration of the South Heemskirk Tin Field, in *"Mineral Exploration and Tectonic Processes in Tasmania",* Symposium of the Geological Society of Tasmania, Burnie, Tasmania, November.

Skey, E.H., 1983. The Geology of the Zeehan Tin Deposits, Tasmania, in *High Grade Metal Deposits in Eastern Australia,* a Seminar sponsored by New South Wales Department of Mineral Resources, September.

# 2.2 The Cleveland Stratabound Tin Deposits, Tasmania, Australia: A Review of Their Economic Geology, Exploration, Evaluation and Production*

R. Cox[1] and E.V. Dronseika[2]


## Abstract

The tin field of western Tasmania, Australia, contains several large stratabound tin deposits associated with massive iron-sulphides in bedded carbonate host-rocks of Upper Proterozoic to Early Cambrian age. Past and present producers include Mt Bischoff, Renison and Cleveland. They all are closely associated, both spatially and temporally, with altered high-level granitic stocks and/or quartz porphyry dykes, emplaced on major faults/lineaments. These stocks and dykes form a late phase of the Late Devonian Meredith Granite pluton.

The Cleveland deposits were explored during 1962—1964 by geological mapping and an evolving series of diamond drilling programmes. A detailed evaluation followed during 1964–66, involving underground exploratory development, comparative sampling, bulk sampling and metallurgical testing, and further diamond drilling. The deposits were brought into production in February 1968 with ore reserves (indicated plus inferred) totalling 2.85 Mt averaging 1.02% Sn, 0.43% Cu.

Since 1968, the Cleveland Mine has been Australia's second largest tin producer. Production to June 1984 has totalled 5.5 Mt ore yielding 24,000 t Sn and 9,592 t Cu. Continued exploration has led to the delineation of new reserves, such that the original reserves (measured plus indicated) are now estimated to have been 10.3 Mt averaging 0.78% Sn, 0.33% Cu (using a cut-off grade of 0.35% Sn), of which 5.7 Mt averaging 0.71% Sn, 0.32% Cu remained at May 1983. Mining is by sub-level open-stoping using tackless diesel equipment with decline access to all levels. Milling is by crushing, heavy-medium separation, grinding, sulphide flotation, gravity separation and tin flotation.

The Cleveland tin-copper deposits occur within the Tasman Fold Belt of Eastern Australia in a N-S trending eugeosynclinal trough (the Dundas Trough) which developed in western Tasmania during the Cambrian. Regionally the unfossiliferous metasedimentary/volcanic sequence in the mine area occurs at the base of and in the western portion of the Dundas Trough. Following strong deformation during the Early-Middle Devonian Tabberabberan orogeny, a number of high-level granitic plu-


[1] Consultant Economic Geologist, Roy Cox & Associates Pty. Ltd., Brigalow Place, Westleigh, N.S.W. 2120, Australia

[2] Project Geologist, Cleveland Tin Ltd., Luina, Tasmania, 7321, Australia

* This chapter was prepared in October 1984; the Cleveland Mine has been closed since 1987

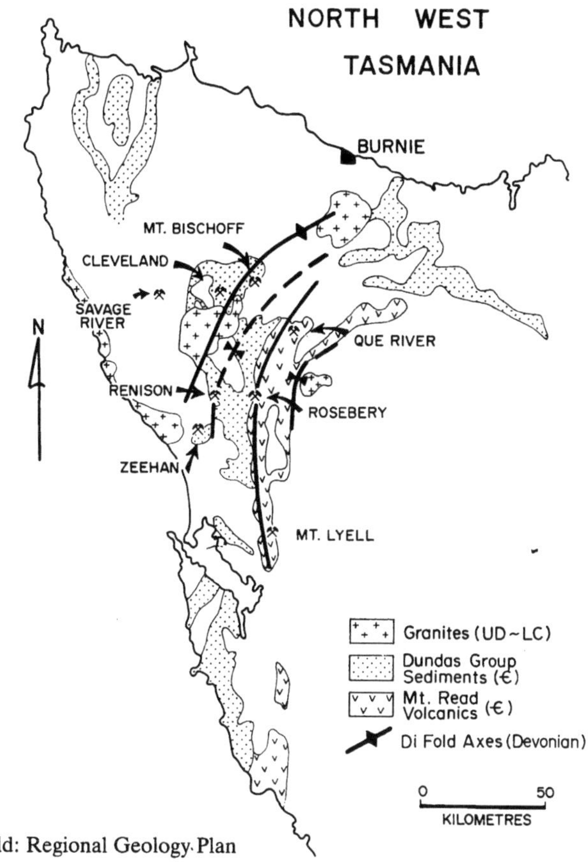


**Fig. 1.** Western Tasmanian Tinfield: Regional Geology Plan

tons were emplaced during the Late Devonian. The Cleveland Mine is located 4 km north of the northern margin of the largest of these plutons, the Meredith Granite.

The Cleveland deposits are stratabound, occuring as a series of subparallel tabular orebodies of pyrrhotite-pyrite-cassiterite-stannite-chalcopyrite mineralisation within a well-layered carbonate-chert-shale unit (Halls Formation). This unit conformably overlies a turbidite greywacke-shale unit (Crescent Spur Mica Sandstone) and is in turn overlain by a spilitic basalt-pyroclastic unit (Deep Creek Basic Volcanics). The sub-vertical orebodies, which strike NE, attain a maximum thickness of 30 m and a maximum strike length of 550 m. Along strike the tin-sulphide mineralisation passes into unaltered carbonate beds. These stratabound deposits are closely associated, both spatially and temporally, with a plumbing system, the locus of which is occupied by a pipe-like body of quartz porphyry surrounded by a W-Mo-Bi quartz vein stockwork deep in the central footwall (Crescent Spur Mica Sandstone) portion of the mine sequence.

The metal zoning and wall-rock alteration patterns indicate the mineralization is essentially post-deformation and emanated from a central plumbing system. The stratabound deposits formed by metasomatic replacement of carbonate beds by hydrothermal solutions derived from a late phase of the Meredith Granite.

## 1 Introduction

The Cleveland Mine (latitude 41° 28'S, longitude 145° 24'E) is situated at Luina in the mountainous rain-forest of northwestern Tasmania (Fig. 1).

Although the gossans of the outcropping Cleveland deposits were discovered in 1898, seventy years elapsed before the deposits were successfully brought into production in 1968. For the past sixteen years they have been Australia's second largest tin producer.

This paper reviews their exploration, evaluation and production, together with their economic and mining geology as presently understood.

The authors thank Aberfoyle Limited and Cleveland Tin Ltd. for permission to present this paper, Ms Janice Mackie for typing the text and Mr Greg Cayzer for drafting the figures.

## 2 Economic Geology

### 2.1 Regional Setting

The western Tasmanian tinfield contains several large stratabound sulphide-cassiterite deposits (Renison, Cleveland, Mt Bischoff) in finely-bedded carbonate host-rocks. These deposits occur in a narrow (25—45 km wide) north-trending eugeosynclinal trough, the Dundas Trough (Fig. 1), which developed during the Cambrian period by crustal displacement between two areas of Proterozoic basement. During the Early to Middle Devonian the Lower Palaeozoic sediments and volcanics infilling this trough were folded and faulted as a result of the Tabberabberan Orogeny. A number of high-level, late-tectonic granite plutons were emplaced as a final phase during the Late Devonian (Solomon, 1981).

The major stratabound sulphide-cassiterite deposits are all closely associated, both spatially and temporally, with altered granitic stocks, cupolas and/or quartz-porphyry dykes (Groves et al., 1972; Collins, 1981; Newnham, 1975; Patterson et al., 1981), which were emplaced along major faults/lineaments.

At the Cleveland Mine, stratabound sulphide-cassiterite bodies and stockwork W-Mo-Bi mineralisation are closely associated with a sub-vertical, pipe-like body of altered stanniferous quartz-porphyry located 4 km north (in plan) of the northern margin of the Meredith Granite, the largest granite pluton in western Tasmania (Fig. 1).

The post-Tabberabberan has been mainly a period of denudation, leading to the erosion and deep-dissection of the pre-Carboniferous rocks.

### 2.2 Mine Stratigraphy

The stratabound tin deposits at Cleveland Mine (Figs. 2 and 3) occur as a series of sub-parallel tabular bodies of sulphide (pyrrhotite-pyrite-chalcopyrite-stannite) – cassiterite within a 100—200 m thick, well-layered carbonate-chert-shale unit known as Halls Formation (Cox, 1968a; Cox and Glasson, 1971). Eleven lenses of mineralisa-

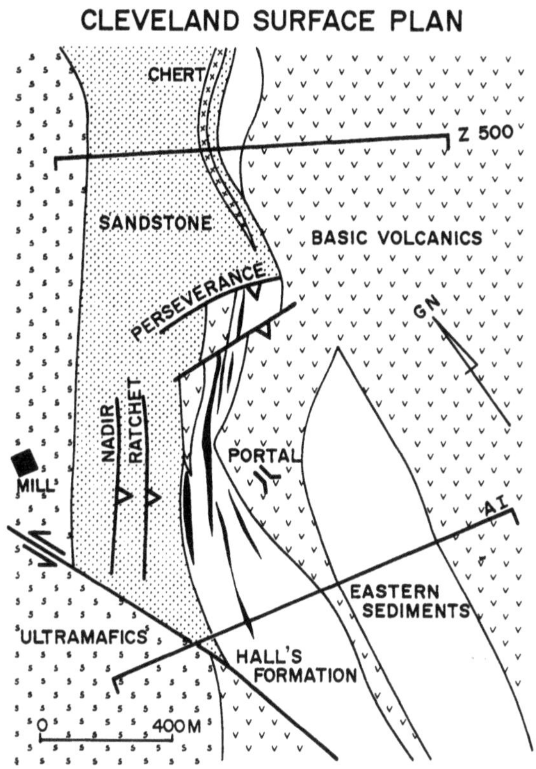


**Fig. 2.** Cleveland Mine: Mine Geology Plan

tion are presently known. Along strike some of these lenses have been observed to pass into pure, laminated micritic limestone composed mainly of calcite, authigenic quartz, chert and clay. Unlike the Renison and Mt Bischoff stratabound deposits, this limestone has low iron, magnesium and manganese contents; it is not dolomitic (Collins, 1981).

Halls Formation conformably overlies a +350 m thick turbidite greywacke-shale unit, known as the Crescent Spur Mica Sandstone Formation (Cox and Glasson, 1971), composed of poorly-sorted, fine to medium grained micaceous greywacke with interbedded siltstone, mudstone and shale. The greywacke beds, generally 10–20 cm thick, show graded bedding.

Halls Formation is overlain by a +200 m thick spilitic unit, the Deep Creek Basic Volcanics Formation (Cox and Glasson, 1971), composed predominantly of massive and locally pillowed spilitic basalt with interbedded pyroclastics, volcaniclastic greywacke, tuffaceous shale and chert.

Evidence from graded bedding in the greywackes and indentation marks in the pyroclastics in the mine area (Cox, 1968a; Cox and Glasson, 1971) clearly shows the way-up of the stratigraphy with the volcanics being younger than the greywackes. This contradicts Collins (1981).

The unfossiliferous sedimentary-volcanic sequence of the mine area is probably equivalent to the Crimson Creek Formation of Early Cambrian age in the Renison area (Newnham, 1975; Patterson et al., 1981).

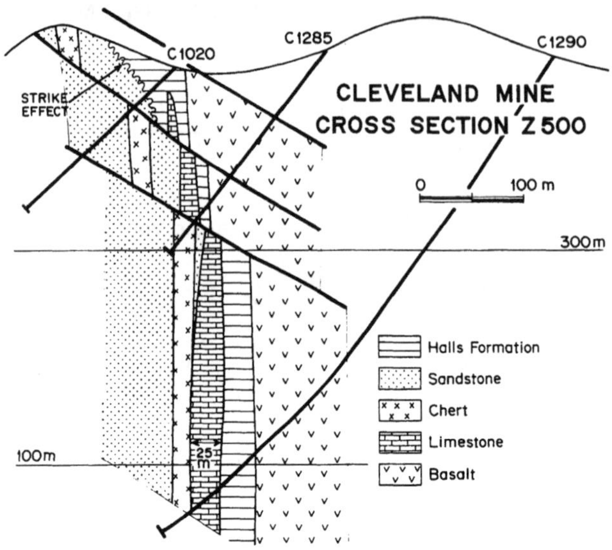


**Fig. 3.** Cleveland Mine: Mine Geology Cross-Section Z 500

## 2.3 Mine Structures

Within the Cleveland Mine area the stratigraphic sequence strikes northeast (Fig. 2), dips subvertical (Fig. 3) and faces east. The oldest unit, the Crescent Spur Mica Sandstone, occupies the core of a regional anticline whose subhorizontal axis is located to the northwest of the mine. Further to the northwest the Crescent Spur Mica Sandstone is in faulted contact with serpentinised ultramafics of the Cambrian (?) Heazlewood River Complex (Figs. 1 and 2).

An analysis of the macroscopic and microscopic structures reveals a complex structural history of folding and faulting related to the Tabberabberan Orogeny (Cox and Glasson, 1971; Ransome and Hunt, 1975; Collins, 1981).

Small-scale folds are well-developed in the finely-layered rocks of Halls Formation, particularly the mineralised beds, shales and cherts, and in tuff/chert horizons within the Deep Creek Basic Volcanics (Cox, 1968a).

The mine sequence is cut by three major series of faults:

(1) a series of sub-parallel axial plane faults, striking northeast and dipping steeply northwest, particularly evident in the upper levels of the mine (Cox and Glasson, 1971),
(2) a series of sub-parallel reverse faults, striking northeast and dipping 40° –50° southeast, and
(3) numerous post-mineralisation normal faults, striking northwest and dipping subvertically.

## 2.4 Intrusives

The Meredith Granite, located 4 km south of Cleveland Mine (Fig. 1), is a large pluton of equigranular and porphyritic biotite adamellite/granite outcropping over approximately 300 $km^2$. A contact metamorphic aureole, composed mainly of albite-, epidote-, and hornblende-hornfels, extends up to 2.5 km (in plan) from the contact.

Primary tin mineralisation occurs as cassiterite in quartz-tourmaline greisen veins within the granite and in hydrothermally-altered quartz-feldspar porphyry dykes intruding the country rocks at Cleveland, Mt Bischoff and Renison.

## 2.5 Mineralisation

### *2.5.1 Orebodies*

Eleven lenses of stratabound sulphide-cassiterite mineralisation are presently known at Cleveland Mine, of which eight have been stoped. The tabular lenses, up to 550 m in strike length, 30 m in width and 800 m dip length, occur sub-parallel and *'en echelon'* to each other, increasing in depth to the southeast (Figs. 2 and 3). They are subvertical and have been divided into two groups (Ransome and Hunt, 1975). The footwall group comprise three relatively thick lenses (Henry's, Lucks, Khaki) which have strike lengths of less than 200 m. The hangingwall group (Halls Lenses A, B, C, D, etc.) are thinner but persistent over greater strike lengths.

Each lens is conformable with the bedding of the host-rock, and consists of interlayered mineralised beds and chert. The footwall and hangingwall of each lens, as well as contacts between interlayered chert and mineralised beds, are sharp. The mineralised beds display fine compositional banding representing original bedding (Cox and Glasson, 1971; Collins, 1981).

In addition to the stratabound deposits, tin also occurs at Cleveland Mine in an altered cassiterite-bearing quartz-feldspar porphyry plug or dyke surrounded by a tungsten-molybdenum-bismuth mineralised quartz-fluorite vein stockwork. This mineralisation, known as the Foley Zone (Dronseika, 1983) is presently not economic to mine.

### *2.5.2 Ore Mineralography*

The mineralised beds of the stratabound deposits consist mainly of fine to medium grained quartz, siderite, tourmaline, fluorite, chlorite, amphibole and sulphides (10–40% by volume). The chief sulphide minerals are pyrrhotite, pyrite and marcasite, with lesser amounts of chalcopyrite, arsenopyrite, stannite, sphalerite, tetrahedrite, chalcocite and covellite (Cox and Glasson, 1971).

Tin is present chiefly as cassiterite (about 90% of the total tin), which occurs as subhedral to euhedral grains 0.02–0.7 mm across, with minor amounts of stannite/iso-stannite which occurs mainly as exsolution rims (up to 0.2 mm thick) around earlier sulphides.

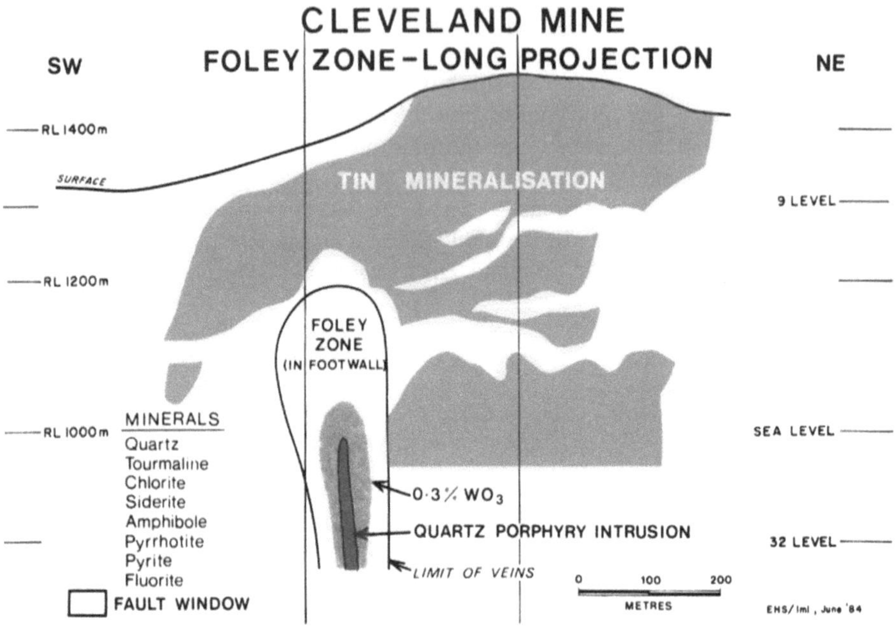


**Fig. 4.** Cleveland Mine: Longitudinal Section, Lens B and Foley Zone

The fine compositional layering of the mineralised beds shows a strong between-laminae variation in mineral composition. Cassiterite is preferentially developed in quartz-rich laminae, and chalcopyrite in sulphide-quartz-rich laminae (Cox and Glasson, 1971).

### *2.5.3 Spatial Distribution*

The constituent minerals of the mineralised beds within each orebody show distinct spatial distribution patterns centred about the top of the Foley Zone, as exemplified by Halls Lens B (Fig. 4). As a result of these mineral patterns, each orebody also shows distinct metal zoning for tin, copper and zinc. The mineralised beds pass laterally into unaltered carbonate beds.

### *2.5.4 Controls*

The stratabound tin-copper deposits at Cleveland Mine show strong stratigraphic controls of mineralisation on a wide range of scales, from macroscopic to microscopic.

The metal zoning and spatial distribution patterns of ore and gangue minerals indicate strong structural controls of mineralisation related to a central plumbing system, the locus of which is the Foley Zone in the footwall Crescent Spur Mica Sandstone. The core of this zone is occupied by a cassiterite-bearing quartz-feldspar porphyry body (Dronseika, 1983).

#### *2.5.5 Age*

The stratabound porphyry and stockwork mineralisation at Cleveland Mine are closely associated, both spatially and temporally, with a high-level, granitic body forming a late phase of the Meredith Granite. Although the mineralisation and quartz-feldspar porphyry dyke have not been directly age dated, the Meredith Granite has been dated by both the K-Ar and Rb-Sr methods at 350 Ma (McDougall and Leggo, 1965; Brooks, 1966).

### 2.6 Wallrock Alteration

An elliptical zone of sericitization of wallrocks, elongated northeastwards and measuring approximately 1,000 m by 500 m in plan, surrounds the Cleveland deposits (Cox, 1968a; Cox and Glasson, 1971). Within this zone the detrital feldspars of the footwall greywackes and the albites of the hangingwall spilitic lavas show progressive alteration from the margin inwards. The locus of this alteration halo is the plumbing system for the hydrothermal solution which formed the Foley Zone and the mineralised beds of the stratabound sulphide-cassiterite deposits.

Biotite alteration of the greywackes in the mine area increases with depth (Collins, 1981) but it is not certain whether this is related to the mineralisation or is a contact metamorphic effect related to the underlying Meredith Granite.

### 2.7 Genesis

A detailed analysis of all the presently available factual data relating to the economic and mining geology of the Cleveland Mine indicates that the stratabound sulphide-cassiterite deposits most probably originated from the metasomatic replacement of limestone beds by hydrothermal solutions, of low to moderate salinity and initially low $CO_2$ content (Collins, 1981), derived from a high-level, late-tectonic granitic source related to the Late Devonian Meredith Granite.

## 3 Exploration

### 3.1 Geology

The Aberfoyle group secured the Cleveland leases in September 1961 and commenced an initial phase of exploration comprising detailed geological mapping of surface outcrop and old open-cuts at 1:480 scale. The main objective of this programme was to define the lithologic/structural framework within which the outcropping sulphide-cassiterite mineralisation is localised, and to outline drill-targets.

Mapping and surveying were aided in the rugged, heavily-forested terrain by cutting grid lines across the prospect area at 30 m line centres. Gridding was subsequently extended, during 1964—1966, at 150 m line centres, for a total strike length of 6 km to facilitate regional mapping at 1:1,200 scale (Cox, 1968a). Outcrop is generally poor and vegetation cover thick. Soils are thin and soil profiles poorly developed on the youthful topography, characterised by 25°—35° slopes.

During the exploration period, 1962—1966, a total of 67 km of traverse line were cut, pegged, surveyed and geologically mapped. Of these 34 km were traversed with a magnetometer to aid geological mapping. In the prospect area detailed geochemical soil sampling and self-potential (S.P.) surveying were undertaken, on 6 km and 13 km of traverse line respectively, to delineate mineralisation and assist the definition of initial drill targets (Cox, 1968a).

### 3.2 Drilling

Surface diamond drilling commenced in March 1962 to test the stratabound sulphide-cassiterite deposits in depth and along strike from the shallow old workings (of the 1908—1914 period). The initial drill holes confirmed the geological interpretation and stratigraphic/structural controls of mineralisation, so additional drilling was undertaken. This first drilling programme, consisting of 46 holes totalling 3,690 m over a strike length of 610 m and to a maximum vertical depth of 120 m, was completed by July 1963 (Cox and Glasson, 1969).

Although encouraging tin-copper values had been intersected in 44 of the 46 drill holes, the results indicated that the geological structure was complex and correlation of drill intersections was in many cases speculative. Thus the inferred resource, tentatively estimated at approximately 1.5 Mt averaging 1% Sn, 0.3% Cu, had a relatively low order of reliability.

## 4 Evaluation

### 4.1 Underground Testing

In order to more fully evaluate the complex geology of the stratabound deposits and to upgrade the confidence limits that could be assigned to the tin and copper grades obtained from sampling diamond drill core (AXT-size), a programme of underground exploratory development and comparative sampling was undertaken during the period April 1964 to October 1965 (Cox, 1968a; Cox and Glasson, 1969).

An adit was driven into the steep hillside at 396 m above sea-level to intersect the stratabound deposits which were then tested by 908 m driving and cross-cutting over a strike length of 410 m, and by 36 underground diamond drill holes totalling 950 m. Drives were developed along the footwall contacts of Halls Lens A and Henry's Lens. Eight cross-cuts were driven through Halls Lenses A and B at 37 m intervals along strike for the comparative sampling programme (Cox, 1968b) which involved:

336 groove samples over 249 m in 8 cross-cuts
318 channel samples over 230 m in 8 cross-cuts
72 bulk samples over 96 m in 8 cross-cuts
346 split core samples over 215 m on 21 cross-sections
756 groove samples from development drives.

The prime objectives of this programme were achieved:

(1) the nature of the complex geological structures was better appreciated, thus allowing geological interpretation of drilling results and ore reserve tonnage estimates to be completed with a high degree of confidence, and
(2) the use of comparative sampling methods yielded closely checking average results from groove, channel, AXT-size split core and bulk sampling for both tin and copper content of ore. These results (Cox, 1968b) indicated that ore reserve grade estimates based on assays from AXT-size exploration drill holes and/or groove sampling would be reliable within acceptable, practical limits.

In addition, the underground exploratory development provided valuable opportunities to obtain material for metallurgical research and testing; to investigate the nature of ground conditions for mine planning; to determine density factors for ore and waste; and to study and analyse variations in width, grade, mineralogy and internal waste (chert content) of the orebodies.

Following completion of a revised geological interpretation, in January 1965, total indicated ore reserves and inferred resources were upgraded to 1.95 Mt averaging 1.01% Sn, 0.43% Cu in April 1965.

### 4.2 Deep Drilling

Between February 1965 and March 1966 a programme of deep diamond drilling was undertaken to further test the deposits in depth and to increase ore reserves. Sixteen additional holes totalling 3043 m were completed to a maximum vertical depth of 300 m below surface outcrop. This programme confirmed the geological interpretation and allowed inferred resources to be increased.

### 4.3 Final Feasibility

Final feasibility studies commenced in March 1966 based upon total indicated reserves and inferred resources of 2.85 Mt averaging 1.02% Sn, 0.43% Cu. These estimates were based upon 135 mineralised intersections in 98 diamond drill holes (totalling 7683 m) and 908 m underground exploratory development; they comprised 1.495 Mt indicated ore and 1.355 Mt inferred ore (Cox and Glasson 1969).

Metallurical design and operating studies were completed, utilising results obtained from the bulk sampling and pilot-scale testing. The mill was designed to treat 250,000 tonnes per year (t/y) by crushing, heavy-medium separation (to remove internal waste and wallrock dilution), grinding, classification, sulphide flotation and gravity concentration. Cassiterite recovery from the complex sulphide ores was estimated at 60%. Copper recovery was estimated at 70%, producing a copper concentrate by-product containing 21% Cu and 6% Sn (as stannite).

Various mining methods and design layouts were extensively investigated before selecting mechanised sub-level open stoping by overhand benching, with access by adits and decline to all working faces. Wallrock dilution was estimated at 12% (by weight), giving mining reserves of 3.2 Mt averaging 0.91% Sn, 0.38% Cu, sufficient for 12.8 years at the planned mining and milling rates.

Cleveland Tin N.L. was incorporated on 8 June 1966 to develop the deposits for production. A new town, Luina, was built to house the work-force. The mine was brought into production during February 1968, and officially opened on 29 March 1968.

## 5 Production

Since the commencement of production in 1968 the Cleveland Mine has been Australia's second largest tin producer. Total mill throughput to June 1984 has been 5,504,000 t for the production of 24,600 t Sn and 9,592 t byproduct copper. During this period ore reserves have been depleted by 4.6 Mt averaging 0.86% Sn, 0.35% Cu, the balance of 0.9 Mt reflecting wallrock dilution in the mill-feed.

Continuing exploration during the production period has resulted in further increases in ore reserves for the stratabound deposits. The original undiluted resource, using a 0.35% Sn cut-off grade, is now estimated to have been approximately 10.3 Mt averaging 0.78% Sn, 0.33% Cu of which 5.7 Mt averaging 0.71% Sn, 0.32% Cu remain. In addition, recent drilling of the Foley Zone porphyry and stockwork deposits has delineated further (presently subeconomic) resources of 3.0 Mt averaging 0.28% $WO_3$, 0.05% Sn, 0.02% $MoS_2$ using a 0.2% $WO_3$ cut-off grade (Dronseika 1983).

### 5.1 Mining

Mining is by open stoping using trackless diesel equipment with access above level 7 by adits into the hillside and to all other levels (no's. 8–26) by a 3.5 km long decline (at 1 in 9 grade) in the footwall Crescent Spur Mica Sandstone (Buckland 1980). The vertical interval between levels is approximately 20 m.

Narrow high-grade orebodies of 1–3 m width are extracted by shrinkage stoping.

Ore and waste rocks are generally strong and competent and require little ground support. Rock bolts, with or without steel-mesh, are used as required. Stope-walls are self-supporting and stope-fill has not been used to date. A surface crown-pillar and stope rib-pillars have enhanced general mine and stope stability.

Mine ventilation is downcast fresh air through the decline, with upcast through the stopes aided by exhaust fans.

Mine equipment consists of three 3-boom jumbos for development, three 2-boom jumbos for stope drilling, five front-end loaders, seven 40 t trucks and two 25 t trucks.

A programme of grade control and blending of ore from three stockpiles helps reduce the amount of heavily-diluted ore going to the mill to achieve a consistent mill-feed.

## 5.2 Milling

Installed mill capacity was increased to 450,000 $t\,y^{-1}$ during the 1970's but International Tin Council production quotas have reduced the current throughput to about 210,000 $t\,y^{-1}$ yielding 950 t Sn in concentrates. Throughput has been reduced further to 100,000 $t\,y^{-1}$.

Milling is by crushing (primary jaw crusher and secondary cone crusher), heavy-medium separation (ferrosilicon at S.G. 2.95), grinding (by rod mill), sulphide flotation, gravity concentration and tin flotation (Robinson 1980).

After grinding the circuit feed contains 10–20% sulphides which must be removed to facilitate gravity separation of cassiterite from lower density gangue minerals. Chalcopyrite and stannite are removed by differential flotation to produce a copper concentrate containing 20–25% Cu and 2–4% Sn.

The gravity separation involves cyclone, spiral classifier, hydrosizer, tables and vanners. The slime fraction (minus 106 micron) passes to a tin flotation section. The final flotation concentrate at 35–40% Sn is combined with the gravity concentrates before filtering and bagging for sale. The saleable product contains 50–55% Sn.

## References


Brooks, C., 1966. The rubidium-strontium ages of some Tasmanian igneous rocks. *Geol. Soc. Aust. Journ.*, vol. 13, 457–469.

Buckland, K. G., 1980. Tin-copper ore mining at Cleveland Tin Limited, Luina, Tasmania. *In "Mining and Metallurgical Practices in Australasia"*, 405–407. Monograph 10, Australas Inst. Min. Metall., Melbourne.

Collins, P.L.F., 1981. The geology and genesis of the Cleveland tin deposit, western Tasmania: fluid inclusion and stable isotope studies. *Econ. Geol.*, vol. 76, 365–392.

Cox, R., 1968a. The economic geology of the Cleveland and Magnet mines, Tasmania. *Unpubl. PhD thesis,* Univ. Sydney, 454 p.

Cox, R., 1968b. The use of comparative sampling methods at Cleveland Mine, Tasmania. *Proc. Australas. Inst. Min. Metall.*, no. 226, pt. 1, 17–27.

Cox, R. and Glasson, K.R., 1969. The exploration and evaluation of a sulphide-cassiterite deposit at Cleveland mine, Tasmania. *A Second Technical Conference on Tin, Bangkok,* vol. 1, 341–354. International Tin Council, London.

Cox, R. and Glasson, K.R., 1971. Economic geology of the Cleveland mine, Tasmania. *Econ. Geol.*, vol. 66, 861–878.

Dronseika, E.V., 1983. Geological assessment of the Foley Zone mineralisation at Cleveland mine, Tasmania. *Unpubl. Cleveland Tin Limited report,* 65 p.

Groves, D.I.; Martin, E.L.; Murchie, H. and Welington, H.K., 1972. A century of tin mining at Mt Bischoff, 1871–1971. *Tasmania Geol. Surv. Bull.* 54, 310 p.

McDougall, I. and Leggo, P.J., 1965. Isotopic age determinations on granitic rocks from Tasmania. *Geol. Soc. Aust. Journ.*, vol. 12, 295–332.

Newnham, L.A., 1975. Renison Bell tinfield. *In: "Economic Geology of Australia and Papua New Guinea"*, 1, *Metals,* 581–584. Monograph 5, Australas. Inst. Min. Metall., Melbourne.

Patterson, D.J., Ohmoto, H. and Solomon M., 1981. Geologic setting and genesis of cassiterite-sulphide mineralisation at Renison Bell, western Tasmania. *Econ. Geol.*, vol. 76, 393–438.

Ransome, D.M. and Hunt, F.L., 1975. The Cleveland tin mine. *In "Economic Geology of Australia and Papua New Guinea"*, 1, *Metals,* 584–591. Monograph 5, Australas. Inst. Min. Metall., Melbourne.

Robinson, R.J., 1980. Tin-copper ore treatment at Cleveland Tin Limited, Luina, Tasmania. *In "Mining and Metallurgical Practices in Australasia"*, 418–422. Monograph 10, Australas. Inst. Min. Metal., Melbourne.

Solomon, M., 1981. An introduction to the geology and metallic ore deposits of Tasmania. *Econ. Geol.*, vol. 76, 194–208.

# 3 Canada

# 3.1 An Evaluation of Reconnaissance and Follow-up Geochemical Surveys to Delineate Favourable Areas for Tin Mineralization in the Northern Canadian Cordillera

S.B. BALLANTYNE and D.J. ELLWOOD[1]


## Abstract

The application of regional reconnaissance multi-element geochemistry is evaluated for delineating areas with tin potential. This approach is supported by a detailed follow-up geochemical exploration strategy which led to the discovery of previously unknown granite-hosted tin mineralization.

In the Canadian Cordillera of northern British Columbia and adjacent Yukon, stream water and sediments were sampled at a regional scale as part of the Uranium Reconnaissance Program. During this helicopter-supported survey, 4,400 locations were sampled at a density of approximately one site per 13 km$^2$. Computer-generated contoured maps of U, F, Mo, W, Zn and Sn were produced at 1:1,000,000 scale to investigate their distribution and abundance in the 62,000 km$^2$ area. This suite of elements is considered indicative of Sn-W mineralization hosted in or associated with granites. The map highlighted stream geochemistry patterns which delineated terrains of "specialized" granites. The multi-element regional geochemical anomalies not only outlined known occurrences of Mo, W, Zn and Sn in the Omineca Crystalline belt, but also demonstrated that portions of the Surprise Lake batholith were greatly enriched in this suite.

Using a subset of the regional data (362 samples), two anomalous catchment basins within the batholith were selected for assessment of their tin potential. One drainage system showed a Sn-W-F elemental association while the other displayed an atypical Sn-Zn-F relationship. A detailed follow-up mixed-method sampling programme was conducted. Stream waters and sediments were sampled, heavy-mineral concentrates (HMC) were panned from stream sediments, and size fraction analyses of these media were performed. Stream water and sediment results confirmed the magnitude of the regional anomalies and the elemental associations. The areal distribution of anomalous HMC coincided with the results for various fractions of unpanned sediment and showed greater contrasts over background. For example, in the stream drainage system anomalous in Sn, W and F, the Sn contents in the -500, -250 and -177 micron mesh fraction of HMC ranged up to two times, five times and eleven times greater than in the same fractions of unpanned sediments. The W content contrast was comparable, ranging up to 21 times greater in the -177 micron mesh fraction. Compared with samples collected from background areas, Sn in HMC from the stream drainage anomalous in Sn, Zn and F showed greater contrast in the finer frac-


[1] Geological Survey of Canada, 601 Booth Street, Ottawa, Ontario, K1A OE8, Canada

tions ranging from four times in the -2 mm mesh fraction to nine times in the -250 micron mesh fraction. No increase in Zn contents was observed in HMC compared to unpanned stream sediments. Analyses of different size fractions helped define prospecting areas within the catchment areas. Magnetite, cassiterite and gadolinite were identified in HMC.

Within the Sn-W-F anomalous drainage system, fractured and/or shear-related mineralization was discovered. It consisted of cassiterite-wolframite in quartz veins and veinlets, quartz-arsenopyrite veins and quartz-fluorite veins in altered granite. The mineralization within the Sn-Zn-F enriched drainage system is more complex. It consists of a network of fractures, veins and replacement bodies thought to have formed near the apex of the granitic body. Cassiterite was found in two mineralized zones. It occurs in massive quartz-sericite-cassiterite-beryl greisen and underlying replacement bodies of biotite-magnetite-sphalerite with minor cassiterite. Fluorite and kasolite were found in the peripheral alteration zones which could account for the consistent F-U anomalies measured in the streams.

The careful evaluation of regional reconnaissance stream geochemical data and subsequent detailed follow-up survey results have proven that a multi-method, multi-element approach to exploration can facilitate the discovery of tin mineralization.

## Purpose

Tin, now a so-called "strategic element", has until recently not been actively explored for in Canada. As relative newcomers in the search for tin deposits, we present an evaluation of multi-element, multi-method regional reconnaissance geochemical data for an area of the northern Canadian Cordillera. This approach delineated areas with tin potential. It is supported by an example of a detailed follow-up geochemical exploration strategy which led to the discovery of previously unknown granite-hosted tin-tungsten mineralization.

Specific objectives of this investigation are:

(1) to examine the regional geology in terms of tin metallogeny and current lithotectonic evolutionary models for the Canadian Cordillera,
(2) to consolidate regional reconnaissance stream sediment and water data for the 4,400 sample sites in the 62,000 km$^2$ area in the form of computer-generated element distribution maps to facilitate interpretation,
(3) to evaluate the distribution and abundance of the suite of elements (U, Mo, W, U, Sn, F) thought to be indicative of "specialized" tin granites in terms of known mineral occurrences or deposits,
(4) to enhance anomalies without known sources by producing computer generated elemental-associated maps at a greater scale to define drainage system anomalies and pathfinder elements for specific targets,
(5) to access this strategy by choosing two anomalous drainage systems within the regional data subset and complete a practical detailed follow-up stream survey,
(6) to use the methods developed in the follow-up and interpret studies to define specific prospecting targets,

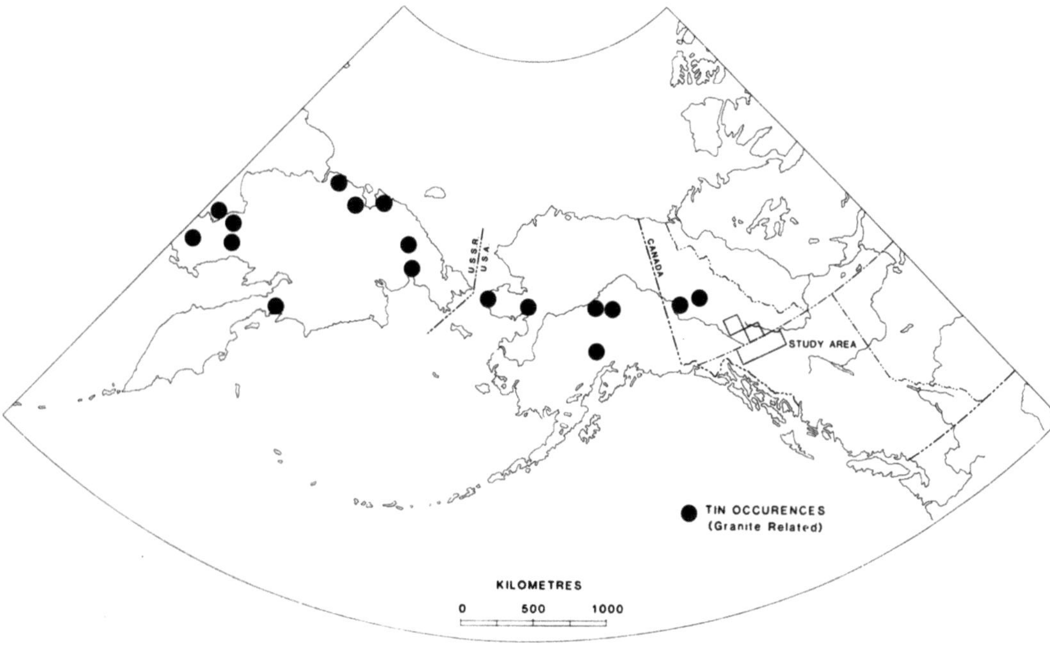


**Fig. 1.** Location of study area and some major tin occurrences

(7) to discover the mineralization and/or alteration which have given rise to the specific anomalies,
(8) to document this overall approach and to suggest that it is a viable exploration strategy in the northern Canadian Cordillera.

## Location

In the northern Canadian Cordillera, approximately 1,200 km NNW of Vancouver and centered about latitude 60° N, lies the 62,000 km$^2$ survey area (Fig. 1, 2). It is covered by 1:250,000 scale NTS (National Topographic Series) map sheets 104N, 104O and 104P in northern British Columbia and NTS map sheets 105B and 105F in southern Yukon Territory (Fig. 3).

Situated on the western edge of the survey area and approximately 180 km south of Whitehorse (the main service center and capital of the Yukon) lies the town of Atlin (population 400). Juneau, Alaska is situated approximately 150 km southwest of Atlin, British Columbia (Fig. 2).

Maintained gravel roads (Alaska highway, South Canol Road, Stewart-Cassiar highway, Atlin, B.C., highway # 7) provide the major access to and transect the survey area. Light aircraft use the three gravel airstrips in the area. Float-equipped light aircraft provide access to the numerous lakes, however, most of the region is only accessable via helicopter.

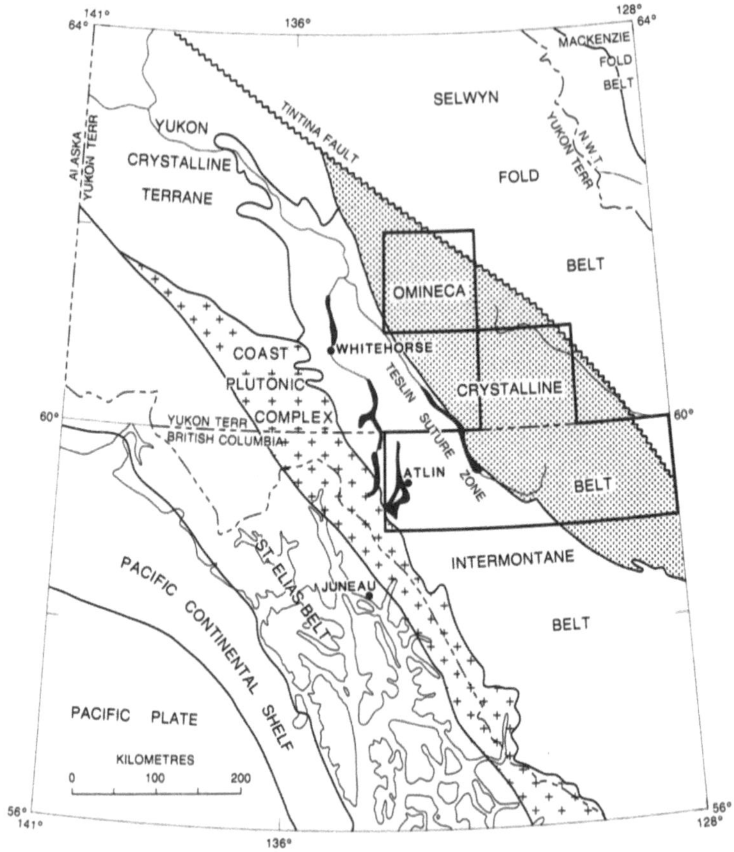


**Fig. 2.** Tectonic belts of the Cordilleran Orogen. Study area is outlined

## Physiography and Climate

The study area is characterized by long northerly to northwesterly trending high mountain ranges often containing icefields, snowfields, and glaciers. Broad alluvial valleys or rolling upland plateaus separate the ranges. Similar-trending major lake systems are outlined in black in Fig. 2 and 3. Run-off from the survey area eventually reaches the oceans via the Yukon River to the Bering Sea, the Liard River to the Mackenzie River and the Beaufort Sea, and the Nakina-Taku Rivers to the Pacific. Most streams and creeks selected for sampling during the regional survey reconnaissance have headwaters in alpine elevations and flow rapidly over a short distance to lakes or rivers in the wooded low-lying areas. Tree-line is generally below 1,067 metres but forest cover is often interrupted by patches of swamp and muskeg in poorly drained areas. Higher elevations usually exhibit good bedrock exposure which also can be seen in incised streams, stream valleys and canyons.

In NTS map sheet 104N elevations, above sea level, range from 683 m at Atlin to 2,100 m at Paradise Peak south of Atlin (Fig. 3). Similarly, in NTS map sheet 104P,

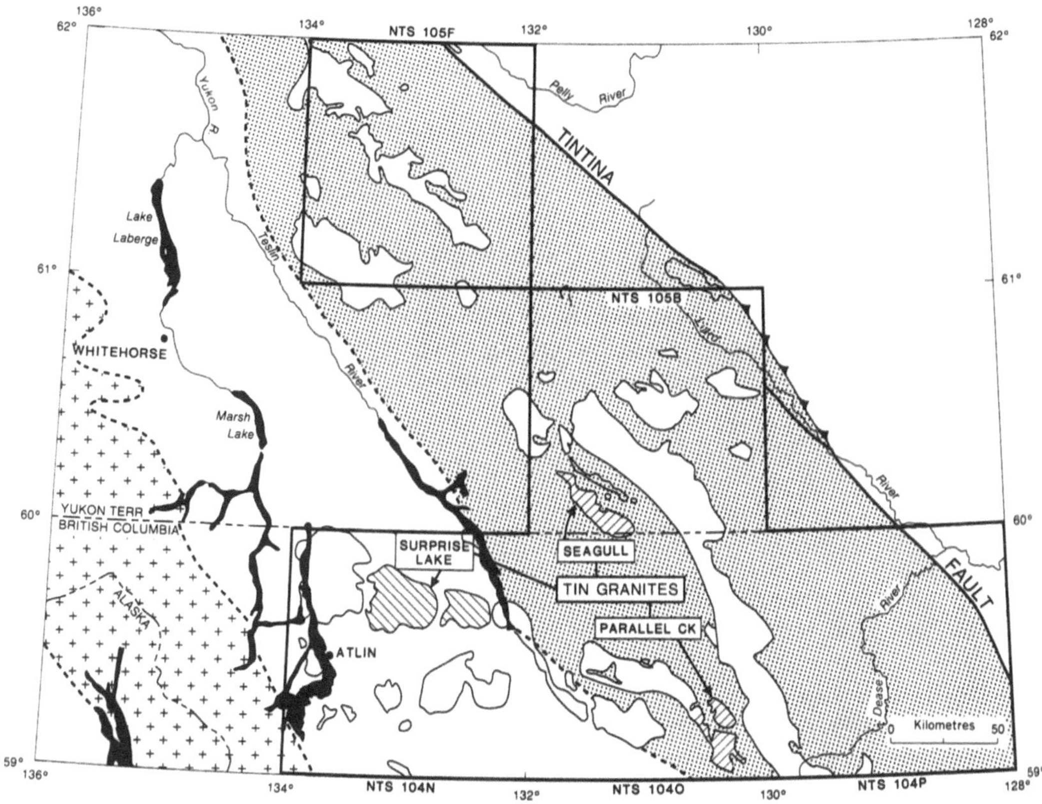


**Fig. 3.** Simplified distribution of Cretaceous-Tertiary granitoid intrusions within the study area (after Tipper et al., 1981). Shading is the same as on Fig. 2. Stanniferous plutons are cross-hatched. Study area is outlined

1,500 m maximum relief from plateau or plain to mountain range can be found. In NTS map sheet 105F in the Pelly Mountains, peaks reaching 2,400 m elevations yield the maximum elevations.

Precipitation ranges in the western position of the study area from about 28 cm annually at Atlin to about 76 cm annually in the Cassiar Mountain Ranges in the east. About one half of the precipitation falls as snow which may persist in the mountains until the middle of June and after mid-September.

Glacial, fluvioglacial and post-glacial erosion and deposition modify local topography often resulting in spectacular features. These include eskers and esker complexes, kames and kettles, shorelines of glacial lakes, terraces, abandoned deltas, drumlinized till, hummocky terrain, moraines and outwash deposits, hanging valleys, and eratics. Most valleys are drift-covered to some extent. Examples of thick boulder and/or basal till (50 m) are evident in the north-western portions of NTS 104N (Aitken, 1959) and in the north-eastern portion of NTS 104P (Gabrielse, 1963) on the low-lying Liard Plain (Fig. 2b). All of these examples of the effects of continental glaciation are due to the extensive advances of the Cordillera ice sheet, Cassiar Lobe, during the McConnel advance or are remnants of at least two older glaciations (Hughes et al., 1969). In the study area, the main ice-stream flowed northwest along the Teslin Suture zone – Teslin River system and the Atlin Lake valley, while in the eastern portion of the survey area an eastwardly flow is noted (Aitken, 1959; Gabrielse, 1963, Hughes et al., 1969). However, because of the complexity of these Wisconsin age glacial deposits, advances, stagnation and retreats have occurred in many directions.

Alpine glaciation has also altered the topography. In the southwestern corner of NTS map sheet 104N, Atlin, British Columbia, glaciers are still active today. Alpine glacial features as evidenced by deep cirques, cirque lakes, cleaver ridges, spurs and horns, have had a pronounced effect on increasing bedrock exposure throughout the region. Much of the alpine glaciation is thought to have taken place prior to the last continental ice sheet advance (Aitken, 1959; Gabrielse, 1963, 1969).

## Previous Work

### Geology

In the area covered by the regional reconnaissance geochemical survey, general geology investigations and descriptions of the mineral deposits may be found in reports and 1:250,000 scale maps by Aitken (1959), Gabrielse (1963, 1969), Mulligan (1969), Poole et al. (1960), and Tempelman-Kluit (1977). Recent geological compilation maps at 1:1,000,000 scale and the tectonic assemblage map of the Canadian Cordillera at 1:2,000,000 scale have been modified and used as the geologic base in this paper (Gabrielse et al., 1980; Souther et al., 1979; Tipper et al., 1981). Important new regional geologic-tectonic concepts relevant to the survey area have been described by Abbot (1981), Monger (1977), Monger and Price (1979), Monger et al. (1982, 1983) and Tempelman-Kluit (1979).

### Geochemistry

As part of the Uranium Reconnaissance Program, geochemical orientation surveys were initiated in the Atlin area (Fig. 2) in 1976 (Ballantyne et al., 1978). Reconnaissance stream water and stream sediment sampling surveys were then conducted as part of the continuing Regional Geochemical Reconnaissance Program. The Federal Government Uranium Reconnaissance Program provided funding for the survey area within the Yukon Territory. For the area within adjoining British Columbia the survey was jointly sponsored and funded by the Province of British Columbia and the government of Canada under an agreement within the Uranium Reconnaissance Program.

Within a year from sample collection the survey results were published as single element maps at a scale of 1:250,000. They were accompanied by computer-generated raw data and summary statistics files for each of the map sheets. The data were also made available on magnetic tape.

Geological Survey of Canada Open File Reports 517 (1978) and 561, 562, 563, 564 (1979) originally contained stream sediment results for a routine element suite (Zn, Cu, Pb, Ni, Co, Ag, Mn, Fe, Mo, U) and in some cases the additional elements (Hg,

W). Also included in the survey were stream water data (U, F and pH determinations).

All reconnaissance sediment samples are archived at the Geological Survey of Canada. Interpretation of the data (W, Mo, U, F) indicated that the area could also have tin potential. Archived samples for NTS map sheet 104N were subsequently analysed for tin and the data published in a revised addition of Geological Survey of Canada Open File report 517 (1979). Recently archived samples for NTS map sheets 105F and 105B have also been analysed for tin but results are not yet published at 1:250,000 scale. However, the authors have included in this paper the newly acquired tin data at a scale of 1:3,000,000.

The mineral industry is continuing to use the reconnaissance geochemical survey data in follow-up exploration programmes. As a result, many new mineral occurrences have been found or known mineral occurrences have been re-assessed within the context of current mineral deposit genetic models. Result of their work may be found in mineral claim assessment reports or in various publications by the British Columbia Ministry of Energy, Mines and Petroleum Resources or the Yukon Department of Indian and Northern Affairs, Geology Section.

## Methods of Study – Sampling, Analytical Procedures and Computer Processing

### Regional Geochemical Reconnaissance

The survey was conducted in 1977 and 1978 during the months of July and August: the snow in the upper elevations had melted by this time and high water levels in the streams had returned to normal. The survey area is rugged and remote, and most of the stream sample sites were visited by helicopter. In areas with road and lake access sampling was conducted by ground crews. Silt- and clayrich sediment and water samples were routinely collected by the contract crews at each of the 4,409 sites in the approximately 62,000 $km^2$ area. The sample density was approximately one sample per 13 $km^2$ or on average one sample every 2 km of stream drainage. The sediment samples were air dried in the field and then sieved to obtain the minus 117 micron mesh fraction. This fraction was then ball-milled to about 60 microns and analysed for Zn, Cu, Pb, Ni, Co, Ag, Mn, Fe, Ba, Mo, W, Hg, U and Sn. The water samples were analysed for U and F and pH was determined.

More information about survey methodology, analytical methods, quality control and monitoring of data may be found in the GSC Open File reports (1978–1979).

### Tin Analytical Method

In map sheet 104N (Atlin), tin was determined in regional stream sediment samples by a method which was only slightly modified from that described by Welsch and Chao (1976). A 1.0 g sample is heated with $NH_2I$ in a modified pyropot furnace for 15 minutes at 500°C. This reaction, which converts the tin in the sample to $SnI_4$, is done in a 25 by 200 mm test tube. After cooling, the residue is leached with 20 ml of a solution which is 5% V/V in HCl and 6% W/V in ascorbic acid. After leaching, the sample is cooled to room temperature and 5 ml of 4% W/V trioctyloposphine oxide in methyl isobutyl ketone (MIBK) are added. The test tube is capped and shaken for 60 seconds. The solvent layer is then transferred to a small tube and centrifuged. The tin in the solvent layer is then determined by atomic absorption spectroscopy using a nitrous-oxide acetylene flame at 2,863 Å.

For map sheets 105B (Wolf Lake) and 105F (Quiet Lake) the regional stream sediment samples were analysed for tin by an XRF method. Approximately 2–5 g of about 60 micron powder was pressed into a pellet by a hydraulic press. The pellet was then analysed for tin using an X-ray fluorescence spectrometer.

Detailed stream sediment samples were analysed for tin by both methods of analysis and the results were found to be reasonably comparable. Tin values from detailed survey samples quoted in this paper were all determined by the XRF method.

### Detailed Survey

The follow-up survey included stream water sampling, stream sediment sampling and heavy mineral panned concentrate collection. Unfiltered waters were analysed for uranium and fluorine by the same methods used for the regional survey. Results were similar to those of the regional survey and all the resampled regional anomalies were confirmed. Detailed stream sediment samples were sieved into three size fractions: minus 500 to plus 250 microns, minus 250 to plus 177 microns and minus 177 microns. Each fraction was then ball-milled to approximately 60 microns thus ensuring a homogenous sample for the analyses. Bulk 4–5 kg composite sediment samples were collected from accumulations in potholes and from behind boulder and cobble rubble in the stream bed. By volume, they evenly filled a conventional gold pan. After screening to remove gravel and after careful washing to remove the clay and silt, the residue was separated by conventional gold panning techniques into heavy mineral (approximately 70%) and light mineral concentrates by drying and sieving to completion the minus 2 mm to plus 500 micron fraction, the minus 500 to plus 177 micron fraction and the minus 177 micron fraction. Examination and mineral identification of the important minerals was completed before each fraction was ball-milled to approximately 60 microns and analysed.

## Computer Processing

### Reconnaissance: Element Distribution Maps

Computer generated contoured maps assisted in the interpretation of the regional reconnaissance analytical data. Various element values were contoured and plotted at a scale of 1:1,000,000 on an Applicon colour plotter. In the production of these maps, a regular grid, 1,600 m square, was laid over the irregularly spaced data points. Grid values were then computed in two steps. In step one, a value was assigned to the center of each grid cell containing one or more data points. Grid cells containing more than one data point were assigned the average of the contained analytical values; grid cells containing one data point were assigned the analytical value of that point. In step two, a new value was computed for each grid cell by averaging the

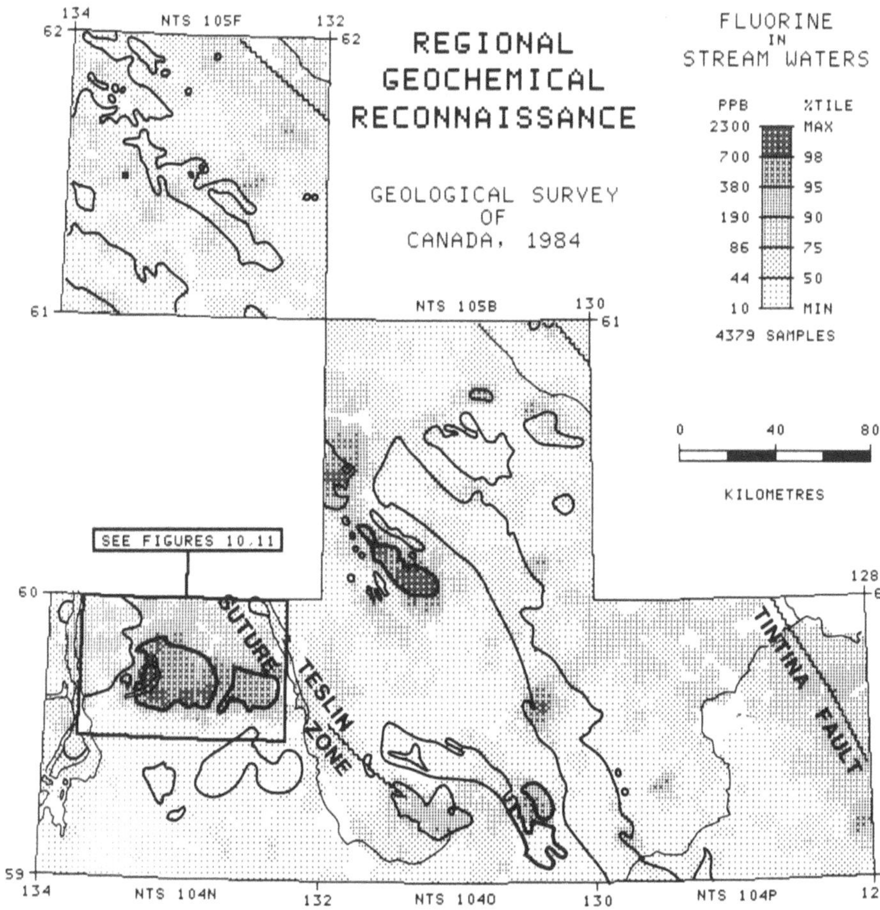


**Fig. 4.** Distribution of fluorine in stream waters. Cretaceous-Tertiary granitoids are outlined; stanniferous plutons are outlined in bold. White areas indicate no data

weighted values from the values of the five nearest cells assigned in step one. Weighting was by an inverse distance function 1/d cubed. Finally, the maps were produced by contouring this regular grid using contours levels based on percentiles from the original data set. Contours were not plotted in areas greater than four km from a data point. Figures 4, 5, 6, 7, 8 and 9 are black and white versions of the original 1:3,000,000 scale colour maps. They were produced from grids prepared in the same way as the colour maps and plotted on film using an Optonics laser plotter. This method of contouring regional data filters out minor irregularities and allows ready appraisal of major regional geochemical trends and patterns. The maps give a clear regional overview by emphasizing particular geochemical signatures which may be associated with productive terrains.

The element distribution maps portray data for molybdenum, tungsten, zinc and tin in stream sediments and uranium and fluorine in stream waters. This suite of elements is used to define areas which clearly show that the single element distribution maps reflect bedrock geochemistry related to "specialized granitoids".

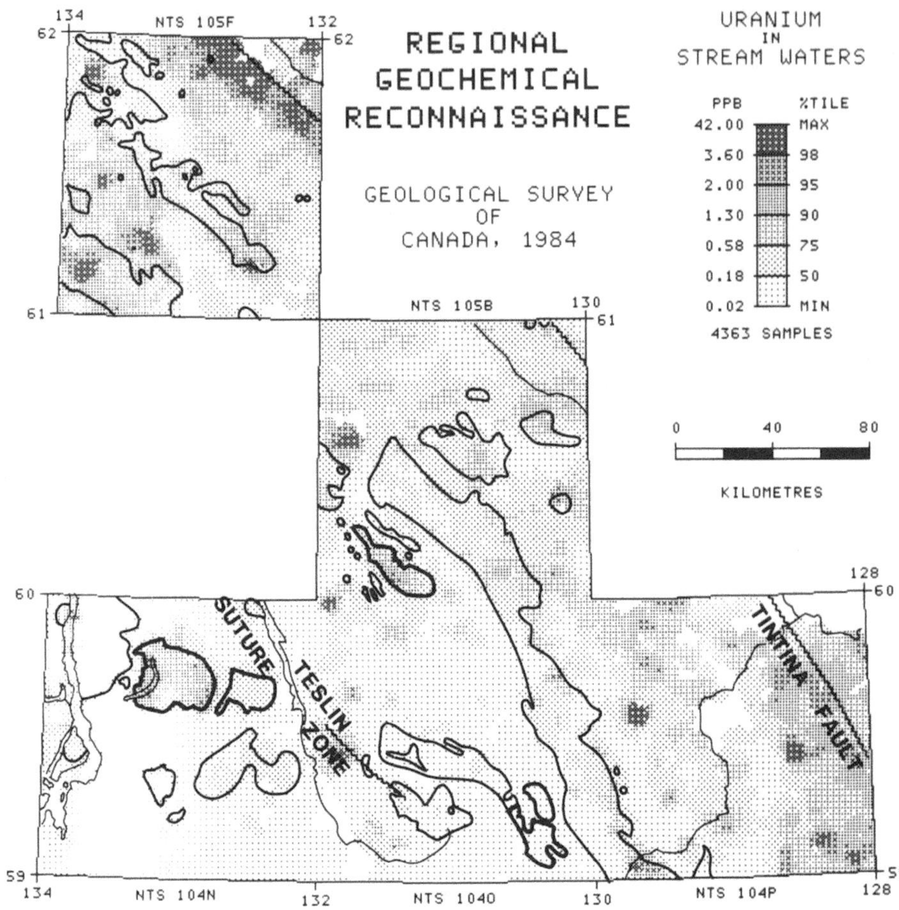


**Fig. 5.** Distribution of uranium in stream waters. Granitoids are outlined as on Fig. 4. White areas indicate no data

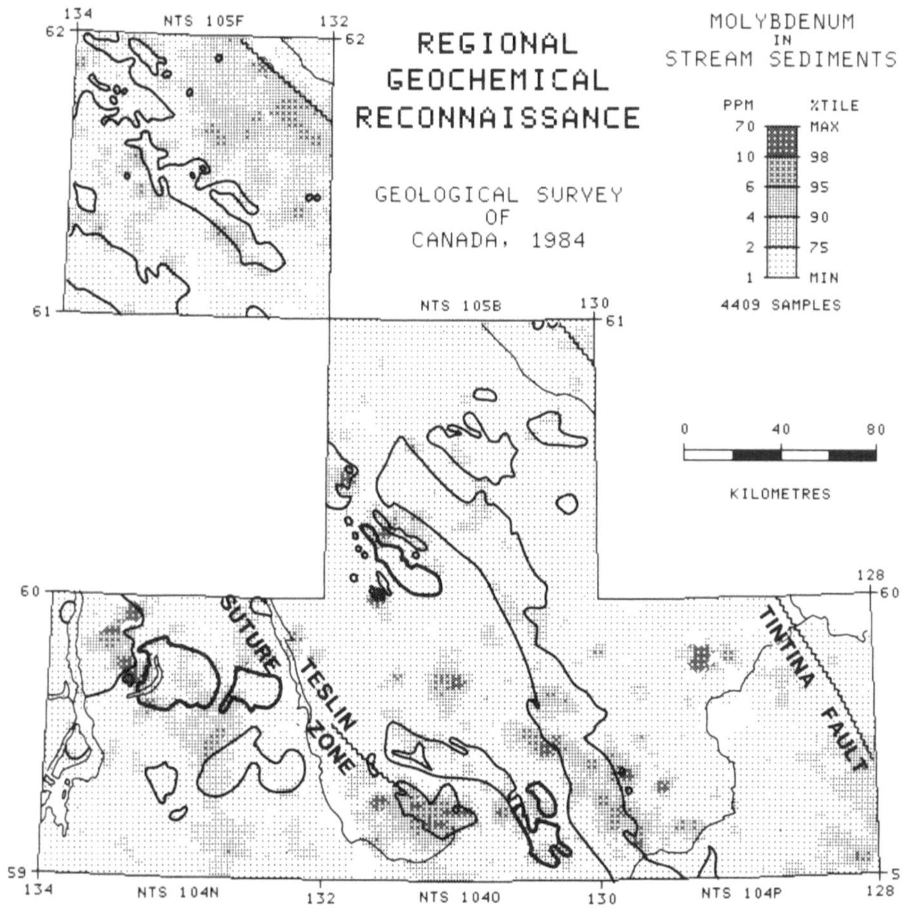


**Fig. 6.** Distribution of molybdenum in stream sediments. Granitoids are outlined as on Fig. 4. White areas indicate no data

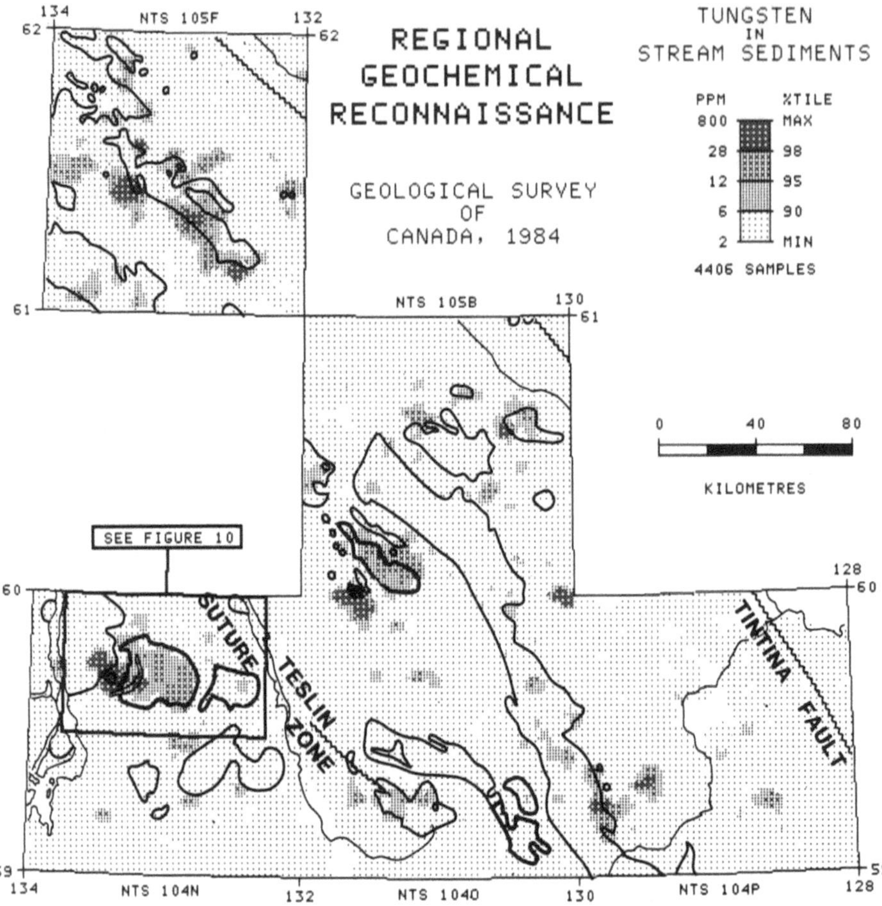


Fig. 7. Distribution of tungsten in stream sediments. Granitoids are outlined in Fig. 4. White areas indicate no data

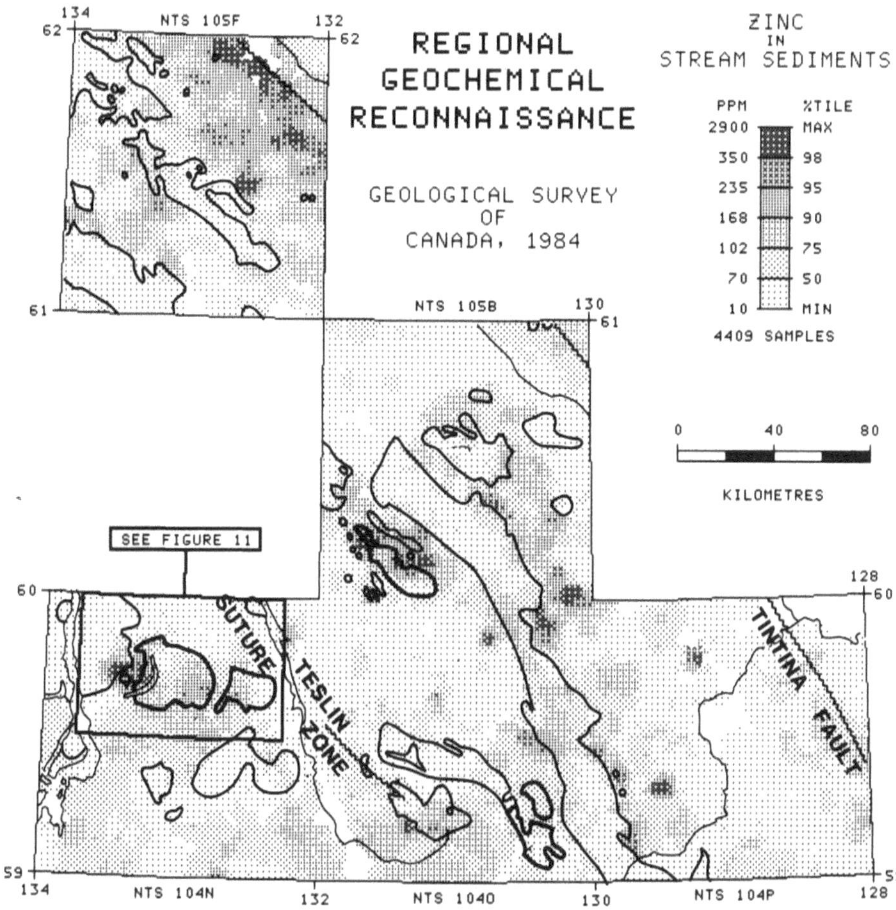


Fig. 8. Distribution of zinc in stream sediments. Granitoids are outlined in Fig. 4. White areas indicate no data

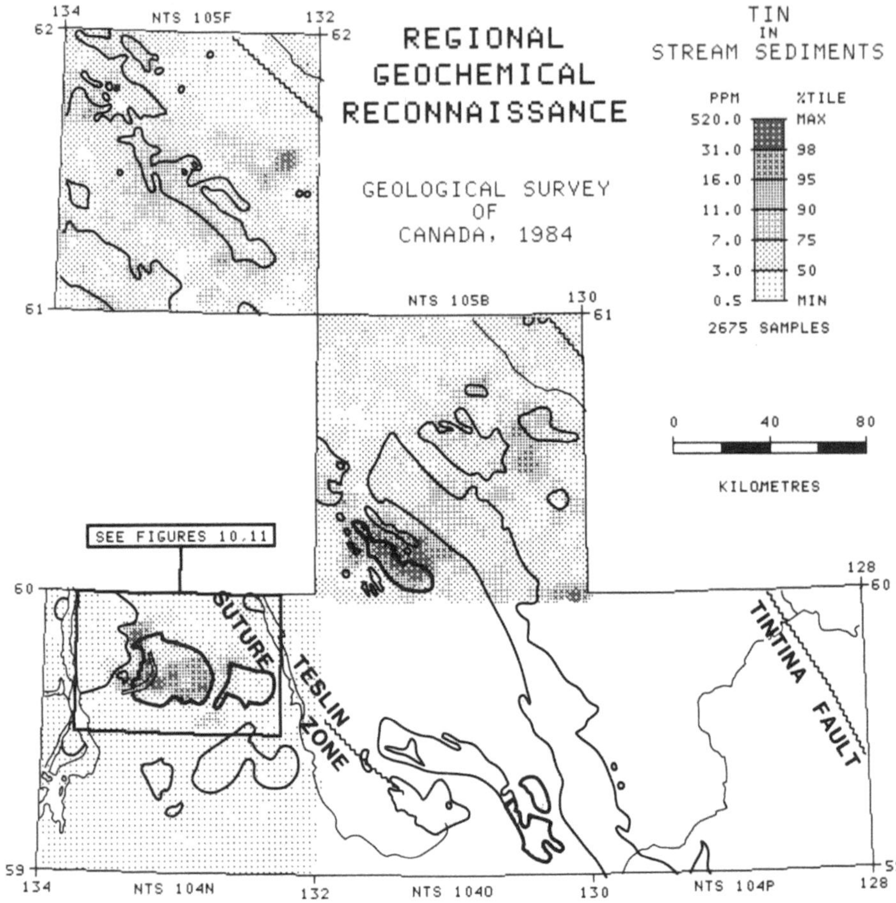


**Fig. 9.** Distribution of tin in stream sediments. Granitoids are outlined in Fig. 4. White areas indicate no data

## Regional Three-Element-Maps

When outlining targets worth prospecting within a reconnaissance survey area, it is useful to look at the associations between the various analytical variables. The authors found that associations between three elements could be shown by plotting them all on the same map, two as symbols and one contoured. These three-element-maps were plotted in colour at a scale of 1:250,000 with one element plotted as blue squares, another as red circles and a third as coloured contours. The size of the symbol was proportional to the data value. The point at which the symbol changed size was based on percentiles of the data being plotted. The contours were again done from a regular grid, which was prepared from the analytical variable as for the reconnaissance maps. Figures 10 and 11 are enlarged black and white versions of a portion of the original three element colour map of NTS 104N, Atlin, British Columbia. The

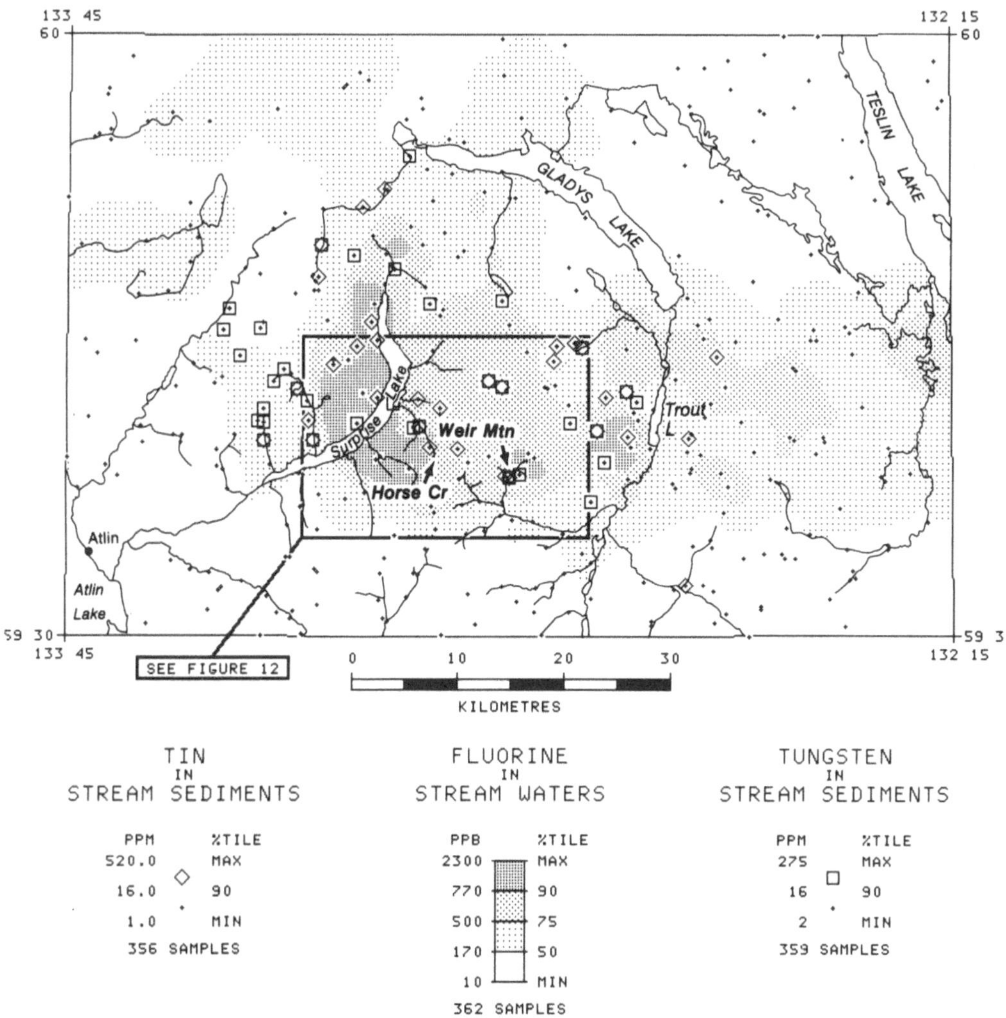


**Fig. 10.** Distribution of tin and tungsten in stream sediments and fluorine in stream waters in the Surprise Lake batholith area, Atlin, British Columbia

number of symbol sizes and contour levels have been reduced to improve readability. Figure 12 shows an enlarged portion of Figures 10 and 11. It contains almost as many symbol sizes and levels as the original colour version. The combination of the circles and squares at each sample site along with the underlying contouring allows easy visual comparison of the three elements at and between each sample site. Many users do not have access to sophisticated data processing. However, this application of multi-element plotting is shown to be an effective way to manipulate sample site data and summary statistics to define element associations and anomalies. Examples of enrichment, depletion and apparent zoning are shown by the distinctive stream geochemistry for the two illustrated drainage systems located within the Surprise Lake batholith, Atlin, British Columbia.

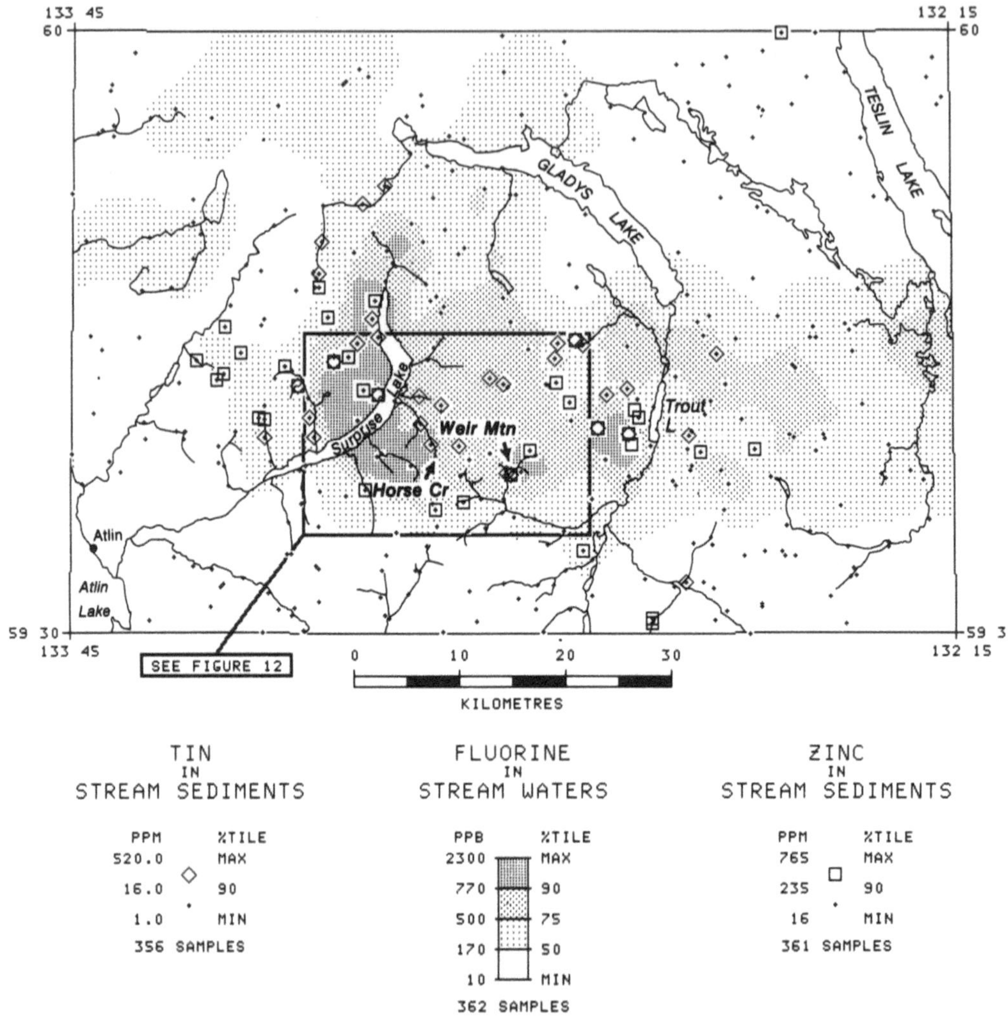


**Fig. 11.** Distribution of tin and zinc in stream sediments and fluorine in stream waters in the Surprise Lake batholith area, Atlin, British Columbia

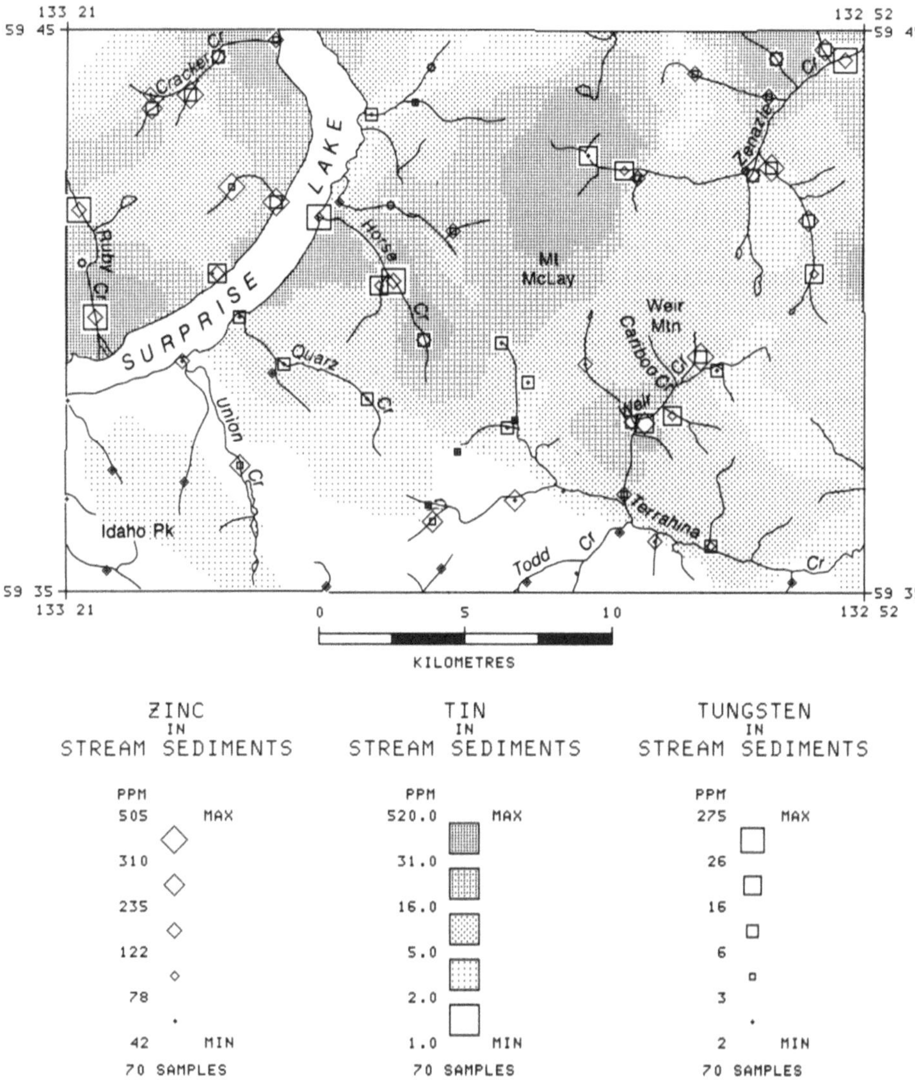


**Fig. 12.** Detailed enlargement of part of Figures 10 and 11, showing the distribution of zinc, tin and tungsten in stream sediments

## Geology

### Regional Tectonics and Structural Setting

The study area lies within the Cordilleran Orogen of Western Canada and is outlined in terms of the major regional geologic belts in Figure 2.

Current tectono-stratigraphic views for this region emphasize its complex nature in terms of a tectonic mosaic composed of separate structural blocks perhaps far-travelled exotic terrains and fragments that accreted by a variety of processes to the margins of North America during Mesozoic and early Cenozoic time (Monger and Price, 1979; Monger and Davis, 1982; Monger et al., 1982, 1983; Jones and Silverling, 1982). Within the study area the dominant structural style produced during accretion is thrust faulting at the Teslin Suture zone (Fig. 2) and later modification by strike-slip faulting along the Tintina fault (Fig. 2). In Late Cretaceous and/or Early Tertiary time approximately 450 km of dextral movement occurred along the Tintina fault (Tempelman-Kluit, 1980).

Stanniferous plutonism in relation to the concept of a "tin belt" (Sainsbury et al., 1969) can now best be descriped in terms of a collision model. It is important to understand the granitic intrusive events in terms of lithotectonic terrains.

It has been proposed that in the early Mesozoic, suspect terrains were amalgamated and subsequently accreted to the ancient continental margins of North America during an initial collisional event in Jurassic time (Coney et al., 1980). The continent had moved westwards into offshore Paleozoic arc systems (Tempelman-Kluit, 1979). The back-arc marginal basin collapsed and was driven up and over the leading edge of the continent at the Teslin Suture zone (Fig. 2). As a result of tectonic overlap, loading by allochthonous terrains and compressional thickening east of the Teslin Suture zone, the Omineca Crystalline Belt was formed as a major metamorphic-plutonic welt (Monger et al., 1982, 1983; Tempelman-Kluit, 1979) (Figs. 2, 3). Farther west, amalgamation of other allochthonous terrains in late Jurassic time led to a second collision with the new continental margin in Cretaceous time. This resulted in a second high-grade metamorphic and plutonic belt, the Coast Plutonic Complex (Monger et al., 1982, 1983) (Figs. 2, 3).

West of the Teslin Suture zone and east of the Coast Plutonic Complex, lies the Intermontane Belt (Figs. 2, 3) which in the study area is predominantly made up of Cache Creek Group Upper Paleozoic rocks of the Atlin Terrain (Monger, 1975). These sub-greenschist metamorphic facies rocks are separated by the Teslin Suture zone from the more highly metamorphosed tectonic melange of sheared and tectonically interleaved cataclastic, volcanic, intrusive and carbonate rocks of the Omineca Crystalline Belt (Tempelman-Kluit, 1979; Abbot, 1981). The Atlin Terrain is generally comprised of Upper Paleozoic chert, pelite, carbonate, volcanic and ultramafic rocks, Mesozoic granitic intrusions such as the Surprise Lake batholith, and minor late Mesozoic or early Tertiary and Pleistocene volcanic rocks (Monger, 1975).

**Geology and Tin Metallogeny of the Survey Area**

In the evaluation of the tin potential, our major focus must be the Cretaceous andTertiary granitoids as outlined in Figure 3. As a result of the collisions, post tectonic plutonism "welded" the tectonic belts and allochthonous terrains together. Although granitoid age dates are few, some conclusions concerning depths and times of emplacement can be drawn from the available data.

Omineca Crystalline Belt

In NTS 105F (Fig. 3), the post tectonic Quiet Lake batholith situated in the southwestern corner of the map sheet is composed of medium- to coarse-grained equigranular, biotite quartz monzonite. The Big Salmon batholith and Nisutlin batholith, lying in a northwest to southeast trend across the map sheet, are composed of porphyritic, medium-grained, biotite quartz monzonite (Tempelman-Kluit, 1977). These three granitic bodies have yielded K/Ar age determinations of about 95 Ma, mid-Cretaceous (Tempelman-Kluit, 1977; Tempelman-Kluit and Currie, 1978).

From 130° west longitude and straddling the border of NTS map sheets 104O and 104P (Fig. 3), the Cassiar batholith and related stocks trend north-northwest transecting NTS map sheet 105B. Since the Cassiar batholith is not deformed it provides evidence for prior to mid-Cretaceous northeastward thrusting from the Teslin Suture zone (Monger et al., 1982). It is a cross-cutting, post-tectonic, generally coarse-grained equigranular, biotite quartz monzonite yielding K/Ar ages of about 100 Ma (Abbot, 1981). In contrast, related stocks show more textural variation and give K/Ar ages of 72.4 ± 2.5 Ma (Cooke and Godwin, 1984; Panteleyev, 1980).

To the south and west of the Cassiar batholith, in NTS map sheet 105B, lies the Seagull batholith, a tin granite (Fig. 3), dated at 100 ± 2.8 Ma (Mato et al., 1983). The Seagull batholith commonly contains miarolitic cavities and tourmaline in veins, veinlets and spots. It shows compositional variation from biotite granite to quartz monzonite with textural variations from coarse-grained, equigranular, to porphyritic phases. Abbot (1981, p. 42—43) has suggested that the coarse-grained phase of the Seagull batholith is essentially identical to biotite quartz monzonite of the Cassiar batholith lying to the east, but it has been emplaced at a higher level.

The Seagull granites and related outliers host tin occurrences (Abbot, 1981, p. 43 and 145—146; Mato et al., 1983) but perhaps of greater importance to the tin metallogeny of the area are the related skarns (Dick, 1980; Layne and Spooner, 1983). Numerous and varied tin-bearing silicate mineral assemblages are found within the skarns. These include andradite garnet, malayaite and/or cassiterite-bearing skarn and the rare occurrence of the Ca-Sn-B mineral, nordenskiöldine (Dick, 1980; Layne and Spooner, 1983).

Also of note is the Parallel Creek batholith situated in the south-eastern corner of NTS map sheet 104O (Fig. 3). Like the Seagull batholith to the north, it lies east of the Teslin Suture zone and west of the margin of the main granitic body of the area, the Cassiar batholith (Fig. 2, 3). It has been K/Ar dated at 78 ± 4 Ma (Wanless et al., 1970).

The Parallel Creek batholith ranges in composition from biotite granite to quartz monzonite and texturally from equigranular, medium-grained to coarsely megacrystic phases. Similar to the Seagull batholith, it too is locally tourmaline-bearing and miarolitic.

The Parallel Creek batholith is classified as a tin granite because of its relationship to tin-bearing garnet skarns (Mulligan, 1969, 1975). In an associated roof pendant the senior author and L.A. Dick discovered and recognized (Dick, 1980) "wrigglite" skarn (Kwak and Askins, 1981; Taylor, 1979) or "ribbon rock" (Sainsbury, 1969; Dobson, 1982) as described for deposits in Australia and Alaska respectively. In thin section, the skarn contains rare cassiterite, disseminated arsenopyrite and mainly, rhythmically alternating layers of vesuvianite, magnetite, fluorite and white mica. The only sample analysed contained 13.8% F and 0.082% Sn.

The Omineca Crystalline Belt contains other Cretaceous to Tertiary batholiths and stocks (Fig. 3).

## The Intermontane Belt

In the study area, between the Coast Plutonic Complex and the Omineca Crystalline Belt (the two major metamorphic-plutonic welts of the Canadian Cordillera) lies the relatively tectonically undisturbed Intermontane Belt (Figs. 2, 3). However, in NTS map sheet 104N situated to the west of the Teslin Suture zone and east of the Coast Plutonic Complex, the Atlin region is also underlain by Cretaceous to Tertiary granitoids (Fig. 3).

In the context of this paper the most important granitic complex is the Surprise Lake batholith situated northeast of the town of Atlin (Fig. 3). The batholith crosscuts the regional trend and occupies about 800 $km^2$. It is cut in two by a down faulted block of Atlin terrain comprised of Paleozoic Cache Creek rocks.

A sample from within the batholith yielded a K/Ar age of 75.4 ± 2.5 Ma and six samples collected from granitic phases relating to the Adanac molybdenum deposit on the western edge of the batholith gave an average age of 70.6 ± 3.8 Ma (Christopher and Pinsent, 1982).

The Surprise Lake batholith is a complex, multi-textured, biotite granite to leucocratic granite. The textures include quartz-feldspar porphyries to crowded porphyry, medium- to coarse-grained inequigranular, coarse-grained equigranular and fine-grained equigranular types. Schlieren and miarolitic cavities are locally numerous, however, tourmaline is notably absent. It was originally designated as a tin granite because of the cassiterite reported in bulk gold placer concentrates and the high tin content (up to 41 ppm Sn) in granitic material (Mulligan, 1975). More recent work by Ballantyne and Littlejohn (1982) identified new granite-hosted occurrences associated with uranium mineralization and compared whole rock and trace element lithogeochemical data with other tin-granites in the world. These findings (Littlejohn and Ballantyne, 1983) and results reported in this paper clearly define the Surprise Lake batholith as a tin granite.

North of the town of Atlin and west of the Surprise Lake batholith (Fig. 3) is the Fourth of July Creek batholith designated by Aitken (1959) as being part of the Coast intrusions. It is composed mainly of granodiorite and quartz monzonite and appears to be zoned. In places it is a fairly uniform, medium-grained, biotite-hornblende diorite. The Coast intrusions have been assigned a Jurassic age by Aitken (1959) but hornblende from granodiorite of the Fourth of July Creek batholith yields a K/Ar age of 110 ± 4 Ma while biotite gives an age of 73.3 ± 2.6 Ma (Christopher and Pinsent, 1982). Clearly more dating is required for the granites in the Atlin district.

South of the Surprise Lake batholith are situated the Mount McMaster stock and the Mount Llangorse batholith, both of quartz diorite. Aitken (1959) also grouped these bodies with the Coast intrusions.

### Coast Plutonic Complex

To the west of the Intermontane Belt lies the Cretaceous and early Tertiary granitoid rocks of the Coast Plutonic Complex. These predominantly granodioritic to quartz monzodioritic plutons mark the second suture zone formed during the second collision of North America with outboard allochthonous terrains (Monger et al., 1982).

Since the Coast Plutonic Complex generally lies outside the study area (Figs. 2, 3) it is not necessary to discuss it further here. It should be noted however that in the Coast Plutonic Complex most deposits associated with intrusions are copper skarn deposits or porphyry copper-molybdenum deposits (Sinclair, 1986).

## Economic Geology and Mineral Deposits

Placer gold was discovered in 1874 on McDame Creek in NTS map sheet 104P (Gabrielse, 1963), but near Atlin (Fig. 3), in NTS map sheet 104N, even more important placer gold discoveries were made in 1898 (Aitken, 1959).

Within the study area, mineral exploration increased after 1942 when the construction of the Alaska highway transected the region.

Several gold quartz veins are presently being mined near the town of Cassiar, in NTS map sheet 104P. Except for continued placer mining at Atlin, the Cassiar Asbestos deposit, discovered in 1950, is the only active mine in the region.

However, a wide range of genetic types of mineral deposits and occurrences have been found and recently a renaissance of mineral exploration for Mo, W, U, Sn, Pb-Zn-Ag and Au has taken place. Some major discoveries have been made since the release of results from regional reconnaissance geochemical surveys (GSC Open File, 1978 and 1979).

Many of the region's important but presently uneconomic deposits are granite-related or -hosted. A recent review by Sinclair (1986) discusses northern Canadian Cordillera Mo, W and Sn deposits but we will only draw attention to those lying within the study area. Molybdenum porphyry deposits of note include Adanac (NTS 104N) (Christopher and Pinsent, 1982; White et al., 1976) and the Casmo and Cassiar molybdenum deposits (NTS 104P) (Panteleyev, 1980; Saydam, 1983; Woodcock, 1980).

Important tungsten-molybdenum deposits comprised of stockworks, skarn and hornfels-hosted mineralization include Logtung (NTS 105B, 104O) (Noble et al., 1984) and Mount Haskins – Mount Reed (NTS 104P) (Woodcock, 1980). Tungsten skarns are also known to occur in NTS map sheets 105F and 104N.

Granite-related tin mineralization in the district is generally of two main types namely, quartz and/or greisen veins and tin-bearing skarns. Descriptions of the tin mineralization may be found in Abbot (1981), Dick (1980), Layne and Spooner (1983), Littlejohn and Ballantyne (1983), Mato et al. (1983), Mulligan (1969, 1975), Mulligan and Jambor (1968) and Sinclair (1984). Some minor tin as stannite and cassiterite is also present in the Cassiar District lead-zinc-silver replacement/vein deposits in the area (Mulligan, 1969, 1975).

In Figures 2 and 3 we have illustrated that the Seagull and Parallel Creek tin granites have been emplaced east of the Teslin Suture zone in the Omineca Crystalline belt. Our and other workers' investigations have shown that these granites are what we will now define as "tin-fluorine-boron" geochemical types.

Lying west of the Teslin Suture zone in the Intermontane Belt, the Surprise Lake batholith tin-granite can be defined as a "tin-fluorine" geochemical type where boron is notably absent. The tin-granites of this part of the Canadian Cordillera have many similarities with the tin-granites of the Seward Peninsula, Alaska (Hudson and Arth, 1983). The role of a co-existing volatile phase in more highly evolved residual granitic magma has been considered in Ballantyne and Littlejohn (1982), Ballantyne et al. (1983), Littlejohn and Ballantyne (1983) and Mato et al. (1983). The enriched volatile nature of the Seagull and Surprise Lake batholiths begs comparison with the Seward granites but as yet models seeking to explain the difference between the two different geochemical types of tin-granites (Sn-F-B versus Sn-F) have not been resolved.

Uranium mineralization is also granite-hosted in the region. The Surprise Lake batholith is greatly enriched in radioactive elements (Ballantyne and Littlejohn, 1982). It also shows a high heat generation (Jessop et al., 1984) and thus exhibits features comparable to the high heat production tin-granites of southwest England (Durrance et al., 1982).

Occurring in all map sheets in the study area are base metal skarns and lead-zinc-silver veins and/or replacement deposits. They are often spatially related to intrusions with which they are probably genetically related (Abbot, 1984; Mulligan, 1969). Important examples are the Atlin-Ruffner deposit (NTS 104N) located northeast of Atlin and hosted within the Fourth of July Creek batholith (Aitken, 1959) (Fig. 3) and the Silvertip-Rancheria deposit (NTS 104O) hosted in Paleozoic limestone and argillites on the west side of the Cassiar batholith (Mulligan, 1969). Also in the locality and probably of the same type is the newly discovered Midway deposit situated approximately at the corner of NTS 105B, 104O and 104P.

We have given only a brief description of the economic mineral potential of the study area. However, the region hosts a variety of mineral occurrences many of which are related to granitic intrusions. This will become more evident in the following interpretation of the regional reconnaissance geochemical survey.

## Reconnaissance Geochemistry

### Fluorine

The map for fluorine in stream water (Fig. 4) shows that reconnaissance hydrogeochemistry can be used to outline fluorine-enriched granites. The maximum fluorine content measured for the 4,379 water samples is 2,300 ppb. Since high fluorine values have been shown to be indicative of geochemically "specialized" granites (Tischendorf, 1977; Tauson et al., 1978; Tischendorf et al., 1978), elemental distribution maps of this type can be used to distinguish those granitoids which are geochemically distinct from similar age "normal" granites of the region.

Anomalous fluorine patterns for tin-granites occur in streams draining Surprise Lake, Seagull and Parallel Creek batholiths. Other fluorine anomalies related to granites are usually associated with W-Mo porphyry deposits (Logtung) or Mo-porphyry stocks which intrude the Cassiar batholith (Casmo).

## Uranium

The distribution of uranium in stream water (Fig. 5) demonstrates that hydrogeochemistry is an effective tool when assessing relative uranium enrichment in granites. The maximum uranium concentration measured for the 4,363 water samples is 42 ppm.

Strong (1981) discussed the fact that more differentiated felsic plutons often form granophile Sn-W-U-Mo ore deposits. The uranium elemental distribution map (Fig. 5) can be used to indicate the geochemical "specialization" of granite batholiths or stocks in the region.

From measurements of stream water the tin-granites (Surprise Lake and Seagull batholiths) are anomalous in uranium, whereas, the Parallel Creek batholith has low uranium contents. Uranium in stream water anomalies also overlap some of the Mo-porphyry stocks.

Boyle (1982) noted, from an experiment to determine the amount of labile uranium in intrusive rocks, that samples from the Adanac Mo-porphyry deposit and rocks of the Surprise Lake batholith contained by far the most readily available labile uranium as compared to other Canadian Cordillera intrusive rocks measured. Therefore, the intense uranium anomaly associated with the Surprise Lake tin-granite is directly related to the proportion of labile uranium (Boyle, 1982) and the uranium mineralization and high radioactive element contents of the host-rocks (Ballantyne and Littlejohn, 1982). Our investigations have shown that within the Surprise Lake batholith the uranium minerals kasolite, metazeunerite, and beta-uranophane are often spatially associated with zones of hydrothermal alteration and tin mineralization.

## Molybdenum

The molybdenum in stream sediment (Fig. 6) shows distinctive patterns which are in part due to its more limited dispersion in the surficial environment as compared to the hydrogeochemical patterns displayed for fluorine and uranium in stream waters. The maximum concentration for molybdenum in 4,409 stream sediment samples is 70 ppm. However, most molybdenum anomalies are found to be above 10 ppm.

The Surprise Lake batholith is drained by streams highly anomalous in Mo. Anomalies are associated with known molybdenum occurrences. For example, west of the Surprise Lake batholith, the small anomaly associated with a stock marks the location of the Adanac Mo-porphyry deposits. In comparison, the other two tin-granites east of the Teslin Suture zone are reflected by subdued molybdenum contents in stream sediment.

The distinct anomaly relating to the Logtung W-Mo deposit is associated with the small stock west of the Seagull batholith on the border of NTS map sheets 104O and 105B.

Other granite-related molybdenum anomalies are either associated with known porphyry Mo systems or the source is yet unknown. Most anomalies east of the Cassiar batholith and west of the Tintina fault are underlain by carbonaceous black shales.

## Tungsten

The tungsten stream sediment (Fig. 7) shows limited dispersion in the surficial environment. This is partially due to the heavy and resistant characteristics of the minerals scheelite and wolframite but perhaps more importantly to the relatively short distance they travel from source. Tungsten minerals have been "dumped" at slope breaks and thus spatially mark specific drainages as being enriched in tungsten. The maximum tungsten concentration measured for the 4,406 sample sites is 800 ppm; most samples, however, contained less than 6 ppm (Fig. 7).

The tungsten anomalies are in most cases associated with known tungsten skarns (NTS 105F) or W-Mo porphyry deposits such as Logtung (NTS 105F and NTS 104O boundary). But not all geochemical anomalies are exclusively attributable to previously known occurrences. For example, anomalies found within the Surprise Lake tin-granite are due to newly discovered wolframite mineralization which is often spatially associated with tin occurrences. To the west, the distinct anomaly near the Adanac Mo-porphyry deposit is partially derived from the Black Diamond Tungsten (quartz-wolframite) deposit (Aitken, 1959, p. 72). In this area scheelite-cassiterite-bearing skarns have also been discovered by the senior author. These occurrences help to explain the high concentrations of wolframite, scheelite and cassiterite in the placer gold concentrates recovered from the Atlin placers.

The Seagull batholith tin-granite also shows tungsten enrichment but the Parallel Creek batholith does not indicate tungsten "specialization".

## Zinc

The zinc in stream sediment (Fig. 8) also shows a distribution pattern which in some cases depicts anomalies which are granite-related. The maximum zinc concentration measured for the 4,409 stream sediment samples is 2,900 ppm. Most granite-related anomalies generally are above 168 ppm zinc, the 90th percentile of the total data set.

The elements U, Mo, W, and F may be considered as pathfinder or ore elements associated with "specialized" or "rare metal" granitoids (Strong, 1981; Tischendorf, 1977; Tauson et al., 1978; Tischendorf et al., 1978). Zinc, on the other hand, is not commonly considered as being enriched in leuco-granites or felsic plutons. As a result, the origin of the zinc anomalies within the Surprise Lake tin-granite will be considered further.

Anomalous levels for zinc in stream sediments associated with the Seagull batholith tin-granite may generally be explained by zinc-rich skarns which are often found far from the intrusive contact (Abbot, 1981, p. 43; Dick, 1980). The Parallel Creek batholith tin-granite also has zinc-rich skarns spatially related to the "wrigglite" skarn which could account for the moderate anomalies in the area.

Other zinc anomalies related to the Cretaceous-Tertiary granites in the area may be explained by lead-zinc-quartz veins hosted within the granites (Abbot, 1984; Mulligan, 1969). No base metal-tin lode deposits are known to occur in the study area.

## Tin

In the study area, the combined stream sediment data for tin are not as complete as for the previously discussed elements (Figs. 4, 5, 6, 7, 8). The elemental distribution map for tin (Fig. 9) shows distinctive patterns and anomalous concentrations confined to specific granitic bodies. The maximum tin concentration measured for the 2,675 stream sediment samples is 520 ppm. This sample which was collected within the Surprise Lake batholith drainage system deserves a closer examination.

It is important to note that although numerous tungsten anomalies were outlined in NTS map sheet 105F (Fig. 7) the tin levels are generally much more subdued (less than 11 ppm Sn).

In NTS map sheet 105B distinct tin enrichment and specific anomalies are associated with the Seagull batholith. These anomalies are readily explained by the tin mineralization described by Abbot (1981), Dick (1980), Layne and Spooner (1983), and Mato and others (1983). Other tin anomalies east of the Cassiar batholith, which generally do not coincide with mapped granitic bodies, have not yet been explained.

The western portion of the Surprise Lake batholith shows definite tin enrichment as compared to the eastern portion. It also has distinct anomalies within it. Our findings and interpretation of two of these anomalies are presented in this paper.

## Conclusions

The single element distribution maps (Figs. 4, 5, 6, 7, 8, 9) have shown how multi-method (stream water and stream sediment) reconnaissance geochemical data can characterize geochemical enrichment and depletion, define background concentrations, and establish anomalous patterns and zonation over a vast area within the Canadian Cordillera.

As compared with other granites of similar age in the region, the tin-granites outlined by surficial geochemical methods exhibit varying degrees of "specialization" as indicated by associated anomalous results. The reconnaissance survey not only identified "specialized" granites and known occurrences of Mo, W, Zn and Sn from the Omineca Crystalline Belt, but also demonstrated that portions of the Surprise Lake batholith in the Intermontane Belt are also greatly enriched in this suite of elements, particularly tin.

Coincident multi-element anomalies over a granitic body probably reflect the mineralizing processes within the granite or at least its geochemical "specialization" in these elements. Therefore, these maps have indicated that the Surprise Lake batholith should be assessed for its tin potential.

## Regional Interpretation: Three Element Maps

A subset, of reconnaissance water and sediment data, from streams which drain the Surprise Lake batholith and proximal country rocks, was chosen for closer evaluation. Reconnaissance survey data had demonstrated that anomalous U, F, Mo, W, Zn and Sn was located in drainages within the selected area. Figures 10 and 11 illustrate elemental association maps for up to 362 sample sites. The elemental association Sn-F-W (Fig. 10) and Sn-F-Zn (Fig. 11) are plotted to show three different geochemical measurements and/or anomalous concentrations at each sample site. This allows easy visual comparison of data from other regional sample sites and drainages.

The content of fluorine in stream water less than the 50th percentile or 170 ppb are shown as white areas on both maps. Distinct fluorine in water anomalies are restricted to specific drainages (Figs. 10, 11).

Tin in stream sediments greater than the 90th percentile are illustrated by diamond symbols, and tungsten and zinc are depicted by square shaped symbols on Figures 10 and 11 respectively.

These elemental association maps define drainages and sample sites which have coincident element anomalies. Figure 10 illustrates that the Horse Creek drainage has coincident Sn-F-W anomalies at most sample sites. By comparison, Horse Creek shows no Zn anomalies (Fig. 11).

The second drainage chosen for closer examination is situated south of Weir Mountain. This drainage displays a greater than 75th percentile fluorine in water anomaly and minor Sn-W and Sn-Zn anomalies at some sites. To the north of Weir Mountain a coincident Sn-W with no Zn-F anomaly is found (Figs. 10, 11). Explanation for anomalous Sn-F-W (Fig. 10) and Sn-F-Zn located west of Trout Lake (Fig. 11) can be found in Ballantyne and Littlejohn (1982).

Element anomalies situated on the west side of Surprise Lake are not discussed in this paper because of contamination and disrupted drainage pattern due to placer gold mining activity. However, cassiterite, wolframite and scheelite concentrations are found in the placer heavies from these drainages.

Figure 12 is a detailed enlargement of the area outlined around Surprise Lake, Horse Creek and Weir Mountain as shown in Figures 10 and 11. It contains 70 reconnaissance stream sediment sample sites. The concentration levels chosen for Zn-Sn-W are based on the total data used to construct Figures 10 and 11. On this elemental association map, Zn and W are depicted as symbols and Sn is contoured.

The Horse Creek drainage system shows a distinct underlying tin anomaly and increasing anomalous tungsten concentrations downstream. Zinc concentrations are relatively minor.

The Weir Creek drainage system, south of Weir Mountain, shows a strong but spacially minor tin anomaly centered at a confluence. A zinc anomalous sample is coincident with the tin enrichment and tunsten and zinc anomalies surround the periphery of the anomaly (Fig. 12).

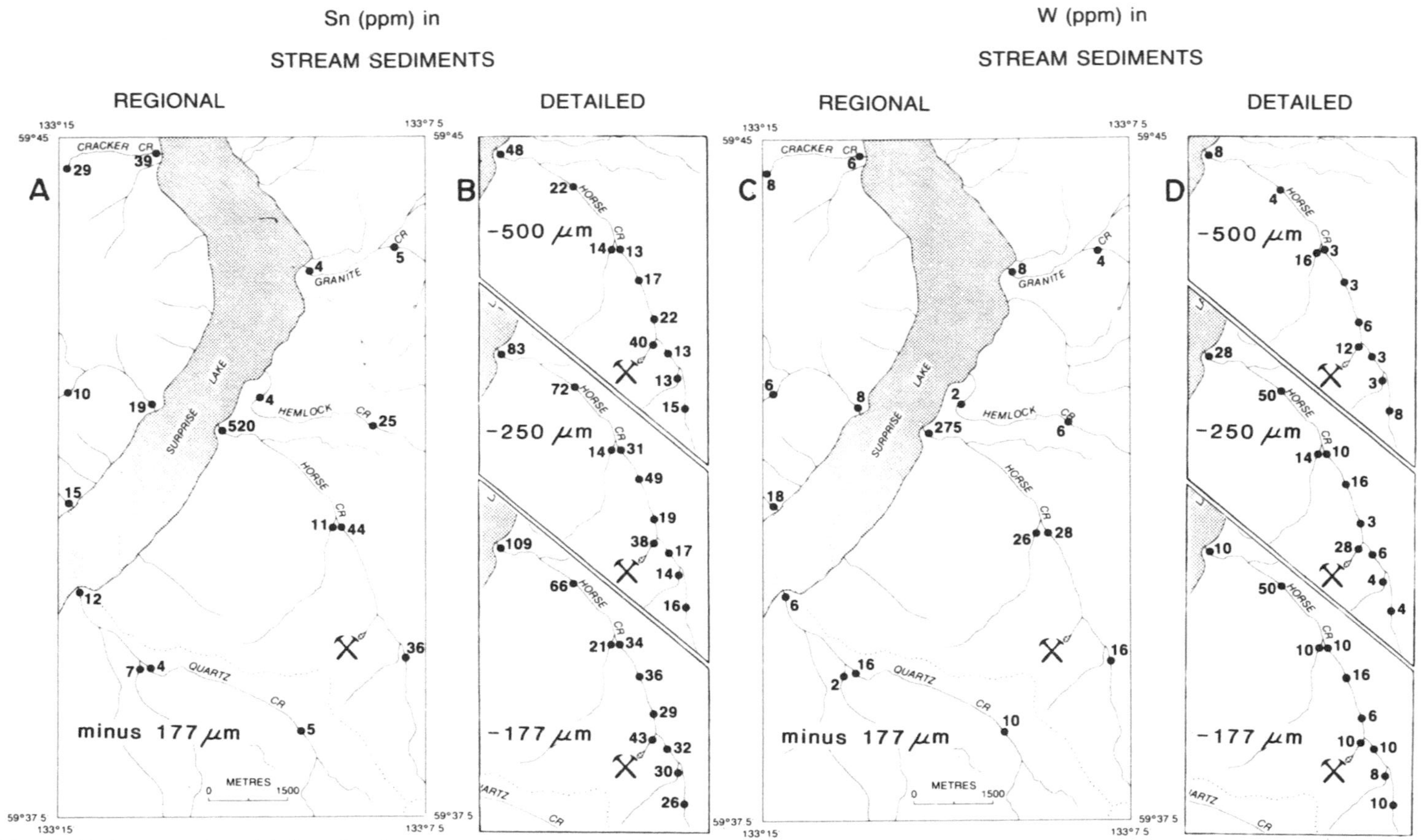


**Fig. 13. A)** Concentration of tin in regional stream sediment samples, Horse Creek study area; **B)** Concentration of tin in three size fractions of detailed stream sediment samples, Horse Creek study area; **C)** Concentration of tungsten in regional stream sediment samples, Horse Creek study area; **D)** Concentration of tungsten in three size fractions of detailed stream sediment samples, Horse Creek study area

## Regional Conclusions

These computer-generated maps have utilized the reconnaissance survey stream data to define elemental associations and to target anomalous drainages for further exploration. On the basis of anomaly contrasts (Sn-F-W vs Sn-F-Zn), Horse Creek and Weir Creek drainages were selected for detailed follow-up geochemical sampling surveys.

## Detailed Follow-Up Survey

### Horse Creek Study Area

Figures 13a and 13c show the stream sediment concentrations of tin and tungsten at the regional sample sites. The detailed follow-up survey included size fraction analyses of stream sediment material; these data are presented for tin and tungsten in Fig. 13b and 13d respectively.

By comparing the tin contents in the minus 177 micron fraction of regional (Fig. 13a) and detailed (Fig. 13b) stream sediments, the greatest enrichment within the Horse Creek drainage is located where it meets Surprise Lake. Horse Creek is relatively fast flowing and it is at the mouth at the major break in slope that tin would appear to be dumped in anomalous concentrations. Examination of Figures 13c and 13d shows that the magnitude of the anomalous regional tungsten sample at the mouth of Horse Creek was not repeated in the detailed stream sediment minus 177 micron fraction.

The crossed hammer symbol in the diagrams marks the location of mineralization discovered to date. Tin and tungsten levels reported for the coarser sized fractions in Figures 13b and 13d indicate that samples for the two tributaries draining the mineralized zone have generally higher contents of Sn and W than samples on the main Horse Creek drainage. They more specifically show the prospect area especially in the case of the Sn concentration.

Heavy mineral panned concentrate sampling of the stream sediment material was also conducted at suitable locations within the drainage system. Wherever possible sites were selected so that direct comparisons with an unpanned sediment sample could be made. The contents of tin and tungsten in four size-fractions of detailed heavy mineral concentrates (HMC) are reported in Figure 14. The areal distributions of anomalous HMC are coincident with the results obtained for the various size fractions previously reported for the unpanned sediments (Fig. 13) although they show greater contrast over background.

The most anomalous tin HMC sample is located on the main Horse Creek drainage immediately below the tributary draining the mineralized zone (29, 31, 252, 422 ppm Sn). The tin contents in the 500, 250 and 177 micron fractions of HMC ranged up to 2 times, 5 times and 11 times greater than in the same fractions of unpanned sediment collected at this site. The tungsten contrast was equally large, ranging up to 21 times greater in the minus 177 micron fraction at the same site (340 ppm vs 16 ppm).

Evaluation and interpretation of the detailed follow-up stream sediment data confirmed the elemental association of Sn-W and generally the magnitude of the anomalies of the regional data. Analysis of different size fractions helped to define

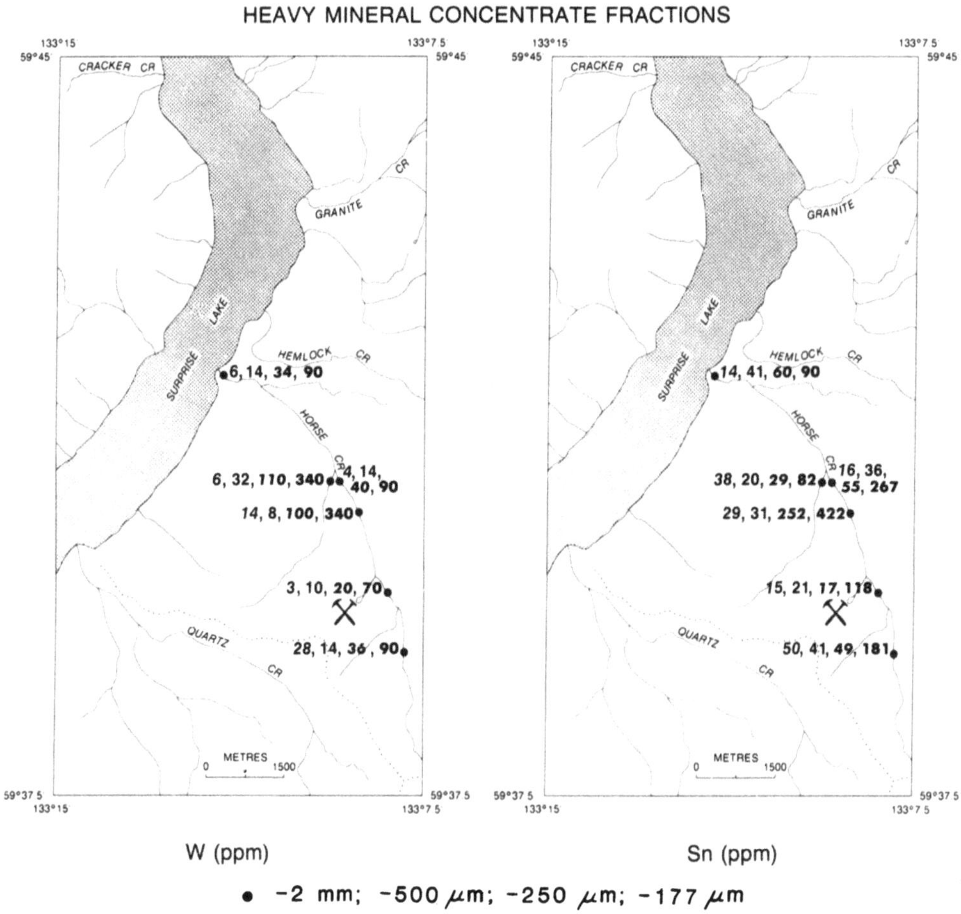


**Fig. 14.** Concentration of tin and tungsten in four size fractions of detailed heavy mineral panned concentrates, Horse Creek study area. The values at each site are in order of their grain size from coarse to fine

prospecting areas within the catchment area. Since all sites are not suitable for HMC sampling, supplementary unpanned sediment sampling helped to define and confirm anomalous results.

Further assessment of the Horse Creek drainage system has led to the discovery of previously unknown tin-tungsten mineralization which may in part explain the anomalous data reported for this study. Mineralization relating to fractures and shears or both consists of (a) quartz veins and veinlets, (b) quartz-fluorite veins, (c) cassiterite-wolframite in quartz veins and veinlets, and (d) quartz-arsenopyrite replacements and veins. The cassiterite and wolframite were found to be generally fine-grained and restricted to the borders of quartz veins in altered, bleached to rusty granite or fractures. The extent and economic significance of the occurrences have not been thoroughly investigated.

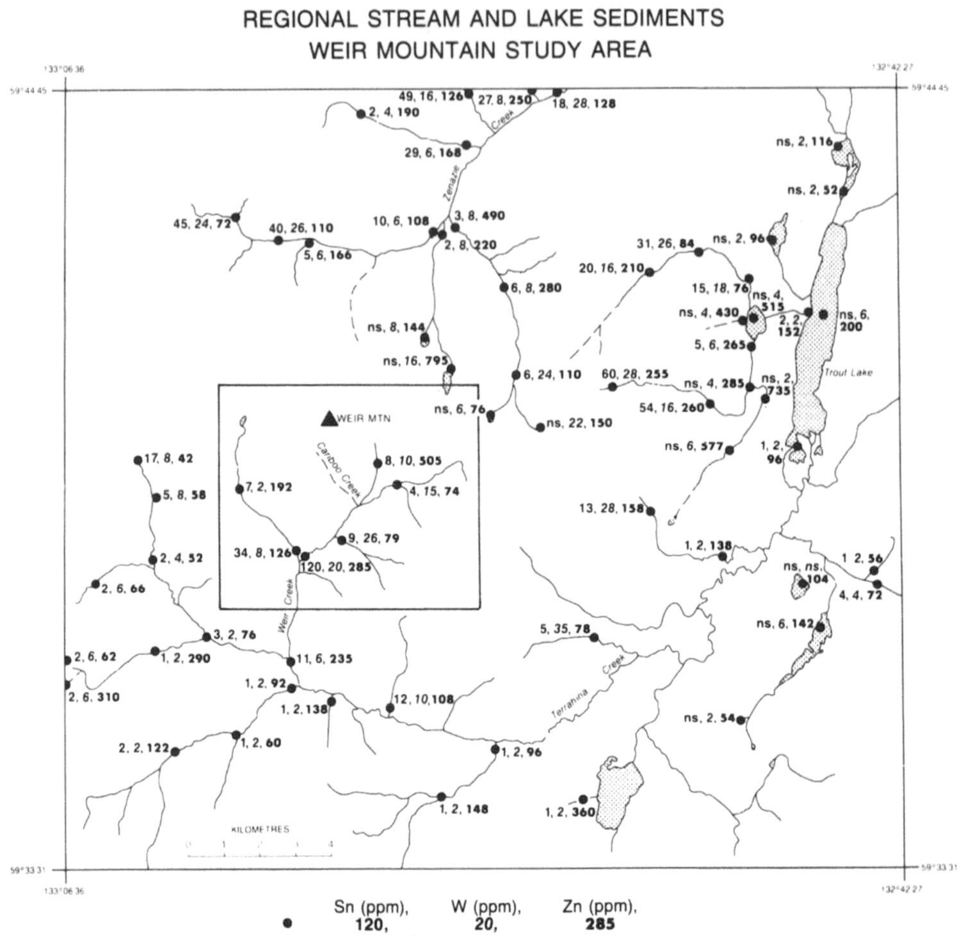


**Fig. 15.** Concentration of tin, tungsten and zinc in regional stream and lake sediment samples, Weir Mountain study area. Values at each site represent tin, tungsten and zinc in that order. Insufficient material for analysis is represented by "ns". Area covered by Fig. 16 is outlined

## Weir Mountain Study Area

Figure 15 shows the stream sediment – lake sediment contents of Sn, W, Zn at regional sample sites. The greatest enrichment of Sn on Weir Creek is at a confluence downstream from Cariboo Creek. The sample site yielded concentrations of 120 ppm Sn, 20 ppm W, and 285 ppm Zn in sediments. Within the Weir Mountain study area outlined in Figure 15 the next highest tin valve is 34 ppm. It should also be noted that the W anomalies reported on Figure 15 are of a much lower order than the response shown for Horse Creek (Fig. 13c). Samples anomalous in Sn and Zn are relatively restricted and only a few tributaries would appear to be responsible for the results. Therefore, during the detailed follow-up survey, HMC samples were collected upstream from the enriched Sn-Zn regional sample on Weir Creek.

Sn (ppm) IN HEAVY MINERAL CONCENTRATE FRACTIONS

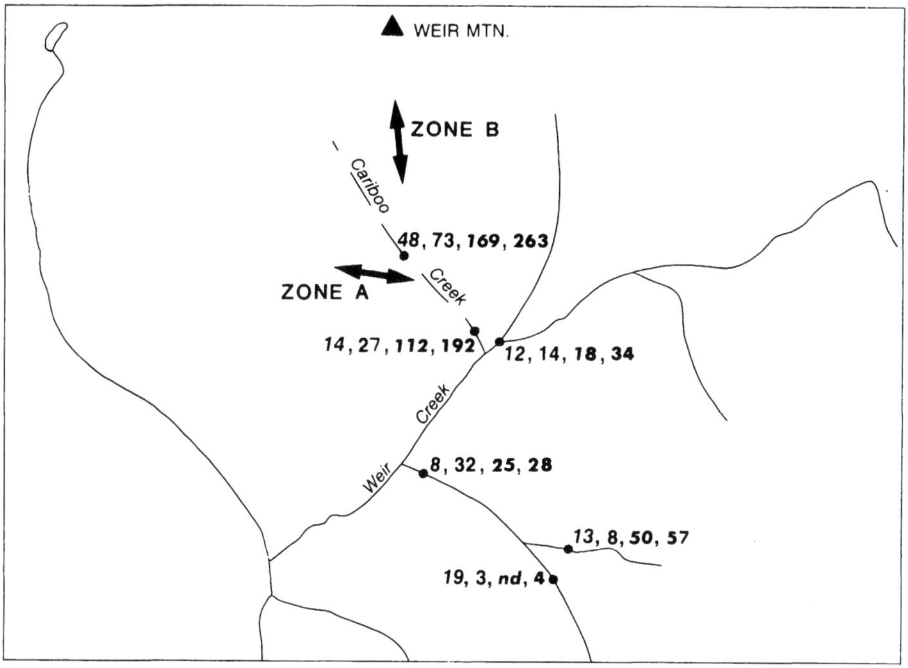


• −2 mm; −500 μm; −250 μm; −177 μm

**Fig. 16.** Concentration of tin in four size fractions of detailed heavy mineral panned concentrates, Weir Mountain study area. The values at each site are in order of their grain size from coarse to fine. Insufficient material for analysis is represented by "ns". Mineralized zones are indicated

The contents of tin in four size fractions of detailed heavy mineral concentrates are reported in Figure 16. This sampling clearly shows Cariboo Creek to be the source of the tin in the Weir Creek drainage.

Using the sample from Weir Creek collected just above the confluence of Cariboo Creek the contrast is 4, 5, 9, and 8 times greater for the various fractions. The finer fractions appear to yield the greatest contrast. Compared to the anomalous regional unpanned sediment sample containing 120 ppm Sn and 285 ppm Zn, Zn contents of the various size fractions of HMC collected showed very little enrichment ranging from 191–333 ppm.

Generally, the magnitude of the tin levels in anomalous samples collected in the Weir Creek study area are comparable to those reported at Horse Creek.

Mineral identification of the HMC from Cariboo Creek showed the presence of magnetite, two different colours of cassiterite and gadolinite, a yttrium-iron-beryllium silicate which has a specific gravity similar to cassiterite.

Mineralized zones A and B (Fig. 16), located on both sides of Creek and on Weir Mountain, were discovered as a result of the detailed follow-up investigations. The minerals identified in HMC samples were important in the identification of previously unknown tin-bearing zones.

On Weir Mountain, which can be considered to lie near the apex of the batholith, a system of fractures and vein-like bodies occurs within the sheared and/or silicified highly altered granite. Zone A is the best exposed and most altered part of the hydrothermal system which consists of an *upper zone* of (a) barren quartz veins and veinlets, (b) vuggy and honey-combed massive quartz, (c) quartz greisen with purple fluorite; a *middle zone* of (a) massive quartz-sericite-cassiterite-beryl greisen, (b) greisen sheeted veins with and without fluorite and dasolite; and finally a *lower zone* of magnetite-sphalerite-galena-biotite replacement veins with minor cassiterite. Some pods of sphalerite contain up to 26% Zn in hand specimen. It is the greisen zones and replacement veins which account for the two different colours of cassiterite found in the "heavies" of Cariboo Creek.

Zone B which leads up to Weir Mountain consists of altered and sheared granite containing fluorite, kasolite and wulfenite, minor quartz veins as well as magnetite-sphalerite-galena-biotite replacement veins with minor cassiterite.

Mineralized zones outside of the Weir Mountain system are characterized by quartz-fluorite veins, quartz-wolframite veins and classic quartz-wolframite greisen bordered veins.

The mineralization found to date at Weir Mountain and its perimeter readily explains the Sn-Zn-F and minor W elemental associations found in the regional stream sediment and water data, and the heavy mineral response of the detailed follow-up study.

## Implications for Future Work

This report has documented the use of multi-element, multi-media regional reconnaissance geochemical stream data to identify potentially tin metalliferous areas in the northern Canadian Cordillera.

Single element reconnaissance maps of U, F, Mo, W, Zn and Sn were used as an interpretive aid to better geochemically define "specialized", enriched, or mineralized granitoids. Regional elemental association geochemical maps (three elements) were used to aid in the more detailed examination of potential exploration areas or drainage systems. Anomalies worthy of detailed follow-up surveys were defined within the Surprise Lake batholith, Atlin, British Columbia.

A complementary stream sediment and heavy mineral concentrate survey utilizing various size fractions of the two methods was conducted in an anomalous Sn-F-W drainage net. Interpretation of results and ground investigations led to the discovery of previously unknown tin-tungsten mineralization.

A heavy mineral concentrate survey utilizing varying size fractions helped to define a Sn-F-Zn anomaly found on another drainage system. Geochemical and mineralogical analyses of the heavy mineral concentrates led to the discovery of tin and tin-zinc mineralization.

The economic significance of these two different tin occurences awaits further investigations. In the district, the many other geochemical anomalies outlined within the regional reconnaissance survey should also be assessed and evaluated. The approach illustrated in this paper was a successful geochemical exploration strategy. However, as we in Canada are relative new-comers in the search for tin deposits the authors feel that other successful case histories will be forthcoming as the true potential of this metalliferous region of the Cordillera is explored in the future.

*Acknowledgements.* The data presented in this paper were obtained by the cooperation of many colleagues at the Geological Survey of Canada. We acknowledge the excellent contributions of A.L. Littlejohn and G. Martin during follow-up field investigations and sample collection. We gratefully acknowledge the expertise of P.J. Lavergne in sample preparation and mineral separation, and of H.G. Ansell and A.C. Roberts for mineral identification. Staff of the Analytical Laboratories Section of the Resource Geochemistry Subdivision and of the Analytical Chemistry Section of the Economic Geology Division are thanked for their contributions during sample analyses. Skillful drafting was supplied by staff of Cartographic Services and J.C. Belec is acklowledged for her excellent typing.

The autors fully acknowledge the contribution of the sponsors of the Regional Geochemical Reconnaissance Program which provided the excellent data base without which our evaluations and follow-up surveys would not have been possible.

## References

Abbott, J.G., 1981. Geology of Seagull Tin District; *in Yukon Geology and Exploration 1979–80,* Geology Section, Department of Indian and Northern Affairs, Whitehorse, p. 33–44.

Abbott, J.G., 1984. Siver-bearing veins and replacement deposits of the Rancheria district; *in Yukon Exploration and Geology, 1983,* Department of Indian and Northern Affairs, Whitehorse, Yukon, p. 34–44.

Aitken, J.D., 1959, Atlin map-area British Columbia (104N); *Geological Survey of Canada, Memoir* 307, p. 89.

Ballantyne, S.B., and Littlejohn, A.L., 1982. Uranium mineralization and lithogeochemistry of the Surprise Lake Batholith, Atlin, British Columbia; *in Uranium in Granites,* ed. Y.T. Maurice, Geological Survey of Canada, Paper 81–23, p. 145–155.

Ballantyne, S.B., Boyle, D.R., Goodfellow, W.D., Jonasson, I.R., and Smee, B.W., 1978. Some orientation surveys for uranium mineralization in parts of the Atlin area, British Columbia; *Geological Survey of Canada, Paper* 78–1A, p. 467–471.

Ballantyne, S.B., Littlejohn, A.L., and Ellwood, D.J., 1983. A tin-tungsten multimedia geochemical case history: Surprise Lake batholith, British Columbia; *Geological Association of Canada Annual Meeting, Victoria, Program with Abstracts,* v. 8, p. 3.

Berg, H.C., Jones, D.L. and Coney, P.J., 1978. Map showing pre-Cenozoic tectono-stratigraphic terrane of southeastern Alaska and adjacent areas; United States *Geological Survey Open File Report,* 78–1085.

Boyle, D.R., 1982. Characteristics of the Okanagan Highlands Intrusive Complex as a source for basal-type uranium deposits, south central British Columbia; *in Uranium in Granites,* ed. Y.T. Maurice, Geological Survey of Canada, Paper 81–23, p.37–47.

Christopher, P.A., and Pinsent, R.H., 1982. Geology of the Ruby Creek and Boulder Creek area near Atlin (104N/11W); British Columbia Ministry of Energy, Mines and Petroleum Resources, notes to accompany Preliminary Map 52.

Coney, P.J., Jones, D.L., Monger, J.W.H., 1980. Cordilleran suspect terranes; *Nature,* v. 288, p. 329–333.

Dick, L.A., 1980, A. comparative study of the geology, mineralogy and conditions of formation of contact metasomatic mineral deposits in the northeastern Canadian Cordillera; unpub. Ph.D. thesis, Queen's University, Kingston, 473 p.

Dobson, D.C., 1982. Geology and alteration of the Lost River tin-tungsten-fluorine deposit, Alaska; *Economic Geology,* v. 77, p. 1033–1052.

Durrance, E.M., Bromley, A.V., Bristow, C.M., Heath, M.J., and Penman, J.M., 1982. Hydrothermal circulation and post-magmatic changes in granites of south-west England; *Proceedings of the Ussher Society,* no. 5, p. 304–320.

Gabrielse, H., 1963. McDame map-area, Cassiar district, British Columbia; *Geological Survey of Canada, Memoir* 319, 138 p.

Gabrielse, H., 1969. Geology of Jenning River map-area, British Columbia (104O); *Geological Survey of Canada, Paper* 68—55, 37 p.

Gabrielse, H., Tempelman-Kluit, D.J., Blusson, S.L., and Campbell, R.B., 1980. MacMillan River, Yukon - District of Mackenzie - Alaska, Sheet 105, 115; Geological Survey of Canada, Map 1398A, 1:1,000,000, Geological Atlas.

Geological Survey of Canada/British Columbia Ministry of Energy, Mines and Petroleum Resources, 1978. Regional stream sediment and water geochemical reconnaissance data, northwestern British Columbia (NTS 104N); Geological Survey of Canada, Open File 517 (National Geochemical Reconnaissance Release NGR 28—177) (Revised and re-released 1979).

Geological Survey of Canada/British Columbia Ministry of Energy, Mines and Petroleum Resources, 1979a. Regional stream sediment and water geochemical reconnaissance data, northern British Columbia (NTS 104O). Geological Survey of Canada, Open File 561 (National Geochemical Reconnaissance Release NGR 41—78) (Revised and re-released in 1980).

Geological Survey of Canada/British Columbia Ministry of Energy, Mines and Petroleum Resources, 1979b. Regional stream sediment and water geochemical reconnaissance data, northern British Columbia (NTS 104P). Geological Survey of Canada, Open File 562 (National Geochemical Reconnaissance Release NGR 42—78) (Revised and re-released in 1980).

Geological Survey of Canada, 1979c. Regional stream sediment and water geochemical reconnaissance data, southern Yukon Territory (NTS 105B); Geological Survey of Canada, Open File 563 (National Geochemical Reconnaissance Release NGR 43—78) (Revised and re-released in 1980).

Geological Survey of Canada, 1979d. Regional stream sediment and water geochemical reconnaissance data, southern Yukon Territory (NTS 105F); Geological Survey of Canada, Open File 564 (National Geochemical Reconnaissance Release NGR 44—78) (Revised and re-released in 1980).

Gordey, S.P., Gabrielse, H., and Orchard, M.J., 1982. Stratigraphy and structure of Sylvester Allochthon, southwest McDame map area, northern British Columbia; *in* Current Research, Part B, *Geological Survey of Canada, Paper* 82—1B, p. 101—106.

Hosking, K.F.G., 1979. Tin distribution patterns; *in* Geology of tin Deposits, ed. C.H. Yeap, *Bulletin of the Geological Society of Malaysia,* no. 11, p. 1—71.

Hudson, T., and Arth, J.G., 1983. Tin granites of Seward Peninsula, Alaska; *Geological Society of America Bulletin,* v. 94, p. 768—790.

Jessop, A.M., Souther, J.G., Lewis, T.J. and Judge, A.S., 1984. Geothermal measurements in northern British Columbia and southern Yukon Territory; *Canadian Journal of Earth Science,* v. 21, p. 599—608.

Kwak, T.A.P., and Askins, P.W., 1981. Geology and genesis of the F-Sn-W (-Be-Zn) skarn (wrigglite) at Moina, Tasmania; *Economic Geology,* v. 76, p. 439—467.

Layne, G.D. and Spooner, E.T.C., 1983. The JC Sn skarn, S. Yukon: A contact metasomatic Sn-Fe-F (-Be-B-As) deposit related to the Seagull batholith; *Geological Association of Canada/Mineralogical Association of Canada/Canadian Geophysical Union Joint Annual Meeting, Program with abstracts,* v. 8, p. 41.

Layne, G.D. and Spooner, E.T.C., 1983. The JC Sn-Fe-F (-Be-B-As) skarn, Wolf Lake area, Yukon Territory; *in* Mineral Deposits of Northern Cordillera; *The Canadian Institute of Mining and Metallurgy, Special Volume,* p. 38.

Littlejohn, A.L., and Ballantyne, S.B., 1983. Paragenesis of cassiterite and associated minerals in the Surprise Lake Batholith, Atlin, British Columbia; *Geological Association of Canada Annual Meeting, Victoria, Program with Abstracts,* v. 8, p. 43.

Mato, G., Diston, G., and Godwin, C.I., 1983. Geology and geochronometry of tin mineralization associated with the Seagull batholith, south-central Yukon Territory; *Canadian Institute of Mining and Metallurgy, Bulletin,* v. 76, no. 854, p. 43—49.

Monger, J.W.H., 1975. Upper Paleozoic rocks of the Atlin terrane, northwestern British Columbia and south-central Yukon; *Geological Survey of Canada, Paper* 74—47, 63 p.

Monger, J.W.H., 1977. Upper Paleozoic rocks of the western Canadian Cordillera and their bearing on Cordilleran evolution; *Canadian Journal of Earth Sciences,* v. 14, no. 8, p. 1832—1859.

Monger, J.W.H. and Davis, G.A., 1982. Evolving concepts of the tectonics of the North American Cordillera; *in Frontiers of geological exploration of western North America,* eds. A.E. Leviton, P.U. Rodda, E. Yochelson, M.L. Aldrich, Pacific American Association for the Advancement of Science, p. 215—248.

Monger, J.W.H. and Price, R.A., 1979. Geodynamic evolution of the Canadian Cordillera - Progress and problems; *Canadian Journal of Earth Sciences,* v. 16, p. 770—791.

Monger, J.W.H., Price, R.A. and Tempelman-Kluit, D.J., 1982. Tectonic accretion and the origin of the two major metamorphic and plutonic welts in the Canadian Cordillera; *Geology,* v. 10, p. 70—75.

Monger, J.W.H., Price, R.A. and Tempelman-Kluit, D.H., 1983. Comment and reply on tectonic accretion and the origin of the two major metamorphic and plutonic welts in the Canadian Cordillera; *Geology,* v. 11, no. 7, p. 428—429.

Mulligan, R., 1969. Metallogeny of the region adjacent to the northern part of the Cassiar batholith, Yukon Territory and British Columbia; *Geological Survey of Canada, Paper* 68—70, 13 p.

Mulligan, R., 1975. Geology of Canadian tin occurrences; *Geological Survey of Canada, Economic Geology Report* No. 28, 155 p.

Mulligan, R., 1984. Geology of Canadian tungsten deposits; *Geological Survey of Canada, Economic Geology Report* 32, 121 p.

Mulligan, R., and Jambor, J.L., 1968. Tin-bearing silicates from skarn in the Cassiar District, British Columbia; *Canadian Mineralogist,* v. 9, p. 358—370.

Noble, S.R., Spooner, E.T.C. and Harris, F.R., 1984. The Logtung large tonnage, low-grade W (scheelite) - Mo porphyry deposit, south-central Yukon Territory; *Economic Geology,* v. 79, p. 848—868.

Panteleyev, A., 1980. Cassiar map-area (104P); *in Geological Fieldwork 1979,* British Columbia Ministry of Energy, Mines and Petroleum Resources, Paper 1980-1, p. 80—88.

Poole, W.H., Roddick, J.A., Green, L.H., 1960. Geology, Wolf Lake map-area, Yukon Territory; *Geological Survey of Canada, Map* 10-1960.

Sainsbury, C.L., 1969. Geology and ore deposits of the central York Mountains, western Seward Peninsula, Alaska; *United States Geological Survey, Bulletin* 1287, 101 p.

Sainsbury, C.L., Mulligan, R.R. and Smith, W.C., 1969. The Circum-Pacific "tin belt" in North America; *in A second technical conference on tin,* ed. W. Fox, Bangkok, Thailand, v. 1, p. 123—148.

Saydam, A.S., 1983. Ground geophysical investigations over the Casmo molybdenite deposit; *Canadian Institute of Mining and Metallurgy, Bulletin,* v. 76, no. 852, p. 80—88.

Sinclair, W.D., 1986. Molybdenum, tungsten and tin deposits and associated granitoid intrusions in the northern Canadian Cordillera and adjacent parts of Alaska, *in Canadian Institute of Mining and Metallurgy, Special Volume,* No. 37. Conference information: Mineral deposits of northern cordillera symposium, p. 169—177.

Souther, J.G., Brew, D.A., and Okulitch, A.V., 1979. Iskut River, British Columbia - Alaska, Sheet 104, 114; *Geological Survey of Canada, Map* 1418A, 1:1,000,000, Geological Atlas.

Strong, D.F., 1981. Ore deposit models - 5. A model for granophile mineral deposits; *Geoscience Canada,* v. 8, p. 155—161.

Tauson, L.V., Cambel, B., Kozlov, V.D. and Kamenicky, L., 1978. The composition of tin-bearing granites of the Krusne hory (Bohemian massif), Spis-Gemer (Rodohorie) Ore Mts. (Western Carpathians) of Czechoslovakia and the Transbaikalian Provinces of the U.S.S.R.; *in Metallization Association with Acid Magmatism,* eds. M. Stemprok, L. Burnol, G. Tischendorf, Geological Survey, Prague, v. 3, p. 297—303.

Taylor, R.G., 1979. *Geology of tin deposits;* Elsevier, New York, 543 p.

Tempelman-Kluit, D.J., 1977. Geology of Quiet Lake (105F) and Finlayson Lake (105G) map-areas, Yukon Territory; *Geological Survey of Canada, Open File* 486.

Tempelman-Kluit, D.J., 1979. Transported cataclasite, ophiolite and granodiorite in Yukon: Evidence of arc-continent collision; *Geological Survey of Canda, Paper* 79—14, 27 p.

Tempelman-Kluit, D.J., 1980. Evolution of physiography and drainage in southern Yukon; *Canadian Journal Earth Science,* v. 17, p. 1189—1203.

Tempelman-Kluit, D.J. and Currie, R., 1978. Reconnaissance rock geochemistry of Aishihik Lake, Snag and Steward River map-areas in the Yukon Crystalline terrane; *Geological Survey of Canada, Paper* 77-8, p. 72.

Tipper, H.W., Woodsworth, G.J. and Gabrielse, H., 1981. Tectonic assemblage map of the Canadian Cordillera and adjacent parts of the United States of America; *Geological Survey of Canada, Map* 1505A.

Tischendorf, G., 1977. Geochemical and petrographic characteristics of silicic magmatic rocks associated with rare-earth mineralization; *in Metallization Associated with Acid Magmatism,* eds. Stemprok, M., Burnol, L., and Tischendorf, G., International Geological Correlation Program, v. 2, p. 41–96.

Tischendorf, G., Schust, F. and Lange, H., 1978. Relation between granites and tin deposits in the Erzgebirge, G.D.R.; *in Metallization Associated with Acid Magmatism,* eds. Stemprok, M., Burnol, L. and Tischendorf, G., Geological Survey of Prague, p. 123–138.

Wanless, R.K., Stevens, R.D., Lachance, G.R. and Delabio, R.N., 1970. Age determinations and geological studies, K-Ar isotopic ages, Report 9; *Geological Survey of Canada, Paper* 69-2A, 78 p.

Welsch, E.P. and Chao, T.T., 1976. Determination of trace amounts of tin in geological materials by atomic absorption spectrometry; *Analytical Chimica Acta,* v. 82, p. 337–342.

White, W.H., Stewart, D.R. and Ganster, M.W., 1976. Adanac (Ruby Creek); *in* Porphyry Deposits of the Canadian Cordillera, *Canadian Institute of Mining and Metallurgy, Special Volume* 15, p. 476–483.

Woodcock, J.R., 1980. Molybdenum exploration and development in western Canada; *Western Miner,* February, p. 55–60.

# 3.2 The Geology and Mineralogy of the JC Tin Skarn, Yukon Territory, Canada

G.D. Layne and E.T.C. Spooner[1]

## Abstract

The JC deposit is a stanniferous skarn related to a ridge-like lobe of the mid-Cretaceous Seagull Batholith. Skarn replaces Mississippian carbonate sediments of the Yukon Cataclastic Complex along an elongate contact zone with a known strike length of 850 m. Tin mineralization was discovered at JC in 1977 and subsequent drilling has outlined estimated geologically inferred reserves of 1,250,000 t grading 0.54% Sn with a cut-off grade of 0.30%.

The skarn displays a complex zoning in stable mineral assemblages which can be divided into several stages: (I) andradite + hedenbergite ± calcite, (IIA) Fe-amphibole + fluorite ± quartz ± magnetite ± cassiterite, (IIB) Fe-amphibole + fluorite ± quartz ± pyrrhotite ± chalcopyrite ± sphalerite, (III) epidote + quartz + calcite ± axinite ± malayaite, (IV) biotite + fluorite ± quartz ± cassiterite ± arsenopyrite with associated vein minerals including beryl, tourmaline and danalite, (V) Late stage chlorite + calcite + pyrite/marcasite.

The predominant mineral residence of tin in the early skarn stages is as tin-bearing silicates, stanniferous andradite (stage I), stanniferous Fe-amphibole (stages IIA and IIB) and malayaite (stage III), although tin as cassiterite occurs with quartz-magnetite-Fe-amphibole rocks in stage IIA. In stage IV tin occurs predominantly as cassiterite, both in narrow (1–2 cm) veins and in massive biotite-fluorite-quartz zones. The stage IV tin mineralization, in particular, is localized above the actual contact between skarn and intrusion.

Both JC and stanniferous skarns in Tasmania, Australia (Moina and Mt. Lindsay) show a well-evolved mineral assemblage zonation and an association with a porphyritic, partially altered lobe of the related intrusion.

## Introduction

Very few detailed descriptions of tin-bearing skarns are available in the literature. The best studies include those on Moina (Kwak and Askins, 1981) and Mt. Lindsay (Kwak, 1983) in Tasmania and Dobson's (1982) study of the skarn at Lost River, Alaska. Recently a number of stanniferous skarns, including the JC skarn, have been

[1] Dept. of Geology, University of Toronto, Toronto, Ontario, Canada M5S 1A1

discovered in the Seagull Batholith area of the south-central Yukon Territory [Dick (1980), Yukon Geology and Exploration 1979–80 (1981), Layne and Spooner (1983)].

The JC deposit is located on the southwest flank of the Seagull Batholith approximately 29 km N of the Alaska highway and 35 km NW of Swift River, British Columbia (Fig. 1A). Several vein-type tin deposits are also associated with the batholith and four of these have been described by Mato et al. (1983).

Cominco Ltd. and Dome Mines Ltd. have been joint venture owners and explorers of the property since 1977. Diamond drilling, totalling 3,822 m in 33 holes on the JC skarn zone between 1979 and 1982 has produced estimated inferred reserves of 1,250,000 t, grading 0.54% Sn with a cut-off grade of 0.30% Sn.

## Regional Geology

Mato et al. (1983) divide the mid-Cretaceous leucogranites of the Seagull Batholith into two textural types. The first is coarse-grained (3–10 mm) and seriate textured, and the second fine-grained and porphyritic.

The porphyritic type characteristically contains large phenocrysts of rounded quartz and angular Alk.-feldspar up to 10 mm in length. Both leucogranites are commonly fluoritic and the batholith frequently contains miarolitic cavities or spheroidal clots of black tourmaline. The fine-grained porphyry is more common to the northwest, where the batholith appears to be more shallowly eroded. Contacts between the porphyry and the coarse-grained leucogranite appear gradational over several m and grain-size variations are especially noticeable within 50 m of the upper or outer contacts of the batholith (Mato et al., 1983). It would appear, from these observations, that the fine-grained porphyritic phase represents a marginal or contact zone of the batholith.

**Fig. 1a.** The Seagull Batholith area of the south-central Yukon territory. Shaded portion corresponds to the map area of Fig. 1b

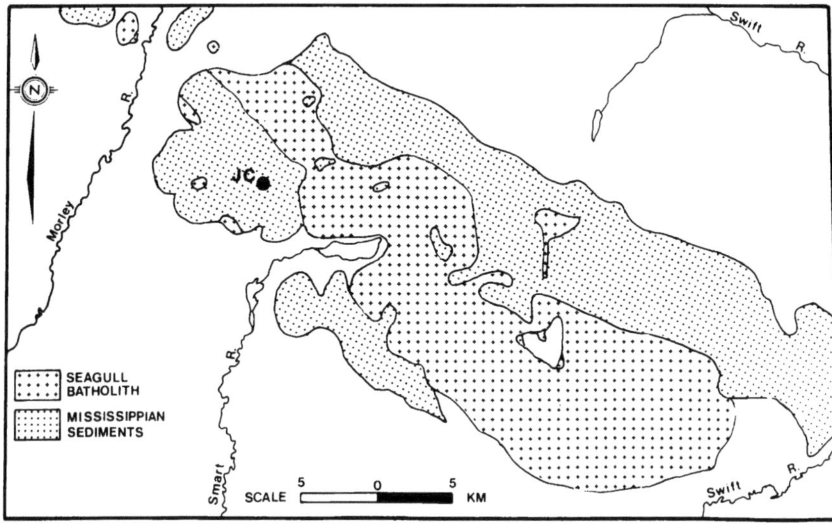


**Fig. 1b.** Simplified regional geology of the Seagull Batholith area (after Poole, 1956) showing the location of the JC prospect

The batholith as a whole displays pervasive jointing, comprising N-S and E-W trending vertical fractures, and less pronounced sub-horizontal fractures. In many cases this jointing is also evident in the overlying country rock.

Both the northeast and southwest flanks of the Seagull Batholith are largely overlain by Mississippian rocks of the Yukon Cataclastic complex (Fig. 1B). Abbott (1981) gives the following description of the Cataclastic complex in this area:

"Most mappable units appear to be part of a uniformly dipping stratigraphic sequence with few folds, but rock units lack lateral continuity and have knife-sharp tectonic contacts. The complex is an intensely sheared sequence of lenses or "fish" with little stratigraphic integrity."

As a result, stratigraphic relations between the many sedimentary and lesser volcanic units of Mississippian age in the Seagull Batholith area are largely unknown.

Many locally mappable carbonate units are present within the sequence. Most are less than 50 m thick and are comprised of white siliceous marble and interlayered, often highly sheared, nodules and bands of chert. Coral and crinoid fossils from several of these units have tentatively been dated as Early to Middle Carboniferous (Abbott, 1981). As with other units, the lateral continuity of the carbonate horizons is difficult to establish and the actual number of discrete carbonate horizons is uncertain.

## Local Geological Relationships

At JC a lobe of the Seagull batholith granite has intruded a thick sequence of Mississippian quartzite containing a 30–35 m thick carbonate horizon. The carbonate horizon strikes ESE (110–120°), dips shallowly (23°) to the south and is in contact with

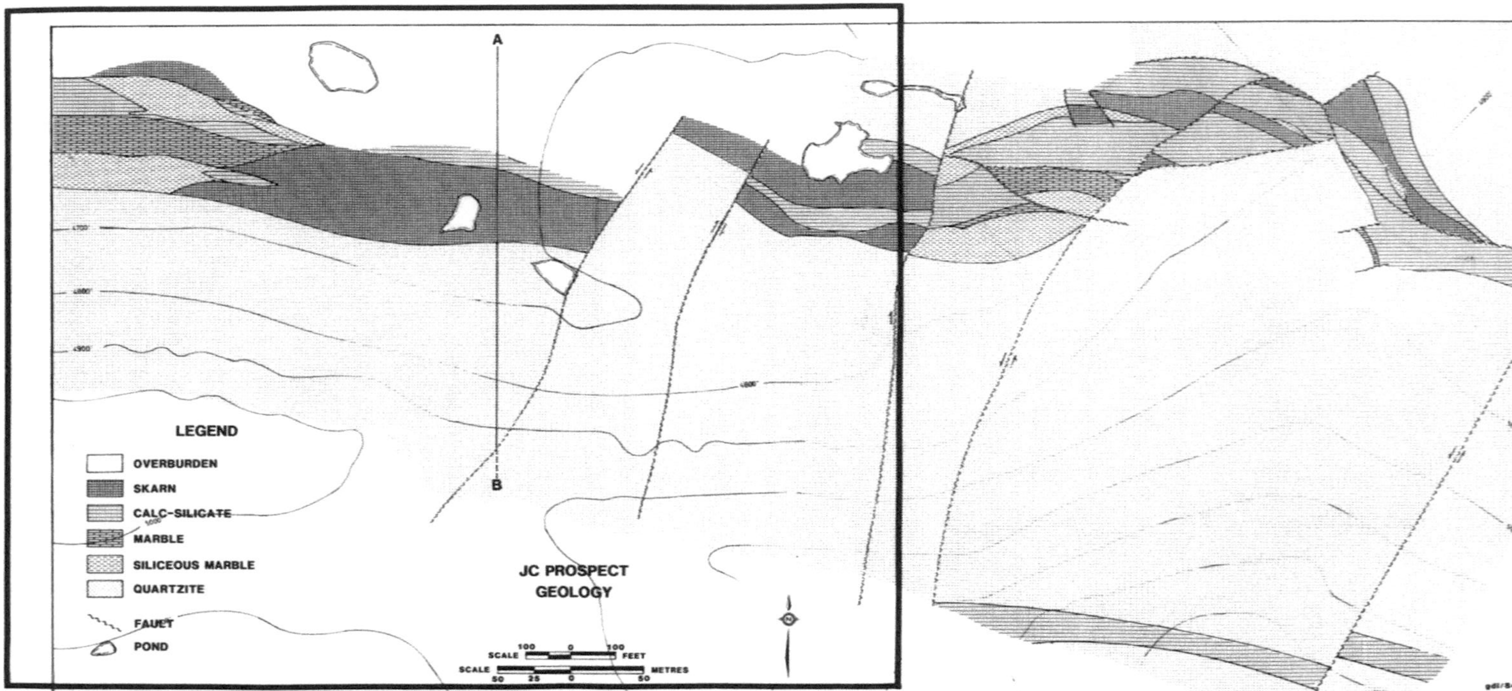


**Fig. 2.** Geological map of the JC prospect. The strike of the sediments is 110−120° and the dip is almost uniformly 20−25° to the south, despite disruption by faults. The trace of the cross-section (Fig. 3) is shown as A-B. Topographical contour interval is 100 feet. Boxed area is that shown in Figs. 4, 5, 12−14

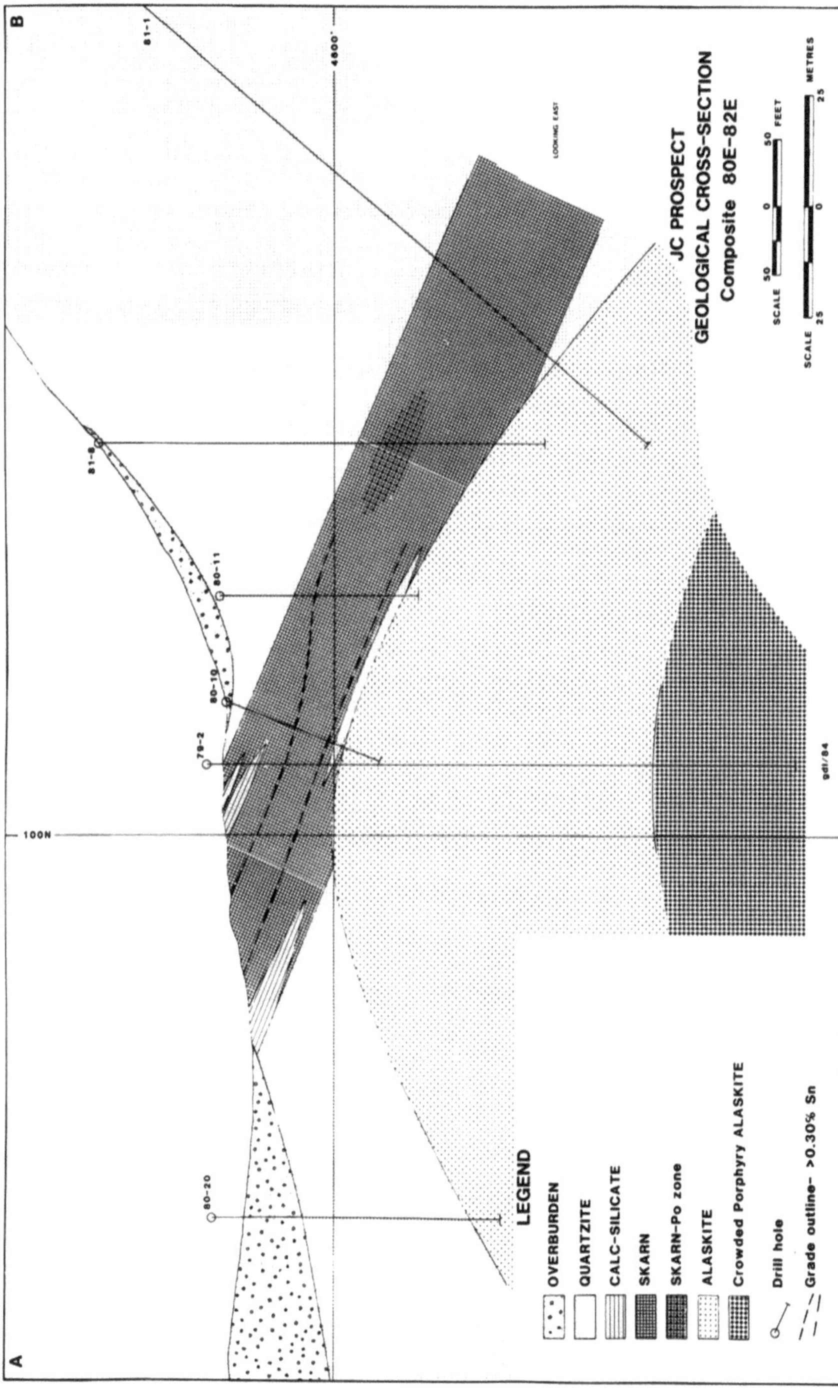


**Fig. 3.** Composite geological cross section, looking east, of the JC prospect. Some drill collars do not appear on ground surface due to projection. The lateral extent of the skarn pyrrhotite zone and its relationship to the closure of the 0.30% Sn grade contour remains uncertain

the granite along a narrow elongate zone, also striking approximately ESE. It is this contact zone which appears, from diamond drilling, to be the locus of skarn formation in the protolithic carbonate horizon. Figure 2 is a map of the surface geology of the JC deposit and Figure 3 shows the general nature of the skarn to granite contact along the A-B cross-section.

The skarn horizon and the underlying intrusion are both disrupted by a series of at least five near-vertical faults, trending approximately 30° and dipping steeply (85–90°) to the east or west. This fault set manifests itself as deep gulleys in the quartzite cliffs in the south part of the map area and faults are commonly filled with dykes of amygdaloidal basalt, the youngest rock in the JC area. A second set of faults, trending roughly E-W, is recognized in the eastern half of the map area (Fig. 2).

The larger E-W fault in the extreme northeast corner of the map (Fig. 2) is part of a major lineament which is visible on air photographs for several km to the east and west of JC. A major 30° fault several hundred m to the east of the JC map area forms a similar, N-S trending lineament.

Estimation of the direction and distance of movement along these faults has been hampered by a lack of sufficient geological markers. The relative age of the two fault sets is also indistinct, although both post-date the skarn formation. The intrusion itself does not outcrop in the JC map area but has been intersected by a series of 23 drill holes within the boxed area of Figure 2. The drill information is contoured in Figure 4 to show the top of the granite. The intrusion appears as a fairly uniform ridge-like body with a domal cross-section. The disruption of the intrusion by several of the 30° faults is also quite evident, as broad kinks or bends in the contours. In Figure 5 empirical corrections have been applied to remove the effects of fault movement and the data have been slightly smoothed. The granite now appears as a continuous ESE-striking ridge with a domal cross-section similar to that in Figure 3. This appearance is emphasized in three-dimensional transect plots of the same data in Figure 6 and 7.

The rock of the intrusive ridge seems typical of the fine-grained porphyritic leucogranite described from the rest of the Seagull Batholith by Mato et al. (1983) and Abbott (1981).

It commonly contains rounded quartz augen and Alk.-feldspar euhedra up to 10–15 mm in diameter in a finer-grained (1mm) groundmass of quartz and Alk.-feldspar, occasionally as a granophyric intergrowth, with lesser plagioclase and about 5% black biotite. Below the JC skarn, the intrusion almost invariably contains narrow subvertical fractures filled with tourmaline and accessory quartz, fluorite, pyrite and arsenopyrite. These fractures have a yellowish alteration selvedge of variable width (usually a few cm). The colouration appears to be the result of sericitic alteration of the biotite and feldspars. Remnant Alk.-feldspar phenocrysts often appear replaced by a yellow talcose material. Concentrations of fluorite from 2–25% and more intensely sericitized rock are frequently encountered in the upper few m of the intrusion.

**Fig. 4. A** Contour map of the top of the JC granite. Contour interval is 25 feet. This and subsequent contour and transect plots produced using the SURFACE II plotting programme (Sampson, 1978). Marginal grid units are in feet. **B** Posting of drill collar locations for holes used to produce Fig. 4A. Marginal grid units are in feet

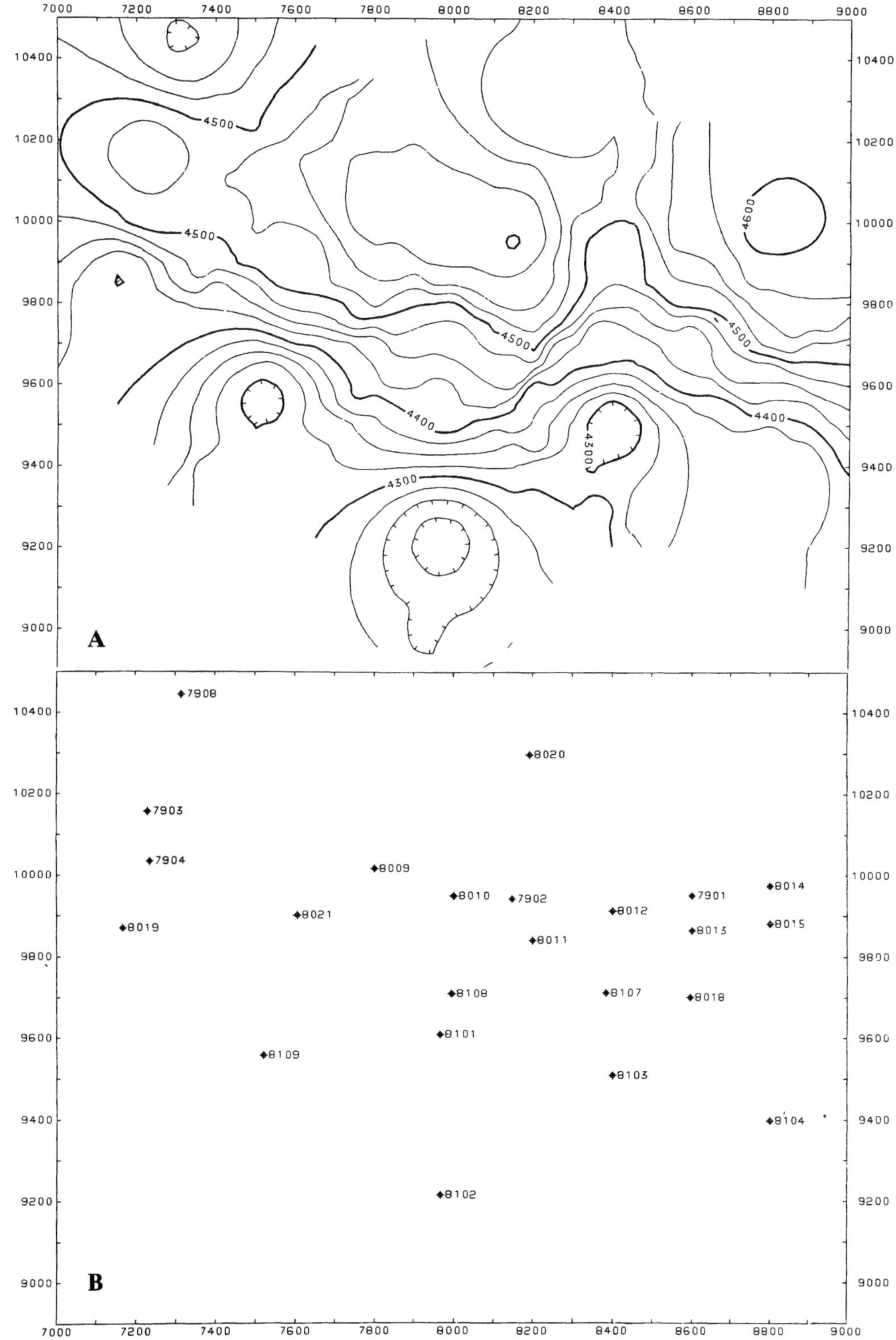
7000
7200
7400
7600
7800
8000
8200
8400
8600
8800
9000
10400
10200
10000
9800
9600
9400
9200
9000
4500
4500
4500
4500
4600
4400
4400
4300
4300
A
B
7908
8020
7903
7904
8009
8010
7902
8012
7901
8014
8021
8019
8011
8013
8015
8108
8107
8018
8101
8109
8103
8104
8102

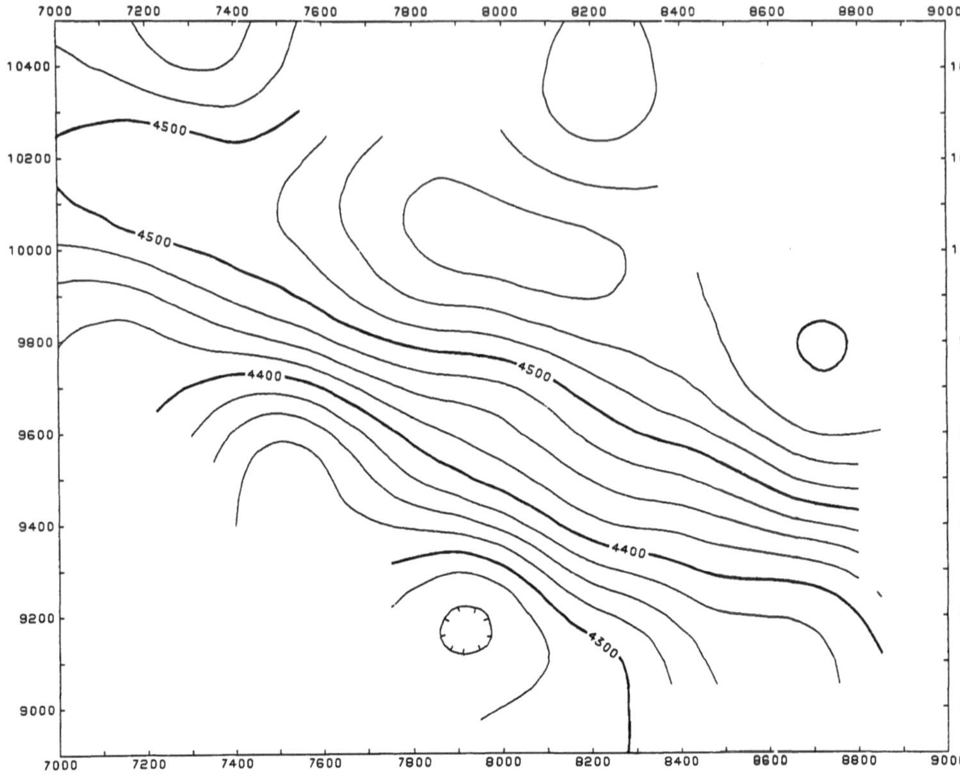


**Fig. 5.** Contour map of the top of the JC granite. Same data used for Fig. 4A, but corrected for the effect of faulting and smoothed to make contours more uniform. Marginal grid units are in feet

A single deep drill hole showed that, at about 73 m below the upper contact of the intrusion, the overall grain size and the number of quartz and feldspar phenocrysts began to increase gradually until, at the bottom of the hole (138 m), the rock comprised a crowded porphyry, with greater than 40% phenocrysts in a groundmass of 2–3 mm grain size.

The contact between skarn and intrusion is generally quite sharp. Only one short intersection of "endoskarn" (calc-silicate alteration minerals within the intrusion) has thus far been encountered by drilling.

The unreplaced carbonate protolith comprises both sparry crystalline marble and a siliceous marble. The latter contains narrow irregular interbeds of chert or fine-grained quartz. These interbeds, generally 1–2 cm in width, produce a characteristic differentiated weathering pattern in outcrop (Fig. 8A).

The sparry marble is quite massive, seldom displaying any remnant sedimentary structures, and occurs as uninterrupted beds, several m thick, or as narrow irregular interbeds in the siliceous marble. Similarly, the siliceous marble appears either as discrete beds or as interbeds or lenses within the sparry marble. Due to the discontinuous exposure of the carbonate protolith, it is not apparent whether these two lithological subdivisions of the horizon occur as stratigraphically continuous beds, or represent intercalated lenses.

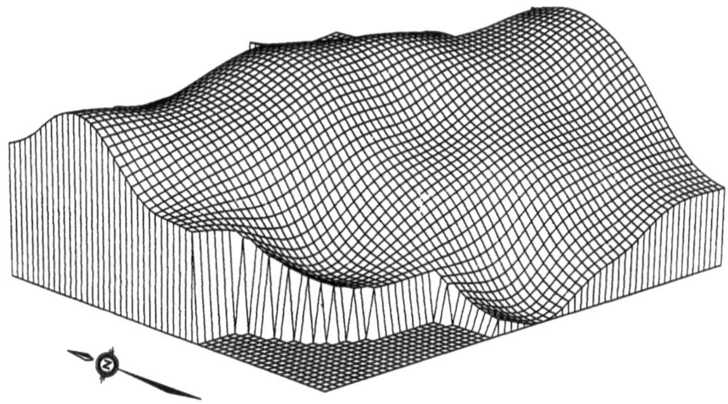

**Fig. 6.** Three-dimensional transect plot of granite top for same data as Fig. 5. Viewed from 225° azimuth, 15° elevation

Both overlying and underlying the carbonate horizon is a thick sequence of very fine-grained quartzite and argillaceous quartzite. The overlying quartzite is known to be at least 200 m thick at JC and is pervasively fractured in the same manner as the roof of the batholith.

Contact metamorphism of the carbonate protolith has, in part, produced a siliceous, fine-grained, finely-banded rock termed "calc-silicate" (Fig. 8B). The bands often appear pale pink or green in colour due to the presence of tiny (10 $\mu$) crystals of garnet (grossularite) or pyroxene (diopside) in parts of the rock.

Some of the calc-silicate rock may also contain quite large (5—10 mm) grossular garnets in irregular elongate bands, which appear to mimic original argillaceous bands in the siliceous marble (Fig. 8C). Wollastonite is also present locally. Remnant pods of siliceous marble within the calc-silicate rock and the banded texture strongly suggest that it is the contact metamorphosed equivalent of the siliceous marble.

The major effect of the contact metamorphism on the purer calcite marble appears to have been simple recrystallization to a 0.5 to 1 mm grain size.

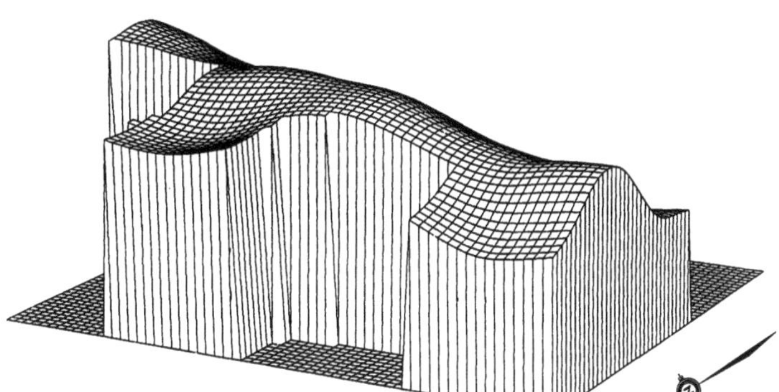

**Fig. 7.** Three-dimensional transect plot of granite top for same data as Fig. 5. Viewed from 330° azimuth, 15° elevation

**Fig. 8.** (**A**) Siliceous marble showing the differentiated weathering pattern of the cherty layers; (**B**) Calc-silicate rock with banding after original siliceous layers; (**C**) Garnetiferous calc-silicate with elongate garnets mimicking original siliceous or argillaceous layers in siliceous marble; (**D**) Polished slab showing garnet-rich stage I skarn (pale grey) partially replaced by Fe-amphibole of stage II (dark grey). Both stages are traversed by a quartz-beryl-cassiterite vein of stage IV

Contact metamorphosed carbonate rocks containing calc-silicate minerals, particularly if they are relatively coarse-grained, are often referred to in the literature as "skarnoids". This term is generally used to describe rocks that, although they are skarn-like in appearance and mineralogy, are contact metamorphic, in the classical sense, rather than truly metasomatic in origin (Einaudi et al., 1981). The calc-silicate rocks at JC would correctly be termed a skarnoid unit.

None of the calc-silicate which remains unreplaced by later skarn contains appreciable tin values.

## Skarn Development

Metasomatism of the carbonate protolith and parts of the calc-silicate unit has produced a skarn which has been traced along the narrow contact between intrusion and carbonate for a distance of 850 m and extends up-dip from the contact for a distance of 70—100 m. The maximum down-dip extension of the skarn is unknown, but appears to be greater than 90 m on the basis of diamond drilling (Fig. 3). Skarn may replace the entire 33 m thickness of the original carbonate horizon. Alteration of the surrounding quartzite, however, is not evident except where tourmaline-bearing veinlets appear in restricted parts of the footwall.

The skarn itself can be broken down into several spatial zones, based on stable mineral assemblages. These zones seem sequential in time, with each subsequent zone wholly or partially replacing its predecessors. Each zone may also replace previously unmetasomatized rock.

It appears that the earlier contact metamorphism of siliceous marble to calc-silicate has greatly reduced the permeability and reactivity of these rocks to metasomatic fluids. As a result, the initial skarn stages often preferentially replace sparry marble horizons, although replacement of calc-silicate horizons is also quite evident, particularly by the later vein-like stages of skarn.

The paragenetic sequence of the JC skarn is summarized in Table 1.

### Stage I

Stage I skarn consists largely of coarse-grained (up to 10 mm) andraditic garnet ($Ad_{50}$-$Ad_{99}$) and lesser hedenbergite pyroxene. The pyroxene occurs largely as inclu-

**Table 1.** Summary of mineral paragenesis

| Stage | Stable Mineral Assemblage |
|---|---|
| Stage I | andradite + hedenbergite ± calcite |
| Stage IIA | Fe-amphibole + fluorite ± quartz ± magnetite ± cassiterite |
| Stage IIB | Fe-amphibole + fluorite ± quartz ± pyrrhotite ± chalcopyrite ± sphalerite |
| Stage III | epidote + quartz + calcite ± axinite ± malayaite |
| Stage IV | biotite + fluorite ± quartz ± cassiterite ± arsenopyrite and associated vein minerals including beryl, tourmaline and danalite |
| Stage V | chlorite ± calcite ± pyrite/marcasite |

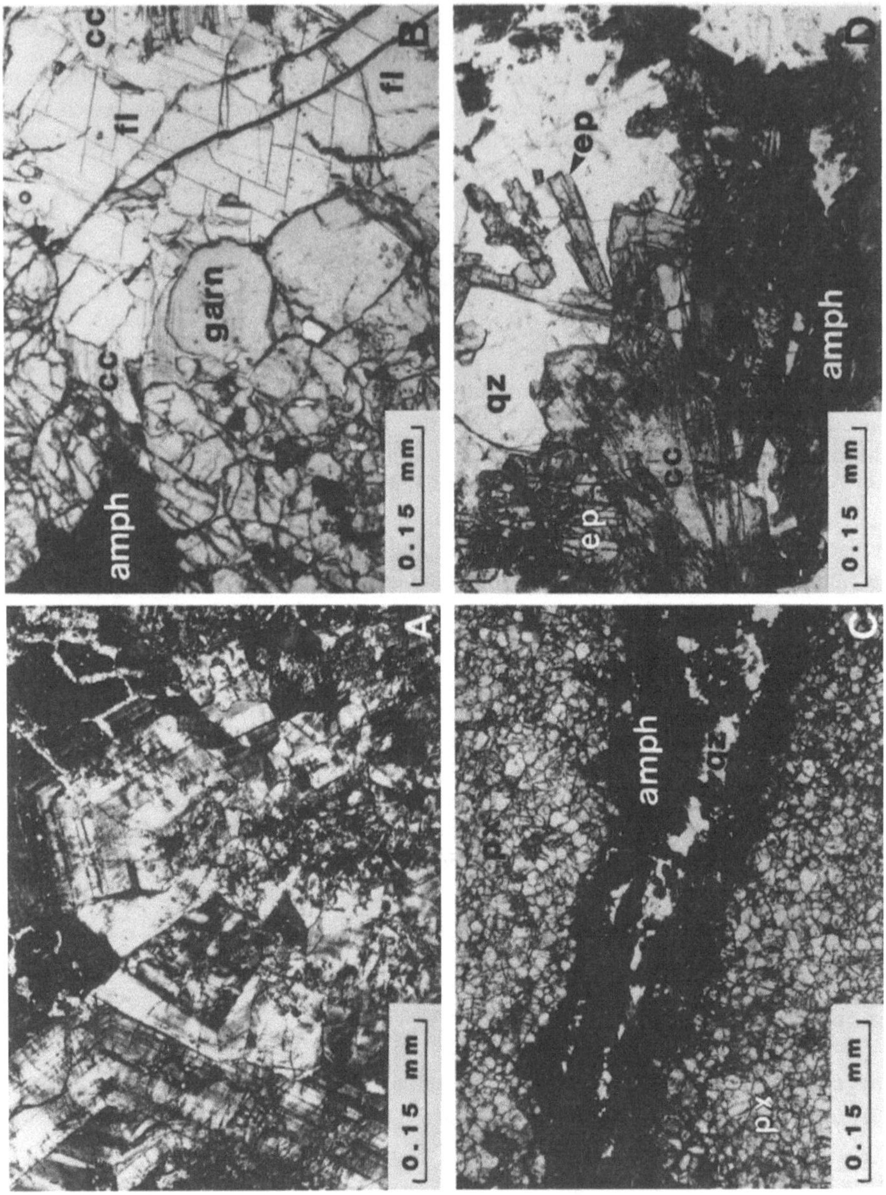


**Fig. 9.** (**A**) Concentrically zoned anisotropic andradite garnet from stage I skarn (cross-polars); (**B**) Amphibole (amph) and fluorite (fl) of stage I calcite (cc) is replaced by fluorite and appears to be a remnant of interstitial calcite from stage I; (**C**) vein-like replacement of pyroxene (px)-rich skarn of stage I by amphibole and quartz of stage II; (**D**) Amphibole (dk. grey) of stage II partially replaced by prismatic epidote (ep) with accompanying calcite and quartz (stage III)

sions in the garnet and as a surrounding groundmass. The garnets are generally anisotropic and typically show both concentric and sector zoning in thin section (Fig. 9A).

These garnets invariably contain appreciable Sn as a structural component. There is a definite positive correlation between $Fe^{3+}$ content (Ad component) and tin content, which ranges from 0.15% to 0.38% Sn. Garnets are the major residence of tin in this early stage and stage I skarns generally assay between 0.10% and 0.25% Sn depending on the modal percentage and composition of the constituent garnets. Earlier grossularitic skarnoid garnets ($Ad_{15}$-$Ad_{40}$) are generally tin-free.

Calcite is also a common mineral in stage I, as grains interstitial to garnet and pyroxene. Quartz is generally absent.

## Stage II

Stage II is typified by the replacement of the pyroxene (Fig. 9C) and interstitial calcite of stage I by Fe-amphibole.

The Fe-amphibole is accompanied by fluorite and quartz and forms a distinctive texture where it surrounds original garnet euhedra (Figs. 8D and 9B). Stage II skarn also occurs in manto-like extensions, replacing previously unmetasomatized carbonate.

This stage has been subdivided into substages IIA and IIB, depending on whether the predominant opaque phase is magnetite or pyrrhotite, respectively. Magnetite is the most common opaque mineral in Fe-amphibole-bearing assemblages but pyrrhotite-rich Fe-amphibole rocks with lesser chalcopyrite and sphalerite appear in a restricted zone above the intrusive contact.

It is not yet clear whether the pyrrhotite-rich zones represent a sulphide facies of stage II or a slightly later replacement of magnetite-bearing rocks. There is, however, sporadic textural evidence of stage IIB sulphides replacing or cross-cutting stage IIA magnetite. The exact geometry and distribution of the stage IIB skarn is still unclear.

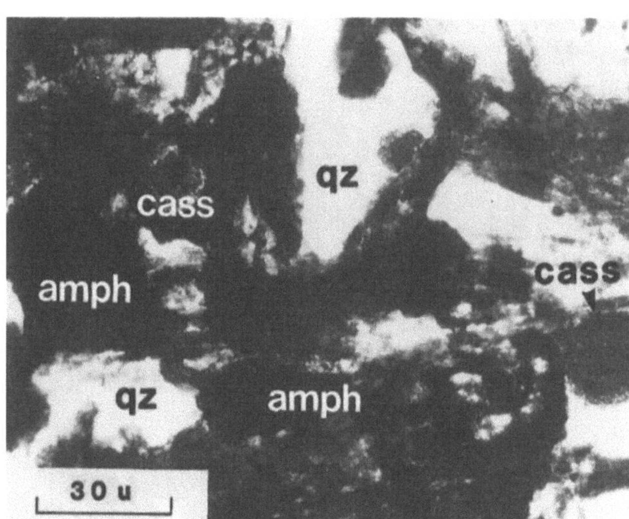


**Fig. 10.** Fine-grained cassiterite (cass), as apparently spherical grains, with amphibole, magnetite and quartz of stage IIA

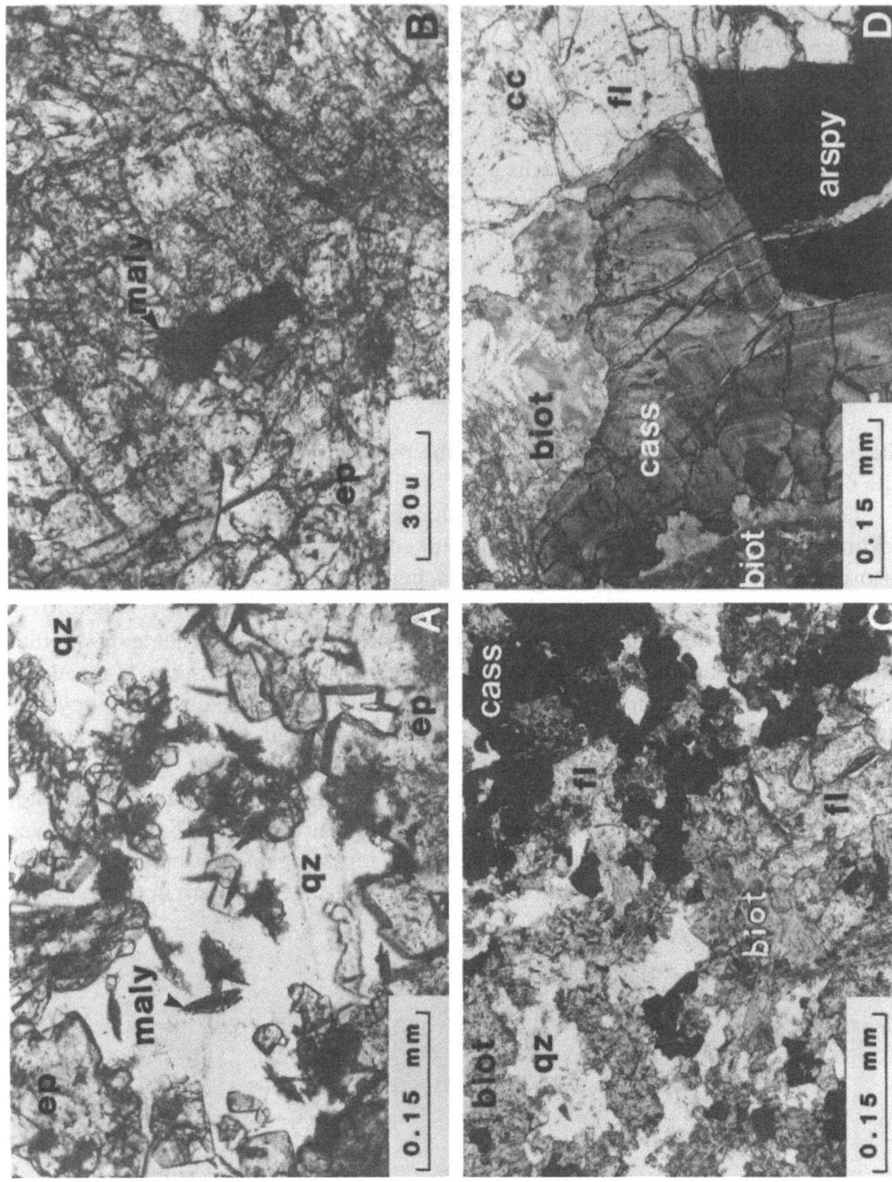


**Fig. 11.** (**A**) Sphenoidal grains of malayaite (maly) occurring with quartz and prismatic malayaite with epidote (stage III); (**C**) Subhedral cassiterite with biotite (biot), quartz and fluorite (stage IV); (**D**) Coarse-grained, colour zoned cassiterite in stage IV vein. The vein contains biotite, fluorite and arsenopyrite (arspy). The vein selvedge (left margin of photograph) is almost entirely comprised of biotite. Minor calcite of stage V forms late-stage veinlets cutting stage IV vein minerals

The Fe-amphibole itself may contain up to approximately 0.3% Sn. Some Fe-amphibole-quartz-magnetite rocks also contain cassiterite as tiny (25 $\mu$) dark brown grains, often displaying a peculiar rounded shape (Fig. 10). Sections of this type may assay greater than 0.30% Sn, predominantly contained in cassiterite.

## Stage III

The typical stage III assemblage is epidote, calcite and quartz (Fig. 9D) with accessory axinite ($((Ca,Mn,Fe)_3Al_2BO_3(Si_4O_{12})OH)$). The assemblage replaces previous stages most often as broad (up to 2 m) diffuse vertical zones or as small (several cm) gash-like discontinuous veins.

The major tin residence in this stage is malayaite ($CaSnSiO_5$) which occurs as tiny (30—50 $\mu$) prisms, often in sheaves, scattered throughout epidote-bearing rocks (Figs. 11A and 11B). Malayaite is most easily identified by its bright green fluorescence under ultraviolet illumination. Scheelite, which occurs only sporadically in the JC skarn, also seems to be associated with this stage.

On the whole, this stage contains low tin grades except where local concentrations of malayaite are encountered.

## Stage IV

This is the principal cassiterite-bearing stage. It is characterized by annitic (Fe-rich) biotite and accompanying fluorite replacing previous assemblages along large (many m wide) diffuse zones or as discrete narrow (1—2 cm) veins (Fig. 8D). Veining may follow subvertical fractures or subhorizontal planes (remnant bedding [?]) parallel to the dip of the original carbonate horizon.

Accessory mineral commonly recognized in the veins include quartz, beryl, tourmaline and danalite [$(Fe,Mn,Zn)_4(Be_3Si_3O_{12})S$]. A very common accessory phase is arsenopyrite as coarse (1—5 mm) euhedra. Bismuth is sporadically associated with arsenopyrite, as very small (10 $\mu$) blebs. Cassiterite appears in this stage as relatively large (up to 2 mm) reddish brown subhedra which display irregular colour zoning in thin section. Cassiterite, which appears in discrete narrow veins (Fig. 11D), is generally coarser-grained and has more well-defined colour zoning than the cassiterite which occurs in more massive, biotite-fluorite zones (Fig. 11C).

Sections which contain predominantly stage IV assemblages have assayed as much as 2—3% Sn.

## Stage V

Late stage pyrite and chlorite with accompanying calcite locally replace previous assemblages, generally as narrow veinlets. Pyrite is almost invariably partially altered to marcasite. This stage also includes several zones of highly altered calcite-filled breccia.

## Grade Distribution

The strike direction of the axis of the granite ridge is slightly different from that of the carbonate protolith horizon and the ridge axis also plunges to the west. As a result the exact attitude and extent of the skarn relative to the granite contact is not immediately obvious. In order to produce a model of the contact zone the granite ridge has been approximated by a simple second order trend surface (Fig. 12).

This is by no means an ideal representation but enables further manipulation of the data in modelling of the contact zone. The ridge axis strikes 105° and is shown in Figure 12 and in subsequent figures for reference.

Tin values from diamond drilling have been represented as grade multiplied by true width of section data, compiled from assays of all skarn intersections in each hole (Fig. 13).

The data have been normalized to a 33 mm true width of skarn in each hole. This normalization de-emphasizes the effect of dilution by later basalt dykes and unreplaced lenses or pods of calc-silicate or quartzite. In essence the contours in Fig. 13 represent the average tin grade of all skarn intersections over the underlying width of the replaced carbonate horizon. It can be seen in Fig. 13 that the highest tin grades (> 0.35 %) form a restricted elongate zone subparallel to the ridge axis.

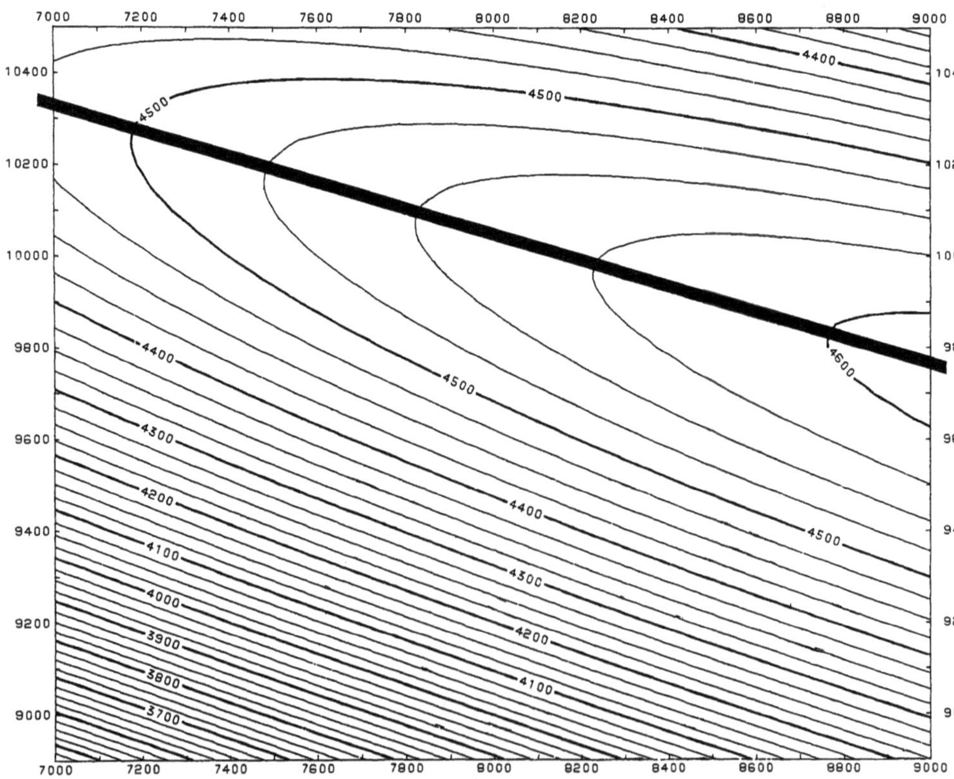


**Fig. 12.** Contour map of second order trend surface fitted to granite top data in Fig. 5. Ridge axis is shown as bold line. Contour interval is 25 feet. Grid units are in feet

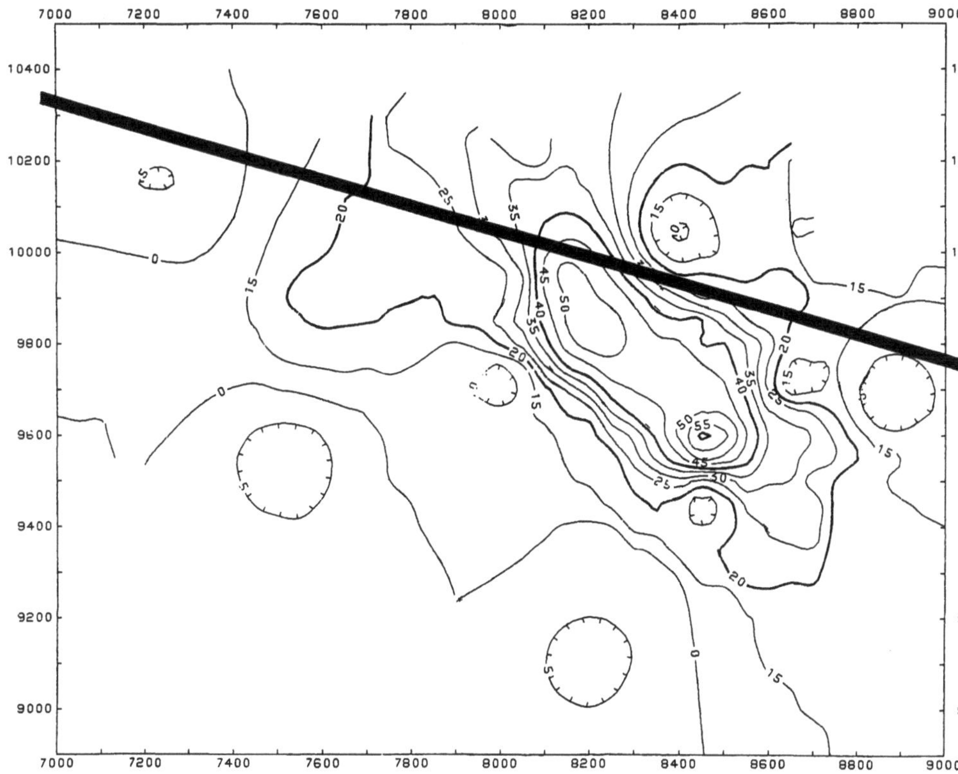


**Fig. 13.** Tin assay data contoured in average percent tin x 100 for drill intersections of skarn (see text). Contour interval is 5 (i.e. 0.05 Sn x 100). Superimposed is granite ridge axis from Fig. 12. Marginal grid units are in feet

The footwall of the original carbonate horizon, before faulting, may be quite accurately represented by a first order (planar) trend surface striking 114° and dipping 23° to the south. This representation of the skarn footwall has been subtracted from the granite trend surface (Fig. 12) to produce a plot of the resulting residual surface (Fig. 14). The zero contours in Fig. 14 thus outline the actual zone of contact between granite and skarn.

The contact zone is seen to be relatively flat-lying with a long axis almost parallel to the granite ridge axis. The zone is approximately 105 m wide with a maximum penetration of the carbonate footwall of about 10 m.

Higher tin grades form a well defined elongate body centred over the northern half of the contact zone, overlying the south flank of the granite ridge. The area of > 0.35% Sn in Figure 15 corresponds closely to the maximum extent of stage IV cassiterite mineralization. The relatively tin-barren area around the central apex of the contact zone corresponds to drill intersections of highly pyrrhotitic stage IIB mineral assemblages. The grade contours of 0.35% or more are, in reality, open to the east in a direction parallel to the strike of the granite ridge and contact zone.

This model of the relationship of skarn to contact zone is, of course, highly oversimplified. The contact zone is likely to be somewhat more irregular than shown in

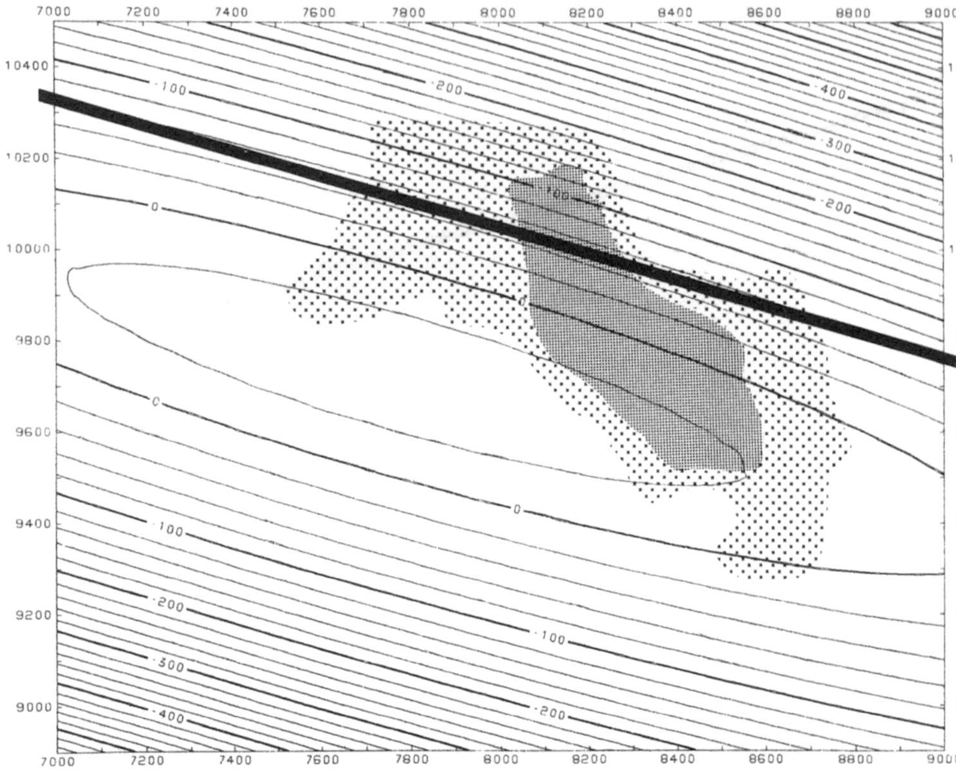


**Fig. 14.** Residual surface produced by subtracting trend surface representation of protolithic carbonate footwall from trend surface representation of granite top (Fig. 12) (see text). Contour interval is 20 feet. Superimposed are grade data from Fig. 13. Small crosses, > 0.20% Sn, dense spotted area; > 0.35% Sn. Also shown is ridge axis from Fig. 12. Marginal grid units are in feet

Figure 14, although the direction of elongation and general cross-sectional shape appear reasonably constant where the deposit has been well-defined by diamond drilling. The model also neglects any more localized structural controls which may have affected the development of vein-like stage IV mineralization. However, this simplified model of grade distribution serves to demonstrate the proximal nature of JC skarn and its relationship to one flank of a ridge-like intrusion. The fault-corrected projections of the high grade tin contours and contact zone provide a working exploration model. It is hoped that this type of treatment can be extended eventually to the modelling of mineral assemblage zonation in the JC skarn.

## Discussion

The JC skarn shows a broad evolution in the geometry and texture of mineral assemblage zones as each subsequent assemblage replaces the earlier skarn minerals. There is a general progression from massive, hornfels-textured andradite-hedenbergite skarn (stage I) to "manto-like" amphibole skarn of stage II, in places forming long shoots into previously unreplaced carbonate, to the more vein-like biotite-fluorite skarn of stage IV. This progression to a more vein-like geometry may be seen as a natural result of the replacement of highly reactive carbonate during stages I and II. Calcite-filled breccias represent the latest stage (V) of skarn development and apparently have formed along fracture zones.

The Mt. Lindsay skarn has been represented as having an orderly outward progression of mineral zones (youngest to oldest) from the intrusive contact (Kwak, 1983). Throughout the skarn, however, the later amphibole and biotite-bearing assemblages have locally replaced earlier skarn, often in a patchy or vein-like manner. This geometric evolution bears a great resemblance to that of the JC skarn.

The Lost River skarn (Dobson, 1982), on the other hand, is a complex association of greisen, largely vein-like skarn and late stage breccia pipes. The stockwork vein-type skarn overlies, and is largely separated from the intrusion. There is little or no development of massive skarns such as those at JC or Mt. Lindsay.

The mineral residence of tin in the JC skarn, as discussed above, is quite varied. Tin-bearing garnets from skarns are well documented and Einaudi et al. (1981) tabulate compositional data from several studies showing maximum tin contents ranging from 0.5 to 5.8 wt.%. The tin-contents of other silicates, including amphiboles and Sn-sphenes, are reviewed by Taylor (1979) and Nekrasov (1971).

Cassiterite is the most important in ore mineral in almost all deposits. At JC, cassiterite occurs in two distinct generations: fine-grained, dark-brown cassiterite with amphibole-quartz-magnetite rocks of stage IIA, and coarse-grained, colour-zoned cassiterite with biotite-fluorite and associated vein minerals of stage IV. It is sections of these cassiterite-bearing assemblages which provide the most likely source of economically recoverable tin in the JC deposit.

In the case of stage IV skarn it is apparent that biotite alteration of earlier assemblages is important in breaking down Sn-bearing silicates and fixing the tin as cassiterite. However, stage IV zones in the skarn have overall tin grades which are generally higher than those of the earlier stages which they replace. This grade distribution suggests that tin has also been added to the skarn by the metasomatic fluid at this stage. The $> 0.35\%$ grade area for tin in Figure 13 is, in fact, a fairly accurate reflection of the extent of the stage IV mineral assemblage.

In other tin skarns, however, this process of enrichment in cassiterite and overall tin grades can be reversed. For example, the earliest apparent stage at Mt. Lindsay is a cassiterite-magnetite-ilmenite-Alk.-feldspar-siderite zone grading up to 0.8% Sn. In subsequent stages cassiterite has reacted to form Sn-bearing amphibole, sphene and ilvaite. The tin grade in the latest stage biotite-fluorite-quartz skarns is essentially zero (Kwak, 1983). At Moina (Kwak and Askins, 1981), where primary wrigglite (fluorite-magnetite-vesuvianite-cassiterite-scheelite-adularia) skarn is altered, tin is essentially lost from those zones in the skarn.

However, Dobson (1982) notes that, at Lost River, a hydrous stage of alteration, which produced an assemblage of fluorite-biotite ± hornblende with accompanying cassiterite and sulphides, was an important process in liberating tin from earlier silicates. Late retrograde chlorite-carbonate alteration at Lost River also appears to have played a part in liberating and concentrating tin as cassiterite.

At JC, low tin values are associated with the pyrrhotite-rich zone developed in stage IIB, although the accompanying magnetite-bearing skarn (stage IIA) often contains appreciable tin grades. A similar type of pyrrhotite zone, also relatively poor in tin, has been observed at Mt. Lindsay (Kwak, 1983). This antipathetic relationship between tin and sulphide concentrations brings up the question of a genetic link between tin skarns and large sulphide-rich carbonate replacement tin deposits such as Renison, Tasmania.

The Renison deposit (Patterson et al., 1981) is a pyrrhotite-rich replacement of several carbonate horizons with very limited development of calc-silicate gangue minerals. Tin occurs largely as cassiterite and the deposit contains on the order of 24 Mt grading 1% Sn (Patterson et al., 1981).

It has been suggested by some authors that magnetite-rich tin skarns and sulphide-rich carbonate replacements represent metasomatic deposits which are proximal and distal, respectively, to the associated intrusion (Kwak and Askins, 1981; Kwak, 1983). At Renison, mineralizing fluids appear to have risen along steep fault zones from an intrusion, as yet unlocated, beneath the replaced dolomite horizons. The inferred distance of transport would have been at least several hundred meters (Patterson et al., 1981). Skarns, such as JC and Mt. Lindsay, on the other hand, are often found immediately at the contact between an intrusion and a carbonate horizon. However, this division of skarn from replacement deposit is by no means clear-cut. For example, at Moina, the skarn appears separated from the underlying granite by at least 200 m of sandstone. Metasomatic fluids appear to have reached the carbonate horizon through a set of subparallel tension fractures in the sandstone which now display Sn-W mineralization (Kwak and Askins, 1981). The distal/proximal division of metasomatic tin deposits by itself does little to explain the existence of tin-poor sulphide zones in skarns as opposed to tin-rich sulphide assemblages in replacement deposits.

The leucogranite at JC appears to comprise a fine-grained, porphyritic marginal zone of the generally coarser-grained Seagull Batholith. A similar case appears at Mt. Lindsay (Kwak, 1983) where the associated granite becomes finer-grained and more porphyritic (quartz and Alk.-feldspar) as the skarn contact is approached. Both deposits also appear to have a similar zone of sericitic alteration near the contact. Kwak (1983) also notes veinlets of quartz-tourmaline-fluorite with accessory sulphides in this sericitized zone, similar to those observed at JC.

Poole (1956) suggested that the roof of the Seagull Batholith may be undulatory in nature. The ride-like geometry of the JC intrusive ridge, its elongate shape and shallow westward plunge roughly mimicking the suggested northwestward plunge of the Seagull Batholith, may represent a single "undulation", possibly one of a series. Perhaps it is this juxtaposition of positive relief in the roof of the batholith with overlying calcareous sediments which was the major control of the formation of proximal skarns in the Seagull district.

## Conclusions

The JC and other stanniferous skarns display a wide variety of mineral residences of tin. In the earlier stages of the JC skarn tin is contained largely in garnet and amphibole while later stages contain tin predominantly as cassiterite. However, for stanniferous skarns in general, the cassiterite-bearing assemblages are not readily predictable on the basis of relative age or mineral association. Progressive stages of skarn development may alter earlier tin-bearing silicates to produce cassiterite, or cassiterite may be replaced by tin-bearing silicates. Metasomatic or hydrothermal fluids may also either enrich or deplete the total tin content of the skarn at any given stage.

The JC skarn, and in particular the late biotite-fluorite-cassiterite mineralization, are concentrated directly above the elongate zone of contact between the flank of the ridge-like intrusion and what was the original carbonate horizon. This represents a rather simple relationship between skarn and intrusion (cf. Mt. Lindsay, Kwak, 1983) as opposed to a more complex, distal relationship (e.g. Moina, Kwak and Askins, 1981).

The recognition of the cassiterite-bearing assemblages within a stanniferous skarn and their spatial relationship to the associated intrusion would appear essential to the efficient exploration of such deposits.

*Acknowledgements.* Cominco Ltd. and Dome Mines Ltd. kindly provided access to the JC deposit and Dome Ltd. and Cominco Ltd. employed one of the authors (G.D.L.) for two field seasons (1981 and 1982) as site geologist for the exploration programme at JC. Special thanks are extended to W.J. Wolfe, D. Cooke and L.J. Nagy (Cominco Ltd.) and G.S.W. Bruce (Dome Mines Ltd.) for their continuing assistance and encouragement in the study of the JC skarn. Many thanks also to Cam Stephen of J.C. Stephen Explorations Ltd., Vancouver, for access to the JC drill core and for assistance in the field.

Financial support for the initial part of this study was provided by E.M.R. grants 87-4-82 awarded to E.T.C.S. Subsequent support has been provided by N.S.E.R.C. operating grant A6114 (E.T.C.S.). Financial assistance to G.D.L. in the form of an N.S.E.R.C. Postgraduate Scholarship is gratefully acknowledged.

We also thank Brian O'Donovan (photography) and fax (typing services) for their indispensable assistance.

## References

Abbott, J.G., 1981. *Geology of the Seagull tin deposit,* in Yukon Geology and Exploration 1979–80, Geology section, Dept. of Indian Affairs and Northern Development, Whitehorse, 3–44.

Dick, L.A., 1980. A comparative study of the geology, mineralogy and conditions of formation of contact metasomatic mineral deposits in the northeastern Canadian cordillera; *Unpub. Ph.D. thesis, Queen's University, Kingston,* 471 p.

Dobson, D.C., 1982. Geology and alteration of the Lost River tin-tungsten-fluorite deposit, Alaska. *Econ. Geol.,* V. 77, 1033–1052.

Einaudi, M.T., Meinert, L.D., and Newberry, R.J., 1981. Skarn deposits. *Econ. Geol.,* 75th Aniv. Volume, 317–391.

Kwak, T.A.P., 1983. The geology and geochemistry of the zoned, Sn-W-F-Be skarns at Mt. Lindsay, Tasmania, Australia. *Econ. Geol.* v. 78, 1440–1466.

Kwak, T.A.P. and Askins, P.W., 1981. Geology and genesis of the laminar F-Sn-W (-Be-Zn) skarn (wrigglite) at Moina, Tasmania. *Econ. Geol.,* v. 76, 439–467.

Layne, G.D. and Spooner, E.T.C., 1983. The JC Sn skarn: A contact metasomatic Sn-Fe-F (-Be-B-As) deposit related to the Seagull Batholith. *GAC Abs. with prog.:* v. 8, p. A41.

Mato, G., Ditson, G. and Goodwin, C., 1983. Geology and geochemistry of the tin mineralization associated with the Seagull Batholith, south-central Yukon Territory. *Canadian Instit. Min. Metall. Bulletin,* v. 76, 43–49.

Nekrasov, I. Ya, 1971. Features of tin mineralization in carbonate deposits, as in eastern Siberia; *Int. Geological Review,* v. 13, 1531–1542.

Patterson, D.J., Ohmoto, H. and Solomon, M., 1981. Geological setting and genesis of cassiterite-sulphide mineralization of Renison Bell, western Tasmania; *Econ. Geol.,* v. 76, 393–438.

Poole, W.H., 1956. Geology of the Cassiar Mountains in the vicinity of the Yukon-British Columbia boundary. *Unpub. Ph.D. thesis, Princeton University,* 247 p.

Poole, W.H., Roddick, J.A., and Green, L.H., 1960. Geology of Wolf Lake, Yukon Territory; *Geol. Surv. Canada, Map* 10-1960.

Sampson, R.J., 1978. Surface II Graphics System, Series on spatial analysis, no. 1, J.C. Davis ed., *Kansas Geological Survey,* Kansas.

Taylor, R.G., 1979. *Geology of tin deposits;* Elsevier, Amsterdam, 543 p.

Yukon Geology and Exploration 1979—80. *Summaries of assessment work, descriptions of mineral properties, and mineral claims staked in 1980:* Wolf Lake area; Geology section, Department of Indian Affairs and Northern Development, Whitehorse, 142—160.

# 4 Europe (Iberian Peninsula)

# 4.1 The Geochemistry of Granitoid-Related Deposits of Tin and Tungsten in Orogenic Belts

M.G. Oosterom[1]


## Abstract

Trace element geochemistry combined with multivariate statistical analysis are applied to lithogeochemical investigation of mineralized and barren Hercynian granites in the NW trending belt of the Iberian peninsula. Different phases and facies within the granitoids of this tin-tungsten province can be distinguished by their geochemical specialization in trace elements. Geochemically defined facies within single granite bodies can be mapped by contouring the anomalies. The applied procedure proves an effective and rapid aid in recognizing mineralized situations on the local scale. A preliminary survey of samples from Bangka and Belitung shows similar geochemical trends to be present in the Triassic-Jurassic-Cretaceous granite belt of Southeast Asia.


## Introduction

Multi-element lithogeochemical studies of granitic rocks have been applied in different situations and on various scales, to elucidate the relation between mineralizations involving tin, tungsten and other elements such as tantalum and their parent rocks (Ivanova, 1963; Tischendorf, 1977; Oosterom et al., 1982). So-called specialized granites show a high content of specific trace elements in comparison with non-mineralized occurrences. However, the relationship between this geochemical specialization and ore depositon is complex (Oosterom et al., 1982).

High content of an ore element, or an appreciable variance in its distribution, is regarded as a favourable indication of the ore producing potential of the parent rock (Beus and Gregorian, 1977). Tin, tungsten, molybdenum, and in certain cases uranium, are among the economically important metals occurring in primary ore deposits within granitoid rock of felsic peraluminous nature. Ore deposition is nearly always limited to late-stage (auto)metasomatic processes, such as albitization and muscovitization, these causing enhanced dispersion of the ore elements and accompanying alkaline and volatile elements (Barsukov, 1975; Taylor, 1979).

A tin and tungsten province related to allochthonous granites of Hercynian age is found in the northern and western part of the Iberian peninsula. A very large deposit

[1] Department of Geochemistry, Institute of Earth Sciences, University of Utrecht, P.O. Box 80021, 3508 TA Utrecht, The Netherlands

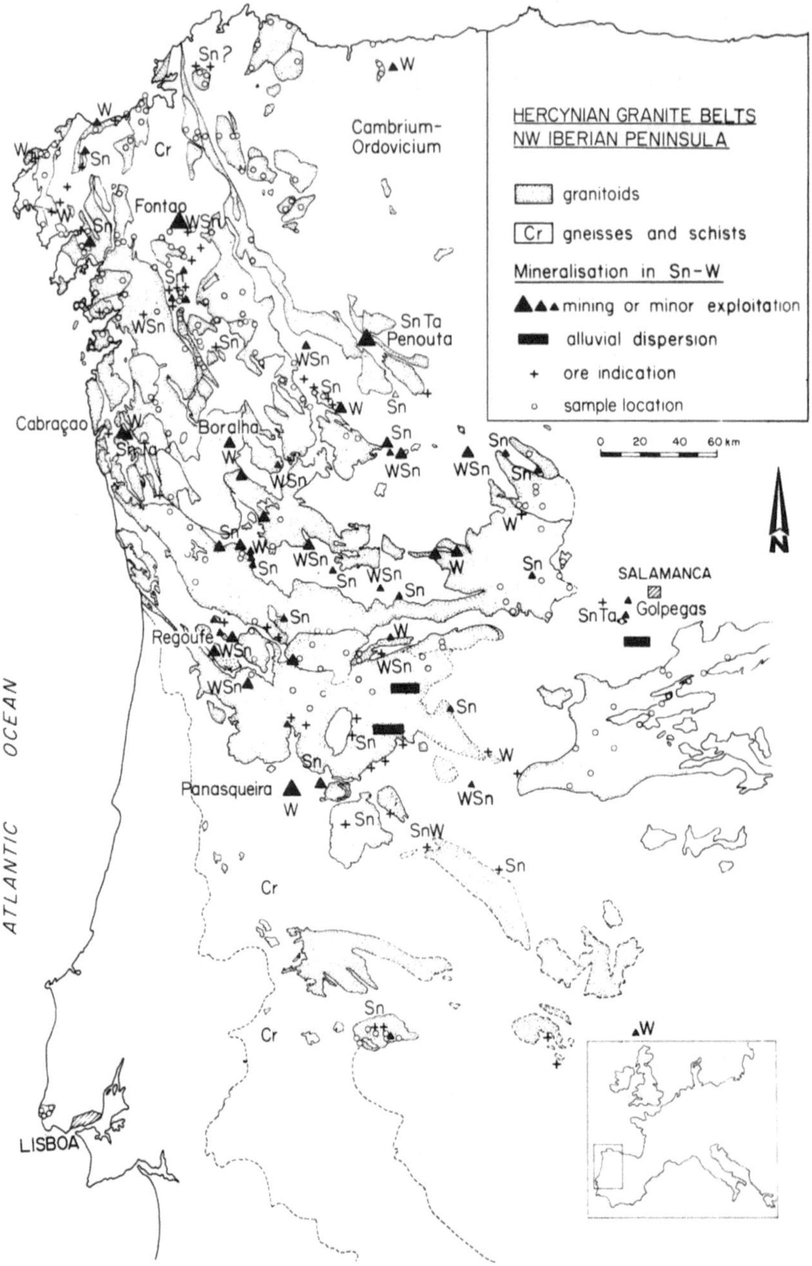


**Fig. 1.** Sketchmap of NW Iberian peninsula with tin-tungsten mineralized occurrences and extension of Hercynian granite areas

of wolframite occurs at Panasqueira, smaller mineable deposits of tungsten and tin at Regoufe, Cabraçao, Fontao and Penouta. Tantalum also occurs occasionally in this ore province.

By ages determined with the Rb-Sr whole rock isochron method, the granitoids in this province can be divided into Older granites of about 330–320 Ma and Younger granites of about 320–280 Ma. Some of the more important mineralized granites date at 287 ± 3 Ma (Priem et al., 1984).

The mineralized source rocks in the Hercynian fold belt in Portugal and Spain are mainly the younger biotite granites, although some of the older two mica leucogranites may also produce mineralization in tin and tungsten. The mineralized granites generally show a higher degree of intrusive complexity and have a hypabyssal level of intrusion. A sketchmap of the granite belts and the mineralization in the NW section of the Iberian peninsula is given in Fig. 1. A major part of the mineralization occurs within granites, emplaced in a syn- to post-tectonic phases of orogeny. Larger and smaller ore indications are found mainly along the border zones of the major granite batholiths, although some of the economically important deposits, e.g. Panasqueira and Regoufe, are related to the roof zones of smaller dome-like granite structures outside the major granite areas.

A spatial linkage of Sn-W occurrences and certain types of granites is assumed. The predominant mineralization is tungsten with subordinate tin, although both elements occur also independently in mineralized veins. Due to the prevailing conditions of relief and weathering, few placer deposits were developed in this otherwise extensively mineralized ore province.

## Lithogeochemical Trends in Hercynian Granites

Several attempts have been made to study the relationship between the chemical characteristics of the enclosing granites and the presence of tin and tungsten deposits (e.g. Burnol, 1974; Tischendorf, 1977; Bussink, 1984). Primary concentration trends have been proposed by Tauson (1974) for a number of granitophile elements. However, the pattern of enrichment is commonly complicated by metasomatic alteration processes, that have modified the original magmatic imprint. Next to an increase in Sn, and less abundantly in W, the following metal elements are detected generally in elevated amounts: Be, Nb, Ta and the alkali elements Li, Rb and Cs. Concurrent with the metals, there is quite commonly a strong increase in F, Cl, B and P, which act as mobilizers and carriers of the metal elements (Barsukov, 1975; Taylor, 1979; Oosterom et al., 1982). More comprehensive work is still required; dealing fully with this subject is beyond the scope of this paper.

In investigation covering this problem in the province of Galicia (Spain), it is shown that granites with ore indications have a higher degree of geochemical specialization. In Fig. 2 the trends in element distribution of about 230 samples of various Hercynian granites are depicted, and means, standard deviation and range of concentration are shown. It demonstrates the arbitrary significance of calculated average content of elements in granite types considering the range of values which usually prevails. Two trends of mutually correlating elements can be distinguished in the magma evolution of the granites of the NW belt in the Iberian peninsula. Sn and its associated

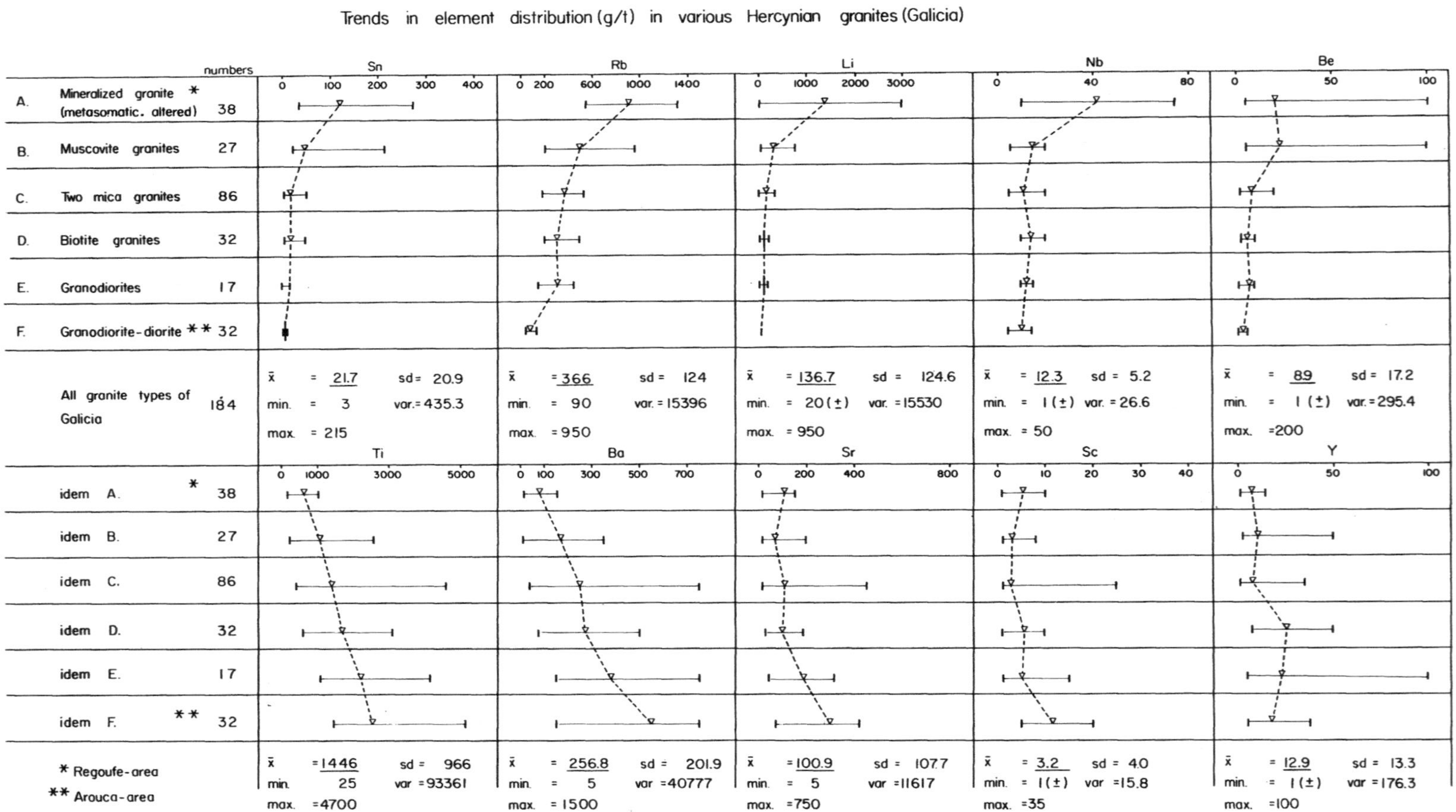


**Fig. 2.** Trends in trace element content in ppm of successive granite phases of Hercynian belts, NW Spain-Portugal. Average content is indicated by small triangle

alkali- and volatile elements, show a considerable increase in the process of specialization of the granite. Ti, Ba, and Sr stongly decrease during the evolution of granitic magmas towards two-mica leucogranites; Sc and Y show more irregular trends. Surprisingly enough, a decrease of some elements during evolution is hardly mentioned in the literature, although Rb/Sr and Ti/Ta ratios have been advocated as suitable indicators of the numeral potential of granites; e.g. the Ti/Ta parameter has been suggested for tin ore specialization of granites (Lehmann, 1982).

The interrelation of the trace elements in the granite evolution exhibits a wide range in element content; they show trends as shown in Fig. 2. Multivariate statistical methods can be applied to deal with these trends in element enrichment or depletion. Such statistical programmes are readily available, e.g., in SPSS, the Statistical Package for the Social Sciences (Nie et al., 1975). Factor analysis, provided that it is correctly applied, is a powerful tool in the interpretation of these types of data sets. Factor analysis is aimed at the explanation of the interrelations of the variables (Vriend et al., 1985), in cases when the elements are active in late-stage metasomatic processes and ore deposition. The factor loadings resulting from the factor analysis may be used to correlate the factors with geochemical processes, while the factor scores can be used to contour the spatial effect of these processes.

## Lithogeochemical Patterns on the Local Scale

In the preceding section lithogeochemical dispersions of the relevant elements in granitoids were discussed in general. The following section of the investigation is concerned with possible geochemical trends within or outside the mineralized granites themselves. Two areas of geochemical as well as economic interest have been selected for more detailed exploration. One case is the vein-related mineralization within the granite, or outside in the contact-aureole of the leucogranite dome of Regoufe; formerly mined for tin and tungsten. The other case is the well known ore occurrence of Panasqueira, with extensive mining of tungsten-bearing quartz-veins surrounding a greisenized granite cupola.

## The Regoufe Case

The Regoufe granite in the Douro Litoral province, about 250 km north of Lisboa, is situated in the centre of the Douro synclinal zone and is intrusive into metasedimentary rocks of the Beira schist complex. Around the dome a well developed contact-aureole of spotted phyllites is present (Sluyk, 1963). The main intrusive rock type is a porphyritic two-mica granite, its age has been determined at 280 ± 9 Ma. Towards the roof zone at the NE contact, this granite grades into a very specialized muscovite-albite granite. The latter is also rich in sulphide minerals, mainly arsenopyrite (Vriend et al., 1985). Within and surrounding the granite there are numerous small tungsten bearing quartz veins. However, only two underground mines of any importance have ever operated. Mineralization consists mainly of wolframite with additional cassiterite, the latter mineral occurring also in greisen zones along the quartz veins.

For the discovery of possible geochemical trends 55 samples of the different types of granite of Regoufe were collected in an area of about 6 km$^2$. For each location representative sampling of the coarse-grained granite requires about 10 to 15 chips of rocks, 1.5 to 2 kg in weight, over an area of about 250 m$^2$. Rb, Sn, W, Sr, Ta, Cs, Nb, P, Ti Zr, and U were determined by XRF on pressed powder pellets of rock. Cu, Zn and Li were determined in solution using A.A.S.-photospectrometry, and F with the ion-selective electrode technique after fusion with $Na_2CO_3$ (Vriend et al., 1985). The distributions of the elements are shown by SYMAP contour plots in Fig. 3. The Sn-plot is given as an example of a single element distribution, this shows very high Sn-content of up to 100 ppm and concentration along the NE border of the granite occurrence. Factor analysis has been applied to reveal the multi-element relations. In the resulting three factor-model, factor 1 shows high positive loadings for Rb, P, Nb and Ta and high negative loadings for Ti and Zr. Factor 2 is controlled by the ore

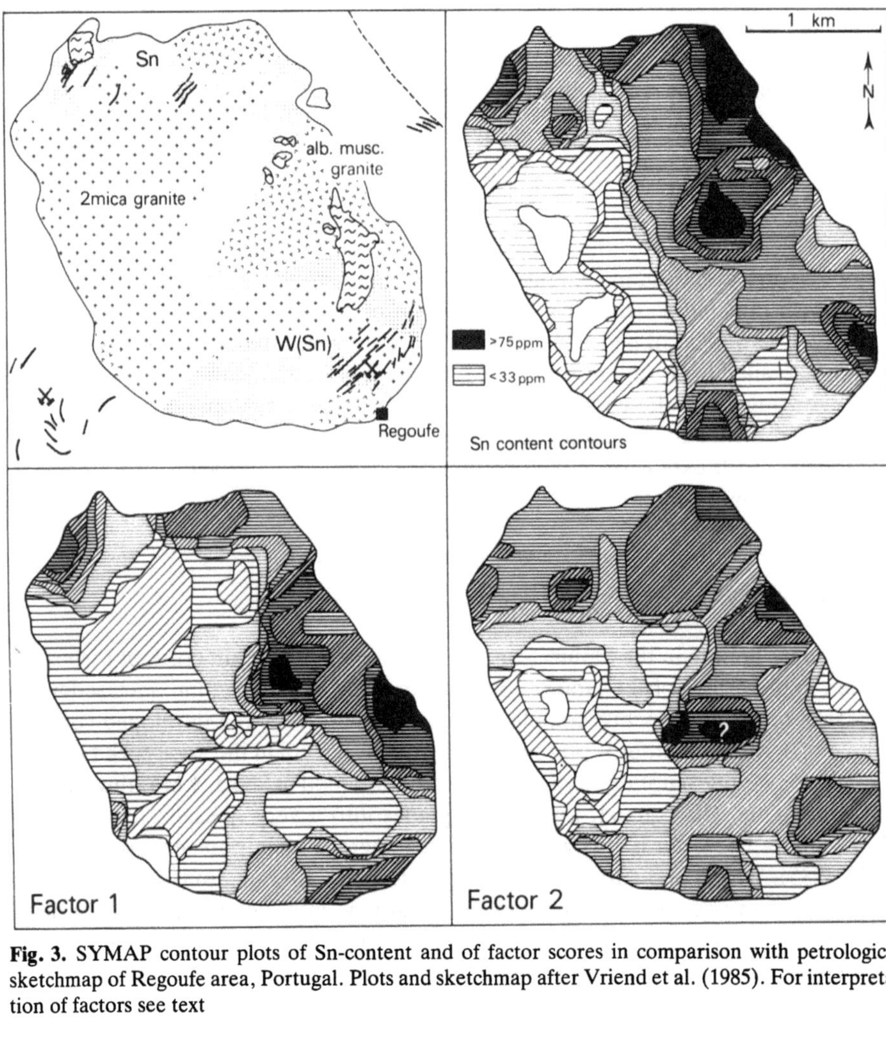


**Fig. 3.** SYMAP contour plots of Sn-content and of factor scores in comparison with petrological sketchmap of Regoufe area, Portugal. Plots and sketchmap after Vriend et al. (1985). For interpretation of factors see text

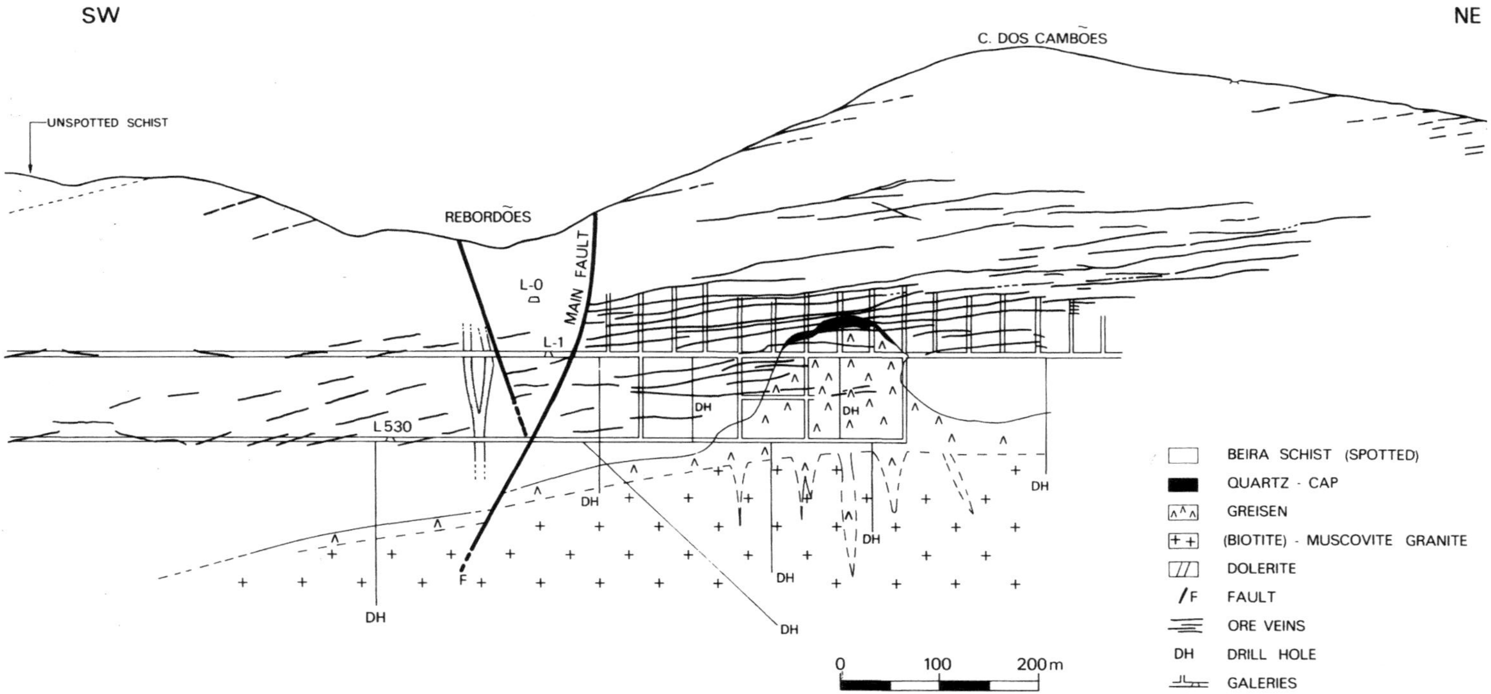


**Fig. 4.** Longitudinal section through Panasqueira granite-greisen cupola and vein-system according to underground maps B.T.W.-company

related elements Sn, Zn, Cu and W with contributions also from Rb and F. Both factors are plotted in Fig. 3. Vriend et al. (1985) interpreted the contour plots of factor 1 to be due to alibitization and greisenization (compare with sketchmap); factor 2 appears to be connected to ore deposition. When considered in relation to the geology of the area, these composite geochemical patterns point to successive processes of metasomatism and ore deposition at Regoufe, and may very well be used for exploration purposes.

## The Panasqueira Case

The Panasqueira ore deposit, near the town of Covilha in the Beira Beixa province of northern Portugal, consists of a flat-lying swarm of parallel quartz veins surrounding a hidden greisenized cupola (Fig. 4). The Beira schists encasing the ore deposits are metasedimentary rocks of marine origin. They are mainly composed of shales with intercalations of greywackes and fine-grained sandstones (Thadeu, 1973). Due to the influence of contact metamorphism by the granite intrusions, the Beira shales are converted into spotted schists over large areas. The vein system may have contributed to the high temperature metasomatic effect surrounding the cupola as the ore fluids reached temperatures of 325° to 250°C (Kelly and Rye, 1979). Shale and schist samples were collected from surface outcrops covering the ore deposits, but also at distances of up to several km from known ore zones. About 175 samples of surface schists were collected in an area of roughly 30 $km^2$ with a sample density of about six samples per $km^2$. Analyses of metal elements, alkalis and fluorine were performed by the same techniques as used in the Regoufe case.

It appears that elements such as Sn, W, Rb, Li and F, which are remarkably enriched in the greisenized granite of the Panasqueira cupola, also occur in high amounts in the schist, forming extensive aureoles surrounding the cupola and its related ore fields. Fluorine contents go up to 0.85% (see Fig. 5), and also Sn- and W-contents in contact schists are as high as 100 ppm and 55 ppm respectively (Oosterom, 1984). In this case a tin anomaly surrounds a tungsten deposit.

The lithogeochemical contour maps of the area of investigated surface samples were produced with the help of the SYMAP computer package (Dougenik and Sheehan, 1976). As an example the plot of fluorine is shown in Fig. 5. Fluorine appears to be a good indicator element for veins mineralized in tin and tungsten.

A final evaluation of the multi-element geochemical contouring method can be made by comparing the outlines of the underground extension of the ore-bodies with the contour patterns of factor scores in Fig. 5. The factor score-anomalies appear to reflect the dispersion of the elements Rb, F, Sn, Li and W through leakages in the joint system of the schists. The fault system in the area influences the anomalies, since contours follow the direction of the major faults in the orefield.

The presentation of geochemical anomalies by multi-element parameters clearly offers advantages compared with the single-element distribution patterns, e.g. of fluorine, since it corrects possible irregularities in the single element contour plots. From the lithogeochemical investigations in the test cases of Regoufe and Panasqueira, it may be concluded that trace element-plotting in combination with multivariate statistical data treatment, is an effective and rapid aid in locating late-stage granite metasomatism and related vein mineralization.

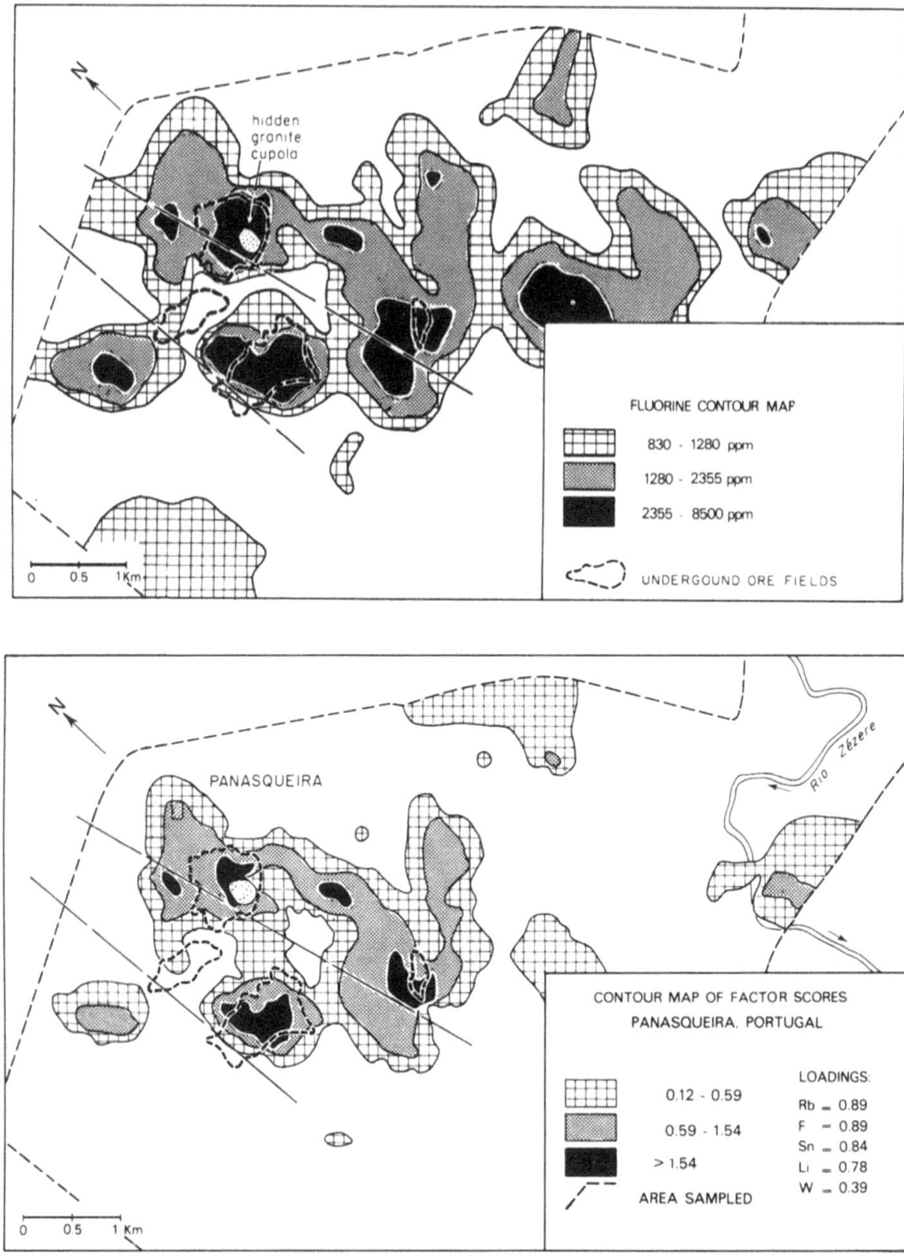


**Fig. 5.** SYMAP contour plots showing distribution of fluorine and of factor scores. Dashed areas represent subhorizontal underground vein deposits at the Panasqueira tin-tungsten deposit, Portugal (Oosterom, 1984)

Other tin-tungsten ore provinces related to granite belts or shields have also revealed the presence of geochemically specialized granites. According to Taylor (1979) there are few exceptions of highly specialized granitoids which are not directly or indirectly associated with tin deposits. An intriguing aspect of the metallogeny of tin is its occurrence in major and minor ore occurrences in global scale linear tin belts, e.g. occurring on the continents around the Atlantic Ocean and in the Gondwana continent (Schuiling, 1967, 1983). Here it may be added that tin mineralized belts, on the scale of the Iberian province, are usually also tungsten bearing. This is also the case in the S.E. Asian belt, e.g. the primary tungsten deposits on Belitung (Bemmelen, van, 1970).

A worldwide comparison of the trace element geochemistry of granitoids may point to general tendencies for specialized granites in containing anomalous amounts of one or more of the metals Sn, Be, Li, Rb, Cs and other elements such as F, Cl and B (Taylor, 1979). However, when investigated in detail many granitoid complexes in orogenic belts reveal more complicated geochemical relations. More work at different scales is still required; a preliminary effort is presented here.

The Southeast Asian orogenic zone is intruded by granitoids which occur as a series of linear batholiths over a distance of more than 3,000 km from Indonesia to

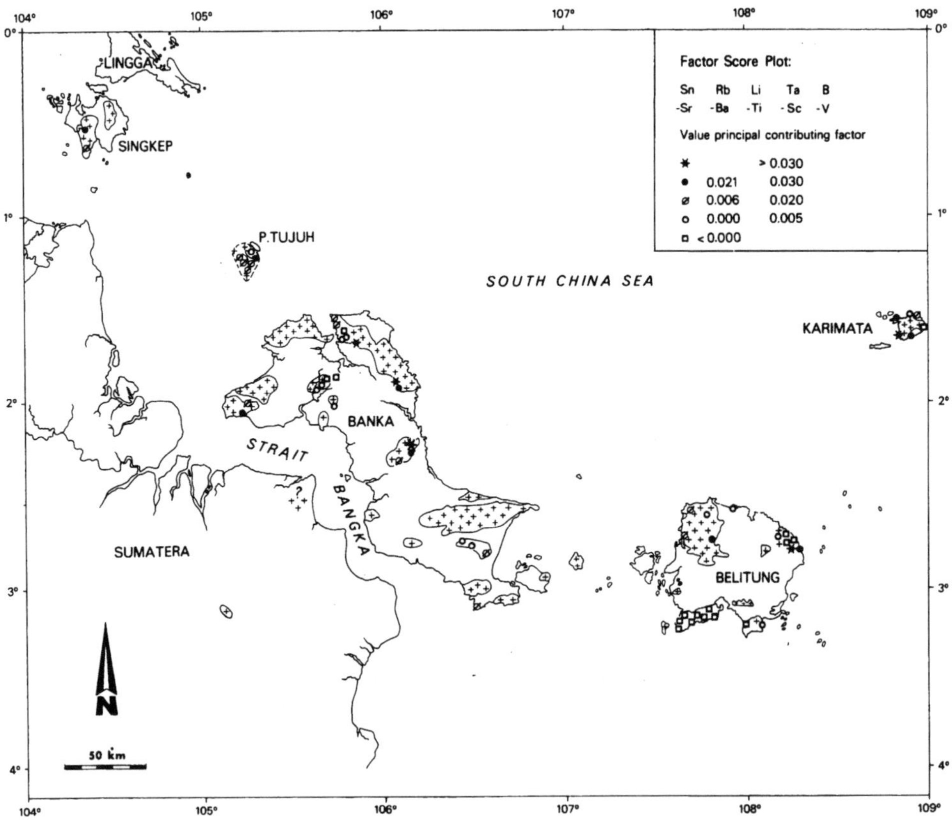


**Fig. 6.** Trace element-geochemistry of granites in the Belitung-Bangka section of the Southeast Asian tinbelt (preliminary plot)

Yunnan in China. Numerous primary tin ore occurrences are found in a variety of types (Taylor, 1979). In these aspects it resembles the mineralized Hercynian belt in SW Europe. Therefore, a comparison of the geochemistry of mineralized granitoids of the two orogenic systems seems quite justified and a preliminary geochemical investigation is presented here. Lithogeochemical analysis by the Department of Geochemistry, Utrecht University, on 50 samples of granodiorites, adamellites and diorites from Bangka and Belitung, Indonesia give trends of trace element distribution which are very similar to that of the Hercynian granites in Fig. 2. The resulting geochemical parameter is a factor composed of elements with positive loading for Sn, Rb, Li, Ta, and B; and negative ladings for Sr, Ba, Ti, Sc, and V (see the legend of Fig. 6). It appears that by multi-element data treatment the granitoids may be divided into different groups with mineralized granites showing higher factor scores (Fig. 6). It is noteworthy that the element Ta takes the place of Nb as a characteristic trace element in the Southeast Asian granites.

From this set of analyses the preliminary conclusion is drawn that granitoids of the Triassic to Cretaceous belts of Southeast Asia can be classified geochemically by similar analytical and statistical methods proposed and applied to the Hercynian mineralized granites of Europe. As is demonstrated by the Regoufe and Panasqueira test cases, this multi-element geochemical procedure can be used as a practical tool for geochemical prospecting for various granite related mineralizations on the local scale, and with modifications may apply as well to granite areas under conditions of weathered bed rock surfaces or poor exposure, for example Bangka and Belitung.

*Acknowledgement.* This review is the result of intermittent field and laboratory work by staff, students and analysts of the Department of Geochemistry, Utrecht. It was partly financed by the European Community under contract no. 025.79 MPP NL.

## References

Barsukov, V.L., 1975. The source of ore material. In: A.J. Tugarinov (ed.). *Recent contributions to Geochemistry and Analytical Chemistry*. Israel Program for Scientific Translations, Keterpress Enterprises, Jerusalem, 303–317.

Bemmelen, R.W. van (1970). *The Geology of Indonesia, 2nd Ed., vol. 2: Economic Geology*. Government Printing Office, The Hague.

Beus, A.A., Gregorian, S.V., 1977. *Geochemical exploration methods for mineral deposits*. Applied Publishing Ltd. Wilmette Ill., U.S.A., 207 pp.

Burnol, L., 1974. Géochemie du beryllium et types de concentration dans les leucogranites du Massif Central Français. *Mém. du Bur. Rech. Géol. Minières,* Orléans, no. 85, 170 pp.

Bussink, R.W., 1984. Geochemistry of the Panasqueira Tungsten-Tin deposit, Portugal. *Geologica Ultraiectina,* No. 33, Dr. thesis State University of Utrecht, 170 pp.

Dougenik, J.A., Sheehan, D.E., 1976. *SYMAP User's Reference Manual*. Harvard University, Mass., 5th Ed.

Ivanova, G.F., 1963. Content of tin, tungsten and molybdenum in granites enclosing tin-tungsten deposits. *Geochem. Int.*, 492–500.

Kelly, W.C., Rye, R.O., 1979. Geologic, fluid inclusion and stable isotope studies on the tin-tungsten deposits of Panasqueira, Portugal. *Econ. Geol.* 74, 1721–1822.

Lehmann, B., 1982. Metallogeny of Tin: Magmatic Differentiation versus Geochemical Heritage. *Economic Geology,* 77, 50–59.

Nie, N.H., Hull, C.H., Jenkins, J.G., Steinbrenner, K. and Bent, D.H., 1975. *Statistical Package for the Social Sciences,* 2nd Ed. McGraw Hill Inc., New York, NY, 675 pp.

Priem, H.N.A., Schermerhorn, L.J.G., Boelrijk, N.A.I.M., Hebeda, E.H., 1984. Rb-Sr Geochronology of Variscan granitoids in the Tin-Tungsten province of northern Portugal: A progress report. *Terra Cognita. The Journal of the European Union of Geosciences,* 4, no. 2, 212—213.

Oosterom, M.G., 1984. Lithogeochemical studies of Aureoles around the Panasqueira Tin-Tungsten deposit, Portugal. *Mineral Deposita* 19, 283—288.

Oosterom, M.G., Vriend, S.P., Bussink, R.W., Moura, M., 1982. Geochemistry of tin, tungsten and related elements such as tantalum and niobium. *E.E.C. Research Programme on Primary Raw Materials, Brussels; open file report,* contract no. 02579 MPPNL, 58 pp.

Schuiling, R.D., 1967. Tin belts on the continents around the Atlantic Ocean. *Economic Geology* 62, 540–550.

Schuiling, R.D., 1983. The position of Indian tin occurrences in the tin belts of Gondwana. *Journal of the Geological Society of India* vol. 24, no. 2, 101—105.

Sluyk, D., 1963. Geology and tin-tungsten deposits of the Regoufe area, northern Portugal. *Thesis, University of Amsterdam,* 123 pp.

Tauson, L.V., 1968. Distribution regularities of trace elements in granitoid intrusions of the batholith and hypabyssal types. In: L.H. Ahrens (ed.). *Origin and the distribution of the elements.* Pergamon Press, Oxford, 629—639.

Taylor, R.G., 1979. *Geology of tin deposits.* Developments in Economic Geology II. Elsevier Scientific Publishing Company Amsterdam, 543 pp.

Thadeu, D.C., 1973. Les gisements stanno-wolframitiques du Portugal. *Ann. Soc. Geol. Belg.,* 96, 5—30.

Tischendorf, G., 1977. Geochemical and petrographic characteristics of silicic magmatic rocks associated with rare-element mineralizations. In: Stemprok, M., Burnol, L., and Tischendorf, G. (eds.). *MAWAM (Metallization Associated with Acid Magmatism).* Vol. 2, Schweizerbart's Verlag, Stuttgart, 41—96.

Vriend, S.P., Oosterom, M.G., Bussink, R.W. and Jansen, J.B.H., 1985. Trace element behaviour in the W-Sn granite of Regoufe, Portugal. *Journal of Geochemical Exploration,* Elsevier Amsterdam, 23, 13—25.

# 5 South America

# 5.1 The Tin Ore Deposits of Bolivia

A. VILLALPANDO B.[1]


## Abstract

This paper contains some general observations about Bolivian tin deposits and their mineralization. Diverse opinions exist about their origins; these theories are discussed and synthesized.

The deposits comprise four genetic groups according to their geotectonic setting, kinds of paragenesis, and the shape of ore bodies (Schneider and Lehmann, 1977).

Recent geochemical research on the metallogenic provinces (Schneider et al., 1978; Dulski et al., 1982) results in several hypotheses concerning the origin of the mineralization, in which, the "Metallogenetic heritage" explains the occurrence of the tin mineral deposits during the geotectonic development from the Precambrian up to Tertiary in the Bolivian crustal segment.


## 1 Introduction

Tin was first exploited in Bolivian territory by the Incas, who made rustic bronze with tin, probably from the deposits of the Cordillera Real. However, its exploitation in reality began at the end of the XIX century, an epoch in which tin began to become important in the international market and also in which the production of silver decreased considerably.

The production of Bolivian tin, since that period, increased gradually from a few hundred tons per year, reaching 47,000 t in 1929, a figure which represents the highest level ever of our production. Since then, it has decreased considerably, reaching between 25,000 and 30,000 t a year (Fig. 1).

About three fourths of the Bolivian tin reserves belong to the Bolivian Mining Corporation (COMIBOL) and estimates of the potential positive and probable reserves are around 200,000 t. Possible and prospective reserves are estimated at about 500,000 t with an additional 300,000 found in placers, dumps and tailings. Mining in general and tin in particular is still today the most important aspect of the Bolivian economy.

[1] National Mining Exploration Fund, Cassila 3080, La Paz, Bolivia

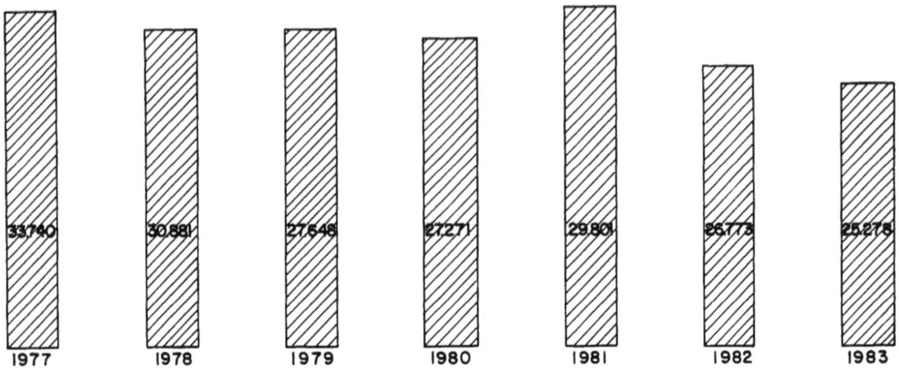


**Fig. 1.** Bolivian tin production (tonnes)

## 2 The Bolivian Tin Deposits

The Bolivian tin deposits belong to the province of Rondonia located in the Brazilian shield area found in the north eastern border of Bolivia with Brazil and in the Andean province, located in the western part of Bolivia.

The Andean province is the only one which is significant for Bolivia at the moment, because in it are found the tin deposits currently being exploited. However, several important deposits are present in the Rondonia province. These have enabled Brazil to increase its production considerably (16,000 t in 1983).

For these reasons, I refer to the Bolivian tin yielding Andean province, pointing out its noteworthiness because it comprises a relatively narrow and well defined arc zone of approximately 900 km in length. It is located in the highest western part of the eastern Andes and extends from the edge of Peru (Condoriquiña) to northeast Argentina (Pirquitas) (Fig. 2).

The eastern Andean Cordillera changes in direction from NW-SE to N-S, a feature known as the Codo de Arica (Arica's Elbow), located northeast of Santa Cruz. This weak zone divides the Easter Cordillera into two parts: the first known as the Septentrional and the other as Meridional, whose morphological and mineralogical characteristics differ.

The tin deposits appear predominantly in the form of veins, although stockworks, impregnations and strata-bound deposits of the "manto" or "bedding" type also occur. The host rock is generally of Lower Paleozoic sedimentary or meta-sedimentary rocks (lutites, sands, quartzites, etc.), and of magmatic bodies. The absence of limestone is noticeable and it could be one of the reasons for the presence of tin deposits only in Bolivian territory (Petersen, 1970).

The formation temperatures of the Sn-W deposits, determined by thermometry of fluid inclusions, range from above 500°C in the initial stages of mineralization to below 100°C in the later stages (Kelly and Turneaure, 1970).

In the northern part of the tin province, the deposits are intimately related with the batholithic granites of the Cordillera Real and are characterized by a relatively simple mineralogy, a low content of silver, and essentially do not present an individual zonation (although as a group, they are zoned around the batholiths).

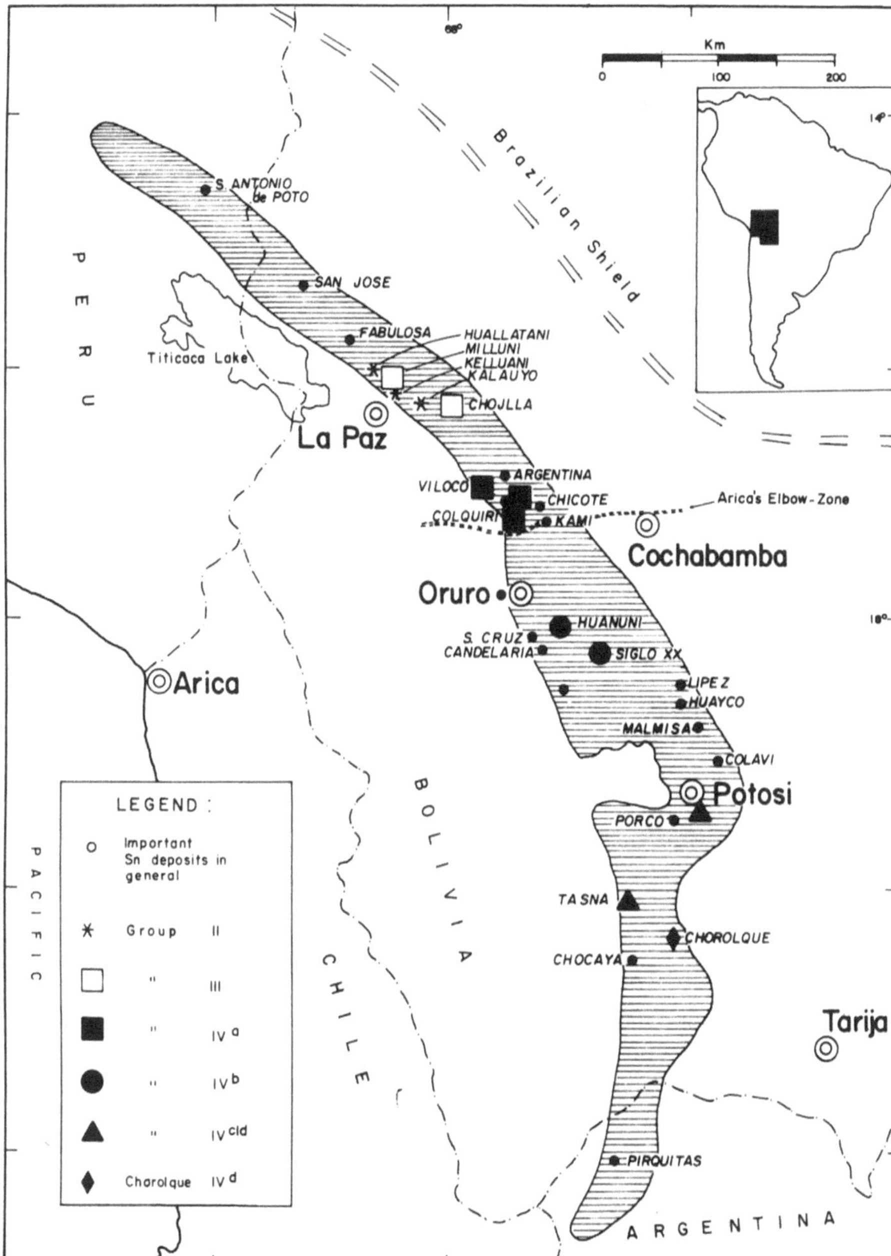


**Fig. 2.** The Andean tin province of Bolivia and the main Sn deposits (after Dulski et al., 1982)

In contrast, in the southern part of the belt, the deposits are more commonly associated with subvolcanic individual stocks and are characterized by their more complex mineralogy which consists of a high silver content, occurrence of sulphosalts, and a well developed zonation. Tin deposits associated with plutonic bodies are an exception, with an example in the Kari-Kari Batholith in Potosí.

## 3 Genesis of the Deposits

Diverse opinions have been expressed regarding the genesis of these deposits and several theories have been elaborated that explain in one way or another their origin; however, no consensus has been reached on this topic.

The granitic rocks that appear in the Septentrional part of the Bolivian tin-yielding province have a NW-SE trend and have been considered since the times of Steinmann as Tertiary in age. Consequently this age was assumed for all the deposits in the province. Kozlowski concurred with this conclusion, based on the similarity between the tin mineralization of the plutonic and the volcanic deposits of the Cordillera Oriental. Based on the notion of a single mineralization cycle, corresponding to the mid-Tertiary, Ahfeld (1937) explained the paragenetic variety of the Sn-W-Bi-Ag-Pb-Zn-Sb deposits as resulting from a magmatic differentiation process. However, this notion had to be changed beginning with the first radiometric dating performed on igneous rocks of the Cordillera Real by Evernden in 1961, which have an age ranging from 180 to 199 Ma.

Based on these results, Schneider-Scherbina (1962) postulated the existence of another magmatic cycle during the Lower Mesozoic at the Triassic-Jurassic boundary, and which had a deep plutonic character in contrast to the igneous activity of the Upper Tertiary that occurred at low depths and was accompanied by intense volcanic activity. Consequently, Schneider-Scherbina (1962) stated that, in the tin-yielding province, two distinct cycles of mineralization are evident which are directly related to these magmatic events. The first, which occurred in the Early Mesozoic, was associated with deep plutons that evolved throughout the folding Axes of the Cordillera Oriental and whose mineralization comprises three phases: a) Auriferous quartz; b) W, Sb, Hg and (Se) mineralization; c) Sn, Bi with polymetallic aureoles of Cu, Zn and Pb mineralization.

The second mineralization epoch of the Upper Tertiary (7.8—19 Ma) is characteristically polymetallic, rich in silver, and of a subvolcanic type. The Tertiary deposits are superimposed on those of Mesozoic age, which according to Schneider-Scherbina (1962), caused a geochemical contamination of the Tertiary hydrothermal solutions, giving way to the formation of hybrid deposits such as those of the Cerro Rico in Potosí, San Jose in Oruro, Colquechaca, etc., all of which belong to the well known "Tin-argentiferous formation" (designated as the Potosí type by Stelzner, 1898).

Subsequently, other more extensive geochronological studies performed by Cordani, 1967 (in Clark and Farrar, 1973); Clark and Farrar, 1973; Robertson, 1974; Clark et al., 1976, confirmed the existence of the two great magmatic cycles.

Ahlfeld (1967) performed a detailed analysis of the Bolivian tin deposits and tried to clarify the relationship between tectonics, magmatism and the formation of these deposits. He was of the opinion that the mineralization of tin is related to a mountain system with fault blocks and folding only of secondary importance. Vertical movements of the faulted blocks predominate. The arc of the province is divided into two different parts as far as its historical development is concerned, by an older geotectonic weakness zone of an approximate E-W direction (Fig. 2). The septentrional part is elevated with respect to the meridional part. In the northern part, granitic plutons flourish which are related to block and fold tectonics that occurred in the Lower Hercynian and which intrude into Lower Paleozoic sediments. The possibility that the

Lower Hercynian ulterior crustal movements of the Hercynic (pulsation) caused the rise of granitic magmas becomes important (Radelli, 1966). Synorogenic mineralization of Au and Sb and post-orogenic deposits of W, Sn, Bi, Zn and Pb are associated with these granites. Tin mineralization is the youngest.

A great number of cassiterite veins are found in the eastern flank of the Cordillera Real. These are intensely eroding and are related to aplitic granites rich in potassium (Chojlla mine). In this type of mineralization one can point out the deposits of Fabulosa (pegmatic veins with Li), Chacaltaya-Milluni-Carmen, Amutara-Chicote-Kami, Colquiri (sulphurous deposit) and the district of Conde Auqui (Fig. 2).

The majority of the tin deposits are found in the southern part (47 districts with several hundred mines). In it, Ahfeld (1967) differentiated two mineralization epochs: a strong plutonic phase in the Miocene that provoked a relatively weak mineralization (Kari-Kari, Andacaba with Sn, W, Bi, Zn, Pb) and another subvolcanic phase in the Pliocene which was of greater intensity than all other older metallogenetic epochs and which is characterized by an abundant silver content. This latter phase is associated with subvolcanic latitic-rhyolitic stocks or is of a cryptovolcanic character. All deposits were more or less disturbed by fault block crustal movements during the Upper Pliocene. There is a pronounced vertical and lateral zonation. In contrast to the deposits in the northern part of the province, the mines are composed in a great part by sulphides; cassiterite is present in varied forms. High temperature paragenesis with tourmaline, to low temperature paragenesis with colloidal wood-tin and "needle tin" (Santa Fe, Colavi, etc.). The abundant presence of silver-yielding tin sulphosalts, among which franckeite, cylindrite, and teallite can be cited, is typical of this type of deposit.

With reference to the hybridization of the tin-argentiferous formation (Schneider-Scherbina, 1962), Ahlfeld (1967) points out that it is more likely that the youngest phase of mineralization, which is rich in silver, also originates in the same source that fed the older mineralizing phases. Its rarity could be explained by pointing out that such formations, which are close to the surface, could have disappeared through erosion in other older metallogenetic provinces that contain tin.

In the Bolivian tin-yielding province in general, successive phases of younger mineralization are observable. These range from the Permian-Triassic to the Pliocene-Pleistocene and constitute an excellent example of the compatible nature of this mineralization with various metallogenetic epochs within the same region (Ahlfeld, 1967).

On the other hand, Halls (1975) relates the formation of the Bolivian tin deposits to plate tectonics. According to the seismic data for Bolivia, tin would have a tendency to concentrate when the subducting plate has reached a depth of 300 km below the edge of the subjacent continental shelf. In this manner, according to Halls (1975), the confused distribution of the tin deposits can be explained. He considers that within this type of deposit, the Tertiary igneous province of Southeast Bolivia is a good example. Avila (1975) shares this opinion by associating the calc-alkaline magmatism of the Cordillera Real (plutonic) and Southern Bolivia (subvolcanic), and the mineralization of tin, with the subduction of the Nazca plate which has a tendency of introducing lithospheric materials to those originated in the mantle.

In summary, many authors point out that the development of the Bolvian tin-yielding province took place in multiple, successive and different magmatic stages (Schneider-Scherbina, 1962; Ljunggren, 1962; Ahlfeld, 1967; Turneaure, 1971).

The discovery of tin deposits in the Rondonia, Brazil, region (Kloostermann, 1967) and other research indicate that the Brazilian shield extends more to the west in Bolivia (Schlatter and Nederlof, 1966; Botello et al., 1972) and therefore displays a fundamental influence on the development of the Andes. This has given credence to the notion of the existence of a relationship between the Precambrian mineralization of tin (Rondonia, northeast Bolivia) and those belonging to the Mesozoic and Tertiary of the Bolivian tin province.

Following the concept developed by Schuilling (1967) of tin belts in South America, the Bolivian tin-yielding province would be located at the intersection of the Precambrian Rondonian and the Cenozoic Andean tin belts. A geochemical reactivation of tin occurred (Fig. 3). The notion of a close relationship between a first Precambrian mineralization and several subsequent magmatic mineralization epochs, is also proposed by Stoll (1964, 1976) for the crustal segment of Argentina and Bolivia. This author establishes a magmatic remobilization of tin and tungsten which extends from the Precambrian deposits of the Sierras Pampeanas in Northern Argentina to the Recent occurrences of wood-tin in Pliocene volcanic rocks in Southern Bolivia.

Schneider and Lehmann (1977) take these criteria into account, and consider the presence of Silurian synsedimentary tin deposits of the "manto" type as a conclusive model of the metallogenetic development of the Bolivian tin-yielding province. They also provide the missing data to establish the relationship of the "heredity of tin" from

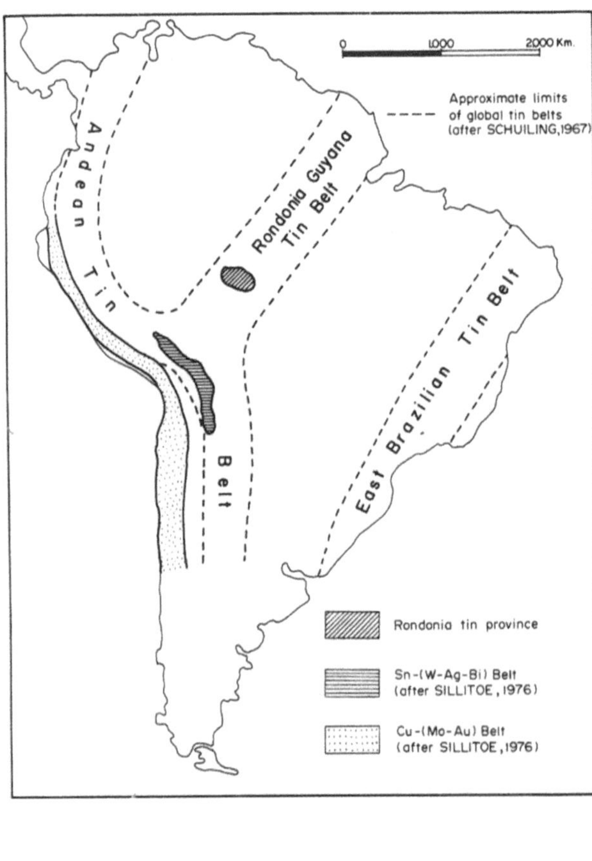


**Fig. 3.** Tin belts and tin provinces in South America (according to Schuilling, 1967, and Sillitoe, 1976, in Schneider and Lehmann, 1977)

the Precambrian to the Cenozoic and elaborate a new and more detailed genetic concept which indicates that tin heritage within the core is not only the result of successive and different magmatic-volcanic events from the Precambrian to the Neogene, but also a regional remobilization caused either by a magmatic process (anatexis) or by a process of re-sedimentation cycles.

## 4 The New Metallogenetic Concept of the Bolivian Tin Deposits

If the metallogenesis of the Bolivian tin-yielding province is considered as a geological sequence of different formation processes, it then becomes a genetic grouping which fits perfectly within the geotectonic development of this part of South America.

Based upon Ahlfeld's (1967) concept, this new genetic classification is the result of the research of a working group from the Free University and the Hahn Meitner Institute of Berlin and the Universidad Mayor de San Andres in La Paz, Bolivia and of which the author forms a part. The following geological criteria were developed:

a) Geotectonic position (for example age and genesis of the host rock).
b) Geometric shape of the deposit (for example veins, stocks, greisens, mantos, etc.).
c) Mineral Paragenesis and typical mineralogical traits.
d) Absolute age.

From this, the following groups of deposits can be differentiated (Schneider and Lehmann, 1977; Villalpando et al., 1977; Schneider et al., 1978; Dulski et al., 1982) (Fig. 4):

| Genetic Classification | Paragenesis | 1 Precambrian | 2 Lower Paleozoic | 3 Mesozoic Triassic-Jurassic | 4a Tertiary Upper Oligocene-Lower Miocene | 4b,c,d Tertiary Miocene-Pliocene |
|---|---|---|---|---|---|---|
| Strata-bound deposits (Mantos) (In part paleo_placers) Diagenesis | Oxidic Sn (Fe) | | Kelluani Huallatani Area of Poopo | | | |
| Hydrothermal Deposits – Epithermal: Subvolcanic "breccia pipes" Vein swarms "Tin porphyries" | Oxidic Sn-W Sulphidic Sn-Sb Ag-Pb | | | | | Potosi Oruro Chorolque Llallagua-Siglo XX Huanuni |
| Hydrothermal Deposits – Hypothermal: Vein deposits (High-temperature) | Oxidic Sn-W Sulphidic Sn-Bi-As Zn-Pb-Cu | | | Fabulosa Milluni Chojlla (Cordillera Real) | Viloco Argentina Colquiri (Quimsa Cruz) | |
| Pneumatolytic and pegmatitic-Deposits | Oxidic Sn-Nb-Ta | Pegmatites and Greisens (Area of Rondonia and NE from Bolivia) | | | | |

**Fig. 4.** Classification of Bolivian Tin Deposits (after Schneider et al., 1978)

## Group I

In Rondonia and NE Bolivia on the Brazilian shield region, tin-yielding granites (greisen) and Precambrian pegmatites are present. Their paragenesis is of cassiterite and wolframite. Columbite also occurs locally as is the case in the Precambrian Nigerian deposits where it is very common (Schuiling, 1967; Kloostermann, 1967). These granites are of at least two generations. The most recent ones generally present anular structures. Dating on two Rondonia samples by A-Ar and Rb-Sr yielded an age close to 950 Ma, probably belonging to the younger formations (Priem et al., 1966). Among these deposits the following can be cited: Santa Barbara, San Francisco, Igarape Preto, Sao Domingo, Macangana and others in Brazil and the small deposit at Cachuela Carmen in Bolivia.

## Group II

These are deposits of the strata-bound ("manto") type which are intercalated within Silurian strata of the Catavi and Llallagua Formations. The deposits of the "manto" type are widely distributed. Examples are: Kelluani, Huallatani, Kalauyo, as well as San José, Monte Blanco, Candelaria, Ocuri, Poopó, etc.

In the Cordillera Real zone, the mineralization shows a simple paragenesis of quartz, tourmaline, fine granular cassiterite, and zircon. In this case, an origin associated with a reworked fossil placer and in some cases additional pyrite is not discounted. In contrast, in the southern zone of the Andes, a polymetallic paragenesis with Ag, Pb, Zn, Sb, Sn is present. Locally, this mineralization is also observed associated with discordant veins (Lehmann, 1979).

## Group III

These are discordant quartz veins directly related to the Mesozoic granitic intrusions whose mineralization consists of cassiterite, base metal sulphides, and locally wolframite and bismuthinite. The deposits are restricted to the northern portion of the Andean metallogenic province: that is to the Cordillera Real. A few examples include Chojlla, Milluni, Fabulosa and others.

## Group IV

These are polymetallic deposits associated with Tertiary plutonic and volcanic phases. This group can be subdivided into four subgroups:

### *Subgroup IVa*

Predominantly nets of parallel and discordant veins related to granitic intrusives of the Upper Oligocene to the Lower Miocene. These are located in the Cordillera of

Quimsa Cruz and Santa Veracruz. The paragenesis is of oxides and sulphides, similar to that of Group III. Among the deposits of this subgroup, Viloco, Mina Argentina, Colquiri, etc. can be cited.

*Subgroup IVb*

Vein and disseminated deposits associated with subvolcanic bodies of quartz rhyodacitic composition which show an intense hydrothermal alteration and brecciation. Mineralization is disseminated in igneous bodies constituting the so-called "tin porphyries" in the Sillitoe et al. (1975) sense, on the one hand, and in vein swarms on the other. Recent K-Ar dating performed by Evernden et al. (1977) and Grant et al. (1977) indicate an age of about 20–21 Ma for these intrusions. Examples of these deposits are Siglo XX-Llallagua and Huanuni.

*Subgroups IVc and d*

Breccia-pipes and vein swarms of a subvolcanic character whose characteristics are of a silver-rich paragenesis and an intense "telescoping" can be observed in the southern part of the Bolivian Andes. However, several accentuated differences are obvious in mineralization (predominance of Ag: Potosí, Oruro types; predominance of Bi: Tasna type) and in age as indicated by the K:Ar data. The age of the Tasna deposit is 15.8 to 16.1 Ma. The age of the Cerro Rico in Potosí lies between 12.9 and 13.8 Ma on the other hand. Taking into account that the tin from Tasna is from small veins situated in the peripheral part of the bismuth deposits, this can be considered a genetic subgroup which would belong to category IVc.

The most recent subvolcanic mineralization phase, characterized by intravolcanic breccia-pipes and vein swarms with a complex paragenesis of oxides and sulphides, constitute subgroup IVd. Such is the case of the deposits at Cerro Rico, Potosí.

A genetic problem which has not yet been resolved has to do with the Chorolque deposits. The latest K-Ar dating indicates an age of approximately 2.7 Ma. This would appear to indicate the existence of an intense hydrothermal alteration late phase from the nucleus of the rhyodacitic breccia. This is corroborated by the presence of hydrocassiterite (varlamoffite) (Ahlfeld and Schneider-Scherbina, 1964).

## 5 The Contribution of Geochemistry to the New Concept

Based on the new metallogenetic concept, elaborated according to classical geologic criteria, the working group from the Hahn-Meitner Institute and the Free University in Berlin and the University in La Paz, under the direction of Prof. H.-J. Schneider and Dr. P. Moeller, carried out their research in an attempt to determine congruent geochemical criteria. One of their objective was to try to clarify certain aspects of the genesis of the Bolivian tin deposits. The concept of a remobilization of Precambrian tin in the younger Mesozoic and Tertiary deposits was taken into consideration.

A model of trace element fractionation in cassiterites was utilized. The fractionation of a trace element, both in the primary formation of minerals as well as in the processes of remobilization or hybridation of parent solutions, has been confirmed in practice and theory through the use of several models (e.g. in fluorite, apatite, calcite, dolomite, etc.), the results of which are found in several publications (Schneider et al., 1975; Moeller et al., 1976; Parekh and Moeller, 1977). The results, which are presented here in brief form, are from 120 cassiterite samples from 15 deposits repressentative of the Bolivian tin-yielding province.

The analytic methods employed are Neutron Activation Analysis (NAA) and Spark Source Mass Spectrometry (SSMS). With the combination of these methods, 24 elements were routinely analyzed. The evaluation was performed through the grouped analysis method (Dulski, 1980).

Of the numerous interpretations that were performed, only a few significant results which are associated with the elements W, Ta, Zr and Hf are now presented. Their fractionation and geochemical distribution is related with the four genetic groups and with the metallogenetic concept of the "inheritance" of tin.

Figure 5 shows a clear fractionation of tungsten during the development of geologic history from the Precambrian to the Tertiary, This also indicates that the

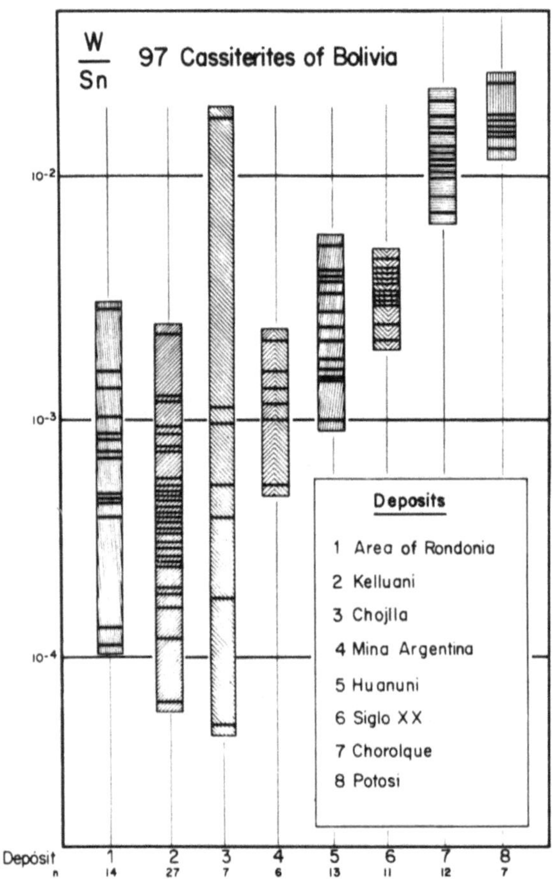


**Fig. 5.** W/Sn ratios of cassiterites from various Bolivian tin deposits (after Schneider et al., 1978)

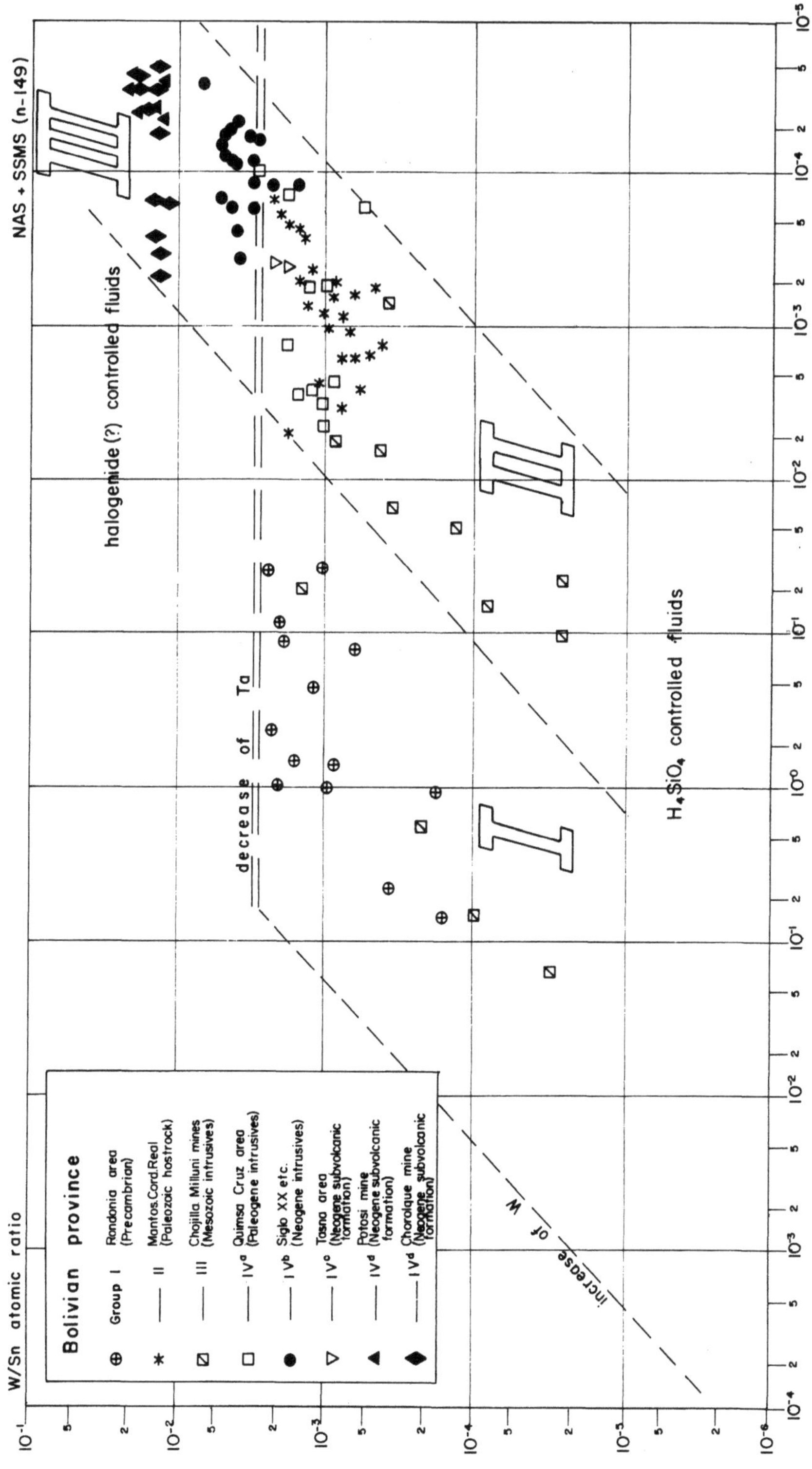


**Fig. 6.** W/Sn vs W/Ta atomic ratios of cassiterites from the Brazilian-Bolivian tin province (after Dulski et al., 1982)

geotectonic grouping proposed is reflected more or less in a marked manner in the groupings of the geochemical results. The most interesting diagram is the one which shows a W/Sn and W/Ta variation (Fig. 6). In it a noticeable grouping of the geochemical results can be observed. This also demonstrates a significant correlation of the cassiterite geochemistry with the geotectonic-genetic grouping of the deposits according to Schneider and Lehmann (1977).

In this figure it is also evident that the cassiterite from the older deposits is poorer in W and richer in Ta than the younger deposits.

Considering the paragenesis of the analyzed cassiterites, three principal groups can be recognized: I, II and III, which are separated by the dotted lines and by the double line (horizontal). Group I includes cassiterites whose paragenesis is preferably associated with oxides (e.g. pegmatites, greisens, metaquartzites, etc.). Those in Group II meanwhile display a sulphidic paragenesis.

The horizontal double line could limit the maximum content of W in cassiterites formed from fluids rich in silicic acid and with a minimum influence from halogen elements. Above this line is Group III, whose higher content of W would obey a different control process, possibly in relation to high concentrations of halogens. This is present in the subvolcanic deposits, especially in Potosí and Chorolque where the high relationship of W/Sn is probably due to the increase in the solubility of tungsten in hydrothermal solutions.

Additionally, the middle dotted line separates the two mineralization stages suggested for Mina Chojlla. This discordant vein-type deposit is genetically associated with a Mesozoic granite body belonging to the Cordillera Real. Crystal-chemical investigations on wolframite (H/F coefficient), such as the fractionation of trace elements, show a significant change in the results for the zone between the 4th and 5th levels. That is to say, the levels which are below this zone show an older paragenesis formed by a "primary fractionation" of the cassiterite crystallization. On the other hand, the paragenesis above level 5 was formed by a younger process of recrystallization (remobilization). Finally, it is interesting to note that the cassiterites of the "manto" or bedded-type deposits, and of the depositions associated with Mesozoic plutons, show more or less the same border relation W/Sn. This could mean, although not necessarily, that both groups of deposits were enriched by inheritance beginning with ancient pegmatites or paleoplacers. Rejuvenation would have occurred in similar formation conditions.

The Zr vs. Hf system can be observed in Figure 7. From its fractionation, clear differences between cassiterites of a pegmatitic and/or greisen origin and those of a hydrothermal origin can be seen.

The data points in this diagram show that the relationship Zr/Hf varies over the order of one magnitude. That is to say, that the Precambrian samples show a Zr/Hf relationship of around 10. The younger samples show a relationship of around 100. The cassiterites of the "manto" type deposits show intermediate proportions. This could be taken as an important reinforcement of the conclusions of the previous diagram. These depositions are derived from older pegmatitic deposits which would constitute – at least in part – the product of a remobilization in multiple stages of the tin content of the Paleozoic sediments. The great extension of strata-bound deposits of 800 km could be explained in this manner (Dulski et al., 1982).

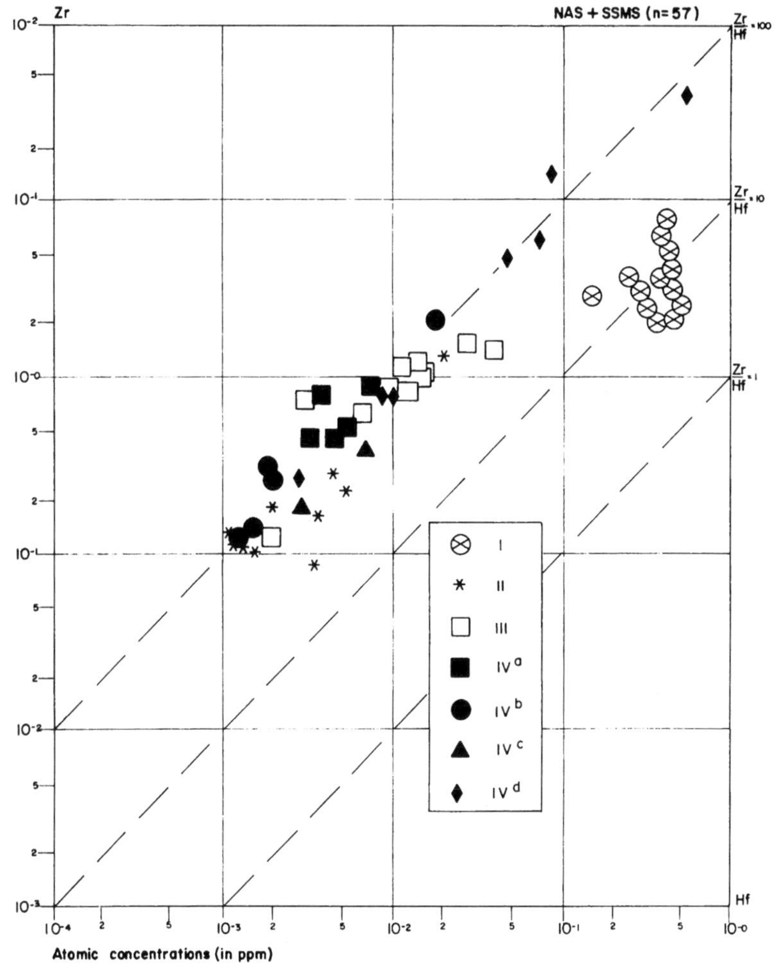


**Fig. 7.** Zr/Hf ratios of cassiterites from the Brazilian-Bolivian tin province. For grouping see Fig. 6 (after Dulski et al., 1982)

The extreme values of the Zr/Hf relationship of the Chorolque cassiterites confirm what was observed in Fig. 6. The unexpected increase of Zr in these samples is probably due to the interaction of the mineralizing hydrothermal solutions with the host rock and with meteoric water during the alteration phase (Grant et al., 1979, cited in Dulski et al., 1982).

## Conclusions

From the above discussion the following conclusions can be reached: the deposits in the Bolivian-Brazilian tin province, according to radiometric dating, belong to three distinct epochs: Precambrian (Rondonia and NE Bolivia), Early Mesozoic and Mid to

Late Tertiary (Provincia Andina). The latter two epochs are in a direct relationship to the Hercynian (deep magmatism) and Andina Orogenies (subvolcanic magmatism), respectively. Its development, according to the majority opinion of the authors, occurred in multiple, successive, and different stages.

- Clear differences can be observed in the Andean province between the tin deposits of the Septentrional part and the deposits of the south Andes delineated by the weakness zone of the "Codo de Arica" (Arica's Elbow). These differences are principally that the deposits in the north are associated with deep-seated granitic-plutonic rocks characterized by a relatively simple paragenesis and very low silver content. The deposits in the south are of a subvolcanic character related to dacitic-rhyodacitic stocks, have a complex paragenesis which includes sulphosalts with a high silver content, and possess a marked individual zonation, telescoping and intense hydrothermal alteration.
- Based on the work of Ahlfeld (1967), a new genetic classification of the Bolivian tin deposits has been established. This classification divides them into four greater groups, based on the geotectonic setting, the geometric shape, the mineral paragenesis, and the age of the deposits. They occur within the geotectonic development of this part of South America (Schneider and Lehmann, 1977; Dulski et al., 1982).
- Finally, the notion of an association between the initial Precambrian metallogenesis and other subsequent magmatic mineralization epochs (Schuiling, 1967; Stoll, 1964, 1976; Schneider and Lehmann, 1977), becomes strengthened through geochemical results which demonstrate a systematic fractionation of certain elements in cassiterites such as W, Zr, Ta, Hf, etc. These results also indicate an interrelationship between the different genetic groups of the Bolivian tin deposits ranging from the oldest to the youngest. That is to say, that tin became enriched by "heredity" during the geotectonic evolution of the Bolivian crustal segment from the Precambrian to the Tertiary.

## References


Ahlfeld, F., 1937. Ueber das Alter der zinnbringenden Magmengesteine Boliviens. *Zbl. Miner. Geol. Palaont.* Abt. A, 34–38.

Ahlfeld, F., 1967. "Metallogenetic epochs and provinces of Bolivia", *Mineral. Deposita*, 2, 291–311.

Ahlfeld, F. and Schneider-Scherbina, A., 1964. Los yacimientos minerales y de hidrocarburos de Bolivia. *Dep. Nl. de Geologia, Bol.* 5, 338 pp., La Paz.

Avila, W., 1975. Un modelo de Tectónica de Placas para el origen del Cinturón Estañífero Boliviano. *Bol.* 21, *S. G. B.*, 31–50.

Botello, R., Martinez, C., Subieta, T. and Tomasi, P., 1972. La carte tectonique de Bolivie, *Cah. ORSTOM, sér. Géolo.*, 4, 149–152.

Clark, A.H. and Farrar, E., 1973. The Bolivian tin province: Notes on the available geochronological data. *Geology,* 68, 102–116.

Clark, A.H., Farrar, E., Caelles, J.C., Hajnes, S.J., Lortie, R.B., McBride, S.L., Quirt, G.S., Robertson, R.C.R. and Zentilli, M., 1976. Longitudinal variations in the metallogenic evolution of the Central Andes; a progress report, *Geol. Assoc. Canada Spec. Pap.*, 14, 23–58.

Dulski, P., 1980. Spurenanalyse als Hilfsmittel bei der geochemischen Untersuchung der Genese bolivianischer Sn-Lagerstaetten. *Berliner geowiss. Abh.* (A), 28, Reimer, Berlin.

Dulski, P., Moeller, P., Villalpando, A. and Schneider, H.-J., 1982. Correlation of Trace Element Fractionation in Cassiterites with the Genesis of the Bolivian Metallotect. *Metallization Associated with Acid Magmatism,* Ed. by A.M. Evans, J. Wiley & Sons Ltd., 71–83.
Evernden, J.F., Kriz, S.J., and Cherroni, C., 1977. Potassiumargon ages of some Bolivian rocks, *Econ. Geol.,* 72, 1042–1061.
Grant, N., Halls, C., Avila, W., Avial, G., 1977. Igneous geology and the evolution of hydrothermal systems in some sub-volcanic tin deposits of Bolivia. *Geol. Soc. (London), Spc.* vol. 7, 117–126.
Halls, Ch., 1975. Dépositos de estaño y su relación con la Tectónica Global. (Trad. de "Tin deposits and their relation to global tectonics" - 1974, por W. Avila), *Bol.* 21, *S.G.B.,* 25–30.
Kelly, W.C. and Turneaure, F.S., 1970. Mineralogy, Paragenesis and Geothermometry of the Tin and Tungsten Deposits of the Eastern Andes, Bolivia, *Econ. Geol.* 65, No. 6, 609–680.
Kloostermann, J.B., 1967. A tin province of the Nigerian type in Southern Amazonia. In: Fox, W. (ed.): *A technical Conference of Tin,* Vol. 2, London. Intern. Tin Council, 383–398.
Lehmann, B., 1979. Schichtgebundene Sn-Lagerstaetten in der Cordillera Real/Bolivien. *Berliner geowiss. Abh.* (A), 14, Reimer, Berlin.
Ljunggren, P., 1962. Bolivian Tin Mineralization and Orogenic Evolution. *Econ. Geol.,* 57, 978–981.
Moeller, P., Parekh, P.P. and Schneider, H.-J., 1976. The application of Tb/Ca-Tb/La abundance ratios to problems of fluorspar genesis. *Mineral. Deposita,* 12, 247–262.
Parekh, P.P. and Moeller, P., 1977. Revelation of the genesis of minerals in paragenesis with fluorites, calcites and phosphates via rare earths fractionation. In: *Nuclear Techniques and Mineral Resources* (ed. by IAEA). IAEA, Vienna, 356–359.
Petersen, U., 1970. Metallogenic Provinces in South America. *Geol. Rundschau,* B. 1, 174–183.
Priem, H.N.A., Boelrijk, N.M., Hebeda, E.H., Verschure, R.H., and Bon, E.H., 1966. Isotopic age of tin granites in Rondonia, N.W. Brazil, *Geol. Mijnbouw,* 45, 191–192.
Radelli, L., 1966. New data on tectonics of Bolivian Andes from a photograph, by Gemini 5 and field knowledges. *Trav. Lab. Geol. Grenoble* 42, 237–261.
Robertson, R.C.R., 1974. Notas sobre el método del K/Ar de datación de edades obtenidas hasta ahora. *Serv. Geol. de Bolivia,* Proy. Plutonismo, Inf. Prel., 2.
Schlatter, L.E. and Nederlof, M.H., 1966. *Bosquejo de la Geología y Paleogegrafía de Bolivia,* La Paz, Geobol, 49 pp.
Schneider, H.-J. and Lehmann, B., 1977. Contribution to a new genetical concept on the Bolivian tin province, in D.D. Klemm and H.-J. Schneider (eds.) *Time- and Strata-bound Ore Deposits,* 153–168, Springer, Berlin.
Schneider, H.-J., Moeller, P. and Parekh, P.P., 1975. Rare earth elements distribution in fluorites and carbonate sediments of the East-Alpine mid-Triassic sequences in the Nördliche Kalkalpen. *Mineral. Deposita,* 10, 330–344.
Schneider, H.-J., Dulski, P., Luck, J., Moeller, P. and Villalpando, A., 1978. Correlation of Trace Element distribution in cassiterites and geotectonic position of their deposits in Bolivia. *Mineral. Deposita,* 13, 119–122.
Schneider-Scherbina, A., 1962. Ueber metallogenetische Epochen Boliviens und den hybriden Charakter der sogenannten Zinn-Silber-Formation. *Geol. Jahrb.,* Hannover, 81, 157–170.
Shuilling, R.D., 1967. Tin belts on the continents around the Atlantic Ocean. *Econ. Geol.* 62, 540–550.
Sillitoe, R.H., Halls, C and Grant, J.N., 1975. Porphyry tin deposits in Bolivia. *Econ. Geol.* 70, 913–927.
Stoll, W.C., 1964. Sn-W-Bi provinces and epochs in Argentina and Bolivia and their genetic interrelationship. *Geol. Soc. Am., Abstr.,* Meeting Miami Beach, Fl., 197–198.
Stoll, W.C., 1976. Provincias metalogenéticas en Argentina, Bolivia y Chile: Aspectos de una teoría evolutiva de la Metalogénia. *Bol. R. Soc. Española Hist. Nat.* (Geol.) 74, 171–189.
Turneaure, F.S., 1971. The Bolivian tin-silver province. *Econ. Geol.* 66 (2), 215–225.
Villalpando, A., 1977. Los yacimientos de estaño de Bolivia. Rev. Científica No. 3, *Univ. Tomas Frías,* Potosí, 37–51.
Villalpando, A., Dulski, P., Luck, P., Moeller, P. and Schneider, H.-J., 1977. Nuevos resultados sobre la geoquímica de depósitos de casiterita de Bolivia. *II Simposio Internacional del Estaño,* La Paz (unpubl.).

# 5.2 Geology of the Brazilian Tin Deposits

E.C. DAMASCENO[1]

## Abstract

This paper presents a brief summary of the geology, types of deposits and economic figures on Brazilian tin deposits. The data were collected from the available literature, government agencies and from the reports of private mining companies.

## Introduction

The Amazonian Craton and Goiás are considered the most favourable regions for containing tin deposits. However there are other places in Brazil where tin shows and prospects are known, as at Amapá, Northeast Brazil, Bahia, Minas Gerais, Mato Grosso and Rio Grande do Sul. During the last few years extensive field exploration programmes were conducted and new mines will be put in operation in the near future (Bettencourt et al., 1981).

Alluvial placers have been mined in Amazonia and primary deposits are the main tin sources in Goiás. In both regions is observed the association of cassiterite with many lithologies such as alluvium, greisen, pegmatites, aplites, veins in granites and albitized granites.

## 2 Amazonian Tin Deposits

Primary tin mineralization in Amazonia are associated with three generations of anorogenic granitoids of Proterozoic age, ranging from 977 to 1800 Ma. The economic deposits are mostly of placer type (Fig. 1).

About 20 mines operate in the region, which have provided an accumulated production of about 75,000 t of tin. Several tin deposits have been discovered during the last ten years. However, the real tin potential in Amazonia is still unknown and it is believed that the known reserves should be increased in the next few years.

The main Amazonian tin fields are located at Mapuera (Pitinga), Tapajós, Surucucus, Abonari, Xingú, Teles Pires and Rondônia. Rondônia and Pitinga are the main tin producing districts from large and high grade tin placer deposits.

[1] Departamento de Engenharia de Minas, Escola Politécnica da USP, Caixa Postal 61548, São Paulo, Brazil

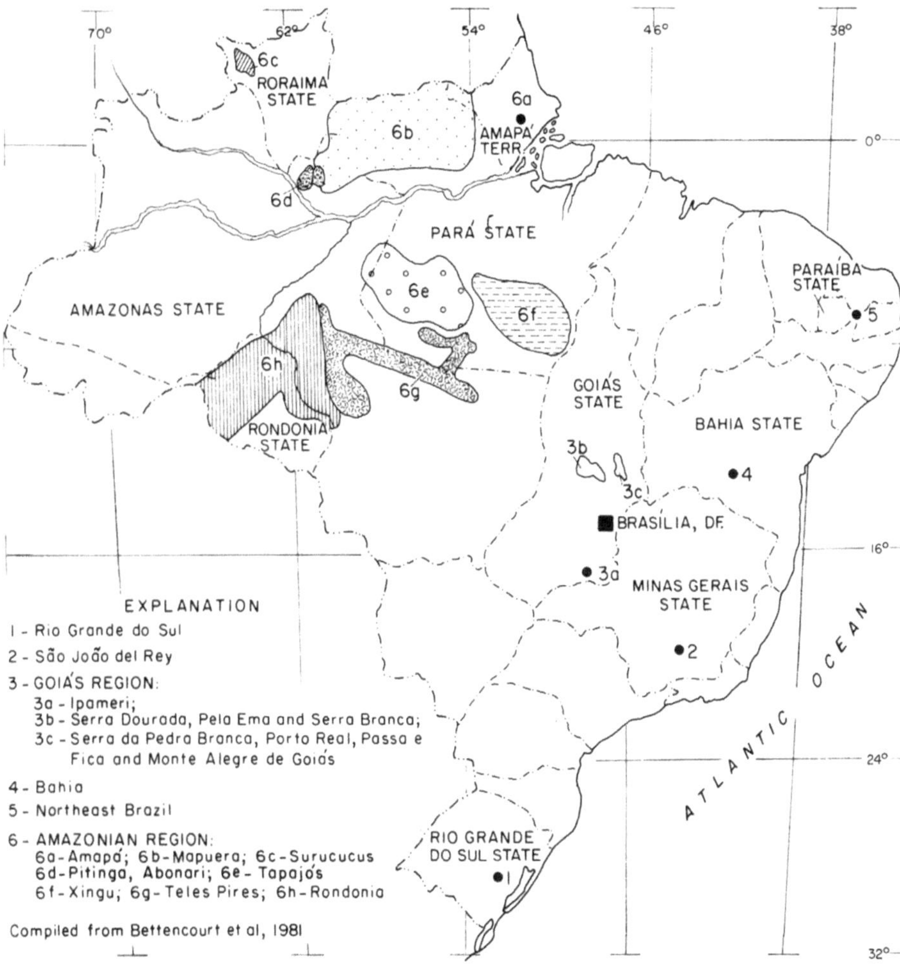


**Fig. 1.** General reference map – distribution of tin deposits and tin potential regions in Brazil

Primary mineralization in Rondônia is represented by veins, stockworks associated with greisen, possible diapiric breccia pipes, and topaz veinlets related to granites (980 to 1000 Ma). The Potosi Ore Body is the largest primary tin deposit presently known in Rondônia. At Pitinga the primary tin is related to albitized granites.

Rondônian tin placer deposits are associated with braided streams, colluvial-eluvial and alluvial placers of Pleistocene-Holocene age. The richest bonanzas are related to recently exposed granites. The deepest deposits have large ore volume but poorer grades.

The alluvial placer deposits can be classified into two types:

a) placers associated with paleovalley fill, with bottom deposits situated at 60 to 70 m depth, with an extention of about 3—4 km, and more than 500 m width. The mineralized gravel – coarse sand seams may be 1 to 6 m thick. The average grade is about 0.3 to 1.2 kg Sn $m^3$.

b) placers associated with present day valley fill, with gravel-coarse sand mineralized seam of about one meter thick and length from 3–4 km. In general the volume of the present valley fill placers is smaller than the paleovalley fill, but they present very high grades, about several kg Sn $m^3$.

## 3 Goiás Tin Deposits

In Goiás state there are more than 30 tin (and tin/columbite – tantalite) occurrences, many of them are exploited as "garimpos" (diggers). The most important tin deposits are situated in Serra Dourada, Serra Branca, Serra da Pedra Branca, Porto Real-Passa e Fica and Monte Alegre de Goiás. There are three small operating mines located at Pela Ema (Serra Dourada), Serra Branca and Passa e Fica. Cassiterite occurs in pegmatites, aplites, greisen, albitized granites, veins in schists and/or granites and in its weathered eluvium.

The Goiás accumulated production is estimated at more than 7,000 t of tin. This production, the geological environment and the expected reserves are encouraging signs for the establishment of many small tin mines for exploitation of both the eluvial and the primary hard-rock deposits.

## 4 Economic Figures

During the last few years, tin production has contributed to the improvement of Brazilian metal and mineral economy. Tin metal production and tin ore reserves are rising annually. In 1981 the reserves were of about 250,000 t of tin. However exploration conduced during the last three years have enlarged the Amazonian tin reserves (Pitinga mainly) and the total amount of Brazilian reserves are now estimated at about 500,000 t of tin (Empresas Brumadinho, 1983).

The production of cassiterite concentrates rose to 15,900 t during 1982 (11,300 t or 72% were produced in Rondônia). For 1986 the projected production is 41,000 t of cassiterite concentrates (or 27,000 t of tin metal).

## References

Bettencourt, J.S., Damasceno, E.C., Fontanelli, W.S., Franco, J.R.M. and Pereira, N.M., 1981. Brazilian tin deposits and potential. Presented at the Fifth World Conference on Tin, Kuala Lumpur, Malaysia.

Empresas Brumadinho 1983. A industria do estanho (1983/1984), ano II, n°. 2, São Paulo, Brazil.

# 5.3 The Andean Batholith and the Southeast Asian Tin-Belt Granites Compared

E.J. Cobbing[1]


## Abstract

The Andean Batholith of Peru consists of about 1,000 interlocking plutons and can be subdivided into plutons, super-units, and batholithic segments.

The primary level is that of the pluton which may be internally zoned by pulses and surges of related magma. Chains of plutons of identical lithology, which have consistent intrusive relationships to other members of the batholith, are referred to as super-units. The batholithic segment consists of a very large area of the batholith which is comprised of a specific number of super-units. Five segments have been recognised but only two have been mapped in detail: the Lima segment 400 km long and the Arequipa segment 1,000 km long. Although the lithological composition of the two segments is similar and ranges from gabbro to monzogranite, the plutons of the Arequipa segment are relatively enriched in $K_2O$ and the segment is also more intensely mineralised with copper and base metals.

The granites of the Main Range in Malaysia provide a total contrast to those of the Andes in that they do not form super-units, they are invariably true granites and are mineralised with tin.

The granitoids of the Eastern Belt in Malaysia are intermediate in character, having a lithological range from diorite to granite with a slight tendency to form super-units. They are mineralised both with tin and with base metals.


## Introduction

The essential element of all granitoid belts is the pluton. This is invariably a geographically singular body located in one particular place and having a certain limited size. In its simplest form it consists of a uniform body of granite contained within one continuous outer contact which separates it from the country rocks, or envelope into which it is emplaced. Plutons however are commonly more complex and may be made up of a series of annular shells which may vary in composition; if the components of those shells are related to one another, the pluton may be called complex and if unrelated composite. In granitic terrains granitoid batholiths are made up of a large number of plutons which appear to have coalesced together, however in such a terrain it is always possible to identify the component plutons and it is also desirable to estab-

[1] British Geological Survey, Keyworth, Nottingham, N4R 5GG, England

lish the relative age relationships so that the geological history may be established. Geochemical and geochronological studies depend upon the correct identification of the plutons within a batholith.

The simplest plan view that a pluton may present is of a circular stock. Plutons of this form may occur up to a diameter of 15 km but are usually smaller. Those of larger size are generally elliptical in plan and are commonly elongated in the direction of regional strike. Other shapes such as square, rectangular or elongate are also common and may reflect the geological history and mode of emplacement, while irregular shapes result from the process of later plutons intruding earlier ones. Other bodies, such as dykes and sills are commonly associated with plutons and are an essential part of the plutonic structure, being small plutons themselves. There is in fact a continuous sequence of phenomena ranging down from plutons through dykes and sills to veins and joint plane coatings, all of which have to be considered in relation to individual plutons or pluton assemblages.

## The Andean Batholith

Studies of the Andean Batholith in Peru have shown that it is made up of a great number of interlocking plutons but, whereas each pluton is a geographically singular body as defined above, there are other relationships which enable an overall classification of the plutons in that batholith to be made. It has been established that plutons of a particular, precise lithology and texture are arranged in chains which structurally interfere with other chains of different lithologies and textures. The relative age relationships between those chains of plutons is always the same and it has subsequently been established that their geochemical and isotopic signature is always specific for a particular chain or super-unit as it is now called.

The Andean Batholith however is not only made up of granitoid super-units; it has also been established that it is divided into batholithic segments, each of which is several hundred km in length and which is characterised by an assemblage of super-units specific to that segment. Modern work in the Circum Pacific region suggests that this pattern of batholithic formation is widespread in Mesozoic and Tertiary granitoid terrains of predominantly tonalitic or granodioritic character which have been generated by processes related to the subduction of oceanic lithosphere beneath continents. These are the classical I type granitoids of Chappell and White (1974) or, more properly the Cordilleran I type granitoids of Pitcher (1983). Granitoids of this type are generally associated with a base metal type of mineralisation.

It has been known for many years that certain orogenic belts are characterised by granitoid batholiths which are different in character from those of the Circum Pacific, among them being the granitoids of the tin belts of Southeast Asia and Australasia, but although much detailed work has been undertaken in these areas, regional studies of the granitoid batholiths as a whole have not been done and consequently the geology and structure of the batholiths have not been properly established except in localised areas. It was with these factors in mind that a research project was undertaken in Southeast Asia which had the specific objective of identifying the constituent plutons of the various batholiths and of establishing whether the hierarchial pattern of pluton – super-unit – batholithic segment, which characterises Andean batholiths, was developed in the granitoids of that region.

## Southeast Asia

The granitoids of Peninsular Malaysia were selected for the first experimental part of the study since it was already known that granite belts of contrasted type existed in that country (Hutchison, 1977) which are known as the Eastern and Western Belt Granites respectively. Field work has now been completed and geochemical and isotopic studies are continuing, and on the basis of the work completed to date it is possible to come to some preliminary conclusions.

**Table 1.** Comparison of Some Features of the Granitoid Bodies of Malaysia and Peru

| | Malaysia | | Peru |
|---|---|---|---|
| | Western Belt Granites | Eastern Belt Granites | |
| Range of lithology | granite – granodiorite | granite – gabbro | granite – gabbro |
| Predominant lithology | granite | granite | tonalite |
| Textural variation | wide range within each rock unit | narrow range within each rock unit | narrow range within each rock unit |
| Hornblende | absent | commonly present | always present |
| Muscovite | abundant in some rocks | very rare | absent |
| Average or range of pluton size | 40 x 10 km | 15 x 15 km | 70 x 35 km to 5 x 5 km |
| Type of pluton | complex, commonly with metasomatic overprint | simple | complex and simple |
| Type of batholith | aggregate of large complex plutons | aggregate of small simple plutons | aggregate of large and small complex and simple plutons |
| Granite type in repeated plutons or unique for each pluton | unique | unique | repeated plutons |
| Magnetic susceptibility | low | generally low | high |
| Ilmenite/Magnetite series | Ilmenite | Ilmenite | Magnetite |
| Granite typology | S type | I type | I type |
| Associated metals | Sn, Wo, Ta, Nb | Sn, Wo, Fe, Au, Cu, Pb, Zn, Ag | Cu, Mo, Au, Pb, Zn, Ag, Hg, Bi |

Whereas in the Andean Batholith, variations in the mafic constituents dominate the texture, in Southeast Asia the salic minerals are of greater importance, although some of the Eastern Belt granitoids are characterised by the nature of their mafic minerals. The size, shape, abundance, spacing and distribution of alkali feldspar megacrysts have proved to be critical factors, particularly in the Western Belt where it has been found that the textures developed within a single pluton are usually sufficiently distinctive to enable its distinction from every other pluton examined to date. Thus, in the case of the Western Belt granitoid batholiths, individual plutons can be mapped but they are not related to one another in the manner of the super-units of the Andean Batholith: each pluton is comprised of its own distinctive granite unit. The distinguishing features of the granitoid belts are summarised in Table 1.

In the Eastern Belt, plutons can also be distinguished and mapped and although, as in the Western Belt most plutons are of distinctive and unique granite units, there are petrographic similarities between certain plutons belonging to one or more batholiths, suggesting that they may be related to one another in the manner of the super-units of the Andean Batholith. However, neither the scale of development nor the range of lithologies is analogous to that seen in the Peruvian Andes. Thus, multi-pluton granite units as developed in the Andes are the exception rather than the rule in Southeast Asia.

## Mineralisation

The differences between the granitoids of the various belts, when considered together with other known geological factors, suggest that the granitoids are associated with different kinds of mineralisation. Thus the Western Belt is a province of tin, tungsten, tantalum and nobium, whereas the Eastern Belt is one of tin, tungsten, iron gold, copper, lead, zinc and silver. The Andean belt is a province of copper, molybdenum, gold, lead, zinc, silver and other metals. Thus the pattern of mineralisation within the belts reflects the pattern of differences between the granitoids with the Eastern Belt of Peninsular Malaysia providing an intermediate case both in the characteristics of the granitoids and the associated metals.

It is hoped that those preliminary results will provide a realistic basis for future investigations.

## References

Chappell, B.W., and White, A.J.R., 1974. Two contrasting granite types. *Pacific Geology* v. 8, p. 173–174.

Hutchison, C.S., 1977. Granite emplacement and tectonic subdivision of Peninsular Malaysia. *Geological Society of Malaysia, Bulletin* 9, p. 187–207.

Pitcher, W.S., 1983. Granite type and tectonic environment. *Mountain Building processes.* Ed. K. Hsu, Academic Press, London, 19–39.

# 6 Southeast and East Asia

## 6.1 General and Regional

# 6.1.1 The Tin Metallogenic Provinces of S.E. Asia and China: A Gondwanaland Inheritance

C.S. Hutchison[1]


## Abstract

Southeast Asia is interpreted as a composite of stable continental blocks which rifted from the northern margin of Australia and drifted northwards. The intervening highly deformed mobile and suture belts are the remains of the Paleotethys ocean when the blocks collided. The first blocks coalesced to form a Cathaysian East Asian Continent which occupied equatorial latitudes in the Permian. Other blocks rifted later from Australia and collided in Mesozoic and Cenozoic times with the East Asian continent to form the Eurasian Plate.

The tin was carried in the continental infrastructure of these blocks, which are all of Gondwanaland ancestry. Tectonic events which have the greatest continental crustal involvement are the most important in mobilizing tin into economic concentrations.

The foremost tin metallogenic event is a Malayan-type collision between two continental blocks, resulting in crustal thickening and S-type granite batholiths. Any subsequent igneous event superimposed on a Malayan-type belt will also have the ability to mobilize tin. The two important Malayan-type tin belts are (a) the Mesozoic belt formed by the collision of Sinoburmalaya, the Burma Plate, the Qantang-Tangla, and the Lhasa-Gandise blocks with the East Asian Continent, (b) the Caledonian East China belt.

Pre-rift thermal reworking of continental crust produces roughly concentric areas of subalkaline anorogenic granite with a high potential for tin mobilization, commonly with a tantalum-niobium association. The most important areas are centred in South China – North Vietnam, with isolated examples in Peninsular Malaysia.


## Introduction

The tin metallogenic belts of Southeast Asia were drawn by shading areas on maps linking up all the known tinfields and deposits. Commonly these maps were constructed without due consideration to the tectonic framework of the region. The first, by Sainsbury (1969), even made a mistake in the location of Nam Phatene in Laos. The shaded areas recognized a belt extending north from Billiton, through the Malay Peninsula, towards Mawchi in Burma. Hosking (1970) drew a similar pattern, with a

[1] Department of Geology, University of Malaya, Kuala Lumpur, Malaysia

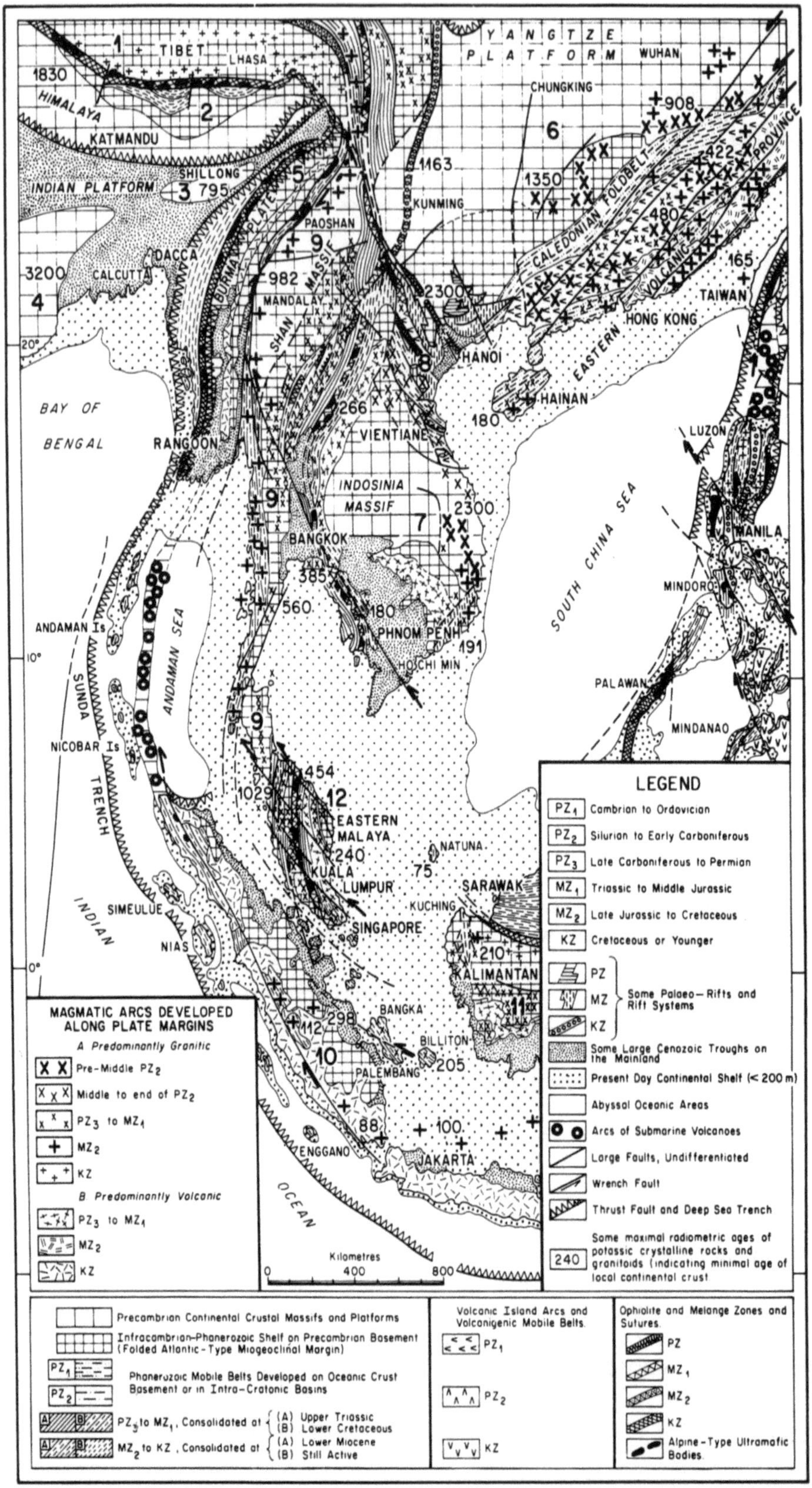
TIBET
LHASA
YANGTZE PLATFORM
WUHAN
CHUNGKING
HIMALAYA
KATMANDU
SHILLONG
INDIAN PLATFORM
KUNMING
PAOSHAN
DACCA
CALCUTTA
BURMA PLATEAU
SHAN MASSIF
MANDALAY
CALEDONIAN FOLDBELT
VOLCANIC PROVINCE
TAIWAN
HONG KONG
EASTERN
HANOI
HAINAN
VIENTIANE
BAY OF BENGAL
RANGOON
INDOSINIA MASSIF
BANGKOK
LUZON
SOUTH CHINA SEA
MANILA
MINDORO
PHNOM PENH
HO CHI MIN
ANDAMAN Is
ANDAMAN SEA
SUNDA TRENCH
NICOBAR Is
PALAWAN
MINDANAO
EASTERN MALAYA
KUALA LUMPUR
NATUNA
SARAWAK
KUCHING
SINGAPORE
SIMEULUE
NIAS
INDIAN OCEAN
KALIMANTAN
BANGKA
BILLITON
PALEMBANG
ENGGANO
JAKARTA
1830
795
3200
982
1163
1350
908
422
480
165
2300
266
180
2300
385
560
180
191
454
1029
240
75
210
298
112
205
88
100
20°
10°
0°
1 2 3 4 5 6 7 8 9 10 11 12
Kilometres
0 400 800
LEGEND
PZ1 Cambrian to Ordovician
PZ2 Silurian to Early Carboniferous
PZ3 Late Carboniferous to Permian
MZ1 Triassic to Middle Jurassic
MZ2 Late Jurassic to Cretaceous
KZ Cretaceous or Younger
PZ
MZ
KZ
Some Palaeo-Rifts and Rift Systems
Some Large Cenozoic Troughs on the Mainland
Present Day Continental Shelf (< 200 m)
Abyssal Oceanic Areas
Arcs of Submarine Volcanoes
Large Faults, Undifferentiated
Wrench Fault
Thrust Fault and Deep Sea Trench
240 Some maximal radiometric ages of potassic crystalline rocks and granitoids (indicating minimal age of local continental crust
MAGMATIC ARCS DEVELOPED ALONG PLATE MARGINS
A Predominantly Granitic
Pre-Middle PZ2
Middle to end of PZ2
PZ3 to MZ1
MZ2
KZ
B Predominantly Volcanic
PZ3 to MZ1
MZ2
KZ
Precambrian Continental Crustal Massifs and Platforms
Infracambrian-Phanerozoic Shelf on Precambrian Basement (Folded Atlantic-Type Miogeoclinal Margin)
PZ1
PZ2
Phanerozoic Mobile Belts Developed on Oceanic Crust Basement or in Intra-Cratonic Basins
A B PZ3 to MZ1, Consolidated at (A) Upper Triassic (B) Lower Cretaceous
A B MZ2 to KZ, Consolidated at (A) Lower Miocene (B) Still Active
Volcanic Island Arcs and Volcanigenic Mobile Belts
PZ1
PZ2
KZ
Ophiolite and Melange Zones and Sutures
PZ
MZ1
MZ2
KZ
Alpine-Type Ultramafic Bodies

suggestion that the main tin belt from Billiton to Mawchi may bifurcate through the Gulf of Thailand towards Rayong into Indochina, heading towards Southeast China.

The second generation of such metallogenic maps paid due recognition to the differences between granitoid belts in the delineation of the tin metallogenic provinces. Thus, Garson et al. (1975) showed two parallel tin belts, a western along the Main Range, which they equated with the Phuket-Tavoi-Mawchi belt, displaced by the Ranong-Khlong Marui faults; and a parallel eastern belt through Rayong, offset in north Thailand by the same faults. Hosking (1977) portrayed a similar pattern, and documented the differences between the western and eastern belts.

The exercise of drawing tin metallogenic belts could proceed no further until the region had been tectonically analysed. Hutchison and Taylor (1978) and Taylor and Hutchison (1979) showed the tin deposits and their scale of importance without attempting to draw in belts.

## The Tectonic Analysis

The region is now known to be composed of rigid Precambrian continental lithospheric blocks, sutured together by narrow highly deformed mobile belts (Fig. 1). The continental blocks are composed of Precambrian crustal massifs and platforms, which extend laterally beneath Infracambrian to Phanerozoic folded Atlantic-type miogeoclinal rifted margins (Gatinsky et al., 1984).

These blocks are divided into two major categories by the suture zone which extends from south to north along medial Malaya (the Bentong-Raub Line), beneath the Gulf of Thailand, extending between the Indosinia and Shan massifs as the Uttaradit – Luang Prabang Line, to link up with the Song Ma suture of North Vietnam. The dividing suture extends northwestwards into northern Tibet as the Litian-Junsha Jiang suture.

### Blocks of Cathaysian Affinity

All the blocks lying east and north of this suture have Permian Cathaysian floral affinities (Gatinsky and Hutchison, in press). The large East Asian Continent was formed of the Yangtze Platform (6) and its southeast Caledonian collision belt, Indosinia massif (7), Eastern Malaya (12), South Sumatra (10), and the West Borneo Basement (11) (see Fig. 1). The West Borneo Basement is thought to have been formerly attached to the east coast of Vietnam, and to have been displaced southwards along the straight faulted continental shelf only in Late Mesozoic times. The collision of Indosinia with the Yangtze Platform along the Song Ma Suture was completed in Late Devonian or Early Carboniferous, and the whole East Asian Continent lay in equatorial latitudes during Permian times.

**Fig. 1.** Tectonic map of Southeast Asia (after Gatinsky and Hutchison, in press). The continental blocks are: 1 = Lhasa-Gandise, 2 = Himalaya, 3 = Shillong, 4 = Indian Platform, 5 = Burma Plate, 6 = Yangtze Platform (Cathaysia), 7 = Indosinia, 8 = Phu Hoat microcontinent, 9 = Sinoburmalaya, 10 = Sumatra, 11 = West Borneo Basement, 12 = Eastern Malaya

The Infracambrian and Cambrian stratigraphy, which includes glacial and phosphorite deposits, and the paleomagnetic data suggest that the Yangtze Platform may have been attached to the northern margin of Australia during Cambrian time.

Thus, all the continental blocks of Cathaysian floral affinities may have rifted from northern Australia in earliest Paleozoic times, and arrived in equatorial latitudes to coalesce to form the East Asian Continent by Permian times.

## Blocks of Gondwana Affinity

Several continental blocks of Gondwana floral affinity have been identified (Fig. 1). They are characterized by Gondwana Permian and Triassic flora and important Carbo-Permian glacial deposits.

The largest block has been named Sinoburmalaya (Gatinsky and Hutchison, in press). It is a narrow elongated continental block (9), extending south from China through the Shan States, Peninsular Thailand and into the western part of Peninsular Malaysia. The northern part of Sumatra has to be included, for it contains characteristic glacial formations identical to that found in the Malay Peninsula.

Sinoburmalaya is interpreted as having rifted from Australia in Carboniferous times and to have drifted northwards to collide with the East Asian Continent. The collision is well dated at 220 to 200 Ma ago (Rhaetic to Lias) by the impressive collision-related Main Range granite batholith of Peninsular Malaysia (Liew, 1983).

The Qantang-Tangla block of northern Tibet collided with the East Asian Continent in Late Triassic-Early Jurassic times. It may therefore have been an integral part of Sinoburmalaya at that time, subsequently disrupted by collisions of other Gondwana blocks behind it.

The Burma Plate (5, Fig. 1) is defined along its eastern margin by the strike-slip Sagaing-Namyin Fault, along which it has moved about 450 km northwards (Mitchell, 1981). I interpret the Late Cretaceous plutonic arc, which extends northwards from Phuket, as a result of collision of the Burma Plate with Sinoburmalaya. The western margin of the Burma Plate is defined as the Indo-Burman Ranges.

The Lhasa-Gandise Block (1, Fig. 1) collided with Eurasia in Middle Cretaceous times to form the Pangong-Nu Jiang suture with the Qantang-Tangla Block.

The Middle Miocene collision of the Indian-Shillong Platform (3, 4, Fig. 1) with Eurasia resulted in the complex crustal thickening and overthrusting in the Himalaya (2, Fig. 1). Its collision on the east with the Burma Plate resulted in the uplift of the Indo-Burman Ranges.

The complex collisions between the continental blocks have resulted in important wrench faults (Fig. 1), some of which have followed ancient lines of weakness. The continental crust has locally been subjected to rifting at various times, preceded by an episode of pre-rift thermal reworking of the continental crust. The area north and east of the Red River and the ancient Song Ma suture has been prone to such extensional tectonics and related igneous events.

**The Major Tin Metallogenic Processes**

There is compelling evidence that continental crust is a necessary requirement for economic tin concentration (Hutchison and Chakraborty, 1979). Continental crust becomes progressively differentiated with respect to tin by polycyclic events involving metamorphism, anatexis and related magmatic-tectonic processes. As a result, the distribution of tin becomes progressively more concentrated in regions which have had the maximum involvement of continental crust in the magmatic-tectonic events (Fig. 2). Maximum continental crustal involvement may be obtained both in convergent and extensional tectonics.

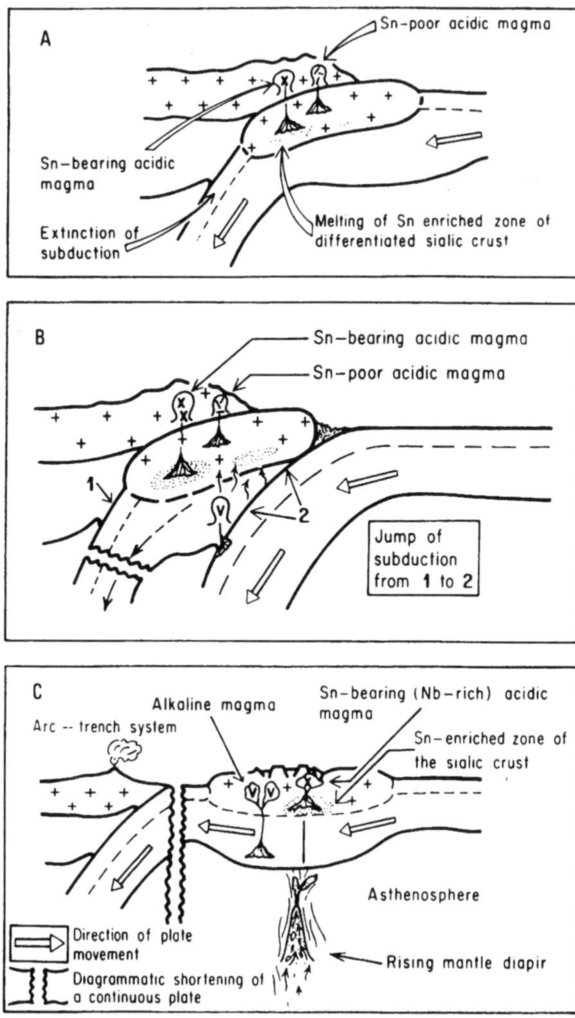


**Fig. 2 A–C.** The major tectonic events which can mobilize tin from the Precambrian continental basement (after Hutchison and Chakraborty, 1979). **A** = Malayan-type collision causing crustal thickening and S-type granites. **B** = Subduction-related magma generation from collision belts. **C** = Pre-rift thermal reworking of continental crust giving subalkaline granites of crustal origin

### a) Convergent Tectonics

The tectonic process which gives the maximum continental crustal involvement in magmatic-tectonic events is a collision between two continental blocks (Fig. 2A). The collision causes continental crustal thickening and S-type granite anatexis in the underthrust plate, rising up to be emplaced in the overthrust plate. The overlying strata are compressed into a linear foldbelt which characteristically contains overthrust zones. Because of its importance in the Malay Peninsula, I have called this a Malayan-type collision belt.

Superimposed tectono-magmatic events on an old Malayan-type collision belt lead to an ideal situation for mobilizing tin, and the successively younger granitic episodes are likely to become more tin specialized (Fig. 2B).

### b) Extensional Tectonics

An important precursor to the rifting of continental crust is a pre-rift cycle of updoming resulting from a high heat flow from the mantle. The igneous events accompanying this episode are tholeiitic basaltic dykes and trap volcanism, in the initial stages, followed by more extensive thermal reworking of the continental crust characteristically giving subalkaline biotite granites. Such granites do not form belts, but have near equidimensional geographical distributions resulting from the localized underlying mantle plume (Fig. 2C). It is only at the later stages of continental rifting that distinctly alkaline magmatism, such as shoshonite, syenite and A-type granites are emplaced. There is maximum involvement of the continental crust at the pre-rift stage when the plutonism is subalkaline.

An illustration of the pre-rift crustal reworking is offered by the tectonic evolution of the offshore Tertiary basins of the South China Sea (Hutchison, 1984). The first stage is the emplacement of tholeiitic basaltic dykes dated at 110 to 100 Ma (Albian). This is followed by extensive pre-rift reworking of the continental crust forming a wide geographic province of 75 to 65 Ma old subalkaline granites (Late Cretaceous to Palaeocene), encountered locally in the Malay Peninsula, Natuna, and in the drilled basement of the offshore basins. The main subsidence and dilation of the basins came in Oligocene times. These pre-rift subalkaline granites are not as potassic as the S-type granites of the Late Triassic-Early Jurassic Malayan-type collision belt (Hutchison, 1983), and they would not obviously be classified into the A-type of White and Chappell (1983). Therefore, I call them subalkaline.

**Fig. 3.** Metallogenic-tectono-magmatic provinces of Southeast Asia (after Gatinsky and Hutchison, in press). The two of importance for tin mobilization are: Malayan-type collision belts; and the pre-rifting districts of crustal reworking associated with subalkaline granites

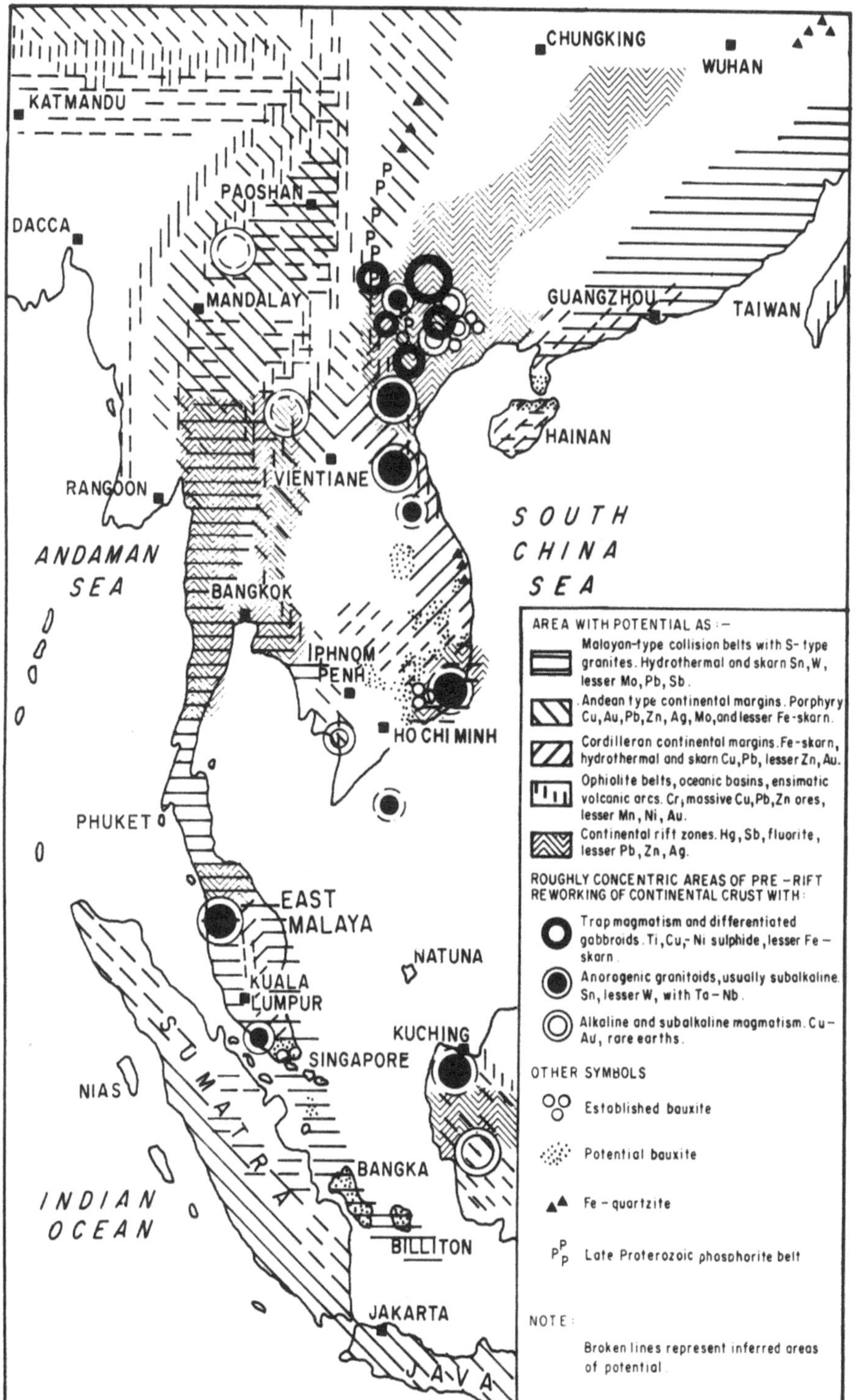
CHUNGKING
WUHAN
KATMANDU
PAOSHAN
DACCA
MANDALAY
GUANGZHOU
TAIWAN
HAINAN
VIENTIANE
RANGOON
ANDAMAN SEA
SOUTH CHINA SEA
BANGKOK
PHNOM PENH
HO CHI MINH
PHUKET
EAST MALAYA
NATUNA
KUALA LUMPUR
KUCHING
SINGAPORE
NIAS
SUMATRA
BANGKA
INDIAN OCEAN
BILLITON
JAKARTA
JAVA
AREA WITH POTENTIAL AS :-
Malayan-type collision belts with S-type granites. Hydrothermal and skarn Sn, W, lesser Mo, Pb, Sb.
Andean type continental margins. Porphyry Cu, Au, Pb, Zn, Ag, Mo, and lesser Fe-skarn.
Cordilleran continental margins. Fe-skarn, hydrothermal and skarn Cu, Pb, lesser Zn, Au.
Ophiolite belts, oceanic basins, ensimatic volcanic arcs. Cr, massive Cu, Pb, Zn ores, lesser Mn, Ni, Au.
Continental rift zones. Hg, Sb, fluorite, lesser Pb, Zn, Ag.
ROUGHLY CONCENTRIC AREAS OF PRE-RIFT REWORKING OF CONTINENTAL CRUST WITH:
Trap magmatism and differentiated gabbroids. Ti, Cu,- Ni sulphide, lesser Fe-skarn.
Anorogenic granitoids, usually subalkaline. Sn, lesser W, with Ta-Nb.
Alkaline and subalkaline magmatism. Cu-Au, rare earths.
OTHER SYMBOLS
Established bauxite
Potential bauxite
Fe-quartzite
Late Proterozoic phosphorite belt
NOTE:
Broken lines represent inferred areas of potential.

## S.E. Asian Tectono-Magmatic-Metallogenic Provinces

The distribution of the tectono-agmatic-metallogenic provinces in Southeast Asia is given in Fig. 3. The tin metallogenic provinces may be classified as either Malayan-type collision belts with S-type granites, or roughly concentric areas of pre-rift reworking of continental crust associated with anorogenic granitoids.

### a) Malayan-Type Collision Belts

The Caledonian foldbelt of S.E. China (Fig. 1) is an important tin metallogenic province. Pre-Devonian collision resulted in thickening of the continental crust and the formation of Caledonian granites. Xu et al. (1982) have emphasized that the foldbelt has been subsequently affected by subduction-related magmatic-tectonic events to produce polycyclic granites of Hercynian, Indosinian, and Yanshanian age. The situation is shown diagramatically in Fig. 2B, and is an ideal setting for mobilizing tin. This tinbelt has no extension into S.E. Asia and terminates at the Song Ma suture in north Vietnam.

The major Malayan-type belt, which extends from Billiton, along the Malay Peninsula, bending westwards into the Himalaya, can be shown to result from a succession of collisions (Gatinsky and Hutchison, in press). The first collision is interpreted as that of the narrow central Malayan platform with the East Asian Continent, giving crustal thickening in East Malaya near the mining district of Gambang and producing S-type Permian granites of 240 to 270 Ma age in a belt along the eastern part of the Malay Peninsula. Liew (1983) has shown that the granites resulted from anatexis of a 1000 to 1300 Ma old underlying continental basement which does not outcrop here.

This relatively minor event was followed in Latest Triassic-Earliest Jurassic times by the most spectacular S.E. Asian (Indonesian) event – the collision of Sinoburmalaya with the East Asian Continent, giving rise to crustal thickening and the impressive 200 to 220 Ma old Main Range S-type granite batholith. Liew (1983) has shown that this granite resulted from anatexis of a 1500 to 1700 Ma old underlying basement which does not outcrop here.

The next collision event was a Late Cretaceous collision of the Burma Plate with Eurasia. The collision took place when the Burma Plate lay at least 450 km farther south than its present position, for it has been slipping northwards relative to Sinoburmalaya along the Sagaing-Namyin Fault since Late Cretaceous times. The collision gave rise to the Late Cretaceous S-type Phuket-Tavoi granite belt which borders the Andaman Sea.

The Malayan-type collision belt extends westwards from north Burma and south China into the Himalayas and Tibet, where successive collisions have also occurred. The tin metallogenic province therefore extends through the magmatic arcs of the Himalayas and of the Lhasa-Gandise blocks.

The belts shaded as Malayan-type collision zones on Fig. 3 do not therefore imply a single magmatic cycle. Both the East China and the Malay Peninsular-Himalayan belts are polycyclic, and this is why both are important in tin metallogeny. The Himalayan and Tibetan parts of this belt should have good potential for tin, and it is felt that they warrant more consideration.

b) Pre-Rift Anorogenic Areas

There are a number of localities scattered throughout the region which do not belong to any belts. These tin provinces result from the early stages of extensional tectonics.

South China-North Vietnam has experienced several episodes of continuing rifting from the Paleozoic to the Cenozoic. The tin centres of Gejiu in Yunnan, North Vietnam, and Nam Patain in Laos are considered to result from pre-rift reworking of the Proterozoic continental basement.

The Yanshanian granites of SE China were emplaced predominantly and extensively in the back-arc region of the Middle Jurassic to Late Cretaceous East China volcanic arc. To describe them as back-arc granites is nothing more than a geographic description. The Yanshanian "orogeny" witnessed the greatest development of extensional basins in China, both in front of and behind the Yanshanian volcanic arc (Li, 1984). Most of the basins were well established in Cretaceous times and many had their extensional tectonics, which affected most of eastern China. The granites are predominantly S-type (Xu et al., 1982) but this does not mean that they have the same tectonic setting as the S-type Main Range Batholith of the Malayan-type collision zone. It appears necessary to consider that the Yanshanian granites of China result from pre-rift or early-rift reworking of the older continental crust. Thus on Fig. 3, the SE China tin province is shown as a Malayan-type collision belt. This classification refers to the Caledonian collision episode. The province, however, has become preeminent for tin mobilization because of the superimposed Yanshanian Jurassic-Cretaceous pre-rift or early-rift anorogenic (extensional) reworking of the previously thickened continental crust.

In the Malay Peninsula, the localities at Kedah Peak and Bakri (see Taylor and Hutchison, 1979) appear to be of this type because of their tantalum-niobium association.

Areas in the West Borneo Basement and in S.E. Vietnam have been suggested as having potential, but they have not yet been proven to be tin-bearing.

*Acknowledgements.* I am grateful to Mr S. Srinivass for drafting the final figures. I also gratefully acknowledge the financial help provided by R.M.R.D.C. to allow me to attend the conference. Finally I am grateful to the Ministry of Geology of the People's Republic of China for the invitation and opportunity to see at first hand the remarkable tin deposits of Dachang.

## References

Garson, M.S., Young, B., Mitchell, A.H.G., Tait, B.A.R., 1975. The geology of the tin belt in Peninsular Thailand around Phuket, Phangnga, and Takua Pa. *Instit. Geol. Sciences, London, Overseas Memoir 1.*

Gatinsky, T.G. and Hutchison, C.S. (in press). Cathaysia, Gondwanaland and the Paleotethys in the evolution of continental Southeast Asia. *Proceedings GEOSEA V. Geol. Soc. Malaysia, Kuala Lumpur.*

Gatinsky, T.G., Hutchison, C.S., Minh, N.N., and Tri, T.V., 1984. Tectonic evolution of Southeast Asia. *27th Internat. Geol. Congr. colloquium 05 Tectonics of Asia. Reports Vol. 5, Nauka, Moscow, 225–240.*

Hosking, K.F.G., 1970. The primary tin deposits of Southeast Asia. *Minerals Science and Engineering 2, no. 4, 24–50.*

Hosking, K.F.G., 1977. Known relationships between the 'hardrock' tin deposits and the granites of Southeast Asia. *Geol. Soc. Malaysia Bull. 9, 141–157.*

Hutchison, C.S., 1983. Multiple Mesozoic Sn-W-Sb granitoids of Southeast Asia. Roddick, J.A. (ed.) Circum Pacific plutonic terranes. *Geol. Soc. America Memoir 159, 35–60.*

Hutchison, C.S., 1984. A layman's view of the Malay Basin. *Geol. Soc. Malaysia Petroleum meeting December, Kuala Lumpur (Abstract).*

Hutchison, C.S. and Chakraborty, K.R., 1979. Tin: a mantle or crustal source? *Geol. Soc. Malaysia Bull.* 11, 71–79.

Hutchison, C.S. and Taylor, D., 1978. Metallogenesis in S.E. Asia. *Journ. Geol. Soc. London,* 135, 407–428.

Li, Desheng, 1984. Geologic evolution of petroliferous basins on continental shelf of China. *Amer. Assoc. Petrol. Geol. Bull.* 68, 993–1003.

Liew, T.C., 1983. Petrogenesis of the Peninsular Malaysian granitoid batholiths. *Unpubl. Ph.D. thesis, Australian National Univ., Canberra.*

Mitchell, A.H.G., 1981. Phanerozoic plate boundaries in mainland S.E. Asia, the Himalayas, and Tibet. *Journ. Geol. Soc. London,* 138, 109–122.

Sainsbury, C.L., 1969. Tin resources of the world. *U.S. Geol. Survey Bull.* 1301.

Taylor, D. and Hutchison, C.S., 1979. Patterns of mineralization in Southeast Asia, their relationship to broad-scale geological features and the relevance of plate-tectonic concepts to their understanding. Jones, M.J. (ed.) *Proc. 11th Commonwealth Mining and Metall. Congr. Hong Kong 1978. Instit. Min. and Metall. London,* 93–107.

White, A.J.R. and Chappell, B.W., 1983. Granitoid types and their distribution in the Lachlan Fold Belt, Southeastern Australia, Roddick, J.A. (ed.) Circum-Pacific Plutonic terranes. *Geol. Soc. America Memoir* 159, 21–34.

Xu Keqin, Hu Shouci, Sun Minzhi, Zhang Jingrong, Ye Jun and Li Huiping, 1982. Regional factors controlling the formation of tungsten deposits in South China. Hepworth, J.V. and Yu Hong Zhang (eds.) *Tungsten geology Jiangxi, China. ESCAP/RMRDC, Bandung, Indonesia,* 473–488.

# 6.1.2 Distribution of Tin Deposits in China and Their Metallogenic Conditions*

CHEN XIN and WANG ZHITAI[1]

**Abstract**

The distribution of tin deposits in China is conditioned by the regional geological settings. The significant tin deposits are limited to only a few metallogenic regions or belts, which correspond to certain geological-tectonic units. In this paper, four tin metallogenic regions are outlined. They are: the south China fold system, the Daxinganling and the Jilin-Heilungjiang fold system, the Sanjiang fold system, and the Yangtze paraplatform. The metallogenic conditions of tin deposits in China are also briefly discussed.

China is one of the first countries to have used tin metal. Copper, tin and lead were used in the manufacture of the binary and ternary alloy-ancient bronze. The Shang dynasty established an outstanding bronze culture in the middle and lower reaches of the Yellow River as early as 1,000 B.C. or earlier.

The presently mined tin deposits all occur in south China. In the early days after the founding of the People's Republic, geological work on tin deposits was confined to a few areas in south China. With the advances of regional geological surveys and mineral prospecting and extensive application of geophysical and geochemical exploration techniques, not only have new deposits been continuously discovered and reserves been increased in some old mining districts and in their surrounding areas, but some new metallogenic provinces and a group of tin deposits of industrial value have also been identified.

The tin deposits of China consist of primary and placer types, with the former predominating. China's tin placers are mainly distributed in areas where primary tin deposits are concentrated in the South China tin province, mostly near the Tropic of Cancer. This paper deals mainly with the distribution or primary tin deposits and their metallogenic conditions.

The primary tin deposits occur mainly in geosynclinal fold belts. Examples are the tin province of the South China fold system, the tin province of the Daxinganling fold system and the Jilin-Heilungjiang fold system, and the tin province of the Sanjiang fold system [in the middle and upper reaches of the Jinsha, Lancang (Mekong) and Nujiang (Salween) Rivers]. Tin deposits are also distributed in the Tianshan, Qilian,

[1] Department of Regional Geology and Mineral Resources, Ministry of Geology and Mineral Resources, Beijing

* This paper has used relevant data collected and processed by Shi Lin, Research Institute of the Bureau of Geology and Mineral Resources of Yunnan Province, and Ma Linqing, Bureau of Geology and Mineral Resources of the Zhuang Autonomous Region of Guangxi.

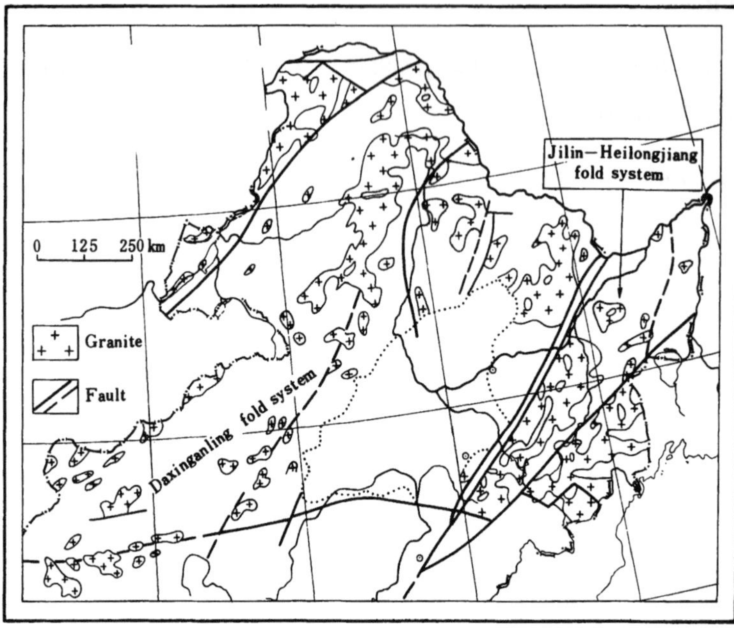


**Fig. 1.** Distribution of tin provinces in China

Kunlun and Qinling fold system. As for platform regions, tin deposits mainly occur in the Yangtze paraplatform. Several tin anomalies have also been found in the Sino-Korean paraplatform. Based on the results of geological investigations conducted in recent years, this paper introduces the distribution of tin deposits of China and discusses briefly their regional metallogenic conditions (Fig. 1).

## 1 Tin Province of the South China Fold System

It is located in southern China, extending from southeastern Yunnan in the west to the coastal areas in the east. Geographically the province is called the Nanling Mountains, belonging to the marginal-Pacific tectonic domain. The Nanling Mountains represent a late Caledonian geosynclinal fold belt, with an upper Paleozoic and Mesozoic cover. The Zhejiang-Fujian-Guangdong coastal areas in the east represent a Mesozoic volcanic belt. Intense Yenshanian tectono-magmatism and mineralization gave rise to the western Pacific metallogenic belt. This province is the most important tin province of China as well as the maintin-producing region at present. According to the regional metallogenic conditions, the province may be divided into the western, middle and eastern parts (Fig. 2).

*Western part.* This part is located in the Youjiang fold belt in southeastern Yunnan and western Guangxi. The late Caledonian movements transformed it into a platform. Still rather mobile, it subsided and a major depression was formed in which a considerable thickness of Devonian to Triassic sediments was deposited. At the beginning of the late Triassic Indosinian movements, it entered again into an active development

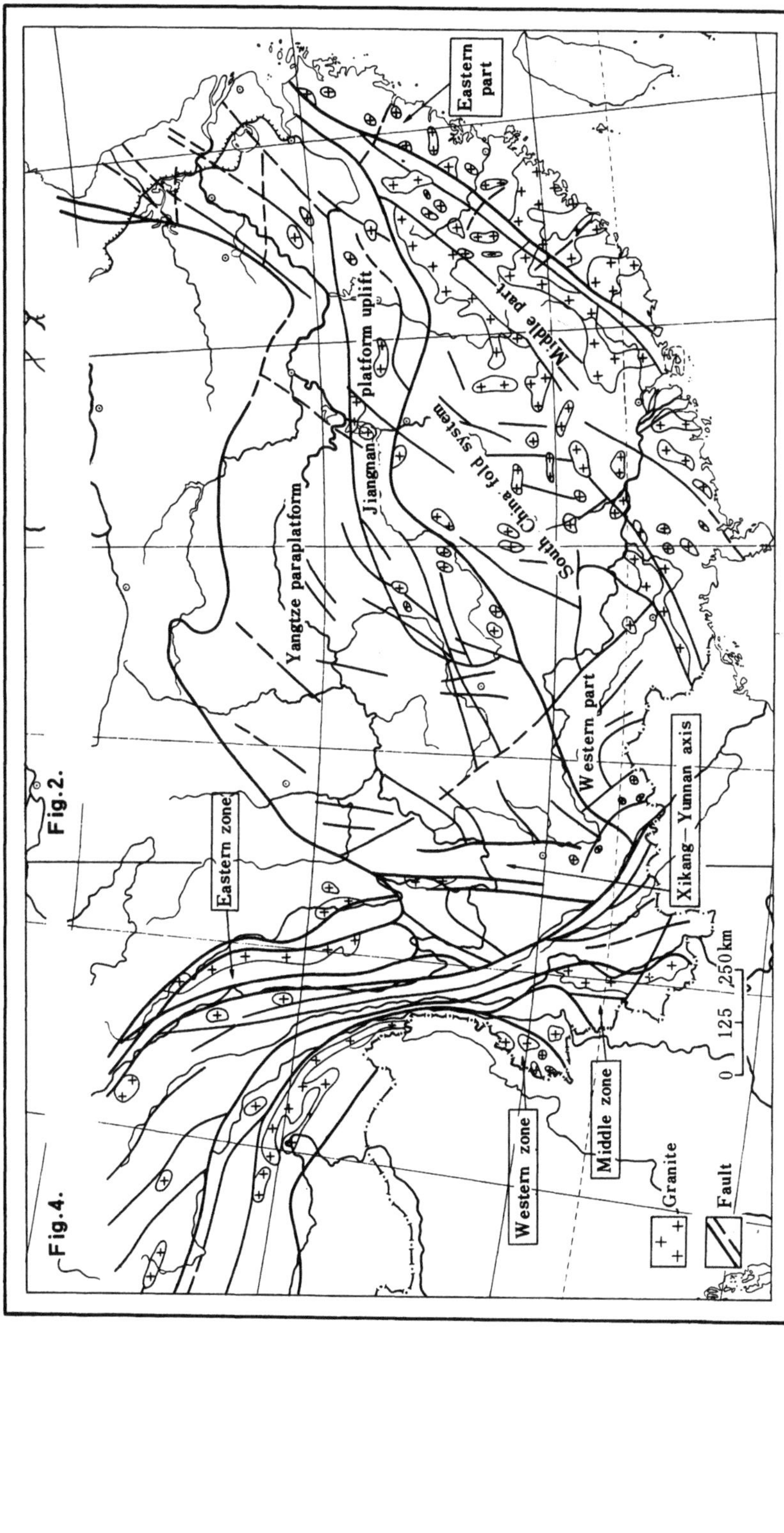


**Fig. 4.** Tin province of the Sanjiang fold system

**Fig. 2.** Tin provinces of the South China fold system and the Yangtze paraplatform

stage. Granites in this area are not widespread; some of them are deep seated. According to the isotopic ages, the granites associated with tin mineralization may be assigned to the Caledonian, Variscan-Indosinian and Yenshanian ages. Tin mineralization related to the Yenshanian granites is the most important. China's proved and demonstrated tin ore reserves are mainly concentrated in this region and the famous Gejiu and Dachang tin mines are situated here.

The Gejiu tin mine is located at the southwestern end of the Nanling latitudinal structural belt, bounded on the southwest by the Ailaoshan uplift of the Sanjiang fold system, on the northwest by the Xikang-Yunnan axis and on the southeast by the North Vietnamese massif. The sediments in the area are very thick. The exposed strata consist mainly of Triassic carbonate rocks. During the Indosinian movements this area was uplifted and basic magmatism took place. The Yenshanian movements generated intense folding and large-scale intrusion of acid and slightly alkaline granites, thus forming tin-polymetallic deposits in the area. The granites related to tin deposits include porphyritic biotite and equigranular granite, which are considered to be of late Yenshanian age. Deposits are mainly of the sulphide-cassiterite skarn type, and mostly occur in contact zones between granites and Triassic limestones. The metals are mainly tin, copper, lead, zinc and tungsten, associated with molybdenum, bismuth, beryllium, indium, silver, arsenic, sulphur and fluorine. Within the contact zones lie concordant interformational orebodies as well as disconcordant vein and stockwork orebodies. Most of the ores have been oxidized. There are also tourmaline-cassiterite veinlet zones and cassiterite greisens.

The Dachang tin mine lies in an Upper Paleozoic downfaulted basin in the southwestern trend of the Jiangnan platform uplift. The mineralization is related to late Yenshanian hidden biotite granite stocks controlled by a major NNW-trending fracture. The stocks are intruded into Devonian carbonate strata with organic reefs. Orebodies mainly occur in the exocontact zones. There are large vein orebodies, veinlet-zone orebodies and veinlet-disseminated (stockwork-disseminated) stratoid orebodies. The types of ore minerals are complex. Metallic minerals include cassiterite, marmatite, jamesonite, galena, pyrrhotite, scheelite, arsenopyrite, pyrite, chalcopyrite, stannite, franckeite, boulangerite, marcasite, pyritogelite, magnetite and stibnite, and gangue minerals include quartz, calcite, tourmaline, fluorite, dolomite and siderite. Many kinds of complex sulphosalt minerals also occur.

In addition to the sulphide-cassiterite type deposits, there are many other types in this area. The important ones are deposits of tantalum- and niobium-bearing cassiterite occurring on the tops of small granite stocks, skarn-type deposits, greisen-type deposits and tungsten-tin quartz vein-type deposits.

*Middle part.* This part extends from eastern Guangxi eastwards and is separated by a major fracture from the Zhejian-Guangdong coastal volcanic belt. The basement of the area is represented by the late Caledonian geosynclinal fold system, covered mainly by Upper Paleozoic strata. Starting from the late Triassic, the Indosinian, Yenshanian and Himalayan movements made this area an important component of the continental margin mobile belt west of the Pacific. Magmatic rocks are widespread. The granites are Caledonian, Variscan-Indosinian and Yenshanian in age. Granites intimately associated with mineralization are mainly of Yenshanian age. There is a wide diversity of types of tin deposit, including sulphide-cassiterite, quartz veins (large-vein or veinlet zones, with tin mostly associated with tungsten), greisens, skarns (skarn tin-tungsten deposits, skarn tin-copper deposits and skarn stanniferous magnetite deposits), pegmatites and granites in which tin is associated with tantalum and niobium. In recent years porphyry-type tin deposits have also been found. Tin ores of all types are often associated with various kinds of non-ferrous and rare metals (tungsten, bismuth, molybdenum, tantalum, niobium, copper, lead, zinc., etc.).

*Eastern part.* The southeastern coastal tin zone is located in the coastal areas of Zheijiang, Fujian and Guangdong, representing a Mesozoic volcanic belt. Tin mineralization is mainly associated with late Yenshanian granites. Types of deposits include sulphide-cassiterite, quartz veins, greisens, skarns and porphyries.

## 2 Tin Province of the Daxinganling Fold System and the Jilin-Heilungjiang Fold System

This province lies in the eastern part of the Caledonian Variscan Tianshan-Xingan latitudinal geosynclinal fold region in northern China. The Mesozoic intense tectonomagmatism caused it to evolve into a component part of the marginal-Pacific tectonometallogenic domain (Fig. 3).

The Daxinganling Mountains tectonically lie in the compounding zone of the NNE-trending structural system and the Tianshan-Xingan latitudinal structural belt. Exposed strata include mainly Permian clastic, carbonate and volcanic rocks. Yenshanian volcanic-intrusive rocks are well developed. The granites are assigned to both the Variscan and Yenshanian ages. Tin mineralization began to occur in late Variscan time, but mainly during Yenshanian time. Magnetite-cassiterite skarns and sulphide-cassiterite assemblages are the main types. Next comes the greisen type. Mineralization zones of tin have been found in Late Jurassic volcanic rocks.

In the Jilin-Heilungjiang fold system, some iron-polymetallic ores contain tin. The deposits are related to Variscan granites and granodiorites intruding the Upper Paleozoic strata.

## 3 Tin Province of the Sanjiang Fold System

This is one of the new tin provinces discovered since the 1960s. It is mainly distributed in Sichuan and western Yunnan, and may extend northwestwards into Qinghai and Tibet, and southwards to be linked with the tin belt of Southeast Asia. According to the investigations made in recent years by geologists in Sichuan and Yunnan, it may be divided into three zones in terms of the regional metallogenic conditions (Fig. 4).

*Eastern zone.* It extends from northwestern Yunnan in the south along the eastern side of the Jinsha River up to western Sichuan, and lies in a zone east of the Jinshajiang deep fracture zone in the westernmost part of the Indosinian Sonpan-Garze fold system and west of the Garze-Litang deep fracture zone. Exposed strata are dominated by a strongly folded Upper Triassic volcanic rock series. They were intruded by granites with multiple phases, thus forming a composite granite zone. The granites are of Yenshanian-Himalayan age. The late Yanshanian granite masses are

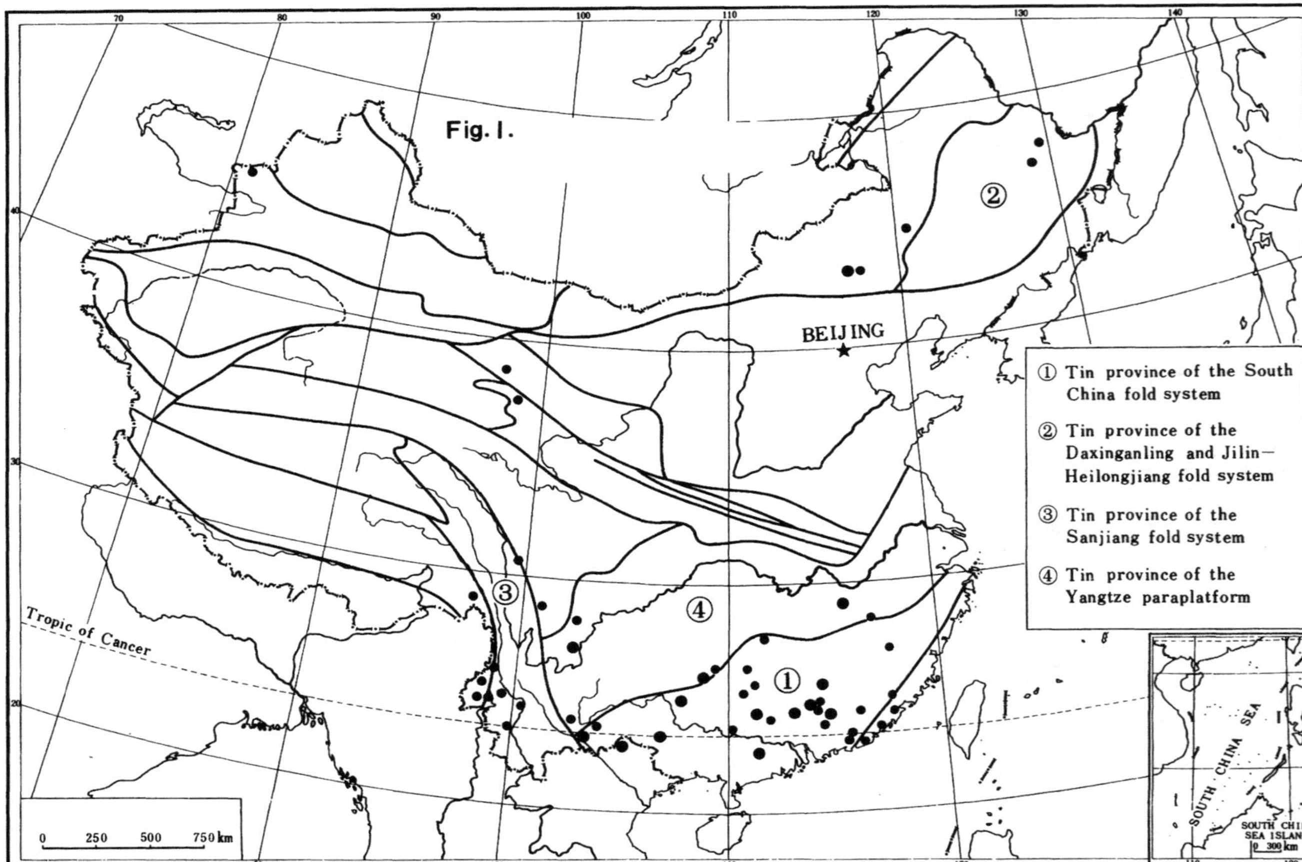


**Fig. 3.** Tin province of the Daxinganling fold system and the Jilin-Heilungjiang fold system

more closely related to tin deposits. Northwestern Yunnan contains tin-tungsten-polymetallic sulphide deposits, while western Sichuan contains mainly cassiterite-polymetallic skarn and cassiterite skarn deposits.

*Middle zone.* It is mainly distributed along the western side of the Lancang River, belonging to the southern sector of the Indosinian Sanjiang fold system. This zone extends from the western bank of the Lancang River in the east to the Kejie fracture in the west. It may be further divided into the eastern and western subzones according to the characteristics of the structures and magmatic rocks. The two subzones converge and merge in the north.

The eastern subzone is largely distributed on the western side of the Lincang granite batholith and at its southward and northward plunging apices. The Lincang granite batholith is composed of Indosinian synorogenic granites. Tin mineralization is closely related to the granite formed at the late stage of evolution and differentiation of the Lincang granite. Country rocks are mainly represented by the Upper Proterozoic Lancang Group. The northern sector is partly represented by the Congshan Group, exhibiting the character of a eugeosynclinal origin. Deposits are mainly of the quartz-cassiterite type. Ore minerals include cassiterite, quartz and tourmaline ($\pm$ sulphide) assemblage and cassiterite and quartz (greisen) assemblage; they are characterized by a rich content of tourmaline.

The strata of the western subzone consist mainly of a late Proterozoic metamorphic series. The southern sector is represented by the Ximeng Group; while the northern sector, i.e. the Mengtong Group, which is composed of clastic and carbonate formations, exhibits the character of a miogeosynclinal origin. The Indosinian-early Yenshanian two-mica granite was intruded into them, thus resulting in the formation of tin deposits. Deposits are dominated by the quartz-cassiterite type. The ore mineral assemblage includes mainly cassiterite, tourmaline, topaz, quartz (greisen) and sulphides. Stanniferous pegmatites also occur.

Between the eastern and western subzones there is a late Paleozoic fracture zone, in which mylonitic and metasomatic granite appears intermittently. Studies show that this kind of granite was probably formed by mylonitization and silica-alkali-metasomatism of primary rocks (volcaniclastic and carbonate rocks), in which tin ores were deposited. Deposits are of the quartz-cassiterite type, enriched in tourmaline.

*Western zone.* The western zone is distributed in the vicinity of Tengchong and Lianghe in westernmost Yunnan. Tectonically it is located at the southern end of the late Yanshanian Gangdise-Nyainqentanglha fold system. Exposed strata consist mainly of Carbo-Permian carbonate rocks and Carboniferous sandy slates. Through recent work it has been considered that tin mineralization in this zone is closely related to lithium- and fluorine-bearing biotite granite, two-mica granite and alkali-feldspar granite fromed at the late stage of magmatic differentiation. According to the characteristics of structures and magmatic rocks, this zone may be divided into three subzones.

Tin deposits in the eastern subzone are associated with minor high-level anatectic granite masses in the northern sector of the early Yenshanian Menglian granite belt. The granite masses were intruded into Carbo-Permian carbonate rocks and Carboniferous sandy slates. Geochemical anomalies reveal a copper, lead, zinc, and tin assemblage, often associated with a high concentration of silver and arsenic. In the carbonate rocks of the exocontact zones there occur stanniferous, silver-rich, lead and zinc (or copper) skarns and stanniferous magnetite ores. Tin is present as an associated component.

Tin deposits in the middle subzone occur both in the endo- and exocontact zones of the late Yenshanian Guyong granite batholith and the minor stocks nearby. Country rocks are Carboniferous sandy slate and conglomeratic slate. Geochemical anomalies indicate a tin, tungsten, boron, copper and lead assemblage. Two types of deposit are distinguished: one is greisenized skarn deposits occurring in exocontact zones, always associated with magnetite and scheelite; and the other is cassiterite greisen veins.

Tin deposits in the western subzone are intimately associated with the early Himalayan Binlangjiang granite belt. Granite masses were intruded into Carboniferous sandy slate or Paleozoic metamorphics. Geochemical anomalies suggest a tin, tungsten and bismuth assemblage, with indications of tantalum and nobium. Deposits are dominated by the cassiterite-greisen (topaz)-sulphide type; however, the cassiterite-wolframite-quartz (greisen) vein type is also present.

There are some tin-rich weathering crusts of granite masses, with a tin grade meeting or approaching the requirements of industrial mining.

## 4 Tin Province of the Yangtze Paraplatform

This tin province lies in the central-southern part of China, between the Qinling and the Nanling Mountains. The Yangtze paraplatform was formed in the late Proterozoic. The Proterozoic metamorphic series that makes up the basement is exposed in the marginal zones of the paraplatform. There occur granite intrusives of various ages from the late Proterozoic to Yenshanian. In this province tin deposits of industrial value are distributed in two zones: the tin zone of the Xikang-Yunnan axis and the tin zone of the Jiangnan platform uplift.

The tin zone of the Xikang-Yunnan axis is situated in the western part of the paraplatform. Exposed strata are mainly the Proterozoic metamorphic series. There occurred multiple phases of tectono-magmatism. The biotite granite associated with tin mineralization was intruded into the Proterozoic metamorphics, belonging to the Jinning-Chengjiang period (1050—700 Ma). Sulphide-cassiterite skarns are the main type of tin deposits; however, there are also stanniferous magnetite skarns and tungsten-tin-quartz veins.

The tin zone of the Jiangnan platform uplift is a long-active uplift zone on the southeastern margin of the paraplatform. Exposed strata include the Proterozoic metamorphic series, intruded by late Proterozoic to Yenshanian granites. Tin deposits are mainly distributed in northern Guangxi and northern Jiangxi. Late Proterozoic granites in northern Guanxi fall into two phases: the first-phase (Sibao-phase) granite is of plagioclase granite-granodiorite (with a zircon U-Pb age of 1100 Ma, and a whole-rock Rb-Sr age of 1063 ± 95 Ma), which is spatially closely associated with the basic-ultrabasic rocks; the second-phase (Xuefeng-phase) granite is of biotite granite (with a zircon U-Pb age of 760 Ma) and a whole rock Rb-Sr age of 730 Ma. Tin mineralization took place during both phases. According to recent data, the granite mainly associated with tin mineralization may be assigned to the Xuefeng phase. Orebodies mainly occur in the basic-ultrabasic rocks in the exocontact zones of the granite masses. Cassiterite-tourmaline-quartz-sulphide-type deposits predominate. The granite associated with tin mineralization in the platform uplift region in northern Jiangxi is of Yenshanian age. The granite was intruded into Sinian strata, forming sulphide-cassiterite skarn-type deposits.

## Discussion

The distribution of China's tin deposits is obviously controlled by regional geological conditions. Important tin deposits of industrial value are concentrated in a few individual tin provinces or zones. These provinces or zones coincide in the main with certain tectonic units. The tin province of the South China fold system and of the Daxinganling and Jilin-Heilungjiang fold systems are both located in the marginal-Pacific tectonic domain. The latter province is continuous to the Far East Asian tin province of the Soviet Union. The tin province of the Sanjiang fold system is located in the Tethyan-Himalayan tectonic domain. Southwards it adjoins the Southeast Asia tin province. Most of China's tin deposits of industrial value and demonstrated tin ore reserves are concentrated in the tin province of the South China fold system. This province is also the region where the world's foremost tungsten deposits are mostly concentrated. The Youjiang fold system lies in the compounding zone of the marginal-Pacific tectonic domain and the Tethyan-Himalayan tectonic domain. More than half of China's proved and demonstrated tin ore reserves and two major tin mines – Gejiu and Dachang – are situated in this region. Late Proterozoic granites are often exposed at the peripheries of the Yangtze paraplatform. The data of the regional geological surveys demonstrate that this kind of granite generally has a relatively high content of tin, which is closely related to the distribution of the tin deposits of the Yangtze paraplatform region.

China's tin provinces are characterized by polycyclic tectono-magmatic activity, and the formation of the stanniferous granites and tin deposits exhibits a polyphase character. In the tin provinces of the geosynclinal fold belts, there often occur stanniferous granites of different ages, some of which have formed composite rock masses and evolved into late-stage granites. Their tin abundance tends to increase markedly, thus forming important tin deposits. The tectonic polycyclicity and polyphase tectonomagmatism in China's tin provinces may constitute important conditions for the formation of stanniferous granites and tin provinces.

There has been much discussion in the world about the regional geological conditions for the formation of tin enriched zones. Some people advocate an original crustal and mantle enrichment in tin, while others suggest the important role that the formation and evolution of specific magmas have played. These ideas are mainly based on the analysis of some geological conditions of the known tin enriched zones, and are still unlikely to be confirmed with certainty. In accordance with the analysis of some regional geological conditions of China's tin provinces, the two above-mentioned factors are most likely to provide a reasonable explanation for the important geological conditions for the formation of the tin provinces. In recent years, Chinese geologists have analysed the conditions for the formation of tin provinces in terms of the concept of plate tectonics and proposed new ideas.

The formation ages of the granites associated with primary tin deposits are late Proterozoic, Caledonian, Variscan-Indosinian, Yenshanian and Himalayan. The tin mineralization of Yenshanian granites is the most important. The level of research on granites varies from province to province. Of these provinces, the South China province is best studied. In recent years, on the basis of regional geological surveys, granites in the Sanjiang region have also been studied more intensively. The research level for other tin provinces is relatively low.

Tin deposits are associated with granites of crustal anatectic derivation. Such granites are mostly biotite granites, rich in silica and alkalis ($K_2O > Na_2O$), alumina-oversaturated and poor in magnesium and calcium. For example, all granites generating tin mineralization in most parts of the South China tin province, the Sanjiang tin province, the Daxinganling and the Xikang-Yunnan axis, belong to this category. Some granites associated with tin mineralization exhibit the character of syntexis. For example, granites generating some tungsten-tin molybdenum-bismuth deposits, some porphyry tin deposits and porphyry tungsten-tin deposits in the South China province belong to this category. Tin-producing areas associated with magmatitic granite include the western Guangdong, eastern Guangxi and the Jiangxi-Fujian border area. In the magmatite zone of the western Yunnan, there also occur tin-, tantalum- and nobium-bearing pegmatites. Mylonitic metasomatic granite in western Yunnan may possibly also belong to this category.

The formation of different types of tin deposit depends on such factors as the geological-structural conditions, the nature of stanniferous granites as well as characteristics of their generation and evolution, and the stratigraphical conditions of metallogenic provinces. In recent years, with the increasingly extensive and intensive geological investigations of tin deposits, new understanding concerning the mechanisms of mineralization and regularities of distribution of different types of tin deposit has been gained and tin perspectives have been continuously broadened.

## References

Cheng Yuqi et al., 1983. Further discussion on the problems of minerogenetic series of mineral deposits. *Geological Review,* Vol. 29, No. 2 (in Chinese).

Li Chunyu et al., 1980. China's plate tectonic framework, *Bulletin of Chinese Academy of Geological Sciences,* Vol. 2, No. 1 (in Chinese).

Li Chunyu et al., 1983. Some problems on subdivision of palaeoplates in Asia. *Acta Geological Sinica,* Vol. 57, No. 1 (in Chinese).

Liu Zenggian et al., 1983. A preliminary study on the north boundary and the evolution of Gondwana and Tethys in light of the new data on Qinghai-Xizang (Tibet) plateau. *Contribution to the Geology of the Qinghai-Xisang (Tibet) Plateau,* No. 12. Geological Publishing House (in Chinese).

Wang Zhifen, 1983. Some problems on the mineralization of tin deposits in Gejiu; Yunnan. *Acta Geological Sinica,* Vol. 57, No. 2 (in Chinese).

Xu Keqin et al., 1982. On the origin and metallogeny of the granites in South China. *Proceeding of Symposium on Geology of Granites and Their Metallogenetic Relations.*

Xu Keqin et al., 1983. On the genetic series of granites as exemplified by the Mesozoic granites of South China. *Acta Geologica Sinica,* Vol. 57, No. 2 (in Chinese).

Yang Chaoqua, 1982. Genetic types of the granitoids in South China. *Proceeding of Symposium on Geology of Granites and Their Metallogenetic Relations.*

# 6.1.3 Tectonic Zoning and Genetic Types of Tin-Bearing Granites in Western Yunnan and Their Relationship with Tin Deposits

SHI LIN, CHEN JICHEN, ZHANG WEILI, and FAN YUCHUN[1]

**Abstract**

The tin-bearing granites of western Yunnan have an obvious zoned distribution, composed of eastern, western and central tectonic-magmatic zones with different characters. The tectonic, petrological and mineralization characters of these three zones fundamentally agree with the corresponding zones of Southeast Asia.

The granites of western Yunnan may be classified into five kinds: 1. mesozonal S-type; 2. epizonal S-type; 3. gneissic fault-zone mylonitic granite; 4. calc-alkaline hypabyssal granite; and A-type; 5. alkaline-granite. Each type has its own tin mineralization characteristics.

West Yunnan is the northward extension of the tin belts of Southeast Asia, where granitoids are strongly developed, and a large number of tin deposits have been found in recent years. The distribution of granites and tin deposits are closely related to plate tectonic activities. As in Southeast Asia, the granitoids may be divided into east, middle and west tin belts. The granites may be grouped into five types, which have different genetic connections with individual regional tin mineralizations. The known commercial tin deposits normally have an immediate genetic connection with anatectic granite of high level emplacement, except that a few are closely associated with fractured and metamorphosed mylonite granites.

## 1 Tectonic Zoning of Tin-Bearing Granites in Western Yunnan (Fig. 1)

The region west of the Lancang River in Yunnan forms a specific crustal tectonic zone connected with Burma, Thailand, Peninsular Malaysia, Sumatra, Bangka and Billiton, called the Yunnan-Burma terrain in this paper. The terrain is bounded on the northeast by the line of Bangong Lake–North Nu River–Lancang River–Uttaradit–Chantaburi, and on the west by the well-known Yalu Zanbo River–Naga mountains suture line. It is a massive intermediate between the Old and Young Tethyan geosynclines. The data show that the massif might have been part of Gondwanaland, and drifted north from a lower latitude. From Late Proterozoic to Early Permian, westwards subduction of oceanic crust occurred along the Lancang River (Fig. 2), forming the Lincang island arc and Shuangjiang-Menglian back arc basin. Metamorphism and magmatic activities round the island arc and along the Ximeng-

[1] Yunnan Institute of Geological Sciences, Kunming, Yunnan

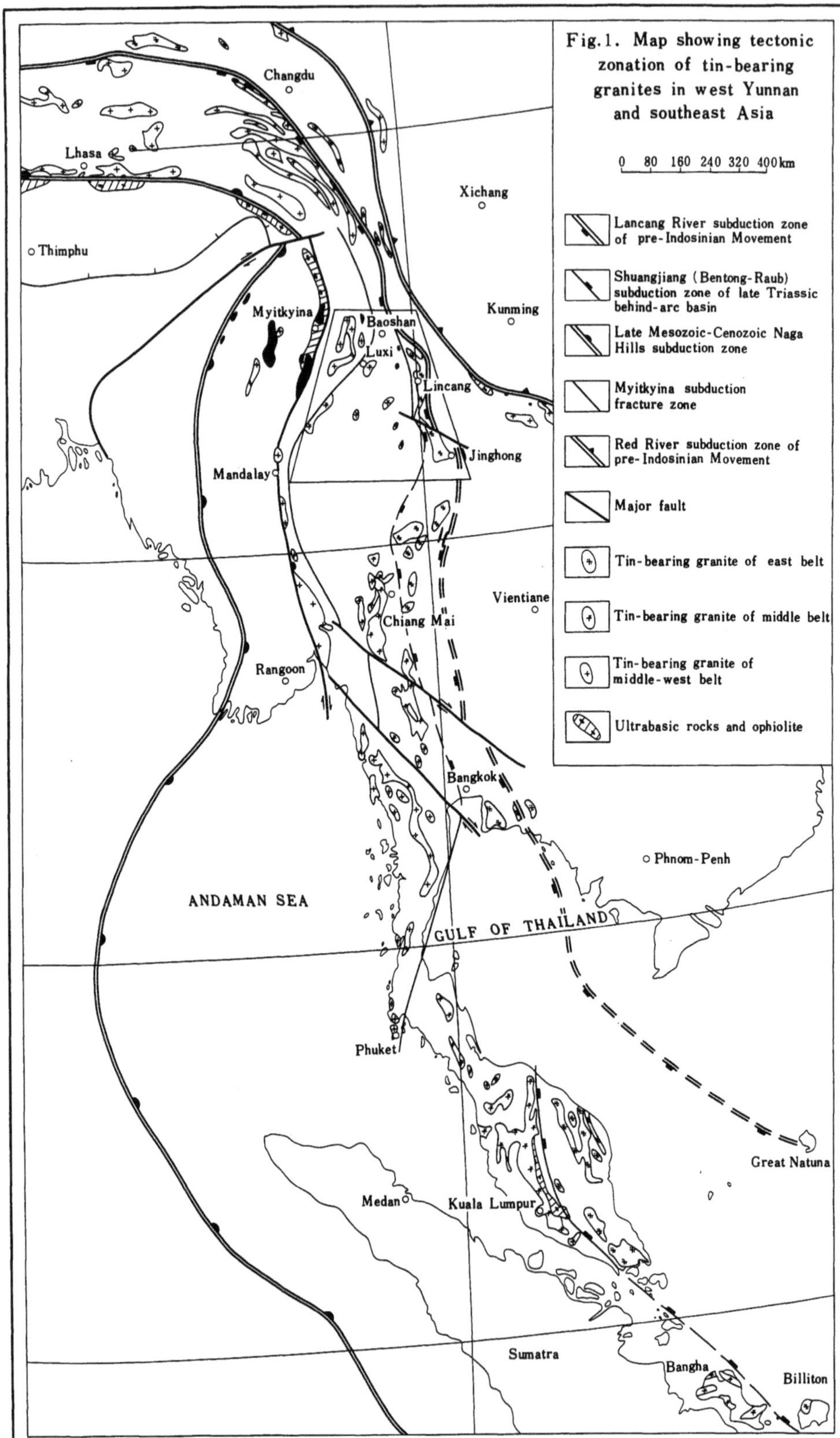
Fig.1. Map showing tectonic zonation of tin-bearing granites in west Yunnan and southeast Asia
0 80 160 240 320 400km
Lancang River subduction zone of pre-Indosinian Movement
Shuangjiang (Bentong-Raub) subduction zone of late Triassic behind-arc basin
Late Mesozoic-Cenozoic Naga Hills subduction zone
Myitkyina subduction fracture zone
Red River subduction zone of pre-Indosinian Movement
Major fault
Tin-bearing granite of east belt
Tin-bearing granite of middle belt
Tin-bearing granite of middle-west belt
Ultrabasic rocks and ophiolite
Changdu
Lhasa
Xichang
Thimphu
Kunming
Myitkyina
Baoshan
Luxi
Lincang
Jinghong
Mandalay
Vientiane
Chiang Mai
Rangoon
Bangkok
Phnom-Penh
ANDAMAN SEA
GULF OF THAILAND
Phuket
Great Natuna
Medan
Kuala Lumpur
Sumatra
Bangha
Billiton

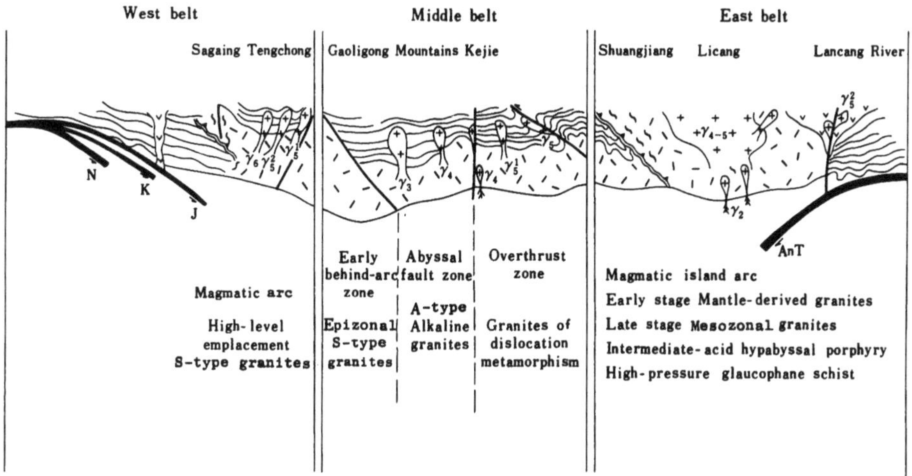


**Fig. 2.** Tectonic setting of different granite belts in western Yunnan

Gengma-Longling back-arc epicontinental zone led to the formation of early granites of both the east and middle magmatic zones. This geological event is verified by isotopic ages of the granites (715, 456, and 347 Ma) and of the metamorphic rocks (645, and 581 Ma). The back arc basin is similar to that of the Benton-Raub zone in Malaysia. From Late Permian to Early Triassic, the basement of the basin subducted towards the magmatic arc, causing closure of the basin and the collision between the magmatic arc and the back-arc continent. As a result, a high-pressure glaucophane schist belt came into existence on the overthrust side of the magmatic arc, and an extensive high-temperature metamorphosed belt of granites occurred, finally completing the evolution process of the tin-bearing granites of the east belt in west Yunnan. The age of the glaucophane schist is indicated by the phenite dated at 260—217 Ma and the east-belt granite is dated at 236—210 Ma.

The west margin of the landmass was active during the Jurassic-Eocene. The subduction there shifted thrice from east to west. Corresponding to the three subduction positions, three parallel back-arc granite subzones from E to W are dated 120—190 Ma, 77—71 Ma, and 58—53 Ma respectively to the west of the Gaoligong-mountain back-arc fracture. Together they constitute the west tin-bearing granite belt of western Yunnan.

The tectonic belts of tin-bearing granites of western Yunnan can be fully correlated with those of Southeast Asia. The east belt correlates with the tin-bearing granite belt, mainly of Variscan-Indosinian age along the east coast of Trengganu and Pahang of Peninsular Malaysia. The west belt is the back-arc granite belt produced by subduction and collision during the convergence of the Indian and Eurasian plates in Yenshanian–Early Himalayan times, corresponding to the granite belt along the west coast north of Phuket in Peninsular Thailand.

**Fig. 1.** Map showing tectonic zonation of tin-bearing granites in west Yunnan and Southeast Asia

The middle belt lies in the back-arc zone and has been superimposed by successive subductions and collisions on the eastern side, then on the western side, corresponding to the Main Range granite belt of the western part of the Peninsula south of Phuket.

## 2 Genetic Types of Granites in Western Yunnan

The granitoids of western Yunnan may be subdivided into five types: mesozonal S-type granites; epizonal S-type granites; fractured-metamorphosed mylonite granites; calc-alkaline hypabyssal granites; and A-type alkali granites.

1) Mesozonal S-type granites occur in the high-temperature metamorphosed zone of the active belt along the margin of the upthrust plate. They are the products of the crustal materials. They form batholiths concordant with the regional tectonic structures. Most of them contain metamorphosed enclaves; they form an association of peraluminous porphyritic biotite adamellite, cordierite granite and granodiorite. Their rock-forming minerals project in the right half of the adamellite area on the Q-A-Pl diagram, showing no distinct trend of differentiation (Fig. 3). Their micas are ferriferous-magnesian biotites. The porphyroblastic potash feldspars are mostly microclines which contain many mineral inclusions, whereas the plagioclases are predominantly andesine (An 32—45); the accessory mineral association is ilmenite-monazite. The lithochemical compositions are homogeneous, K/Na = 1.5, K/Rb = 161. The patterns of rare earth elements generally show a slight deficiency in Eu.

2) Epizonal S-type granites generally occur in the back-arc magmatic belt of the continental-crust plate, in the form of batholiths or stocks. They are characterized by consanguineous, contemporaneous and multi-stage differentiation facies belts. The complete evolution of a rock mass may be represented by a series composed of porphyritic amphibole-biotite adamellite, granular biotite granite and muscovite-biotite alkali-feldspar granite. They appear as a linear path in the Q-A-Pl diagram from the left half of the adamellite area to the alkali-feldspar granite area (Fig. 3). The biotites are mainly lepidomelane and siderophyllite; the growth sequence of the potash feldspars are orthoclase → intermediate-microcline → maximum microcline; the plagioclase is correspondingly in the sequence of An 32—40 → 24—26 → 7—9. Similarly, the accessory minerals change gradually from the association of magnetite, sphene and allanite to ilmenite, monazite and topaz. The lithochemical compositions are high in Si, rich in K and Na, K/Na = 7.66—8.38, K/Rb = 145—61. The pattern of rare earth elements varies from a slight Eu deficiency in the early lithofacies to an intense Eu deficiency in the late lithofacies; the heavy rare earth elements were obviously enriched during the late stage.

3) Fractured-metamorphosed mylonite granites are a unique kind of small-sized granitic rock bodies formed by the replacement of mylonitization and granitization of the epicrustal rocks along the overthrust belt near the collision zone of the back-arc basin. Remnant bedded structures and layers are still preserved in some of the rock bodies. Mylonitic texture and oriented structure can be observed. In hand specimen both the K-feldspar and the plagiodase contents vary greathy. K-feldspars are mostly of very

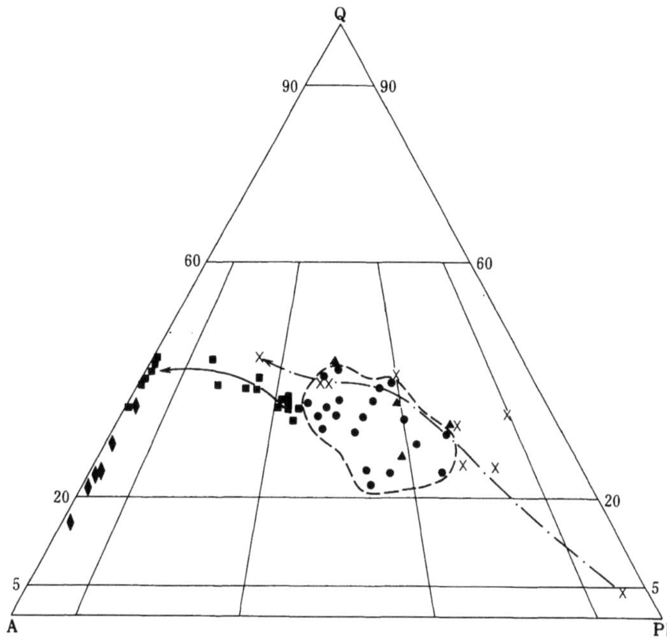


**Fig. 3.** Q-A-Pl diagram of various granites in west Yunnan

large microclines, presenting optical anomalies. These rocks are scattered in Q-A-Pl diagram, indicating a lack of homogeneity. Their composition varies widely, with an average corresponding to tonalite, which is low in silica, poor in alkali, and with an extremely low Rb/Sr value and very high Ba/Rb value. A slight deficiency of Eu exists in their pattern of rare earth elements.

4) Calc-alkaline hypabyssal granitic intrusives occur along the deep-fracture zone. They are of andesitic magma mixed with upper crustal material, forming a series composed of tonalite, quartz diorite, plagiogranite and granite-porphyry. In the Q-A-Pl diagram, an evolutionary line is shown from the lower right to the upper left. $K_2O<Na_2O$, with no deficiency of Eu and enrichment of heavy rare earth elements. The pattern of rare earth elements shows a curve declining more sharply to the right.

5) Alkali granites occur as stock-like bodies in tensional fracture belts. They are peraluminous with graphic textures, containing alkaline minerals such as riebeckite. Their compositions project into the lower left area of the alkali-feldspar granite in the Q-A-Pl diagram (Fig. 3). Nearly all of the feldspar of the granites are K-feldspars, with only 3—5% albites, characteristic of A-type granite.

Mesozonal S-type, the epizonal S-type and the fractured-metamorphosed mylonite granites fall on the evolutional line of the "granite series of crustal transformation" of L.V. Tauson in the Ternary diagram between F (Ba+Sr), and (Rb+Li) (Fig. 4). Each type occupies a definite segment and is transitional to the next. This indicates that the three types of rocks are the products of the transformation of continental-crust in different maturity stages.

Because the fractured-metamorphosed mylonite granites were not homogenized, the whole-rock Rb and Sr isotopes are distributed as scattered points in the isochron diagram. The mesozonal S-type granites do not give strict isochrons but show a band-shaped distribution of obvious trend (Fig. 5). The slope represents the approximate

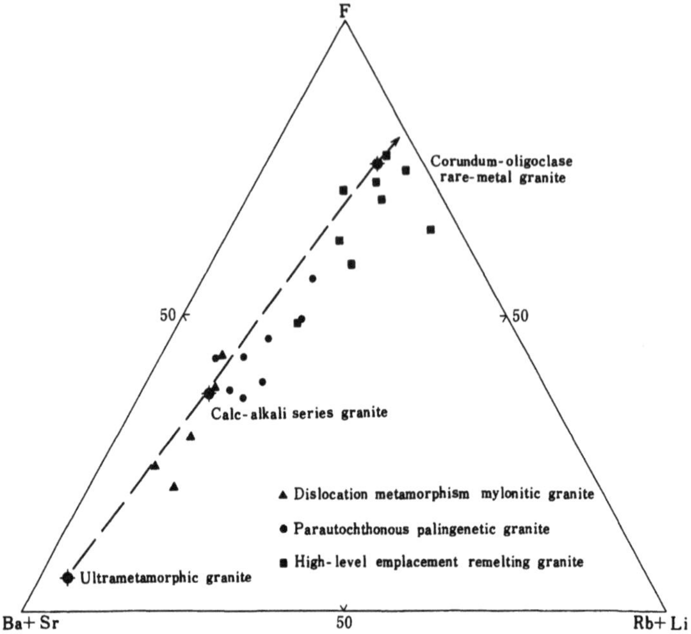


**Fig. 4.** F-(Ba+Sr)-(Rb+Li) diagram of granites in west Yunnan

age of the rock body, while the varying interval of the initial values of $Sr^{87}/Sr^{86}$ might indicate the inhomogeneity of the constitutents of the source rock. The epizonal S-type granites have an excellent whole-rock Rb/Sr isochron in their lithofacies of individual stages (Fig. 6). Their age values vary by about 10 Ma and their initial $Sr^{87}/Sr^{86}$ valves approximate each other. All these features characterize the formation of consanguineous and contemporaneous rocks at various stages.

## 3 The Relations Between Different Types of Granites and Tin Deposits in Western Yunnan

In western Yunnan, tin mineralization of variable intensity and kinds can be found in nearly all varieties of granites, except for the alkali granites.

Tin mineralization of the mesozonal S-type granites occurs mainly in the hydrothermal alteration zone along the mylonite zone within the rock body, most appearing as small and dispersed cassiterite-quartz veins or greisen veins, locally as clusterd veins or greisenized granites, forming small-sized tin deposits genetically related to hydrothermal solution of partially remelted magma of the mesozonal granites.

Most of the major tin deposits have genetic relations with the epizonal anatectic granites. Their mineralization process is evidently controlled by the differentiation and the hydrothermal activity of the magma. The isotope determination indicates a span of only four million years from the early stage of magmatic condensation to the postmagmatic hydrothermal mineralization stage (cassiterite-greisen stage). This proves that there exists a continnuous evolution process from magma crystallization to mineralization.

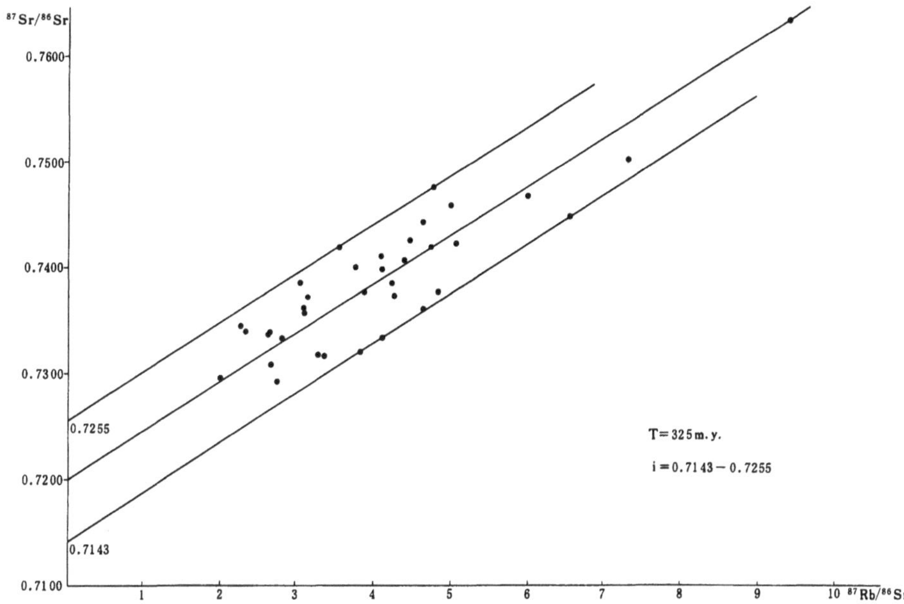


**Fig. 5.** Whole-rock Rb-Sr isochron diagram of mesozonal S-type granite (Licang rock body)

The types of tin deposit related to the anatectic granites are as follows:

1) Pegmatite-cassiterite type occurs mostly in the cupolas of plutons or penetrates the country rocks, forming large veins or vein swarms. Cassiterite appears as black coarse-grained aggregates in Na-metasomatic parts of pegmatites, distributed unevenly and associated with topaz, microlite, columbite, beryl, etc.
2) Cassiterite-rare metal granitic type occurs in fine-grained alkali-feldspar granite containing Li-muscovite in the late stage of differentiation. The contents of Li, F, and Rb are relatively high in ore-bearing rocks. The cassiterite is associated with the columbite, ilmenorutile, tapiolite, microlite, etc. The existence of large but low grade rare metal-cassiterite deposits is quite probable.

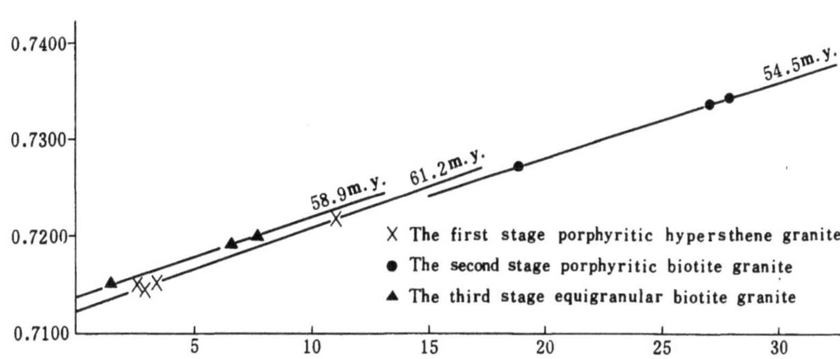


**Fig. 6.** Whole rock Rb-Sr isochron diagram of consanguineous, contemporaneous and multistage anatectic complex rock bodies

3) Cassiterite-greisen type is the most predominant type of tin mineralization in western Yunnan; usually it occurs in the endo- or exo-contact zone between the muscovite-biotite alkali-feldspar granite and the country rocks. The greisen-cassiterite veins or the cassiterite-sulphide-greisen type formed through remineralization with dominant pyrites have been observed. In the vicinity of the remnant cap of the country rock above the cupolas of plutons, plane-type mineralization often occurs, with cassiterite-greisenized granite in the lower part and tin-bearing skarn or hornfels superposed by greisenization in the upper part.
4) Quartz-cassiterite and quartz-tourmaline-cassiterite vein type occurs in the crushed zones of the rock body and its country rocks. The former is preferentially developed in the west zone, occurring generally in association with greisen or in the form of compound veins. The latter represents the predoninant pattern of tin mineralization in both the east and middle zones, mostly associated with large amounts of sulphide such as pyrite and pyrrhotite in addition to tourmaline.
5) Tin-bearing polymetallic skarn type is mainly distributed in the eastern subzone of the west zone, occurring at the contact of Early Yenshanian epizonal granites. The tin occurs in the metallic or calc-silicate minerals as micro-grain cassiterite or impurity in the crystalline structures and may locally form small cassiterite-bearing greisenized and sericitized orebodies after undergoing the late hydrothermal process.

The fractured-metamorphosed mylonite granite has an extremely close spatial association with tin deposits. Its mineralization occurs along fissures or in the interlayer crushed zone of the country rocks near the granite. It is a type of quartz-tourmaline-cassiterite deposit containing sulphides such as pyrite, arsenopyrite, etc. The mylonitic granite associated closely with the deposit has neither the characteristics of anatectic magma nor intrusive characteristics, and lacks contact metamorphic phenomena. The cassiterite shares similar cataclastic characteristics with the rock-forming minerals in the mylonite granite. The ore has a negative $^{34}S$ value, and its mineralization may be caused by metamorphism-hydrothermal activity simultaneously with the alteration of the mylonite granite. Some granite porphyries and quartz-porphyries within the calc-alkaline hypabyssal rocks show special alteration phenomena characteristic of tin porphyry. Such altered porphyry not only forms considerable geochemical anomalies but also contains cassiterite, wolframite, molybdenite and pyrite, occurring in fine disseminated, spotted or veinlet forms in the rock. The content of Sn and Mo may locally reach commercial grades.

In western Yunnan, a variety of granites and tin deposits are formed through long and complex crustal orogeny. The authors have made a preliminary approach to the regularities of the formation of granites and tin deposits, and have attempted to correlate in general with the tin belts of Southeast Asia.

# 6.1.4 An Approach to Ore-Forming Characteristics and Metallogenic Model of the Granites Emplaced Through Tectonic Remelting in the Yunlong Tin Belt, Western Yunnan

Zou Shu, Lin Yongcai and Gao Zepei[1]


## Abstract

This paper deals with the multi-phase and ore-containing setting of the granite in western Yunnan from the viewpoint of plate tectonic activity. A discussion of the genesis of the tin-bearing granite terrain is made on the basis of the time of intrusion, mode of occurrence, petrochemistry, trace elements, rare-earth elements, strontium isotopes and formation temperature of the terrain. In combination with the analysis of the mineralization characteristics, the authors conclude that the host rock for mineralization represents a remelted emplaced granite whose metallogenic model lies in the concept that the plate tectonic activity gave rise to the segregation of tin-containing material to form an initial tin-rich belt, which through remelting gave birth to tin-bearing magma that converted into ore-forming magma by means of emplacement and further evolution, and finally orebodies were formed in the favourable space at the contact with the granites through transportation and precipitation.


## 1 General Geology

The Sanjiang tin ore belt of western Yunnan is strictly controlled by three deep fractures – Jinsha River, Lancang River and Nujiang River. It is connected with the Thailand-Burma tin belt to the south, and extends in a WNW direction beyond the territory of China through Tibet and Xinjiang. It falls within the same category as the famous Malaysian metallogenetic zone, as shown by its similarities to the latter in respect of tectonics, magmatic activities, and mineralization characteristics (Fig. 1).

The Yunlong tin belt, an important mineralization region, is located at the northern end of the Lancang tecto-magmato-metamorphosed zone within the Sanjiang tin belt. Its areal extent (Fig. 2) starts from Guangshan in the south, running through Tiechang, and ends round the terrain of Thibenshan and Shiganghe, with an extension of more than 30 m. It has a width of 5—15 km, bounded on the east by Caujian-Luyintang fault, and on the west by the Baoshan back-arc basin which is split by the Qiangjie-Wenjuan fracture. The ore belt, striking 330—340° parallel to the regional lineament, occurs in a triangular configuration between the Lancang River (Lancangjiang) deep fracture and the Wenquan fracture.

[1] No. 301 Geological Party of Southwest China, Geological Exploration Corporation, CNNC

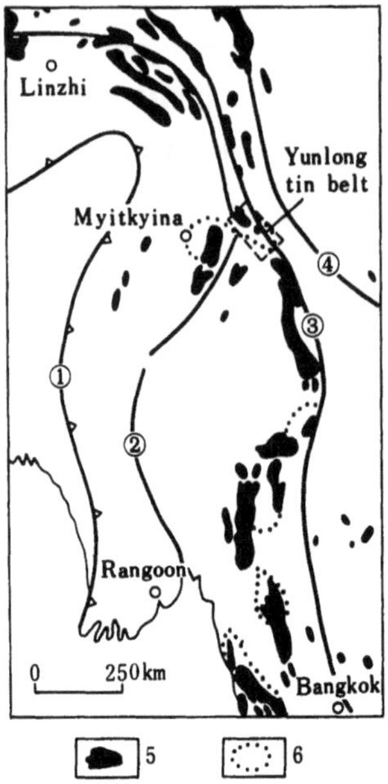


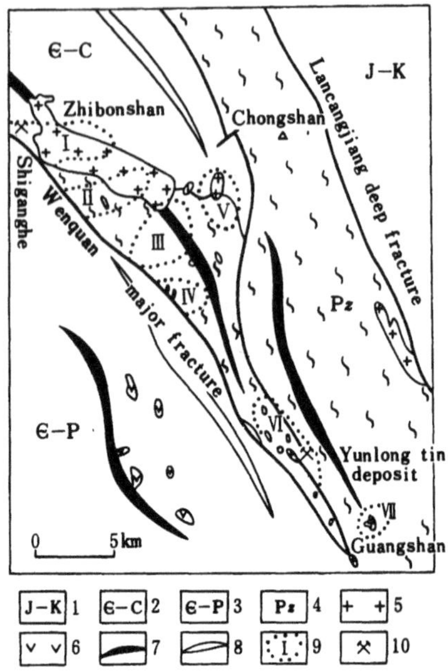


**Fig. 1.** Tectonic and magmatic distribution of the western Yunnan-Thailand-Burma tin belt. 1-Plate suture line; 2-Nujiang great fracture; 3-Lancang-River deep fracture; 4-Honghe River suture line; 5-Granite; 6-Tin ore field (zone)

**Fig. 2.** Sketch map of geological structures of the Yunlong tin belt, western Yunnan. 1-Jurassic-Cretaceous; 2-Cambrian-Carboniferous; 3-Cambrian-Permian; 4-Palaeozoic metamorphic rocks; 5-Granite; 6-Basic rock; 7-Anticline; 8-Syncline; 9-Heavy mineral cassiterite anomaly and its serial number; 10-Orebody location

Seven anomalous (mainly Sn) areas have been detected in the region, among them the priority of evaluation was given to anomalies No. 1 and No. 4 for which economic results were obtained.

## 2 Plate-Tectonic Activity and Tin Mineralization Setting

On the western side of the Yunlong tin belt lies the suture between the Indian and the Eurasian plates, and on the eastern side the Lancang River fracture.

The western Yunnan subplate, with the Lancang River fracture as the subduction zone, started to be subjected to multi-stage subduction and compression at the Jinning period 1,000 ± 0.5 Ma ago, and ended at the late stage of the Indosinian movements 195 ± 5 Ma ago. This led to the extinction of oceanic crust and the collision of continental crust, the increasing maturity of the Tengchong volcanic island-arc zone, and the formation of the Baoshan back-arc basin (Fig. 3).

Accompanying the plate activity was the emplacement of granitic magma, for instance the composite Lincang granitic batholith gives dates in the range 715—194 Ma (Fig. 4). The porphyritic granite (585.8 Ma) and gneissic granite (236—194 Ma) represent thermal events in the plate activity.

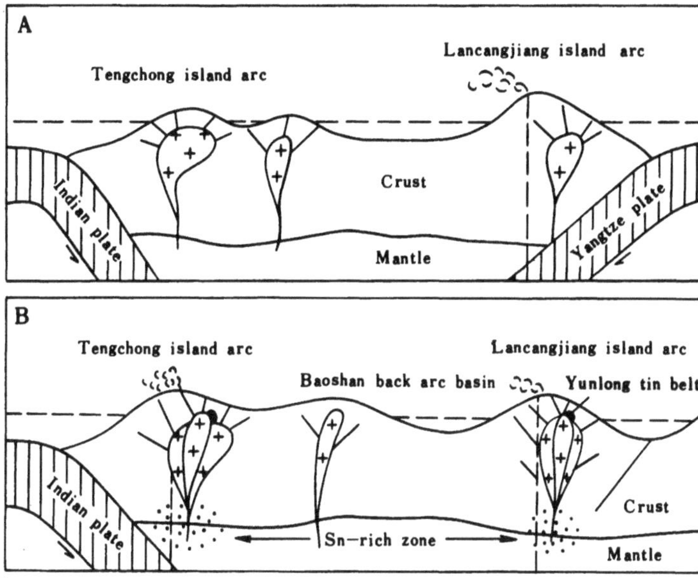


**Fig. 3.** **A** Diagrammatic cross section sketch of subplate activity in western Yunnan during the Jinning-Indosinian stage; **B** Relationship between Yenshanian tin-bearing granite and plate activity

During the middle and late stages of the Yenshanian movements (137 ± 3 to 67 ± 3 Ma) under the impact of subduction of the Indian Plate, which turned from northward to eastward on the western Yunnan continental crust (Fig. 3-b), a series of tin-bearing granitic bodies came into being along the intercontinental rift. Examples of

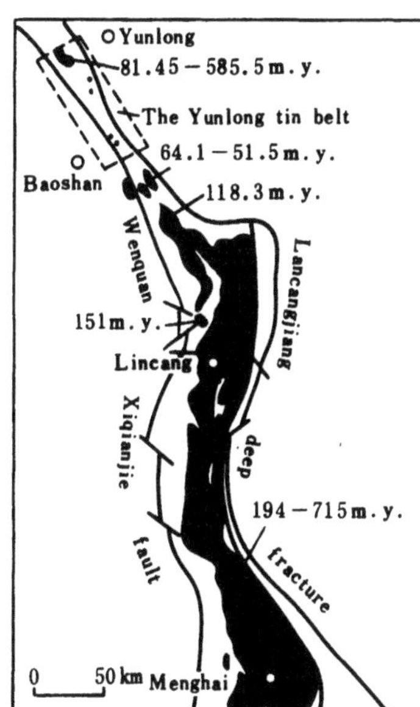


**Fig. 4.** Distribution of granitic bodies along the Lancang-River tectonic belt

these are the Yangtoyen rock body (151 Ma) and the Pinghejie rock body (118.3 Ma) occurring at the middle segment of the Lancang-River structural-magmatic-metamorphic zone, and the Zhibenshan, Zhenggoushan and Tiechang-Guanshan rock bodies in this region (112.9—81.45 Ma) (Fig. 4).

In the Himalayian period (less than 80 Ma), emplacement of a tin-bearing granite took place (64.1—51.5 Ma).

It is quite obvious that plate activity led to the formation of a volcanic arc with relatively high background tin values, causing the formation of an initial tin-rich belt, and also the emplacement of tin-bearing granite. This has established the material basis for the tin deposits.

## 3 Features of Ore-Forming Rock Bodies

1. The Zhibenshan and Chenggoushan rock bodies are in the form of stocks, having intrusive contacts with the country rocks, which produced hornstone aureoles several meters to 80 m wide when in contact with sandstone and slate (Fig. 5-a). The emplacement of the Tiechang-Guangshan rock body along the Caojian fault extends in a belt-shape in plane section, with evident protrusion at the top (Fig. 5-b). Three late-stage vein clusters exhibit a close relation with tin deposits (Fig. 5-c).
2. The ore-forming rock bodies are mostly composite and multi-staged, such as the Zhibenshan rock body, which is known to have three age values, 585.8, 112.9, and 81.45 Ma, and has marked contemporaneous differentiation. Belonging to the late stage of the Yenshanian movements, the main part of the rock is of biotite granite, the transitional part of two-mica monzonitic granite, and the periphery of fine-grained muscovite granite. Formed along with the magmatic differentiation are pegmatite, aplite, felsic, albite, and quartz veins, among which the aplite and quartz veins have a close relation with tin deposits.
3. From Fig. 6 it can be seen that the $SiO_2$ of the rock bodies of the region is 74.37—74.65%, higher than that of normal granite of the world (70.18%), and approaches that of the Laochang rock body of Gejiu (74.11%) and that of the tin-bearing granites of the world (73.10%). The content of $K_2O + Na_2O$ is 7.24—7.61%, approaching that of tin-bearing granites of the world (7.59%). Meanwhile, the $K_2O/Na_2O$ ratio ranges between 1.27 and 1.5, also very close to that of the Laochang (1.44) and that of the

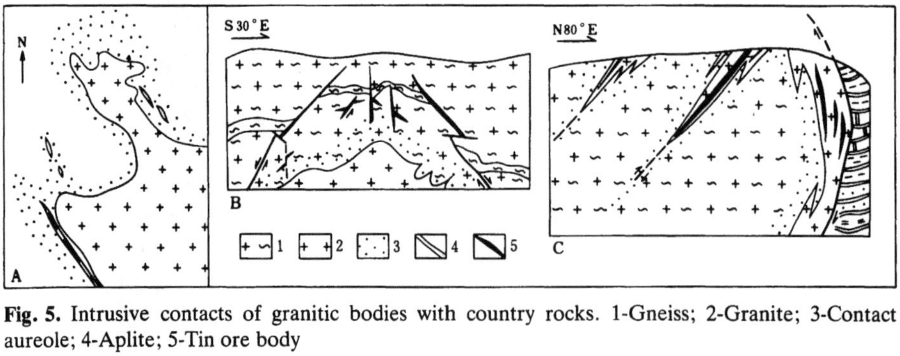


**Fig. 5.** Intrusive contacts of granitic bodies with country rocks. 1-Gneiss; 2-Granite; 3-Contact aureole; 4-Aplite; 5-Tin ore body

| Rock body | Lithology | Sample number | Chemical composition (%) | | | | | | | | | | | | | Parameters | |
|---|---|---|---|---|---|---|---|---|---|---|---|---|---|---|---|---|---|
| | | | $SiO_2$ | $TiO_2$ | $Al_2O_3$ | $Fe_2O_3$ | FeO | MnO | MgO | CaO | $Na_2O$ | $K_2O$ | $P_2O_5$ | $H_2O$ | Total | $K_2O/Na_2O$ | A/CNK |
| Zhibenshan rock body | Two-mica monzonitic granite | 20 | 74.65 | 0.07 | 13.24 | 1.39 | 0.97 | 0.06 | 0.26 | 0.81 | 2.90 | 4.34 | 0.17 | 0.57 | 99.43 | 1.50 | 1.28 |
| Tiechang rock bodies | Striped granite | 48 | 74.37 | 0.21 | 12.41 | 0.69 | 1.27 | 0.02 | 0.69 | 0.93 | 3.35 | 4.26 | 0.25 | 0.40 | 98.95 | 1.27 | 1.04 |
| Average | | 68 | 74.51 | 0.14 | 12.83 | 1.02 | 1.12 | 0.04 | 0.48 | 0.87 | 3.13 | 4.30 | 0.21 | 0.49 | 99.14 | 1.37 | 1.13 |
| Gejiu-Laochang rock body | Biotite granite ① | | 74.11 | 0.38 | 12.81 | 0.79 | 1.28 | 0.05 | 0.22 | 0.90 | 3.49 | 5.04 | 0.23 | | | 1.44 | 1.00 |
| Tin-bearing granites in the world ② | | 185 | 73.10 | 0.21 | 13.96 | 0.91 | 1.44 | 0.05 | 0.55 | 1.21 | 3.01 | 4.58 | 0.20 | | | 1.52 | 1.16 |
| Normal granites in the world ③ | | 546 | 70.18 | 0.39 | 14.47 | 1.57 | 1.78 | 0.12 | 0.88 | 1.99 | 3.48 | 4.11 | 0.19 | | | 1.18 | 1.04 |

| Zhibenshan rock body: Locality \ Parameters | Ca | Na | K | Q | Ab | Or |
|---|---|---|---|---|---|---|
| Central phase | 14 | 34 | 52 | 36 | 33 | 31 |
| Transitional phase | 13 | 35 | 52 | 43 | 28 | 29 |
| Marginal phase | 20 | 55 | 25 | 43 | 45 | 12 |

① From Wu Qinsheng

②③ From Stempron and Skvor

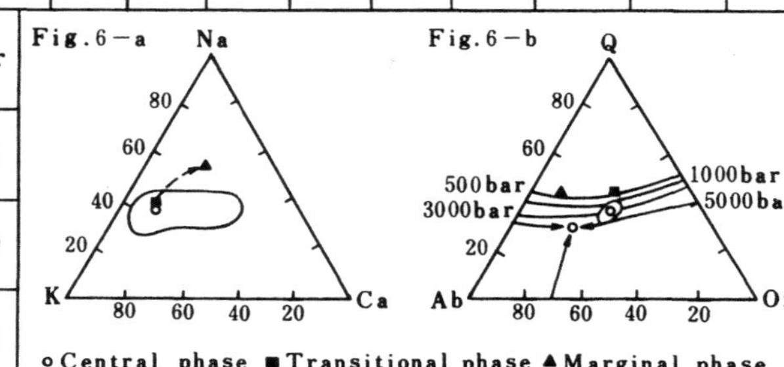


**Fig. 6.** Lithochemistry of tin-bearing granite in the Yunlong tin belt

tin-bearing granites of the world (1.50), higher than that of normal granites of the world (1.18). The value of the A/CNK ranges from 1.04 to 1.28, nearly the same as that of normal granite (1.04) and tin-bearing granite (1.16) of the world and the Laochang rock body of China (1.00). The rock is obviously poor in Ca, Fe, and Mg, which indicates the highly acidic and alkali-rich peraluminous nature.

When the Ca, Na, and K values of the 25 samples of the Zhibenshan rock body are projected on triangular diagrams, they all fall into the magmatic granite area (Fig. 6-a). What is more interesting is that when the six normative minerals of the biotite granite samples from the main part of the Zhibenshan rock body are projected on Tuttle's Q-Or-Ab triangular diagram, they all fall within the low-temperature depression. Around the depression are the 19 samples of the transitional part. Samples from the peripherial rock body, affected by the hydrothermal fluids, appear far beyond the depression (Fig. 6-b). In describing the normative $Q + Or + Ab \geqslant 80$ falling within and around the low-temperature depression, the Chinese geologist Cong Loling has said: "this indicates that the granite was crystallized from magma or fluids produced by melting of rocks (mainly sedimentary and metamorphic rocks) at depth. Such regularity would never have come into existence if the formation of granite had been caused by gaseous or liquid metasomatism of the original rocks". The Japanese scientist Yagishozo holds that the result of the study on the low-temperature depression has put an end to the long dispute amongst petrologists whether granite is crystallized from magma or formed by granitization of the original rock. The authors of this paper consider that during the formation of the Zhibenshan rock body there did exist a magmatic evolution stage characterized by the lowest Ab-Or-Q three element value.

4. As listed in Table 1, the contents of minor elements such as Sn, Bi, Mo, B, and Nb in the rock body seem by far higher than those in the crustal granite. The concentration coefficient of Sn is higher than 16.67, whereas the contents of iron group elements V, Cr, Co, and Ni, and the medium-low temperature ore-forming elements Cu, Pb, and Zn are lower than those of the crustal granite. This is identical with the lithochemical compositions of highly-acidic and alkali-rich rocks.

5. In Table 2, the total REE content of the rock bodies of Yunlong tin belt is 134.92 ppm, 25.5 times that of chondrite (5.3 ppm). Light and heavy REE ratio $\Sigma Ce/\Sigma Y$ gives a value of 2.44, which is 2.9 times that of chondrite ($\Sigma Ce/\Sigma Y=0.84$). The chondrite-normalized REE pattern of the rock bodies exhibits intense deficiency of Eu, forming a V pattern (Fig. 7)

6. The initial $Sr^{87}/Sr^{86}$ values range 0.7085—0.7208, higher than that of mantle materials (0.69898 ± 0.00003), upper mantle (0.7015—0.710), submarine mantle (0.702 ±

**Table 1.** Minor elements in metallogenetic rock mass, Yunlong tin belt (ppm)

| Name of rock body \ Element | V | Cr | Co | Ni | Cu | Pb | Zn | Sn | W | Bi | Mo | B | Nb |
|---|---|---|---|---|---|---|---|---|---|---|---|---|---|
| Ore-forming rock body, Yunlong belt | 16 | 15 | 10 | 1 | 10 | 15 | 100 | 50–200 | 20–60 | 85–211 | 1–6 | 476–956 | 50–90 |
| Ca-rich crustal granite | 44 | 41 | 1 | 15 | 30 | 15 | 60 | 15 | 1–3 | | 1 | 9 | 20 |
| Ca-poor crustal granite | 30 | 2 | 1 | 45 | 10 | 19 | 39 | 3 | 22 | 0.01 | 73 | 10 | 21 |

**Table 2.** REE in metallogenetic rock mass, Yunlong tin belt

| Name of rock body / REE | La | Ce | Pr | Nd | Sm | Eu | Gd | Tb | Dy | Hd | Er | Tm | Yb | Lu | Y | Σ | Rock Σ / Chondrite Σ | Σ Ce / ΣY | Rock Σ / Chondrite Σ |
|---|---|---|---|---|---|---|---|---|---|---|---|---|---|---|---|---|---|---|---|
| Zhibenshan (2) | 8.47 | 18.80 | 2.16 | 8.03 | 2.59 | 0.23 | 2.77 | 0.56 | 3.76 | 0.3 | 1.87 | 0.29 | 1.92 | 0.34 | 20.94 | 13.66 | | 1.41 | |
| Tiechang (2) | 33.84 | 67.35 | 8.07 | 26.36 | 6.03 | 0.54 | 5.97 | 1.10 | 6.14 | 1.5 | 2.66 | 0.57 | 2.69 | 0.38 | 33.00 | 196.20 | | 3.09 | |
| Average (4) | 21.16 | 43.08 | 5.12 | 17.20 | 4.31 | 0.41 | 4.37 | 0.83 | 4.95 | 1.15 | 2.27 | 0.43 | 2.31 | 0.36 | 26.97 | 134.91 | 25.5 | 2.44 | 2.9 |
| Chondrite (9) | 0.33 | 0.88 | 0.112 | 0.60 | 0.181 | 0.068 | 0.244 | 0.047 | 0.338 | 0.070 | 0.200 | 0.030 | 0.200 | 0.034 | 1.96 | 5.300 | | | |

* From A.G. Herman

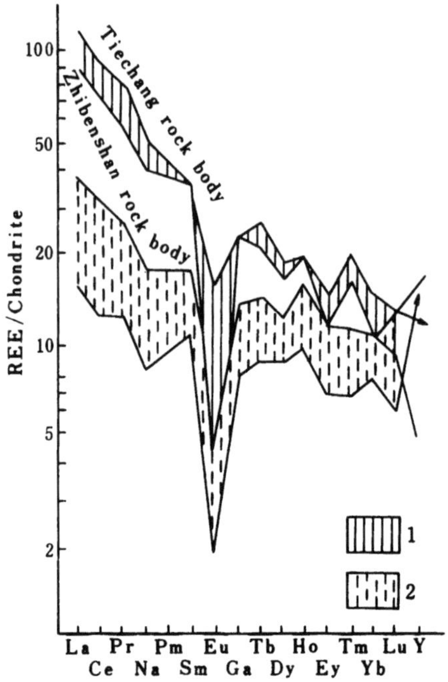


**Fig. 7.** Chondrite-normalized REE pattern of the Yunlong tin belt. 1-Tiechang; 2-Zhibenshan

0.705), and island basalt (0.7037 ± 0.0001). They fall mainly within the area of crustal substances (Fig. 8).

7. The rock body is mainly characterized by magmatic crystalline and vitreous inclusions for its major part, the homogenization temperature being higher than 800°C, which represents the diagenetic temperature. The peripheral part of the rock is mainly of gaseous and fluid inclusions, the homogenization temperature ranging between 437 and 196°C, which represents the temperature for the formation of residual magma and ore fluid.

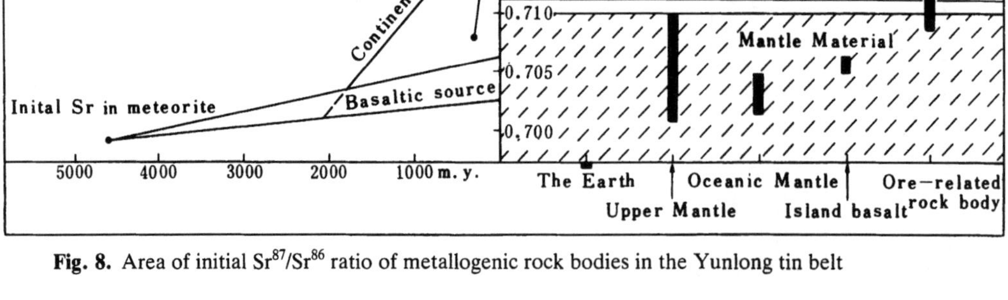


**Fig. 8.** Area of initial $Sr^{87}/Sr^{86}$ ratio of metallogenic rock bodies in the Yunlong tin belt

## 4 Mineralization Features

From Fig. 2 one can see that the Yunlong tin belt can be divided into two mineralization zones.

The mineralization of the southern zone is immediately controlled by the late intrusive fine-grained granitic veins. Three rock vein clusters have three corresponding groups of veined tin orebodies (Fig. 5-c). The middle rock vein cluster is largest in size, and the mineralization is consequently more intense, forming large concentrated orebodies. Several dozens of orebodies in this part appear to be concealed or blind underlying a depth of 7—130 m or more. The orebodies are in right-handed oblique orientation from west to east in en echelon arrangement in plane. A similar feature can also be observed on the profile, with most of the main orebodies inclining towards the west and laterally plunging towards the north.

The large concentrated orebodies in the southern zone are characterized by superimposed multistage mineralization as shown in Fig. 9. The mineralization in the upper part is of cassiterite-quartz type and cassiterite-tourmaline type at an elevation above 2,300 m. At 2,240—2,300 m above sea level we find the cassiterite-sulphide type; however, below the height of 2,240 m quartz-cassiterite type is observed. Thus a vertical zoning of different types of mineralization is formed (Fig. 9).

Intense mineralization sites are found to lie along the semiclosed compresso-shear faults and nearby fissure belts, or at the joining part of two or more fractures. Rich tin ore pockets are very common.

The dominant mineralizations are of the cassiterite-tourmaline-quartz and cassiterite-sulphide types.

The metallic minerals include cassiterite, arsenopyrite, pyrrhotite, pyrite, minor chalcopyrite, scheelite, bismuthinite, sphalerite and galena. The gangue minerals are mainly quartz, magnesium tourmaline, iron tourmaline, etc.

The early-stage cassiterite is of light colour, mostly crushed into brecciated forms which indicates the structural compression and shearing in the later stage of ore-formation. Such cassiterite contains 0.01—0.06% of Nb, but no Ta. The temperature for its formation ranges between 230 and 600°C. However, the late-stage cassiterite is of deep colour. It has perfect crystal forms and inclusions of early-stage cassiterite. Its formation temperature is 185—205°C.

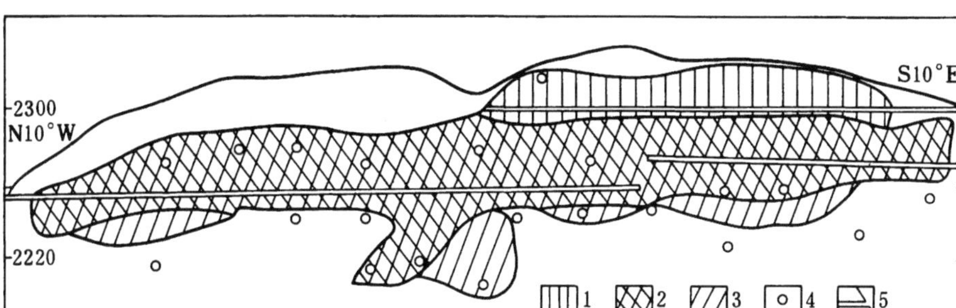


**Fig. 9.** Vertical zoning of ore types of the large rich orebodies of the Yunlong tin deposit, western Yunnan. 1-Cassiterite-tourmaline-quartz vein type; 2-Cassiterite-sulphide type; 3-Cassiterite-quartz type; 4-Borehole; 5-Tunnel

The pyrite which has close connection with the tin deposit can be grouped into two stages. The first-stage pyrite contains 0.1% Sn, formed at the temperature of 270–290°C, while the second-stage pyrite contains 0.01–0.005% Sn, formed at the temperature of 220–230°C.

The content of Sn in pyrrhotite is 0.1%. The formation temperature is greater than 290°C.

The arsenopyrite contains 0.4% Sn, whereas that far away from the orebodies contains only 0.05%. The temperature for its mineralization ranges 175–345°C.

The quartz closely related to cassiterite may fall into two stages. The first-stage quartz is associated with tourmaline and has wavy extinction. The formation temperature was 350–450°C. The second-stage quartz is associated with sulphides such as pyrite, formed at a temperature of 200–345°C.

The magnesium tourmaline closely related to tin deposit contains 0.4–1.9% Sn. Its mineralization temperature was 200–290°C. Iron tourmaline is the product of early mineralization.

The mineralization of the northern zone is strictly controlled by the Zhibenshan rock body. The mineralization took place along the endo- and exocontact zones of the rock body. As shown in Fig. 10, the Zhibenshan rock body occurs in the shape of a hat brim, which has two mineralizations above and below the elevation of 2,300 m respectively. Above this level occurs the Zhibenshan Sn ore, where the mineralization is controlled by the cold tensile fissure group, while below is the Shiganghe tin deposit where the mineralization is controlled by the tongue of the rock body, i.e. the front of the emplacement of the rock body, forming three prospective mineralization zones stretching round the tongue.

The most remarkable feature of the northern zone is characterized by the minerogenic combination of three components: ore-localizing structure, aplite veins, and cassiterite-quartz veins. Without any one of the three there will be no possibility of mineralization. Another feature is the zoning of ore-forming elements: on a plane section mainly Sn at the center, gradually transitioned to Sn, W, and mainly W in the outer part. The cross section shows a format of W in the upper and Sn in the lower.

The intensely mineralized portions are the front of the emplacement of the rock body (the tongue of the rock); the concave part caused by the contact between the rock body and the country rock; and the place where the minerogenic rock veins are overlain by a sandstone or slate bed as a screening layer. All the rich and thick orebodies occur in these positions.

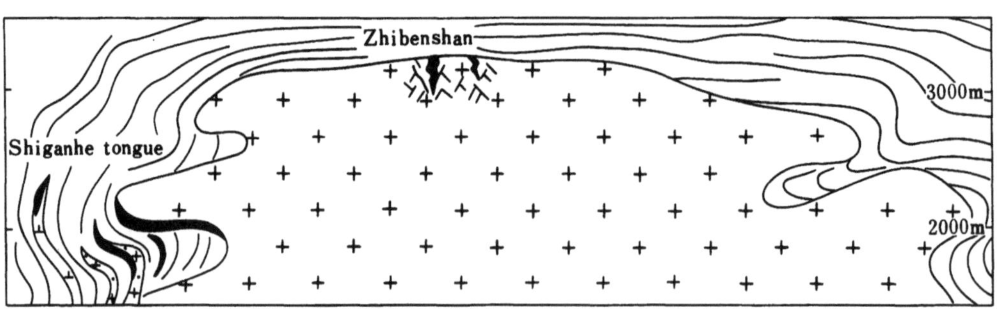


**Fig. 10.** Longitudinal projection of the Zhibenshan granitic body

The predominant mineralization is of cassiterite-quartz type.

The metallic minerals include cassiterite, wolframite, scheelite, pyrite, arsenopyrite, etc. The gangue minerals include magnesium tourmaline, iron tourmaline, quartz, etc.

The cassiterite of the early stage is of brownish colour, containing about 0.01—0.06% Nb, its mineralization temperature is higher than 350°C. That of the late stage is of rose colour, with a mineralization temperature of 280—340°C.

The earlier quartz is associated with iron tourmaline, with a mineralization temperature 280—330°C, whereas the late quartz with magnesium tourmaline and rosy cassiterite had a temperature of 185—276°C.

## 5 Metallogenesis of the Tin Deposit

Under the conditions of complex geological settings and the long duration and extensive tectonic activities around the Yunlong tin belt, the tin minerogenic rock body and its country rock had been metamorphosed to different extents. Therefore it is not surprising that this has aroused all kinds of arguments among the researchers on the metallogenesis of the tin belt.

The authors have concluded from their field and microscopic studies of many years that the occurrence of the deposits of this tin belt is on the whole controlled by the rock body and geologic structures. This can be verified by the following:

1. The uniformity of the spatial situation of the rock body, structures and mineral deposit.
2. The evolutionary inheritance relationship between rock body and ore body.
3. The consanguinity relationship between rock body and mineral deposit in their inclusion types and compositions.

It can be confirmed that the anatectic granite emplaced during the late Yenshanian movements should be the parent rock, and the genesis of the mineral deposit should be of postmagmatic high-temperature pneumato-hydrothermal to medium-temperature hydrothermal type. This conclusion is based on: a) the intrusive contact between the rock body and the country rock; b) magmatic inclusions and vitreous inclusions in metallogenetic rock body representative of undergoing the magmatic stage; c) similar lithochemical composition and minor element association to that of tin-bearing granite in other places of China and foreign countries; d) serious Eu deficiency and V pattern of the REE, indicating the crustal source and the complete evolution; d) initial Sr isotope ratio representing the melting of the crust; and f) rocks of the main part falling in the magmatic granite area of the Ca-Na-K triangular diagram of Raju and in the low-temperature depression of the Q-Or-Ab triangular diagram of Tuttle.

## 6 Metallogenic Pattern

Based on the foregoing, the metallogenic pattern of the Yunlong tin belt is shown in Fig. 11. The concept is that the multi-staged acitivity of the subplate formed the initial

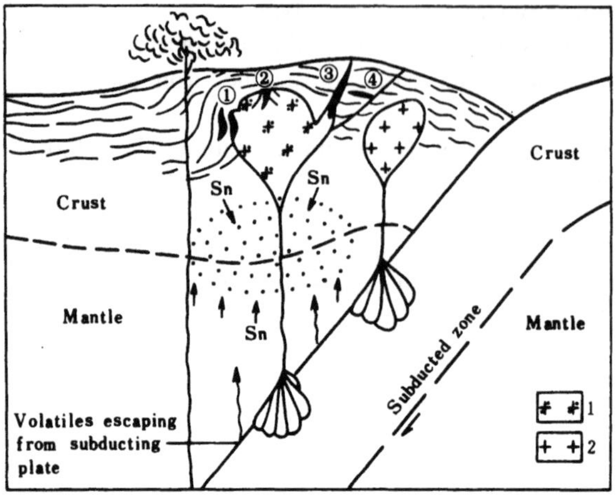


**Fig. 11.** Mineralization model for the Yunlong tin belt. 1. Tin-bearing granite; 2. Tin-deficient granite. (1) = Shiganghe pattern (controlled by the shape of the rock body and the interstratal structures of the country rock); (2) = Zhibenshan pattern (controlled by the tensile fractures at the top of the rock body); (3) = Tiechang pattern (controlled by rock veins and compresso-torsional faults); and (4) = Liziping pattern (controlled by interstratal crushed zones at intersections with ore-guiding structures)

tin-rich belt, and its anatectic melting resulted in the formation of tin-bearing magma whose emplacement, differentiation and evolution led in turn to the high concentration of tin and volatile constituents until the eventual precipitation of cassiterite in favourable ore-storing spaces along the inner and outer contact zones of the rock body. As a result, the following ore deposits came into existence:

1. The Zhibenshan tin deposit controlled by the tensile fractures at the top of the rock body;
2. The Shiganghe tin deposit controlled by the shape of the rock body and the interstratal structures of the country rock;
3. The Tiechang tin deposit controlled by rock veins and compresso-torsional faults; and
4. The Liziping tin deposit controlled by interstratal crushed zones at intersections with ore-guiding structures.

These four mineralization patterns can be found similarly in other tin depositional locations in west Yunnan. Guided by this concept, we have made ore-forecasting and evaluation of the ore potential in the northern part of the tin belt, whose correctness has later been fully substantiated.

## 7 Conclusions

1. The tin deposits within the belt are closely related in time and space to magmatic granite;
2. The tin-bearing magma can only be formed through the anatectic melting of an initially tin-concentrated belt;
3. The thorough differentiation and evolution of the emplaced tin-bearing magma is an indispensable factor for the concentration of tin; and
4. Semi-enclosed tectonic fracture systems are the optimum sites for the occurrence of tin deposits.

# 6.1.5 Time-Space Distribution of Tin/Tungsten Deposits in South China and Controlling Factors of Mineralization

XU KEQIN and ZHU JINCHU[1]


## Abstract

The abundant tin and tungsten deposits in South China were formed in different geologic ages. They began to appear in the late Protereozoic, and reached their paramount development during the Yenshanian period. In regional distribution the W deposits tend to occur in the eastern part of this region, mostly in post-Caledonian uplifts, while the Sn deposits tend to occur in the western part, mostly in Hercynian-Indosinian depressions. Both of them are found either within the middle-late Proterozoic and lower Paleozoic basemental structure, or adjacent to the basement rocks. Genetically, they are mainly associated with continental crustal anatectic granitoids. The following factors are considered to be most important in controlling the time-space distribution of Sn/W deposits: regional existence of ore source beds of middle-late Proterozoic and early Paleozoic ages; crustal mechanism in the generation of multi-cyclic granitoids and their evolution; activities of ore-bearing hydrothermal solutions; and localization of Sn/W mineralization in close relation to structural environment and wall rock properties.


## Introduction

This paper concerns an extensive region of South China, from the Yangtze River in the north to the Red River in the west and reaching coastal areas to the southeast. The core part of this region is South China's Caledonian fold belt: to the north and northwest is the Precambrian Donganian-Xuefengian fold belt; and to the east is the Zhejiang-Fujian-Guangdong Yenshanian volcanic belt and the Taiwan Himalayan fold system (Fig. 1).

South China has a prolonged and complicated history in its geologic development, characterized by multi-cyclic tectonic movements of the earth crust and multi-cyclic generations of granitic rocks. It has undergone intermittently Donganian, Xuefengian, Caledonian, Hercynian-Indosinian, Yenshanian and Himalayan movements. Furthermore, directly related to these orogenic movements, repeated generation of granitoids and Sn/W deposits of corresponding ages took place.

[1] Department of Geology, Nanjing University, Jiangsu

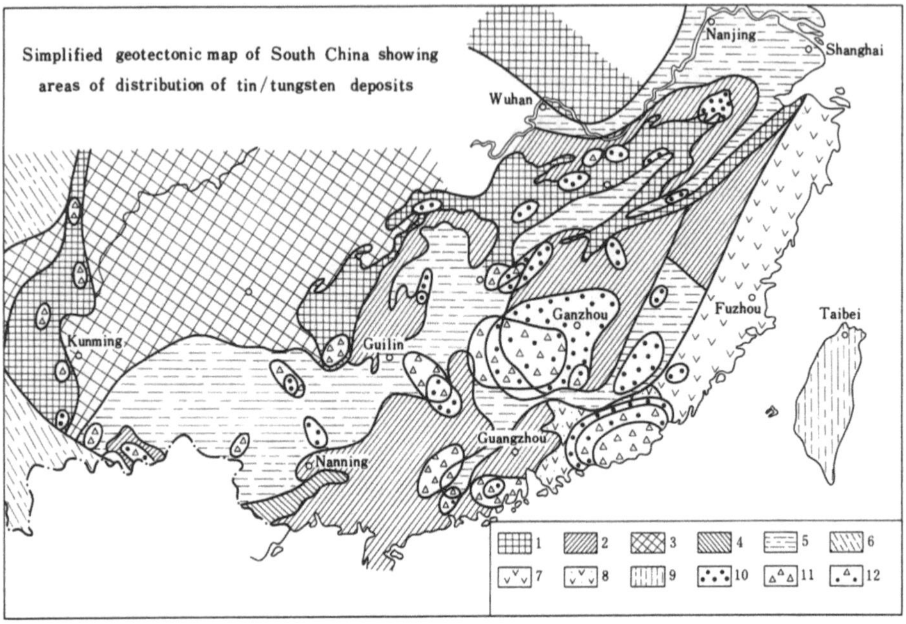


**Fig. 1.** Simplified geotectonic map of South China showing areas of distribution of tin/tungsten deposits

South China is one of the most important regions for tungsten and tin deposits in the world. The present paper deals particularly with the time-space distribution of Sn/W deposits, their relationship with ore source beds and granites, and localization of Sn/W mineralizations.

## Formation Ages of Tin/Tungsten Deposits

Most of the primary Sn/W deposits in South China are found to be genetically related to granitoids. Based on field evidence and isotopic determinations, a histogram of isotopic ages for 52 Sn/W-bearing granite bodies was plotted (Fig. 2). It is evident that the formation of Sn/W deposits in South China is multi-aged and bears some features of inheritance. The formation of Sn/W deposits began during the late Proterozoic,

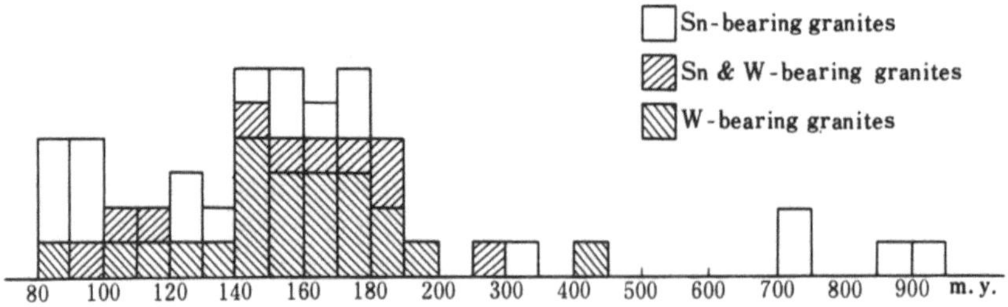


**Fig. 2.** Histogram of isotopic ages of 52 Sn/W-bearing granite bodies in South China

intermittently continued in the Paleozoic, and reached a culmination in the Mesozoic. Nevertheless, there also exist some differences in formation ages between tin and tungsten deposits.

Several late Proterozoic tin deposits with considerable commercial value have been found within the old Precambrian cratonic massifs. No economic Sn deposits of the Caledonian time have ever been observed. However, some cassiterite mineralization of placer character in Devonian sandstone of south Jiangxi was reported. Therefore, there is no reason to exclude the possibility of Caledonian Sn-bearing granites and Sn deposits. A few Hercynian-Indosinian tin deposits were discovered in Hunan and west Guangxi. The early Yenshanian tin deposits are usually accompanied by tungsten. The most important tin deposits in South China are late Yenshanian or Cretaceous with isotopic ages of 120—80 Ma. Especially significant are the late Cretaceous deposits with isotopic ages concentrated in the range of 80—95 Ma, which are mainly distributed in southeast Yunnan, northwest Guangxi, south Hunan and west Guangdong.

Independent tungsten deposits began to develop also in the late Proterozoic, but those with economic importance are few. The Caledonian W deposits occur occasionally in southwest Hunan. The ages of formation of stratiform and stratabound scheelite deposits and W-Sb-Au deposits found in late Proterozoic volcano-sedimentary sequences of northwest Hunan and Jiangxi are still not clear. It is quite probable that these deposits were formed through superimposed transformation of the W-rich late Proterozoic sequences by metamorphic hydrothermal solutions. The Hercynian-Indosinian tin deposits are usually accompanied by a limited amount of tungsten. The paramount development of commercial W deposits was reached somewhat earlier than that of Sn-deposits, i.e., it was reached during the early Yenshanian, or Jurassic, with the isotopic ages of 190—137 Ma. These W deposits are extensively distributed in south Jiangxi, south Hunan, north Guangdong and northeast Guangxi. Several early Yenshanian W deposits also contain considerable amounts of Sn, which are visible in the histogram. The late Yenshanian W deposits are far less important than early Yenshanian ones. They are mostly small-sized low temperature deposits.

The kind of multi-aged formation and inherited development of Sn/W deposits within a large tectonic region, and the tendency for their concentration associated with progressively younger aged granitoids, are also reported to occur in Southeast Asia and South America. This evidently shows that repeated transformation and differentiation-evolution of the continental crust are favourable conditions for the formation of tin and tungsten deposits.

## Spatial Distribution of Tin/Tungsten Deposits

Both tin and tungsten are typical granitophile elements. In geological processes, especially during the formation of granites and concentration of metals, these two elements have much in common and are therefore frequently associated with each other. On the other hand, due to some important differences of geochemical behavior, ore-formation mechanisms and features of ore source beds, these two elements show considerable separation.

In the spatial distribution of Sn/W deposits in South China the following points are worth mentioning:

1) South China as a whole is not only the most important region for tungsten deposits in the world, but is also one of the outstanding regions for tin deposits.
2) Within the territory of South China there is a tendency for concentration of W deposits in the eastern part, especially in Jiangxi, Hunan, Guangdong and east Guangxi, and Sn deposits in the western part, especially in east Yunnan and Guangxi (Fig. 1). The discovery of some Sn/W deposits in west Yunnan indicates that the Sanjiang Indosinian fold system is a northward extension of the Southeast Asia tin belt.
3) Within South China's Caledonian geosynclinal fold belt, the most important commercial W deposits are mainly distributed in the south Jiangxi post-Caledonian uplift and adjacent peripheral part of the Hercynian-Indosinian depression, while the most important economic Sn deposits are mainly distributed in the south Jiangxi post-Caledonian uplift and adjacent peripheral part of the Hercynian-Indosinian depression, while the most important economic Sn deposits are mainly distributed in the Hercynian-Indosinian depression. If one goes from Yudu, south Jiangxi, westward Linwu, south Hunan, the following regional variation in types of Sn/W deposits can be observed: W deposits of wolframite-quartz vein type→W and Sn deposits of wolframite-quartz vein type→skarn type scheelite deposits and cassiterite deposits→cassiterite-sulphide type Sn deposits.
4) Most of the W and Sn deposits in South China were formed either in the basement structural layer consisting of middle-late Proterozoic or early Paleozoic eugeoclinal volcano-sedimentary sequences, or in the peripheral part of the Hercycian-Indosinian depression adjacent to the folded basement.
5) Regional deep-seated faults and faults of second order as well as intensive and extensive shear zones or sheeted zones, related to some regional stresses, are important factors controlling the spatial distribution of Sn/W-bearing granites and deposits.

## Preliminary Concentration of Tin/Tungsten and Ore Source Beds

Why do the Sn/W deposits of South China have this time-space distribution? To answer this question, great attention has been paid to the detection of the preliminary concentration of Sn/W in the stratigraphic sequences on a regional scale.

The research work done by some faculty members and graduate students of Nanjing University in recent years has shown that the Sinian and Cambrian strata in Dayu, south Jiangxi, contain W as high as 7–80 ppm, and in Yudu area 16–63 ppm. However, in some other localities of south Jiangxi, which lack tungsten deposits, the stratigraphic sequences of the same age contain only 2 ppm W.

The research Group on Tungsten Deposits, Ministry of Metallurgy (1981) analysed more than 1,000 stratigraphic samples for ore-forming elements and found that on the Jiangnan Geanticline belt lying in north Guangxi, east Guizhou, west Hunan and north Jiangxi, the middle-late Proterozoic volcano-sedimentary sequences over 20,000 m thick contain 7.8–11.3 ppm W on the average, and 7.7–22.0 ppm Sn. In the

Caledonian uplift regions, the Sinian, Cambrian and Ordovician volcano-sedimentary sequences over 10,000 m thick contain W and Sn respectively: Sinian, 9.8 and 12.0 ppm, Cambrian 10.1 and 13.1 ppm, Ordovician 9.0 and 6.5 ppm. In south Jiangxi and north Guangdong, an area with topmost concentration of W deposits, the Cambrian sequences contain 17.5 ppm W on the average.

The Sn-W contents of the sediments of the Hercynian-Indosinian structural layers of the platform regime are usually lower than those of the older geosynclinal accumulation of the same region. However, in comparison with the Sn/W contents of the regions other than South China they are still somewhat high. For example, the Devonian sediments in Dongpu, south Hunan, contain 1.3—18 ppm W. Some Devonian sandstone in Yaogangxian, south Hunan, contain 0.1—0.01 % W. The Devonian clastic rocks of northeast Guangxi contain 7—12 ppm Sn. Middle to Upper Devonian sandstone and shale in Lufu, northwest Guangxi, contain 3—101 ppm Sn.

The preliminary concentration of some deep-source magmatic materials of South China reflects the hetereogeneity of the upper mantle. The mafic dykes of east Hunan and north Guangdong contain 1.5—0.10 ppm W; some Carboniferous pyroclastic rocks and lavas of dacite composition at the Fengling W-Cu deposit, north Jiangxi, are rich in Fe and W. The Jurassic-Cretaceous volcanic-sedimentary rocks of the Yenshanian regenerated geosyncline in east Guangdong contain 15—106 ppm Sn and 2.1—3.3 ppm W (Shou Junian, 1982).

One of the most fascinating problems in recent years is the discovery of considerable concentrations of tin in the middle-late Proterozoic mantle rocks in north Guangxi and Sichuan-Yunnan area, where the Sn contents of ultramafic or mafic intrusive bodies reach 17—18 ppm and 20—45 ppm respectively. Furthermore some economic tin deposits were also found near or within these rocks. This fact sheds important light on the problem of preliminary concentration of tin.

These relatively high concentrations of W/Sn in the stratigraphic sequences of different ages and in deep-source materials reveals that the upper crust, lower crust and upper mantle of South China probably bear considerably higher Sn/W abundances than other parts of the world, indicating that South China is an anomalous region for W and Sn. This accounts for the extremely widespread W/Sn deposits in this region.

## Tin/Tungsten-Bearing Granites

The great majority of primary tin and tungsten deposits in South China are genetically and spatially related to the crustal anatectic (or S-type) granitoids which are mainly distributed on the continent of South China and are formed by granitization, partial melting and remelting of crustal materials. Only a small proportion of Sn/W deposits are found in association with I-type granitoids. In conformity with the tectonic movements, multi-cyclic and multi-stage processes of granitization, partial melting and remelting have repeatedly transformed, readjusted and mobilized the basement rocks rich in Sn/W, leading to the generation of Sn/W-bearing granites.

The Sn/W-bearing S-type granites have some features in common: They are usually late-stage small bodies of composite intrusions or occupy the top and peripheral parts of large intrusions. They are medium to fine grained, acidic to ultra-acidic in composition, Al-oversaturated, rich in such components as alkalies, rare alkalies, (Li,

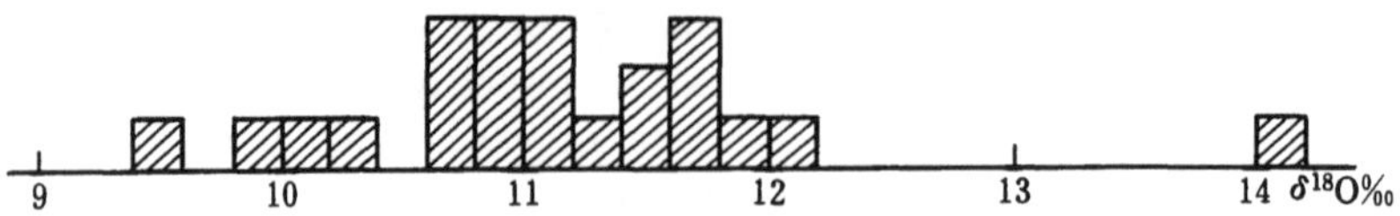


**Fig. 3.** Oxygen isotope composition of 22 Sn/W-bearing granite bodies in South China (whole rock samples)

Rb, Cs), ore-forming elements (W, Sn, Be, Nb, Ta, etc.) and volatiles (F, B, etc.), poor in Sr and Ba, and associated with intense hydrothermal alteration (especially K-feldspathization, albitization, greisenization and tourmalinization), and with relatively high initial Sr ratios and $\delta^{18}0$ values.

## Sr and O Isotopes

According to Wu Qinsheng (1982), Zhao Zijie et al. (1982), Wang Liankui et al., (1982), Zhang Jinrong (1982) and Fujian Bureau of Geology (1981), the initial Sr ratios $(^{87}Sr/^{86}Sr)_i$ of 10 Sn/W-bearing S-type granite bodies range from 0.710 to 0.735, averaging 0.720.

According to Zhang Ligang et al. (1982), Zhang Chongzhen and this paper, the $\delta^{18}0$ values of 23 Sn/W-bearing granites in South China vary from 9.3 to 14.0‰, averaging 11.0‰ (Fig. 3).

The relatively high values of $(^{87}Sr/^{86}Sr)_i$ and $\delta^{18}0$ indicate that these Sn/W-bearing S-type granites were derived from the geosynclinal sediments which had undergone surficial sedimentation and weathering processes and were thereby enriched in $\delta^{18}0$ and radiogenic $^{87}Sr$.

**Fig. 4**

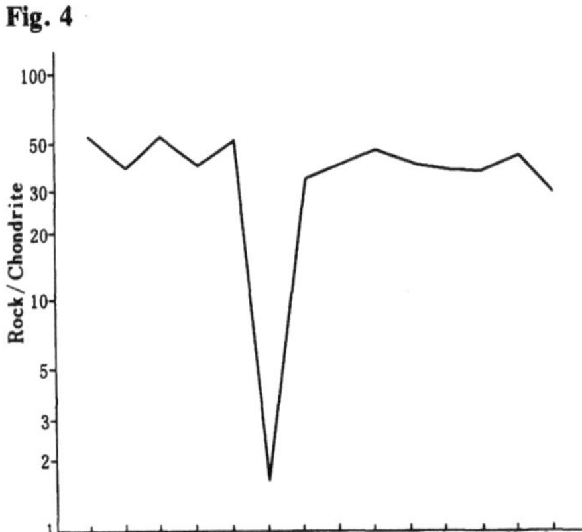


**Fig. 5**

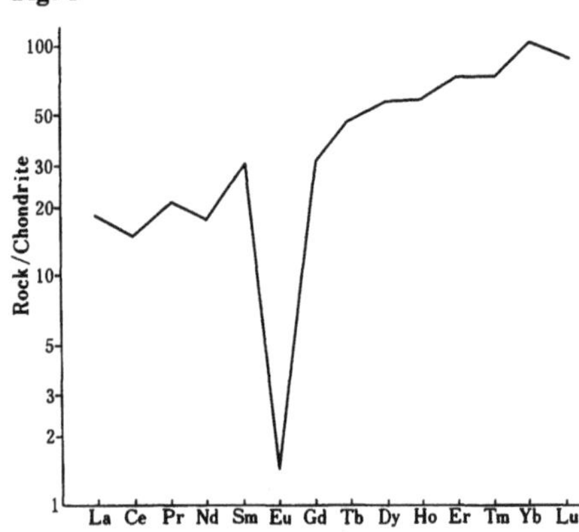


**Fig. 4.** Chondrite-normalized REE pattern of the Gejiu tin-bearing granite

**Fig. 5.** Chondrite-normalized REE pattern of the Xihuashan tungsten-bearing granite

**REE Distribution Patterns**

Studies on the REE composition of some 20 Sn/W-bearing S-type granites show that, in most cases, the REE contents of Sn/W-related granites (250–100 ppm) are usually lower than those of average granites (290 ppm, Emermann, 1970). During intense hydrothermal alteration, especially albitization, greisenization and tourmalinization, REE are commonly selectively leached out together with Ca, resulting in a significant decrease of REE contents. The heavy REE are comparatively enriched, with LREE/HREE ratios usually close to 1 or <1. Eu depletion is sharp and $\delta$ Eu usually <0.3. The chondrite-normalized REE curves are commonly symmetrically V-shaped (Fig. 4). A few granites, such as the Xihuashan W-bearing biotite-granite, are extremely enriched in HREE-containing gadolinite and some other HREE minerals, which results in REE curves with distinct positive slope (Fig. 5).

**Trace Elements**

According to the studies of this paper, Liu Yinjun et al. (1982), Xu Jinfang (1981), Wang Weiyu (1983), and Sun Chengyuan et al. (1983), the Sn/W-bearing S-type granites in South China contain F>2000 ppm, sometimes over 3000–5000 ppm; rare alkalies $\Sigma$ (Li+Rb+Cs)>600 ppm, averaging 1000 ppm; $\Sigma$ (Sr+Ba)<100 ppm, averaging 50 ppm; (Li+Rb+Cs)/$\Sigma$ (Sr+Ba)>10, averaging 20; and the Rb/Sr ratio is usually >25.

**Sn/W Contents**

The Sn/W contents of 9 Sn/W-bearing S-type granites in South China are listed in Table 1. They are several times or tens of times those in average granites.

**Table 1.** Average W and Sn contents of selected ore-bearing granites in South China

| Locality | W (ppm) | Sn (ppm) |
|---|---|---|
| Xihuashan, Jiangxi | 7– 70 | 32–52 |
| Huangsha, Jiangxi | 16– 32 | 37–42 |
| Hongshuizhai, Jiangxi | | 12.4 |
| Qianmutian, Zhejiang | 8 | |
| Denpuxian, Hunan | 50–103.6 | 28.7 |
| Qianlishan, Hunan | 18– 39 | 43–71 |
| Yaogangxian, Hunan | 10.4 | |
| Gejiu, Yunnan | | 16–25 |
| Dachang, Guangxi | | 15–43 |

**Discrimination of Ore-Bearing Potential**

Some geochemical parameters, such as the metal and trace element contents, the element-pair ratios, and different combinations of these figures, can be used to discrimi-

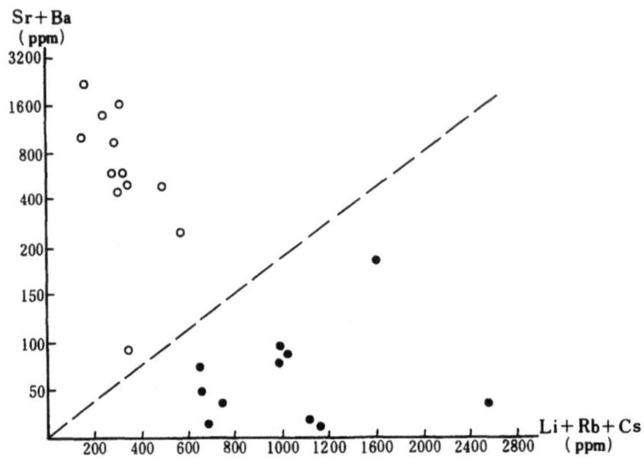


**Fig. 6.** Sr + Ba versus Li+ Rb + Cs diagram of 11 Sn/W-bearing (solid circles) and 12 Sn/W-barren granite (open circles) bodies in South China

nate the ore-bearing potential of granites. The drawing of different diagrams, such as Σ (Sr-Ba) versus Σ (Li+Rb+Cs), W versus F, Li/Mg versus K/Rb, and Sn versus Rb/Sr diagrams (Figs. 6—9) can give excellent results.

## Localization of Tin/Tungsten Deposits

The existence of ore source beds with preliminary Sn/W concentration and generation of Sn/W bearing granites seem to be prerequisites for the formation of Sn/W deposits. However, where and how can the Sn/W deposits be formed? This is, to a large extent, dependent on the structural environment of the region, the evolution and alteration of the ore-bearing granites, as well as on the physicochemical properties and attitude of the country rocks. The tungsten and tin deposits are to be dealt with separately below.

As mentioned above, most W deposits in South China are found in post-Caledonian uplift areas, especially in the south Jiangxi post-Caledonian uplift, where the exposed stratigraphic sequences are mostly pre-Devonian or pre-Cambrian phyllite, schist, slate, greywacke and metavolcanics. The distribution of the Yenshanian granites is commonly controlled by deep regional faults. The distribution of small ore-bearing granite bodies and W deposits is controlled by the faults of second order, and the

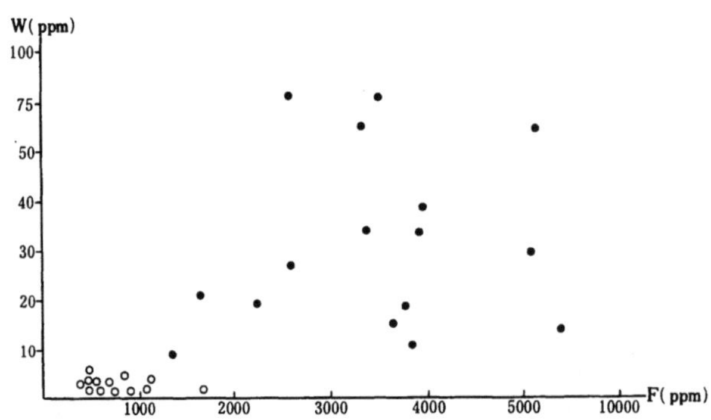


**Fig. 7.** W versus F diagram of 16 Sn/W-bearing and 13 Sn/W-barren granite bodies in South China (solid circles refer to ore-bearing granites, open circles to ore-barren ones)

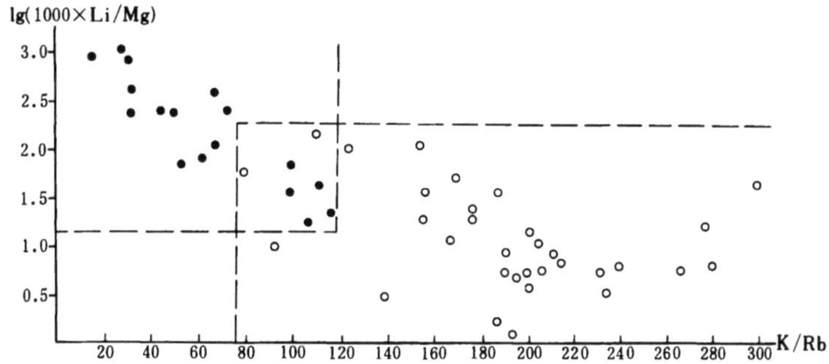


**Fig. 8.** Li/Mg versus K/Rb diagram of 17 Sn/W-bearing (solid circles) and 32 Sn/W-barren (open circles) granite bodies in South China

distribution of wolframite-quartz vein bodies is controlled by sheared fractures. This kind of structural control over wolframite deposits has universal significance. For instance, the Chijiang fault in the area between Dayu and Nankang, south Jiangxi, is a deep fault of regional scale. The Yenshanian granites are developed on the NW side or hanging side of NE trending Chijiang fault. Most of the tungsten deposits and W-related small granite bodies are found along the NNE trending second order faults such as the fault from Xihuashan to Zongshukeng. In a wide area of south Jiangxi and its neighbourhood more than 90% of the wolframite-quartz veins are developed subparallel to the nearly E-W trending fractures. This kind of orientated sheared fractures usually cut across the pre-Devonian metasediments and Yenshanian granites. The fractures were even still active after formation of the W-filled veins. Obviously, this was caused by the prolonged action of a regional stress field.

When the ore-bearing granites are emplaced in compact, plastic, impermeable and chemically inert rocks such as phyllite and slate, screening of country rocks confine the ore-bearing hydrothermal solutions to a relatively closed system, which results in a more complete differential evolution and produces intense pervasive metasomatic alterations expressed in the upward vertical zoning of K-

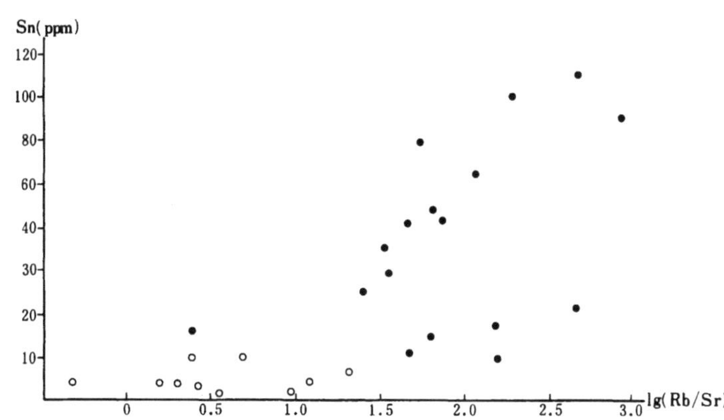


**Fig. 9.** Sn versus Rb/Sr diagram of 17 Sn/W-bearing and 10 Sn/W-barren granitic bodies in South China (solid circles refer to ore-bearing granites, open circles to ore-barren ones)

feldspathization→albitization→greisenization and pegmatoid layer. At the stage of opening of fractures in granitic rocks, the hydrothermal alteration of pervasive type is developed into fissure type. The formation of alteration zones and pegmatoid layer may be considered as a very important criterion for Sn/W mineralization in granites.

The high temperature wolframite-quartz veins are all formed in non-calcareous chiefly pre-Devonian metasediments, Yenshanian granites and, locally, Devonian shale and sandstone. The scheelite-bearing skarn deposits are usually formed in Devonian or Cambrian calcareous beds intercalated with shale and sandstone near the W-related small granite bodies. These deposits are found mostly in the transitional zone between post-Caledonian uplift and Hercynian-Indosinian depression of southeast Hunan, but are not observed in the central part of the depression with very thick carbonate sequences.

As noted above, the most important tin deposits in South China appear in the Hercynian-Indosinian depression, where the upper Paleozoic and lower Mesozoic sediments (chiefly carbonates) are extensively developed, with a total thickness of 3000–5000 m. The tin deposits are distributed adjacent to the pre-Cambrian or pre-Devonian basement. The main factors controlling localization of tin deposits can be summarized as follows: (1) places near deep regional faults or in fault depressions; (2) anticlinal domes consisting of thick-bedded carbonate rocks; (3) Sn-bearing granite intrusions beneath the anticlinal domes. For instance, the Gejiu, Dachang, Xianhualin and Limu deposits are all provided with these controlling factors.

In addition to occurrence of tin deposits in the apical parts of granite stocks, tin deposits tend also to be controlled by the contact zones of granite bodies with thick carbonate sequences. The existence of thick-bedded carbonate strata near tin-bearing granite bodies seems to be especially favourable for precipitation of tin from the hydrothermal solutions.

**Fig. 10.** Stratiform sulphide body of Dachang cassiterite-sulphide deposit

**Fig. 11.** Cassiterite-quartz veinlets cut across stratiform sulphide body. Dachang deposit

In addition, in some localities of Hercynian-Indosinian depressions, especially near fault depressions, some Devonian or Triassic stratiform sulphide beds (sulphides of Fe, Zn, Pb, Sb, etc.) of sedimentary or exhalative-sedimentary origin are developed. These sulphide sequences seem to be extremely favourable for precipitation of tin from hydrothermal solutions.

Under such an environment was formed the most important economic type of tin deposit of China, i.e., the cassiterite-sulphide ores in the Dachang area, Guangxi, and the Gejiu area, Yunnan. These cassiterite-sulphide bodies are usually stratiform within carbonate country rocks (Fig. 10); laminated, rhythmic and convolute structures are common; $\delta\ ^{34}S$ varies in a wide range, mostly of negative value, with the character of sedimentary sulphur. The S, Pb, Zn, and Sb contents of host carbonate are siliceous rocks and high. The Sn content is also higher than normal. The flat or sub-horizontal stratiform sulphide bodies are frequently cut across by vertical or steeply-dipping cassiterite-quartz veinlets (Fig. 11). All these phenomena have evidently demonstrated that most of the Fe, Zn, Pb, Sb and possibly a part of Sn in the cassiterite-sulphide ores, were derived from the sedimentary formations. The tin-bearing hydrothermal solutions released from granites have leached considerable amounts of Fe, Zn, Pb, S and a certain amount of Sn from the country rocks, and some favourable sites (especially where sulphide beds were dominant), precipitated them and thus formed commercial concentrations of tin.

## Brief Conclusions

The formation mechanism and controlling factors of tin/tungsten mineralization can be briefly summarized as follows:

1) The Sn/W distribution in the upper mantle is far from homogeneous. South China as a whole is characterized by a high Sn/W background and is therefore an important anomalous Sn/W province of the world. In the course of a prolonged and complicated geological history, through different endogenic and exogenic processes, the ore source beds of different ages were formed. Among them, the most important are the middle-late Proterozoic and lower Paleozoic volcano-sedimentary sequences.
2) Through magmatization, granitization, partial melting and remelting processes these ore source beds were transformed and the Sn/W-bearing granite bodies (mainly middle-late Mesozoic) were formed. Differentiation and evolution of these Sn/W-bearing granitic magmas led to the generation of Sn/W-bearing hydrothermal solutions.
3) It was through metasomatic alteration by hydrothermal activities, fissure-filling and other forms, as well as under favourable structural and wall rock conditions that the economic tin/tungsten deposits were formed.

## References


Cheng Xianyao, Huang Yude, 1984. On the primary enrichment of tin. *Paper for International Symposium on Geology of Tin deposits* (in Chinese).

Department of Geology, Nanjing University, 1980. Investigation on the time and spatial distribution of the granitic rocks of Southeastern China, their petrographic evolution, petrogenetic types and metallogenetic relations. *Journal of Nanjing University*, special issue (in Chinese).

Hu Shouxi, 1984. An important model of ore formation related to granites. *Paper for 27th IGC, Moscow.*

Liu Yingjun et al., 1982. The geochemical characteristics of trace elements in granitic rocks of South China. *Proceedings of Nanjing Symposium on Geology of Granites and Their Metallogenetic Relations* (in Chinese).

Sun Chengyuan et al., 1983. The geochemistry of lithium, rubidium and cesium in granitic rocks of South China. *Journal of Nanjing University,* No. 4 (in Chinese).

Xian Baiqi, 1984. A discussion on the ore-forming conditions and distributional regularity of the tin ore deposits of Guangxi. *Acta Geologica Sinica,* Vol. 58, No. 1 (in Chinese).

Xu Keqin et al., 1981. Regional controlling factors of tungsten deposits in South China. *Proceedings of Symposium on Tungsten Deposits,* Geological Publishing House, Beijing (in Chinese).

Xu Keqin et al., 1982. Petrogenesis of the granitoids and their metallogenetic relations in South China. *Proceedings of Nanjing Symposium on Geology of Granites and Their Metallogenetic Relations* (in Chinese).

Ye Jun et al., 1984. On the mechanism of ore-formation of some cassiterite-sulfide deposits in South China. Proceedings of International Symposium on Geology of Tin Deposits.

Zhang Chongshen, 1982. Evolution and mineralization of the Mesozoic granitoids in South China. *Proceedings of Nanjing Symposium on Geology of Granites and Their Metallogenetic Relations* (in Chinese).
Zhang Ligang et al., 1982. A study on the oxygen isotopes of some Mesozoic granitoids in southeastern China. *The same proceedings* (in Chinese).
Zhu Jinchu, Zhang Chenhua, 1981. On the occurrence of the Carboniferous volcanics and origin of the copper and tungsten deposits of the Fenglin district, Dongxiang, Jiangxi. *Journal of Nanjing University,* No. 2 (in Chinese).

# 6.1.6 On the Ore-Forming Mechanism of Some Cassiterite-Sulphide Deposits in South China

YE JUN, ZHOU HUAIYANG, and CHEN ZHUGI[1]

## Abstract

Cassiterite-sulphide deposits represent one of the most important types of primary tin deposit in China. The granites related to tin deposits of this type are mainly of late Yenshanian age. A number of characteristics of these granites indicate that they belong to typical continental crustal (S) type. These deposits are located in fault depressions, and the ore-bodies have many features of stratabound hydrothermal superimposition. Magmatic activities not only provided ore-forming materials, but also served as the necessary thermal source in the formation of the ore deposits. The ore-forming materials came partially from the crust, and partially from the upper mantle. These deposits are characterized by dual features, i.e., they are related to both granites and stratabound hydrothermal epigenetic deposition in fault depressions.

## Introduction

In East and South China, granitic rocks occur extensively, which are accompanied by rich tin deposits. The primary tin deposits are mainly cassiterite-sulphide type, cassiterite-wolframite-quartz vein type, greisen type, cassiterite skarn type and porphyry tin deposits. Among these, the cassiterite-sulphide deposit type seems to be most important. The famous Dachang tin deposit in Guangxi and the Gejiu tin deposit in Yunnan belong to this type. We have made some field investigations and laboratory work on cassiterite-sulphide deposits under Prof. Xu Keqin's guidance. This paper mainly deals with the ore-forming mechanism, based on the characteristics of the granitoids and ore deposits.

## 1 Characteristics of the Granitic Rocks Related to Cassiterite-Sulphide Deposits

The granitoids related to cassiterite-sulphide deposits are mainly late Yenshanian biotite granite dated by the K-Ar method at 67–115 Ma. They commonly occur as batholiths, stocks and cupolas, with some hypabyssal-suprahypabyssal dykes of gran-

[1] Department of Geology, Nanjing University

**Fig. 1.** Phenocrysts of alkali feldspar with sub-parallel orientation in the porphyritic biotite granite of Dachang (530 m level, Lamo)

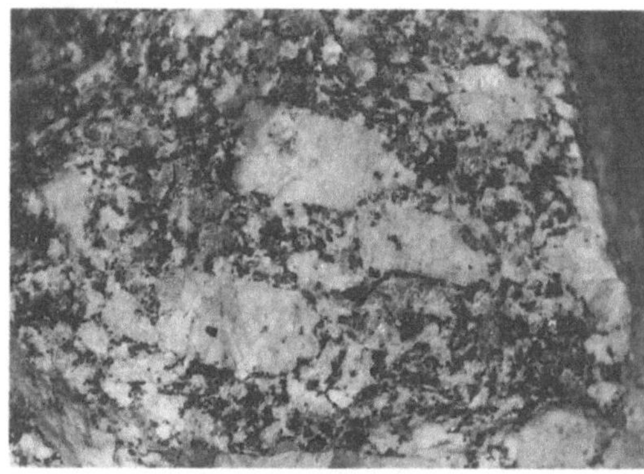

**Fig. 2.** Porphyritic biotite granite of Dachang (hand specimen)

**Fig. 3.** Large phenocrysts of alkali feldspar in porphyritic biotite granite (hand specimen from 530 m level, Lamo)

ite porphyry, dacite-porphyry and diorite-porphyrite in the mining areas. Magmatic activities are obviously multi-staged.

(a) Petrographic Characteristics

The granitic rocks are characterized both by a porphyritic and equigranular texture (Fig. 1, 2, 3). The major mineral constituents are alkali feldspar, plagioclase, quartz and biotite. Alkali feldspar includes microcline, perthite and orthoclase, while the plagioclase ranges from oligoclase to andesine ($An_{20}-An_{38}$). Biotite is rich in Fe and

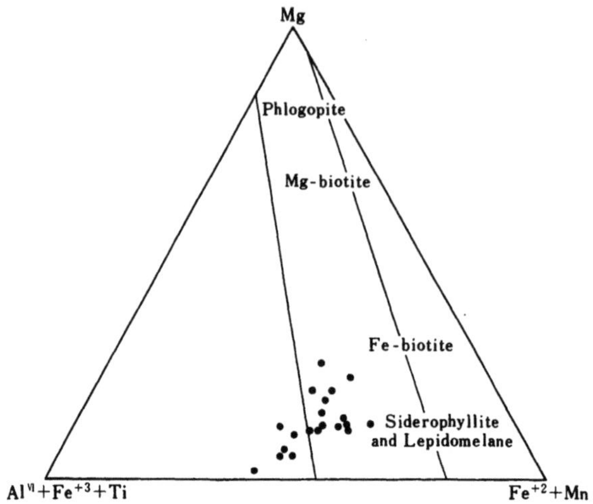


**Fig. 4.** Mg-($Al^{VI}$+$Fe^{3+}$+Ti) – ($Fe^{2+}$+Mn) diagram of biotites from granites in the Dachang area

poor in Mg. The Fe/(Fe+Mg) ratio of biotite in the Dachang granite varies from 0.80 to 0.87, averaging 0.85, whereas its Mg/Fe ratio varies from 0.03 to 0.25, averaging 0.19. The Mg – ($Al^{VI}$+$Fe^{3+}$+Ti) – ($Fe^{2+}$+Mn) triangular diagram shows that the biotite from the Dachang granite is Fe-biotite and siderophyllite (Fig. 4), as is also the case for the biotite from the Gejiu granite. This variety of biotite is rich in Li, ranging from 0.4 to 1.5% with an average content of 0.71%, and relatively rich in Al and K. The K/(K+Na+Ca) ratio varies from 0.84 to 0.97, averaging 0.90. Al/Σ Me is somewhat higher (0.24). It can be clearly seen from a diagram of Mg versus Al/Σ Me of biotite from granites (Fig. 5) that the content of Mg in the biotite of the crustal anatectic (or S) type granitoids is less than that of the I-type.

The accessory minerals are dominated by tourmaline, fluorite, topaz, apatite, etc., accompanied by a small amount of zircon, titanite, allanite, rutile, xenotime, ilmenite, uranium minerals, etc.

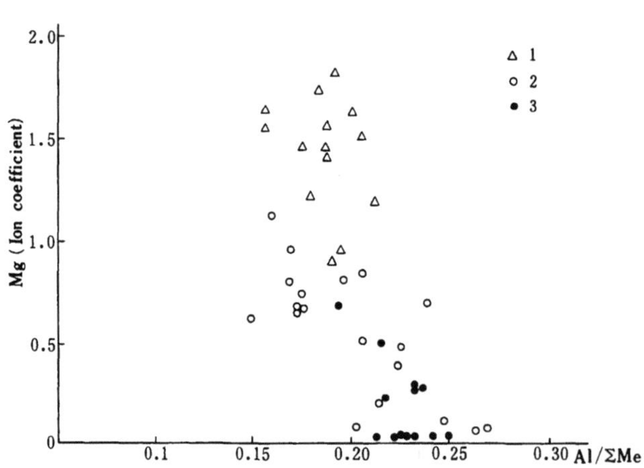


**Fig. 5.** Correlation between Mg and Al/Σ Me (total amount of metallic ions) of biotites from granites. Symbols as in Fig. 7

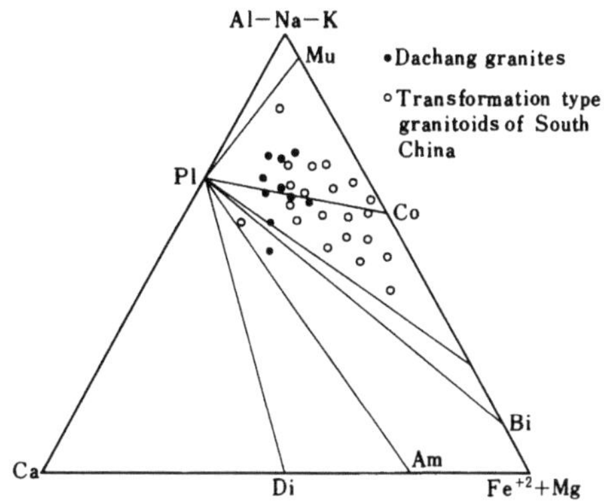


**Fig. 6.** (Al-Na-K)-Ca-(Fe+Mg) diagram of granites in the Dachang area

(b) Petrochemical Characteristics

These granitoids are rich in $SiO_2$ (>70%, S=80) and alkalies ($K_2O + Na_2O > 7\%$, $K_2O > Na_2O$), with $K_2O$ generally around 4%, and even reaching 6% for the highest. They are poor in CaO (varying within 0.70—1.80), $Al_2O_3 = 14\% \pm$. The Al'value (Al' = Al – K – Na – 2Ca = 38 ±) suggests that the granitoids are peraluminous. All the points of the granites projected on the triangular diagram (Al-Na-K) – Ca – ($Mg+Fe^{2+}$) are concentrated in the plagioclase-biotite-cordierite (muscovite) region (Fig. 6).

These granitoids are poor in $Fe_2O_3$, FeO, MgO and $TiO_2$, ($Fe_2O_3$+FeO)<2.5%, sometimes about 3%, resulting in rarity or absence of amphibole and only a small amount of biotite. They are rich in Li (averaging 247 ppm), and volatile components such as F (averaging 3700 ppm) and B (averaging 766 ppm).

(c) Characteristics of Trace Elements

These granitoids are rich in Rb, with the average Rb value of 18 whole rock samples from the Dachang granite being 810 ppm and the highest reaching 1402 ppm. Obvi-

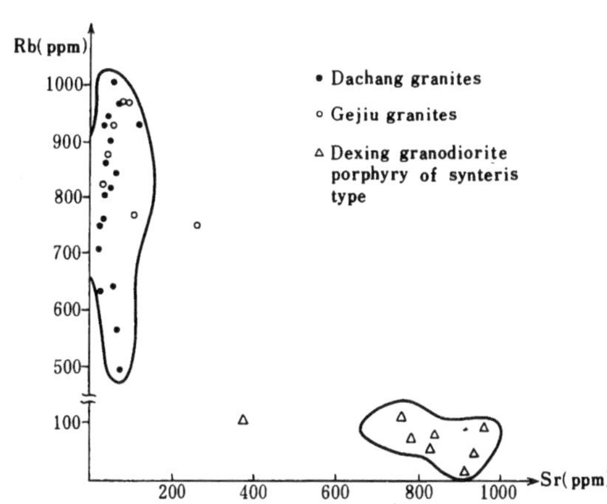


**Fig. 7.** Diagram showing Rb-Sr relationship of granites

ously, the Rb content of these granitoids is much higher than that of common granitic rocks 200 ppm ±).

These granitoids are poor in Sr, with the average Sr content being 37.7 ppm and the lowest value 16 ppm. The Rb/Sr ratio in the Dachang granite varies from 8 to 87.6, averaging 26.2. The Rb-Sr ratio in the Gejiu granite is generally higher than 20. The Rb-Sr ratio in the I-type granodiorite porphyry of the Dexing Porphyry Copper Mine is low, varying within 0.025—0.26.

The Rb/Sr ratios of the S-type granitoids and the I-type granitoids are obviously different. The S-type is rich in Rb, poor in Sr, while the I-type is just the opposite (Fig. 7).

In addition, these granitoids are rich in Sn. The Sn content of granitic bodies is over 10 ppm, and that of biotite is over 100 ppm. They are also rich in U, Nb and Ta, the contents of Nb and Ta being three to nine times as high as those in common granitic rocks.

### (d) REE Characteristics

The ΣREE of these granitoids varies considerably, with the average value being 150—200 ppm and some even higher than 300 ppm. The LREE/HREE ratio ranges from 3.8 to 4, δ Eu=0.3 ±. The REE distribution patterns of the Dachang granitoid and the Gejiu granitoid have a distinct negative Eu anomaly, and their chondrite-normalized REE pattern is a "seagull" shaped curve (Fig. 8).

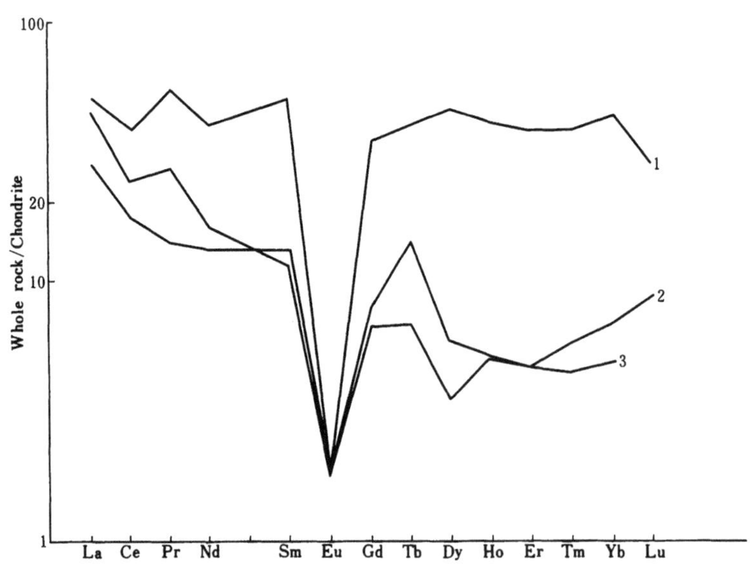


**Fig. 8.** Chondrite-normalized REE pattern of the granites from Dachang and Gejiu

(e) Isotope Characteristics

Their initial $^{87}Sr/^{86}Sr$ ratio is high, generally over 0.710. The initial $^{87}Sr/^{86}Sr$ ratio of the apatite from fine-grained biotite granite in Dachang is 0.7223 ± 0.0007. The initial $^{87}Sr/^{86}Sr$ ratio of whole-rock samples of biotite granites in Gejiu range from 0.7102 ± 0.0007 to 0.7142 ± 0.0008 (Wu Qinsheng et al., 1982). These granitoids are rich in $^{18}O$, with the $\delta\ ^{28}O$ of two quartz samples in the Dachang granite being 12.07‰ and 11.48‰. All these characteristics indicate the reasonableness of relating the granitoids associated with cassiterite-sulphide deposits to the typical S-type granitoids derived from the continental crust (Xu Keqin et al., 1982, 1983).

## 2 Characteristics of Ore Deposits

It is obvious that different mineral assemblages of these cassiterite-sulphide deposits are zoned horizontally and vertically around the biotite granites. The vertical zoning of orebodies above the Laochang No. 1021 granite cupola in the Gejiu area is shown in Fig. 9. The different mineral assemblages are zonally distributed from the contact zone to wallrocks in order of Cu→ Sn, Cu→ Sn, Pb. In the Dachang tin deposit, from the contact zone outwards, there successively occur skarn Zn, Cu type (V) → stratiform-substratiform Sn-Zn sulphide metasomatic type (IV) → stratiform-substratiform Sn-Zn sulphide type (III) → cassiterite-sulphide veinlet type (II) → cassiterite-sulphide large vein type (I) (Fig. 10) (Hu Shouxi et al., 1982).

Hydrothermal activities in these deposits are intense, and the constituents and textures of ores are complicated. It can be seen from Fig. 11, that stockwork and stratiform or substratiform orebodies in the Dachang tin deposit consist of bedded

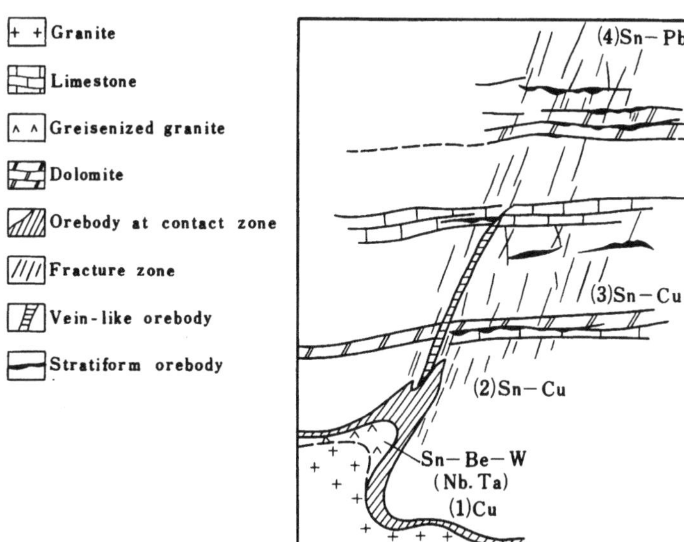


**Fig. 9.** Diagram illustrating the vertical zoning of orebodies above the Laochang 1021 granites cupola in the Gejiu area

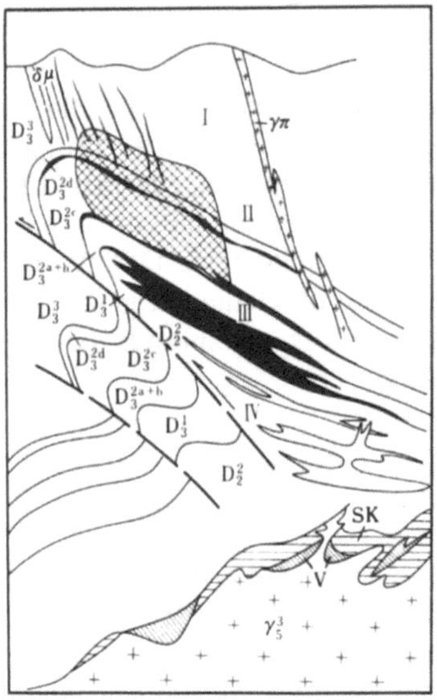


**Fig. 10.** Diagram illustrating the mineral zonation of the Dachang tin deposit

sulphides and cassiterite-quartz veins. In places, the cassiterite-sulphide vein cuts across the bedded sulphide orebodies at the Changpo Mine, Dachang (Fig. 12).

Because these deposits are located in fault depressions, the orebodies have many features in common with the stratiform or substratiform hydrothermal epigenetic type:

**Fig. 11**

**Fig. 12**

**Fig. 11.** NE-trending cassiterite-quartz vein cutting across the bedded sulphide orebodies in siliceous rocks ($D_3$) (level No. 6, Changpo Mine)

**Fig. 12.** Cassiterite-quartz vein cutting across the bedded sulphide orebodies in laminated limestone (level No. 6, Changpo Mine)

(1) The main orebodies are strictly controlled by stratigraphic horizons, and their wall rocks consist mainly of Late Paleozoic and Middle Triassic marine carbonates, siliceous and clastic rock formations, and carbonate reefs.
(2) The main orebodies occur as stratiform bodies remaining stable in thickness and extending along strike and dip for a long distance. They were folded synchronously with the stratigraphic sequences (Figs. 13, 14, 15).
(3) The ores may retain some sedimentary features, such as lamination (Fig. 16), banding and synchronous folds. In the Dachang tin deposit, for example, the dark bands of banded ore consist of pyrite and pyrrhotite (Fig. 17), the laminated ore consists of pyrite, pyrrhotite, arsenopyrite, and newboldite, and the ore with bedding convolution texture consists of newboldite, arsenopyrite, pyrite and cassiterite (Fig. 18). In addition, the ores may retain early framboidal pyrite and residual organic structures.
(4) Volcanic rocks can be found in ore-bearing strata of the mining area and its vicinity. It seems that the ore-forming materials were partially derived from volcanic activities. In the Dachang mining area, there are extensive occurrences of siliceous rocks, probably related to volcanic activities, especially to hot springs. Scanning electron microscopy of siliceous rocks in the Dachang mining area (Fig. 19) shows that the quartz crystals are even-grained or mosaically aggregated with clean and smooth crystal faces and edges; these characteristics are similar to those of quartz crystals related to volcanic activities in Tibet (Li Kang et al., 1982). Scanning electron photomicrographs (Fig. 20) show that siliceous rocks from the Upper Devonian in the Yelan mercury deposit of the Dachang mining area may contain crystal debris of quartz and tuffaceous materials.
(5) The pyrite from stratiform or substratiform orebodies is different from that from vein type orebodies in Co/Ni ratio, being <1 and >1 respectively (Fig. 21). This

**Fig. 13**

**Fig. 14**

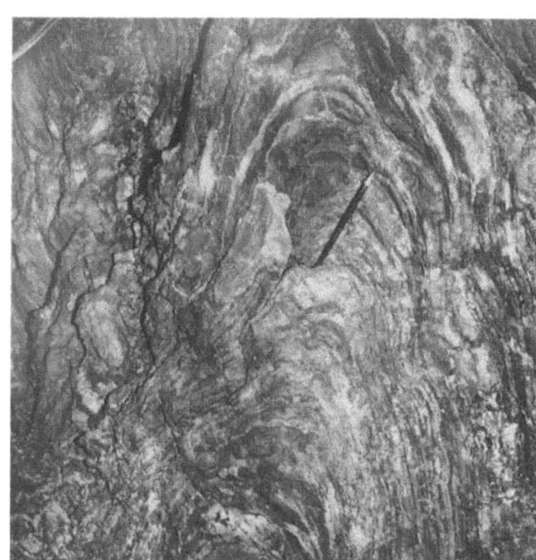

**Fig. 13.** Bedded sulphide orebodies folded synchronously with the strata (level No. 6, Changpo Mine)

**Fig. 14.** Bedded sulphide orebodies folded synchronously with the strata (level No. 6, Changpo Mine)

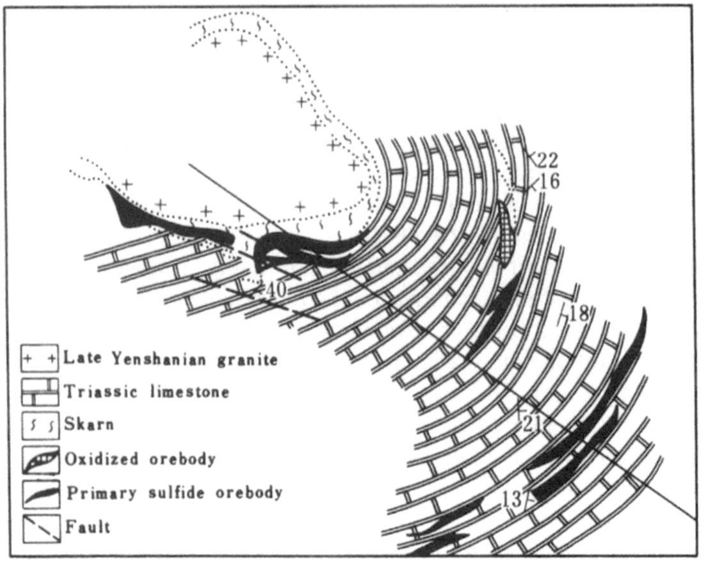


**Fig. 15.** Schematic geological map of a tin deposit in the Gejiu area

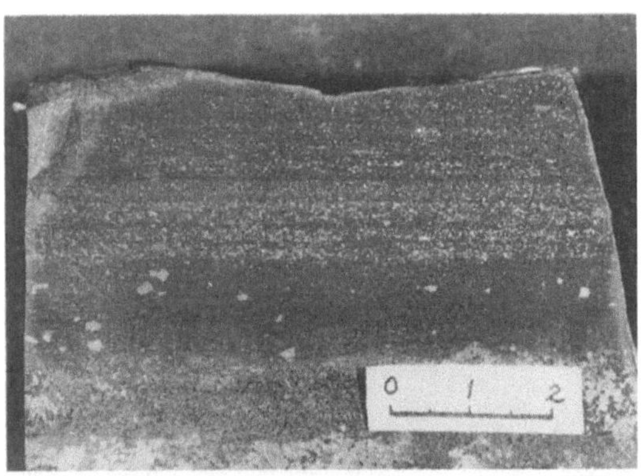


**Fig. 16.** Lamination structure of pyrite in siliceous rocks (hand specimen from level No. 7, Changpo Mine)

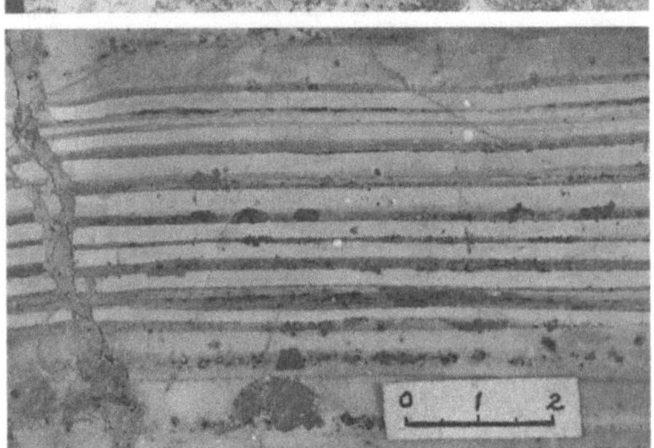


**Fig. 17.** Banded ore. Dark bands consist of pyrite and pyrrhotite (hand specimen from level No. 7, Changpo Mine)

**Fig. 18.** Ore with bedded convolution texture, consisting of newboldite, arsenopyrite, pyrite and cassiterite in siliceous rocks (hand specimen from level No. 9, Toncang)

difference indicates that the stratiform or substratiform orebodies may be of sedimentary origin. The pyrite in stratiform or substratiform and disseminated orebodies is rich in Ni, and poor in Co, while the opposite holds for the pyrite in cassiterite-quartz veins.

(6) The ore-controlling source beds related to cassiterite-sulphide deposits have high contents of Cu, Zn, Pb and Sn. For example, the Late and Middle Devonian siliceous rocks, carbonate rocks and arenaceous shale from the representative stratigraphic section of Lufu in the vicinity of the Dachang Mine are rich in Cu, Zn, Pb and Sn. The average content is 19.2 ppm, 52.6 ppm, 25 ppm, and 3.6 ppm for Cu, Zn, Pb and Sn respectively.

**Fig. 19.** Scanning electron photomicrograph of siliceous rocks of the Toncang Mine ($D_3^1$), Dachang

**Fig. 20.** Scanning electron photomicrograph of crystal debris of quartz in tuffaceous siliceous rocks of the Yelan area

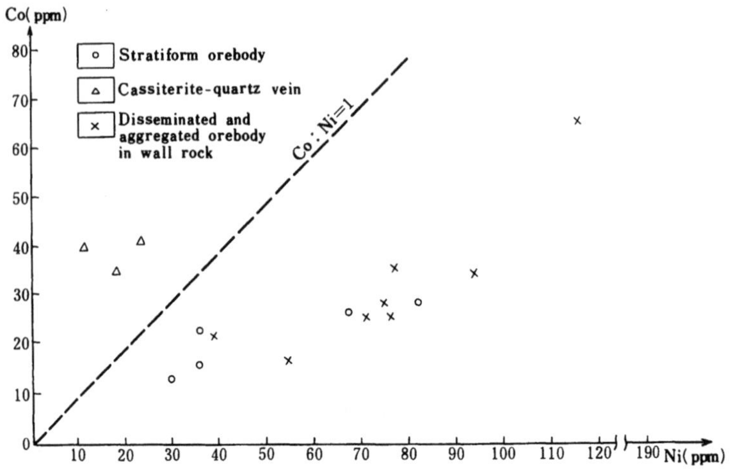


**Fig. 21.** Diagram showing the variations in Co and Ni contents of pyrite from differently shaped orebodies in the Dachang tin deposit

These features suggest that the cassiterite-sulphide deposit bears the obvious characteristics of stratabound hydrothermal epigenetic type, quite similar to some iron and copper deposits occurring in fault depressions in Southeast China (Xu Keqin et al., 1980, Ruan Huichu et al., 1982).

## 3 The Ore-Forming Mechanism

The formation of cassiterite-sulphide deposits in South China may be represented by the following simplified scheme:

Geosynclinal accumulation
(old folded basement)
↓
volcanism, exhalative hydrothermal process
↑ ↑
Late Paleozoic-Middle Triassic fault depressions
(Multiple-aged source beds or strata of Sn, Pb, Zn, Fe, S, etc.)
↓
multi-cyclic granites
↓
favourable structural conditions
↓
Late Yenshanian biotite granite
↓
metasomatism, fissure filling
hydrothermal superimposition
economic cassiterite-sulphide deposit

It can be supposed that these deposits occur in fault depressions on the old folded basement. The multiple-aged source-beds or strata of tin and polymetals underwent long processes of multi-staged magmatic activities, tectonic movements and different kinds of mineralization. In consequence, large-sized economic tin deposits, such as Dachang and Gejiu, were formed. Magmatic activities, therefore, not only provided ore-forming materials, but also served as the necessary thermal source. The ore-forming materials came partially from the crust, and partially from the upper mantle. These deposits are characterized by dual features, i.e. they are related to both granites and stratabound hydrothermal superimposition in fault depressions.

*Acknowledgements.* We would like to thank Mr. Chen Shanyou (No. 215 Party of Dachang, Guangxi) for his help during the field work. We also thank the Dachang Administration of Mines and the Lamo Zinc Mine of Dachang for their support. Professor Hu Shouxi and Professor Zhu Jinchu (Department of Geology, Nanjing University) are acknowledged for their critical review of the manuscript.

## References

Hu Shouxi et al., 1982. *Ore Deposits.* Geological Publishing House, p. 132, p. 217.

Li Kang et al., 1982. Application of SEM to geological research. In: *Research of Geology* (1), Institute of Geology, Academia Sinica (ed.), Cultural Relics Publishing House, 251–254.

Ruan Huichu, Gu Lianxing, Zheng Sujuan and Ye Jun, 1982. On the origin of the Middle Triassic stratiform and stratabound iron ore deposits in the lower reaches of the Yangzi River. *Journal of Nanjing University,* No. 3.

Wu Qinsheng et al., 1982. Application of apatite to the study of strontium isotope geology. *Proceedings of Symposium on Geology of Granites and Their Metallogenetic Relations* (Abstracts), Nanjing, 78–81.

Xu Keqin (Hsu Kechin), Zhu Jinchu, and Ren Chijiang, 1980. On the origin of some iron and copper deposits in several marine fault depressions of southeastern China. *Scientific Papers on Geology for International Exchange, Prepared for the 26th International Geological Congress,* 3. Geological Publishing House, 49–57.

Xu Keqin et al., 1982. On the origin and metallogeny of the granites in South China. *Proceedings of Symposium on Geology of Granites and Their Metallogenetic Relation* (Abstracts), 1–3.

Xu Keqin, Hu Shouxi, Sun Mingzhi, Zhang Jingrong, Ye Jun and Li Huiping, 1982. *Regional factors controlling the formation of tungsten deposits in South China,* Geological Publishing House.

Xu Keqin, Hu Shouxi, Sun Mingzhi and Ye Jun, 1982. On the two genetic series of granites in southeastern China and their metallogenetic characteristics. *Mineral Deposits* Vol. 1, No. 2.

Xu Keqin, Hu Shouxi, Sun Mingzhi, Zhang Jingrong and Ye Jun, 1983. On the genetic series of granites, as exemplified by the Mesozoic granites of South China. *Acta Geologica Sinica,* No. 2.

## 6.2 China: Exploration Methods

# 6.2.1 Some Results and Prospects in the Application of Geophysical and Geochemical Methods to the Search for Tin Deposits

WU GONGJIAN and GAO RUI[1]

## Abstract

This paper deals with some results in the application of geophysical and geochemical methods to the search for tin deposits based on some examples of prospecting in some areas on China. In addition, the ore potential for different types of tin deposit has been estimated in combination with the study of regional geological structures.

## Introduction

The effectiveness of application of geophysical and geochemical approaches to the search for tin deposits in China is very remarkable, for they can rapidly delineate potential metallogenic areas. Because of their stable chemical properties, mechanical dispersion haloes and trains of cassiterite are commonly formed in soils and stream systems around orebodies. Tin anomalies, therefore, can be found rapidly through soil, stream sediment, and heavy mineral surveys. Judging from the geochemical anomalies of the known commercial tin deposits in China, evident tin concentration centres with values exceeding 20 ppm always exist over the tin deposits. Tin is an important ore-forming element in granite; and in the surrounding country rocks there often occurs a zonation of element associations. The effectiveness of geophysical methods lies essentially in the delineation of orebodies and in the study of structures, rock types and alteration zones. As a result, different approaches in the search for tin deposits and the most appropriate methods can be established, and potential in prospecting for tin is predicted.

### Applicability of Geochemical Method in Rapid Delineation of Prospective Metallogenic Areas

In 1964, two localities with tin anomalies were found in an area of Sichuan Province through stream sediment survey, with concentration of 100 and 200 ppm respectively (Fig. 1). These two localities occur in the vincinity of the contact of the Jinning-period granite. As early as 1959, a number of cassiterite heavy mineral anomalies were found in streams of this particular region, with the best anomalies located at the contact, and characterized by a short-distance transportation. In 1964, pyritization and skarni-

[1] Institute of Mineral Deposits, Chinese Academy of Geological Sciences, Beijing

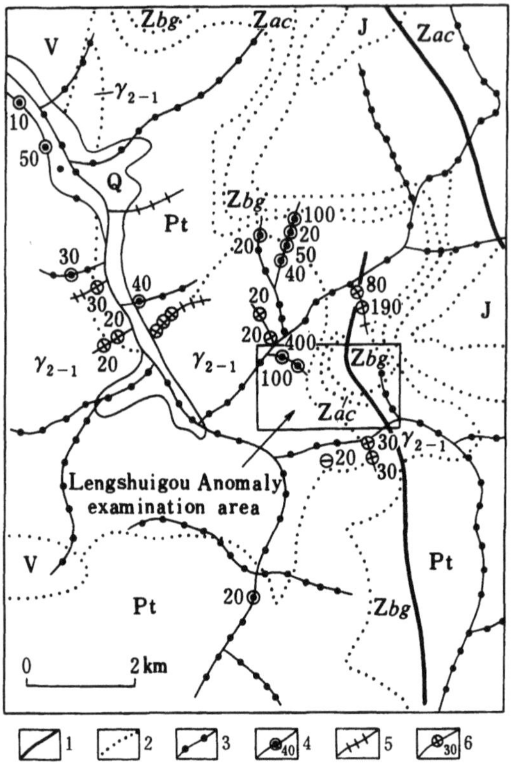


**Fig. 1.** Diagrammatic map showing the two localities of tin anomalies found through stream sediment survey in Lengshuigou region and its vicinity, Sichuan Province. (after the Geophysical Survey Party of the Sichuan Geological Bureau, 1974)

zation were discovered, indicating the presence of primary cassiterite. Later, by using arsenic as a indicator element, a good number of new anomalies were found (Fig. 2), subsequently proved to be economic by examining the geochemical anomalies with prospecting trenches exposing an alteration zone about 1,000 m in length, consistent with the extension of the anomalies.

In 1965, a soil survey was conducted on the scale of 1:50,000 and 1:10,000 in the areas surrounding the Gejiu Tin Mine, as a result of which several primary and placer tin deposits were found, thus expanding the ore reserves of the mine.

Within only three months an area of 3,300 $km^2$ was covered by geochemical prospecting, resulting in the discovery of 46 anomalies, showing the high effectiveness, low expense and rapid achievement of results in applying geochemical approaches.

During the prospecting for iron ores with magnetic methods in a region of Yunnan Province in 1958, a geochemical survey was also conducted. At first no attention was paid to the tin anomalies discovered in the course of prospecting for iron ore. These already discovered geochemical anomalies were later re-investigated only when economic tin deposits had been found in this particular region. They were proved to form a polymetallic ore deposit with tin as the main economic element, in combination with anomalous contents of W, Mo, Cu, Pb and Zn (Fig. 3).

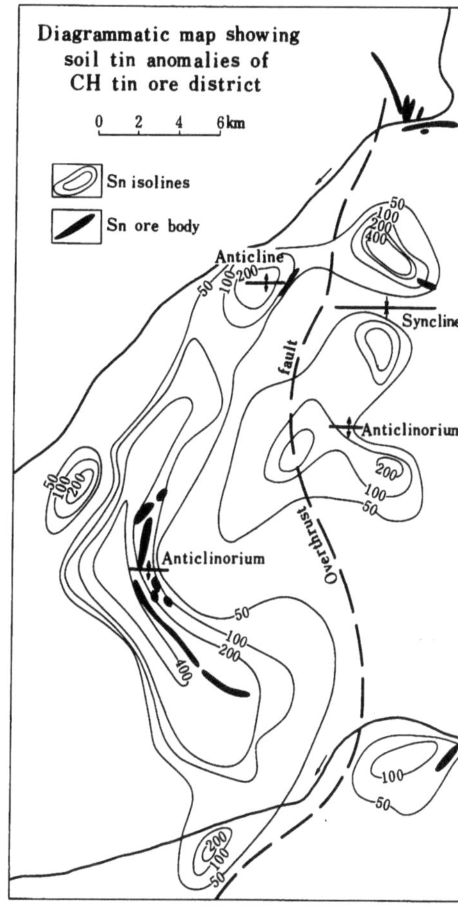


**Fig. 2.** Schematic map showing the distribution of geochemical anomalies in a region of Sichuan Province. (after the Geological Party of Sichuan Geological Bureau, 1980)

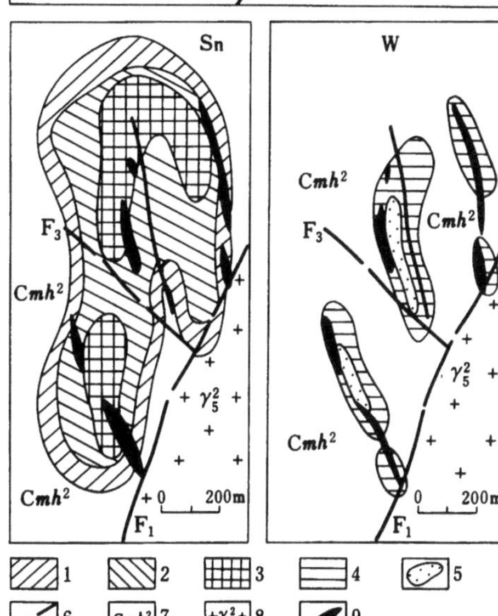


**Fig. 3.** The distribution of Sn and W anomalies in soils of the TYS tin (tungsten) deposits

The Usefulness of Geophysical Approaches to the Direct Delineation of Orebodies and to the Investigation of Alteration Zones

During prospecting for tin mineralization in a region of Hunan Province, the main attention was originally paid to both the axial part and the western flank of an overturned anticline. However, the anomalies proved by induced polarization and the geochemical survey were found to be distributed along the eastern flank of the overturned anticline, thus changing the site for tin prospecting (Fig. 4). The original plan to look for Pb and Zn mineralization had to be changed to prospecting for tin deposits following the induced polarization and geochemical investigations, which showed that the particular region might be potential for cassiterite-sulphide mineralization. The inference made, based on the geophysical and geochemical investigations, was proved by consequent drilling and chemical analysis of the cores within the area of anomalies. Commercial ore was proved in six out of ten boreholes.

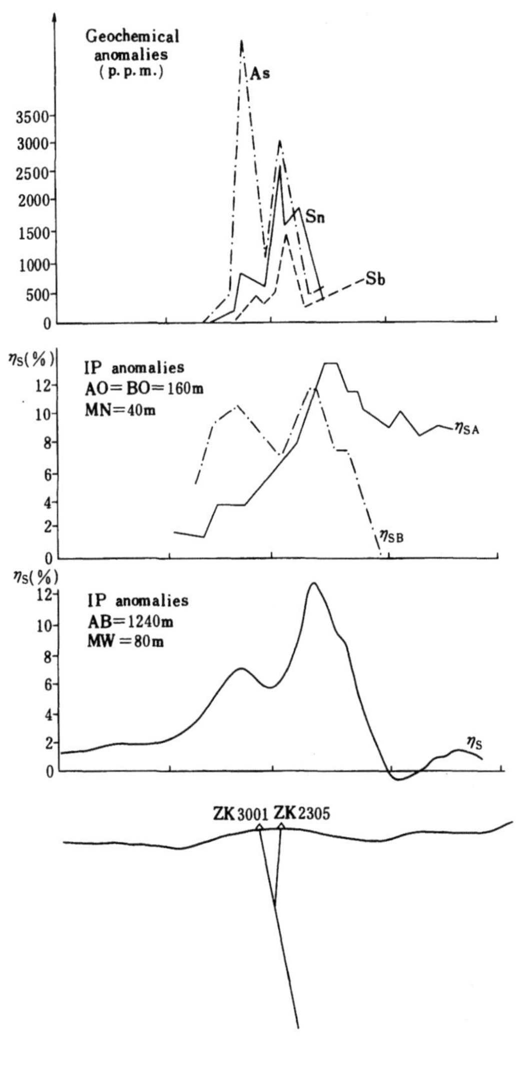


**Fig. 4.** Profile showing geophysical and geochemical anomalies in HNSSW area

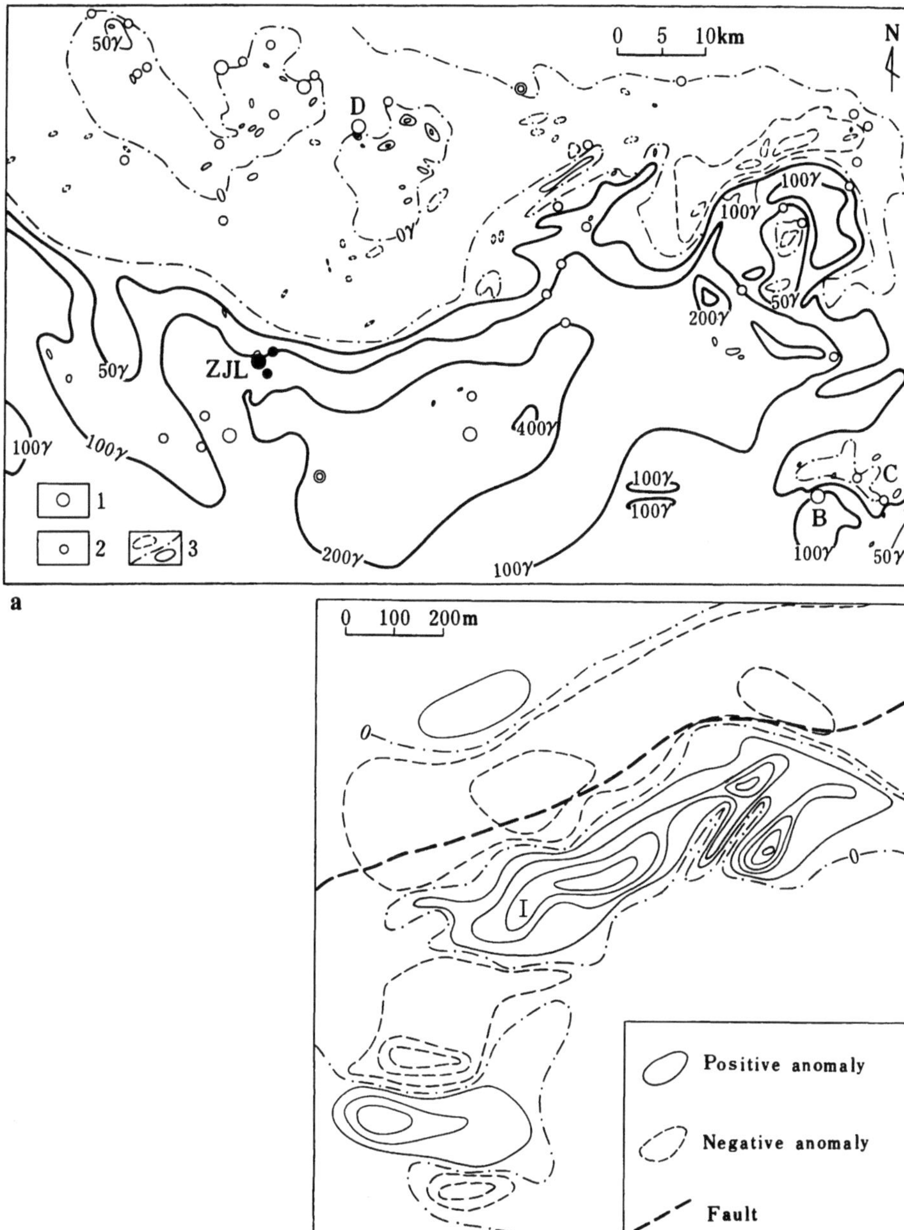


**Fig. 5. a** Schematic map showing the regional aeromagnetic anomalies and Cu-polymetallic deposits in northern Jiangxi; **b** Schematic map showing magnetic anomalies of the ZJL tin mine area

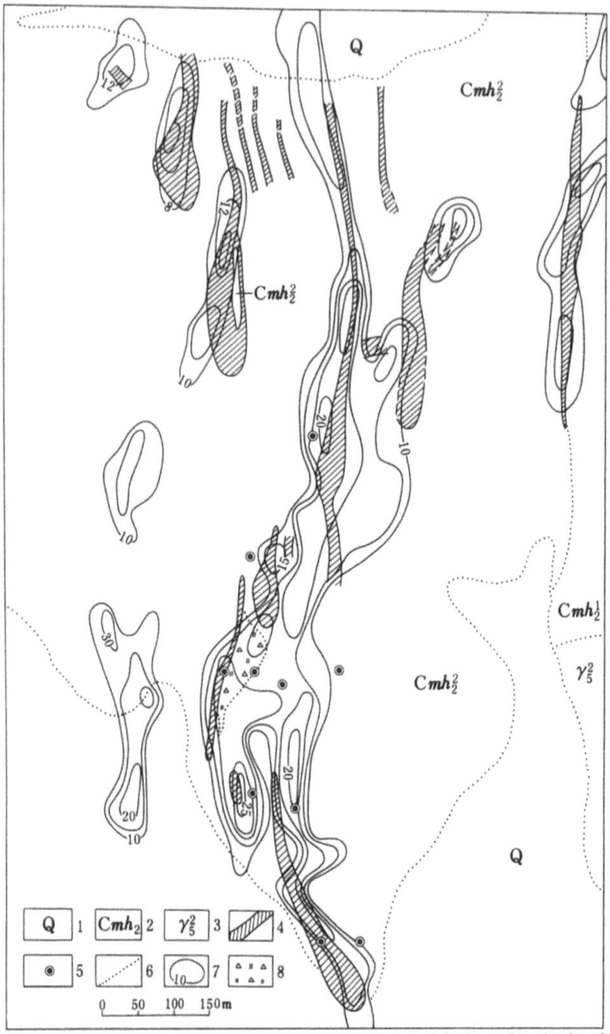


**Fig. 6.** Schematic map showing the geology and induced polarization anomalies in the TIS area

In 1966, magnetic and geochemical surveys on the scale of 1:50,000 were conducted in a mineralized district and its vicinity, which resulted in the discovery of an arcuate magnetic anomaly zone consisting of a series of small-sized local anomalies occurring on the southeastern side of a NE-trending fracture zone, with an intensity of 500–2000γ (Fig. 5). This particular magnetic anomaly zone has been checked by drilling, proving it to be a tin-containing site with strong and large-scale mineralization. Some of the local anomalies represent tin orebodies associated with magnetic minerals. By making use of a geochemical survey, the percentage of accurate drilling will become higher.

In the Tengchong area of Yunnan Province, the distribution of induced polarization anomalies is basically consistent with that of known orebodies, which makes it

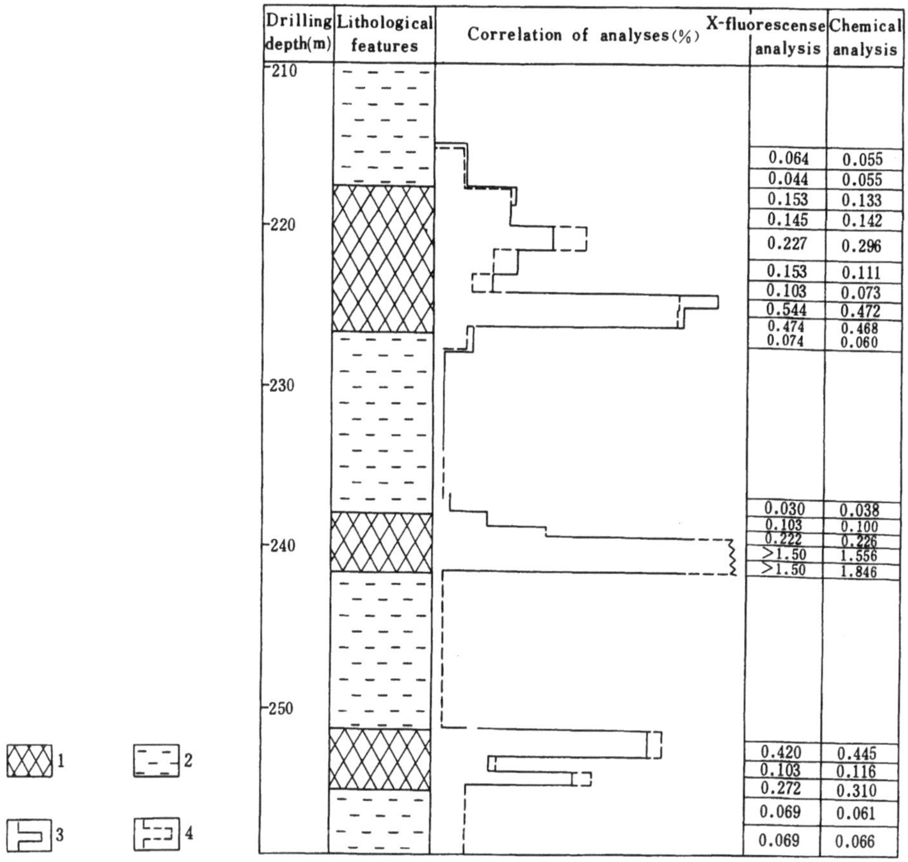


**Fig. 7.** Diagram showing the results of X-ray fluorescent measurement conducted in Guangxi. (after Qin Shuiji, 1982)

possible to delineate orebodies directly by using the anomalies, and the mode of occurrence of orebodies can also be detected by measurements using different sonde spacings (Fig. 6).

When X-rays emitted from radio-isotopes hit the wall of the borehole or rock exposures, they cause the various elements in the rocks and minerals to produce a certain characteristic X-ray spectrum (X-ray fluorescence). The measured fluorescence energy indicates the kind of element, and its intensity indicates the concentration of the element. The results of measurement obtained by using the portable HYX-I XRF spectrometer and of the JXY-I logger are consistent with the chemical analyses. Satisfactory results of measurements have been obtained in tin deposits in both Sichuan Province and Guangxi Zhuang Autonomous Region (Fig. 7).

Fig. 8 shows the results of electrical sounding conducted in Yunnan Province. As the tin mineralization is closely related to the concealed granite body and its configuration, cusps of the granite must be favourable positions for mineralization. The relief of the granite contacts can be determined satisfactorily by electrical sounding.

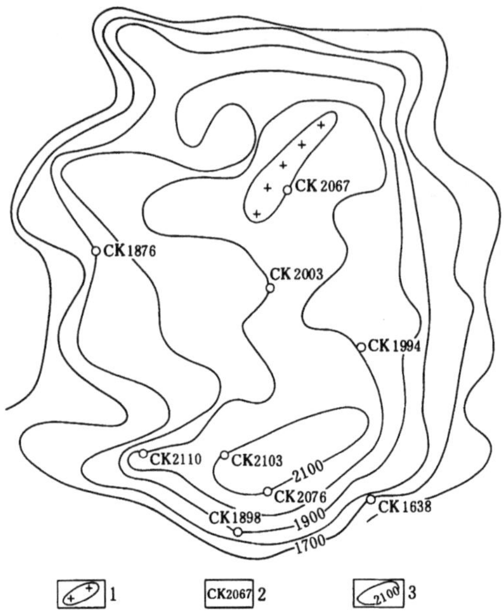


**Fig. 8.** The relief of the interface of granite determined by electrical soundings

## Prospects for Future Investigations to be Conducted in Combination with Tectonic Analysis

This paper deals also with the relations of tin mineralization to the Mohorovicic discontinuity (Fig. 9), deep-seated fractures, and suture lines (Fig. 10). We have found that there exists a definite relationship between the various types of tin deposits and tectonics. Large sulphide tin deposits are commonly distributed around the Qinzhou-Ningbo tectonic line or in this northwestern part, as inferred by geophysical data. Vein-type tin mineralization is mostly found along an E-trending tectonic line, and the tin placer deposits occupy a vast area, so special attention should be paid to the geophysical prospecting for tin placers in the offshore area.

## Investigations into the Relationship Between the Lithology-Controlling and Ore-Controlling Structures of Tin Deposits

The gravity and magnetic anomalies can be used to infer various structures, especially the compounding portions of the structures, which tend to serve as favourable positions for mineralization. Igneous activity is of importance in the region. The igneous rocks formed in the middle and early stages of the Yenshanian orogeny – the host rocks of mineralization, with both hornblende and biotite granite and biotite-granite containing the tin mineralization.

It is evident from Fig. 11 that magnetic anomalies are largely distributed around the granite mass to form a circular or semi-circular zone which serves as a manifestation of magnetitization and pyrrhotitization at both the endo- and exocontacts.

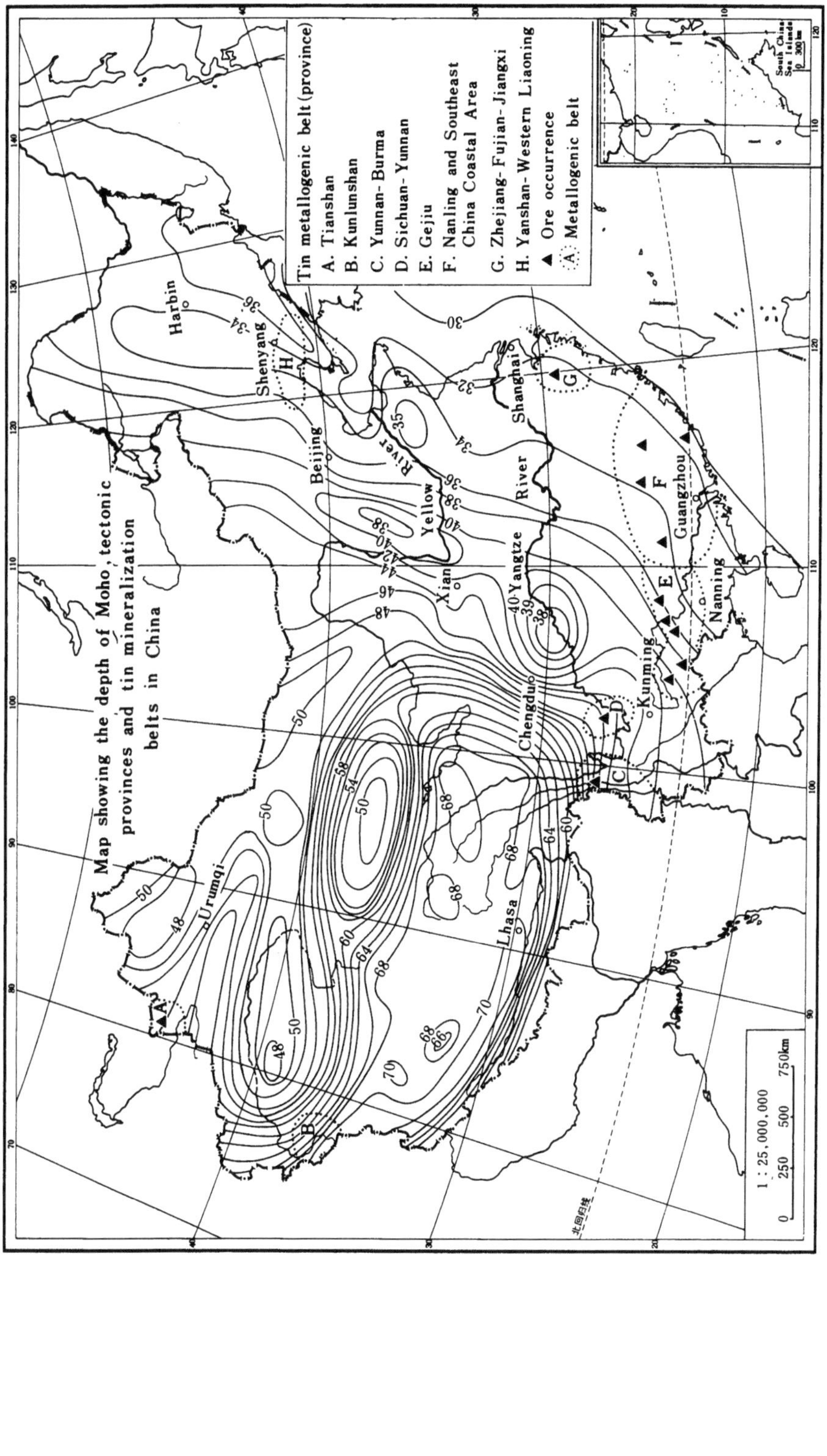


**Fig. 9.** Map showing the depth of the Mohorovicic discontinuity, tectonic provinces and tin mineralization belts in China

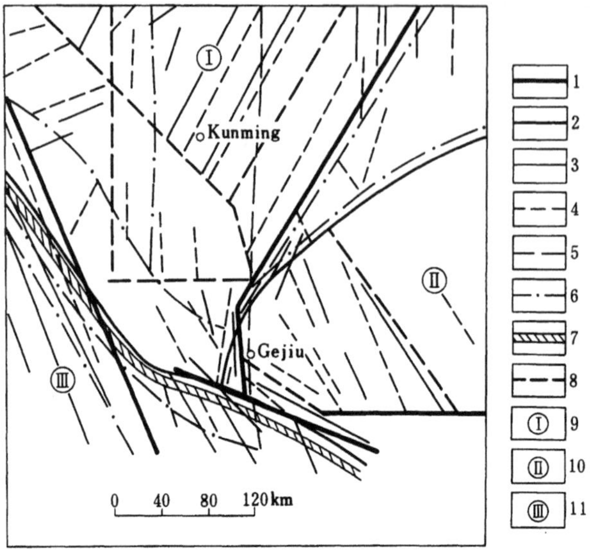


**Fig. 10.** Schematic map showing the distribution of aeromagnetic axial line and tectonics in Gejiu City area. (after Wu Gongjian and Gao Rui, 1983)

Fig. 12 represents the results of gravity measurements conducted above the intrusions. The area of the gravity negative anomaly above the Gejiu Granite is larger than that of the exposed body. Sub-outcropping granite has been proved beneath the anomaly.

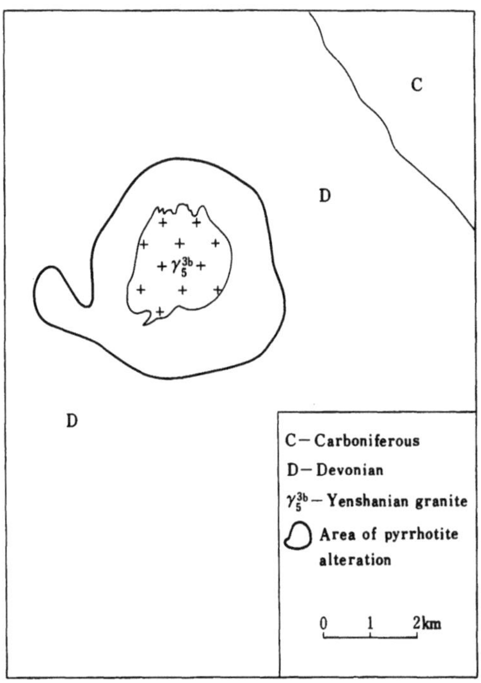


**Fig. 11a.** Schematic map showing the distribution of pyrrhotitization in the DC ore field

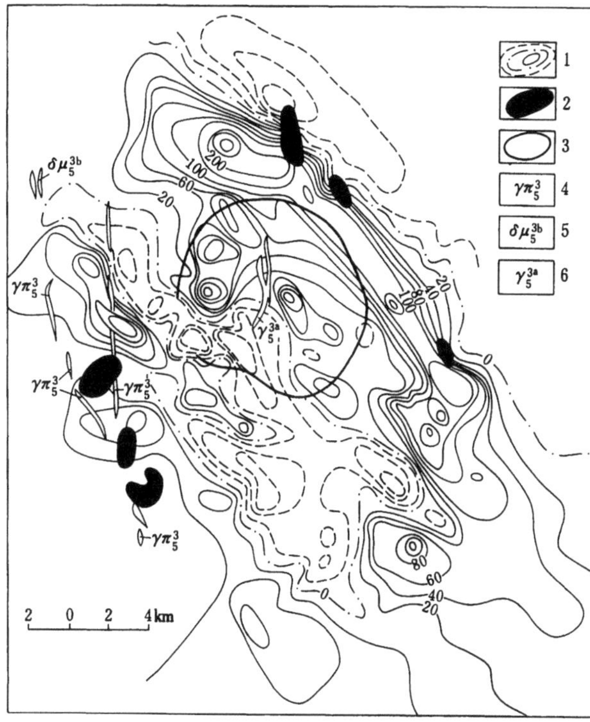


**Fig. 11b.** Diagrammatic map showing the spatial relationship between the DC in field and the aeromagnetic anomalies

A primary dispersion halo in this region is characterized by a zonation of components such as, in outward order, W, Sn, Mo, Cu-Pb, Zn, Ag, and Mn; while in another region the dispersion halo in soil is characterized by an outward zonation as represented by Sn, Cu, Pb. In the same region the vertical zonation of the primary dispersion halo in descending order is Sn, Zn, Cu.

## Selection of Appropriate and Effective Geophysical Methods for Different Types of Ore Deposits

For sulphide-type ore deposits, various electrometric techniques, with the induced polarization method as the main one, should be applied, and magnetic prospecting may usually also provide good results. In prospecting for oxide-type ore deposits, mid-gradient profiling is advisable, while the high-resistance technique is reasonably good for the prospecting of tin-bearing quartz veins and good results have been obtained in other countries by the application of the piezoelectric method. In prospecting for tin placer deposits both on land and near shore, electrical sounding, seismic and sonicseismic techniques have been applied.

Based on our own experience acquired over the past 30 years, we have established a chart of application of appropriate and effective methods with explanation of their aims at various of exploration in any new area (Fig. 13).

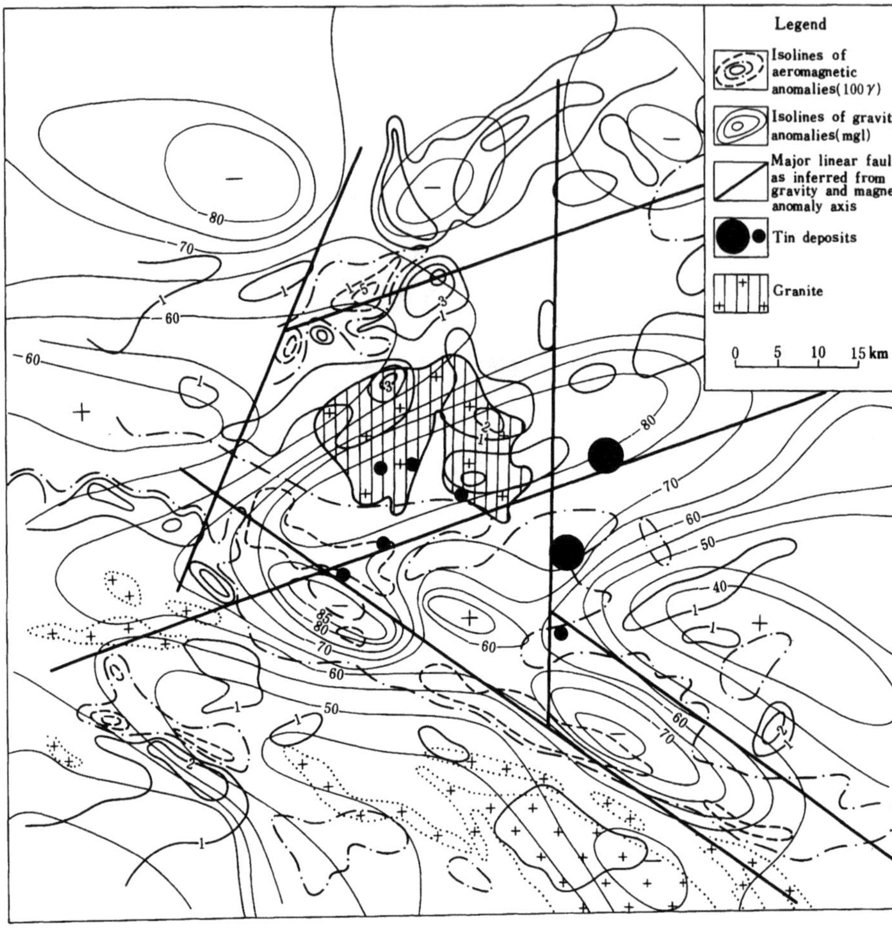


**Fig. 12.** Integrated map showing the gravity and aeromagnetic surveys and geology of Gejiu region. (after Wu Gongjian and Gau Rui)

Geological models for typical tin deposits (or ore field) should be established, which may later be converted into geophysical or geochemical models, and then into mathematic models for prediction of tin resources and for subsequent computer modelling.

*Acknowledgements.* The organizations and individuals whose data have been quoted in the present paper are sincerely acknowledged by the authors.

## References

Wu Gongjian, Gao Rui, 1983. On the relationships between regional aeromagnetic anomalies and deep geological structures in Eastern China. *Regional Geology of China,* No. 6, 83–96.

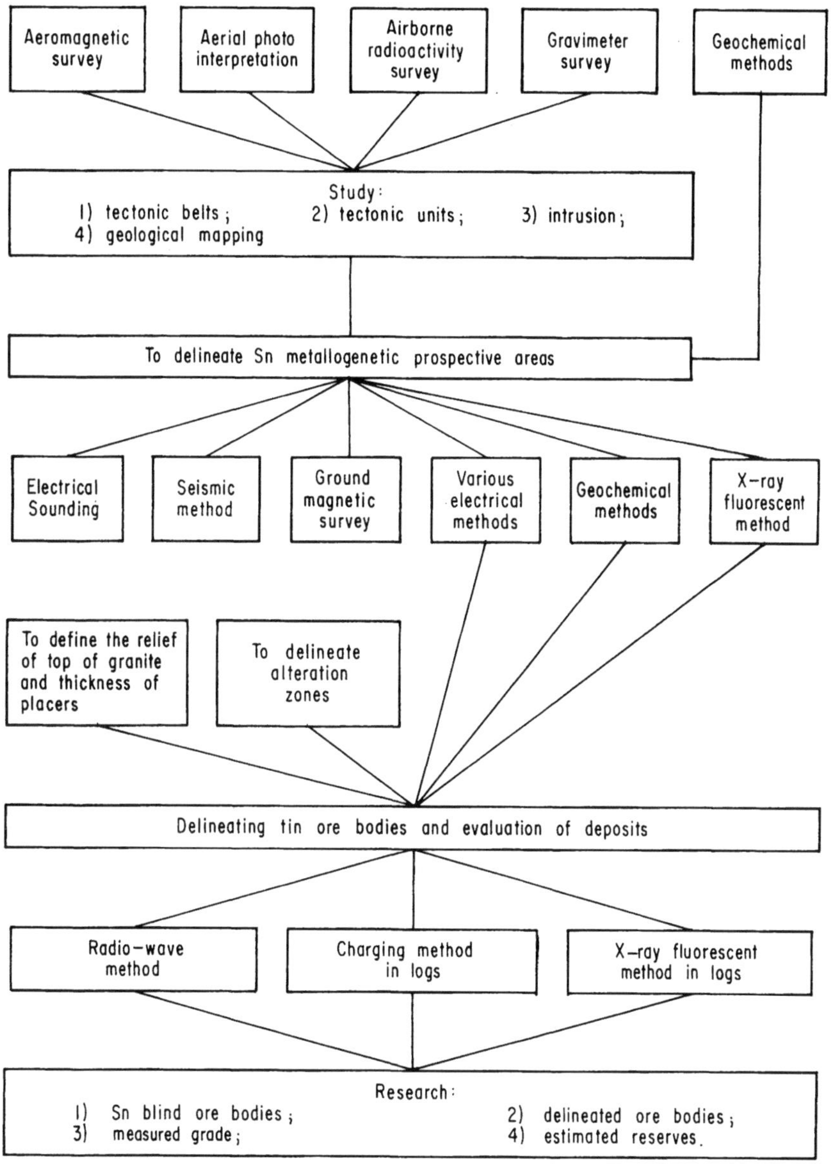


**Fig. 13.** Chart showing application of appropriate and effective methods in prospecting for tin at various stages of development

# 6.2.2 Geological Characteristics of the Tin Deposits of China and the Basic Methods of Prospecting and Exploration

Li Xiji[1]


## Abstract

The tin deposits of China were formed chiefly in the late Mesozoic and occur in marginal belts of the crustal plate or in piedmont troughs of South China where S-type granites are well developed. Primary deposits of cassiterite-sulphide type with Devonian and Triassic carbonate strata as their country rocks are the main source of tin reserves. Placer tin deposits occur chiefly as diluvial and colluvial red clay in karst terrains.

Priority of exploration was given to the tin-placer deposits. Sampling by systematic Bangka drilling, chemical analysis or panning was carried out into the tin or cassiterite contents for different styles of placer deposits. For primary tin deposits, co-ordinated core drilling in a pattern of drifts and crosscuts was the main means of deposit assessment. The accompanying polymetals were comprehensively explored and their utilization was emphasized.

Many new deposits were found chiefly through 1/200,000—1/50,000 regional geological surveying in which heavy mineral and geochemical prospecting methods have proved effective.


## Introduction

This paper first briefly describes the history of exploitation and geological investigation of China's tin deposits. Based on the plentiful data obtained from large scale prospecting and exploration since the founding of the People's Republic in 1949, we have formulated metallogenetic theories and working methods. This has resulted not only in the enlargement of old mines but also in the discovery of many new commercial tin deposits and types, all of which have greatly expanded the tin industry and its future prospects.

## I

China has abundant tin mineral resources. In addition to the well-known Gejiu tin deposit in Yunnan, there are hundreds of recently discovered ones of different sizes, the Dachang tin deposit in Guangxi being the largest.

[1] Bureau of Geology and Mineral Resources of Yunnan Province, Ministry of Geology and Mineral Resources, Kunming, Yunnan

China was one of the earliest bronze-producing countries. As far back as 2,000 years ago in the West Han Dynasty, Gejiu was already known for its tin mining. According to the customs records for the time span of 1889—1983, tin yields totalled about 700,000 t. Guangxi and Hunan also have an early history of tin mining and metallurgy. Early in the Song Dynasty, Dachang began to be mined for silver, and in the early years of the Qing Dynasty, it turned to tin mining. Before 1949, the maximum annual output of tin in Guangxi was 3,419 t and the main mining sites were the Fuchuan-Hexian-Zhongshan tin placer districts.

Before 1949, China's tin deposits had not been explored systematically. Geological surveys were carried out in a few old mines, and Gejiu had been investigated successively by many Chinese and foreign experts since the beginning of this century, while the most detailed work was done by Prof. Meng Xianmin and co-workers in the 1930s. This work was summarized by Meng and Chen (1937). In the 1940s, a regional geological survey, and study of the mineral deposits and mining geology were undertaken by Liu Jinxin, Xiong Bingxin and others (Liu and Lixiji, 1957). All their work laid a good foundation for the geological work to follow. For the Fuchuan-Hexian-Zhongshan tin districts and the Dachang tin deposit and their peripheries, geological work was done by Chang Geng, Meng Xianming, Xie Jiarong, Wu Leibo and others in the 1940s. On the aspects of stratigraphy, intrusive rocks and tin placer deposits, their work furnished a good basis for the work done later. Prof. Xie Jiarong wrote a paper in 1945 entitled "Summary of Fuchuan-Hexian-Zhongshan Tin Placer Deposits in Junan-Guangxi Border Area and Concurrently on the Distribution of Tin Mineralization Belts in China". The Xianghualing tin deposit in Hunan had also been studied by Meng Xianmin and Chang Geng. They wrote a paper in 1936 entitled "Geology of the Tin Deposit in Xianghualing, Lingwu County, Hunan".

Since 1950, to meet the need of construction, large scale geological prospecting and exploration for tin deposits were undertaken by the geological field parties attached to the Ministries of Geology and Metallurgy. Not only has enlargement of the tin deposits been achieved in the old mines and their peripheries, but also many new commercial tin deposits and new ore types have been discovered and form new metallogenic tin areas or belts. By now, we have already identified some medium to large primary tin deposits and tin-bearing stratigraphic horizons such as the upper part of the early Proterozoic Sibao Group represented by the north Guangxi tin deposits, the upper part of the middle Proterozoic Huili (Kunyang) Group represented by the Chahe tin deposit in Sichuan, the late Proterozoic Sinian Dengying Formation represented by the Zengjialong tin deposit in Jiangxi, the middle Cambrian represented by the Dulong tin deposit in Yunnan, the middle to upper Devonian represented by the Dachang tin deposit, the Carboniferous represented by the Lailishan tin deposit in west Yunnan, the Permian represented by the Huanggang tin deposit in Inner Mongolia, the middle Triassic represented by the Gejiu tin deposit, and the Jurassic-Cretaceous represented by some tin deposits in Guangdong and Fujian. Most placer tin deposits occur in unconsolidated Quaternary sediments.

The dates of granite intrusion related to tin mineralization are: Jinning epoch at 860—794 Ma in the tin belt of the Xikang-Yunnan Axis, the Caledonian epoch at 448—382 Ma in the north Guangxi Sn-Cu (Ni) deposit, Hercynian epoch at 326—274 Ma in the SW Guangxi-Sn-Cu deposit, the early Yenshanian epoch at 196—135 Ma in the Nanling W-Sn belts of the Hunan-Guangdong-Guangxi border region, the late

Yenshanian epoch at 124—64 Ma in the Youjiang (Guangxi-Yunnan) Sn-polymetal belts, and the Himalayan epoch with isotopic ages younger than 70 Ma in the W-sub-belt of the tin belt in west Yunnan.

## II

In the practice of ore-searching, we have accumulated geological data and gradually enriched our knowledge of geological characteristics of the tin deposits in China and therefore have preliminarily formulated a series of tin metallogenic theories and working hypotheses. The following are some of our views on the ore forming conditions and regularity of distribution of the tin deposit:

Tin deposits are found in the marginal fold-fault belts of the continental crustal plate or in the piedmont depressions in which there exist regional geochemical tin anomalies and multi-staged ore source beds as well as various tin-bearing rock assemblages.

Most of the tin belts of east China are subduction-related along the Pacific plate margin, whereby the inner zones of the Chinese mainland became an S-type granitoid arc resulting from crustal anatexis. Major tin deposits like Gejiu and Dachang all occurred in fold-fault belts of piedmont depressions. The Gejiu tin district is located in a bay-like trough between the Xikang-Yunnan Axis, the north Vietnam old terrain and the denudated range of the Ailaoshan Precambrian metamorphic terrain (Fig. 1). Some Fe-Cu deposits and strata of the Xikang-Yunnan Axis contain 10—50 ppm Sn (Sun Kexiang et al., 1980). A few samples from the Ailaoshan metamorphic rocks contain 45—150 ppm Sn (Peng Chengdian, 1983). These data suggest that there exists a close connection between the Precambrian beds and their metallogenic environment and the later magmatic activities and mineralization. The Precambrian tin-bearing metamorphic rocks of the Yunnan continental crust might serve as the material base for later forming the tin-bearing granites (Li Xiji et al., 1980).

1. The tin lode bearing horizons in Gejiu are the lower part of Gejiu Formation ($T_2K_1$), of which 105 samples give an average Sn content of 8 ppm by spectral analysis, whilst in the barren middle-upper beds and members, the tin content

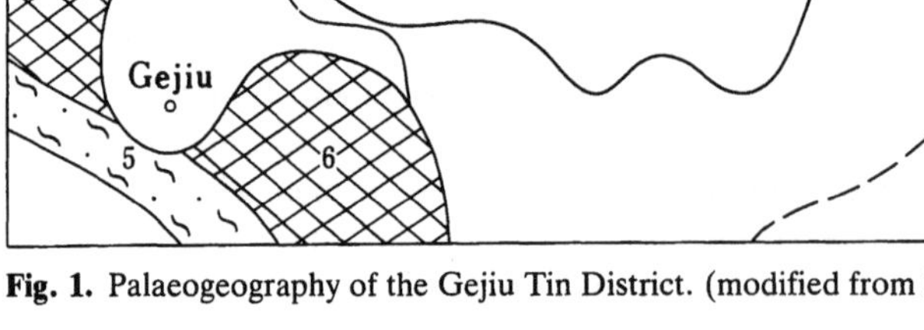


**Fig. 1.** Palaeogeography of the Gejiu Tin District. (modified from Lu Jincai, 1983)

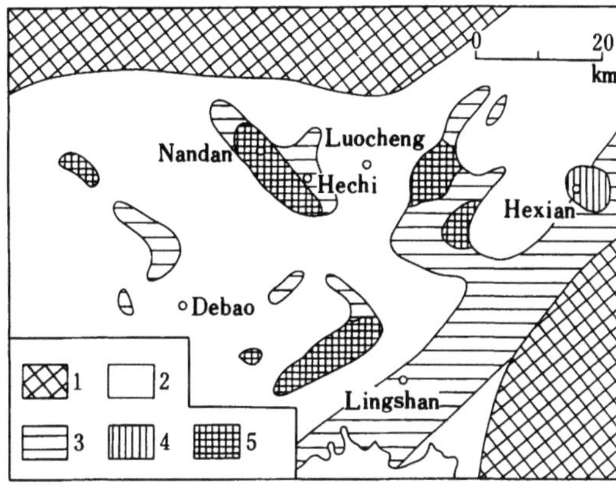


**Fig. 2.** Diagrammatic map showing sedimentary facies of late Devonian and distribution of tin ore source beds in Guangxi. (after Xian Baiqi, 1984)

approaches zero. Gejiu tin-bearing granites contain 10—25 ppm Sn. Guangxi and its neighbouring provinces may be called a geochemical tin metallogenic province with an anomalously high Sn abundance. The average Sn content of the granite in the whole of Guangxi is 9.6 ppm. The Sn content in the early Proterozoic basic-ultrabasic rocks and metamorphic rocks in the north Guangxi tin district is about 10 ppm (Ma Fujun, 1983). Jinning (Xuefeng) cycle granites contain 14 ppm tin. In the chief bedded and stratabound cassiterite-sulphide deposits and their peripheries of Dachang and Fuchuan-Hexian-Zhongshan, the tin-bearing horizons of Devonian carbonates and siliceous beds contain 10—30 ppm tin with local highest contents probably up to hundreds of ppm. The ore source beds with higher Sn content all occur in the narrow and elongated deep trenches along the well-developed regional fractured zones in the flat piedmont depression (Fig. 2) in which element assemblages are complex. Other tin mineralized areas and belts also have similar conditions.

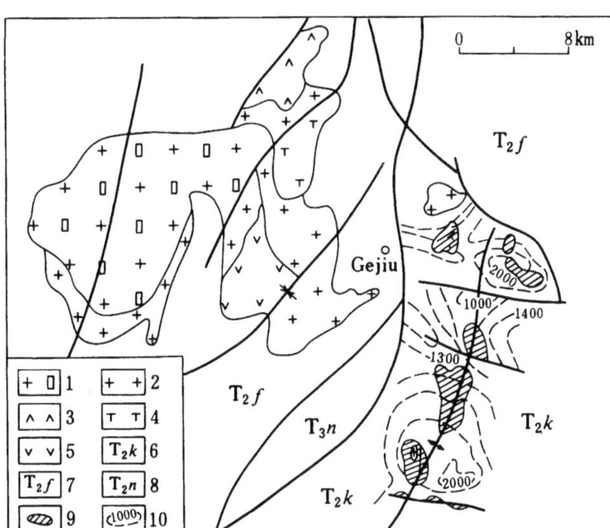


**Fig. 3.** Simplified geological map of the Gejiu tin deposits

2. In the polyclyclic multi-staged orogenic-magmatic-metallogenic activities, there was a tendency for the crustal anatectic tin-bearing magma to become progressively enriched with time and eventually to become rich enough to form commercial tin deposits in the late stages of the Yenshanian cycle. In the Sn-metallogenetic province of South China, from the starting point of ore source beds (mainly Precambrian marine volcanic rocks) the Sn (W)-bearing granites of different ages had inherited their tin (and W) through the poly-cyclic orogenic movements. In this region, the granites from the Jinning cycle to the Indosinian cycle contain 7—9 ppm tin (Xu Keqin et al., 1980), leading to large to intermediate sized tin deposits, chiefly in the late Yenshanian granites. Nevertheless, in some local open deep fracture (rift) zones there was an initial rich mantle source of tin, resulting in the earlier (Proterozoic) tin deposits, e.g. Chahe-type tin deposit. Such deposits even formed in the older basic-ultrabasic rocks and were coexistent with Cu-Ni, e.g. Jiumao-type tin deposit of N-Guangxi, where the role of the Xuefeng-Caledonian granites might have only been thermal activation and superimposition.

3. The tin ore bodies are localized in the outer and inner contact zones of the altered acid to ultra-acid stocks or dykes protruding from composite granite batholiths, controlled by secondary structures and their intersections (Fig. 3). Concealed and semiconcealed granite batholiths which underlie the 1,000 m thick carbonate beds are places favourable for seeking tin fields of cassiterite-sulphide type, e.g. in the east part of Gejiu and the Longxianggai tin field of Dachang.

4. Tin deposits are extensively distributed in some ten provinces or autonomous regions, but their mineralization, though diverse in metallogeny, is concentrated temporally and spatially in the following types: formed in the late Mesozoic; occurring in Devonian and Triassic carbonate beds in South China where S-type granites are well developed; primary deposits of cassiterite-sulphide type with their associated placer deposits chiefly as diluvial and colluvial red clay on mild slopes; and flat terraces and depressions in karst terrains (Fig. 4) derived mainly by chemical weathering of the cassiterite-sulphide deposits. Alluvial tin placer deposits also play important roles in Guangdong, Hunan and some parts of Guangxi.

5. Tin mineralizations always coexist and are accompanied by polymetals in four main mineral assemblages:

a) The Li-Be-Nb-Ta-rare earth-Sn assemblage occurring in deposits of pegmatite type.
b) W-Sn assemblages in deposits of cassiterite-skarn type.
c) Fe-Sn assemblages in deposits of cassiterite-skarn type.
d) Sn-Cu-Pb-Zn-Sb assemblages in deposits of cassiterite-sulphide type.

In the first three assemblages, tin is subordinate and a byproduct of the other main elements, while in the fourth assemblage tin is the chief element co-existing with the other subordinate elements.

In intermediate and large size tin fields, most types of mineral deposits and ore-forming elements exhibit a perfect zonal distribution. From historical records, Gejiu, Dachang and Xianghualing tin mines all began as producers of silver (with Pb, Zn) from the outskirts and superficial parts of the tin mines proper. The main deposits of Sn, Cu and Zn were found later in the mineralization centres.

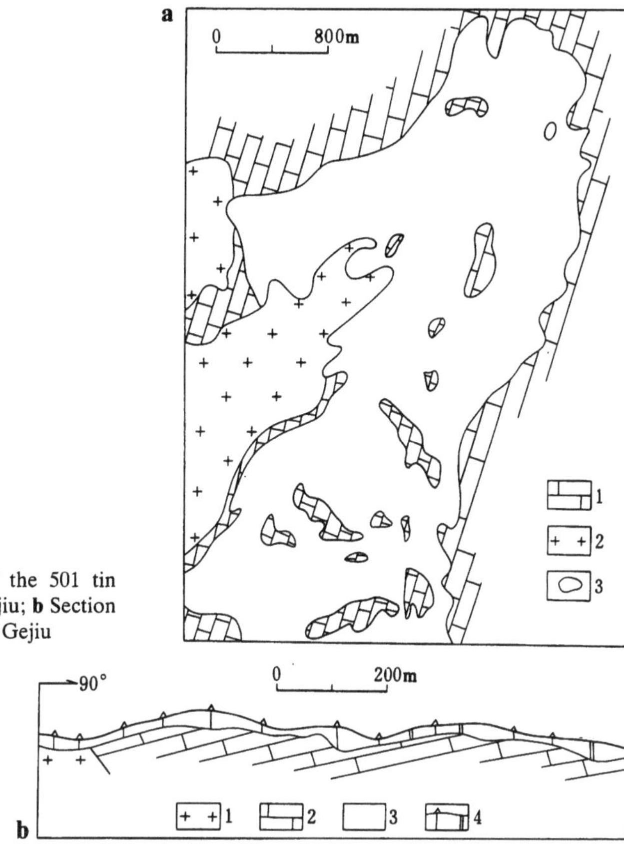


**Fig. 4. a** Simplified plan of the 501 tin placer block of Niushipo, Gejiu; **b** Section of the 501 block of Niushipo, Gejiu

## III

Over the past thirty years, our basic working procedure has been that priority of exploration is given to placer tin deposits. Practice has led to the new idea that the Gejiu red argillaceous tin placers are not merely man-made tailings; they are not "sod ore" or glacial sediments accumulated in some gullies, but are placers characterized by a wide distribution, great thickness (5–30 m, locally 40–70 m), exposed on the ground surface or covered by man-made tailings. The cassiterite is fine grained (0.06–0.25 mm), pretty evenly distributed, with higher content (Sn 0.08–0.30%) and presents easy mining and ore dressing; hence, these placers should be selected first for economic exploitation. From such an understanding, we have concentrated our efforts to systematically explore such tin fields as Niushipo and Laochang by means of banka drilling and some shallow boreholes. We used chemical analysis to substitute for the traditional panning to get more accurate Sn contents of the core and channel samples. In this way, we quickly completed the detailed exploration of the whole placer tin deposit in Gejiu to meet the needs of reconstruction and expansion of the tin industry.

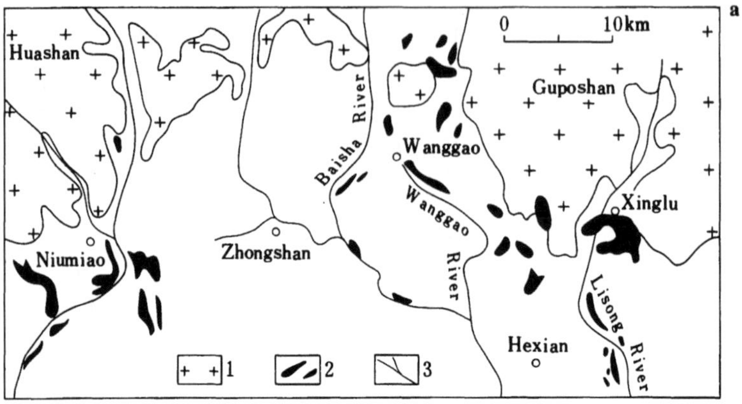


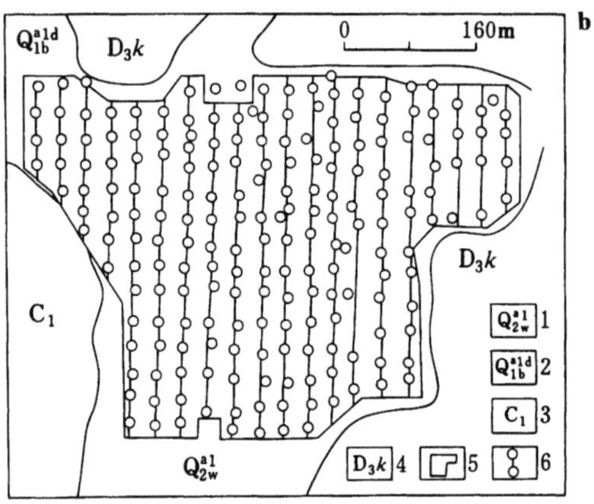


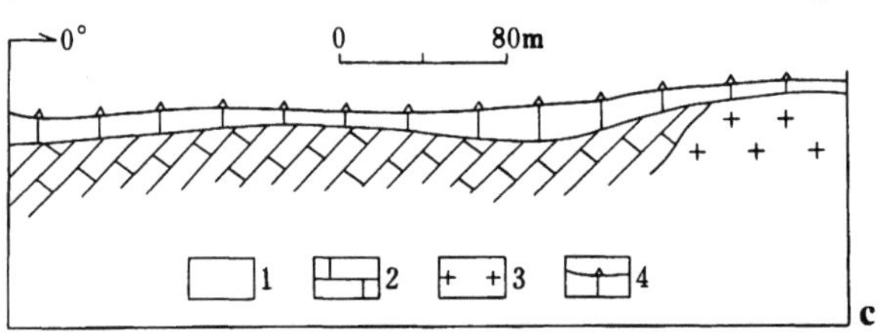


**Fig. 5.** **a** Diagrammatic map of the Fuchuan-Hexian-Zhongshan tin placer district, Guangxi (from Shan Zhenhua, 1983). **b** Plan of alluvial and diluvial tin placer deposit and Banka drill net in Shuiyanba, Guangxi. **c** Geological section of No. O exploration line at Xinggui tin placer deposit in Shuiyanba, Guangxi

The types and working methods of the tin placer deposits in Dachang are similar to those in Gejiu. For all the alluvial and colluvial tin placers in Fuchuan-Hexian-Zhongshan district and the alluvial tin placer deposits in Guangdong and Hunan, banka drilling was used for systematic exploration (Fig. 5-A, -B, -C), and panning was used to measure the content of cassiterite characterized by coarser granularity and lower content as well as to measure and calculate the reserves of all usuable heavy minerals comprehensively.

For the primary tin deposits in old mines such as Gejiu, we coordinated core drilling with tunnel drifts and crosscuts of the mines and use the method of "seeking new orebodies from the known ones". In this way, several old mines such as Laochang, Malage, Songshujiao and Kafang have been rejuvenated into rich and large ones. From some geological, geophysical and geochemical profiles from the surface down-

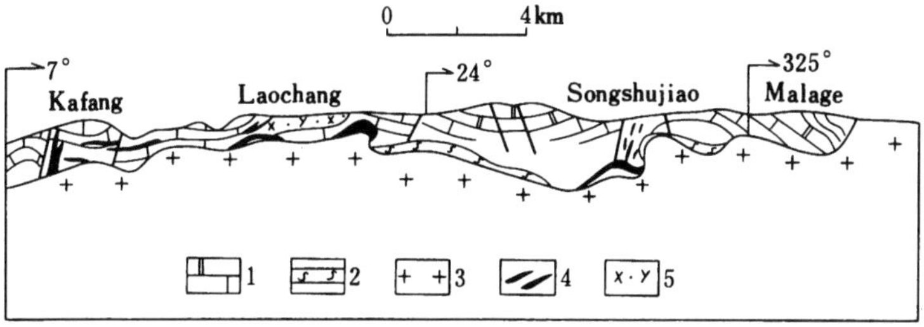


**Fig. 6.** Diagrammatic longitudinal section of Gejiu tin district. (from 308 Geologic Party, 1980)

ward, we can see that, in the upper part, there are the thick carbonate beds with gentle and open folds and well developed fractures, and in the lower part, there is the concealed underlying granite batholith (Fig. 6). Anticline and dome structures seen at the surface always occur in company with concealed diapiric stocks and form the mineralized centres. With these regularities in mind, we used adits as the main with core drilling as supplementary means to explore the upper orebodies of stockworks, vein swarms, interbedded oxide ores and some large ore pockets in intersected zones of ore-bearing fractures. For contact metasomatic skarn and sulphide orebodies, exploration has been done chiefly by drilling from surface and tunnelling in a fanshaped pattern (Fig. 7). The unmineralized areas between the tin fields are synclines and deep-buried batholith depressions. Isodepth contour maps reflecting undulations, depths of burial and forms of the batholith have been drawn resulting from physical prospecting and a few core drills as well as by decouloration and recrystallization of the carbonate beds. Ore-searching indicators such as minor undulations in large synclines, small protrusions in big depressions, comparatively universal tin-bearing beds and members, lithological characters and lithofacies assemblages favourable for mineralization and their contact relations with granite bodies (concordant, discordant), extentions and intersections of tin-bearing fractures, ragged and collapsed

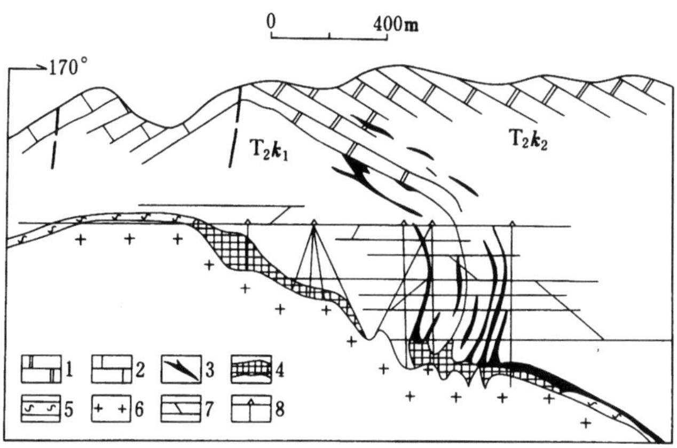


**Fig. 7.** Selected section of Songshujiao tin field showing underground fan shaped drills coordinated in exploring the primary tin deposit of Gejiu. (after 308 Geologic Party, 1980)

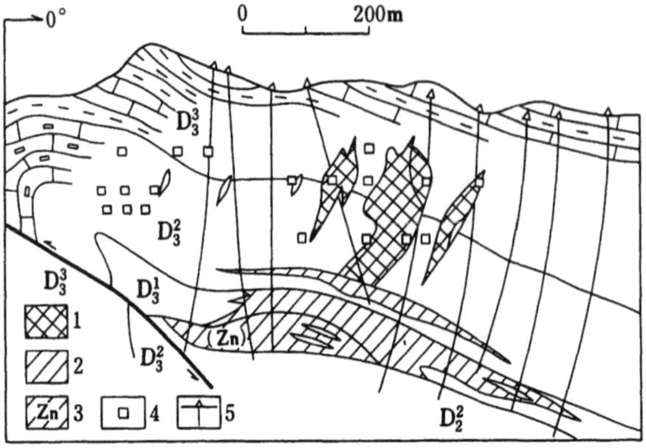


**Fig. 8.** Geological section along No. 203 exploration line of Changpo tin field, Dachang, Guangxi. (from 215 Geological Party)

topography, gossans and other mineralization and alteration features, as well as grades and sizes of tin placers are considered together in predicting deep blind orebodies and determining the sites of prospecting drilling. Ore-forming prognostication based on geology, geophysical and geochemical prospecting, geomathematics and remote-sensing is being made. The success ratio in the initial tests is about 50% (Peng Chengdian, 1984).

For the coexisting and accompanying polymetals in different types of tin deposit, comprehensive exploration and utilization are emphasized. Hard to handle complex ores are under experiment in ore dressing and smelting. Systematic detailed investigation and exploration will not start before satisfactory experimental data are obtained.

Many new deposits have been found chiefly through 1:200,000—1:50,000 regional geological surveys in which heavy mineral and geochemical prospecting methods with stream sediments as the main samples have been especially effective. Geochemical prospecting for primary and secondary haloes quickly delineates target

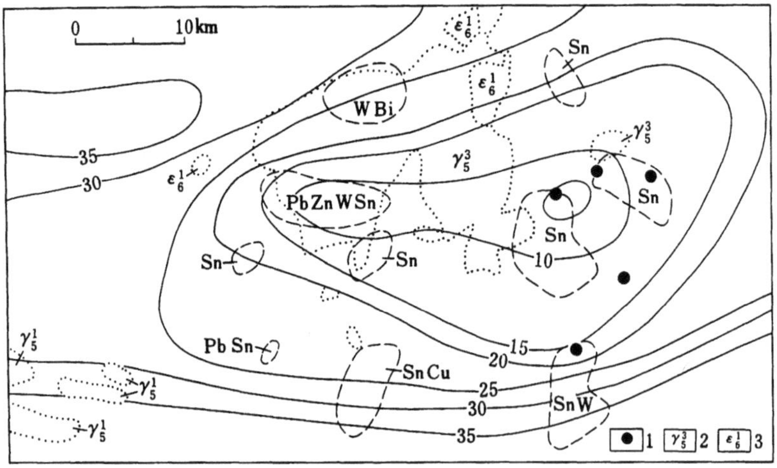


**Fig. 9.** Anomalies of the regional gravity and geochemical prospecting in Gejiu district. (after Geological Field Party of Yunnan Tin Firm, 1980)

areas for geophysical and geological follow-up, and air-borne and ground geophysical methods (gravitational, magnetic and electric) have been effective in discovering favourable ore-forming structures and concealed granite batholiths (Fig. 9) as well as defining magnetic orebodies. Also, by studying the details of the tin-mineralization belts of South China and Thailand–west Yunnan, we have successfully applied the principles to west and southwest Yunnan, west Sichuan, northeast China, Inner Mongolia and the southeast China coastal regions.

In the inland provinces where geological work has been done at more advanced level, ore-searching has entered a new stage. Under the new theory, some known deposits or sites are being re-understood and re-evaluated and a new breakthrough has been achieved. The porphyry tin deposit of Yinyan, Guangdong, has been discovered through new geological work realizing that crust source porphyry deposits might exist. A similar story is said to have happened in the Yejiwei tin deposit in Hunan.

*Acknowledgements.* In the process of writing this paper, valuable support has been given by the relevant units of Yunnan, Guangxi, Guangdong, Hunan and Sichuan. Engineers Yang Zhuang and Yang Yonggeng helped to compile the figures, and Shen Xuezhen undertook the drawing of diagrams and maps. Acknowledgements are due to all the contributors.

## References

Chen Yuchuan, 1965. Primary zonal distribution of certain tin metallogenic belt in Guangxi, *Geological Review,* Vol. 23, No. 1 (in Chinese).

Deng Yishu, 1951. Relationship of tin deposit of Gejiu with structures. *Geological Review,* Vol. 16, No. 2 (in Chinese).

Juang Tingran, 1983. On the geological features of Gejiu type tin placers. *Geological Review,* Vol. 29, No. 2 (in Chinese).

Liu Jinxing and Lixiji, 1957. On the formation and characteristics of mineralization of the Kochiu tin deposits, Yunnan. *Acta Geologica Sinica,* Vol. 37, No. 4 (in Chinese).

Meng Xianming and Chen Kai et al., 1937. Geology of Gejiu tin deposits, Yunnan. *Journal of the Geological Society of China,* Vol. 16 (in Chinese).

Taylor, R.G., 1979. *Geology of Tin deposits* Elsevier, Amsterdam.

Tian Kaiming, 1957. Connection of the Fu-He-Zhong tin placer deposits, Guangxi, with karst. *Acta Geologica Sinica,* Vol. 37, No. 4 (in Chinese).

Wang Zhifeng, 1983. Some problems on the mineralization of tin deposits in Gejiu, Yunnan. *Acta Geologica Sinica,* Vol. 57, No. 2 (in Chinese).

Xian Baiqi, 1984. A discussion on the ore-forming conditions and distributional regularity of the tin deposits of Guangxi. *Acta Geologica Sinica.* Vol. 58, No. 1 (in Chinese).

Xu Keqin et al., 1980. Research on the temporal and spacial distribution, rock evolution, genetic types and ore-forming relations of the granitoids in SE China, *Geological Issue of the Journal of Nanjing University* (in Chinese).

Ye Shisun, 1983. Metallogenetic regularity and prognostication of Dachang Sn-polymetals deposits. *Geology and Exploration,* No. 5 (in Chinese).

## 6.3 China: Mineralogy

# 6.3.1 Spectroscopic Analysis and Genesis of Cassiterite

PENG MINGSHENG, LU WENHUA, and ZOU ZHENGGUANG[1]


## Abstract

This paper deals with the spectral characteristics of cassiterite of four different genetic types by means of its absorption spectra in the ultraviolet – visible – infrared regions.

The ultraviolet spectral study shows that energy gap (Eg) values can be used to indicate the formation temperature of cassiterite, because heterovalent isomorphism prefers to take place at a higher temperature. The crystal-field spectral study shows that the cassiterites from Dachang, Xiling and Yunlong ore deposits contain a small quantity of $Fe^{3+}$, $V^{4+}$, $Mn^{2+}$ and $Fe^{2+}$, thus indicating the physicochemical conditions under which cassiterite was formed.

The variation in infrared spectral shapes and intensity provides information on the crystal structure, composition and origin of cassiterite.


## Introduction

As an indispensable part of quantum mineralogy, mineralogical spectroscopy is not only applied to the study of the composition, structure and physical properties of minerals, but may also be used to explain the essential law of accumulation of elements and the origin of minerals. In earlier papers, a systematic investigation of quantum mineralogy on piemontite was carried out by the authors mainly to study its crystal field spectra and to explain the formation process of the Hainan iron deposit (Peng et al., 1981, 1982, 1984). This paper deals with the spectral characteristics of cassiterite of four genetic types by means of its absorption spectra in the ultra-violet – visible – infrared regions.

The samples were collected from four mines:

(1) The Dachang cassiterite-sulphide hydrothermal deposit:
   A) Cassiterite-quartz association of the early stage of mineralization.
   B) Cassiterite-calcite association of the late stage of mineralization.
(2) The Xiling tin deposit associated with volcanism or sub-volcanism;
(3) The Yunlong tin deposit associated with migmatization:
   A) Cassiterite-quartz-tourmaline association of the early stage of mineralization.
   B) Cassiterite-quartz-sulphide association of the late stage of mineralization.
(4) The Limu tin deposit associated with Nb- and Ta-bearing granite and grano-pegmatite.

[1] Research Group of Quantum Mineralogy, Central-South Institute of Mining and Metallurgy

**Crystal Chemistry**

The crystal structure of cassiterite belongs to $D_{4h}^{14}$—$P4_2/mnm$ and Z=2. Each Sn atom is co-ordinated by six atoms at the corners of an octahedron, which forms chains by sharing edges along the tetragonal axis. Two non-equivalent $Sn^{4+}$ sites with a local symmetry of $D_{2h}$ exist in each unit cell, which can be transformed into each other by rotating through an angle of 90° around the C-axis.

The composition of cassiterite is very complicated. According to Bokiyu (1974), up to 48 elements are present in various forms in cassiterite, particularly $Fe^{3+}$, $Nb^{5+}$, $Ta^{5+}$ substituting for $Sn^{4+}$ ions. Chemical analyses are given in Table 1 for cassiterite of different origins. All of them contain $Fe_2O_3$, $Ta_2O_5$, $Nb_2O_5$, $WO_3$, $TiO_2$ and $SiO_2$. The iron content in cassiterite decreases from the Xiling and Limu deposits to those from the Yunlong deposit. The cassiterite from the Limu deposit has the largest content of Nb and Ta. The cassiterite from the Yunlong deposit has Nb, but does not contain Ta (beyond the precision of the instrument), while Xiling and Dachang cassiterite is poor in Nb and Ta.

Colour changes of these cassiterites are mainly related to the valence states and content of iron, and also to the contents of Nb and Ta. Except for the cassiterite from the Limu deposit which has an apparent pleochroism and feeble ring structures, ring structures are commonly seen in other cassiterites, but only weak pleochroism is seen. Furthermore they show great variation in colour, which indicates the compositional and structural changes of cassiterite formed under different physical and chemical conditions, thus providing information on different origins of cassiterite.

1. Ultra-Violet Absorption Spectra

From the spectra measured with a microspectrometer (300—800 nm), it can be seen that the energy gaps of cassiterites of different associations are restricted mainly to the range of 310—410 nm, which corresponds to the Egs varying within 4—3.2 ev and indicates that cassiterite is a semi-conductor of a wide energy gap. The magnitude of the Eg has much bearing on the colours of cassiterite. In the cassiterite from the Dachang deposit, the colour changes from dark-brown (D), brown (B), to colourless transparent (A), as the mineral association changes from cassiterite-quartz-type to cassiterite-calcite-type, and accordingly Eg increases from 3.26 (D), 3.51 (C), 3.59 (B) to 3.60 ev (A) (Fig. 1a). Cassiterite from the Limu grano-pegmatite has the smallest Eg (3.12 ev), as it is black-brown. Cassiterite from the Xiling deposit is also black-brown, and has small Egs (3.29—3.45 ev). In the cassiterite from the Yunlong deposit, various colours ranging from black-brown ($Y_{2B}$), yellow ($Y_{1B}$) to light grey ($Y_{2A}$) are seen, with its Eg changing correspondently from 3.59, 3.70 to 3.73 ev. The ultraviolet absorption spectra of the cassiterite from the Yunlong deposit vary within a range of 300—350 (Fig. 1a), but in the cassiterite from Dachang they obviously move towards longer wavelengths: 350—400 nm (Fig. 1b).

Cassiterites of different colours in the same mineral association have different Eg values, the white, grey-white or colourless having the largest Eg (3.71—3.75 ev). By measuring the polarized absorption spectra of cassiterite (brown) from a Sn-bearing quartz vein in hornstone, we have obtained the absorption edges or energy gaps in

**Table 1.** Chemical composition of cassiterite of different origins

| Sample number | $SnO_2$ | FeO | $Fe_2O_3$ | $Nb_2O_5$ | $Ta_2O_5$ | $SiO_2$ | $WO_3$ | $TiO_2$ | $VO_2$ | Mode of occurrence |
|---|---|---|---|---|---|---|---|---|---|---|
| $D_A$ | 97.42 | | 0.01 | | | 0.26 | | | | Dachang hydrothermal cassiterite sulphide deposit |
| $D_B$ | 98.75 | 0.08 | 0.06 | 0.0047 | 0.0033 | 0.07 | 0.041 | 0.21 | | |
| $D_C$ | 98.43 | 0.08 | 0.04 | 0.0012 | 0.0049 | 0.33 | 0.085 | 0.36 | | |
| $D_D$ | 99.06 | 0.06 | 0.41 | | 0.0055 | 0.17 | 0.12 | 0.37 | | |
| $Y_{2Al}$ | 99.40 | | 0.172 | 0.043 | 0 | 0.043 | 0.06 | 0 | 0 | Yunlong migmatization related tin deposits |
| $Y_{2B}$ | 99.46 | | 0.172 | 0 | 0 | 0.021 | 0.11 | 0.033 | 0 | |
| $X_E$ | 97.35 | 0.17 | 0.82 | 0.023 | 0.0067 | 0.56 | 0.18 | 0.89 | | Xiling tin deposit related to volcanism |
| $X_D$ | 98.50 | 0.08 | 0.64 | 0.0099 | 0.0085 | 0.33 | 0.15 | 0.78 | | |
| $F_2$ | 93.44 | | 0.27 | 0.78 | 0.68 | 0.32 | 0.19 | 0.684 | 0 | Limu tin deposit related to Sn-, Nb-, Ta-bearing granite and pegmatite |
| $N_2$ | 98.43 | | 0.60 | 0.043 | 0.012 | 0 | 0.07 | 1.0 | 0 | |

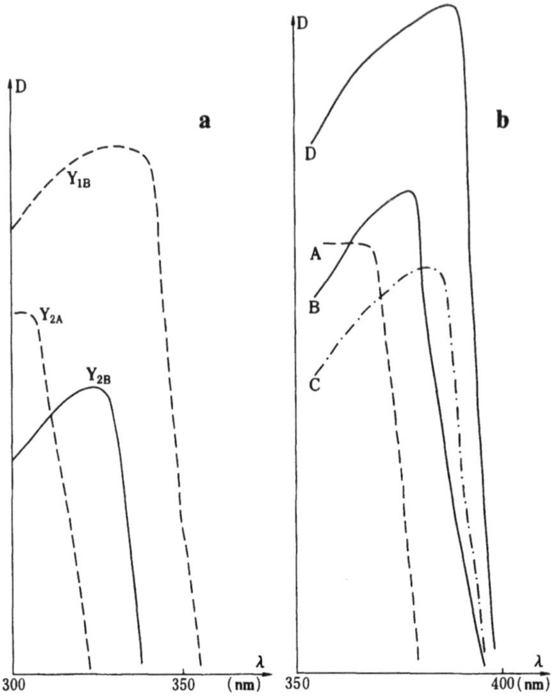


Fig. 1a, b

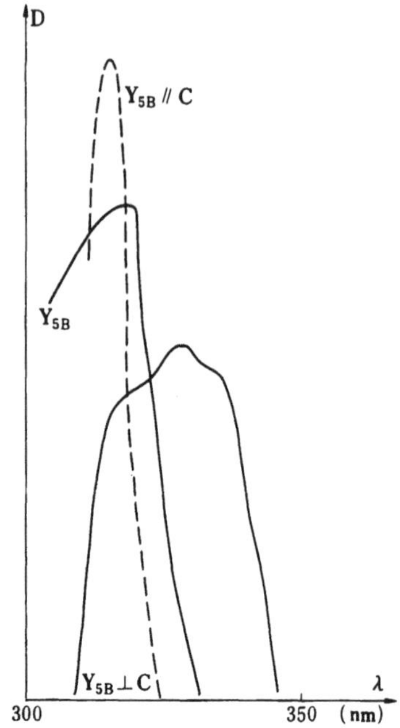


Fig. 2

**Fig. 1 a, b.** UV absorption spectra of cassiterite in Yunlong (**a**) and Dachang (**b**)
**Fig. 2.** UV polarized absorption spectra of cassiterite

two directions: //C-axis, Eg=3.93 ev; C-axis, Eg=3.769 ev (Fig. 2). This result is in agreement with that obtained with the APW method (Arlinghaus, 1974).

The energy gap (Eg) varies inversely with the temperature, the light-coloured or colourless cassiterite has the largest Eg and lower formation temperature than the cassiterite from Yunlong, but the dark one has smaller Egs and higher formation temperature than the cassiterite from Limu. Therefore, the energy gap (Eg) is not only an important parameter of semiconducting properties, but also an indicator of its origin.

The variation in the absorption edges or the Eg of cassiterite is related mainly to the impurity ions. In the energy band structure of $SnO_2$, an intrinsic semiconductor, the 5S orbitals of the Sn atoms constitute the conduction band; and the 2p orbitals of the O atoms, the valence band, with the energy level sequence being plotted in Fig. 3a, according to Nagasaw et al. (1969) and Frohch and Kenklies (1978). It can be seen that there is a large energy gap between the valence band and the conduction band of pure cassiterite, the direct transition from the former ( $_5$) to the latter ( $_1$) (Fig. 3a) only taking place in the ultraviolet region and causing the cassiterite to be colourless in the visible region. Unlike pure cassiterite, which is rarely seen in nature, native cassiterite is usually a non-intrinsic semiconductor containing impurity ions, with the transitional ions, such as $Fe^{3+}$, $Nb^{5+}$, $Ta^{5+}$, etc. substituting for $Sn^{4+}$ as heterovalent isomorphism and causing the energy gap to be sandwiched by the acceptor energy level $Fe^{3+}$ and the donor energy level $Nb^{5+}$, $Ta^{5+}$, and so making it narrowed (Fig. 3b). The addition of the acceptor and donor energy levels to the energy gap prefers higher temperature, thus the energy gap (Eg) decreases and the formation temperature of cassiterite increases successively in a direction from Yunlong, Dachang, up to Xiling.

## 2. Visible-Near-Infrared Absorption Spectra

As shown by the visible-near-infrared absorption spectra of the cassiterite from miarolitic cavities in Dachang, which occurs in black-brown columnar crystals, a strong absorption peak of 460 nm and two weak absorption peaks of 740 and 790 nm are seen along the C-axis (Ne), and a strong peak of 4.10 nm, a less-strong absorption peak of 450 nm and two weak absorption peaks of 730 and 780 nm are present in the direction perpendicular to the C-axis (NO) (Fig. 4); in the former the cassiterite is brown, and in the latter, light-brown. Obviously, the strong absorption peak near the ultra-violet is caused mainly by the substitution of $Fe^{3+}$ for $Sn^{4+}$, thus giving rise to the d-d electron transitions from $^6A_{1g}$ to $^4A_{1g}$ and $^4E_g$ in the distorted octahedron with a site symmetry of $D_{2h}$, and the two weak peaks near infrared were caused by the substitution of $Fe^{2+}$ for $Sn^{4+}$, producing transitions from $^5B_{1g}$ to $^5B_{1g}$ to $^5A_{1g}$. According to the selection rules, the d-d electron transition of $Fe^{3+}$ is spin-forbidden, but that of $Fe^{2+}$ is spin-allowed, so the absorption peaks of the former are weaker than those of the latter, but in this case it is the opposite, as the concentration of $Fe^{3+}$ is greater than that of $Fe^{2+}$. The presence of $Fe^{2+}$ suggests that the cassiterite from Dachang was formed under relatively reducing conditions.

Under the same conditions we measured the polarized absorption spectra of the cassiterite from Xiling. Along the C-axis it is red-brown and two strong absorption peaks of 430 and 470 nm are seen, and perpendicular to the C-axis it is brown-red, with two strong absorption peaks of 420 and 470 nm. There is no absorption peak near 780 nm, i.e. the crystal field spectrum of $Fe^{2+}$, which indicates that the cassiterite from Xiling was formed under relatively oxidizing conditions.

Ring structures are well developed in the above two cassiterites. In the former, for example, colours in the rings change from colourless, yellow blue to brown, as passing through the crystal from the inside to the outside. The absorption spectra of these rings were measured (Fig. 5), with no absorption peak being seen in the colourless part in the visible region, and with a strong peak of 470 nm and a weak peak of

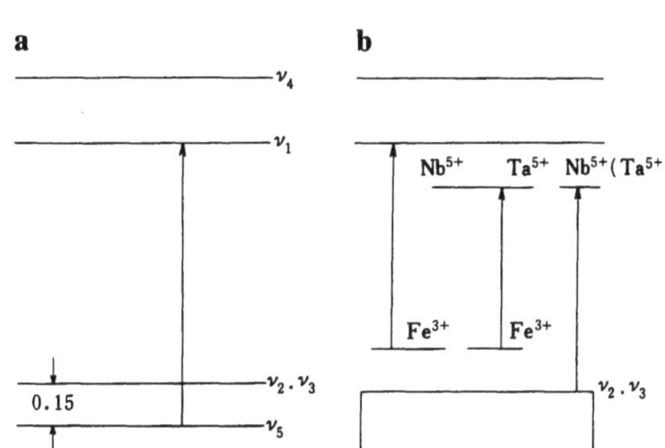


**Fig. 3.** Intrinsic semiconductor (**a**) and non-intrinsic semiconductor (**b**) of cassiterite

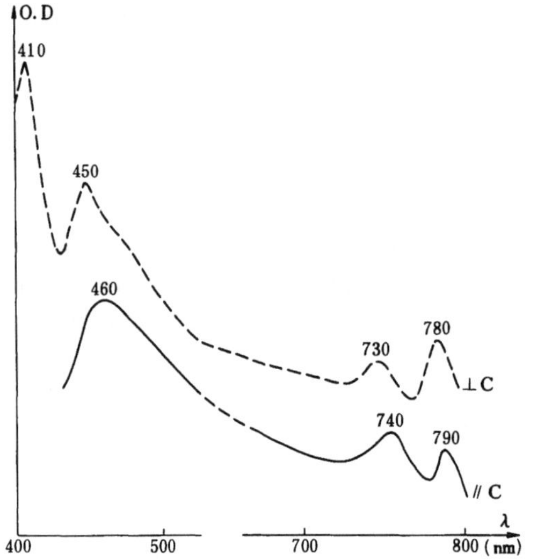


**Fig. 4.** The polarized absorption spectra of miarolitic cavity cassiterite

780 nm appearing in the yellow part, two strong absorption peaks of 430 and 470 nm in the brown part caused by electron transition of $Fe^{3+}$ ions, but the weak absorption peak of about 780 nm caused by ions of $Fe^{2+}$ does not occur, which indicates that the conditions are more oxidizing changing from the inside to the outside.

Various colours are seen in the cassiterite from Yunlong. The polarized absorption spectra are apparently different from those mentioned above, as it has a strong absorption of 760 nm, and a weak one of 625 nm, in addition to an absorption peak of 410–430 nm caused by the d-d electron transition of $Fe^{3+}$ substituting for $Sn^{4+}$ (Fig. 6). These additional spectra are caused neither by the electron transition of $Fe^{3+}$ in the $D_{2h}$ field nor by that of $Fe^{2+}$.

In order to find out the assignment of these spectra we have measured the EPR spectra at room temperature. Eight superfine lines in the spectra indicate the presence of $V^{4+}$ ions substituting for $Sn^{4+}$ in the $D_{2h}$ crystal field, because the electronic configuration of $V^{4+}$ is $3d^1$ with an electron spin $S=1/2$, and a nuclear spin $I=7/2$. The

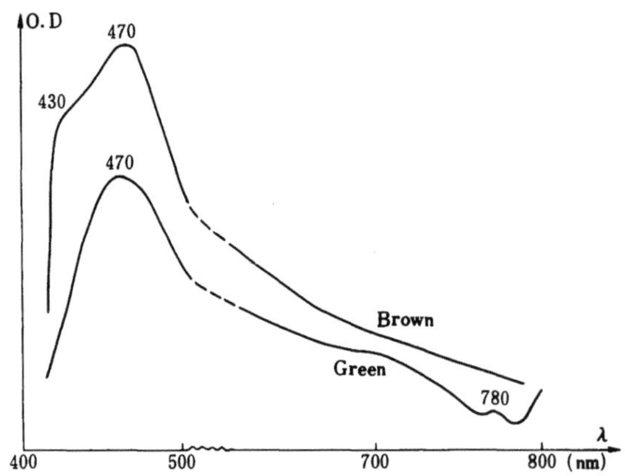


**Fig. 5.** The absorption spectra of the ring structure of cassiterite from Dachang

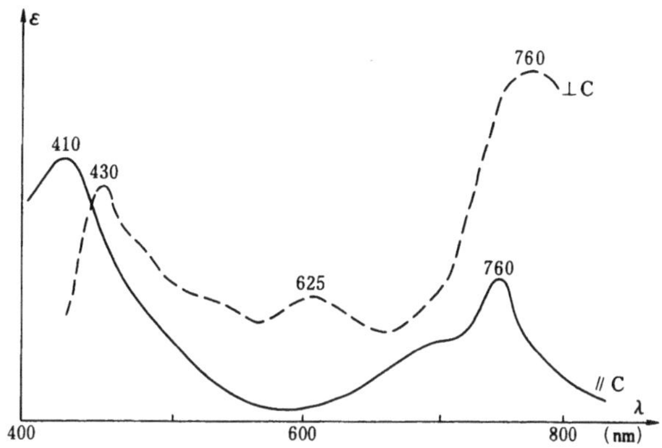


**Fig. 6.** The polarized absorption spectra of cassiterite from Yunlong

mean g-factor of the eight lines is equal to 1.9296 and superfine structure constant at $A=42.46\ 10^{-4}cm^{-1}$.

The substituent $V^{4+}$ ions causes the d-d electron transition in the distorted octahedron with $D_{2h}$ symmetry. Judging from the group theory, the group representations for $d_z2$ and $d_x2_{-y}2$ $d_{zx}$, $d_{yz}$ and $d_{xy}$ orbitals are $^2A_{1g}$, $^2B_{2g}$ and $^2B_{3g}$ respectively. By calculating the crystal field potentials from the formula of crystal field potentials and then the microdisturbance matrix elements: $<Z^2IVIZ^2>$, $<X^2\text{-}Y^2IVIX^2\text{-}Y^2>$, $<Z^2IVIX^2\text{-}Y^2>$ $<XYIVIXY>$, $<XZIVIXZ>$, $<YZIVIYZ>$, we have obtained the sequence of its energy levels as: $^2A_{2g}$ $(d_z2+\lambda d_x2_{-y}2)>^2B_{3g}$ $(d_{xy})>^2B_{1g}(d_{yz})>$ $^2B_{2g}$ $(d_{yz})>^2A_{1g}(d_z2-\lambda d_x2_{-y}2)$. It can be seen from this that the weak absorption peak of 625 nm and the strong absorption peak of 760 nm are caused by the electron transitions from $^2B_{1g}$ to $^2B_{3g}$ and from $^2B_{2g}$ to $^2B_{3g}$ of $V^{4+}$ ion in $D_{2h}$ crystal field. The presence of $V^{4+}$ ion in the cassiterite from Yunlong shows that it was formed under relatively reducing conditions, as $V^{4+}$ can be changed into $V^{5+}$ under high $fo_2$-fugacity conditions.

From the crystal field spectral study on the cassiterites from Dachang, Xiling and Yunlong, it may be seen that the cassiterites of different genetic types contain transition ions in different states to substitute for $Sn^{4+}$, and make the spectra different from each other in their positions and intensities, as they are caused by d-d electron transitions of different energies, thus providing some information on the formation environment of cassiterites of different genetic types.

## 3. Infrared Spectra

In the crystal structure, Sn has a site symmetry of $D_{2h}$ and O of $C_{2V}$. In terms of group theory, four infrared active vibrations are obtained by the correlation analysis between the site groups ($D_{2h}$ and $C_{2v}$) and the factor group ($D_{4h}$), i.e.: $A_{2u}+3E_u$. Therefore, four intrinsic absorption peaks should occur in cassiterite. The IR spectra were measured on a PE-58 infrared spectrophotometer (4000–200 $cm^{-1}$) at room temperature under the same sample-preparation conditions. In more than a hundred IR spectra of cassiterites of different origin, four intrinsic absorption peaks occur at $V_4$ $(A//C)=530-570\ cm^{-1}$, $V_1(Eu'\perp C)=630-670\ cm^{-1}$, $V_2(Eu^2\perp C)=310-340\ cm^{-1}$,

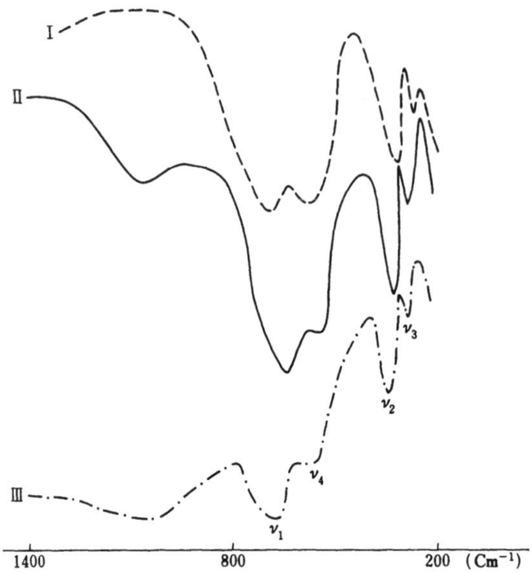


**Fig. 7.** IR spectra of cassiterite; Standard (type I), deformation (type II), and distortion (type III) spectra

$V_3(Eu^3 \perp C)$=265–270 $cm^{-1}$. According to the changes in intensities and spectral shapes, the IR spectra of cassiterites are classified into three types (Fig. 7).

Standard spectra (type I) have intensities $V_1 \approx V_4 > V_2 > V_3$ and symmetrical peaks. This type of spectra is seen only in pure cassiterite.

Deformation spectra (type II) seen in the cassiterites from Dachang and Xiling, have intensities such as $V_1 > V_4 = V_2 > V_3$ and are characterized by symmetrical peaks.

Distortion spectra (type III) have intensities such as $V_1 > V_4 > V_2 > V_3$ and are characterized by asymmetrical peaks, in which the shapes of the $V_4$ peaks are variable, providing some direct information on the origin of cassiterite.

1. Deformation spectra are mainly related to substitution of $Fe^{3+}$ for $Sn^{4+}$, but do not change with the content of Fe, as shown in the cassiterite from cassiterite-sulphide hydrothermal deposits, metasomatic deposits and volcanic hydrothermal deposits.
2. Distortion spectra are mainly related to the substitution of $Nb^{5+}$ ($Ta^{5+}$) and/or $Fe^{3+}$ for $Sn^{4+}$, and vary with the content of $Nb^{5+}$, $Ta^{5+}$ and $Fe^{3+}$ and with the formation temperature. Examples are seen in the cassiterite from Sn, Nb, Ta-bearing granite and grano-pegmatite or from W- and Sn-bearing quartz veins.
3. Occasionally both type II and type III are found to occur in the same sample, which can be used to distinguish metallogenic stages. For example, the spectra of cassiterite from Yunlong in the late-stage hydrothermal cassiterite-sulphides belong to deformation ones, and those in the early-stage cassiterite-oxides belong to distortion ones.

## Conclusions

1. The ultra-violet spectral study shows that cassiterite is a semi-conductor of a wide energy gap (Eg), which is related to the impurity ions, such as $Fe^{3+}$, $Nb^{5+}$, $Ta^{5+}$, etc. Eg is decreased if acceptor or donor levels are added into the energy gap by the presence of $Nb^{5+}$, $Ta^{5+}$ and $Fe^{3+}$ as heterovalent isomorphism in the crystal structure, and the colour becomes darker correspondingly. The values of Eg may be used as an indicator of the formation temperature of cassiterite, because the heterovalent isomorphism tends to take place at a higher temperature.
2. The visible-near-infrared absorption spectral study suggests that the various colours of cassiterite from Xiling are caused mainly by the substitutions of $Fe^{3+}$ and $Fe^{2+}$ for $Sn^{4+}$, giving rise to the d-d electron transition in $D_{2h}$ crystal field, and that the cassiterite from Yunlong contains a small quantity of $Fe^{3+}$ as well as of $V^{4+}$, $Mn^{2+}$, whose form of occurrence in the structure is revealed by their crystal field spectral intensities and positions, and accordingly, the physicochemical conditions under which cassiterite forms may be further inferred.
3. The infrared spectral shapes and intensities may provide information on the crystal structure, composition and origin of cassiterite. The mono-vibration $V_4(A_{2u})$ corresponds to the antisymmetrical stretching vibration of $Sn^{4+}$ and $O^{2-}$ along the C-axis, the spectral shapes obviously varying with the temperature and the content of $Nb^{5+}$, $Ta^{5+}$ (above all $Nb^{5+}$), from which direct information can be obtained on its crystal chemistry.

*Acknowledgements.* The author is very grateful to the Academia Sinica (Chinese Academy) for the financial support. Much gratitude should also go to Mess rs. Zheng Chusheng, Ge Jiugao and Cai Xiuchen and Mrs. He Shuangmei for their kind help in all the spectroscopic measurements.

## References

Arlinghaus, F.J., 1974. Energy in stannic oxide ($SnO_2$). *J. Phys. Chem. Solids,* Vol. 35, 931–935.

Frohich, M. and Kenklies, R., 1978. Band-gap assignment in $SnO_2$ by two-photon spectroscopy. *Phys. Rev. Lett.,* Vol. 41, No. 5, 1750–1751.

Nagasaw, M. et al., 1969. Urbach's rule exhibited in $SnO_2$, *Solid State Comm.,* Vol. 7, 1731–1733.

Peng Mingsheng and Zheng Chusheng, 1980. Quantum mineralogy. *Acta of Central-South Institute of Mining and Metallurgy,* No. 2 (in Chinese).

Peng Mingsheng and Zheng Chusheng, 1981. A study on the optical properties and spectra of piemontite. *Acta of Mineralogy,* No. 3 (in Chinese).

Peng Mingsheng, Xiu Zitou et al., 1982. A study of crystal field theory on piemontite. *Molecular Sciences and Chemical Studies,* No. 2 (in Chinese).

Peng Mingsheng, Zheng Chusheng et al., 1982. A Study on the genesis of Hainan piemontite. *Acta of Mineralogy,* No. 2 (in Chinese).

Peng Mingsheng et al., 1984. A Study on the stability field of Hainan piemontite. *Acta Geologica Sinica,* Vol. 58, No. 1 (in Chinese).

# 6.3.2 Discovery and Study of Titanium-Rich Nigerite

Tan Yansong, Liu Zhengyun, and Zhang Qiuju[1]

**Abstract**

The titanium-rich nigerite found in Shizhuyuan, Hunan Province, is a rare Ti-Sn- and Al-bearing oxide mineral. It occurs in marble of the Late Devonian Shetienqiao Limestone Formation, along the exocontact with an Early Yenshanian biotite-granite stock.

The mineral is disseminated in veinlets composed of fluorite, tourmaline, mica, pyrrhotite and some other minerals, and associated with the nigerite are gahnite and cassiterite.

Chemically, in addition to tin and aluminium, it contains titanium as one of its main components (up to 7.45%), and its ratio of Ti atoms to those of Sn is close to 1:1. Accordingly, the titanium-rich nigerite is close to being the nigerite of the nigerite-hoegbomite series.

A yellow-coloured tabular oxide mineral of tin and alumina, which is rich in titanium, was found during the study of tin occurrence in the tin-bearing marble zone of the polymetallic tungsten-tin deposit of Shizhuyuan. The mineral is preliminarily named titanium-rich nigerite, based on the microprobe and X-ray structural analysis, and on the measurements of its physical and optical characters. Such titanium-rich tin mineral should be the first discovery in China, and has not been reported yet in literature known to the authors.

This discovery furnishes new evidence for the existence of tin in the deposit and indicates the main cause for the low-grade recovery of tin in ore dressing and smelting. Meanwhile, it also provides new information on the geochemistry and mineralogy of tin in the Nanling area of China and on the crystal chemistry of titanium. A preliminary study is carried out on its mineralogy by the authors.

## Occurrence and Associated Minerals

The mineral occurs in marble in the exocontact zone of the limestones of Qiziquiao and Shetianqiao formations of the Middle and Upper Devonian with the biotite granite of the early Yenshanian Qianlishan stock. The upper part of the marble is exposed

[1] Experiment and Test Centre of the Bureau of Geology and Mineral Resources of Hunan Province, Changsha, Hunan

and the lower part is associated with skarns. Ore-forming hydrothermal replacement took place repeatedly during different periods along the fissures in the marble, forming fine and dense stockwork-veinlet marble type tin deposits. The veins are mainly of tourmaline-mica, fluorite-mica-calcite, mica-pyrite-pyrrhotite, etc. The titanium-rich nigerite occurs disseminated and scattered in the veins or in the adjacent marble. The associated minerals include tourmaline, sericite, muscovite, fluorite, nigerite, gahnite, cassiterite, pyrite, pyrrhotite, etc. The titanium-rich nigerite is closely associated and intergrown in mosaic with nigerite and gahnite, forming a mineral association of oxides rich in titanium, tin and aluminium. Considerable amounts of micrograined cassiterite sometimes appear and are enclosed in the titanium-rich nigerite at places where the latter is enriched. Pyrrhotite has also been found to be enclosed, which indicates the relatively late deposition of the mineral.

## Mineralogy

The mineral occurs as hexagonal or tabular plates (Fig. 1). The colour is brownish yellow to light yellow. The lustre is glassy. There is a perfect 0001 cleavage. The mineral tends to thin plates and has a slightly blue fluorescence under the electron beam. The

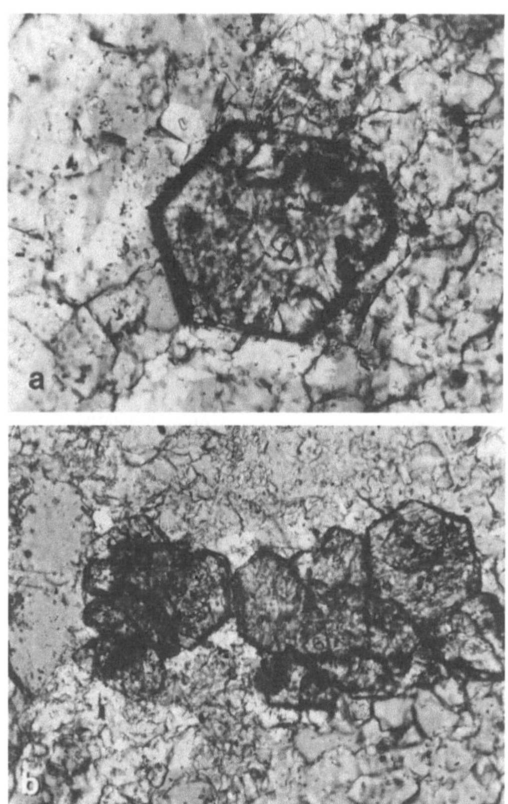


**Fig. 1. a** Single crystal of Ti-rich nigerite occurring in the marble. X 200. linearly polarized light. **b** Aggregate of Ti-rich nigerite occurring in the marble. X 200. linearly polarized light

grain size generally ranges from 0.074 to 0.005 mm. The Vickers hardness is 1280 kg/$mm^2$ (load 50 g). The specific gravity is 4.1 (calculated by the X-ray method). In comparison with 4.4 for common nigerite in other places, the specific gravity of the mineral is low, because of its lower content of tin and iron.

The optical characteristics of the mineral are as follows:

Colour – light yellow to brownish yellow;
Form – hexagonal in cross section and rectangular in section parallel to C;
Ratio of length/width – 2 : 1 to 4 : 1;
Extinction – parallel;
Pleochroism – distinct ($N_O$-yellowish brown, $N_e$-pale yellowish brown;
Birefringence – low;
Interference colour in thin section – first order yellow, with evident zonal structure and anomalous colour;
Refractive indices ($N_0$-larger than 1.78 and less than 1.81);
Uniaxial negative.

According to the data available, nigerite is normally uniaxial positive, except that discovered in Portugal, which is caused by the minor content of titanium. The variation of optical character is thought to depend on the titanium content. Table 1 gives a comparison of this Ti-rich with the normal nigerite.

**Table 1.** Comparison of mineralogical characteristics of nigerite and titanium-rich nigerite

| Property | Ti-rich nigerite | Nigerite |
|---|---|---|
| Colour | Yellowish to yellowish brown | Slightly yellow to slightly yellowish green |
| Cleavage | Perfect | Perfect |
| Transparency | Semitransparent | Semitransparent |
| Electromagnetic property | Medium to weak | Weak to nil |
| Form | Tabulate | Rectangular tabulate |
| Optical sign | Uniaxial negative | Uniaxial negative or positive |
| Pleochrosism | Distinct | Indistinct |
| Anomalous interference colour | Distinct | Indistinct |
| Luminescence | Slightly blue | None |

## Crystal Structure

The X-ray powder data of the titanium-rich nigerite approach those of nigerite. The strongest d-values are: 4.0265 (4); 4.6936 (5); 3.350 (2); 3.140 (3); 2.8376 (9); 2.710 (3); 2.520 (2); 2.4253 (1); 2.1739 (4); 2.009 (5); 1.840 (2); 1.710 (2); 1.6456 (7); 1.440 (6); 1.420 (4).

**Table 2.** Comparison of various nigerites

| 6H Ti-rich nigerite (local) | | | 6H Nigerite (Brazil) | | | 6H Nigerite (Hunan) | | |
|---|---|---|---|---|---|---|---|---|
| hkI | I | d | hkI | I | d | hkI | I | d |
| 302 | 1 | 6.99 | 302 | 60 | 6.88 | 004 | 2 | 6.95 |
| 100 | 1 | 4.95 | 100 | 40 | 4.93 | 100 | 1 | 4.95 |
| 101 | 5 | 4.6936 | 101 | 90 | 4.65 | 102 | 5 | 4.67 |
| | 1 | 4.200 | | | | | | |
| 102 | 4 | 4.0265 | 102 | 90 | 4.00 | 104 | 4 | 4.02 |
| | 1 | 3.5952 | | | | 008 | <1 | 3.50 |
| | 2 | 3.350 | 103 | 40 | 3.346 | 106 | <1 | 3.35 |
| | 3 | 3.14 | | | | | | |
| | 1 | 3.05 | | | | | | |
| 104 | 9 | 2.8376 | 110, 104 | 100 | 2.832 | 110, 108 | 10 | 2.85 |
| | 3 | 2.710 | | | | | | |
| | 2 | 2.520 | | | | | | |
| 113 | 10 | 2.425 | 201, 113 | 100 | 2.420 | 202, 116 | 10 | 2.85 |
| | 1 | 2.319 | | | | | | |
| | 1 | 2.210 | | | | | | |
| 203 | 4 | 2.104 | 203 | 80 | 2.169 | 206 | 3 | 2.180 |
| 106 | 1 | 2.10 | 106 | 10 | 2.134 | 10, 12 | 1 | 2.100 |
| 204 | 5 | 2.009 | 204 | 80 | 2.010 | 208 | 3 | 2.020 |
| | 1 | 1.990 | 115, 007 | 20 | 1.983 | 11, 1000, 14 | 1 | 1.990 |
| | 1 | 1.915 | | | | | | |
| 210 | 2 | 1.840 | 210, 211 | 80 | 1.853 | 212 | 5 | 1.855 |
| 205 | 2 | 1.837 | 205, 107 | | | 2010, 10, 14 | 5 | 1.840 |
| | 1 | 1.80 | 212, 216 | 20 | 1.799 | 214, 11, 12 | 2 | 1.810 |
| | 2 | 1.710 | 213, 008 | 20 | 1.727 | 216, 00, 16 | 2 | 1.730 |
| 214 | 7 | 1.646 | 300, 214 | 90 | 1.643 | 300, 218 | 9 | 1.650 |
| 108 | 1 | 1.620 | 108 | 10 | 1.628 | 10, 16 | <1 | 1.635 |
| | 2 | 1.570 | | | | | | |
| 207 | 7 | 1.545 | 215, 207 | 100 | 1.551 | 21, 10, 20, 14 | 9 | 1.552 |
| | | | 118 | 10 | 1.477 | 11, 16 | 1 | 1.480 |
| | | | 109 | 10 | 1.463 | 10, 18 | 1 | 1.470 |
| 220 | 6 | 1.440 | 220 | 90 | 1.426 | 220 | 9 | 1.430 |
| 208 | 6 | 1.420 | 208, 205 | 90 | 1.412 | 20, 16 | 9 | 1.419 |
| 233 | 2 | 1.360 | 233 | 10 | 1.359 | 226 | 1 | 1.361 |
| 312 | 1 | 1.34 | 312 | 10 | 1.342 | 314 | 1 | 1.345 |
| | 1 | 1.33 | 10, 10 | 10 | 1.329 | 10, 20 | <1 | 1.331 |
| 314 | 3 | 1.271 | 314 | 80 | 1.372 | 318 | 4 | 1.277 |
| 401 | 3 | 1.229 | 401 | 40 | 1.229 | 402 | 3 | 1.233 |
| 402 | 2 | 1.214 | 402 | 40 | 1.217 | 404 | 2 | 1.221 |
| | 1 | 1.198 | 403, 308 | 10d | 1.192 | 406 | <1 | 1.188 |
| | 1 | 1.190 | | | | | 1 | 1.190 |
| 404 | 1 | 1.160 | 404 | 10 | 1.162 | 408 | <1 | 1.168 |
| | 2 | 1.150 | 00, 12, 11, 11 | 10 | 1.149 | 00, 24 | 1 | 1.154 |
| | | | 321, 405 | 10 | 1.127 | 322, 40, 10 | 1 | 1.181 |
| | | | 309 | 10 | 1.119 | 30, 18 | 1 | 1.124 |
| | | | 323 | 10d | 1.099 | 326 | 1 | 1.105 |
| | | | 401, 324 | 10d | 1.077 | 410, 328 | 4 | 1.082 |
| | | | 413, 325 | 40 | 1.048 | 416, 32, 10 | 4 | 1.053 |
| | | | 408 | 40 | 1.004 | 40, 16 | 3 | 1.009 |

**Table 3.** Electron microprobe analyses

| Element Content (%) No. of Sample | $TiO_2$ | $SiO_2$ | MgO | MnO | $SnO_2$ | $Al_2O_3$ | FeO | CaO | ZnO | Total |
|---|---|---|---|---|---|---|---|---|---|---|
| 1 – 1 | 7.476 | 0.099 | 5.046 | 1.206 | 16.972 | 57.366 | 4.508 | 0.21 | 4.620 | 97.498 |
| 1 – 2 | 7.544 | 0.635 | 5.526 | 1.317 | 16.962 | 57.933 | 4.609 | 0.24 | 4.445 | 99.216 |
| 1 – 3 | 7.205 | 0.197 | 5.526 | 1.185 | 14.451 | 55.804 | 5.313 | 0.223 | 5.406 | 95.216 |
| 2 – 1 | 6.797 | 0.070 | 5.403 | 1.096 | 14.871 | 58.667 | 6.192 | 0.152 | 6.061 | 99.308 |
| 2 – 2 | 8.067 | 0.099 | 5.522 | 0.983 | 17.962 | 58.424 | 5.694 | 0.112 | 5.419 | 102.282 |
| 3 – 1 | 7.010 | 0.040 | 5.284 | 1.175 | 17.488 | 59.107 | 4.472 | 0.212 | 5.491 | 100.190 |
| 3 – 2 | 7.086 | 0.059 | 5.323 | 1.175 | 17.532 | 59.584 | 4.850 | 0.170 | 5.588 | 101.366 |
| 4 | 5.037 | 0.277 | 4.823 | 1.139 | 20.334 | 57.846 | 4.609 | 0.098 | 5.009 | 99.172 |
| 5 | 4.485 | 0.159 | 5.861 | 0.888 | 17.392 | 59.771 | 4.911 | 0.096 | 6.203 | 99.767 |
| 6 | 4.996 | 0.049 | 5.187 | 1.218 | 20.462 | 58.735 | 4.803 | 0.100 | 5.235 | 100.785 |
| 7 | 4.022 | 0.199 | 5.838 | 1.091 | 19.979 | 59.353 | 4.905 | 0.208 | 6.172 | 101.767 |
| 8 | 4.732 | 0.409 | 6.969 | 1.140 | 16.610 | 59.304 | 5.383 | 0.116 | 7.788 | 102.452 |
| Mean | 6.205 | 0.191 | 5.525 | 1.134 | 17.585 | 58.491 | 5.021 | 0.161 | 5.619 | 99.932 |

The cell parameters of the mineral were calculated based on X-ray powder data, shown as follows:

A=5.70049 Å $C_o$=13.8699 Å, R-120°, v=390.33 $Å^3$, the mineral analyzed should belong to the 6H polytype. A comparison between the titanium-rich nigerite analyzed in this paper and different 6H polytypes of nigerite by Kloosterman (1974) and Van Tassel (1965) is shown in Table 2.

## Chemical Composition

The chemical composition of the mineral was analyzed by means of the quantitative microprobe method and the semi-quantitative microlaser method. The microlaser analyses showed that the major elements contained are Al, Sn and Fe; the minor ones include Ti, Zn, Mg and Mn, and trace elements are Be and Ca.

The average chemical compositions of the samples obtained by multi-spot analyses are: $TiO_2$ 6.205, $SiO_2$ 0.191, MgO 5.525, MnO 1.134, $SnO_2$ 17.585, $Al_2O_3$ 58.491, FeO 5.021, CaO 0.161, ZnO 5.619.

The content of Ti generally ranges between 4.002 and 7.544, exceptionally up to 8.067%, and some less than 4.022 (Table 3). The total content of Ti varies within the range 0–8.067%, while that of Sn ranges from 14.451 to 20.462%. In comparison with nigerite, the mineral is rich in Ti and Mg and low in Sn, with also a relatively minor content of Zn and Fe (Table 4).

The crystallochemical formula of the Ti-rich nigerite calculated by the oxyatomic method is given as: $Sn_{1.06}Ti_{0.89}Al_{10.70}Fe_{0.55}Zn_{0.52}Mn_{0.18}Ca_{0.04}Mg_{1.29}Si_{0.10}O_{22}(OH)_2$.

For comparison, the crystallochemical formula of common nigerite is:

$Sn_{1.77}Al_{10.52}Fe_{1.96}Zn_{0.57}Mg_{0.33}O_{22.16}(OH)_{1.84}$.

**Table 4.** Comparison of the chemical compositions of nigerite and Ti-rich nigerite

| Locality | Shizhuyuan | | Portugal | Brazil | Nigeria |
|---|---|---|---|---|---|
| Content (%) Composition | Ti-rich nigerite | Nigerite | Nigerite | Nigerite | Nigerite |
| $SnO_2$ | 16.962 | 25.457 | 23.2 | 29.00 | 25.33 |
| $Al_2O_3$ | 57.933 | 57.984 | 55.0 | 51.90 | 50.91 |
| FeO | 4.609 | 5.819 | 10.9 | | 2.65 |
| $Fe_2O_3$ | | | | 7.60 | 11.90 |
| MgO | 5.526 | 1.943 | 0.1 | | 1.28 |
| MnO | 1.317 | 3.124 | 0.1 | | 0.09 |
| ZnO | 4.445 | 4.716 | 5.30 | 9.50 | 4.51 |
| $TiO_2$ | 7.544 | 0.00 | 3.3 | | 0.17 |
| $SiO_2$ | 0.635 | | 0.9 | | 0.48 |
| CaO | 0.24 | | | | |
| PbO | | | 0.1 | | 0.94 |
| $H_2O$ | | | | 0.42 | 1.57 |
| Total | 99.216 | 98.964 | 100.03 | 100.82 | 99.83 |

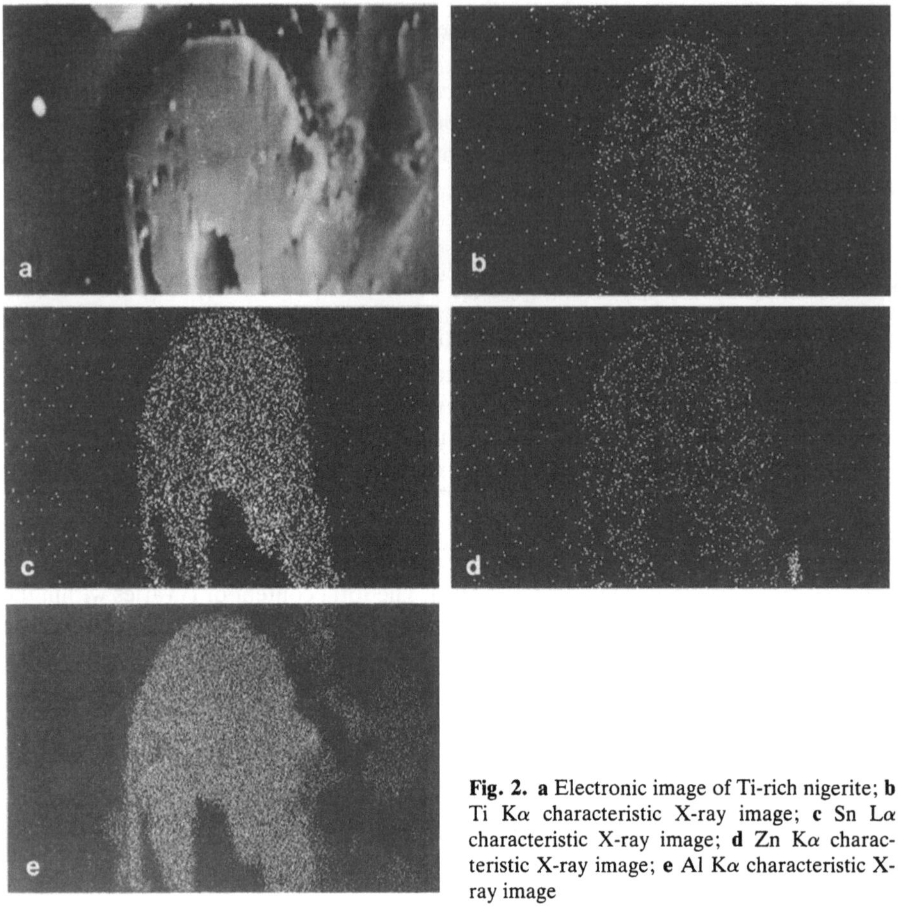


**Fig. 2.** **a** Electronic image of Ti-rich nigerite; **b** Ti K$\alpha$ characteristic X-ray image; **c** Sn L$\alpha$ characteristic X-ray image; **d** Zn K$\alpha$ characteristic X-ray image; **e** Al K$\alpha$ characteristic X-ray image

## Discussion

The K$\alpha$ characteristic X-ray image of Ti shows that Ti is distributed homogeneously throughout the mineral, no distinct concentrated area was observed. So there is little possibility of the existence of Ti inclusions. Sn, Zn and Al are also distributed homogeneously in the mineral (Fig. 2). Some of the crystals are replaced by fine calcite grains due to alteration, and this may be the cause for the varying content of Ti in the mineral.

The microprobe multi-spot analyses of the mineral suggest that there exists a regular variation of Ti and Sn content in the mineral. The Sn content decreases gradually with the increasing Ti content, for example, the content of Sn will be 20–17% when Ti content is 4–5%, and 17–14% when the Ti content is 7%. Obviously the decrease of Sn content is caused by the replacement of Ti (Table 5).

**Table 5.** Microprobe analyses of Sn and Ti

| Sample No. Content Element | | 1 | 2 | 3 | 4 | 5 | 6 | 7 | 8 |
|---|---|---|---|---|---|---|---|---|---|
| $TiO_2$ | % | 4.02 | 4.49 | 4.99 | 5.04 | 7.09 | 7.21 | 7.48 | 7.54 |
| $SnO_2$ | % | 19.98 | 17.40 | 20.46 | 20.33 | 17.53 | 14.45 | 16.97 | 16.93 |

In terms of the geochemistry of Ti and Sn, the relevant parameters are close to each other, for instance, the electric charges of $Ti^{4+}$ and $Sn^{4+}$ give 1.60 and 1.80 respectively and the ionic radii of Ti 0.69 Å approximates to that of Sn in 6-fold coordination (0.77 Å), so the $Ti^{4+}$ might be substituted by $Sn^{4+}$ during postmagmatic crystallization.

According to the available data, the specific feature of the Ti-rich nigerite in this area is its high Ti content, which is one of the main constituents in addition to Sn and Al. Its Ti content is at least twice as much as the Ti-nigerite of Portugal.

The Ti-rich nigerite is considered a new variety close to the end member in the series nigerite-hogbomite-taufeite, because the ratio of the number of atoms of Ti and Sn is close to 1 : 1.

Further precise structure analysis is needed to prove if the Ti-rich nigerite is a new mineral species.

*Acknowledgement.* We thank Professor Guo Zongshan and Peng Zhizhong for checking and approving the paper, and Messrs. Jiang Shengzhang, Luo Shiwei, Zhang Xuerao, Ju Zhonghang, Hang Gongliang and Liu Beyi, and also Mr. Yang Guangming from Wuhang College of Geology for their warm support and useful suggestions.

## References

Bryan M. Gatehouse, 1979. The crystal structure of nigerite-24R, *Amer. Mineral,* Vol. 64, No. 11—12, 1255—1264.

Ding Kuishou, 1982. The discovery and investigation of 6H-nigerite. *Acta Petrologica Mineralogica et Analytica,* Vol. 1, No. 1, pp. 32 (in Chinese).

Kloosterman, J.B., 1974. Nigerite in the tin-tantalum pegmatites of Amapá, Brazil. *Mineral. Mag.,* Vol. 39, 837—846.

Peacor, D.R., 1967. New data on nigerite. *Amer. Mineral.,* Vol. 52, 864—866.

Peng Zhizhong, 1964. The discovery of nigerite in China and a preliminary measurement of its crystalline structure, *Geological Review,* Vol. 22, 476—478 (in Chinese).

Van Tassel, R., 1965. Nigerite from Lixa, near Felgueiras, Douro Litoral Province, Portugal. *Mineral. Mag.,* Vol. 34, 482—486.

## 6.4 China: Dachang and Other Guangxi Deposits

# 6.4.1 Geochemical Characteristics of Indicator Elements and Prospecting Criteria for the Danchi Polymetallic Mineralized Belt of the Dachang Tin Field

YANG JIACHONG[1], LI DADE[1], ZHANG DUOXUN[2], LI SHUIMING[1], LI XINYI[1] and LU XIUFENG[1]

## Abstract

Based on the regional data of heavy minerals and secondary dispersion haloes, a subdivision of geochemical anomalies has been made for the Danchi ore belt. Geochemical features of the Dachang ore field and the Changpo cassiterite-sulphide deposit are discussed and a model of vertical zoning in the Changpo deposit is established. The assessment indexes for anomalies of various types of ore deposit and indicators for cassiterite-sulphide ore deposits are presented.

The study of the geochemical features of the ore fields and deposits, in combination with geochemical prospecting, is evaluated for developing mineral resources in new areas.

## Geological Setting

The NW-trending Danchi polymetallic mineralized belt is in the Danchi depression located on the southern margin of Jiangnan Old Land (Fig. 1). The exposed Devonian to Triassic strata consist of pelites, cherts and carbonates. Most of the polymetallic ores are hosted in the Devonian sediments.

NW-trending linear folds and faults are well developed and distribution of the ore belt is controlled by the Danchi anticlinorium and the major Danchi fault.

Igneous rocks crop out only in the northern and central segments of the belt and consist of quartz porphyry, biotite granite, porphyritic granite and diabase porphyrite, occurring as stocks, dykes or veins, dated radiometrically as late Yenshanian.

The mineralization has clearly taken place in close relation to certain strata, structures and igneous intrusions. There are four polymetallic tin fields in Mangchang, Dachang, Furongchang and the Julan mercury field.

The Dachang field, located in the middle of the Danchi anticlinorium and fault, can be divided into eastern, central and western sub-belts. The Changpo cassiterite-sulphide deposit is in the western sub-belt with cherts ($D_3^1$) and lenticular and striped limestones ($D_3^2$) as its host rocks.

[1] Research Institute of Geology for Mineral Resources, China National Nonferrous Industry Corporation, Guilin, Guangxi

[2] No. 215 Party of Guangxi Geological Exploration Corporation, CNNC

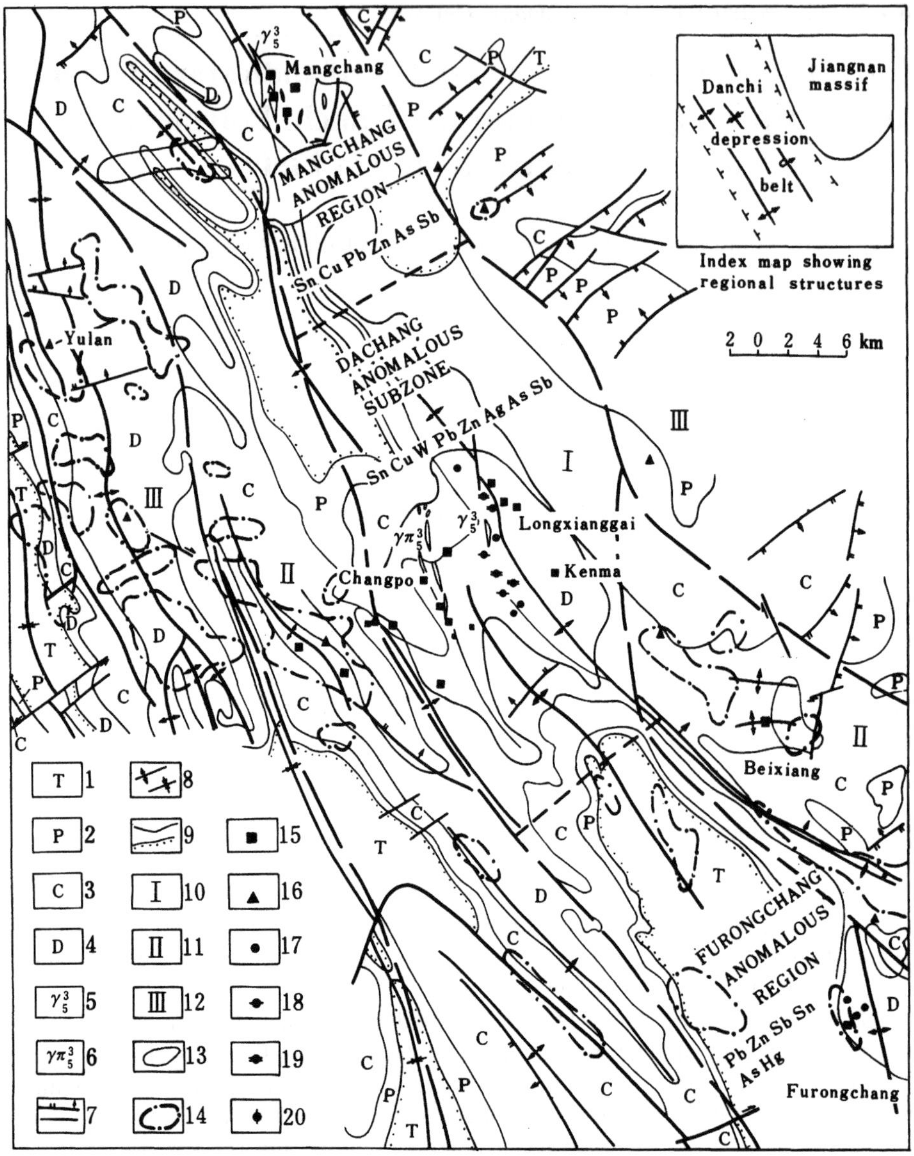


**Fig. 1.** Zonation of geochemical anomalies in Danchi polymetallic belt, Guangxi

## Regional Geochemical Features

### 1. Geochemical Features of the Metallogenic Plutons

Studies show that the plutons which have a close affinity with the mineralization are granite porphyry occurring in the Mangchang ore field and biotite-granite in Dachang. The plutons are rich in $SiO_2$ and volatiles such as P, $H_2O$, F and B, poor in

Ti, Fe, Ca, Mg and characterized by $K_2O>Na_2O$ (Table 1). The diagnostic values of $K_2O/Na_2O$ $(K_2O+Na_2O)/CaO$, $SiO_2/(Fe_2O_3+FeO+MgO+MnO)$, $CaO+Na_2O+K_2O$, in the two plutons are quite different from those in normal granite, but are close to the tin-bearing granites of the world. The Sr/Ca and MgO/FeO ratios of the two plutons are lower than those of normal granites, and the contents of trace elements such as Sn, Cu, Pb, Zn, W in them are 1–66 times that in the same type of rocks (Table 2). The concentration coefficient if Sn is 4.67 for biotite granite, and 11.67 for granitic porphyry. In addition, thge plutons are rich in Li, Be, Nb, Ta etc.

These salient features indicate that the enrichment of Sn, Cu, Pb, Zn and W is usually favoured by the enrichment of Si, alkalic metals (K) and volatiles, and by the depletion of Ti, Fe, Ca and Mg.

## 2. The Zoning of Anomalies

Based on heavy minerals (cassiterite and cinnabar), geochemical soil surveys, distribution of ore deposits and mineral occurrences as well as geochemical properties of the various elements, three anomalous zones have been recognized within the polymetallic belt, i.e. tin-polymetallic zone, Sn-Hg zone, and Hg zone. As the element associations in the tin-polymetallic zone are different from north to south, other three subzones can be identified: (1) Mangchang anomalous subzone, which is located in the Mangchang ore field and is characterized by the association of Sn, Pb,

**Table 1.** Chemical composition and typical ratios of productive plutons

| | Location | Mangchang | Dachang | The world | | China | |
|---|---|---|---|---|---|---|---|
| | Lithology | (1) | (2) | (3) | (4) | (5) | (6) |
| Chemical composition weight % | $SiO_2$ | 74.63 | 74.26 | 73.10 | 70.52 | 71.27 | 71.99 |
| | $TiO_2$ | 0.04 | 0.12 | 0.21 | 0.38 | 0.25 | 0.21 |
| | $Al_2O_3$ | 15.74 | 13.82 | 13.96 | 14.51 | 14.25 | 13.81 |
| | $Fe_2O_3$ | 0.43 | 0.16 | 0.91 | 1.28 | 1.24 | 1.37 |
| | FeO | 0.67 | 1.25 | 1.44 | 1.87 | 1.62 | 1.72 |
| | MnO | 0.07 | 0.07 | 0.05 | 0.07 | 0.08 | 0.12 |
| | MgO | 0.17 | 0.19 | 0.55 | 0.85 | 0.80 | 0.81 |
| | CaO | 0.45 | 0.51 | 1.21 | 2.10 | 1.62 | 1.55 |
| | $Na_2O$ | 0.13 | 3.40 | 3.01 | 3.61 | 3.79 | 3.42 |
| | $K_2O$ | 3.24 | 4.20 | 4.58 | 3.82 | 4.03 | 3.81 |
| | $P_2O_5$ | 0.48 | 0.32 | 0.20 | 0.17 | 0.16 | 0.20 |
| | $H_2O^+$ | 3.93 | 0.71 | 0.73 | 0.62 | 0.56 | 0.64 |
| Typical ratios | $K_2O/Na_2O$ | 24.92 | 1.24 | 1.52 | 1.06 | 1.06 | 1.11 |
| | $(K_2O+Na_2O)/CaO$ | 1.49 | 14.90 | 6.27 | 3.54 | 4.77 | 4.66 |
| | $K_2O+Na_2O+CaO$ | 3.82 | 8.11 | 8.80 | 9.53 | 9.53 | 8.78 |
| | $SiO_2/(Fe_2O_3+FeO+MgO+MnO)$ | 55.49 | 44.47 | 24.77 | 17.33 | 19.06 | 17.91 |
| | Sr/Ca | | 0.001 | | | 0.019 | |
| | MgO/FeO | 0.25 | 0.056 | 0.38 | 0.45 | 0.49 | 0.47 |

(1): Granite-porphyry; (2): Biotite-granite; (3): Tin-bearing granite; (4): Normal granite; (5): Granite; (6): Biotite-granite.

**Table 2.** Trace elements in the productive plutons (ppm)

| Location | Element / Lithology | Sn | Cu | Pb | Zn | Mo | W | Li | Be | Nb | Ta | Rb | Sr | B | F |
|---|---|---|---|---|---|---|---|---|---|---|---|---|---|---|---|
| Mang-chang | Granite porphyry | 35.0 | 20.0 | 20.0 | 150.0 | 5.0 | 100 | 900.0 | 25.0 | 150.0 | | | 50.0 | | |
| | Concentration coefficient | 11.67 | 1.0 | 1.0 | 2.50 | 5.0 | 66.67 | 22.50 | 4.55 | 7.50 | | | 0.17 | | |
| Dachang | Biotite granite | 14.0 | 31.4 | 34.7 | 38.5 | 1.0 | 78.89 | 353.0 | 30.6 | 62.9 | 10.2 | 813.8 | 8.3 | 331.1 | 5850 |
| | Concentration coefficient | 4.67 | 1.57 | 1.73 | 0.64 | 1.0 | 52.39 | 8.83 | 5.56 | 3.12 | 2.91 | 407 | 0.03 | 22.07 | 7.31 |
| A.P. Vi-rogradov | Granite | 3 | 20 | 20 | 60 | 1.0 | 1.5 | 40 | 5.5 | 20 | 3.5 | 200 | 300 | 15 | 800 |

Zn, Sb and As; (2) the Dachang anomalous subzone, which is in the Dachang ore field and is characterized by anomalies of the association of Sn, Cu, Zn, W, Sb, As and Ag; and (3) the Furongchang anomalous subzone, which is characterized by anomalies of the association of Pb, Zn, Sb, Sn, As and Hg at the site of Furongchang ore field.

## Geochemical Features of the Dachang Ore Field

As the cassiterite-sulphide ore deposits mostly occur in the Devonian sediments, it is important to make clear the character of associations of elements and the concentration range of elements in the host rocks and non-host ones for understanding of metallogeny, source of metals, and local settings.

### *1. Associations and Concentration Range of Trace Elements in Host and Non-Host Rocks*

Some unaltered and unmineralized samples were collected from exposed strata in the Dachang ore field for multi-elemental analysis. The comparison of the results with clark values (Vinogradov, 1962) gives the following characteristics (Fig. 2).

(1) The host rocks ($D_1$, $D_2^2$, $D_3^1$, $D_3^2$, $D_3^3$) are enriched in Sn, Sb, As and S, and depleted in Cu and Zn. The contents of such elements as Sn, Sb, As and S are respectively 1.64–5.82 times, 5.6–11 times, 3.82–10.78 times, 5.54–11.17 times higher than their clark values. Furthermore, $D_3^3$ is rich in Mn, Sr, and Ag; $D_1$ in V and Ca; $D_2^2$ in Ag, V and Sr; and $D_3^2$ in Ag and V.

(2) The non-host rocks are characterized by the enrichment of As, Sb, Ag and Bi, and by the depletion of S, Ni, Co, V, Mn and Ba.

(3) Different strata have different background values for certain elements, for example,

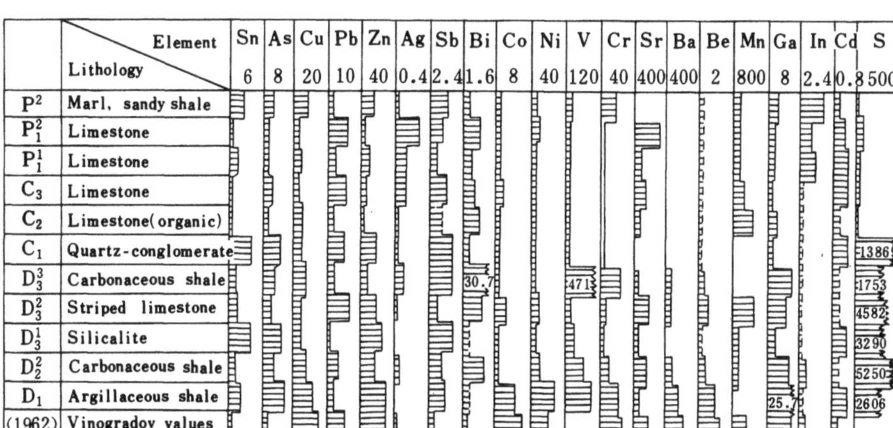


**Fig. 2.** Distribution of trace elements in different sediments of Dachang ore field

$$D_3^1 - D_1 - D_3^2 - D_3^3 - D_2^2$$

$$\longrightarrow$$

Sn reduction

$$D_3^1 - D_3^3 - D_1 - D_2^2 - D_3^2$$

$$\longrightarrow$$

Sb reduction

$$D_1 - D_3^1 - D_3^3 - D_2^2 - D_3^2$$

$$\longrightarrow$$

As reduction

### 2. *Zoning of Composition*

The repeated activities of ore-forming fluids give rise to an overlap of the ore-forming and halo-forming elements in the Dachang ore field. The following zoning can be recognized: the anomalies of W, Bi, Ci and F appear at the centre at Longxianggai, with those of In, Sr, Mn and Ga in the vicinity; and away from the granite body appear the anomalies of Pb, Ag and Sb. Over the whole area, Sn, Cu, Zn, Ba and As occur as common elements.

### 3. *Characteristic Associations of Elements and Ratios for Different Types of Ore Deposits*

Based on the geochemical soil survey and the delineated anomalies, the following elemental associations and ratios can be obtained for different types of ore deposits.

Cassiterite-sulphide deposits are dominated by an association of Sn, Pb, Zn, Sb and As, of which, Sn 60 ppm, Pb or Zn > 120 ppm, and Pb · Zn/Zn · Cu>3.

Zn-Cu deposits are characterized by an association of Zn, Cu, Sn, W, As and Sb, of which Zn>240 ppm, Cu>160 ppm, Pb·Zn/Zn·Cu<2, Zn·Cu·As/Sb and Wx1000>4.

W-Sb deposits are dominated by an association of W, Sb, Zn, Cu and As, with W>60 ppm, Sb>100 ppm, Pb·Zn/Cu·Zn<1, and Zn·Cu·As/Sb·Wx1000<2.

### Geochemical Characteristics of Cassiterite-Sulphide Deposits

The Changpo cassiterite-sulphide deposit is composed, in descending order and in accordance with the mode of occurrence, of the fracture-filling large veins (No. 0 orebody), metasomatic veinlets, stratoid veinlets and disseminated ores (No. 91 orebody), stratoid stockwork and disseminated ores. A study of the minerals and ores has been carried out to find the elemental associations for different types of ore deposits and the metallogenic environments.

## *1. Element Associations in Different Ore Deposits*

The element composition of cassiterite, sphalerite, pyrite, magnetite ores clearly indicates that the No. zero lode is characterized by enrichment in Pb, Zn, Sb, Mn and S, and the depletion of $WO_3$ and Bi; veinlet ores are rich in Ag, Cu, Zn, Fe, $SiO_2$, $TiO_2$ and $Al_2O_3$, and poor in Nb, Ta and Bi; the contents of Sn, Ga, Nb, In, Bi, $WO_3$, $ZrO_2$ and $Cr_2O_3$ are high in No. 91 orebody, with Pb and Sb being depleted; No. 92 orebody is rich in Nb, Ga, Zn, Ag, As and $V_2O_5$ but poor Cu and Mn.

These characteristics suggest that No. zero lode was formed under an environment of a higher content of sulphur but a lower content of oxygen; the veinlets were formed with abundant sulphur and moderate oxygen; No. 91 orebody was formed in an oxidizing environment with less sulphur; and No. 90 orebody in an environment between oxidation and reduction. The relationship between the various types of orebodies is as follows:

No. 91 orebody—No. 92 orebody—veinlets—No. zero orebody

$$\longrightarrow$$

Eh reduction and S increasing

The ore-forming processes are therefore characterized by an enrichment in oxygen at the early stage, and in sulphur at the late stage.

## *2. The Characteristics and Model of Geochemical Anomalies of the Deposit*

Geochemical survey along the two profiles of the deposit and at three different levels shows a regular distribution of positive anomalies of certain elements such as Sn, Cu, Pb, Zn, As, Ag, Mn, Sb, F, I, Hg, Bi and In; and negative anomalies of Sr above the orebodies. Outwards from the orebodies, the zonation of such indicator elements as Sn, Cu, Pb, Zn, Ag, As and Sb is clear (Fig. 3, 4, and 5), and the concentration gradients rise in the direction from granite terrain to orebodies, thus resulting in "bean skin-shaped" anomalies. The study of lateral haloes indicates that most of the elements tend to extend for 20—60 m outwards from orebodies until background values are reached (Fig. 5). The front haloes of certain elements such as Pb, Sb, Zn and Sn can usually reach the surface through fractures. Furthermore, some trace elements and special ratios show a vertical variation, for example, in descending order, the contents of Sn, Cu, Bi, W and In increase while the contents of Ba, Ga, Pb/Sn, Ag/Cu, Sb/Bi, Sb/Sn decrease; and also Mn, Ag and As show high values in the middle portion of orebodies.

Based on the above-mentioned characteristics, a model for vertical zonation of elements has been established (Fig. 6).

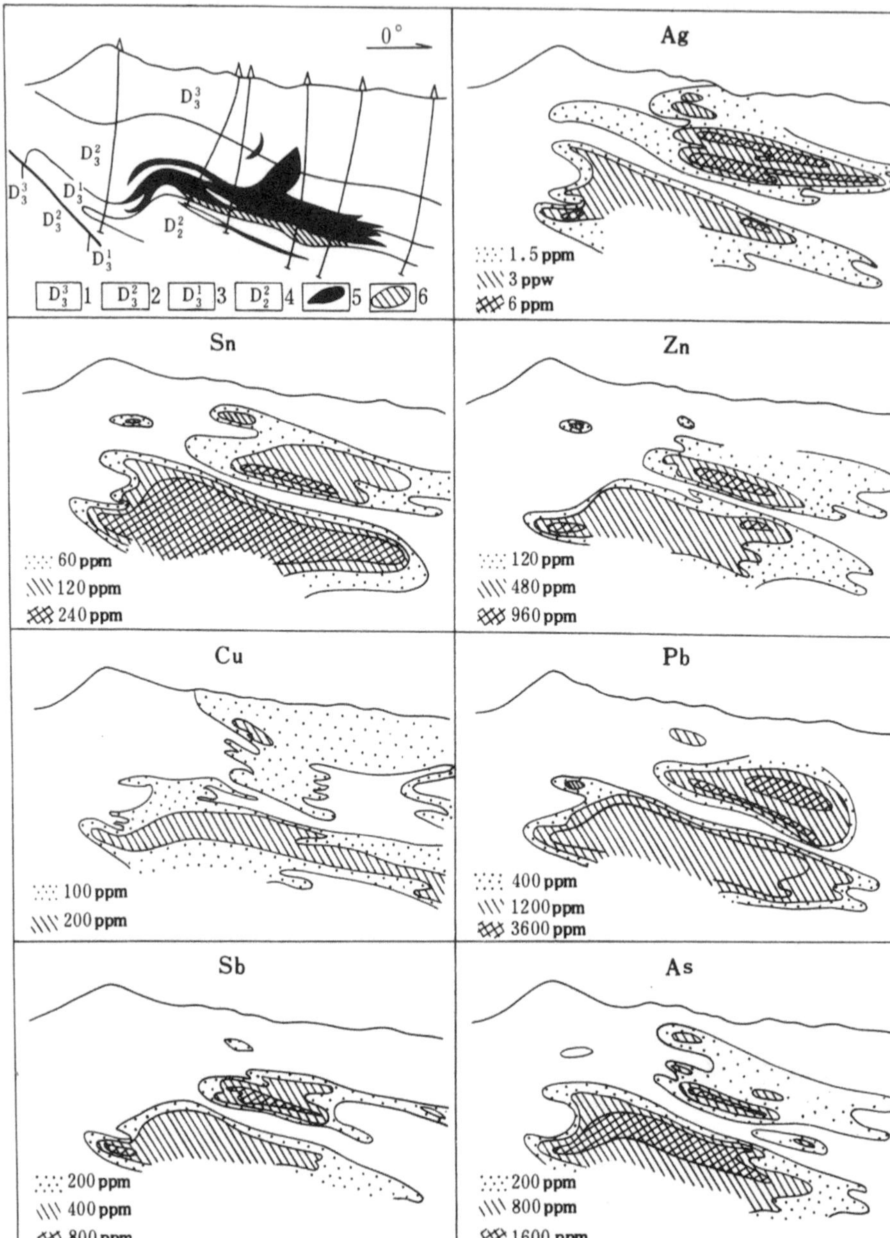


**Fig. 3.** Geochemical anomalies in section No. 202 in Changpo mine, Dachang orefield

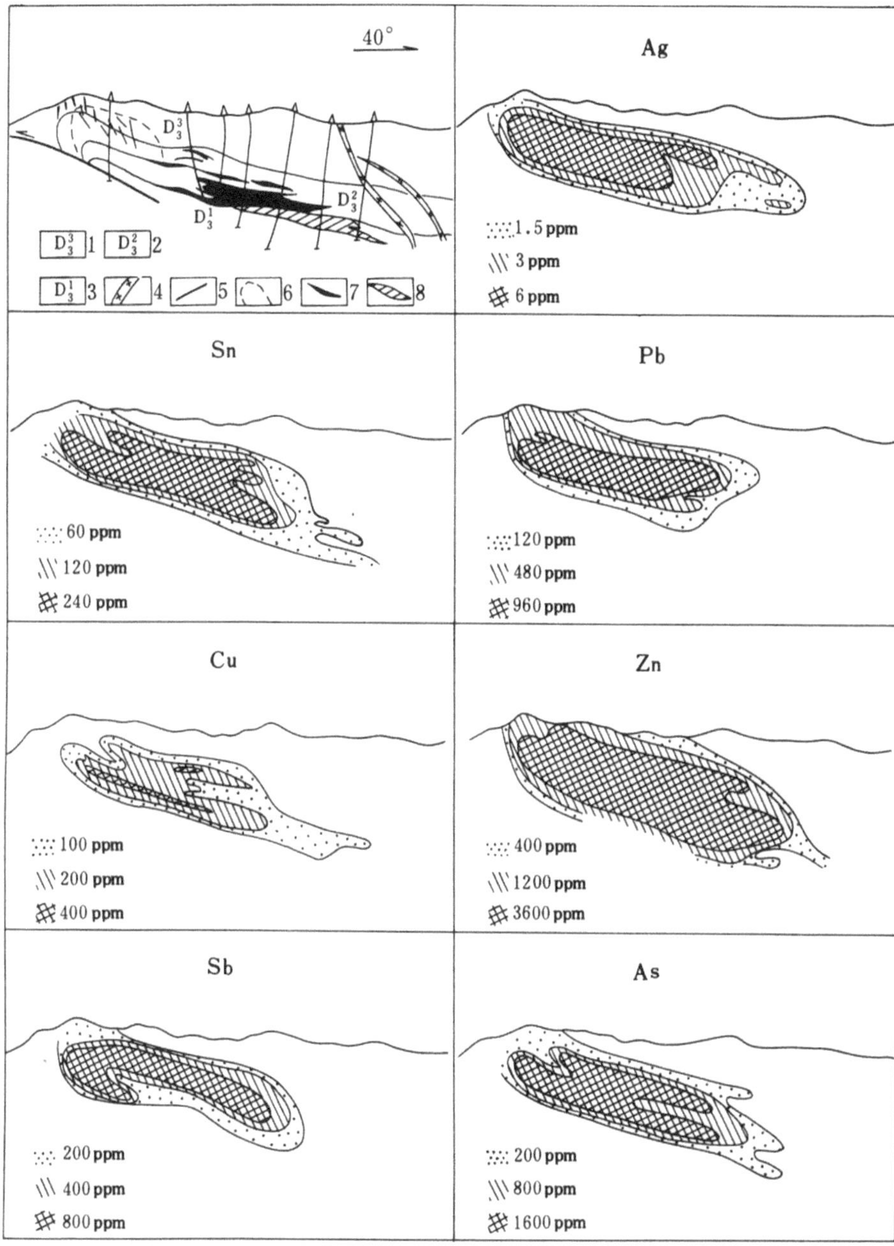


**Fig. 4.** Geochemical anomalies in section I-I in Changpo mine, Dachang orefield

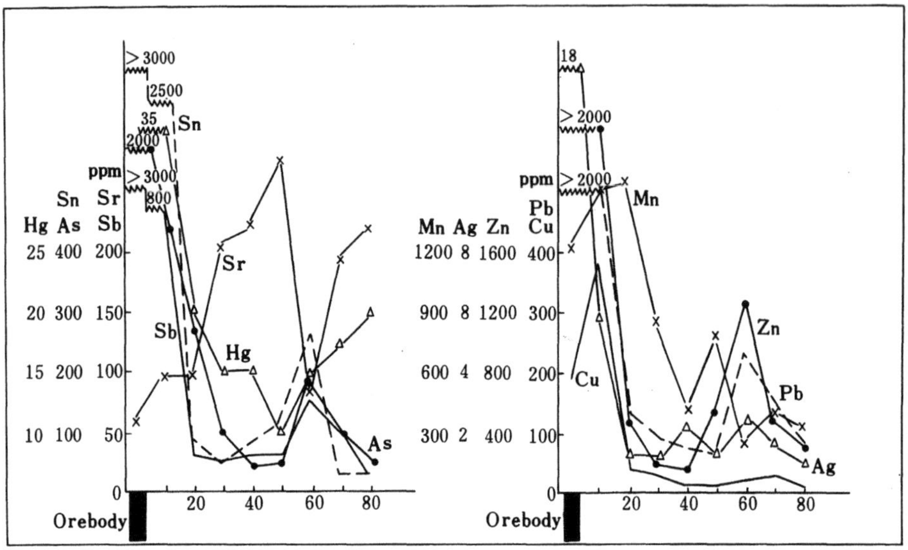


**Fig. 5.** Lateral dispersion distance from orebody

## Prospecting Criteria and Efficiency of Prospecting Techniques

### *1. Prospecting Criteria*

(1) The tin-polymetallic anomaly zone is the most favourable site for the discovery of tin-polymetallic deposits, and the Sn-Hg anomaly zone possesses the second importance. The anomalies of Sn, As, Cu, Pb, Zn, Ag, and Sb occurring in tin-polymetallic anomaly zone are pathfinders for blind ores.

(2) The ore-containing plutons are characterized by oversaturation of $SiO_2$ (>74%), enrichment of P, $H_2O$, F and B, and lack of Ti, Fe, Mg and Ca, with $K_2O/Na_2O>1.2$ $(K_2O+Na_2O)/CaO>7.5$, $CaO+K_2O+Na_2O<4\%$ and $SiO_2/(Fe_2O_3+FeO+MnO+MgO)>44$. Trace elements, such as Sn, Cu, Pb, Zn, and W are several to 10 times higher than their average contents in the same type of rocks.

(3) The anomalies of Sn, Cu, Pb, Zn, Ag, As and Sb tend to occur at the conjunction of NW-trending and NE-trending structures, and are associated closely with the mineralization.

(4) The contents of such trace elements as Sn, Sb, As and S are relatively high in the ore-containing strata.

### *2. The Efficiency of Prospecting Techniques*

In the experimental geochemical soil survey conducted in the Dachang ore field, halogens (F, I), heat-releasing Hg, and $H_2S$ were used. The results show that clear anomalies of heat-releasing Hg occur above the blind orebodies, that halogens occur in weak anomalies, and that the anomalous indication of $H_2S$ is obscure.

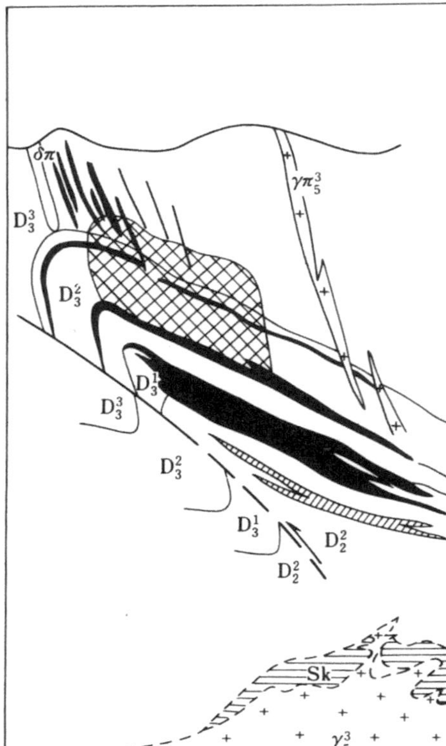


| Type of ore deposit | Location of ore bodies | Associations of elements | Ratios between elements in the anomalies | | | | Zoning of elements |
|---|---|---|---|---|---|---|---|
| | | | Pb/Sn | Sb/Sn | Sb/Bi | Ag/Cu | |
| | Surface | Pb. Zn. Cu. Ag. Sb(Sn) | 2.38 | 4.34 | 82.67 | 0.4 | Pb Sb |
| Cassiterite-sulfide vein | Upper part | Pb*. Ag*. Sb*. Zn. Sn. As. Mn. Cu. (In). Sr(negative) | 1.31 | 3.62 | 71.11 | 0.13 | Pb Sb Ag Zn Sr(−) |
| Cassiterite-sulfide replacement orebody (dense veinlet) | Middle part | Sn. As. Zn. Cu. Pb. Sb. Ag. Mn. F. Sr. (negative) | 1.06 | 0.75 | | 0.08 | SnAsZnCu MnSr(−) |
| Disseminated tin-zinc-sulfide replacement orebody (stratiform ore body, veinlet and stockwork) | Lower part | Sn*. As*. Zn*. Mn. Cu. Sb. In. Ag. Bi. F. Sr(negative) | 0.87 | 0.36 | 20.95 | 0.04 | Sn As Zn In Bi F Sr(−) Sb |

* Strongly anomalous element

$D_3^3$ Limestone with intercalations of shale

$D_3^2$ Lanticular limestone

$D_3^1$ Siliceous shale

$D_2^2$ Calcic shale intercalated with silicalite

$\delta\pi$ Diorite porphyrite dyke

$\gamma\pi_5^3$ Granitic porphyry dyke

Sk Skarn

Fissure vein type cassiterite-sulfide orebody

Veinlet-zone cassiterite-sulfide orebody

Stratoid cassiterite-sulfide orebody

Zinc orebody

**Fig. 6.** Vertical zonation of elements in Changpo cassiterite-sulphide deposit

In order to prove the prospecting efficiency of heat-releasing Hg and halogen, we sampled the soils along three profiles over the Maopingchong ore occurrence in the east of the ore field, resulting in two associated anomalies of heat-releasing Hg, F, and I. These two anomalies, superimposed by those of Sn, Cu, Pb and Zn, were suggested to be produced by a concealed cassiterite-sulphide deposit. Later drilling proved that an economic orebody did exist beneath the anomalies. The result indicates that this prospecting technique is effective in the search for tin-polymetallic deposits.

*Acknowledgements.* This study was conducted with assistance of geologists Chen Sanyou, Ye Xusun and Yang Yanfu from No. 215 Geological Party of Guangxi Nonferrous Metals Industry Corporation, and of Chai Hongyuan, Yang You and Jingboa from the Research Institute of Geology for Mineral Resources, China Nonferrous Metals Industry Corporation.

## 6.4.2 The Sulphosalt Mineral Series and Their Paragenetic Association in the Changpo Cassiterite-Sulphide Deposit, Zhuang Autonomous Region of Guangxi

HUANG MINZHI, CHEN YUCHUAN, TANG SHAOHUA, LI XIANGMING, CHEN KEQIAO, and WANG WENYING[1]

**Abstract**

The ore deposit is dominated by Pb-Sb sulphosalt minerals accompanied by various Sb, Bi sulphosalt minerals of Pb, Ag, Cu, and Sn. The 22 species of sulphosalt minerals and the 9 species of rare sulphides associated with them, as well as the natural elements, have been identified in this paper. Of them, the 23 species including 17 species of sulphosalt minerals and other associated minerals were discovered for the first time in the Dachang ore field. Based on the paragenesis of the minerals, different types of sulphosalt minerals assemblages have been preliminarily determined. Based on the temporal and spatial evolutionary relationship of the mineral assemblages and the ore-forming element assemblages, the zoning pattern of the deposit has been worked out. It exhibited normal characters at the early stage and reverse zoning at the late stage. The sulphosalt minerals were all formed in the late stage and the difference in their species and their paragenesis is a qualitative indicator of the difference in the concentration of the components in the ore-forming fluids.

The Changpo deposit is the largest and the most representative deposit in the well-known Dachang cassiterite-sulphide polymetallic ore field, Zhuang Autonomous Region of Guangxi. In this huge cassiterite-sulphide deposit, tin is the main economic metallic element and a variety of other metals such as zinc, lead, antimony, silver, indium and cadmium can be recovered in varying grades. The ore constituents are complex and mineral associations most varied. In studying the mineralization mechanism, the authors conducted systematic researches on the ore compositions of the Changpo deposit, especially of the sulphosalt minerals. The purpose was to explore the physical-chemical conditions for formation and to find out the mode of occurrence of some useful elements for the integrated utilization of the ore. The study suggests that, in addition to cassiterite and a great quantity of such common sulphide minerals as iron-, zinc-, and arsenic-bearing ones, the deposit is obviously characterized by large amounts of sulphosalt minerals. Such numerous and highly concentrated sulphosalt minerals are rarely seen in the world. Our results are summarized briefly:

[1] Institute of Mineral Deposits, Chinese Academy of Geological Sciences, Beijing

## 1. The Geological Setting of the Deposit

The Dachang ore field is situated on the southwestern margin of the Jiangnan (south of the Yangtze River) massif and within the depression belt of the Danchi axis margin. The Changpo deposit, including Tongkeng, belongs to the western belt of the Dachang ore field. Its main ore-bearing beds are of Upper Devonian silicic rocks ($D_3^1$), striped limestone ($D_3^2$), lenticular limestone, siliceous shale and marl ($D_3^3$). Within the mine field, Rb-Sr whole-rock dating gives an age of 140 Ma for the late Yenshanian granite occurring in the middle ore belt. Within this belt, only two dykes are exposed, the eastern being granitic porphyry and the western dioritic porphyrite. The ore belt is controlled mainly by a compressive overthrust, NW-trending Dachang overturned anticline and sub-unit buried short-axis anticline. The Changpo deposit is located at the junction of the sub-unit latitudinal anticline on the gentle eastern limb of the Dachang overturned anticline, the fissure belt at the axis, and bedding plane slippage. Due to the different country rocks and the structural conditions of the ore body, the most complete zonation of the Changpo deposit can be observed in a vertical direction, i. e. large fissure type→veinlet type→bedding-plane vein→stratoid veinlet type→stratoid network type in downward succession. The mineral constituents within the orebody are complicated; the metallic minerals are abundant, marmatite, pyrite, pyrrhotite, arsenomarcasite, jamesonite and cassiterite, with lesser franckeite and boulangerite, and still less stannite, chalcopyrite, galena, stibnite, accompanied by a great variety of sulphosalt minerals. Gangue minerals are mainly quartz, calcite, and to a lesser amount, tourmaline, sericite, fluorite, gypsum, barite, potash feldspar, chlorite, etc. The mineral associations also show a vertical zonation: from top to bottom, the Fe-sulphides change in order of colloidal pyrite, and marcasite→pyrite→pyrrhotite; in the marmatite, the chalcopyrite and stannite increase as exsolved milky coloured drops; at depth individual chalcopyrite occurs and is mostly associated with pyrrhotite; arsenopyrite and cassiterite are relatively enriched; lead and antimony sulphides and sulphosalts relatively decrease. Franckeite occurs mainly at the margins of the middle and upper orebody, while antimony, bismuth as well as sulphosalt minerals of silver and bismuth are relatively concentrated in the middle and lower parts and the variety of sulphosalt minerals become complicated in the lower part.

## 2. Sulphosalt Minerals and Their Associated Minor Minerals

This deposit is dominated by Pb-Sb sulphosalt minerals and accompanied by various Sb-Bi sulphosalt minerals of metals such as Pb, Ag, Cu and Sn. 22 species of sulphosalt minerals and 9 species of associated less common sulphides and natural elements have been preliminarily identified in this paper. They are:

*1. Pb-Sb sulphosalt mineral series:* (1) geocronite $Pb_{4.85}$ $(Sb_{1.12}As_{0.88})_{\Sigma=2.00}S_8$; (2) boulangerite $Pb_{5.33}$ $(Sb_{3.79}As_{0.25})_{\Sigma=4.04}S_{11}$; (3) Heteromorphite* $Pb_{7.22}Sb_{8.50}S_{19}$; (4) plagionite* $Pb_{5.39}(Sb_{8.04}\ As_{0.17})_{\Sigma\ =\ 8.21}S_{17}$; (5) jamesonite $Pb_{4.32}$ $(Fe_{0.86}Sn_{0.06})_{\Sigma\ =\ 0.92}(Sb_{5.91}As_{0.20}Bi_{0.07})_{\Sigma = 6.11}$ $S_{14}$; (6) zinkenite* $(Pb_{0.95}Sn_{0.05})_{\Sigma = 1.00}$ $(S_{2.22}As_{0.02})_{\Sigma = 2.24}$ $S_4$; (7) fuloppite* $Pb_{3.06}$ $(Sb_{7.65}As_{0.33})_{\Sigma = 7.98}$ $S_3$.

*2. Pb, Cu-Sb sulphosalt mineral series.* (8) meneghinite* $Pb_{13.62}Cu_{0.86}Sb_{7.13}S_{24}$; (9) bournonite $Pb_{1.00}Cu_{1.21}(Sb_{0.93}As_{0.06})_{\Sigma=0.99}S_3$.

*3. Pb, Ag-Sb sulphosalt mineral series.* (10) dadsonite* $Pb_{10.56}Ag_{2.14}(Sb_{11.40}As_{0.51})_{\Sigma=11.91}S_{29}$; (11) andorite* $Pb_{1.27}Ag_{0.96}Sb_{3.03}S_6$; (12) ramdohrite* $(Ag_{3.83}Cu_{0.42})_{\Sigma=4.25}(Pb_{5.65}Zn_{0.80})_{\Sigma=6.45}(Sb_{9.17}Bi_{0.74}As_{0.31})_{\Sigma=10.12}S_{23}$.

*4. Cu, Ag-Sb sulphosalt mineral series.* (13) miargyrite* $(Ag_{0.94}Cu_{0.06})_{\Sigma=1.00}(Sb_{1.02}As_{0.05})_{\Sigma=1.07}S_2$; (14) Ag-tetrahedrite* $(Cu_{7.45}Ag_{2.83}Fe_{1.43}Zn_{0.51}Pb_{0.02})_{\Sigma=12.23}(Sb_{3.73}As_{0.12}Bi_{0.01})_{\Sigma=3.86}S_{13}$; (15) freibergite* $(Ag_{4.23}Cu_{6.12}Fe_{1.68}Zn_{0.49})_{\Sigma=12.52}(Sb_{4.11}As_{0.17})_{\Sigma=4.28}S_{13}$.

*5. Pb-Sb, Bi sulphosalt mineral series.* (16) kobellite* $(Pb_{3.85}Ag_{1.14})_{\Sigma=4.99}(Bi_{4.28}Sb_{4.11})_{\Sigma=8.39}S_{17}$; (17) tintinaite* $Pb_{5.08}(Bi_{2.45}Sb_{4.91}As_{0.39})_{\Sigma=7.75}S_{17}$.

*6. Pb-Bi sulphosalt mineral series.* (18) lillianite* $Pb_{3.42}(Bi_{1.92}Sb_{0.09})_{\Sigma=2.01}S_6$; (19) galenobismutite* $(Pb_{0.75}Ag_{0.26}Cu_{0.03})_{\Sigma=1.04}(Bi_{2.01}Sb_{0.14})_{\Sigma=2.15}S_4$.

*7. Ag-Bi sulphosalt mineral series.* (20) matildite* $Ag_{0.91}(Bi_{1.00}As_{0.06})_{\Sigma=1.06}S_2$; (21) pavonite* $Ag_{1.01}Bi_{2.91}S_5$.

*8. Pb, Sb-Sb sulphosalt minerals and tin sulphides.* (22) franckeite $(Pb_{6.00}Ag_{0.17})_{\Sigma=6.17}Sn_{2.06}Fe_{0.95}Sb_{2.00}S_{14}$; (23) stannite $Cu_{2.06}(Fe_{0.89}Zn_{0.16})_{\Sigma=1.05}Sn_{1.06}S_4$; (24) rhodostannite* $Cu\ Fe_{0.99}(Sn_{2.95}Pb_{0.04})_{\Sigma=2.99}S_8$; (25) herzenbergite* $(Sn_{0.95}Pb_{0.07})_{\Sigma=1.02}S$; (26) teallite* $Pb_{1.00}Sn_{1.04}S_2$.

*9. The paragenetic minerals related with sulphosalts.* (27) bismuth* $Pb_{0.03}Bi$; (28) Pb-bismuth* $Pb_{0.17}Bi_{1.00}$; (29) antimony $As_{0.03}Sb$; (30) galena $(Pb_{0.91}Bi_{0.05}Ag_{0.01})_{\Sigma=1.03}S$ associated with Ag, Bi sulphosalts and $Pb_{1.03}S$ associated with Pb, Sb sulphosalts; (31) stibnite $Sb_{1.95}S_3$; (32) gudmundite $Fe_{1.43}(Sb_{0.95}As_{0.03})_{\Sigma=0.98}S$.

Detailed description and determination data of all the above minerals are available in the paper proper. 23 species of them (with *marks), including 17 species of sulphosalt minerals and other associated minerals, were discovered for the first time in the Dachang ore field, while antimony (with *marks), occurring at the depth of the deposit, was discovered for the first time in the area of this deposit. Furthermore, detailed work has also been conducted on parts of the sulphosalt minerals already examined by previous workers. The discovery of the above minerals and some supplementary researches add new data to the systematic mineralogical study of the Dachang ore field.

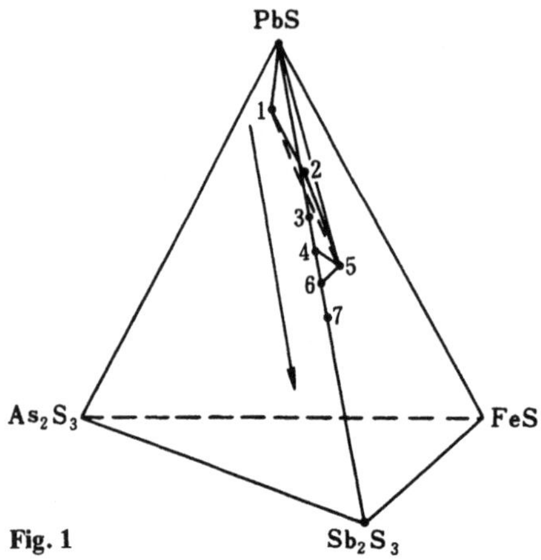


Fig. 1

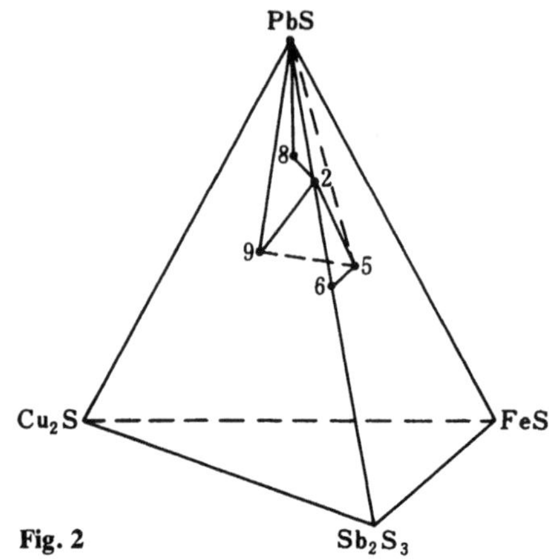


Fig. 2

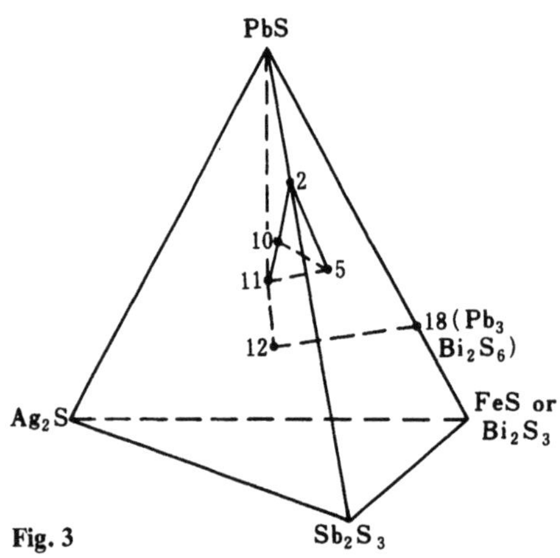


Fig. 3

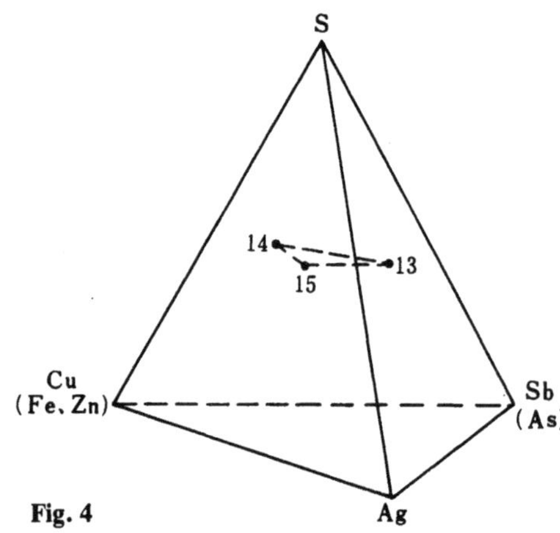


Fig. 4

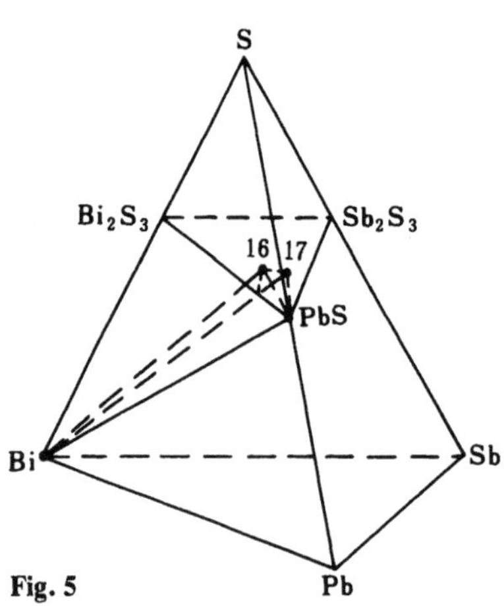


Fig. 5

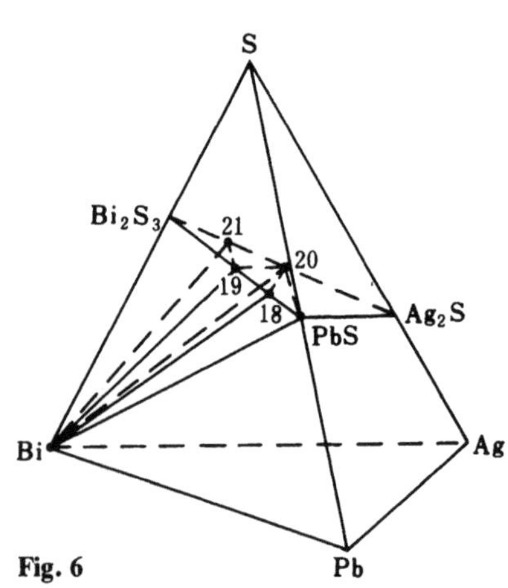


Fig. 6

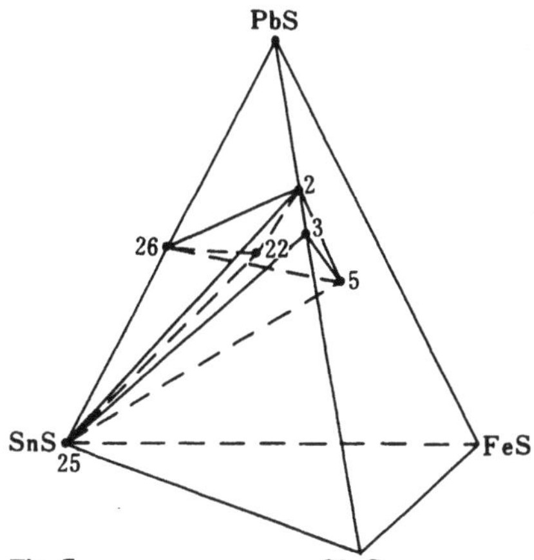


Fig. 7

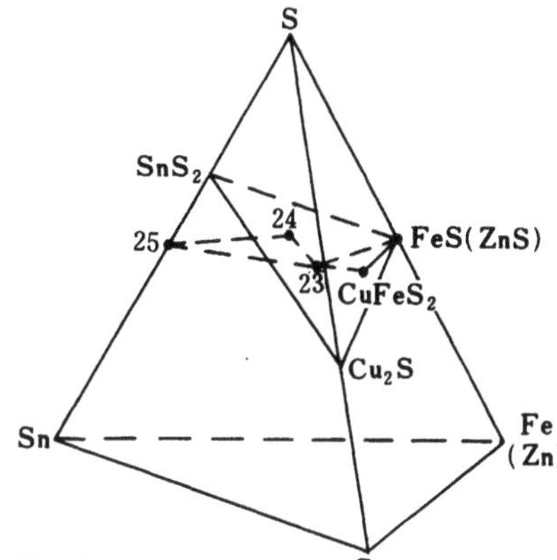


Fig. 8

**Fig. 1.** PbS-$As_2S_3$-$Sb_2S_3$-FeS series mineral parageneses

**Fig. 2.** PbS-$Cu_2S$-$Sb_2S_3$-FeS series mineral parageneses

**Fig. 3.** PbS-$Ag_2S$-$Sb_2S_3$-FeS or PbS-$Ag_2S$-$Sb_2S_3$-$Bi_2S_3$ series mineral parageneses

**Fig. 4.** Cu(Fe.Zn)-Ag-Sb(AS)-S series mineral parageneses

**Fig. 5.** Pb-Bi-Sb-S series mineral parageneses

**Fig. 6.** Pb-Ag-Bi-S series mineral parageneses

**Fig. 7.** PbS-SnS-$Sb_2S_3$-FeS series mineral parageneses

**Fig. 8.** Sn-Cu-Fe(Zn)-S series mineral parageneses

## 3. Based on the Paragenesis of the Mineral, Different Types of Sulphosalt Mineral Assemblages Have been Preliminarily Distinguished

They are listed according to their paragenetic order as follows:

*1. Sn-rich mineral assemblages:* marmatite-chalcopyrite-stannite (23) and stannite (23)-rhodostannite (24)-herzenbergite (25) assemblage, franckeite (22)-teallite (26)-herzenbergite (25) assemblage (Figs. 8 and 7).

*2. Pb, Sb mineral assemblages:* jamesonite (5)-boulangerite (2)-zinkenite (6) assemblage (Fig. 1), belonging to the simple sulphosalt mineral assemblage of Pb, Sb minerals which were deposited earlier from the ore-forming liquids.

*3. Ag, Bi-rich mineral assemblages:* dedsonite (10) or andorite (11)-boulangerite (2)-jamesonite (5) assemblages, (Fig. 3) miargyite (13)-Ag-tetrahedrite (14)-freibergite (15) assemblage (Fig. 4), galena-kobellite (16) or tintinaite (17)-bismuth and galena-lillianite (18) or matildite (20)-bismuth assemblage (Figs. 5 and 6).

*4. Pb-rich mineral assemblages:* galena-bournonite (9)-boulangerite (2) or jamesonite (5) assemblage, galena-meneghinite (8) assemblage (Fig. 2), galena-boulangerite (2)-jamesonite (5) assemblage, galena-geochronite (1)-boulangerite (2) and geocronite (1)-boulangerite (2)-jamesonite (5) assemblage, (Fig. 1) jamesonite (5)-heteromorphite (3)-herzenbergite (25) and jamesonite (5)-boulangerite (2)-teallite (26) assemblage (Fig. 7). These Pb-rich mineral assemblages are mostly the products of late-stage hydrothermal alteration.

*5. Sb-rich mineral assemblages:* fuloppite (7)-stibnite assemblage (Fig. 1).

The above-listed sulphosalt mineral species and their intersecting relationship with the sulphides of the early stage suggest that these minerals are all products of late stage mineralization. At this stage, such metallic elements as Pb, Ag, Cu, Fe, Sn and Sb, Bi, As, especially Pb, Ag, Sn and Sb, Bi were concentrated in the hydrothermal ore-forming fluids. A lot of As was separated out in the early stage and formed arsenomarcasite with the pre-existing rich Fe so that the ore fluids of the late stage were poor in As and rich in Sb and Bi sulphosalt mineral assemblages with abundant Pb and Ag formed.

**4.** The vertical distribution of mineral assemblages in the deposit is not only constrained by the lithology of the country rocks and the structure features, but also closely related to the changes in physiochemical conditions and the evolution stage in time during the mineralization. The mineral assemblages in the ores, the spatial distribution of the metallic elements and other data all indicate that the high-temperature cassiterite-quartz-arsenomarcasite mineral assemblage in the lower part, passing upward to medium and low-temperature sulphosalt mineral association, can be considered as the basic model of normal vertical zoning in the Changpo deposit. However, during the late stage mineralization, abundant elements were accumulated in the ore fluid. In the process of downward recession of the ore fluids, Sb, Bi sulphosalt minerals in association with various medium and low temperature Pb, Cu, Ag – bearing species as well as the final stibnite, calcite association, were superimposed on the high temperature cassiterite, quartz, arsenomarcasite association and cassiterite, sulphide association formed in the early stage at depth. These associations, combined with the simultaneously-formed various metasomatic mineral associations, constitute a great abundance of sulphosalt minerals. Therefore, the evolutionary relationships in time and space of the sulphosalt mineral associations indicate that the general trend of the ore-forming process proceeded first upward from bottom to top in the form of fissure-filling and replacement and then a recession of the ore-fluid took place from top to bottom, i.e., a late-stage reverse zoning was superimposed on an early-stage normal zoning, and the ore fluids were derived from depth. The variation regularity of mineral associations shows that, from early to late stage, ore-forming fluids experienced an evolutionary process from high oxygen through sulphide to carbonate and again the late-stage intense sulphide activities. In this process, oxygen fugacity progressively decreased, sulphur activity continuously increased, partial pressure of carbon dioxide increased, ore-forming temperature and pressure continuously decreased, and ore fluids experienced an evolution from intermediate-acid through neutral to an alkaline reducing environment. The metallic element associations, i.e., from early-stage Sn-Fe-As association through Sn-Fe-Zn-As association to late-stage Sn, Zn-Pb (Ag)-Sb (Bi) association, and their temporal relationship of mutual superposition, imply that the ore fluids were cognate and that their compositions were changing successively and regularly.

**5.** The conditions for the formation of sulphosalt minerals have been tentatively investigated on the basis of the ore mineral associations in the deposit, the characteristics of their spatial and temporal distribution, their textures and other evidence. The sulphosalt minerals were formed at shallow depth (less than 3 km), low pressure (less than 0.9 kilobar), medium-low to low temperature (220°–120°C), and alkaline and reducing environment, and were the final products of the late stage mineralization. The species of the sulphosalt minerals and their depositional sequence reflect the nature of the ore fluids at the late stage. The general trend of their evolution was from Pb-rich to Sb-rich. Therefore, the sulphosalt minerals and their difference in paragenetic relationships may be considered as a quantitative indicator for the differences in the concentrations of the components in the ore fluids. The sulphosalt mineral species and their association types are governed not only by the influence of physiochemical conditions and the geochemistry of elements in the ore-forming processes, but also by the regional geochemistry. The Pb, Sb sulphosalt series minerals are dominant in this deposit and occur extensively in the whole Dachang ore field. Like the Changpo deposit, the Dachang ore field is rich in As, but a lot of this element was separated out at the early stage and formed arsenomarcasite when combined with Fe; therefore, the ore field became poor in As at the late stage mineralization. The Changpo deposit and Lamo Sn-containing skarn-type Zn, Cu deposit contain numerous sulphosalt minerals of Ag, Cu, Bi and native bismuth, jamesonite, fuloppite, etc. Due to high contents of Bi in these minerals, independent minerals of Bi have not yet been found in the Chashan Pb, Zn, Sb deposit and the Furongchang Pb, Zn, Sb deposit, whose ore-forming temperatures are relatively low; nor do they contain Bi in the metallic minerals such as jamesonite. It can be seen that the Dachang ore field and its neighbouring region is a Sb-rich area and is dominated by sulphosalt minerals of Pb, Sb; at the outer area of the ore field, however, the single sulphide of Sb (stibnite) is dominant.

**6.** Work on considerable quantities of samples confirmed the economic value of the sulphosalt minerals in this deposit. The great amount of Pb-Sb sulphosalt minerals is not only the major source of Pb and Sb metals, but also the "carrier minerals" of rare elements such as In and Cd. Furthermore, the occurrence of Sn, Bi and the precious metal Ag indicate that they are of commercial significance. The close genetic relationship of the sulphosalt minerals have also been further understood. The results of this project show that studies on the associated sulphosalt mineral species, their distribution, and the relationship between the associations and geological structures are not only of significance in mineralogy and in providing information for the physicochemical conditions of ore forming process, but also important in ascertaining the mode of occurrence of useful elements in the ores, exploring the possibilities of integrated utilization of minerals resources, and furnishing mineralogical knowledge for ore dressing and smelting experiments.

# 6.4.3 Geological and Metallogenic Features and Model of the Dachang Cassiterite-Sulphide Polymetallic Ore Belt, Guangxi, China

CHEN YUCHUAN, HUANG MINZHI, XU YU, AI YONGDE, LI XIANGMING, TANG SHAOHUA, and MENG LINGKU[1]

**Abstract**

The Dachang ore belt is associated with the Yenshanian epizonal granite in the Zhuang Autonomous Region of Guangxi, and is one of the most important cassiterite-sulphide polymetallic ore belts of China. The relatively complete investigations of the geological features, various ore-controlling factors, and main metallogenic features are described. In addition, a summary of the investigations on the metallogenic patterns and model of the Changpo-Tongkeng ore deposit, and a preliminary discussion on the genesis of the ore deposits are also made.

## 1. Geological Features of the Ore Belt

The Dachang cassiterite-sulphide polymetallic ore belt (Fig. 1) is located on the southwestern margin of the Jiangnan tectonic axis and extends northwesterly for over 100 km, with a width of about 30 km, representing a fault depression formed during the Hercynian and Indosinian movements. Within the depression there are exposed sequences of Devonian to Triassic, of marine flysch and carbonate formation, with a total thickness of over 8000 m. During the Indosinian-Yenshanian orogenies there were intensive tectonic movements which occurred in two manners: one is characterized by NW-trending strong folding and fracturing, as a result of which were formed closely-spaced folds, overturned anticlines and overthrusts in the central part of the depression where the Dachang ore field is located (Fig. 2); weak folds and fractures occur at both ends of the depression, forming a series of brachy-anticlines which contain the Mangchang and Beixiang-Furongchang ore fields. The other manner is marked by N-trending fracturing of the basement, which led to the formation of dextral *en echelon* features expressed as a series of upwarps with intervening downwarps. Within the upwarps Devonian sequences are exposed, while in the downwarps there are Triassic sequences. Along the N-trending fractures there occur lamprophyre and granite-porphyry dykes. The upwarps are commonly characterized by granitic intrusions. The combination of these two major groups of structures and the consequent second-order structures constitute the geotectonic framework of the Dachang ore belt.

[1] Institute of Mineral Deposits, Chinese Academy of Geological Sciences, Beijing

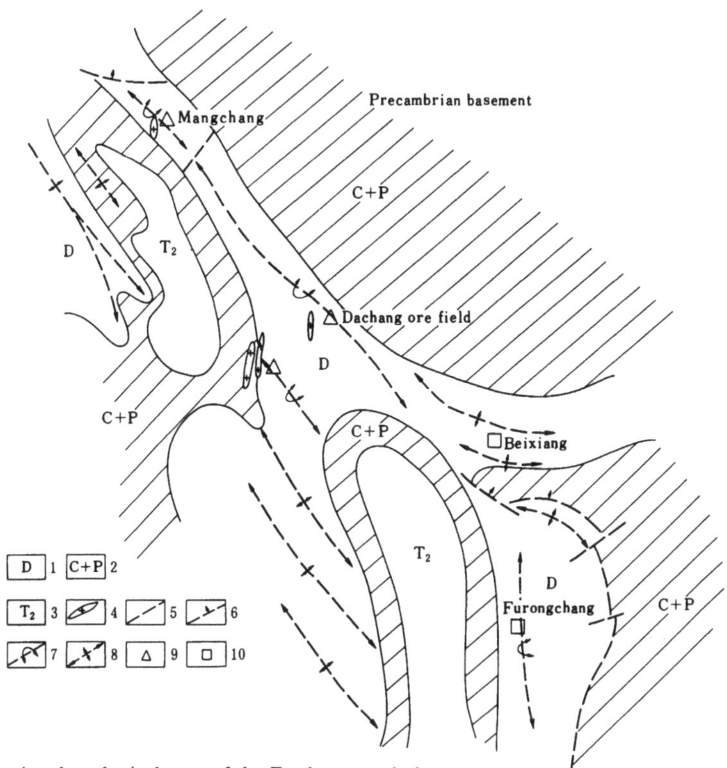


**Fig. 1.** Schematic regional geological map of the Dachang ore belt

The magmatic activity occurred in four phases. The earliest phase, of intermediate and basic porphyrite, is characterized by lamprophyre dykes and amygdaloidal diabase-porphyrite, with the latter occurring only in the granite-porphyry dykes. The second phase is characterized by the formation of granitic stocks at Longxianggai in the Dachang ore belt, composed of three-stage intrusions. The firststage is of porphyritic biotite-granite; the middle or mainstage medium- and fine-grained biotite-granite; and the latestage is of porphyritic biotite-granite. The third phase is of granite-porphyry and quartz-porphyry, which occurred both before and after the mineralization. The fourth and final phase is of tungsten-bearing alaskite, characterized by sills which tend to cut across the skarn-type sulphide-copper and zinc orebodies.

The granitoids are chemically peraluminous and calc-alkaline, while petrographically they are largely alkali-feldspar-granites, with the quartz-porphyry belonging to the quartz-granitoids (Fig. 3). The REE patterns show a certain regularity of evolution (Fig. 4), which lies in the decrease in the total amount of rare earth, the increase in the content of the light rare-earth elements, and a relative rise in the content of heavy rare-earth elements with the staged development of intrusion of magma. The two kinds of porphyrite, intruded at the earliest time, are characterized by an obscure Eu-depletion, indicating another source than that of the granite, i.e., a mantle origin. The isotopic datings are shown in Table 1, with the biotite-granite yielding an age of

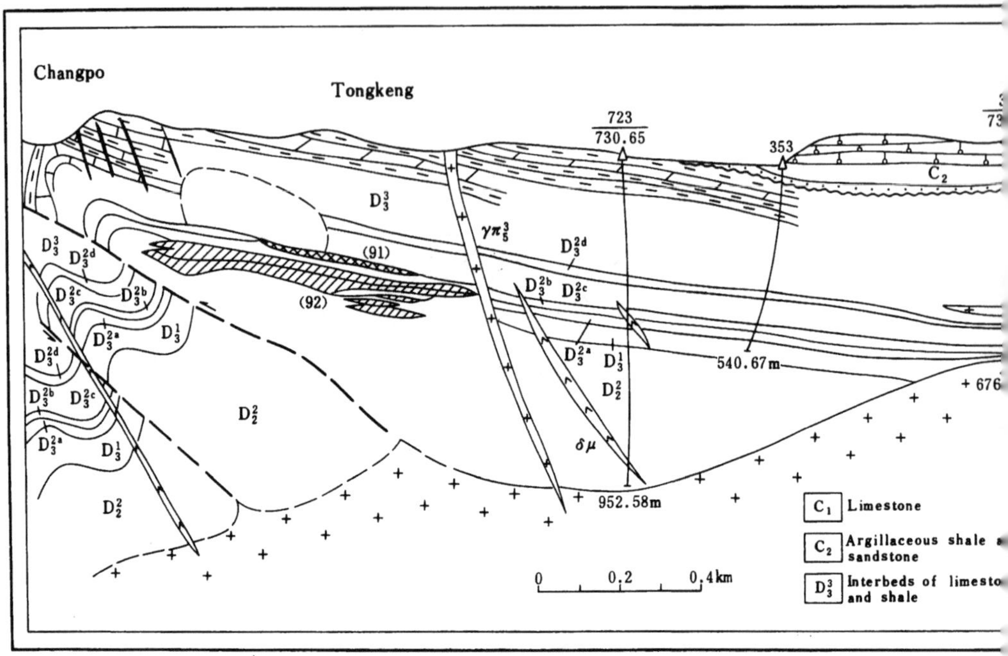


**Fig. 2.** No. 10 geological profile of the Changpo-Longxianggai area in the Dachang ore field, Nandan, Guangxi. (after the No. 215 Geological Party, China National Nonferrous Metals Industry Corporation)

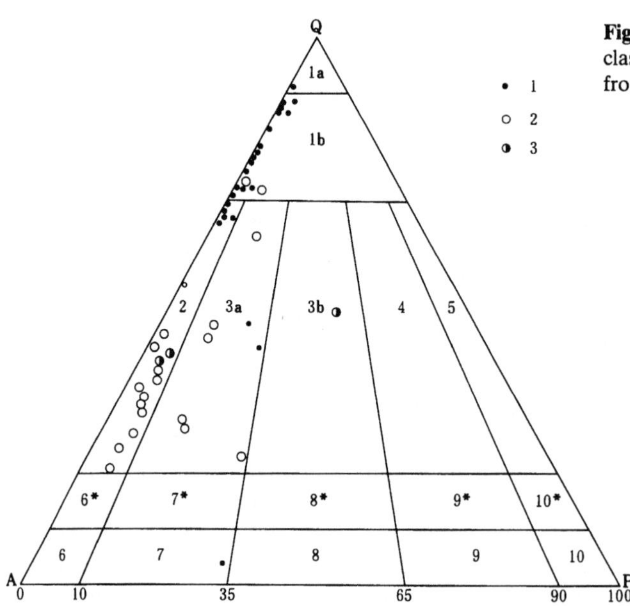


**Fig. 3.** Diagram showing the classification of magmatic rocks from the Dachang ore belt

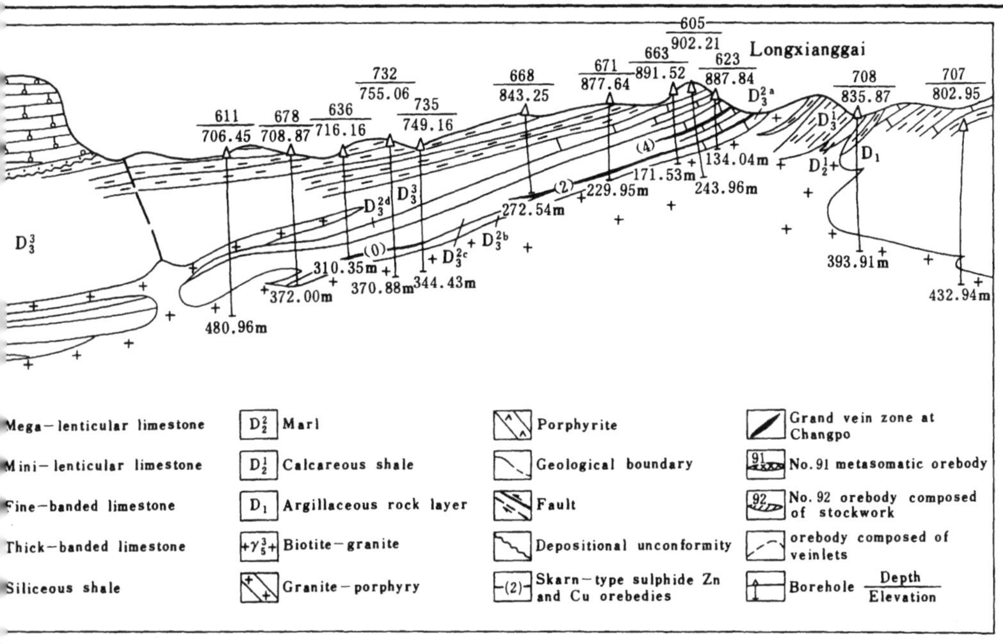


about 130—149 Ma, and the mineralization age later at 120 Ma. Many granite dates are younger (72—89 Ma) because of the effects of alteration. We believe that the magmatic rocks of the Dachang ore belt might have come from the mantle at the initial magmatic stage, and then at a later stage the granitic magma formed by the melting of the continental basement rocks.

## 2. Metallogenetic Series of Sn, Pb, Zn, Cu, Sb, As and Hg (W) Associated with the Yenshanian Hypabyssal Granite in the Dachang Ore Belt

The mineralization is divisible into three stages:

1. The skarn-type mineralization phase, during which zinc and copper sulphide deposits were formed at the exocontact of the granite. The Sn content of the andranite attains several parts per thousand.
2. Cassiterite-sulphide polymetallic mineralization phase, the major mineralization phase, divisible into five stages:

   a) Cassiterite-sulphide;
   b) Tin-bearing sulphide-sulphosalts-carbonate;
   c) Stibnite-quartz-calcite;
   d) Realgar-orpiment-calcite; and
   e) Cinnabar-pyrite-quartz.

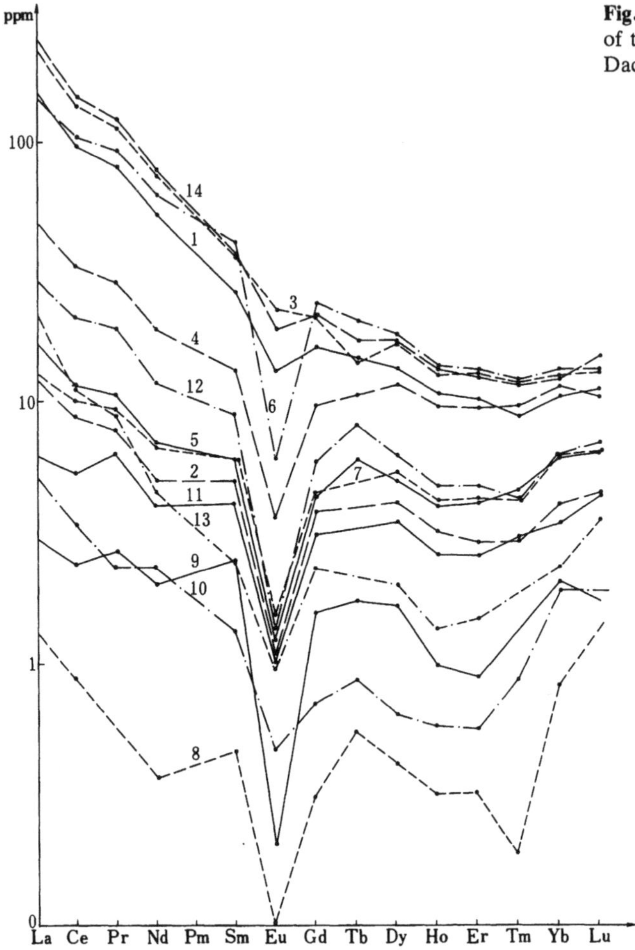


**Fig. 4.** REE distribution pattern of the magmatic rocks from the Dachang ore belt

Spatially, each stage may constitute a single ore deposit, and conversely a single deposit may be formed by superposition of several mineralization stages.

3. Tungsten and antimony phase, whose lodes cut across the cassiterite-sulphide polymetallic mineralization phase.

The salient features of the various mineralization phases are as follows:

Alterations of the country rocks: The skarn-type mineralization is characterized essentially by skarnization, with greisenization occurring on the top of granite, and potash-feldspathization on the edges of the granite. During the cassiterite-sulphide polymetallic mineralization phase, alteration of the country rocks includes tourmalinization, potash-feldspathization, muscovitization, silicification, chloritization, sericitization and sideritization. The alteration of the country rocks during the W- and Sb-mineralization phase is largely of silicification.

Ore-forming temperature: The ore-forming temperatures of the various mineralization phases and stages show a gradual decrease (Table 2).

**Table 1.** Ages of mineralizations and magmatic rocks from the Dachang ore belt

| Rock type | Sample | Age (Ma) | Method |
|---|---|---|---|
| Biotite granite | Whole-rock | 149.5 ± 5 | Rb-Sr($Sr^{87}/Sr^{86}$=0.7) |
| Fine-grained granite | Whole-rock | 138.60 | K-Ar |
| Porphyritic biotite-granite | Whole-rock | 80.30 | K-Ar |
| Porphyritic biotite-granite | Whole-rock | 82.33 | K-Ar |
| Pegmatoid | Potash feldspar | 84.13 | K-Ar |
| Porphyritic biotite-granite | Biotite | 88.22 | K-Ar |
| Porphyritic biotite-granite | Potash feldspar | 85.39 | K-Ar |
| Medium- and fine-grained Biotite-granite | Whole-rock | 77.70 | K-Ar |
| Biotite-granite | Biotite | 86.64 | K-Ar |
| Biotite-granite | Biotite | 89.31 | K-Ar |
| Lamprophyre | Whole-rock | 81.53 | K-Ar |
| Granite-porphyry | Whole-rock | 72.01 | K-Ar |
| Amygdaloidal diabase-porphyrite | Whole-rock | 73.69 | K-Ar |
| Alaskite | Whole-rock | 81.83 | K-Ar |
| Aplite | Whole-rock | 86.92 | K-Ar |
| Tungsten-bearing alaskite | Whole-rock | 84.21 | K-Ar |
| **Mineralization type** | **Sample** | **Age (Ma)** | **Method** |
| Potash-feldspathization occurring at early stage | Whole-rock | 117.89 | K-Ar |
| Sericitized quartz-hornstone | Whole-rock | 104.83 | K-Ar |
| Hydromica formed at late stage of mineralization | Whole-rock | 90.92 | K-Ar |

Characteristics of sulphur isotopes: The variation of sulphur isotopes in the various mineralization phases shows a certain regularity. The $S^{34}S$‰ for the sulphides from the skarn-type mineralization phase is close to zero; the $S^{34}S$‰ for the sulphides from the cassiterite-sulphide polymetallic mineralization phase tends to change from negative to positive values; and the $S^{34}S$‰ for the sulphides from the W- and Sb-mineralization phase again approaches zero (Fig. 5).

Characteristics of the zonation of ore deposits: There is a regular regional zonation of mineralizations within the Dachang ore belt, expressed by the occurrence of a zone of Sn, Cu, polymetals, Sb and W; the Dachang ore field serving as a centre with the polymetallic, antimony and arsenic mineralization zone occurring in its surrounding areas, and the mercury mineralization zone on the margins (Fig. 6).

Around a magmatic centre there exists a positive zonation of ore fields formed at the early stage of mineralization, i. e. in the vicinity of granite terrain there occurs the skarn-type mineralization, while in the surrounding areas are distributed the surrounding hydrothermal cassiterite-quartz-sulphide ore deposits. The late-stage mineralization shows a negative zonation represented by silver-bearing sulphosalt-sulphide veins and the antimony mineralization stages of the cassiterite-sulphide polymetallic mineralization phase and during the W- and Sb-mineralization phase near the granite, but in tectonic belts of different trends (Fig. 7). The ore-containing structures were active synchronously with the mineralization, being developed successively from the granite terrain outwards, and then backward towards the granite terrain, which, in our opinion, was controlled by the upward intrusion and shrinkage of the pluton.

**Table 2.** Homogenization temperatures of the main rock-forming and ore-forming phases in the Dachang ore belt

| Rock-forming or ore-forming phase | Rock or minerals association | Measured mineral | Homogenization temperature (C°) |
|---|---|---|---|
| Granite phase | Biotite-granite | Quartz | 703–743 |
| Potash-feldspathization phase | Potash-feldspathized granite | Quartz | 500–605 |
| Greisenization phase | Greisenized granite | Quartz | 301–342 |
| Skarnization phase | Endoskarn | Andradite | 602–411 |
| | Exoskarn-sulphide mineralization zone | Quartz and fluorite | 275–343 |
| Cassiterite-sulphide mineralization phase | Cassiterite-quartz-sulphide association | Cassiterite and quartz | 350–460 |
| | Cassiterite-sulphide-carbonite association | Calcite, quartz and cassiterite | 270–340 |
| | Silver-bearing Pb-Zn-sulphide association | | 215–290 |
| W- and Sb-mineralization phase | Wolframite-scheelite-quartz vein | Fluorite | 137–178 |
| | Stibnite-fluorite-quartz vein | Fluorite | 150–153 |

We conclude that the metallogenesis is associated essentially with the Yenshanian granite, and the metallogenesis, tectonic movements and magmatic activities were in mutual coordination and polycyclic, thus forming the metallogenic series of Sn, Pb, Zn, Cu, Sb, As and Hg (W). We present a metallogenetic model for the metallogenic series (Fig. 8).

The metallogenic series may be summarized as follows: Within the Hercynian-Indosinian fault-depression and on the margins of the Jiangnan tectonic axis the Indosinian-Yenshanian tectonic movements gave rise to magma intrusion, probably with more basic magma having intruded first from the mantle, then followed by granitic magma from crustal melting.

The repeated upward intrusion and shrinkage of the plutons caused the formation of major ore-containing structures around the granite terrain. The multi-staged magmatic activities and the ore-containing structures operated synchronously with the metallogenic processes. Under such conditions, there was formed an association of cassiterite-sulphide polymetallic ore deposits, characterized spatially by a positive zonation of ore deposits formed at the early stage of mineralization, and a negative zonation of the ore deposits formed at the late stage of mineralization. This particular metallogenic series belongs to the South China Yenshanian granite metallogenetic province of the Nanling region, and also to the multiple tin mineralization activities ranging from the Early Proterozoic to the Yenshanian (Mesozoic) stage in the Nanling tin geochemical field.

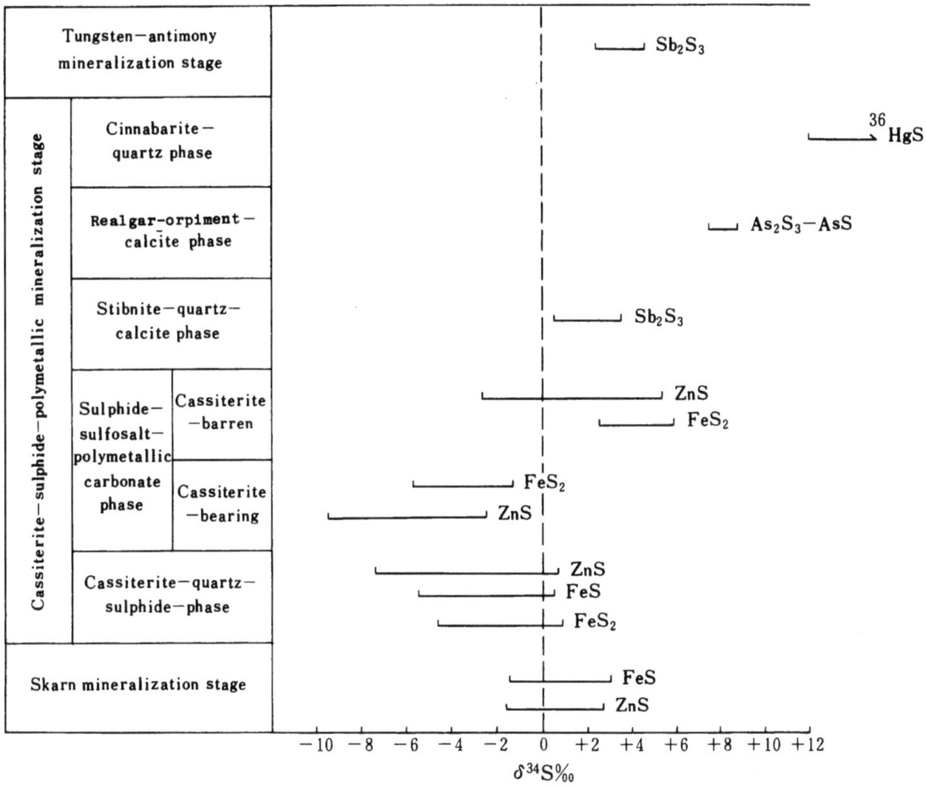


**Fig. 5.** Characteristics of sulphur isotope of some minerals from each mineralization stage in the Dachang ore belt

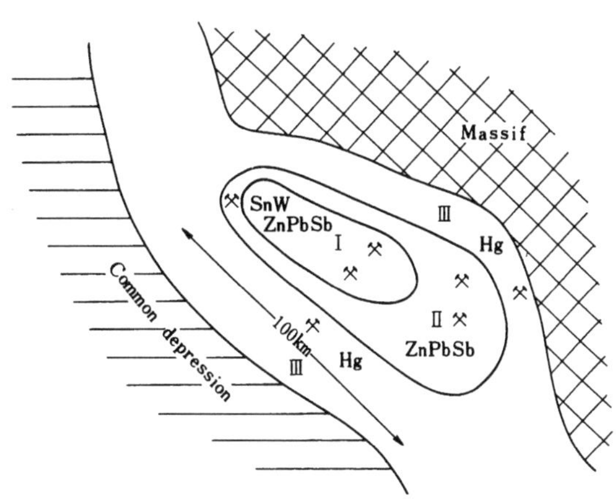


**Fig. 6.** Schematic map showing the distribution of regional mineralization zones in the Dachang ore belt

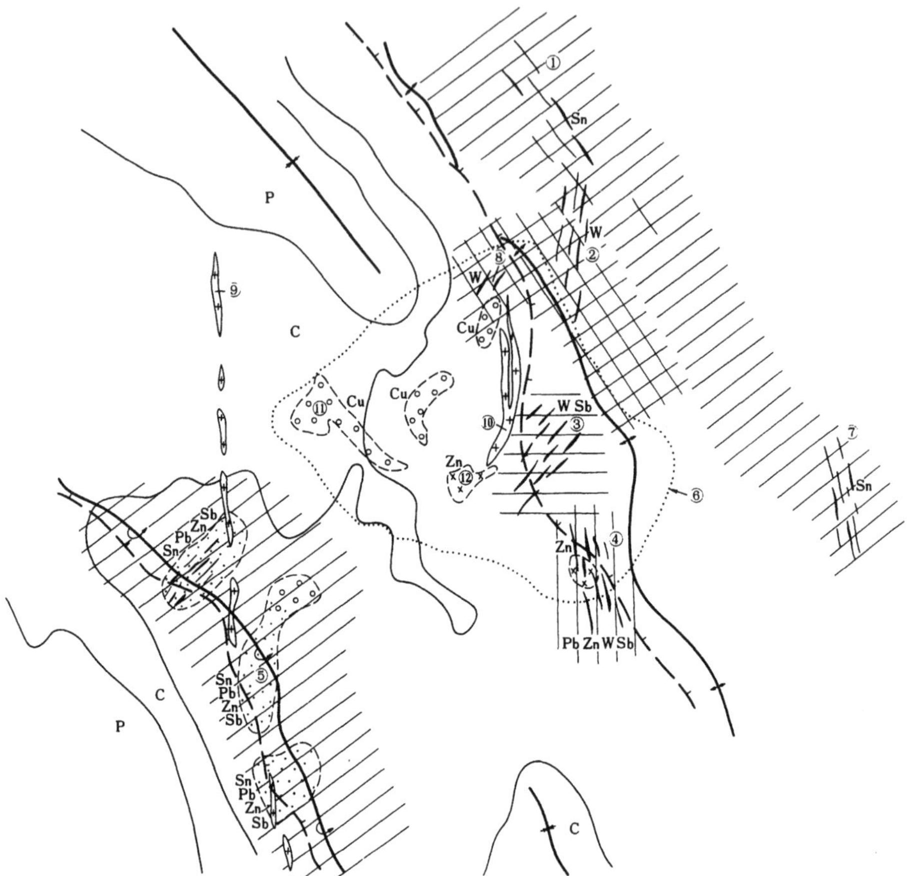


**Fig. 7.** Schematic map showing the distribution of ore deposits within the Dachang ore field

## 3. The Changpo-Tongkeng Ore Deposit and Metallogenic Model

The Changpo-Tongkeng ore deposit is one of the largest and most representative of the Dachang belt. It occurs in the second-order NE-trending transverse anticline at the turning portion of the plunging of the Dachang overturned anticline and is controlled by the tensile fractures, fissures and interstratified dislocation zones in the axial part of the transverse anticline, thus forming lode zones and interstratified metasomatic mineralized layers.

Before the mineralization, there were formed in ascending order the following metamorphic zones around the buried biotite-granite, such as the skarn-hornfels-marble zone, the sericite-marble zone, and the sericite-silicification zone. The mineralization of the cassiterite-sulphide polymetallic phase was superimposed upon the last two metamorphic zones, and takes the form of three stages and six zones in ascending order.

| Geotectonic position | ore-containing strata | Alteration type | Zonation of mineralizations | Mineralizing element associations | Ore-controlling structure | Form and mode of occurrence of orebody | Major mineral associations | Ore-forming temperature (℃) |
|---|---|---|---|---|---|---|---|---|
| Hercynian-Indo-sinian depression on the southwestern margin of the old massif within the Yangtze Paraplatform (the Jiangnan tectonic axis) | Devonian carbonate-flysch formations | Silicification and pyritization | Mercury mineralization zone | Hg | Interstratified displacement zone, fractures, and fissures | Stratoid produced mainly by replacement, fissure-filling-type veins (6) | Cinnabar-pyrite-quartz | |
| | | Silicification, sericitization, and pyritization (carbonatization) | Sulphide-polymetallic mineralization zone | Zn, Pb, Sb, As, Ag (sometimes with Sn and Au) | Fractures-fissures, local stratified displacement zone | Fissure-filling-type veins, with minor metasomatic orebodies (5) | Marmatite-pyrite-jamesonite-boulangrite-galena-carbonate; antimonite-quartz; eolite-arsenblende-calcite | 120-290 |
| | | K-feldspathization-tourmalization-silicification; muscovitization-tourmalitization-silicification; sericitization; chloritization, and sideritization | Cassiterite-sulphide polymetallic mineralization zone | Sn, Zn, Pb, Sb, As, Ag, Bi (sometimes with Au) | Stratified displacement zone, and fracture-fissure zone | Metasomatic stratoid orebodies (4), fissure-filling-type mega-veins (2), veinlet and stockwork zones (3) | Cassiterite-arsenopyrite-pyrite-marmatite-pyrrhotite-quartz; Cassiterite-pyrite-marmatite marmatite-jamesonite-manganocalcite | 270-460 |
| | | Skarnization zone represented essentially by diopside-, garnet-, grammite-, and vesuvianite-skarns | Skarn-stage zinc and copper mineralization zone | Zn and Cu (sometimes with Sn and W) | Exocontact of the granite terrain | Metasomatic stratoid orebodies (1) | Marmatite-pyrrhotite-chalcopyrite (cassiterite) | 275-343 |
| | | K-feldspathization on the top of the granite terrain with local potassium feldspar pegmatoid crust | Early-stage biotite-granite | W and Sb (sometimes with Au) | Fracture-fissure zone | Fissure-filling vein-type orebodies (7) intersecting (1) and (2) orebodies | Scheelite-fluorite; wolframite-antimonite-quartz | 137-203 |

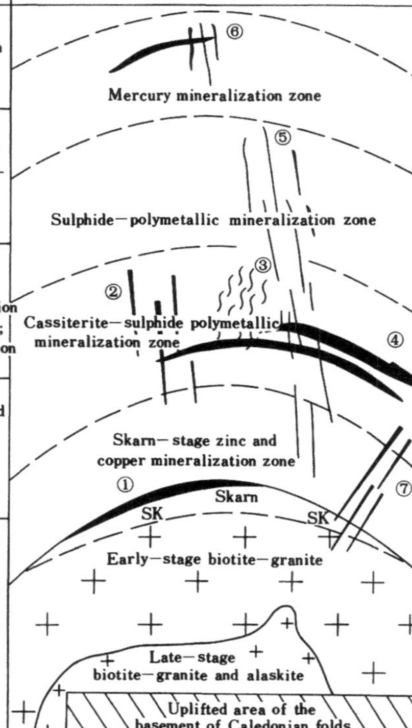


**Fig. 8.** Metallogenic model for cassiterite-sulphide polymetallic mineralization series associated with Yenshanian biotite-granite in the Dachang ore belt

The three mineralization stages are:

1. Cassiterite-quartz-sulphide, with an ore-forming temperature of 330—470°C, divisible into two substages:
   a) cassiterite-quartz-arsenopyrite-pyrite, and
   b) pyrrhotite-marmatite-arsenopyrite-quartz-siderite substage.
2. Cassiterite-sulphide-sulphosalt-carbonate stage, with an ore-forming temperature of 120—300°C, divisible into two substages:
   a) cassiterite-sulphide-jamesonite-manganocalcite, and
   b) cassiterite-franckeite-sulphide-calcite substage.
3. Stibnite-quartz-calcite stage, with an ore-forming temperature of 120—143°C.

The six mineralization zones (Fig. 9) are:

1. The first occurs at the depth of the ore deposit, and is represented by veinlet and disseminated sulphide ores in the strongly skarnized, potash-feldspathized and marblized argillo-calcarous and argillaceous rocks of Middle Devonian age, yielding largely pyrrhotite, marmatite and arsenopyrite, with accompanying chloritization, epidotization and carbonitization. This zone has also been superimposed by jamesonite veinlets.

2. The second zone represents interstratified stockwork and veinlet layers, i.e. No. 92 orebody, within siliceous shale occurring at the base of the Upper Devonian. The stockwork and veinlets are composed mainly of minerals formed during the early cassiterite-quartz-sulphide stage, of the cassiterite-sulphide polymetallic mineralization phase. In addition, there are also veinlets of late stage mineral associations such as cassiterite-sulphosalt-sulphide veinlets which cut across the former veinlets. On both flanks of the veinlets, the cassiterite and sulphide mineralization continues along the rock layers. The siliceous shale has commonly been recrystallized. In the vicinity of cassiterite and sulphide, there occur potash-feldspar and tourmaline, with a siderite zone commonly occurring around the sulphide.

3. The third zone is the stratiform No. 91 orebody, formed by the filling and replacement of the ore-forming fluids along the interstratified dislocation zone within the Upper Devonian striped limestone. It is composed of metasomatic ore bodies, fissure veins, veinlets and stockwork. The metasomatic orebodies themselves consist of two generations, the earlier represented by the cassiterite-arsenopyrite-marmatite-pyrite-quartz-potash feldspar- and tourmaline-association; and the later by the pyrrhotite-marmatite-muscovite-tourmaline- and daphnite association. The stockwork, veinlets and metasomatic ores are of the same composition. Also various Bi-Ag- and Cu-rich sulphide-sulphosalt veinlets of the late-stage mineralization are commonly superimposed upon the stibnite-quartz-calcite veins.

4. The fourth mineralization zone is of veinlet orebodies occurring above the stratiform orebodies. The lodes in the veinlet zone are of different mineral composition, with the first two mineralization stages the most important and superimposed upon each other. On both flanks of the veinlets there are cassiterite and sulphide disseminations into the country rocks, which are commonly tourmalinized and silicified.

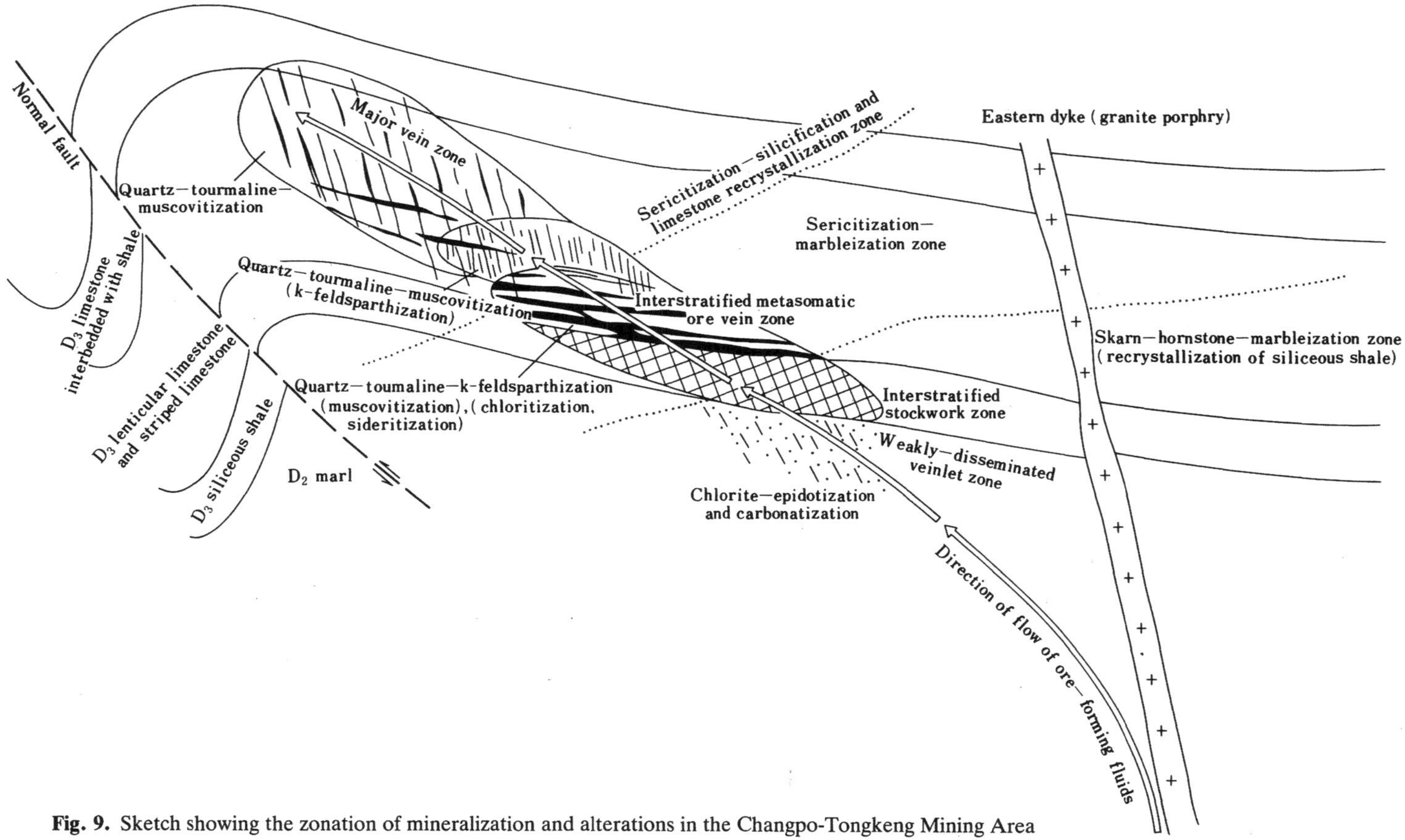


**Fig. 9.** Sketch showing the zonation of mineralization and alterations in the Changpo-Tongkeng Mining Area

On the edges of the veinlets of the early-stage mineralization there also occur potash-feldspathization and muscovitization.

5. The fifth mineralization zone is of large lodes, distributed in the Upper Devonian sequences, being controlled by NE-trending tensile fissures in the axial part of the transverse second-order anticline. The lodes result from the filling by the ore-forming fluid of the second-stage mineralization, and are composed of cassiterite, marmatite, pyrite, pyrrhotite, arsenopyrite, jamesonite, and manganocalcite. Alteration of the rocks surrounding the mineralized veins is mainly by silicification and tourmalinization. In addition, there are also franckeite and stibnite veins which were formed at a still later stage.

6. The sixth mineralization zone occurs at the top of the deposit, located in the upper part of the Upper Devonian. Present here are also minor fissure veins largely of sulphosalt and sulphide formed at the late stage of mineralization. Along the strata occur slight pyrrhotization and pyritization, and the country rocks are characterized by weak silicification and tourmalinization. The mineralization in this particular zone tends to pinch out, commonly not passing through the black shale above the interbeds of the calcareous shales.

It can be seen that the general metallogenic process is continuous and the operating ore-forming liquid is characterized by a regular evolution during the metallogenic process. The ore-forming fluid exhibits an obvious regularity in the spatial activities as shown by its ascending advance to form a positive zonation of high temperature mineralizations at the early stage, and by its descending shrinkage to produce a negative zonation of low temperature late stage mineralizations at the early stage, and by its descending shrinkage to produce a negative zonation of low temperature late stage mineralization, essentially of sulphosalt- and stibnite-mineralization. The oxygen fugacity showed a regular variation, higher at first, lowering later, then rising once again. Exchange of some components between the ore-forming fluid and the country rocks took place in the course of its crystallization. At the early stages the ore-forming fluid replaced the surrounding rocks at depth of the ore deposit leading to the migration of $SiO_2$, CaO, $CO_2$ and $MnO_2$ from the country rocks into the fluid, and while moving upwards, the ore-forming fluid deposited manganocalcite in the fissure veins and large veins at a higher horizon, leading to a universal silicification in the country rocks. Thus a metallogenic model for this particular ore deposit is presented on the basis of the salient features of the Changpo-Tongkeng ore deposit (Fig. 10).

The available data show that the ore-forming metallic elements, such as Sn, W, Cu, Pb, Zn, Sb and Hg, are largely associated with the granite. This does not, of course, exclude the possibility of metallic elements such as Sn, W, Sb, Pb and Zn coming in part from the mineralized basement strata occurring at depth. In the course of partial melting of the latter, the above-mentioned metals would enter the fluid. S, $CO_2$, Ca, Mn and Fe might have come in part from the Devonian strata which have been replaced by the mineralizing fluid. This particular problem awaits further study.

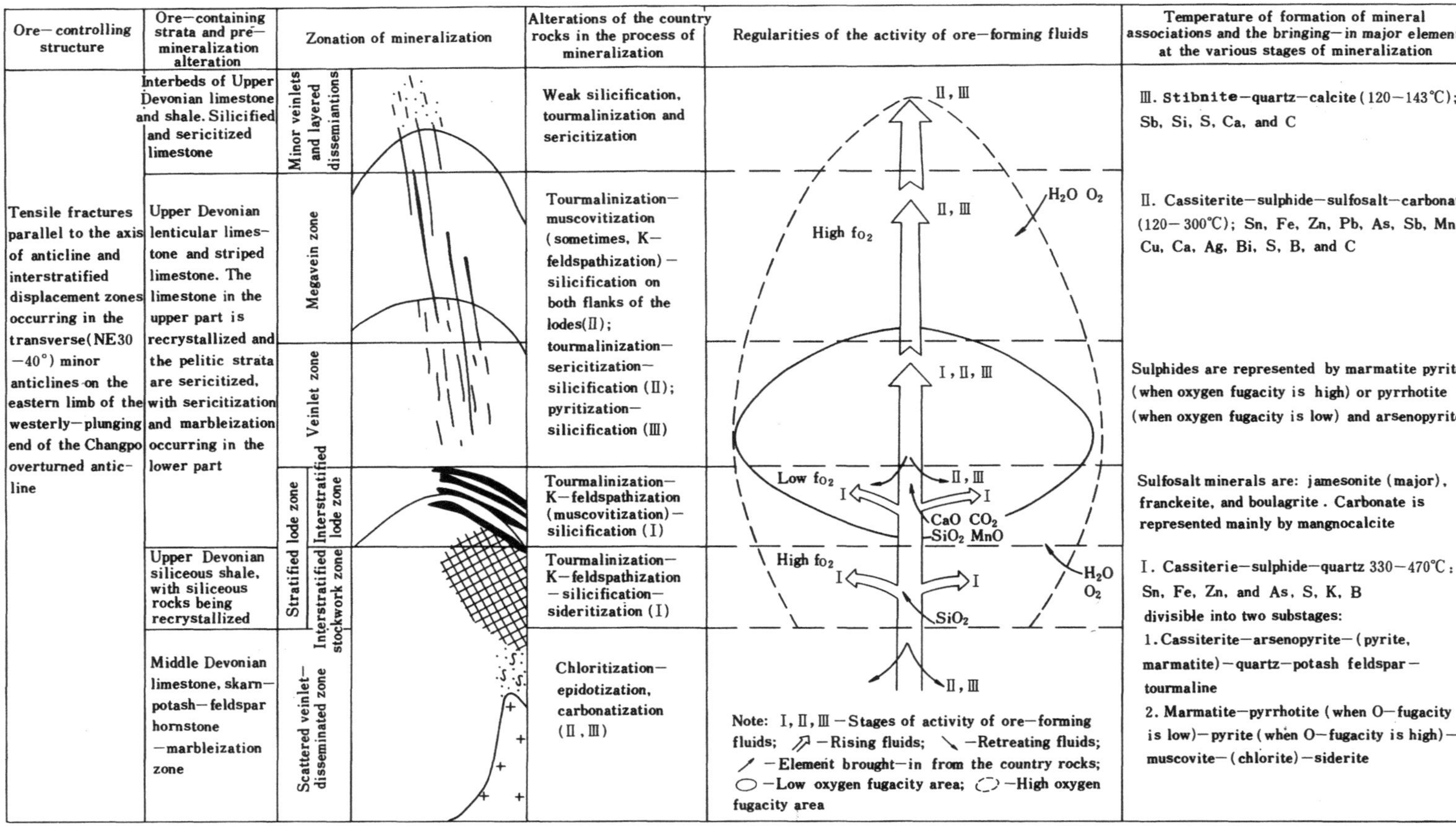


**Fig. 10.** Metallogenic model for the Changpo-Tongkeng ore deposit in the Dachang ore field

## References

Chen Yuchuan, 1964. A Preliminary study on the salient geochemical features in the hydrothermal replacement of the country rock of an ore deposit. *A collection of papers of the first conference on mineralogy, petrology and geochemistry,* organized by the geological society of China, Section on geochemistry (in Chinese).

Chen Yuchuan, 1964. Metallogenic phases and characteristics of a cassiterite-polymetallic ore belt. *Geological Review,* Vol. 22, No. 2, 111–128 (in Chinese).

Chen Yuchuan, 1965. Primary zonal distribution of an ore belt in Guangxi, China. *Geological Review,* Vol. 23, No. 1, 29–41 (in Chinese).

Onihimovskii, V.V., Govrilov, V.I., 1982. Problems of geology and metallogeny. *Science Publishing House* (Izdatelistvo Nauka) (in Russian).

Taylor, R.G., 1978. *Geology of Tin Deposits.* Elsevier, Amsterdam.

Yang Fengjun, Qui Chunyi, Chen Minyang, Wang Liyi, Li Chunsheng, Huo Weiguo, Liu Jieren, and Chai Baoping, 1966. Application of sulphur isotope investigation to the genesis of a certain cassiterite sulphide deposit. *Scientia Geologica Sinica,* No. 3, 217–226 (in Chinese).

Zhang Zhengen, Li Xilin, 1981. Studies on Mineralization and Composition of DC ore field, Guangxi, China, *Geochimica,* No. 1, 74–86 (in Chinese).

# 6.4.4 Experimental Research on the Formation Conditions of the Cassiterite-Sulphide Deposits in the Dachang Tin Ore Field

YANG JIATU, CHEN CHANGYI, ZENG JILIANG, and ZHANG YONGLIN[1]

**Abstract**

According to the characters of mineral distribution in the deposits, the mineralization system of the deposits is divided into several simple systems, each of which has been examined experimentally.

The following minerals have been synthesized experimentally: cassiterite (Ct), monoclinic-pyrrhotite (Mpo), hexagonal-pyrrhotite (Hpo), arsenopyrite (Asp), marmatite (Mmt), jamesonite (Ja), zinkenite (Zin), herzenbergite (Hzb), pyrite (Py), galena (Ga), stibnite (Stb), anhydrite (Ah), ß-hexagonal herzenbergite (Ber), marcasite (Ms), quartz (Q), calcite (Cal), etc.

We counder that: 1) The formation of cassiterite is closely related to monoclinic-pyrrhotite, so good results should be achieved in exploration by the use of magnetic prospecting if the disturbance caused by other magnetic minerals is eliminated. 2) Because the formation temperature of Po-Po+Py-Py falls gradually and the sulphur fugacity rises progressively, the temperature and the sulphur fugacity are the main factors affecting the zoning pattern of minerals in the Dachang cassiterite-sulphide deposit. 3) In the ore-forming hydrothermal solutions, there are elements to form sulphide and sulphate minerals, and there exist the same conditions of sulphur fugacity and temperature which are the main factors to develop large amounts of Ja in the cassiterite-sulphide deposits. 4) After the experiment, the concentrations of metal ions in the medium solution approximately equal the lowest ore-forming boundary concentrations of these principal metal ions in the ore-forming fluids of the ore field. The ore-forming hydrothermal solution is a Sn-Fe-Pb-Sb-S-As-Si-$H_2O$ system high in sulphur and moderately rich in iron. The medium solution is a mixture of alkali metal chlorides and alkali metal fluorides. Ore-forming oxygen fugacity is roughly $10^{-28}$–$10^{-40}$ atm. (350–250°C). 5) It is considered that the metal element zoning with "copper and zinc near the granite, cassiterite farther away" and "cassiterite at the upper part, copper at the lower and zinc at the middle" in the ore field is due to the variation in oxygen fugacity and the difference in sulphophile and lithophile tendencies of the ore-forming metal elements.

[1] Research Institute of Geology for Mineral Resources, China National Nonferrous Metals Industry Corporation, Guilin, Guangxi

From the results of our experiment and the geological features of the deposit, we believe that the biotite granite is the main source of the ore-forming material and metallogenic thermal fluids, whereas part of the sulphur, lead, and antimony is from the wall rocks.

## Introduction

The Dachang tin ore field is situated where the western flank of the frontal arc of the Guangxi type tectonic zone intersects the western end of the Nangling latitudinal tectonic zone. It is also a part of the western Circum-Pacific tin metallogenic zone. Magmatic rocks, folds and faults are well developed in the mining area. Orebodies occur mainly in carbon-rich carbonate formations, silicalite and reef limestone of Mid-Late Devonian. The Longxianggai biotite granite occurs in the centre of the mining area. The zinc-copper skarn deposits occur in the west contact zone of the rock body while tungsten-antimony veins occur in the east side. Cassiterite-sulphide deposits are distributed in both the eastern and western sides, mainly 400—2000 m away from the rock body. Horizontal mineral zoning is obvious: copper and zinc are near the rock body and cassiterite is far away. Vertical mineral zoning is also clear, as suggested by the 215 Geological Party: cassiterite in the upper, zinc in the middle and copper in the lower.

In the cassiterite-sulphide deposits the principal metallic minerals are cassiterite (Ct), monocline-pyrrhotite (Mpo), hexagonal-pyrrhotite (Hpo), arsenopyrite (Asp), marmatite (Mmt), jamesonite (Ja), zinkenite (Zin), herzenbergite (Hzb), pyrite (Py), etc. The gangue minerals are quartz (Q), calcite (Cal) tourmaline, etc. Vertically, the mineral assemblages also show distinct zoning, i.e., in the lower part, arsenopyrite, marmatite, monoclinic-pyrrhotite and cassiterite are relatively concentrated, and in the upper part there are mainly pyrite, sulphosalts and cassiterite.

From the previous data, the cassiterite-sulphide deposits in Dachang can be divided into three main metallogenic stages; the earliest is a cassiterite-quartz stage, the second a cassiterite-sulphide stage, and the third is cassiterite-sulphosalt (calcite) stage.

From the geological characteristics of the deposits, we divide the complex metallogenic system into several simple water-bearing systems so that we can carry out the experimental researches individually. The work is focussed on the simulation experiments of Sn-Fe-S (Si)-$H_2O$ and Sn-Fe-Pb-S (Si)-$H_2O$ systems.

## I. Experimental Conditions and Methods

### 1. Choice of Experimental Conditions

From the data on gas-liquid inclusions from the Dachang tin ore field, the volatile components mainly include $CO_2$, $Cl^-$ with minor $F^-$; the cations include $K^+$, $Na^+$ and minor $Ca^{++}$. The ratio of $K^+/Na^+$ is 1.18, and $Cl^-/F^-$ is 13.04 in fluid inclusions of cassiterite from Bali and Tongkeng. The total salinity, measured by the freezing method, is between 3.64—12.13% NaCl equivalence and the formation temperature of cassiterite ranges from 280 to 393°. From the data of S.D. Scott (1981), the metallogenetic pressure range is calculated from marmatite to be 400—1000 atm. From the data of inclusions and the geological feature of the mining area, we selected the experimental conditions as follows: Temperature range from 400 to 200°C; pressure from 900 to 300 atm, total salinity in liquid from 0.5 to 2 m; ratio of m NaCl/m KCl=1:1; m NaF/KF=1:1; and pH=3—9.

## 2. Experimental Method and Measurement

Before experimentation, a thermodynamic calculation of the mineral assemblage of the cassiterite-sulphide deposits in the Dachang tin ore field was carried out using the available thermodynamic data, and the $fo_2$ versus $fs_2$ figure (Fig. 1) at 300°C and 1 atm was drawn to guide the experiments. From the figure, it can be found that $fs_2$ ranging from $10^{-10}$ to $10^{-15}$ and $fo_2$ from $10^{-32}$ to $10^{-39}$ represent the stable field of the cassiterite-sulphide assemblage. The oxygen fugacities of several buffer pairs are shown in the figure by dotted lines. Among them, the equilibrium lines of Ni and NiO are closest to the oxygen fugacity in the stable coexisting field of the cassiterite-sulphide assemblage.

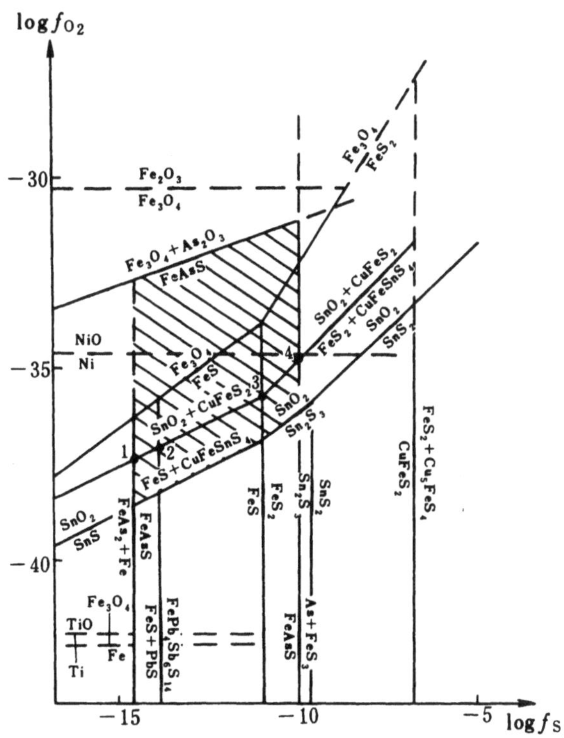


**Fig. 1.** $fs_2$-$fo_2$ diagram of the assemblage of major minerals at 300°C and 1 atm in cassiterite-sulphide deposits of the Dachang ore field

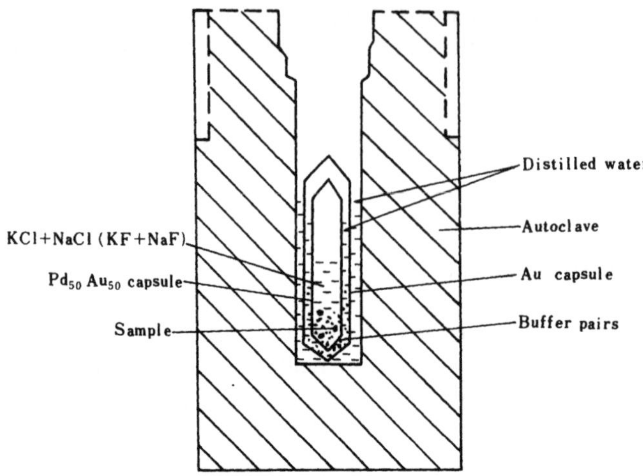


**Fig. 2.** Diagram showing a sample in the autoclave device

The experimental primitive reagents, such as Sn, Fe, Pb, Sb, Zn and S, are compounded by chemical reagents of high purity.

The oxygen fugacity is strictly controlled by double layer capsules ($Pd_{50}Au_{50}$ and Au) and buffer pairs (Ni and NiO, Fe and $Fe_3O_4$ and $Fe_2O_3$, Ti and TiO) in this experiment. A two capsule device of primitive charges and buffers is shown in Fig. 2.

After experimentation, the autoclave is quenched in flowing cold water, and then the solid and liquid facies in the Pd Au capsule are separated for determination and analysis. The liquid facies are analysed by catalytic polarography and atomic absorption spectrophotometry methods, and the solid facies are identified under a Leitz microscope, or by using X-ray diffraction, infra-red absorption spectrometry, or the electron microprobe.

Most of the experimental reactions reach equilibrium which is examined by Gibb's facies law F=n+2-Ø. Some data in the article "Research for the Cu-Sn-S-$H_2O$ system under 300—500°C and the origin of cassiterite-sulphide ores" written by V.R. Hekpacob, were consulted during the experimentation.

The experimental processes were carried on in an autoclave made of $Tc_9$ titanium alloy. The pressure in the autoclave was calculated according to the filling degree of pure water. The temperatures were measured by PtLr—Pt and NiCr-NiSi thermoelectric couples, with an error of ± 10°C. The content of solution was analysed by catalytic polarography with a relative error of 10—15%. The contents of iron, lead, antimony and zinc were analysed by atomic absorption spectrography, with a relative error of 15—20%.

## II. The Results of Experimentation

More than 10 kinds of minerals were synthesized during the experimentation. They are cassiterite (Ct), monoclinic-pyrrhotite (Mpo), hexagonal-pyrrhotite (Po), arsenopyrite (Asp), marmatite (Mmt), jamesonite (Ja), zinkenite (Zin), herzenbergite

(Hzb), pyrite (Py), galena (Ga), stibnite (Stb), anhydrite (Ah), ß-berndtite chalcostibnite (Chal), berthierite (Ber), marcasite (Ms), quartz (Q), calcite (Cal), etc. (Figs. 3 and 4). After the experimentation, Sn, Fe, Pb, Sb and Zn concentrations in the solutions were determined. The experimental results of the Sn-Fe-S-$H_2O$ system under 300°C and 350°C and 500 atm. are shown in Table 3, the representative experimental results of the Sn-Fe-Pb-Sb-S-$H_2O$ system at 300°C and 500 atm. are listed in Table 5.

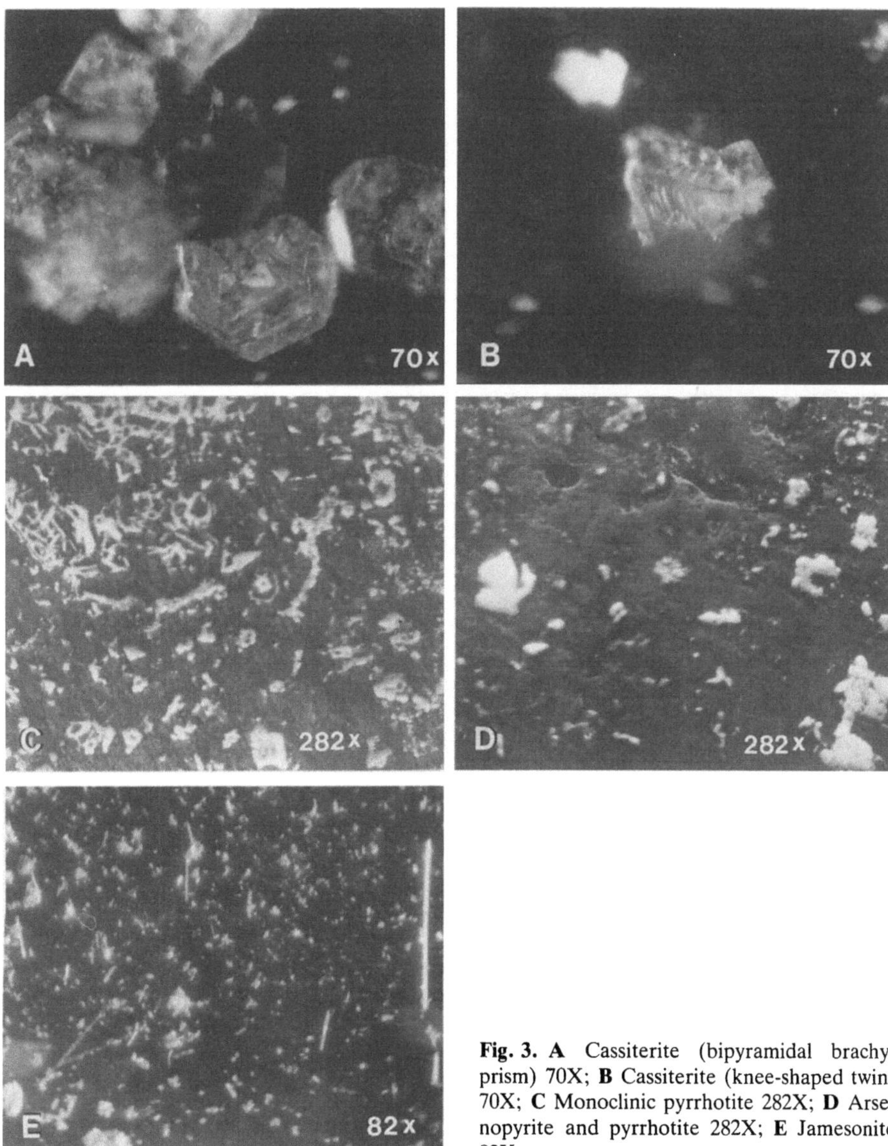


**Fig. 3. A** Cassiterite (bipyramidal brachyprism) 70X; **B** Cassiterite (knee-shaped twin) 70X; **C** Monoclinic pyrrhotite 282X; **D** Arsenopyrite and pyrrhotite 282X; **E** Jamesonite 82X

It can be concluded from the experiments that the principal factors affecting the mineral assemblage are temperature, oxygen fugacity, relative ratio of metallic ion concentration ($Fe^{+2}/S^{-2}$, $Pb^{+2}$, $Pb^{+2}/Sb^{+3}$) in the medium solution; and the secondary factors are type, concentration and pH of the medium solution. The pressure has little effect on the mineral assemblage. The results of experimentation are discussed as follows.

**Fig. 4. A** Zinkenite 82X; **B** ß-hexagonal stannite 282X; **C** Pyrite and hexagonal pyrrhotite 282X; **D** Pyrite and anhydrite 82X; **E** Chalcostibnite 43X

**Table 1.** The relationship between temperature and mineral assemblages

| Experimental temperature | Association of mineral paragenesis (in order of decreasing quantity) |
|---|---|
| 400°C | Asp+Mmt+Hpo+Ct+Q |
| 350°C | Ct+Asp+Mmt+Mpo+Hpo+Q+Ah+Py |
| 300°C | Ct+Mpo+Asp+Ja+Mmt+Hpo+Cal+Py+Ah+Ms |
| 250°C | Mpo+Ja+Ct+Py+Cal+Stb+Zin |
| 200°C | Mpo+Ja+Cal+Ga+Stb |

## 1. The Effects of Temperature on the Mineral Association

According to optimal temperature condition for mineral synthesis, different mineral assemblages could be integratedly distinguished (Table 1).

From Table 1 it can be seen that the formation temperatures of cassiterite range from 400 to 250°C. The temperatures for monoclinic-pyrrhotite, jamesonite, arsenopyrite and marmatite are 350—200°C, 400—300°C and 400—250°C respectively. Iron sulphides successively take the form of Hpo, Mpo and Py with the decrease of temperatures.

## 2. The Effects of Oxygen Fugacity on the Mineral Association

Some experimental results of the Sn-Fe-S (Si)-$H_2O$ system are listed in Table 2, showing that different mineral associations are formed at 350°C when all conditions except the $fo_2$ buffer pairs remain constant.

From Table 2 we can see that the cassiterite-sulphide association can be synthesized only when Ni and NiO or $Fe_2O_3$ and $Fe_3O_4$ buffer pairs are used at 350°C, and that only sulphides of tin and iron can be synthesized when Fe and $Fe_3O_4$, Ti and TiO buffer pairs are used. Thus, according to the formation temperature of the cassiterite-sulphide association (250—400°C), the optimal $fo_2$ is about $10^{-28}$—$10^{-40}$ atm., which is in agreement with the calculation results shown in Fig. 5.

## 3. The Effects of Atomic Ratios ($Fe^{+2}/S^{-2}$, $Pb^{+2}/Sb^{+3}$) of Ore-Forming Elements in the Experimental System on the Mineral Association

In the Sn-Fe-S-$H_2O$ system, mineral associations vary with the change of the $Fe^{+2}/S^{-2}$ ratio (Table 3).

**Table 2.**

| Experiment number | T°C | $fo_2$ buffer pairs | Experimental products (in decreasing quantities) |
|---|---|---|---|
| 83121 | 350 | Ni and NiO ($fo_2=10^{-31}$) | Ct+Ms+Ah+Po |
| 83098 | 350 | $Fe_2O_3$ and $Fe_3O_4$ ($fo_2=10^{-40}$) | Po+Ct+Cal+Py |
| 83032 | 350 | Fe and $Fe_3O_4$ ($fo_2+10^{-37.5}$) | Mpo+Q+Hzb+Py |
| 83080 | 350 | Ti and TiO ($fo_2=10^{-38}$) | Hpo+Hzb+Q+Py |

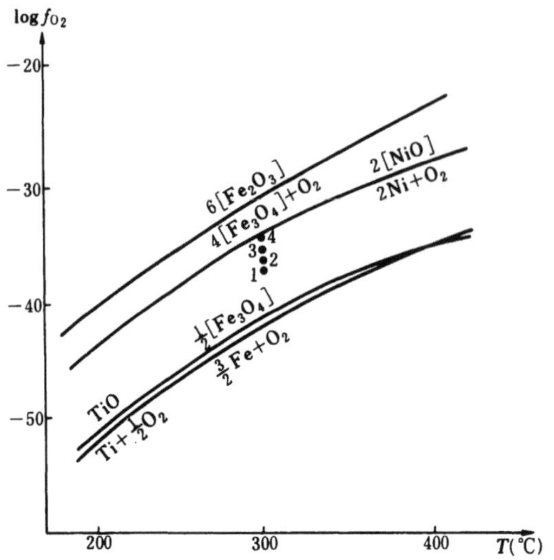


**Fig. 5.** Diagram of log $fo_2$ versus T for some buffer pairs

The formation conditions and the identification characteristics for Mpo and Hpo are listed in Table 4.

In the Fe-Pb-Sb-S-$H_2O$ system, the mineral assemblages are closely related to the ratio of Pb/Sb. Some representative experimental results and some values of theoretical calculation are listed in Table 5.

From Table 5, we can see that jamesonite is present only when the atomic ratios of Pb and Sb are equal or close to 2/3 in iron-bearing solution rich in sulphur. Otherwise only other sulphosalts and simple sulphides of Pb-Sb can be formed.

## 4. The Effects of Composition, Concentration and pH in the Solution

The cassiterite-sulphide association could be synthesised in base metal fluoride solution (0.5 m KF+NaF) or base metal chloride solution (0.5−2 m KCl+NaCl), with the former solution more favourable, pH values of the ore-forming solutions are commonly from 4.5−8.

5. After the experiment the concentrations of metallic ions in the solution (M/1) are as follows: Sn=$nx10^{-6}$ in the solution in equilibrium with cassiterite; Sn=$nx10^{-5}$−$nx10^{-4}$ in the solution in equilibrium with herzenbergite; Fe=$nx10^{-5}$, Pb=$nx10^{-5}$−$nx10^{-4}$, Sb=$nx10^{-5}$−$nx10^{-4}$, Zn=$nx10^{-5}$−$nx10^{-4}$ in the solution in equilibrium with pyrrhotite sulphosalts and marmatite.

**Table 3.**

| $Fe^{+2}/S^{-2}$ ratio | Mineral assemblages (in decreasing quantities) |
|---|---|
| 4−1 | Ct+Mpo* |
| 1−2/3 | Ct+Mpo+Hpo+Py |
| 2/3−1/5 | Ct+Py+Hpo |

* Ct+Mpo+Fe assemblage occurs when the $Fe^{+2}/S^{-2}$ ratio=4.

**Table 4.**

| Name of minerals | Formation temperature | Identification results | | |
|---|---|---|---|---|
| | (°C) | X-ray diffracto gram | * Specific magnetic coefficient C.G.S.M. $cm^3/g$ | Infrared spectrum |
| Hpo | 400—300°C | $d_{102}$ is of simple hump with sharp gradient $d_{102}$=2.0758 | $96.76 \times 10^{-6}$ | Absorption wave number is 430, 350 $cm^{-1}$ |
| Mpo | 350—200°C | $d_{102}$ is of double humps $d_{102} = \frac{2.0668}{2.0897}$ | $5201.04 \times 10^{-6}$ | No absorption wave number |

* Induced electric current is 1 ampere.

**Table 5.**

| Experiment number | Pb/Sb atomic ratio | Mineral assemblage |
|---|---|---|
| | (1.25)* | Bou (boulangerite) |
| 83087 | 1 | Bou+Po+Ga+Ant |
| | (0.875) | het (heteromorphite) |
| 83021 | (0.8333) | Ga+Ant+Stb+Po |
| 83116, 83117, 83118, 83119. | 0.666..(2/3) | Ja+Pla+Ga+Po |
| 83093 | 0.666 | Ja+Zitn+Pla+Po |
| | (0.625) | Pla (plagionite) |
| | (0.5) | Zin (zinkenite) |
| 83127 | 0.4 | Ful+Ant+Stb+Zin |
| | (0.375) | Ful (fuloppite) |
| 83092 | 0.167 | Ant+Stb+Zin+Po |

* The theoretically calculated values are in the brackets.

## III. Some Considerations

1. The optimal physiochemical conditions for the formation of cassiterite-sulphide deposits in the Dachang ore field are as follows: the temperatures are 400—250°C; oxygen fugacities are $10^{-28}$—$10^{-40}$ atm; the hydrothermal solutions may be of the Sn-Pb-Sb-Zn-As-S-Si-$H_2O$ system high in sulphur and moderately rich in iron, in which the ratio of $Fe^{+2}/S^{-2}$ is 4—2/3, Pb/Sb is about 2/3; the medium solution may be a mixed solution of base metal chlorides and fluorides with pH values of 4.5—8.
2. Ferromagnetic monoclinic-pyrrhotite obtained from experimentation is closely associated with the formation of cassiterite. In the Dachang ore field, the Po in the Ct-Po association is usually Mpo, and independent Hpo+Ct association is never found. Therefore provided there is no superimposition of other magnetic minerals or bodies, good results should be obtained by using magnetic methods for mineral prospecting in the Dachang ore field.

3. It can be found from Table 1 that from Po→Po+Py→Py the formation temperature of the iron sulphides falls gradually, and the sulphur fugacities rise progressively. The variation pattern of this kind of association is similar to the vertical zoning of mineral assemblages of the cassiterite-sulphide deposits in Dachang. The temperature and sulphur fugacity could therefore be regarded as the main factors affecting the mineral zoning of cassiterite-sulphide deposits in Dachang.
4. From Table 1 and Figure 5, we can see that the conditions of sulphur fugacity and temperature under which Po and Py could stably coexist are also the stable conditions for Ja, so there are some sulphide- and sulphosalt-forming elements in the ore-forming hydrothermal solution. The similar conditions of sulphur fugacity and temperature are hence the main factors for the formation of jamesonite in large quantities in the Dachang ore field.
5. It has been experimentally shown that during the formation of the deposits the lowest limits for concentration of all major metallic ions in the ore-forming solution are as follows: $Sn=nx10^{-6}$, $Fe=nx10^{-5}$, $Pb=nx10^{-5}-nx10^{-4}$, $Sb=nx10^{-5}-nx10^{-4}$ and $Zn=nx10x^{-5}-nx10^{-4}$.
6. From the experiments and thermodynamic calculations, Sn, Fe, Zn and Cu only form sulphides when the oxygen fugacity is lower ($=10^{-40}$), but when it rises to $10^{-40}-10^{-28}$ atm, the coexistence of Sn oxides and Fe-Zn sulphides occurs. During the migration of magmatic hydrothermal solution from rock body outwards, the oxygen fugacity gradually increases. Because Cu and Zn are sulphophile elements, and Sn is a more lithophile element, Cu and Zn firstly form sulphide deposits near the granite at an earlier stage when the oxygen fugacity is lower, and tin oxides are only formed farther away from the rock body with the gradual increase of oxygen fugacity. So the mineralization zoning "copper and zinc near the granite, cassiterite farther" horizontally, and "cassiterite in the upper, copper in the lower and zinc in the middle" vertically in the Dachang ore field was caused by the relative ratios of ore-forming elements in the metallogenic solution, variation of oxygen fugacity and the different sulphophile and lithophile characters of the ore-forming metallic elements.
7. From the conditions for synthesis of chalcostibnite, we infer that there are probably some Cu-Sb-S minerals in the transitional zone between the middle mineralized belt and the east or west mineralized belts in the Dachang ore field.

Based on the experimental results and the data of fluid inclusions and isotopes, together with the geological characteristics of the deposits, it is thought that the formation of the cassiterite-sulphide deposits is closely associated with the biotite granite, which is regarded as the main source of the metallogenic materials and hydrothermal fluids. The strata of Mid-Upper Devonian age also provide part of metallogenic materials, as revealed by the investigation of lead and sulphur isotopes.

*Acknowledgements.* Many thanks are due to Professor Zhang Benren of Wuhan College of Geology and some colleagues from various sections of my institute for their help and support.

# 6.4.5 The Geological Characteristics and Metallogenic Regularities of Tin Deposits in Guangxi

LIU YUANZHEN, ZHONG KENG, and MA LINQING[1]

**Abstract**

Guangxi, one of the major tin-producing areas in China, lies in the west of the Nanling polymetallogenic belt and is characterized by multicyclic structure–magma–ore-formation. Tin mineralization took place mainly during the Xuefengian, Caledonian and Yenshanian orogenies. According to the genetic relationship between tin deposits and associated metallic deposits as well as the evolution of tin mineralization, the tin ores in Guangxi have been divided into five groups made up of thirteen types. Associated with the structural magmatic zone, five NE-trending tin – polymetallic ore zones have formed. In each tin ore zone, tin-polymetallic ore fields and deposits are frequently distributed in regular zonation around granite bodies. Ore concentration regularities are also discussed in terms of such features as granite, structure and wall rocks. On the basis of ore-forming characteristics and material sources, the tin deposits in Guangxi are grouped into three different metallogenic models of Dachang type, Limu type and Baotan type.

## I. Regional Geological Setting

Guangxi, located in the marginal mobile zone of southeast China and in the western part of the Nanling polymetallogenic belt, is an important part of the Circum-Pacific tectonic and metallogenic belt.

This region, a first order tectonic unit of the South China fold system which is represented by a series of uplifts and downwarps produced by multicyclic tectonic movements, is divided into five tectonic units of second order (Fig. 1), namely, the northern Guangxi uplift, central-eastern Guangxi downwarp, Yunkai uplift, Youjiang folded zone, and Qinzhou folded zone. This region is characterized by active magmatism, and every tectonic cycle is accompanied by magmatic activity. Granitoids are most extensively distributed in this region, consisting of five-NE-trending belts and one NW-trending belt, with the emplacement ages tending to be younger from the northwest to the southeast. The Sibaoian and Xuefengian granitoid bodies occur in the northern Guangxi uplift (i.e., the first zone). The Caledonian granitoid bodies

[1] Bureau of Geology and Mineral Resources of Zhuang Autonomous Region of Guangxi, Nanning, Guangxi

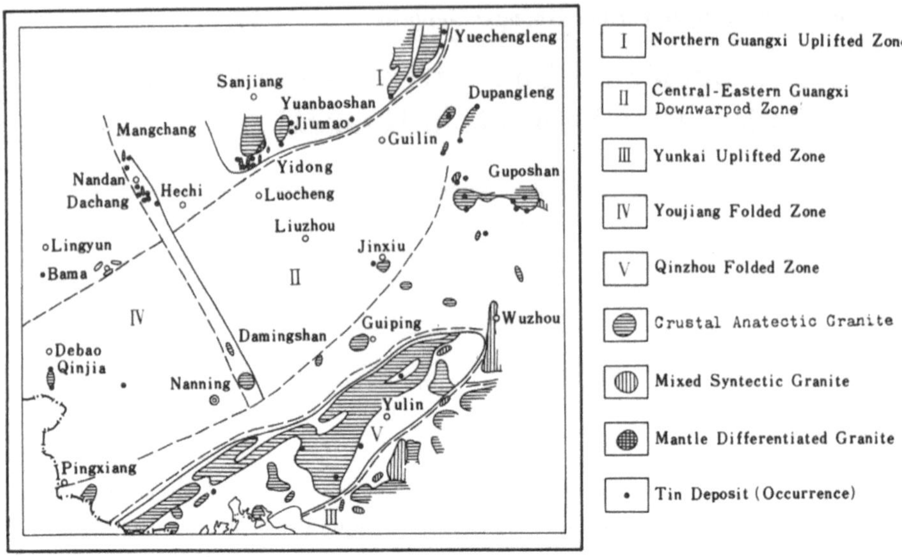


**Fig. 1.** Diagrammatic geological map showing the distribution of tin deposits in Guangxi

generally appear to the southeast, making up the second zone. The Variscian and Indosinian bodies are spread in both the third and fourth zones. The Caledonian bodies are also distributed in the fifth zone. The Yenshanian rocks can be found in all these five zones. The NW-trending sixth zone is mostly occupied by the late Yenshanian granitoids. These tectonic-magmatic activities provided most favourable regional geological conditions for tin mineralization.

## II. Types of Ore Deposit

Tin mineralization is closely related to the tectonism-magmatism in the regional geological history, showing metallogenetic specialization of granite in particular. Though the concepts of polygenetic and polystage metallogenesis are given more attention as the metallogeny is developed, granites are still believed to be the leading factor in the tin mineralization. Guangxi is characterized by various types of ore deposit and diverse mineral associations. On the basis of the genetic relations between tin and other associated metallic ores and their temporal and spatial evolution regularities, and considering the requirements for ore potential assessment, the tin deposits in Guangxi may be classified into five groups and thirteen types:

1) Cassiterite-feldspar-quartz group: composed of deposits of Sn-W-Nb-Ta-bearing granite type and Sn-W (Be)-Nb-Ta-bearing pegmatite type which are respectively related to late stage magmatic differentiation and metasomatism and to postmagmatic differentiation and metasomatism;

2) Cassiterite-quartz group: including greisen type, cassiterite-quartz vein type and cassiterite-tourmaline-quartz type;
3) Cassiterite-skarn group: consisting of cassiterite-hematite-skarn type and cassiterite-chalcopyrite-skarn type;
4) Cassiterite-sulphide group: represented by cassiterite-sulphosalt type, cassiterite-sulphide type and cassiterite-chlorite type;
   The second to fourth groups are related to post-magmatic pneumato-hydrothermal processes.
5) Superficial placer tin group: composed of eluvial-diluvial type, alluvial type and karst type, all related to surface weathering and erosion.

Because the magmatism, tectonism and relevant country rocks varied, different types or assemblages or ore deposit were produced. Consequently, one mineralization zone or field may consist either of deposits of a single type or assemblage of several types. The common assemblages are as follows:

1) Cassiterite-sulphide-skarn type plus cassiterite-sulphide type plus cassiterite-sulphosalt type plus placer type (Fig. 2);
2) Granite type plus greisen type plus cassiterite-magnetite-skarn type plus cassiterite-quartz type plus cassiterite-sulphide type plus placer type (Fig. 3);
3) Granite type plus pegmatite type plus quartz vein type plus placer type (Fig. 4); and
4) Greisen type plus cassiterite-tourmaline-greisen type plus cassiterite-chlorite type plus placer type (Fig. 5).

In terms of commercial potential, the deposits of the greatest importance are of the cassiterite-sulphide group, which are characterized by large size, diverse associated minerals and high value of comprehensive utilization; the placer type, cassiterite-quartz vein type, and Sn-W bearing granite type are of second importance.

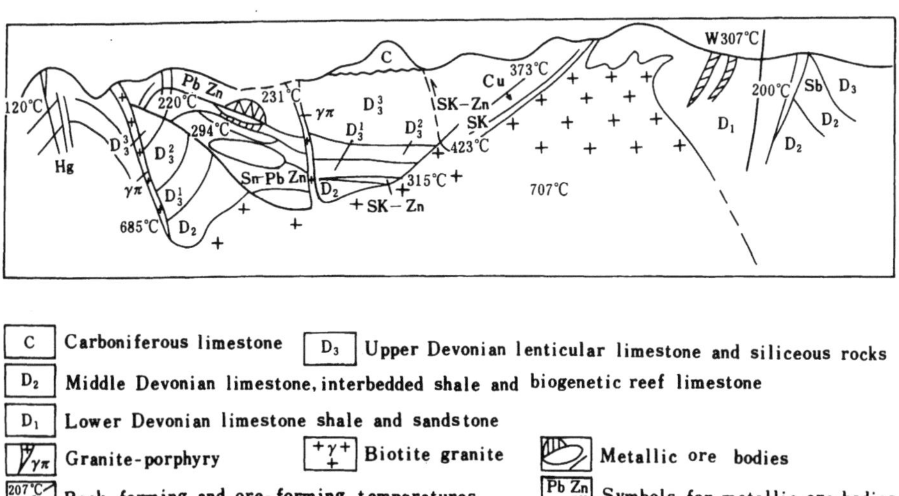


**Fig. 2.** Schematic map showing the metallic-deposit zoning of the Dachang tin field, Guangxi

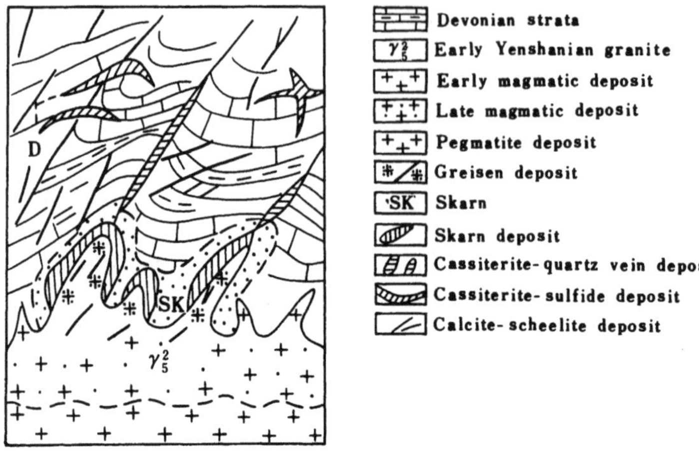


**Fig. 3.** Tin deposit zoning in the Fuhezhong region

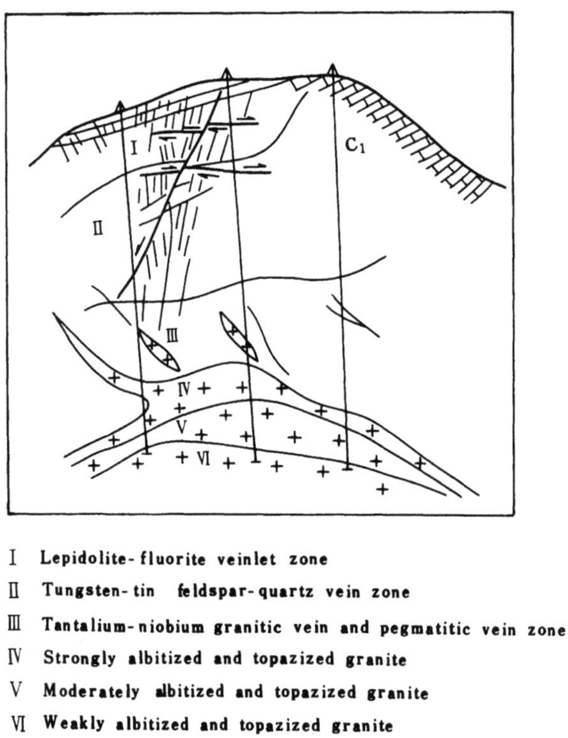


**Fig. 4.** Geological section of the Limu ore field

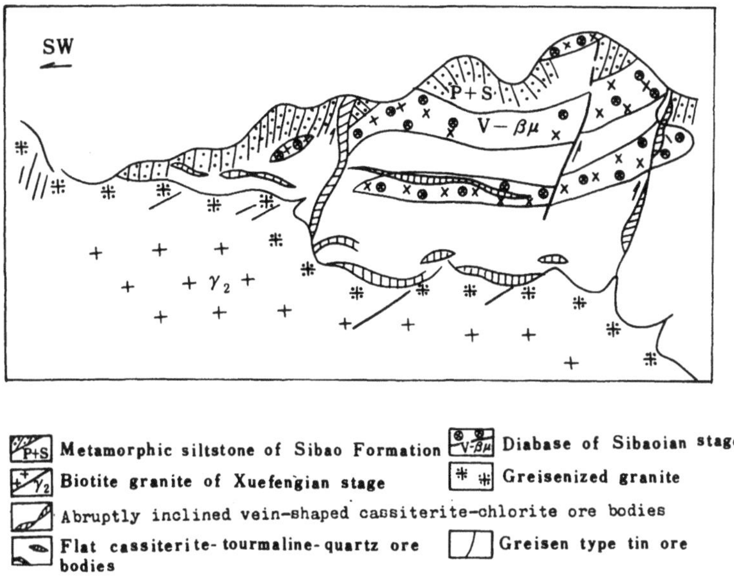


**Fig. 5.** Geological section of the Baotan tin ore deposit, Losheng county

## III. The Spatial Distribution of the Tin Deposits

Controlled essentially by the Nanling E-W tectonic belt, the tin deposits in Guangxi are restricted to two E-W trending tin mineralization zones approximately parallel to each other (Fig. 1). The northern mineralization zone lying across the northern part of Guangxi consists of Dachang, Baotan, Jiumao, Xinlu, and Limu tin polymetallic deposits, dominated by tin with Zn, Pb and Cu associated to the west of Yuechengling and by tin, tungsten and rare metals to the east. The southern mineralization zone is composed of Qinjia, western Damingshan, Xishan, and Rongtang tin polymetallic deposits, predominated by tin with copper associated in the western part and by tin, tungsten, and rare metals in the eastern part.

Corresponding to the tectonic-magmatic zones, five parallel NE-trending tin polymetallogenetic zones may be described, namely, (1) Yuechengling-Danchi, (2) Dupangling-Qinjia, (3) Guposhan-Zhenlongshan, (4) Darongshan-Liuwandashan and (5) Yunkaidashan. In each zone, major ore deposits mostly occur associated with granite in the uplifted zones at the intersections of E-W structures with NE and NW structures or in the secondary uplifts within the depressions, forming a tin polymetallic deposit series generally around a swarm of small rock bodies or a single rock body. In the Yuechengling-Danchi zone, for example, a series of genetically related tin deposits are distributed around the Yuechengling, Yuanbaoshan, Sanfang, Pingying, and Dachang granite bodies, showing a distinct zonation. And in the Guposhan-Zhenlongshan zone, the Keda, Shuiyanba, and Xinlu tin deposits are spread around the Guposhan granite. From the inside of the granite body outwards to the exocontact zone, the mineralization gradually varies, successively represented by granite type,

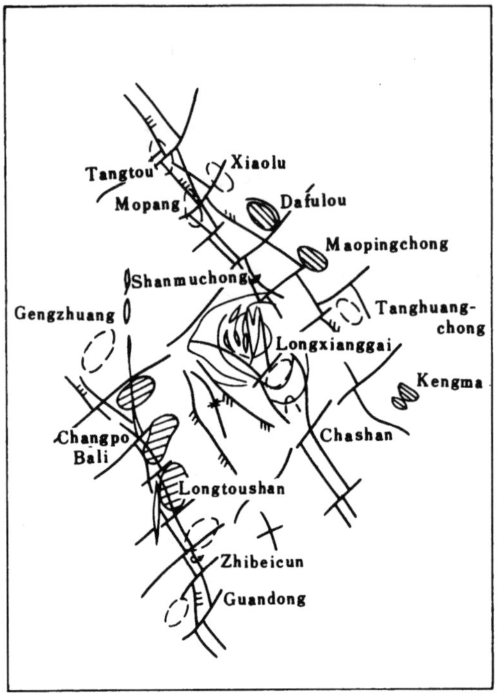


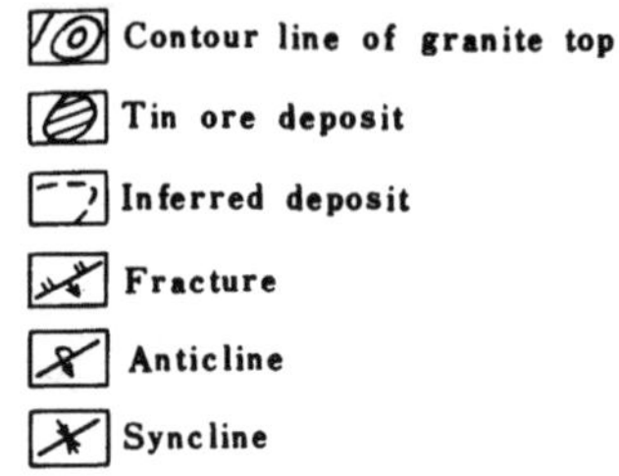


**Fig. 6.** Schematic geological map showing equidistant distribution of various deposits in the Dachang ore field

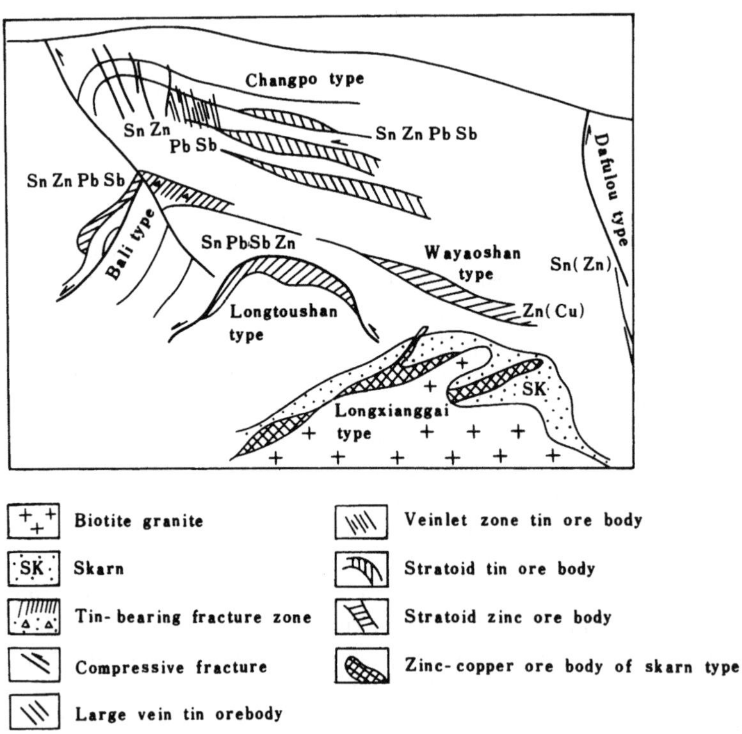


**Fig. 7.** Diagram showing ore-controlling patterns and distribution of major metals in the Dachang ore field

## V. Some Mineralization Regularities

Tin mineralization is not only related to certain magmatic events, but also was conditioned by tectonic settings and physiochemical properties of the country rock.

### 1. Granitoids

The granitoids genetically related to tin mineralization are all of anatectic S-type. Though they are scattered in their distribution and their magmatic stages are not quite identical with the mineralization phases, they are generally characterized by the following features:

1) The granitoids are dominated by medium-fine grained biotite granite with some muscovite-biotite- (or muscovite) granite and are petrochemically peraluminous and oversaturated, containing $SiO_2$=70.38–75.09%, $K_2O+Na_2O$=6.21–7.97%, F-0.055–0.125%, B=0.0015%–0.0022%, K/Rb<100 (genetically ranging 32–85), little Fe and Mg and large amounts of tourmaline, fluorite and topaz as well as minor plagioclase.
2) The granites have a great abundance of Sn, generally 3–30 times higher than the clarke value. On the basis of the statistics, the tin abundance of granite of various stages related to the tin mineralization averages 18.7 ppm for the Xuefengian, 17.6 ppm for Caledonian, 11.1 ppm for Variscan, 8.17 ppm for Indosinian, 21.23 ppm for early Yenshanian and 40.6 ppm for late Yenshanian. The gradual concentration of tin during the differentiation of granitic magma reveals the genetic relationships of tin mineralization in Guangxi to the magmatic activities of the Xuefengian, Caledonian and Yenshanian periods.
3) The mineralized granitic bodies generally occur as stocks and small hidden or semi-hidden batholiths, with outcrop areas generally less than 10 $km^2$. Tin is concentrated at the tops and the endo- and exocontact zones of the plutons. For example, the Dachang and Baotan tin deposits related to the Dachang and Pingying granitic bodies are of this type. The extensively exposed batholiths are mostly compound, in which the tin mineralization is generally related to the supplementary intrusives or late stage small bodies, e.g. Jiumao and Xinlu tin deposits related to the Yuanbaoshan and Guposhan granitic bodies.

### 2. Structural Control of Tin Mineralization

The tin ore fields and deposits mostly occur in the uplifted areas where two structural zones of different orientations intersect and are cut by deep fractures, or in the axial and plunging parts of anticlines and domes. For instance, the Danchi tin ore belt is located in a NW-trending uplifted zone, while Mangchang, Dachang, Beixiang and Wuxu tin polymetallic ore fields are spread in an uplifted area at the intersection of E-W and NE trending structures. In the Dachang tin ore fields, tin deposits occur in zones around the Dachang granite, with Changpo, Bali, and Longtoushan deposits located in the western belt, Dafulou, Maopingchong and Yuanma deposits in the east-

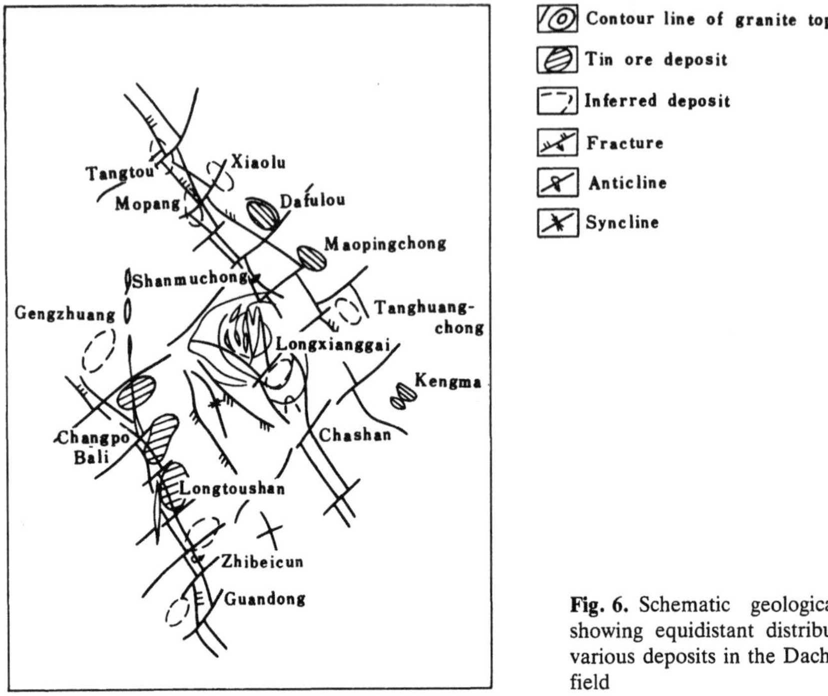


**Fig. 6.** Schematic geological map showing equidistant distribution of various deposits in the Dachang ore field

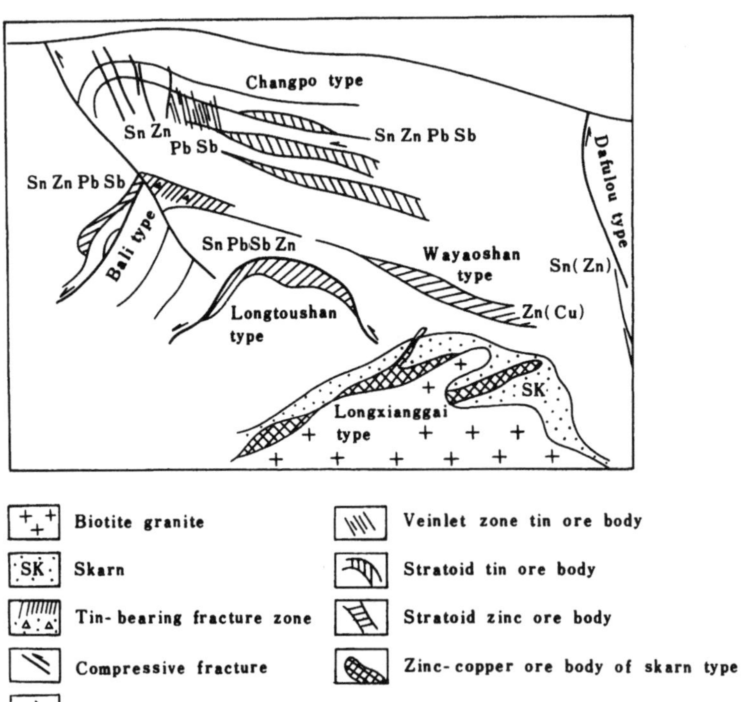


**Fig. 7.** Diagram showing ore-controlling patterns and distribution of major metals in the Dachang ore field

ern belt, and Lamo in the middle, showing a distinct directional, symmetrical, equidistant and zonal arrangement of deposits (Fig. 6). The orebodies occur in the relatively closed fracture zones and fissure zones and in the structures formed by interlayer slipping. The shapes of the orebodies are controlled by structural patterns (Fig. 7). In the upper part, where shearing fissures are well developed, the orebodies generally occur as large veins, veinlets and irregular veins predominated by dense and massive tin ore of high grade. In the middle part, the fissures formed by compression and interlayer slipping are developed and orebodies are generally large and occur as networks of veins in en echelon arrangement. Near the granitic bodies in the deep zone, the orebodies are in layered and striped shapes and controlled by the contact zone. In the Shanhu ore field (Fig. 8), orebodies are concentrated in groups in the secondary fissure zones near the fracture associated with the NE-trending anticline. The major ore veins are in en echelon arrangement both horizontally and vertically.

## 3. The Relationship of Mineralization to Stratigraphy and Lithology

The primary tin deposits in Guangxi mostly occur in the Sibao group (7.4%), Cambrian (3.9%) and Devonian (67%), among which the Devonian (67%) is the most important. The mineralization is closely related to the lithology of the country rocks. The Sibao Group is composed of calcareous clastics and spilite keratophyre, and the Cambrian and Devonian consist of impure carbonate, which are chemically active, forming an alkalic geochemical environment favourable for the separation, metasomatism and precipitation of cassiterite and sulphides. In addition, the Devonian is characterized by a high carbon content and low Devonian is characterized by a high carbon content and low oxygen fugacity, producing a reducing environment, with low oxygen content being advantageous to the precipitation and concentration of

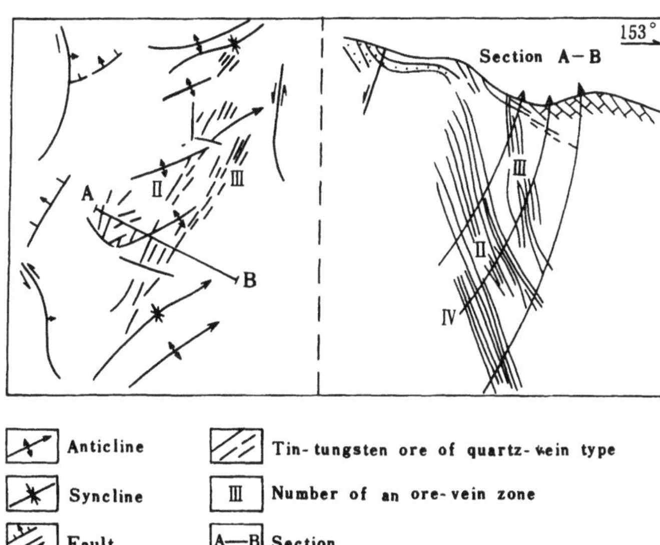


**Fig. 8.** Tin-tungsten ore veins in the en echelon folds and associated fissure zones in the Shanhu ore field

cassiterite and sulphides. The sulphur isotope data obtained from some cassiterite-sulphide deposits show a wide range of $\delta^{34}S$ values (mostly negative), revealing the effects of organic bacterial reduction. It may be inferred, therefore, that part of the sulphur in the deposits may be derived from the sedimentary country rocks. The trace elements and element associations from Devonian rocks of different sedimentary facies have been studied. The analysis of the trace elements Sn, Pb, Zn, Cu, Sb, As and Hg in various rocks from three geological profiles of Xiangzhou type (made up of marl, limestone, dolomite and mudstone intercalated with sandstone which were formed in a shallow water and oxidation environment), Beiliu type (consisting of pure limestone, dolomite, and bioherm limestone formed in an offshore environment characterized by shallow water, rich oxygen, and high energy), and Nandan type (composed of limestone, siliceous rocks, black mudstone with thin-bedded dolomite deposited in pelagic, oxygen-lacking, deep water and quiet environment) show that in the Devonian system (Table 1) the abundances of Pb, Zn, As and Sb are generally higher than their clarke values, especially in Middle-Upper Devonian, where they are several to tens of times higher than the clarke values. The Sn and Cu abundances are close to their clarke values, with Cu being 1—2 times higher only in Middle-Upper Devonian mudstone and marl. Tin abundance is approximately the same in various rocks of different sedimentary facies generally ranging 1—3 ppm, only reaching 7—13 ppm in individual mudstones. In addition, the chemical analysis of some Devonian volcanic rocks also reveals low abundances of tin. For instance, the so-called tuffaceous rocks of the Upper Devonian near Yilan in Nandan contain only 3 ppm Sn. Moreover, many Middle-Upper Devonian volcanic rocks in southwestern Guangxi have no tin at all. Judging from these facts, it may be concluded that the Devonian in Guangxi cannot be the source beds of tin. This can also be indicated by the distribution pattern of tin, copper, lead and zinc deposits. For instance, in the Xiangzhou area there was no granitic activity associated with tin mineralization, only lead, zinc, and copper deposits were formed; while in the Danchi area magmatism associated with tin mineralization was intense and cassiterite-sulphide polymetallic deposits were produced.

## VI. A Discussion of the Metallogenetic Model

On the basis of geological observations combining a large quantity of petrological and mineralogical studies and analysis of stable isotopes sulphur, oxygen, carbon and hydrogen of the typical deposits in Dachang, Limu, and Baotan ore fields, the authors have proposed three metallogenetic models for tin mineralization in Guangxi.

### 1. The Metallogenic Model of Dachang Ore Field Type

The major ore-forming elements such as Sn and Cu are derived from the granitic magma of continental crustal origin, and some ore-forming elements such as Zn, Pb, Sb, As, and S are partly from the sedimentary country rocks (Fig. 9).

**Table 1.** Trace element contents of Devonian strata of different facies in Guangxi

| Strata | | Sn | | Pb | | Zn | | Cu | | Sb | As | Hg |
|---|---|---|---|---|---|---|---|---|---|---|---|---|
| Upper Devonian | Wuzhishan Fm (5) | 0.2–2.8<br>2.1 | | 40–60<br>54 | | 19–74<br>40.2 | | 28–63<br>40.4 | | 3.7<br>(1.6) | 14.7<br>(16) | 0.075<br>(16) |
| | Liujiang Fm (18) | 0.3–7.4<br>3.9 | | 10–100<br>4.6 | | 12–232<br>67.6 | | 25–76<br>57.5 | | | | |
| | Rongxian Fm | | 0.5–4.1<br>1.5 | | 30–70<br>50 | | 26–99<br>62.5 | | 24–30<br>27 | | | |
| Middle Devonian | Luofu Fm (14) | 1–13.2<br>2.5 | | 10–50<br>32.5 | | 11–112<br>37.8 | | 10–80<br>40 | | 49<br>(29) | 18.6<br>(29) | 0.09<br>(29) |
| | Domganling Fm (15) | | 1–4.3<br>1.75 | | 18.9–100<br>63 | | 16–17.5<br>75.3 | | 4.4–288<br>57.8 | | | |
| | Beiliu Fm; Nobiao Fm (8) | 1–4.4<br>2.6 | | 10–50<br>30 | | 15–163<br>78.4 | | 10–100<br>40 | | | | |
| | Yingtang Fm (13) | | 1–8<br>3.4 | | 25–100<br>63 | | 18–221<br>84.2 | | 10–40<br>23.8 | | | |
| | Sipai Fm (3) | | 1–13<br>2.6 | | 63–70<br>65 | | 20–24<br>22.5 | | 1–20<br>15.2 | | | |
| | Xinglou Fm | | | | | | | | | | | |
| | Xiaodong Grp. | | | | | | | | | | | |
| Lower Devonian | Tangding Fm; Moding (5); Ertang Fm; Heeang Fm; Hexian Fm; Qinzhou Grp. | 1–3.9<br>2.1 | 2.98 | 20–64<br>41 | 25.8 | 14–72<br>36.5 | 29.1 | 20–110<br>52.5 | 52.8 | 2.5<br>(10) | 4.4<br>(10) | 0.07<br>(10) |
| | Yijang Fm (65); Yilan Fm (65); Shiqiao Fm | 1.5–5.4<br>3.5 | | 8–54<br>27 | | 20–80<br>48.8 | | 30–90<br>50 | | | | |
| | Nagaoleng Fm (65) | 1.7–5<br>2.82 | | 8–26<br>219 | | 10–30<br>25 | | 20–59.1<br>478 | | | | |
| | Lianhuashan Fm (50) | 3.42 | | 19.6 | | 19.4 | | 14.67 | | | | |

0.2–2.8 = Range of content; 2.1 = Average content; (5) = Number of samples.

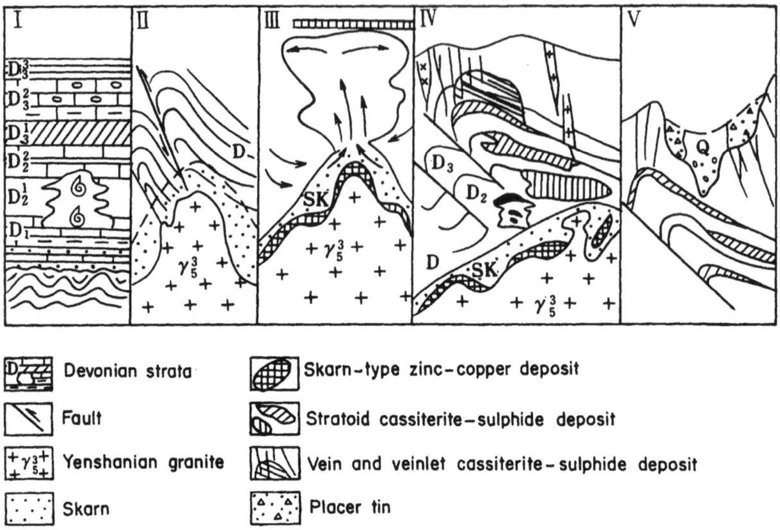


**Fig. 9.** Metallogenic model of the deposits in the Dachang ore field

1) The elements S, Hg, Zn, and As were primarily concentrated in the neritic sedimentary rocks in Devonian times;
2) During the Indosinian-Yenshanian tectonic movements, the structural framework of this region was formed, controlling the later emplacement of granitic bodies and the distribution of ore fields. In the meantime, the crustal temperatures increased to 250°C in Dachang and 235°C in Danchi, high enough to stimulate the activation of magmation and the migration and concentration of Pb, Zn, Sb, As and Hg in the rocks;
3) Biotite granite was intruded during the late Yenshanian stage, producing zinc-copper deposits of tin-bearing skarn type along the contact zones of the sedimentary country rocks. Geochemical analyses of the ores show the values $\delta^{34}S$ 0.03–1.26% for sulphides and $\delta^{18}O$ 5.17–6.31% for cassiterite, indicating the primary magmatic origin of the ore-forming fluids;
4) The postmagmatic pneumatolito-hydrothermal ore-forming solution was mixed with heated underground water and then convected vertically along fracture zones. The repeated mineralization of different stages kept ore-forming solutions continuously concentrated, leading to the formation of cassiterite-sulphide deposits in the exocontact zones. Geochemical analyses for the ore have given the following values: $\delta^{34}S$-8.49- +8.31‰ for sulphides, $\delta^{18}O$ 5.17–6.31‰ for cassiterite, $\delta^{18}O$ 11.99–16.5‰ for quartz, and $\delta^{18}O$ 10.77–22.47‰ and $\delta^{13}C$ (POB) 1.25–7.89‰ for calcite, indicating that the ore-forming solutions were derived partly from magma and partly from country rocks.
5) The tin placer deposits were formed by weathering and erosion processes.

## 2. The Metallogenetic Model of the Baotan Ore Field Type

The major ore-forming elements were derived from granitic magma formed by continental crustal anatexis and some associated elements from basic-ultrabasic complexes (Fig. 10).

1) In the late period of the Middle Proterozoic, the intermediate-ultrabasic complexes were intruded, producing copper-nickel sulphide deposits;
2) The Xuefengian Movements produced the tectonic framework of this field controlling the later granite emplacements and ore deposit distribution;
3) During the Xuefengian Movements, the Pingying biotite granite was intruded and there was postmagmatic pneumatolito-hydrothermal activity. In the early stage, a series of tin deposits of greisen type was formed within or on the margins of granitic bodies; while in the middle and late stages, the postmagmatic pneumatolito-hydrothermal solutions were mixed with heated groundwater, producing cassiterite-tourmaline-quartz type and cassiterite-chlorite type deposits. The $\delta^{34}S$ values are 0.1—22 % for sulphides and $\delta^{18}O$ values are 3.4—7.8 ‰ for cassiterite, indicating double sources.
4) Weathering-produced tin placer deposits.

## 3. Metallogenetic Model of the Limu Ore Field Type

The ore-forming elements were derived from granitic magma formed by crustal anatexis (Fig. 11).

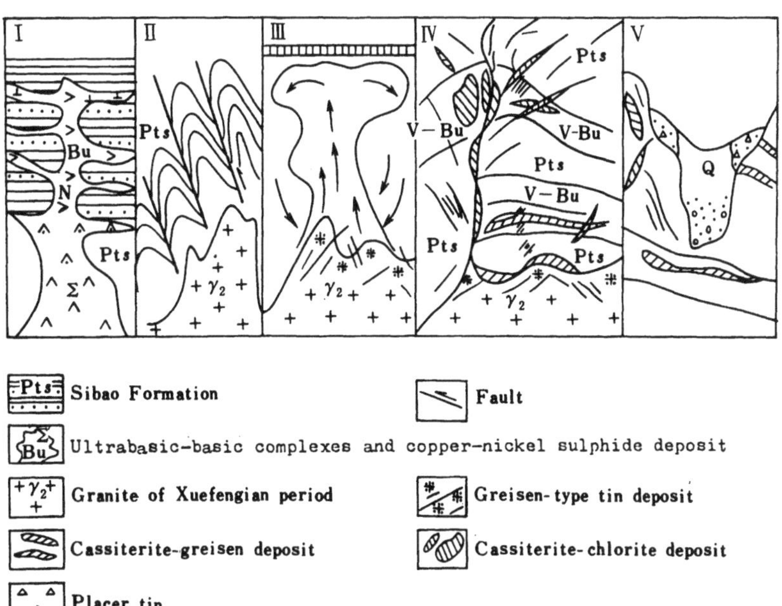


**Fig. 10.** Metallogenic model of the deposits in the Baotan ore field

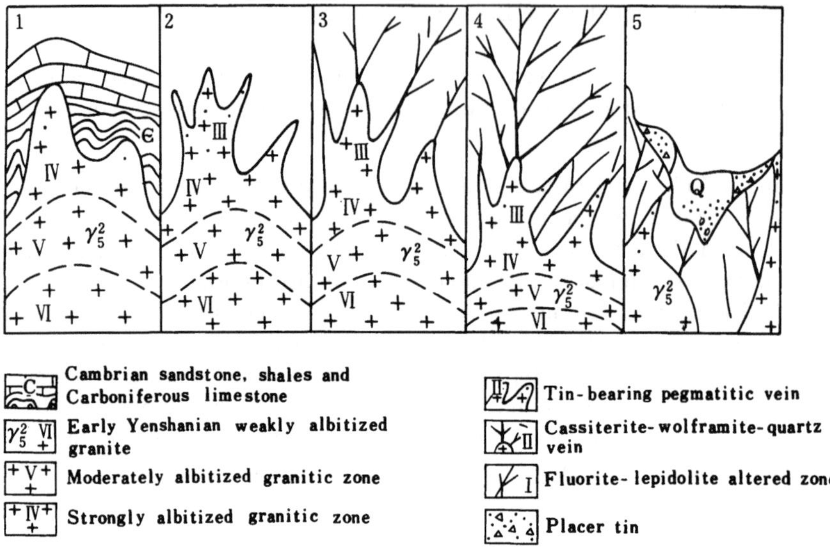


**Fig. 11.** Metallogenic model of the deposit in the Limu ore field

1) In the late stage of the magmatism, three albitized granitic zones (i.e., IV, V, VI) were produced as the result of postmagmatic differentiation-metasomatism. They are, from lower to upper, the weak, intermediate, and strong albitization zones. The contents of Ta, Nb, Sn and W increased as the albitization alteration became stronger. The rock-forming temperature was determined as 500–525°C and ore-forming temperature of wolframite as 315°C. The $\delta^{34}S$ values obtained from ores range from -1.7 to 0.1‰;
2) In the postmagmatic differentiation-metasomatism stage, Sn, Ta and Nb bearing pegmatite veins (Zone III) were also formed, the tin contents of which were higher than in Zone IV;
3) Being the major mineralization stage of cassiterite, the postmagmatic pneumatolito-hydrothermal metasomatism produced many cassiterite-quartz veins and cassiterite-feldspar-quartz veins containing wolframite in the sedimentary country rocks. The $\delta^{34}S$ values of sulphides are -1.7—0.1‰, implying a deep source;
4) The postmagmatic hydrothermal process was responsible for the formation of metallogenetic front-haloes of fluorite-lepidolite alteration (Zone I), where the tin contents were decreased;
5) There also exist tin placer deposits formed by weathering and erosion.

## Conclusions

1) Guangxi is a region tectonically and geochemically very favourable for tin mineralization. A series of genetically related tin deposits was produced as the result of polycyclic tectonism-magmatism. The S-type granites are of tin metal-

logenetic specialization. The major metallogenic epochs are Xuefengian, Caledonian, and Yenshanian.

2) The tin mineralization in Guangxi is characterized by multi-epochs, with multi-stages for each, and by complex metallogenetic processes and diverse composition. The elements Pb, Zn, Sb, As and S associated with tin were derived from both magma and, partly, from sedimentary country rocks. The orebodies generally occur in the Sibao Group and Cambrian and Devonian strata.
3) The tin mineralizations in Guangxi are characterized by obvious zonation, both horizontally and vertically.

*Acknowledgements.* The authors are very grateful to those geological and geophysical prospecting parties of both the Bureau of Geology and Mineral Resources and Geological Exploration Corporation of the National Nonferrous Metals Industry Corporation of China for their primary data on the basis of which this paper is written. The Guiyang Institute of Geochemistry of Academia Sinica and the Institute of Geology and No. 215 and 271 geological parties of Geological Exploration Corporation of CNNC are also acknowledged for their research results and analytical data which are cited by the authors in the paper.

# 6.4.6 Genesis of the Dachang Ore Deposit and the Formation Conditions of Cassiterite-Sulphide Deposits in General

TU GUANGZHI[1]

**Abstract**

The Dachang cassiterite-sulphide ore deposit is controlled by stratigraphy-lithology, acid magmatic activity, and structure. Correlation between the Dachang deposit and the neighbouring stratabound mineral deposits as well as mineralogical, chemical and isotopic investigations of the Dachang deposit reveal the characteristics of multiple source material, multiple stages of formation and multiple genesis of the deposit. It is postulated that some large-scale cassiterite-sulphide deposits with abundant polymetals (not including skarn deposits) may have a similar ore-forming mechanism.

Dachang is an important tin-base deposit in southwest China. For many years the predominant viewpoint on ore genesis has laid emphasis on granite-related postmagmatic hydrothermal fluids. For prospecting purposes, stress has been placed mainly on granite; while the role of stratigraphic, lithological and structural controls have been largely neglected.

In 1976 the author proposed an idea of polygenesis and multiple sources for the ore-forming substances of the Dachang deposit. Three controlling factors were postulated. Recent research work seems to give support to this idea.

The genesis of the Dachang ore deposit should be considered in the background of a regional metallogenic analysis. The Dachang ore deposit occurs in Middle and Upper Devonian carbonate formations composed mainly of limestone, argillaceous limestone, reef limestone, banded and lenticular limestone, siliceous rocks and calcareous mudstones. In an area of 100 km stretching out from Dachang, there occur a number of Pb-Zn, Sb, realgar-orpiment, and Hg stratabound deposits in Devonian strata. No granitoid has ever been found close to these deposits. The Dachang deposit differs from these stratabound deposits in that the former, related to a small biotite granite stock, contains economically important Sn and Cu with some associated W in addition to large amounts of Sb, As, Pb, Zn and Fe sulphides and sulphosalts. Sn, Cu and W are almost totally absent from stratabound deposits near Dachang.

Trace element analysis of the Devonian rocks along the Nandan-Loufu profile far away from ore district shows high contents of As, Sb, Pb, Zn, with As 10 times, Sb 6 times, Pb and Zn 1—2 times the average crustal contents of the respective elements, while the contents of Sn, W, Cu and Mn in Devonian strata are lower than their average contents in the crust (Table 1).

[1] Institute of Geochemistry, Academia Sinica, Guiyang, Guizhou

**Table 1.** Trace element analyses of the Devonian rocks along the Nandan-Loufu profile (in ppm)

| | | Upper Devonian | Middle Devonian | Lower Devonian | Average |
|---|---|---|---|---|---|
| As | range | 0.1–13.33 | 0.4–126 | 6.3–28.4 | 0.4–126 |
| | average | 5.0 | 22.0 | 13.5 | 17.2 |
| | f | 2.9 | 12.9 | 7.9 | 10.1 |
| Sb | range | 2–5 | 2–15 | 2–9 | 2–15 |
| | average | 2.1 | 3.4 | 2.4 | 3.0 |
| | f | 4.2 | 6.8 | 4.8 | 6 |
| PB | range | 10–40 | 10–40 | 10–110 | 10–110 |
| | average | 21 | 25 | 41.0 | 28 |
| | f | 1.3 | 1.6 | 2.6 | 1.8 |
| ZN | range | 10–150 | 10–500 | 20–240 | 10–500 |
| | average | 50 | 116 | 63 | 93 |
| | f | 0.6 | 1.4 | 0.8 | 1.1 |
| W | range | 0.4–2.8 | 0.4–2.4 | 0.4–2.8 | 0.4–2.8 |
| | average | 1.1 | 1.3 | 0.9 | 1.1 |
| | f | 0.9 | 1.0 | 0.7 | 0.9 |
| Cu | range | 10–40 | 10–80 | 10–60 | 10–80 |
| | average | 15.1 | 34 | 28 | 29 |
| | f | 0.3 | 0.7 | 0.6 | 0.6 |
| Sn | range | 0.3–2.8 | 0.3–3.5 | 0.5–4.4 | 0.3–4.4 |
| | average | 0.8 | 1.4 | 1.4 | 1.3 |
| | f | 0.3 | 0.6 | 0.6 | 0.5 |
| Mn | range | 60–2200 | 40–570 | 40–570 | 40–2200 |
| | average | 477 | 164.9 | 95 | 209 |
| | f | 0.5 | 0.2 | 0.1 | 0.2 |

f = average $\frac{\text{content in sediment}}{\text{crustal abundance}}$; After Zhao (1984).

It has been noticed that not only the Devonian rocks in north Guangxi have high contents of As, Sb, Pb, Zn and low contents of W, Cu, Sn, Mn; but the Devonian rocks in the neighbouring Hunan Province also bear similar characteristics in trace element contents. According to Xiao et al. (1984), 50 Sb analyses from the Upper Devonian in central Hunan give an average of more than 10 times its crustal content. Tong et al. (1984) show that the Middle Devonian carbonates along more than 20 profiles in south Hunan give a Pb average at 2.07 times, a Zn average at 3.16 times and a Mn average of 0.30 times their respective crustal contents. This corresponds quite well with the trace element distribution pattern in Devonian rocks in Guangxi. It is worth mentioning that there also occur a number of important stratabound Sb and Pb-Zn deposits in the Devonian rocks of Hunan, including the famous Xikuangshan Sb deposit.

Therefore, Devonian strata, widely distributed in Guangxi, Hunan and Guangdong with a coverage area of more than 200,000 km$^2$, and having high contents of As, Sb, Pb, Zn, form important source beds and deposits might hence have provided such elements as Pb, Zn, Sb and As for the deposits they host, including the Dachang deposit.

There exists a substantial amount of Sb in the Dachang ore, mainly in jamesonite and other Sb-Pb sulphosalt minerals. In contrast, other famous cassiterite-sulphide deposits in the world, such as Gejiu, China and Renison Bell, Australia, contain very little Sb. There is also very little Sb in the Dachang granite. It is thus more reasonable to consider that Sb in the Dachang deposit might have come from the Devonian strata rather than from the granite, since it is the former that have high Sb contents and host independent stratabound Sb deposits.

Arsenic is present in many cassiterite-sulphide ores, in the form of arsenopyrite. However, in the Dachang deposit, the As content is abnormally high, judging from the fact that the ore-hosting Devonian rocks have very high As contents (up to 10 times the crustal abundance), and possess independent realgar-orpiment ore deposits. It is therefore conceivable that part of As in the Dachang ore might have come from Devonian strata.

The characteristics of the sulphur isotope composition of the Dachang ore are:

1. The $\delta^{34}S$ value is around zero in the Lamo Cu-Zn skarn ores;
2. The $\delta^{34}S$ value ranges from 0 to −8‰ in cassiterite-sulphide ores hosted by Upper Devonian siliceous rocks and banded limestone;
3. The $\delta^{34}S$ value ranges from 3 to 10‰ in cassiterite-sulphide ores hosted by Middle Devonian reef limestone (Fig. 1). It seems probable that the S in the Lamo skarn deposit has a granite-related hydrothermal origin while the S isotopic composition in the cassiterite-sulphide ores seems to be controlled by stratigraphic and lithological factors. In the latter case, S in the ore might have been derived mostly from sulphates of the contemporaneous sea water.

Thus, the sulphur in the Dachang ores may have dual sources: mostly from ancient sea water and partly from granitic magma.

The lead isotope composition of the Dachang ores shows distinct characteristics of multiple sources and multiple stages. Part of the lead is abnormal lead, giving negative age values for all the models. This part of the lead is probably of crustal origin. Most lead isotope determinations display normal lead characteristics, giving similar age values from different models. Much of the normal lead shows Pre-Devonian or Devonian ages with some of the lead being related to Mesozoic and Cenozoic events (Table 2). The former can hardly be relevant to the Yenshanian Dachang granite in that it is much older. It is likely to have been a deep-seated Pb coming up and then precipitating in a Devonian basin. The Mesozoic and Cenozoic lead may be genetically related to the Dachang granite.

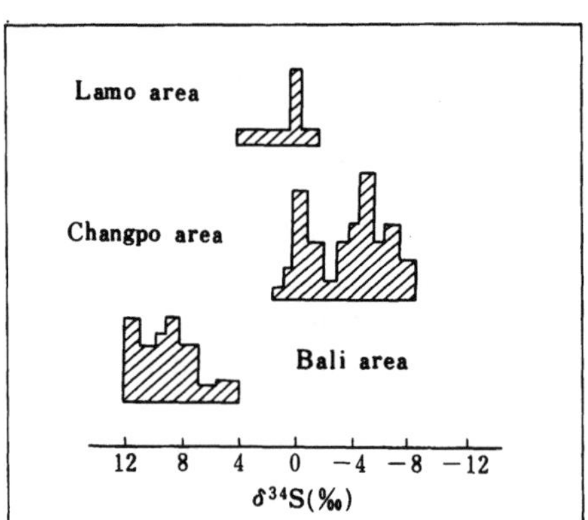


**Fig. 1.** Histogram showing S isotopic composition in Dachang cassiterite-sulphide ores

**Table 2.** Lead isotope composition and model ages in the Dachang ores

| No. | Mineral | $\frac{Pb^{206}}{Pb^{204}}$ | $\frac{Pb^{207}}{Pb^{204}}$ | $\frac{Pb^{208}}{Pb^{204}}$ | Model ages (in Ma) | | | |
|---|---|---|---|---|---|---|---|---|
| | | | | | H-H | R-S-F | Doe | Stacey |
| Pb-9 | jamesonite | 17.786 | 15.589 | 38.094 | 599 | 505 | 597 | 608 |
| Pb-6 | jamesonite | 17.969 | 15.620 | 38.046 | 500 | 387 | 503 | 532 |
| Pb-8 | jamesonite | 17.963 | 15.581 | 38.005 | 398 | 368 | 462 | 460 |
| Pb-7 | jamesonite | 17.969 | 15.573 | 39.957 | 398 | 367 | 449 | 440 |
| A8042 | galena | 18.38 | 15.67 | 39.20 | 268 | 125 | 270 | 327 |
| Pb-30 | galena | 18.37 | 15.66 | 39.20 | 262 | 123 | 266 | 315 |
| Pb-35 | galena | 18.30 | 15.55 | 39.29 | 175 | 78 | 180 | 138 |
| W-739 | galena | 18.41 | 15.54 | 38.57 | 26 | –14 | 86 | 30 |

After S.L. Li and Q.X. Chen.

Therefore, judging from lead isotopes, three probable Pb sources can be distinguished. Much of the Pb was older than or contemporaneous with Devonian sediments. Some lead might have been derived through leaching out from country rocks. Part of the lead was coeval with the biotite granite which, being Cretaceous in age, has long been considered to be the sole source of the cassiterite-sulphide ores. It also has a high $Sr^{87}/Sr^{86}$ ratio (0.7163), as is the case with many Sn-bearing granitic plutons. The Dachang granite is rich in volatiles (Table 3). According to Zhang and Li (1983), the F content may be as high as 5274 ppm (average of 11 analyses) and Cl, 253 ppm. In Table 4 are listed the contents of trace elements in the Dachang granite. It can be seen that the contents of Sn, W, and Bi in the Dachang granite are several times higher than those in ordinary granites.

The Dachang cassiterite-sulphide ore deposit itself and its neighbouring stratabound Pb-Zn, Sb, pyrite and realgar-orpiment deposits are all controlled by structural factors: the orebodies occur in the axial part of an anticline or an overturned anticline or in nearby fracture zones. These structural elements are the product of Late Yenshanian crustal movements.

The main orebodies of the Dachang deposit occur along the axis of the Dachang anticline, which strikes 325° and has steep west and gentle east limbs. In Changpo, the anticline is overturned to the west, with a large-scale reverse fault developed along the anticlinal axis. In addition, secondary latitudinal and longitudinal fracture zones occur in the gentle east limb. All these form important ore-hosting structures (Fig. 2).

Certain ore-hosting structures are confined to certain lithological units. For example, the gigantic stratoid-veinlet orebody (No. 91) occurs mainly in Upper Devonian banded limestone. Another huge stratoid-stockwork orebody (No. 92) is confined to

**Table 3.** Petrochemistry of the Dachang biotite granite (%)

| | W1087 | W1088 | W1091 | W1116 | W1117 |
|---|---|---|---|---|---|
| $SiO_2$ | 73.21 | 74.36 | 73.61 | 74.49 | 74.07 |
| $TiO_2$ | 0.06 | 0.06 | 0.08 | 0.07 | 0.08 |
| $Al_2O_3$ | 13.33 | 12.81 | 13.86 | 12.28 | 12.73 |
| $Fe_2O_3$ | 1.35 | 1.56 | 0.83 | 0.44 | 0.46 |
| FeO | 1.35 | 1.31 | 1.33 | 1.26 | 1.76 |
| MnO | 0.07 | 0.10 | 0.08 | 0.11 | 0.06 |
| CaO | 1.97 | 0.91 | 1.28 | 0.95 | 0.73 |
| MgO | 0.13 | 0.19 | 0.19 | 0.24 | 0.16 |
| $Na_2O$ | 1.65 | 3.45 | 3.60 | 3.65 | 3.45 |
| $K_2O$ | 4.25 | 4.50 | 4.86 | 4.70 | 4.75 |
| $P_2O_5$ | 0.43 | 0.25 | 0.38 | 0.27 | 0.26 |
| $H_2O^+$ | 0.96 | 0.66 | 0.39 | 0.44 | 0.59 |
| $H_2O^-$ | 0.42 | 0.32 | 0.19 | 0.16 | 0.17 |
| Total | 99.18 | 100.48 | 100.68 | 99.06 | 99.17 |

**Table 4.** Trace element contents of Dachang granite (averages of 11 samples) in ppm (from Zhang and Li, 1983)

| Sn | W | Mo | Pb | Zn | Cu | Bi |
|---|---|---|---|---|---|---|
| 25.6 | 52 | 2.2 | 28 | 71 | 25 | 41 |

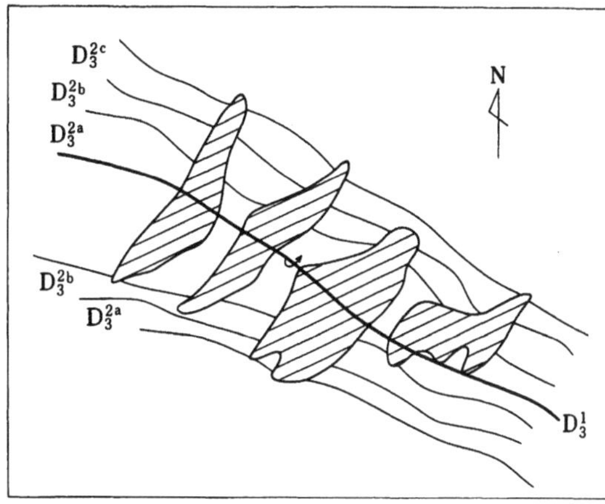


**Fig. 2.** Part of the Dachang orebodies situated along the anticlinal axis

Upper Devonian siliceous rocks. These two orebodies, together with two in Middle Devonian reef limestone, account for most of the Dachang ore reserves.

From the foregoing discussion, a 3-stage mineralization history can be postulated for the Dachang ore deposit (Fig. 3).

1. In Devonian time in north Guangxi, as well as in neighbouring Hunan and Guangdong, carbonate formations were commonly formed in shallow sea environment together with gypsum and salt beds, abundant organic material, siliceous rocks and reef limestones. At the same time, plenty of As, Sb, Pb and Zn were also accumulated in dispersed forms, forming high background values but failing to form independent ore deposits. At this sedimentary-diagenetic stage, important pyritic beds might

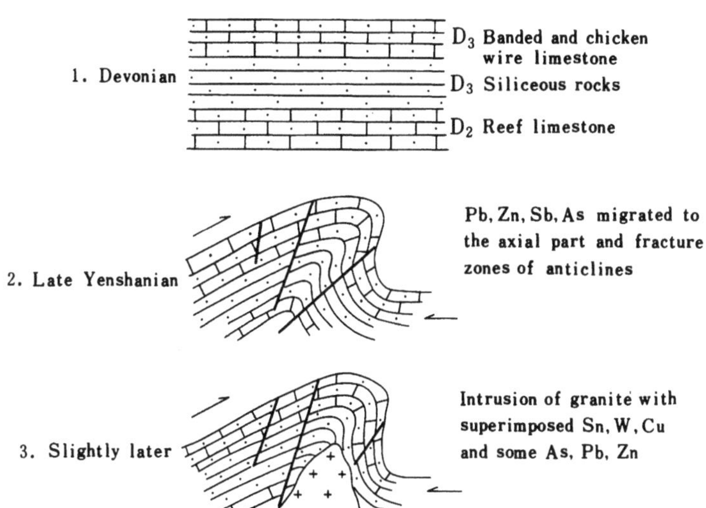


**Fig. 3.** Sketch showing model mineralization history of the Dachang ore deposit

have formed. Deep fracture zones along the southern border of Jiangnan Massif probably played a significant role in bringing up As, Sb, Pb, Zn and other ore substances from the deeper part of the crust with the help of geotherms, hot brines or submarine volcanic activities (though evidence of the latter is lacking at Dachang).

2. During Yenshanian crustal movements, the Late Palaeozoic and Triassic formations were folded and faulted, giving rise to a series of NWW-striking fracture zones, anticlines and synclines, accompanied by secondary NNE structures. The stress thus generated would help to reactivate As, Sb, Pb, Zn and related ore substances. These geochemically active elements tended to migrate towards the crests of anticlines and fracture zones close to them, just as oil and gas would behave under the same conditions. Stratabound ore deposits of Pb-Zn, Sb and realgar-orpiment were thus formed in the axial parts of anticlines and related fracture zones.

3. Simultaneously or slightly later, the emplacement of a biotite granite stock took place also along the axial part of the Dachang anticline. The granitic magma brought its own content of Sn, Cu, W as well as some As, Pb, Zn and superimposed these elements upon the pre-existing stratabound Pb-Zn, Sb, realgar-orpiment deposit, giving rise to the present Dachang ore deposit. The Lamo skarn Cu-Zn deposit was also formed at the same time. Sn in Dachang ore came mostly from granite, with perhaps a small partial contribution from the Devonian rocks.

Therefore, the formation of the Dachang cassiterite-sulphide deposit was subject to lithological (carbonate formation), magmatic (granite) and structural (anticline and related fracture zones) controls. The stress on its sole genetic relationship with granite does not suffice to explain its long lasting and overall complex mineralization history.

An analysis of geological occurrences and geochemical peculiarities of some of the large-scale cassiterite-sulphide ore deposits, both at home and abroad, suggests that they are controlled by similar 3-fold factors. Here large-scale cassiterite-sulphide deposits refer to those deposits with large amounts and massive ores of sulphides of Fe, Pb, Zn, As and cassiterite. In geographic distribution, the cassiterite-sulphide type of tin deposit is much more limited than the quartz-vein or greisen-type of tin deposit. For example, in some major tin-producing countries, such as Thailand and Malaysia, vein and greisen types of tin mineralization are widespread while large-scale cassiterite-sulphide deposits are scarcely observed. The reason may be that the vein and greisen type of tin deposit is controlled mostly by the magmatic factor (Sn-producing granite), with the structural factor playing only a subordinate role, while the large-scale cassiterite-sulphide deposits, not counting those small deposits of skarn type, are controlled by lithological, magmatic, as well as structural factors. The probability of occurrence of the latter type of mineralization is thus more limited than the former.

Furthermore, a correlation between the Dachang deposit and another large scale cassiterite-sulphide deposit (the Gejiu deposit) also demonstrates that the latter is very similar to the former in its controlling factors:

1. Cretaceous biotite granite, high in silica, peraluminous, rich in alkalies and volatiles with abundant content of W, Sn, Be and other trace elements and of a high initial $Sr^{87}/Sr^{86}$ ratio.

2. Anticlines and fracture zones; Triassic carbonate formations containing salt and gypsum. Notwithstanding the fact that there exist significant differences between the two deposits (for example, there is practically no antimony in the Gejiu ore), the two deposits are similar in overall mineralization patterns. The Gejiu deposit, like the Dachang deposit, may have polygenesis and multiple sources of ore-forming substances.

The author now comes to the conclusion that in the search for large-scale cassiterite-sulphide deposits (small skarn deposits are excepted), prospectors should pay attention not only to the tin-bearing granite, usually of S-type, but also to structures (anticlines and fracture zones) and lithology (carbonate formations with gypsum and salt beds, carbonaceous materials, and submarine volcanic intercalations, especially those having high contents of trace elements).

*Acknowledgements.* In preparing this manuscript, the author has had opportunities to discuss the Dachang problem with X.S. Ye, Y.X. Yan, Z.G. Zhang, S.L. Li, Z.H. Zhao and Q.X. Chen. The author wishes to express his gratitude to them, and also to K. Ding, who helped in drawing and calculating.

## References

Beckinsale, R.D., 1979. Granite magmatism in the tin belt of SE Asia. *Origin of granite batholiths: Geochemical evidence,* ed. Atherton, M.P. and Tarney, J., Shiva Publ., England.

Chen, Yuchuan, 1965. Primary zoning in a certain ore deposit in Guangxi. *Geological Review,* Vol. 23, No. 1 (in Chinese).

Tong Quianming et al., 1984. A study on Pb, Zn, Fe and Mn contents of Middle Devonian Qiziqiao Zu in Southern Hunan. *Mineral Deposits,* Vol. 3, No. 3 (in Chinese).

Wu Qinsheng et al., 1983. Characteristics of Sr isotopes in apatite and their application. *Sciences of China,* (Ser. B), No. 7 (in Chinese).

Xia Qiming et al., 1984. An investigation into the genesis of the antimony deposits in Hunan. *Mineral Deposits,* Vol. 3, No. 3 (in Chinese).

Yang Fengyun et al., 1966. Sulphur isotopic analysis, with special application to a cassiterite-sulphide ore deposit. *Geological Science,* No. 3 (in Chinese).

Zhang Zhengen et al., 1981. Studies of mineralization processes and mineralogical composition at the Dachang ore deposit. *Geochemica,* No. 1 (in Chinese).

Zhang, Z.G. and Li, X.L., 1983. Researches on the mineralogical and chemical composition of the Dachang ore deposit, *Geochemica* (in Chinese).

Zhao Zhenhua, 1984. Association of chemical elements in Chinese stratabound deposits. *Geochemistry of Chinese stratabound deposits,* ed. Tu Guangzhi et al., Science Press (in Chinese).

# 6.4.7 Franckeite: Mineralogical Aspects and Genetic Relations to the Dachang and Bolivian Ore Deposits

Li Jiuling[1], Gunther H. Moh,[2] and Naiding Wang[2]


**Abstract**

Synthetic experiments on the mineral franckeite at elevated temperatures revealed an extensive solid solution range due to $Pb^{+2}/Sn^{+2}$ substitution. The derived general formula $(Pb, Sn)^{+2}_{6-x}Fe^{+2}Sb^{+3}_{2}\ Sn^{+4}_{2}S_{14-x}(0 \leq x \leq 1)$ covers the composition of at least three independent minerals: franckeite, incaite, and potosiite. A correlation of the paragenetic patterns of franckeite in the Dachang and Bolivian deposits with the relevant synthetic assemblages gave clues to the formation conditions of the respective natural occurrences. Compared with Bolivia, the sulphosalt assemblages in Dachang are probably formed under relatively high $Sb_2S_3$ concentrations and low sulphur fugacities.


The Dachang polymetallic deposit in China is economically and mineralogically of great importance. Its complicated formation and coexisting patterns display similarities in many respects to the famous Bolivian type deposits. Of the numerous sulphosalts in these two deposits, franckeite has been chosen as the subject of an intensive and systematic investigation. Two main points were taken into account for this choice. First, the composition range, the chemical formula, and the crystal chemistry of this mineral species have not been uniquely determined. Second, the characteristic and locally differentiated parageneses of franckeite in the two deposits permit a deduction of the respective formation condition of the various assemblages.

The investigation includes dry and hydrothermal syntheses, microscopic and microprobe analyses of some 20–30 natural franckeites, Mossbauer analyses and heating experiments of the natural material. On the basis of the experimental results and of information collected on the relevant minerals jamesonite and cylindrite, a tentative correlation with respect to the parageneses and formation conditions of the Dachang and Bolivian deposits was attempted.

The present experiments indicate that franckeite, as a mineral group, can form an extensive solid solution range which covers the compositions of at least three independent minerals (Li, 1984). The multicomponent phase is composed of a complex thioantimonate (III) – thiostannate (IV) with bivalent $Pb^{+2}$, $Sn^{+2}$ and $Fe^{+2}$ cations.

[1] Institute of Mineral Deposits, Chinese Academy of Geological Sciences, Beijing
[2] Heidelberg University

The experimentally confirmed Ag-free formula

$$(Pb, Sn)^{+2}_{6-x} Fe^{+2} Sb^{+3} Sn^{+4}_{2} S_{14-x} \ (0 \leq x \leq 1)$$

involves an extensive substitution of bivalent $Pb^{+2}$ and $Sn^{+2}$. Within the composition range, powder diffraction patterns and microscopic evidence indicate a single phase, with a slight but continuous change of d-spacing. The upper stability temperature of the phase approaches 700°C. Figure 1 shows the solid solution range of dry syntheses at 600°C. Under hydrothermal conditions (300°C/600 bars), this range extends in the direction of the Pb-rich composition. Pb-free franckeite only appeared in the experiments at temperatures higher than 560°C. The analyses of some natural franckeites (Bernhardt, 1984) are also plotted in the figure. Comparing the experimental results with the natural material, the mineral incaite (Makovicky, 1974, 1976) and franckeite may be considered as the Sn-rich and Pb-rich members, respectively, of this solid solution range. The Pb-richest member of the series is the newly reported mineral potosiite (Wolf et al., 1981), which is outside the 600°C range of dry syntheses but inside the range of hydrothermal syntheses (300°C/600 bars). Heating experiments performed at 600°C of the Pb-rich natural franckeites yielded exsolutions of galena. Franckeites with intermediate (Pb, Sn) compositions yielded no exsolutions after heating.

Normal franckeite usually contains some Ag, but not as an essential element. In the franckeite structure Ag may substitute for Pb, which possibly involves partial oxidation of either $Fe^{+2}$ for $Sb^{+3}$, or a change of the relative amount of the formula components. The synthetic Ag-bearing and Ag-free franckeites are optically very similar.

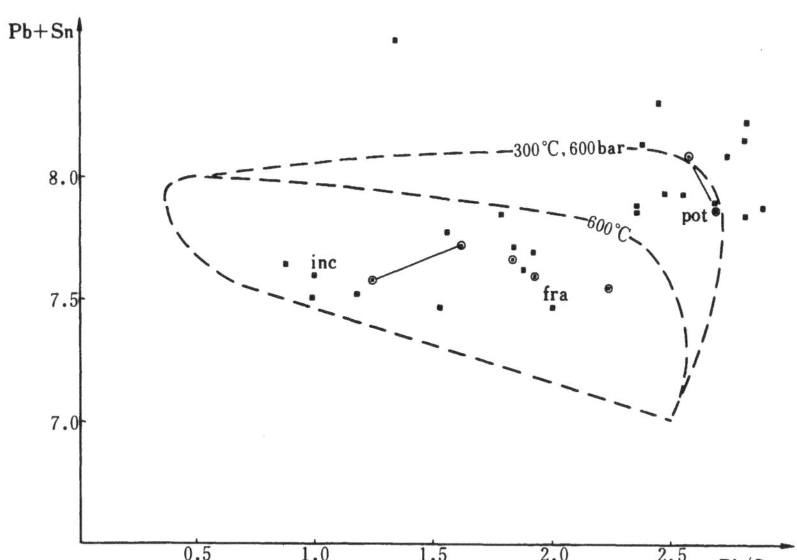


**Fig. 1.** Solid solution range of synthetic franckeite expressed in terms of the $Pb^{+2}/Sn^{+2}$ substitution under dry (600°C) and hydrothermal (300°C/600 bars) conditions, respectively. Small squares represent analyses of natural franckeites plotted against this range (Bernhardt, 1984). Circles are the results obtained from heating experiments. Dashed lines indicate postulated boundaries

In addition to the isomorphic substitution of Pb/Sn or Pb/Ag, variation in the components, such as excess or deficiency of PbS, SnS, FeS, $Sb_2S_3$, or change in the prevailing sulphur fugacity, will result in the formation of different assemblages containing franckeite. Figure 2 shows in addition to the franckeite solid solution, the various assemblages obtained in experimental runs under constant sulphur fugacities and changing proportions of the formula components. Franckeite can form assemblages with galena, teallite, herzenbergite, ottemannite, or cylindrite.

The change in sulphur fugacity is reflected by the change in $Sn^{+4}$ content in the formula. This change may be represented by the third axis running perpendicular to Figure 2. In 3-dimensions, the solid solution range of franckeite having constant $Sn^{+4}$ value appears as a flake-like field, the lateral projection of which is degenerated to a straight line (Fig. 3). This figure also shows the assemblages obtained in the experimental runs due to change in sulphur fugacity. The stability of these assemblages is extremely sensitive to such a change.

At low sulphur fugacities, franckeite tends to coexist with galena or teallite. A slight increase in sulphur fugacity yielded coexisting cylindrite. As an illustration, powder of natural franckeite heated with a small amount of sulphur at 200°C already yielded some cylindrite. Homogeneous cylindrite therefore could be synthesized directly from Sn-rich franckeite by heating with an appropriate amount of sulphur. Further increase in sulphur fugacity to an Sn-rich starting composition resulted in the formation of a cylindrite-berndtite ($SnS_2$) assemblage.

Not included in Figure 3 is the relationship between franckeite and jamesonite ($Pb_4FeSb_6S_{14}$). With increasing $Sb_2S_3$ in the charge, franckeite tends to coexist with jamesonite and disappears under $Sb_2S_3$ excess. Similar to franckeite, an extensive

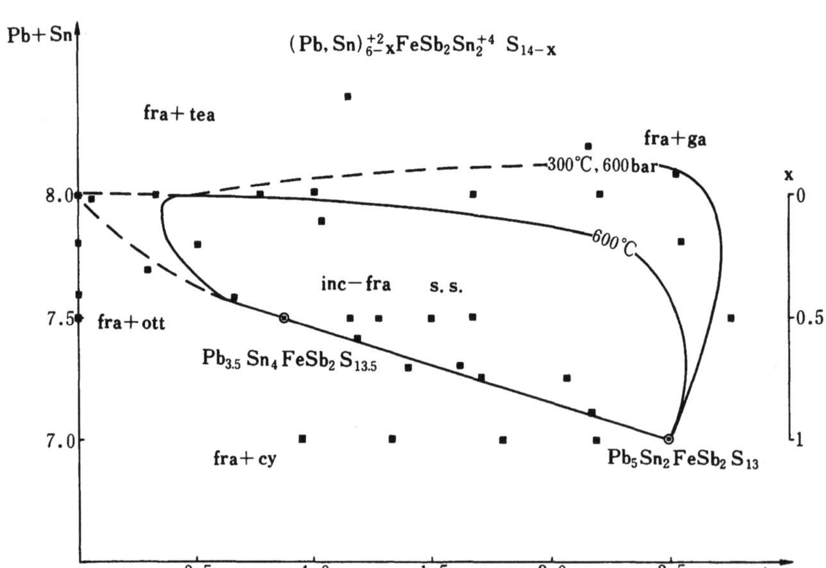


**Fig. 2.** Franckeite solid solution (s.s.) range and the various assemblages containing franckeite obtained experimentally under constant sulphur fugacity and under variation in the relative amounts of the formula components

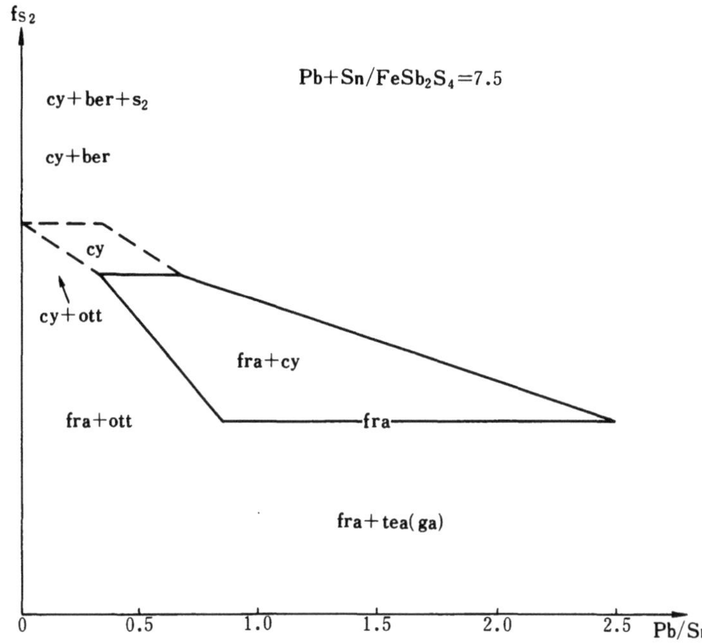


**Fig. 3.** Parageneses of franckeite in dry synthetic experiments plotted against change in sulphur fugacity

$Pb^{+2}/Sn^{+2}$ substitution range has been observed in jamesonite (Wang et al., 1982). At 560° C, jamesonite remains stable under approximately the sulphur fugacity in which franckeite is also stable. In runs with high sulphur fugacities, however, jamesonite was found partially decomposed at 400° C to robinsonite and pyrite.

Brief descriptions of the relevant assemblage observed at Dachang and in Bolivia are as follows: At Dachang, jamesonite is a common mineral while franckeite is relatively seldom found. Franckeite usually coexists with galena, teallite, or boulangerite, occasionally also with traces of stannite residue, exhibiting a characteristic coexisting pattern. Additional coexisting minerals include pyrite, marcasite, sphalerite, geochronite and andorite. The existence of cylindrite or berndtite has not been reported.

In Bolivia, jamesonite shows only restricted distribution and amount, whereas franckeite has been found in 14 mining districts, in two of which franckeite counts as the leading ore mineral. Two different types of franckeite parageneses are recognized in Bolivia. The first type is similar to what occurs in Dachang, with coexisting minerals franckeite-galena/teallite – boulangerite – sphalerite – pyrite – marcasite – arsenopyrite, occasionally also residues of stannite or cassiterite. The second type involves a pattern with cylindrite replacing franckeite in various proportions. With intensive replacement, flakes of franckeite occur only in layers of cylindrite. Other coexisting minerals include traces of boulangerite and residues of stannite.

The observed natural assemblages and the experimental results were correlated with regard to the presence or absence of the relevant sulphosalts and their coexisting patterns. Such correlations yielded valuable, though still preliminary information on the formation conditions of the deposits: the relative abundance of the participating components and the prevailing partial pressure of sulphur.

At Dachang, before franckeite formed, high concentration of $Sb_2S_3$ and relatively low sulphur fugacity favoured the formation of abundant jamesonite. During the franckeite stage, the concentration of $Sb_2S_3$ decreased. An appropriate proportion of the components and the optimal sulphur fugacities required for the formation of franckeite prevailed only locally. As a result franckeite occurs only in very restricted areas and amounts at Dachang. The presence of small amounts of coexisting teallite and galena, and the absence of cylindrite both indicate conditions of relatively low sulphur fugacities.

In Bolivia, on the other hand, the considerably lower $Sb_2S_3$ concentration, and the relatively higher and constant sulphur fugacity during certain mineralization stages, permitted the extensive formation of a considerable amount of franckeite. In some localities, a common feature is the replacement of franckeite by cylindrite, which indicates a further increase in sulphur fugacities of the ore-forming hydrothermal solution in the later stages of mineralization.

In addition, the appearance of berthierite ($FeSb_2S_4$) at Dachang in the later stages of mineralization also reflects conditions of checked sulphur fugacities. Under increasing sulphur fugacity, this mineral tends to decompose to stibnite and pyrite. In Bolivia, no berthierite has so far been reported. Instead, there occurs the mineral berndtite ($SnS_2$), which is only stable under very high sulphur fugacities. The presence or absence of these two minerals reflects to some extent the formation conditions which prevailed during certain mineralization stages of these two deposits.

## References

Bernhardt, H.-J., 1984. The composition of natural franckeites. *N. Jb. Miner. Abh.* 150, 41—45.

Li, Jiuling, 1984. Franckeite syntheses and heating experiments. *N. Jb. Miner. Abh.* 150, 45—50.

Makovicky, E., 1974. Mineralogical data on cylindrite and incaite. *N. Jb. Miner. Mh.* H6, 235—256.

Makovicky, E., 1976. Crystallography of cylindrite. Part 1. Crystal lattices of cylindrite and incaite. *N. Jb. Miner. Abh.* 126, 304—326.

Wang, N., Moh, G.H. and Klein, D., 1982. Synthesis of lead-tin-antimony sulfosalts. *N. Jb. Miner. Abh.* 144, 304—307.

Wolf, M., Hunger, H.-J. and Bewilogua, K., 1981. Potosiit, ein neues Mineral der Kylindrite-Franckeit-Gruppe. *Freib. Forsch. H. C.* 364, 113—133.

# 6.4.8 Isotope Geochemistry of the Dachang Tin-Polymetallic Ore-Field

XU WENKIN, WU QINSHENG, and XU JUNZHEN[1]

**Abstract**

The granite in this area has been dated by Rb-Sr at 99 to 115 Ma, and its genesis is discussed.

The isotope compositions of Pb, S, O, C of the minerals at various ore-forming stages of every belt in the orefield have been measured, and their variation regularities are interpreted. The genesis and source matter of the deposits are also discussed.

## 1. Introduction

The Dachang tin-polymetallic ore-field consists of a middle belt (zinc-copper and antimony deposits), and east and west belts (tin-polymetallic deposits). The exposed igneous rocks are porphyritic and equigranular biotite granite stocks and dykes of porphyritic diabase, porphyritic diorite and granite porphyry. The orebodies' host rocks are striped limestone ($D_3^2$) in the middle belt, striped limestone and siliceous rocks ($D_3^1$—$D_3^2$) in Changpo, and reef limestone ($D_2$) in Longtoushan within the west belt. There is also a bed of gypsum at depth at Longtoushan.

## 2. Rb-Sr Age Dating and REE Contents of Some Magmatic Rocks

Rb-Sr age determinations show that the ages of porphyritic granite and equigranular biotite granite are 115 $\pm$ 3 Ma and 99 $\pm$ 6 Ma respectively with initial strontium isotope ratios of 0.7159 $\pm$ 0.0009 and 0.7163 $\pm$ 0.0061 (Figs. 1 and 2). The fact that the two granite types have similar strontium isotope composition but different ages indicates that they were from the same source but emplaced at different stages. The initial strontium isotope ratios are both greater than 0.7100, implying they are S-type.

The total contents of rare earth elements in the porphyritic granite, equigranular biotite granite and equigranular two-mica granite are 251 ppm, 103 ppm and 66 ppm respectively, with their $\Sigma$ Ce/$\Sigma$ y ratios 4.4, 2.7, and 2.2, and the Eu/Eu* ratios 0.26,

[1] Research Institute of Geology for Mineral Resources, China National Non-ferrous Metals Industry Corporation, Guilin, Guangxi

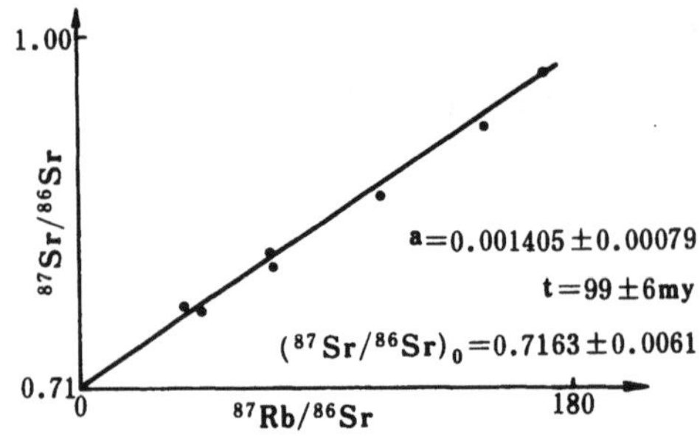


**Fig. 1.** Rb-Sr whole rock isochron diagram of porphyritic biotite granite from Dachang

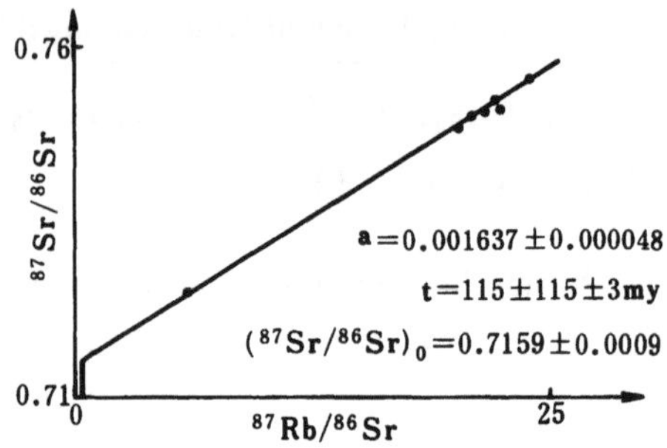


**Fig. 2.** Rb-Sr whole rock isochron diagram of equigranular biotite granite from Dachang

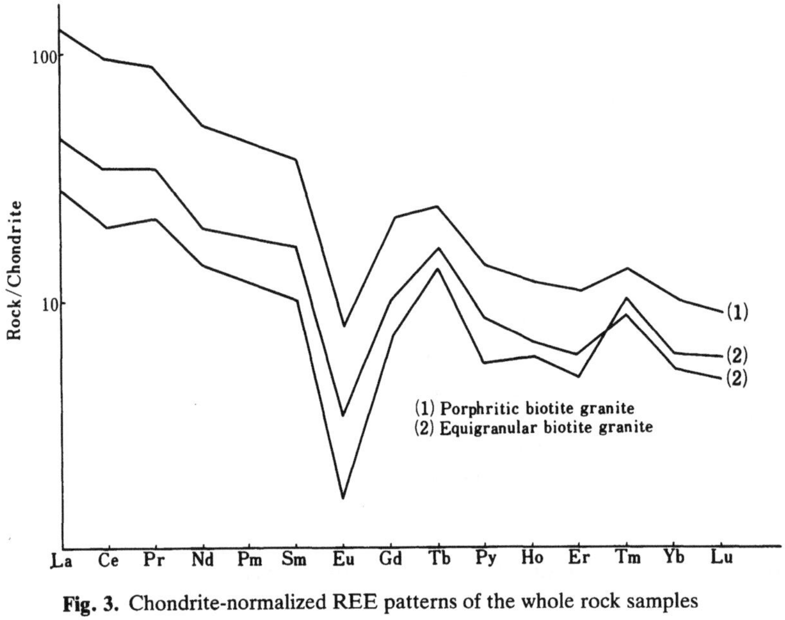


**Fig. 3.** Chondrite-normalized REE patterns of the whole rock samples

0.25 and 0.18 respectively. These differences are closely related to the granite lithological differences. The chondrite-normalized REE patterns of these three rocks are all V-shaped and inclined to the right (Fig. 3); the europium depletions are very obvious. Comparing the above data with the REE data of the granites of South China summarized by Wang Zhougang, we can conclude that the granites in this area are of crustal anatectic type.

**Table 1.** Temperatures, physiochemical conditions and paragenetic associations of minerals at various belts and stages of the Dachang ore-field

| Belt | Stage | Paragenetic association of minerals | Mode of occurrence | Homogenization temperature of mineral inclusion | Average equilibrium temperature | Equilibrium temperature of oxygen isotope | Chosen temperature | $\log f_{O_2}$ | $\log f_{S_2}$ | $\log f_{CO_2}$ | pH | Mole % FeS in the sphalerite |
|---|---|---|---|---|---|---|---|---|---|---|---|---|
| Middle Belt | 1–4 | Skarn minerals | | 560–380°C | | | | | | | | |
| | 5 | Arpy, Py, Mt, Cpy, Fluor, Cal, | bedded | 350°C | 362°C | | 350°C | –28.4** | – 8.5* | 0.4*** | 4.9 | 19.83 |
| (Zn, Cu) | | Spl, Call, Ga, stan, Q, Sec, Po. | | (pressure corrected) | | | | –30.9 | – 9.2 | 2.18 | 5.8 | 22.68 |
| | 6 | Arpy, Py, Chl., Po, Cpy, Fluor, Cal, | vein | 300°C | 319°C | | 300°C | – 31.9** | –10.3 | – 0.44*** | 5.5 | 18.6 |
| | | Mt(Tr), Q, Spl, Ga, Sec, bi, nabi. | | (pressure corrected) | | | | –34.5(?) | –11.5(?) | ? | 5.9 | ? |
| Middle Belt | 1–3 | Sulfides and Q, Fluor, Wf, Sc. | vein | 250–230°C | | | | | | | | |
| (Sb) | 4 | berth, Antim, Py, Q, Fluor | vein | 162–175°C | | 154°C | 150°C | – 46.1 | –17.5 | –2±1 | 3.8 | |
| | | | | (pressure corrected) | | | | –47.9 | | + 1.5 | 5.4 | |
| West Belt (Sn, Zn, Cu) | 1 | Q, Cass, Tl and Arpy (Tr), Py (Tr) | vein | 380–370°C (Boiling inclusion) | | 366°C | 380°C | | | | | |
| | 2 | Kf, Sec, Arpy, Po, Py, Cpy, Mt, | bedded | ? | 335°C | | 350°C | –28.4** | – 8.5* | 0.4*** | 4.9 | 19.6 |
| | | (Tr) Mv, Cal, Q Spl, Cass, Stan. | | | | | | –30.6 | – 9 | 2.48 | 5.8 | 22.0 |
| | 3 | Arpy, Q, Cass, Py. | vein | 350°C (Boiling inclusion) | | | | | | | | |
| | 4 | Arpy, Ja, Po, Py, Spl, Stan, | vein | 300°C (Boiling | 315°C | 280–320°C | 300°C | –31.9** | –10.2* | – 0.8*** | 5.5 | 18.0 |
| | | Cpy, Q, Mt(Tr), Cal, Cas, Ga, bi, nabi, Tl. | | inclusion) | | | | –36.4 | –11.2 | – 0.04 | 5.9 | 23.0 |
| | 5–6 | Sulfosalts, Py, Arpy, Spl, Antim, Call, Mr, O, Q, Sid, cal. | vein | < 270°C | | | | | | | | |

Antim=antimonite, Bi=bismuthinite, Cass=cassiterite, Cpy=chalcopyrite, Chl=chlorite, Ca=calcite, D=dickite, Ga=galena, Arpy=arsenopyrite, Ja=jamesonite, Fluor=fluorite, Mt=magnetite, Kf=K-feldspar, Po=pyrrhotite, Py=pyrite, Mv=muscovite, berth=berthierite, Sec=sericite, Spl=sphalerite, Stan=stannite, Sid=siderite, Q=quartz, nabi=native bismuth, Sc=scheelite, Wf=wolframite, Tl=tourmaline.

Mole % FeS in the sphalerite is from Fu Jinbao (1984). * Sulfur fugacity is calculated by Mole % FeS in the sphalerite. ** Oxygen fugacity is calculated by Mt+Arpy+As reaction equation. *** Carbon dioxide fugacity is calculated by $CaSo_4+CaCO_3$ reaction equation and graphite+$W_2$ reaction equation. pH value is calculated by $CO_2$-$H_2O$-NaCl methods and average value of sericite mineral. Paragenetic associations of minerals are from Yan Yunxiu (1965) and Lai Lairen (1984).

## 3. The Mineral Assemblage and Physico-Chemical Conditions of Various Minerogenic Stages

On the basis of the microscopic studies, sulphur and oxygen isotopic thermometers, filling temperatures of fluid inclusions, and by using the methods of Holland (1959), Scott (1971), and Crerar (1978), as well as the thermodynamics data summarized by Robie (1978), the mineral assemblages, the ore-forming temperatures, and the physico-chemical conditions at each minerogenetic stage of every belt were determined (Table 1). The mineral assemblages at the principal minerogenetic stage of zinc-copper and tin deposits are as follows: sulphide+cassiterite+magnetite (tr.)+calcite+quartz+fluorite+jamesonite (stage 5 at the middle belt and stage 4 at the west belt). However, some stages have only sulphides, but no tin-bearing minerals (e.g. stage 6 at the middle belt and stage 2 at the west belt). On the other hand, some stages have only cassiterite, quartz, arsenopyrite, and tourmaline (e.g. stage 1 and stage 3 at the west belt). The ore forming temperature of the main stage of the zinc-copper orebodies is about 300°C–350°C, and that of the tin orebodies is between 300°C and 380°C. All stages have characteristics of lower oxygen and lower sulphur fugacities. Carbon dioxide fugacities are from 0.16 to 301.9 ($\log f_{CO_2}$=0.8-+2.49). The solutions show acidic and weak basic characteristics; the precipitating temperature of abundant stibnite is 150°C. The physico-chemical conditions for stibnite precipitation are: $\log f_{S_2}$: 17.5; $\log f_{O_2}$: between –46.1 — –47.9; $\log f_{CO_2}$: –2±1 — +1.5; and pH values: 3.8–5.4.

## 4. Studies on Sulphur Isotopes

Due to the effects of magmatic hydrothermal activity, it is difficult to obtain sulphur isotope samples of sedimentary origin from the strata. We have obtained only two samples of sedimentary concretion pyrite (Table 2). The sulphur isotope compositions of these two samples are quite different from each other and are rich in $S^{34}$ in the upper part of the strata. We consider that sedimentation of the sulphides was in a closed system. The sulphide sedimented earlier was rich in $S^{34}$ and the later was rich in $S^{34}$, because there was no other sulphur source to supplement the solution when sulphate was reduced to sulphide by organic or inorganic reduction.

According to the sulphur isotope compositions of more than 200 sulphide and sulphate samples and calculated $\delta\ S^{34}_{\Sigma S}$ values, we can draw the following conclusions (Table 3):

(1) Spatially, $\delta\ S^{34}$ gradually increases from the deeper to the upper part at the 5th stage in the middle belt and at the 2nd stage in Changpo. This trend is the same as for sulphur isotopes of sedimentary rocks, and might be due to the fact that the sulphur

**Table 2.** Sulphur isotope composition of sedimentary concretion pyrite in the Devonian strata

| Strata | Mineral | $S^{34}_{(CPT)}$% | Variation |
|---|---|---|---|
| $D_3^1$ | Pyrite | – 3 | (20.7) |
| $D_3^2$ | Pyrite | +17.7 | |

**Table 3.** Sulphur isotope composition and $\delta S^{34}_{\Sigma s}$ values at various metallogenic stages of the Dachang ore-field

| Belt | Stage | Strata | $\delta S^{34}_{(CPT)}$ ‰ (variation range) | | | | | Galena | $\delta S^{34}$ ± 3% |
|---|---|---|---|---|---|---|---|---|---|
| | | | Sphalerite | Pyrite | Pyrrhotite | Arsenopyrite | Jamesonite | | |
| Middle | 5 | $D_3^2$ | +1.3 + 4.2 | −0.1 +2.4 | +0.6 + 3.6 | +1.2 + 1.8 | | | + 5 |
| Belt | | $D_2^2$ | − 5.4 | −4.1 −4.4 | | | | | |
| (Zn, Cu) | 6 | | −1.1 + 1.2 | −4.4 +1.1 | | | | −0.2 | + 1 |
| | | | | | | | | −1.50 | |
| West | 2 | $D_3^1$ | −3.7 − 1.0 | −2.0 0.0 | − 3.4 | +2.0 | | | + 6 |
| Belt | | $D_3^2$ | +1.1 + 6.6 | −0.7 +6.3 | −1.2 + 3.1 | | | | |
| (Changpo) | 4 | | −4.3 + 1.1 | −4.0 +1.0 | | +0.2 | | | + 1 |
| West Belt | 2 | $D_2^1$ | −37.3 | −4.0 | −19.7 | | +2.4 + 7.6 | ? | |
| (Long-toushan) | 4 | | +4.8 +11.8 | +9.3 | +7.0 +10.2 | | | | |

source of the ore was from Devonian sedimentary rocks. However, their variation range is narrower than that of pyrite in sedimentary rocks and wider than the range (10‰) of sulphur isotope variation predicted from physico-chemical conditions at the same stage. Moreover, the replacement of ores is obvious. So we believe that sulphur isotope variations at stage 5 in the middle belt and at stage 2 in the west belt are caused by the changes in the physico-chemical conditions of ore-liquid, instead of by the sulphur in the strata.

(2) The $\delta\, S^{34}_{\Sigma s}$ values at stage 5 and stage 6 in the middle belt and at stage 2 and stage 4 in Changpo are lower, which indicate that the sulphur source is magmatic.

(3) Sulphur isotope variation (–4‰ to –37.3‰) of sulphides at stage 2 in Longtoushan is larger than the sulphur isotope range predicted from physicochemical conditions at the same stage, suggesting that their sulphur is from more than one source and probably from a mixture of magmatic and biological sources. This idea is confirmed by the fact that framboidal pyrite is widely developed at the same stage and that the ores show obvious replacement characteristics.

(4) The $\delta\, S^{34}_{\Sigma s}$ value at stage 4 in Longtoushan is higher than that at the same stage in Changpo. It shows that sulphur of Longtoushan is a mixture of lower $\delta\, S^{34}_{\Sigma s}$ sulphur and higher $\delta\, S^{34}_{\Sigma s}$ sulphur or was carried by a different hydrothermal liquid. In view of the fact that the products of this stage all occur in veins and that there are gypsum beds at depth in Longtoushan, we consider that sulphur at stage 4 is probably a mixture of magmatic sulphur with a lower $\delta\, S^{34}_{\Sigma s}$ value and sulphur in gypsum with a higher $\delta\, S^{34}_{\Sigma s}$ value. Provided that magmatic sulphur had the same $\delta\, S^{34}$ value as the $\delta\, S^{34}_{\Sigma s}$ value at stage 4 in Changpo and that the sulphur of gypsum had a $\delta\, S^{34}$ value equal to the value of oceanic sulphur in Devonian time (+20‰), using equation (1), the amount of sulphur in the gypsum can be calculated.

$$fi = \frac{\delta S^{34}_{\Sigma s}\text{ mixture value} - \delta S^{34}_{\Sigma s}\text{ gypsum value}}{\delta S^{34}_{\Sigma s}\text{ magmatic value} - \delta S^{34}_{\Sigma s}\text{ gypsum value}} \quad \cdots\cdots\cdots (1)$$

It might have contributed 58% of the sulphur in this stage while magmatic sulphur contributed about 42%.

(5) The mechanisms of sulphur mixing and ore-formation at stage 2 in Longtoushan are different from those of stage 4. This means that they are different hydrothermal liquids. We suggest that the hydrothermal solution of stage 4 was from a deeper magmatic source.

(6) The $\delta\, S^{34}_{\Sigma s}$ value of stage 4 in Changpo is basically identical with that of stage 6 in the middle belt, which indicates that they are from the same source. The mineral association at stage 6, however, is different from that of stage 4. Stage 4 shows boiling characteristics while stage 6 fails to show boiling. We therefore consider that they are probably the products of different periods of magma-hydrothermal activities.

## 5. Carbon and Oxygen Isotope Composition of Calcite

The $\delta\, C^{13}$ and $\delta\, O^{18}$ values of calcite in orebodies from the west belt show negative correlation in the Longtoushan mining area and vertical straight line correlation in the Changpo mining area.

Assuming $\delta C^{13}_{\Sigma CO_2} = \delta C^{13}_{H_2CO_3}$, $\delta O^{18}_{\Sigma CO_2} = \delta O^{18}_{H_2CO_3}$, we can obtain the following equations:

$$\delta C^{13} \text{ calcite} = \delta C^{13}_{\Sigma CO_{2M}} + \Delta^{\text{calcite}}_{CO_2} \text{ ........ (2)}$$

$$\delta O^{18} \text{ calcite} = \delta O^{18}_{\Sigma CO_2} + \Delta^{\text{calcite}}_{CO_2} \text{ ........ (3)}$$

$\Delta^{i}_{\Sigma}CO_2$ is the relative isotope enrichment factor. $\Sigma\ CO_{2M}$ is from a single source, $\Sigma\ CO_{2M} = CO_2(1)$; when $\Sigma\ CO_{2M}$ is from two kinds of source, $\Sigma\ CO_{2M} = CO_2(1) + CO_2(2)$. The correlation modes of the $\delta\ C^{13}$ and $\delta\ O^{18}$ of various kinds of calcite can be built from the above equations.

Comparing the correlation model of the $\delta\ C^{13}$ and $\delta\ O^{18}$ of calcite in Fig. 5 with the data in Fig. 4, we may draw the conclusion that the carbon and oxygen of calcite in the Longtoushan mining area (model C, Fig. 5) are both a mixture of two kinds of source. The carbon in the Changpo mining area is a mixture of two kinds of source but the oxygen is from a single source.

## 6. Studies of Oxygen Isotopes

The studies of the oxygen isotopes of granite and ore-forming solutions at each stage in the mining area show the following characteristics (Fig. 6):

(1) The oxygen isotope composition of water in granite of the middle belt is between +10.9‰ and +11‰, a little higher than that of normal magmatic water (+5.5 — +10‰), implying that the granite was contaminated by a rock with high oxygen isotope values.

(2) The oxygen isotope compositions of ore-forming solutions at stage 1 and stage 3 of Changpo are between +8.6‰ and +9.8‰, basically coincident with normal magmatic water, suggesting that they are from magmatic sources.

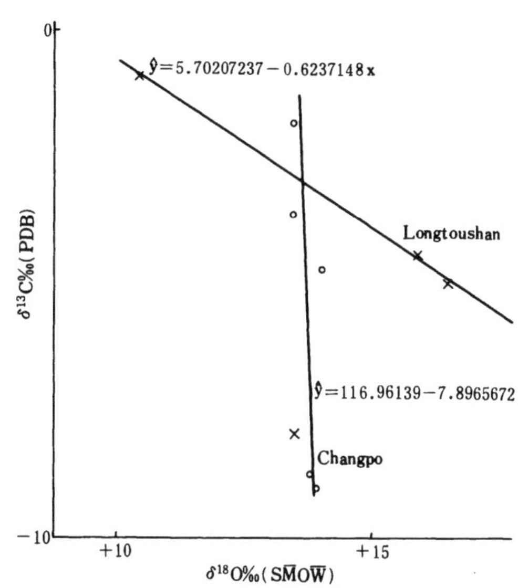


**Fig. 4.** $\delta\ C^{13}$‰ versus $\delta\ O^{18}$‰ diagram of calcite from the Dachang ore field

| Model | | | T°C | $\delta^{13}C-\delta^{18}O$ | |
|---|---|---|---|---|---|
| | $\delta^{13}C_1=\delta^{13}C_1$ | $\delta^{18}O_1=\delta^{18}O_1$ | Fall | Postitve correlation | $\delta^{13}C$ (‰) / $\delta^{18}O$ (‰) |
| $B_1$ | $\delta^{13}C_1+\delta^{13}C_2$ | $\delta^{18}O_1=\delta^{18}O_1$ | Slight variation | 90°C | $\delta^{13}C$ (‰) / $\delta^{18}O$ (‰) |
| $B_2$ | $\delta^{13}C_1+\delta^{13}C_2$ | $\delta^{18}O_1=\delta^{18}O_1$ | Fall | Positive correlation | $\delta^{13}C$ (‰) / $\delta^{18}O$ (‰) |
| C | $\delta^{13}C_1+\delta^{13}C_2$ | $\delta^{18}O_1=\delta^{18}C$<br>$\delta^{18}O_2<\delta^{18}O_1$ | Slight variation | Negative correlation | $\delta^{13}C$ (‰) / $\delta^{18}O$ (‰) |
| O | $\delta^{13}C_1+\delta^{13}C_2$ | $\delta^{18}O_1+\delta^{18}O_2$<br>$\delta^{18}O_1>\delta^{18}O_2$ | Fall | Positive correlation | $\delta^{13}C$ (‰) / $\delta^{18}O$ (‰) |
| $E_1$ | $\delta^{13}C_1=\delta^{18}O_2$ | $\delta^{18}O_1+\delta^{18}O_2$<br>$\delta^{18}O_2<\delta^{18}O_1$ or<br>$\delta^{18}O_2>\delta^{18}O_1$ | Fall | Negative correlation | $\delta^{13}C$ (‰) / $\delta^{18}O$ (‰) |
| $E_2$ | $\delta^{13}C_1=\delta^{13}C_1$ | $\delta^{18}O_1+\delta^{18}O_2$<br>$\delta^{18}O_2<\delta^{18}O_1$ or<br>$\delta^{18}O_2>\delta^{18}O_1$ | Slight variation | 0°C | $\delta^{13}C$ (‰) / $\delta^{18}O$ (‰) |

$\delta^{18}O$ or $\delta^{13}C_1$ = Original $\delta^{18}O_2$ or $\delta^{13}C_2$ = Newly added

**Fig. 5.** Correlation pattern of carbon and oxygen isotopes in calcite from the Dachang ore field

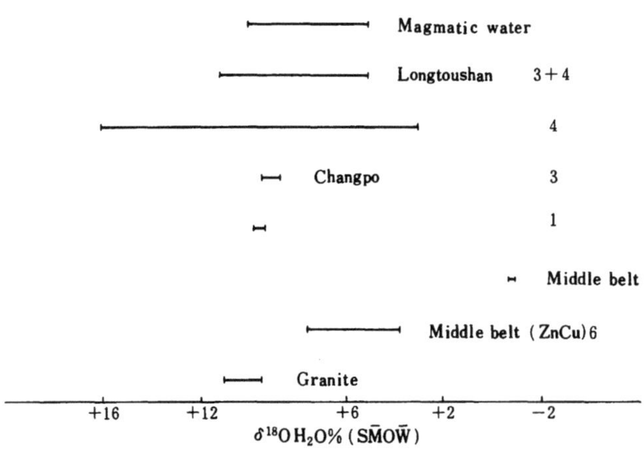


**Fig. 6.** Oxygen isotope composition of coexistent water of cassiterite and quartz from the Dachang ore field

(3) The oxygen isotope compositions of ore-forming solutions at stage 6 in the zinc-copper orebodies of the middle belt ($\delta O^{18}_{H_2O}$ = +3.8% – +7.5%) and at stage 4 of the antimony orebodies of the middle belt ($\delta O^{18}_{H_2O}$ = +1.1 – +1.8%) are different from those of normal magmatic water. They are displaced to values lower than magmatic water. This indicates that these ore-forming solutions were possibly a mixture of meteoric and magmatic water.

(4) The oxygen isotope compositions of ore-forming solutions at stage 4 in Changpo ($\delta O^{18}_{H_2O}$ = +3.0% – +16.2%) and at stages 3 + 4 in Longtoushan ($\delta O^{18}_{H_2O}$ = +5.1% – +11.2%) are also different from those of normal magmatic water. These values are displaced to two directions related to the oxygen isotope composition of magmatic water. This shows that they are possibly mixtures of magmatic with meteoric water or another water which has higher $\delta O^{18}$ values.

## 7. Studies of Lead Isotopes

The lead isotopes of lead-bearing minerals from the middle and the west belts are studied. The results are given in Fig. 7:

(1) The $Pb^{206}/Pb^{204}$ ratios of zinc-copper deposits are 18.147–18.510, their $Pb^{207}/Pb^{204}$ ratios are 15.443–15.730, and their $Pb^{208}/Pb^{204}$ ratios are 38.149–38.830, showing that they are rather constant and vary only in a narrow range. The $Pb^{206}/Pb^{204}$ ratios of tin deposits are 17.491–18.730, their $Pb^{207}/Pb^{204}$ ratios are 14.807–15.810, and the $Pb^{208}/Pb^{204}$ ratios are 36.736–39.300. They are not constant and vary in a wide range.

(2) The lead isotope composition in zinc-copper deposits is relatively constant and shows no linear distribution. Their lead isotope model age values (∅ value of Stacey, 1975) are from negative to 400 Ma. Most of the age values are concentrated in the range of 100 to 300 Ma; $U_1$ values are from 9.57 to 10.09. The (Th/U) ratios are from 3.78 to 3.94. If the source of this lead was from Devonian strata or was the result of mixing between granite and Devonian strata, it would show a wider age range and the lead isotopes should show a linear arrangement. However, this is not the case in the Zn-Cu deposits. We therefore consider that the lead of the zinc-copper deposits was from a granite source. Furthermore, the model age of lead isotopes in the deposit is different from the Rb-Sr age of the granite, indicating that the lead possibly experienced a poly-stage evolution.

(3) The lead isotope ratios of the tin deposits show a linear arrangement (Fig. 8). This suggests that they are mixtures of at least two kinds of sources. Therefore we require to design several kinds of models:

a. mixture of two groups of lead having identical U values;
b. mixture of two groups of lead having different U values;
c. mixture of three groups of lead.

After careful consideration, we chose the model of mixing of two lead groups having the same U value. If the age of tin-forming magma is equal to that of the middle belt granite, and also equal to the age value (80 ± 20 Ma) on the upper intersecting point of the discordant line with the Stacey (1975) growth curve, we can obtain the age value 2,400 ± 120 Ma of the lower intersecting point. The age value of the lower

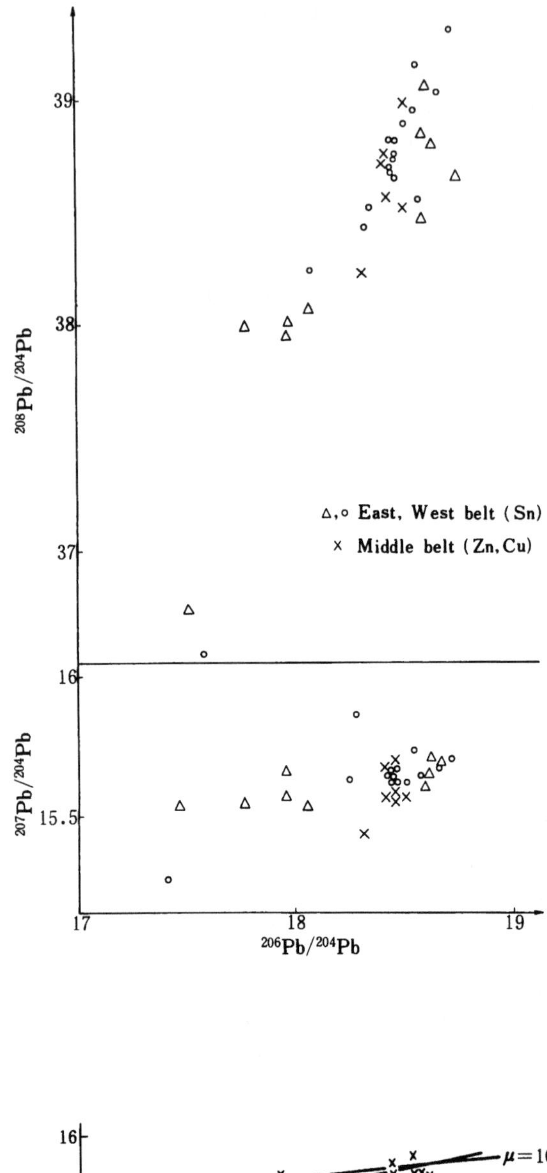


**Fig. 7.** Diagram showing $Pb^{208}/Pb^{204}$ and $Pb^{207}/Pb^{204}$ versus $PB^{206}/Pb^{204}$ of galena and sulphosalts from the Dachang ore field

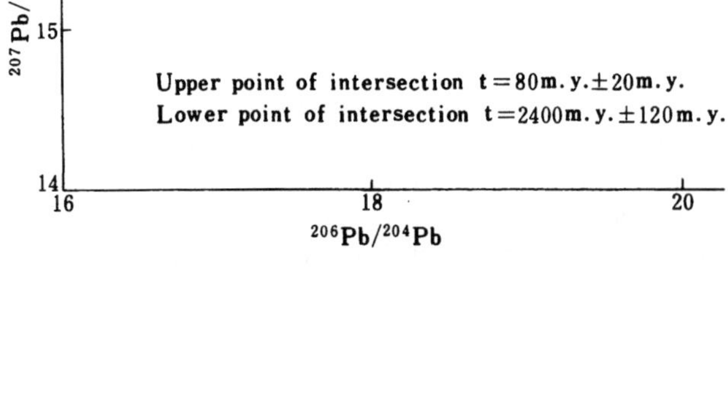


**Fig. 8.** Diagram showing $Pb^{207}/Pb^{204}$ versus $Pb^{206}/Pb^{204}$ of galena and sulphosalts from the Dachang ore field. Galena: a = 12.867, b = 0.154076; Sulphosalts: a = 12.867, b = 0.154076

intersecting point is reasonably identical with the value of the strata of the Shibao Group. We also point out that the lead, and probably tin, represent a mixture from Yenshanian granite and older strata. This is supported by the fact that there is a tin-bearing layer at the unconformity between the Shibao and Banxi groups.

From these isotope compositions, we believe that the granite in the middle belt is of S-type. The metallization of zinc-copper was possibly connected with the granite in the middle belt, and tin mineralization was connected with deep magmatic rock in Changpo. The material sources are different at different belts and different stages. However, the material from a magmatic source was usually predominant in the early stage, and some material from strata were added at the later stage.

*Acknowledgements.* We thanks to Senior Engineer Huang Youde, Engineers Cai Hongyuan, Yang You, Cheng Xianyao, Fu Jinbao, Liu Yuexing, Huang Baozhong, Li Zhifeeng, Lai Lairen, Zhang Wenhua, Yan Yunxiu and ZhuGuitian, who have done a lot of research work to help us.

## References

Chappell, B.W., 1974. Two constrasting granite types: *Pacific Geol.* 8, 173–174.

Crerar, D.A., 1978. Solubility of the buffer assemblage pyrite+pyrrhotite+magnetite in NaCl solutions from 200 to 350°C. *Geochim. et cosmochim. Acta,* V. 42, 1427–1437.

Holland, H.D., 1979. Some applications of thermochemical data to problems of ore deposits: I stability relations among the oxides, sulfides sulfates and carbonate of ore and gangue minerals. *Econ. Geol.,* 54, 184–233.

Ohmoto, H., 1972. Systematics of sulfur and carbon isotopes in hydrothermal ore deposits: *Econ. Geol.,* 67, 551–579.

Robie, R.A., 1978. Thermodynamic properties of minerals and related substances at 298"k and 1 bar ($10^5$ pascals) pressure and at higher temperatures. *U.S. Geol. Survey Bull.,* 1452.

Rye, R.O., and Ohmoto, H., 1974. Sulfur and carbon isotopes and Ore genesis: A review. *Econ. Geol.,* 69, 826–842.

Scott, S.D., 1971. Sphalerite geothermometry and geobarometry. *Econ. Geol.,* 66, 653–669.

Stacey, J.S., 1975. Approximation of terrestrial lead isotope evolution by a two-stage model. *Earth Planet. sci. lett.* 26, 207–221.

Taylor, H.P., 1979. Oxygen and hydrogen isotope relationships in hydrothermal mineral deposits. In: Barnes, H.J. (ed) *Geochemistry of hydrothermal ore deposits,* 2nd ed. New York, John Wiley, 236–277.

Wang Zhonggang, Zhao Zhenhua and Zhao Huilan, 1980. REE Distribution Patterns of Granites in South China. *Geochimica,* No. 1, 1–12 (in Chinese).

# 6.4.9 Discrimination of Sn (W) Metal-Bearing Potential of the Yenshanian Granites in Guangxi with a Discussion of Their Evolution

WANG WEIYU and WEI WENZHUO[1]

**Abstract**

From a comparison of analyses of a number of large samples of Yenshanian granites in Guangxi, some combinations of elements and characters of minerals have been recognized and they may be regarded as discrimination criteria between ore-bearing and barren rock bodies. It is believed through further study that in the evolutionary history of a rock body, the criteria of distinction are almost without exception more obvious in late stage than in early stage and that mineralization is related to late stage development of an intrusive rock body. The more intensely the rock bodies evolved, the better criteria they would show in their late stage, and the more intensive the metallization would be. Therefore, the determination of discriminating criteria and study of the evolutionary degree of granites are of great importance and will provide a useful approach to the exploration for Sn and W deposits. The reason why discriminating criteria are good in the late stage of evolution is also discussed.

In Guangxi, the Yenshanian granites which are distributed in ten areas are of two groups: the ore-bearing granites, to which most Sn (W) deposits in this region are closely related, and ore-free ones. Samples, each weighing over 40 kg, have been collected from about half of the granite bodies in nine areas. To make a comparison between the known ore-bearing and ore-free granites for the purpose of establishing their discrimination criteria, each sample has been analysed for 13 items, namely: the relative contents of rock-forming minerals, petrochemical composition, trace element contents, composition of rock-forming minerals, Sn (W) contents of the micas, proportions of magnetite and ilmenite, semiquantitative analysis of other accessory minerals, Sn (W) contents in bulk samples of heavy minerals, isotopic composition, crystallographic characteristics of zircon, ordering degree of the feldspars, prolongation ratio of plagioclase and the rock-forming temperature. The comparisons indicate that although most methods are not so useful, the following features may be considered as the effective discrimination criteria.

1. Having sifted 27 variables related to petrochemical composition, trace-element geochemistry, Sn (W) contents in micas and others, the correspondence analysis and vector length methods give very similar results, with $TiO_2$, MgO, CaO, FeO, $Fe_2O_3$, Sr and Ba closely associated with the ore-free granite bodies while Sn, W, Ta, Nb, F as well as Sn and W in the micas are associated with the ore-bearing bodies.

[1] Guangxi Institute of Nonferrous Metal and Geology, CNNC, Nanning

Figure 1 is a diagram of the major factor loading of two correspondence analyses. The solid line and circular points represent all the 27 variables while the dotted line and triangles represent the 16 variables of antithetic petrochemical composition. Similar results are obtained from these two analyses. These two combination groups are separated clearly, with a transitional group in between which contains variables of no discriminating effect.

Figure 2 is the total integral curve of the vector length of continuous and dual variables sifted from 27 variables of the ore-bearing and ore-free rock bodies by the vector length method. It distinctly indicates the difference between ore-bearing and ore-free rock bodies. These two methods of calculation adopt the total integral respectively equal to 255 and 430 for the critical values which can effectively discriminate the unknown rock bodies.

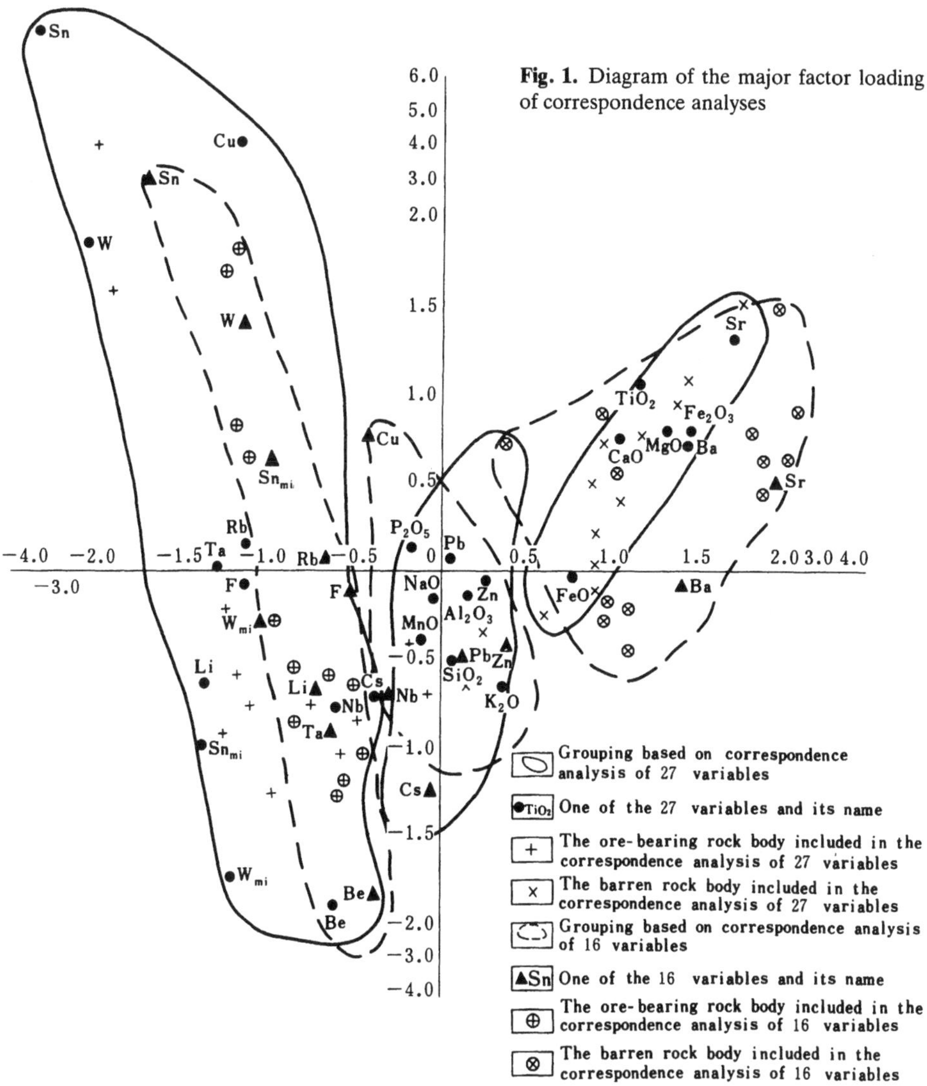


**Fig. 1.** Diagram of the major factor loading of correspondence analyses

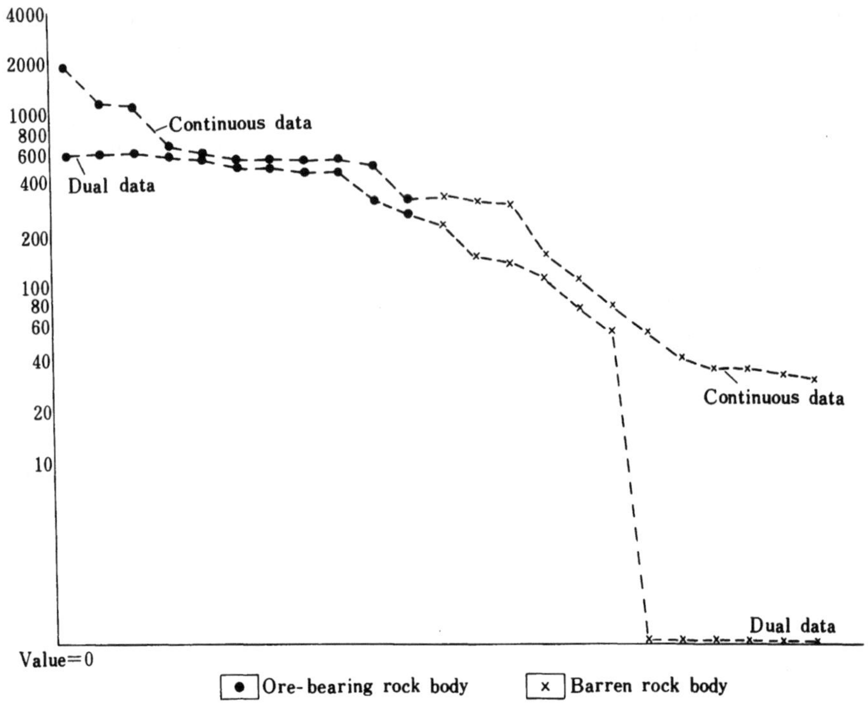


**Fig. 2.** Diagram showing relationship between total integral of the length of two vectors and ore-bearing potential of rock bodies

2. Six ratios recommended previously by geologists, viz. (K+Na)/(Fe+Ti+R"), $10^3$XLi/Mg, Rb/Sr, $TiO_2$/Ta, K/Rb, and (Sr+Ba)/(Li+Rb+Cs), also show very good discriminating effects. The boundary values for the ore-bearing rock bodies are >3.3, >98, >36, <114, <63, and <0.15. In Figure 3 the distinction between the ore-bearing rock bodies and the ore-free ones is very clear. The discriminating effects for previously unknown rock bodies are also very good.

3. The mica varieties and their Sn contents calculated from the chemical composition are also effective criteria. Figure 4 shows the species of the mica group which is commonly seen in granites. Siderophyllite and nearby lepidomelane are the micas characteristic of ore-bearing granites, while magnesium-biotite, biotite and part of nearby lepidomelane are the micas characteristic of ore-free granites. The former group of micas is high in Sn content as shown in Table 1, while the latter group is low in Sn.

4. The contents of magnetite and ilmenite may be used only to perform a one-way discrimination. If the contents of magnetite and ilmenite are high, the granites are surely ore-free. However, their low contents do not necessarily mean the granites are ore-bearing. Table 2 shows that the magnetite and ilmenite contents are quite low in all the ore-bearing granites and although they are generally high in most ore-free granites they could be very low, even approaching zero in some ore-free granites.

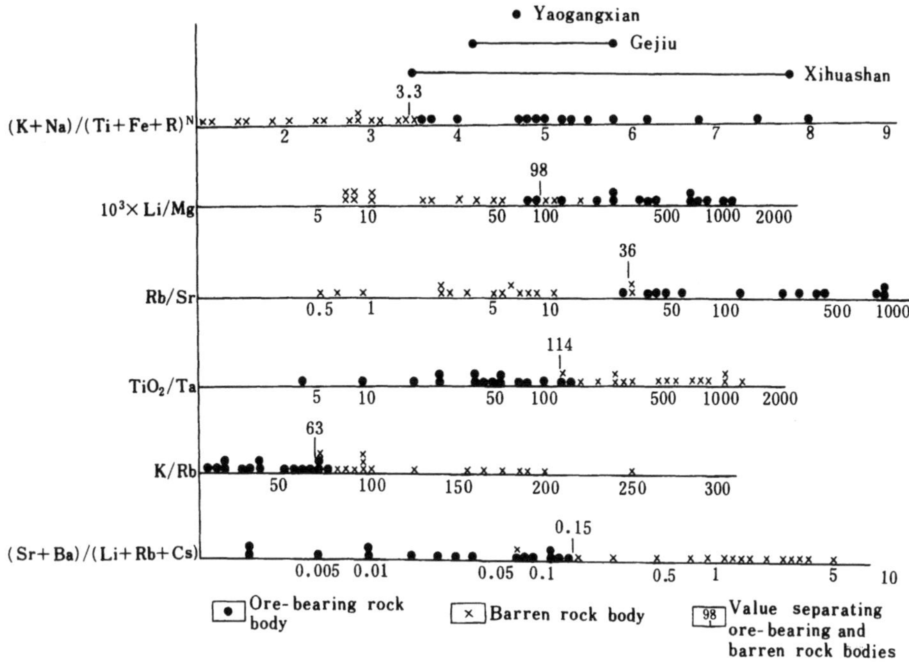


**Fig. 3.** Diagram showing the relation between ratios of six elements and ore-bearing potential of rock bodies

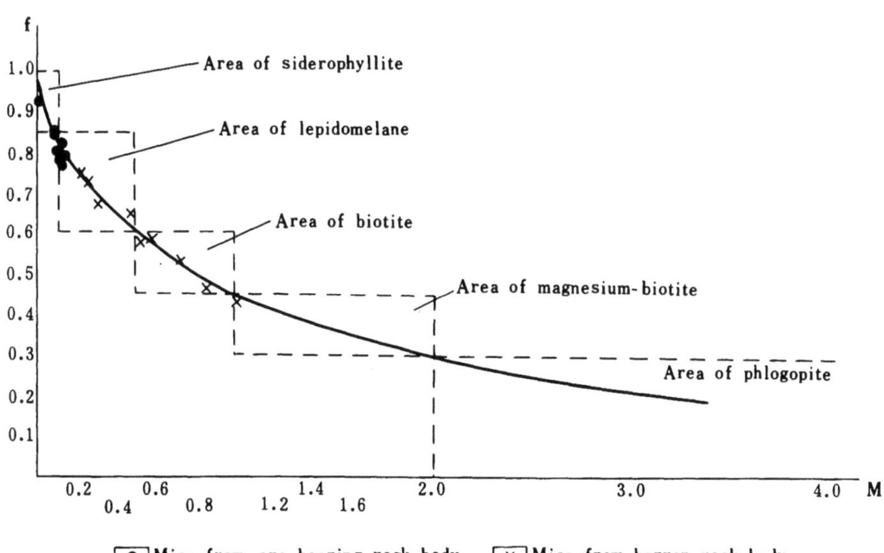


**Fig. 4.** Diagram showing the relation between varieties of mica and ore-bearing potential
$f = Mg/(Mg + Fe)$ $M$ = six co-ordinated $Al^{VI}$

**Table 1.** Sn, W contents of the mica from the ore-bearing and ore-free rock bodies

| Kind | Number | Sn content of the mica (ppm) | W content of the mica (ppm) |
|---|---|---|---|
| Ore-bearing rock bodies | 1 | 183 | 78.2 |
| | 2 | 114 | 14.8 |
| | 3 | 403 | 72.4 |
| | 4 | 381 | 324.2 |
| | 5 | 48 | 110.6 |
| | 6 | 600 | 78.3 |
| | 7 | 379 | 105.9 |
| | 8 | 645 | 76.2 |
| | 9 | 519 | 74.3 |
| | 10 | 115 | 91.8 |
| Average | | 339 | 102.7 |
| Ore-free rock bodies | 1 | 52 | 19.8 |
| | 2 | 101 | 41.6 |
| | 3 | 26 | 7.9 |
| | 4 | 60 | 5.0 |
| | 5 | 11 | 0 |
| | 6 | 55 | 0 |
| | 7 | 249 | 11.8 |
| | 8 | 21 | 2.0 |
| | 9 | 52 | 12.0 |
| Average | | 70 | 11.1 |

**Table 2.** Contents of magnetite and ilmenite in the ore-bearing and ore-free rock bodies

| Kind | Serial number | Content of magnetite (g/T) | Content of ilmenite (g/T) |
|---|---|---|---|
| Ore-bearing rock bodies | 1 | trace | trace |
| | 2 | 2.4 | 0.48 |
| | 3 | 2.0 | trace |
| | 4 | 41.3 | 0 |
| | 5 | 0 | 0 |
| | 6 | trace | trace |
| | 7 | 0 | 0 |
| | 8 | trace | 0 |
| | 9 | trace | trace |
| | 10 | 76.0 | trace |
| | 11 | 52.4 | 0.14 |
| Average | | 15.8 | 0.06 |
| Ore-free rock bodies | 1 | 0.3 | 102.3 |
| | 2 | 219.5 | 12.8 |
| | 3 | 835.0 | 85.5 |
| | 4 | 3.4 | 1.3 |
| | 5 | 4911.0 | 25.2 |
| | 6 | 4651.2 | 456.0 |
| | 7 | 2390.0 | 0 |
| | 8 | 3762.2 | 528.0 |
| | 9 | 81.0 | 123.0 |
| | 10 | 5734.0 | 649.0 |
| | 11 | 3084.0 | 218.0 |
| | 12 | 2810.0 | 413.0 |
| | 13 | 1647.0 | 56.1 |
| | 14 | trace | 0 |
| Average | | 1931.8 | 190.2 |

Using these criteria, the granites studied have been re-discriminated. The lowest percentage of correct discriminations is 82.8%, and the highest reaches 96.6%. The integrated discriminations show better results.

It is considered that the cause of effective discrimination of the Sn (W) metal-bearing potential is related to the evolution of the granite bodies.

The regional evolution of granites takes place as a unit consisting of an evolutionary series within a large tectonic cycle. However, in a particular area the evolutionary series includes a number of multiple-staged intrusions of composite bodies, or a number of granitic bodies in close temporal and derivative relation with each other.

The Yenshanian granites of Guposhan, Huashan, Dachang, Limu, Kunlunguan, Luchuan etc. in Gangxi have been studied in detail. The early, (middle) and late stages of these seven granitic areas reveal regular evolutionary trends by comparing the discriminating criteria. Table 3 shows that of the nine discriminating criteria, the contents of Ti, Fe, Ca, Mg, Sr and Ba are higher in early intrusive bodies while the contents of Sn, W, Ta, Nb, Rb, Cs, K and Na are higher in late intrusive bodies. Almost all the intrusive bodies agree with this trend except the granite-porphyry of the third stage in Dachang, which shows some abnormality due to carbonatization of the feldspar phenocrysts. The contents of Ca, Mg, and their positively correlated Sr and Ba have increased correspondingly. The contents of K, Na and their positively correlated Rb have decreased.

From early to late stage, all the varieties of micas have evolved from high magnesium to low magnesium. Table 4 shows the fall of the variation of ratio Mg/(Mg+Fe) and the rise of six coordinated $Al^{VI}$ while the Sn content of mica is higher at the late than at the early stage.

Prof. Xu Keqin has proposed that the granites in South China show a general trend of evolution from intermediate-acid to acid with high alkali-content. This general trend can only be better shown by statistics of data combinations or the trace element ratios and the properties of some minerals, for the intrusive bodies emplaced at different stages in the same rock body show very little difference, though the evolution of different stages is also consistent with this regularity. It is easier to explain the reason for the evolutionary trend of the individual rock bodies than to explain the evolution of granites on a regional extent, because the evolution of a single rock body is more directly related to the crystallization order of the magma. During the magmatic crystallization Ti, Fe, Ca, Mg, etc. were all concentrated in ilmenite, hornblende, and biotite formed at the earlier stage, while K, Na etc. remained in the residual magma and crystallized at the later stage. Therefore, the ratio (K+Na)/(Fe+Ti+R") can reflect the degree of evolution and may be considered as an effective discrimimating criterion. The trace elements related to these two groups were differentiated correspondingly.

$Rb^+$ is similar to $K^+$ in geochemistry, both being enriched in the late stage of magmatic crystallization. But the $Rb^+$ content in the late stage is higher than that of $K^+$, for the ionic radius of $Rb^+$ is slightly larger. In the pegmatitic stage the amazonite, a potash feldspar rich in Rb, might appear. Therefore, the K/Rb ratio can reflect the degree of magmatic evolution. The behavior of $Sr^{++}$ is very similar to $Ca^{++}$, and they were enriched in the early stage of magmatic crystallization. The Rb/Sr ratio compares two elements representing different directions of evolution, therefore it is a very sensitive and effective discriminating criterion. Some other element ratios of different directions of evolution are also effective criteria.

**Table 3.** The variations of nine parameters in different stages of individual rock bodies

| Rock body | Eastern body of Guposhan | | | Western body of Guposhan | | Huashan rock body | | Dachang rock bodies | | | Limu rock bodies | | Kunlunguan rock body | | Luchuan rock body | |
|---|---|---|---|---|---|---|---|---|---|---|---|---|---|---|---|---|
| Parameter | Early | Mid | Late | Early | Late | Early | Late | Early | Mid | Late | Early | Late | Early | Late | Early | Late |
| (Na+K)/(Fe+Ti+R") | 1.65 | 2.83 | 4.0 | 2.14 | 3.9 | 3.6 | 5.2 | 3.15 | 4.2 | 0.68 | 7.6 | 15.4 | 2.3 | 3.96 | 1.5 | 5.2 |
| X $10^3$ Li/Mg | 10 | 36 | 79 | 31 | 656 | 45 | 74 | 57 | 446 | 220 | 898 | 789 | 10 | 16 | 8 | 23 |
| Rb/Sr | 2.5 | 5.5 | 270 | 2.8 | 288 | 8 | 155 | 12 | 62 | 6.2 | 804 | 1145 | 0.9 | 2.7 | 0.6 | 8.0 |
| $TiO_2$/Ta | 316 | 733 | 113 | 660 | 19 | 275 | 129 | 300 | 37 | 22 | 63 | 4 | 58 | 45 | 1075 | 518 |
| K/Rb | 118 | 76 | 68 | 79 | 43 | 64 | 53 | 42 | 32 | 25 | 16 | 12 | 174 | 136 | 178 | 188 |
| (Sr+Ba)/(Li+Rb+Cs) | 1.20 | 0.83 | 0.02 | 1.36 | 0.068 | 0.58 | 0.03 | 0.65 | 0.09 | 0.14 | 0.017 | 0.002 | 2.27 | 1.11 | 3.18 | 5.74 |
| Total length of vector for the continuous data | | 75.7 | 485.8 | 89.2 | 724.4 | 123.4 | 604.2 | 119.5 | 326.0 | 487.4 | 747.1 | 1924.4 | 58.2 | 93.0 | 35 | 113.2 |
| Total length of vector for the dual data | 0 | 163.3 | 327.2 | 241.2 | 552.6 | 280.7 | 552.2 | 282.1 | 555.2 | 501.8 | 609.2 | 620.9 | 75.4 | 234.4 | 0 | 177.6 |
| Value of major factor in correspondence analysis ($F_1$) | | 0.58 | 0.17 | 0.82 | –0.6 | 0.39 | –1.17 | 0.35 | –0.55 | –0.66 | –1.43 | –2.6 | 0.88 | 0.22 | 1.41 | 0.61 |

**Table 4.** Variation of Mg/(Mg+Fe), 6-fold-coordinated $Al^{VI}$ and Sn contents of muscovite in different stages of rock bodies

| Rock body | Eastern body of Guposhan | | | Western body of Guposhan | | Huashan rock body | | Dachang rock bodies | | Limu rock bodies | | Kunlunguan rock body | | Luchuan rock body | |
|---|---|---|---|---|---|---|---|---|---|---|---|---|---|---|---|
| Parameter | Early | Mid | Late | Early | Late | Early | Late | Early | Mid | Early | Late | Early | Late | Early | Late |
| Mg/(Mg+Fe) | 0.33 | 0.25 | 0.08 | 0.31 | 0.08 | 0.28 | 0.14 | 0.69 | 0.37 | 0.13 | 0.09 | 1.0 | | 0.87 | 0.53 |
| $Al^{VI}$ | 0 | 0.059 | 0.168 | 0 | 0.774 | 0 | 0.002 | 0.367 | 0.65 | 3.029 | 2.16 | 0.46 | | 0 | 0.02 |
| Sn (ppm) | | 52 | 114 | 72 | 116 | 175 | 381 | 89 | 115 | 541 | 579 | 21 | 52 | 249 | 679 |

Both the correspondence analysis and the vector length method are methods of mathematical geology concerning the combination of two elements of contrary evolution, leading to similar results.

It is considered that the geochemical similarity of $Sn^{4+}$ and $Fe^{2+}$ is responsible for the existence of Sn in micas. The micas low in magnesium and high in iron were formed at the late stage of granite evolution, creating a favourable condition for the incorporation of Sn in the mica. Therefore, the content of Sn in the mica is an effective discriminating criterion.

From the discussion above, it may be concluded that in the evolutionary process of whichever rock body, the above-mentioned criteria are more obvious in the late stage than in the early stage. Most intrusive bodies of the early stage fail to reach the ore-bearing criteria. Generally speaking, the more intensely the rock bodies evolved, the better criteria they would reach in the late stage and the stronger the metallization they would encounter; while the simple and poorly evolved rock bodies are unfavourable for mineralization. Therefore, determination of discriminating criteria and study of the evolutionary degree of granites are of great importance and will provide a useful method and approach to the exploration for Sn and W deposits.

# 6.4.10 Mineralization Process and Genesis of Tin Deposits in North Guangxi, China

PENG DALIANG, GUO YURU, and DENG DEGUI[1]

**Abstract**

The tin deposits in north Guangxi probably went through three stages of mineralization: the first was related to Proterozoic Sibuian basic magmatic rocks, the second to Proterozoic Xuefengian granitic intrusions, and the third to Caledonian metamorphism and transformation of the pre-existing tin deposits. The mineralization elements were probably supplied mainly by the mantle-derived basic and ultrabasic rocks, and later during crustal remelting. Therefore, the tin deposits in north Guangxi are multi-staged and of different sources.

The north Guangxi tin deposits can be divided into: 1) deposits repeatedly transformed by magmatic hydrothermal activities, such as the Yidong-Wudi tin deposits; 2) deposits transformed by metamorphic hydrothermal fluids, such as the Jiomao tin deposit; 3) metasedimentary deposits, such as the Xiaodongjiang tin deposit; and 4) greisen Sn-Cu deposits, such as the Wuyongling deposit.

## I. Regional Tectonic Setting

The Baotan and Jiomao tin deposits in north Guangxi are located in the southwestern extreme of the Proterozoic Jiangnan Island-arc (Fig. 1).

Strata exposed in this region are mainly of the Sibu Group, Banxi Group and Sinian System of Proterozoic Era, with Devonian and Carboniferous strata lying locally unconformably above them. Tin deposits occur dominantly in basic-ultrabasic ophiolitic complexes and island-arc flysch sandstone and shale of the Sibu Group. Tin mineralization has not been found in the unconformably overlying Banxi Group and Sinian and Devonian Systems. The Xuefengian granites, which provide an isotopic age of 760–836 Ma, show only locally weak greisen-type Sn-Cu mineralization.

Based on the characteristics of sedimentary formations, ophiolite suites, volcanic rocks and their petrochemistry, it is believed that the Sibu Group might have been developed in an active island-arc environment.

Starting with the Xuefengian event, the activity of the island-arc gradually diminished and a set of miogeoclinal flysch sandstone and shale intercalated with car-

[1] Institute of Geological Sciences, Bureau of Geology and Mineral Resources of Zhuang Autonomous Region of Guangxi, Nanning, Guangxi

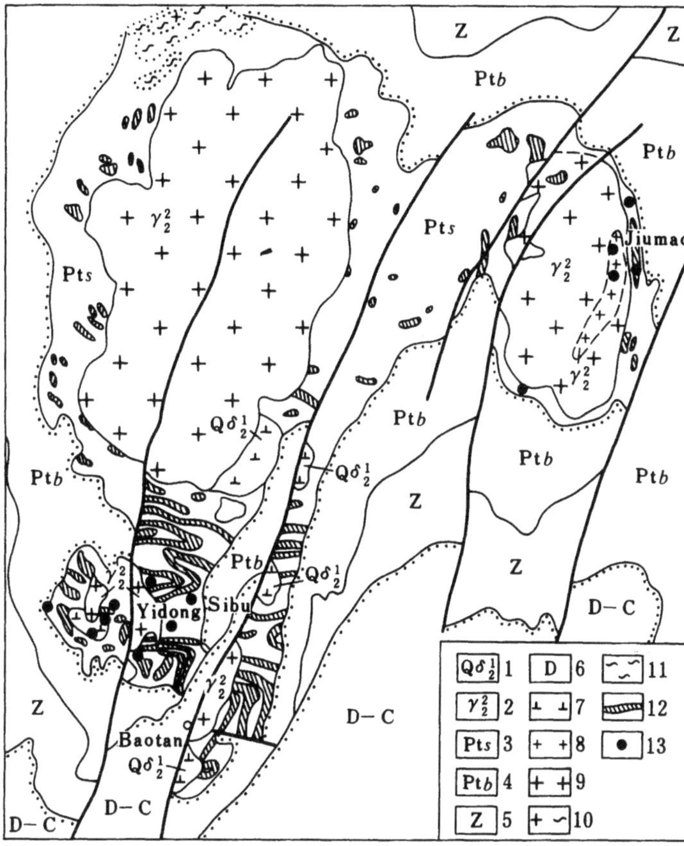


**Fig. 1.** Geological map of tin deposits in north Guangxi

bonate rocks was deposited. Afterwards, this area was subjected to uplift for quite a long time. Paleozoic and Mesozoic rocks and magmatism are poorly developed in this region.

## II. Formation of Tin Deposits

Tin deposits in north Guangxi were formed through a long geological process, which can be subdivided into three stages:

### 1. The Sibuian Basic Magmatic Activity and Associated Tin Mineralization

The Sibuian was not a stage for formation of important tin deposits, but it provided a large amount of ore-forming materials. Tin mineralization at this time was chiefly related to activities of the basic magmas. The main evidence supporting this conclusion is as follows:

(1) Spatial distribution of the tin deposits is obviously associated with the Sibuian basic-ultrabasic complexes. For example, the main bodies of the Jiumao tin deposit occur in the contact zone of ultrabasic rocks, and the attitude of the orebodies varies with that of the contact plane (Fig. 2).

In the Yidong-Wudi tin deposit, most orebodies occur at the intersection between the NNE-trending faults and the EW-trending basic-ultrabasic complexes. Where the faults extend into the metasediments, the orebodies thin out immediately and where the faults enter the basic-ultrabasic rocks again, the orebodies reappear (Fig. 3).

(2) Parts of the basic-ultrabasic complexes are abundant in tin (17–61 ppm). Cheng Xianyao from the Institute of Geology for Mineral Resources, CNNC, has told the authors that there is dispersed tin in basic-ultrabasic complexes, occurring mainly in mafic minerals such as olivine and amphibole, which contain 200–400 ppm Sn, while metamorphic rocks of the Sibu Group and the Xuefengian granites contain only 9.4 ppm and 11.8 ppm tin respectively. So, the orebodies occurring in basic-ultrabasic complexes of the Sibuian are not only larger in size, but also 1–2 orders of magnitude higher in tin than those present in other rocks. The reserves of this kind of deposit rank first in the Zhuang Autonomous Region of Guangxi.

(3) In Yidong-Wudi and Huonggang Mining Areas, ores are in part associated with segregated Sibuian Cu-Ni ore. Some Cu-Ni nodules are surrounded by quartz

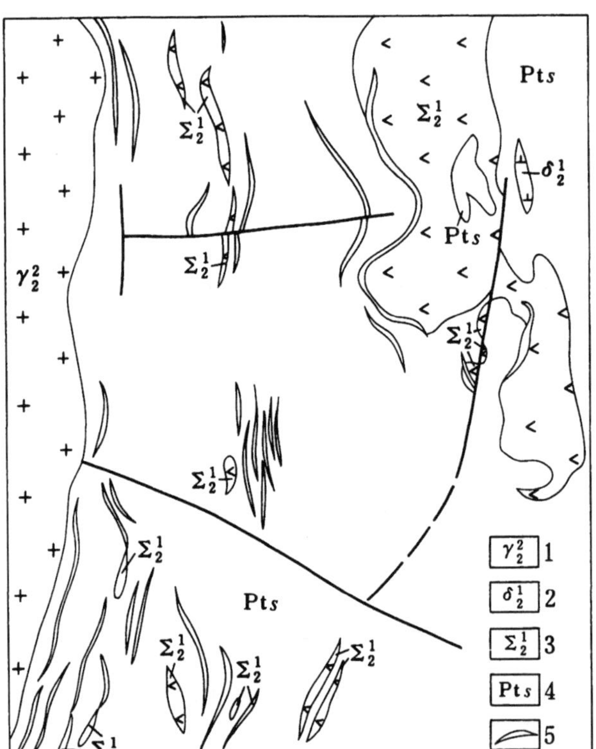


**Fig. 2.** Geological map of the Jiumao tin deposit, Rongshi County

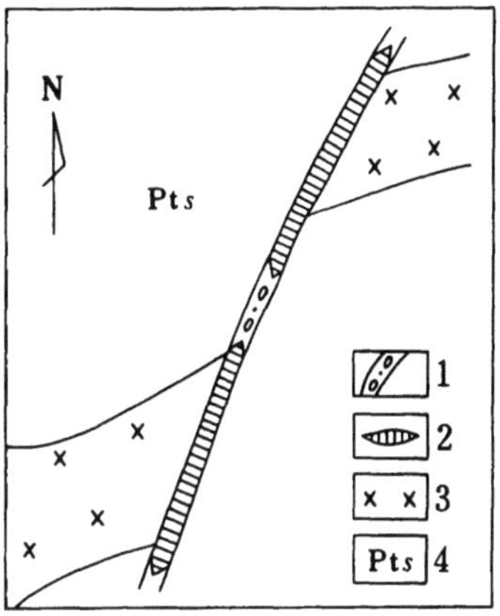


Fig. 3. Distribution of tin orebodies in NNE fault zone in Yidong-Wudi mining area. (from No. 7 Geological Party of Guangxi)

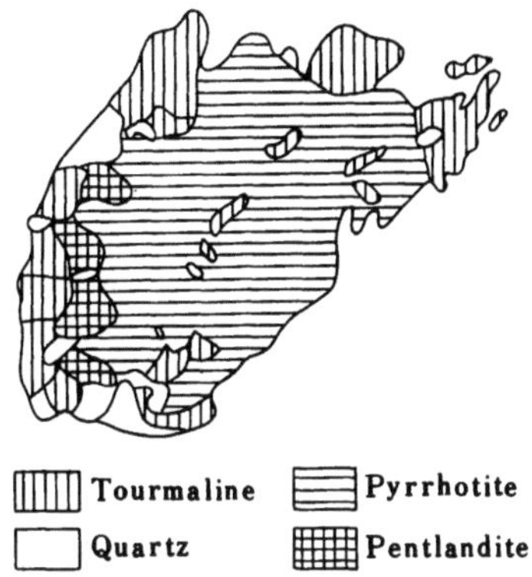


Fig. 4. Sketch showing nodular pyrrhotite and pentlandite surrounded by quartz and tourmaline in Cu-Ni ore of Huonggang

and tourmaline (Fig. 4). The authors found cassiterite inclusions in actinolite from the same area (Fig. 5).

(4) Tin mineralization found in the metamorphosed basal conglomerate of the Proterozoic Banxi Group is convincing evidence for tin mineralization during the Sibuian event. A large number of pebbles of tin-bearing tourmaline rocks and a small number of pebbles of quartz-albite porphyry are found in the basal conglomerate of Banxi Group unconformably overlying the Sibu basic-ultrabasic complexes. Chemical analyses of some of the conglomerate are given in Table 1, from which it can be seen that most gravels and their cement contain tin up to parts per thousand. However, the tourmaline rocks from basal conglomerates found in some parts of Huangma'ao re-

Table 1. Chemical analyses of basal conglomerates of the Banxi Group

| Number of sample | Location | Rock | Contents of elements (%) | |
|---|---|---|---|---|
| | | | Sn | Cr |
| 1001-1-A | Latan, Xiangfen, Rongshui County | sericite-quartz schist (cementing material of conglomerate) | 0.07 | 0.02 |
| 1001-1-B | | tourmaline rock (gravel of conglomerate) | 0.01 | 0.05 |
| 1003-2-A | Dongtian, Rongshui County | tin-bearing quartz-albite porphyry (gravel of conglomerate) | 0.12 | 0.08 |
| 1003-2-B | | biotite-chlorite-quartz schist (cementing material of conglomerate) | 0.07 | 0.13 |
| 1011-1-A | Huangma'ao, Huangjiang County | tin-bearing tourmaline rock (gravel of conglomerate) | 0.38 | 0.01 |
| 1011-1-B | | muscovite-quartz schist (cementing material of conglomerate) | 0.04 | 0.01 |

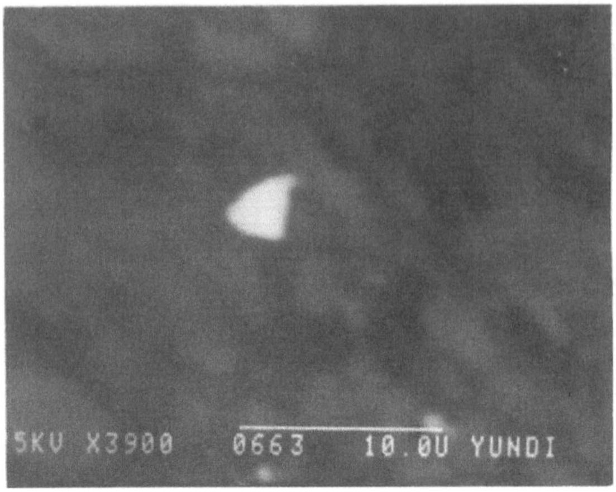


**Fig. 5.** Cassiterite inclusion in actinolite from Cu-Ni ore in the Baotan of Huonggang, Lowcheng

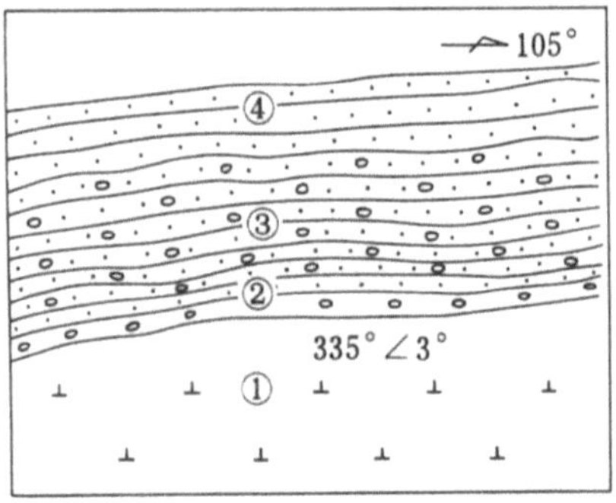


**Fig. 6.** Sedimentary contact relation between the Proterozoic Banxi Group and the Sibuian intermediate-basic rock in Hunagma'ao, Huanjiang County

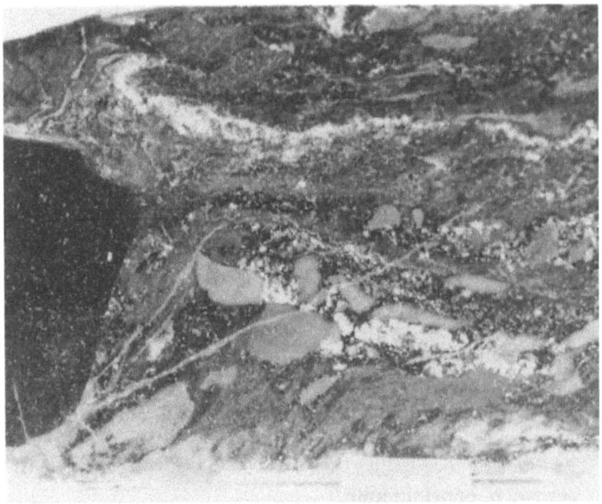

**Fig. 7.** Tourmaline-bearing gravel with tin 0.38% (black) from basal conglomerates of the Proterozoic Banxian Group

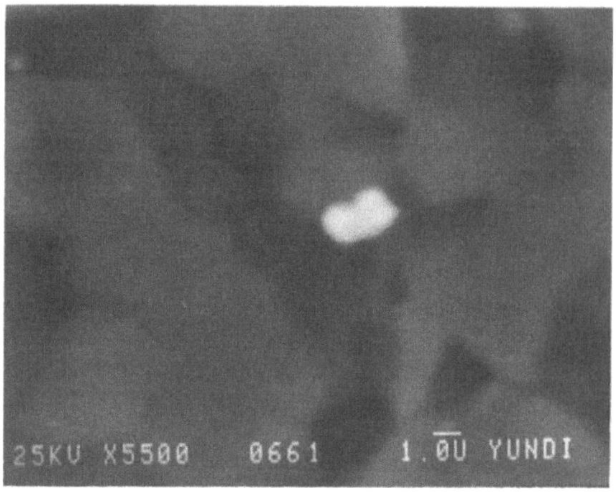


**Fig. 8.** Cassiterite (white) in tourmaline-bearing gravel from basal conglomerate of the Banxian Group in Xiangfen, Rongshui, Lowcheng

gion, Huanjiang county, contain Sn 0.38%, and Cr 0.01%, while their cementing materials contain Sn 0.04%, and Cr 0.01%. Figure 6 is a sketch showing the contact of conglomerates at the sampling site. Figure 7 shows tourmaline-bearing gravel containing Sn 0.38%. Examination of heavy minerals and the scanning electron microscope have demonstrated that there are extremely fine-grained cassiterites (Fig. 8) and chromites in these basal conglomerates.

On the basis of comparison of petrographic textures and distribution patterns of trace elements, it is suggested that the tin-bearing tourmaline-bearing pebbles of the basal conglomerates of the Banxi Group came from the Sibu metasedimentary tin-bearing tourmaline rocks. The tourmaline rocks from the two stratigraphical units are all higher in elements characteristic of basic rocks, such as Mg, Cr, Ni and Co, but lower in those characteristic of granites, such as Li, Be and Ga (Table 2), suggesting that they were derived from the Sibuian basic rocks. Moreover, the fact that the tourmaline of the tourmaline rocks from the basal conglomerates of the Banxi Group is higher in REE, and that its chondrite-normalized abundance is nearly identical with that of the Sibuian keratophyre and tonalite (Fig. 9), shows that they are from a common parent magmatism.

**Table 2.** Comparison of trace elements in various tourmaline rocks from north Guangxi

| Name of rocks | Contents of elements (ppm) | | | | | | | |
|---|---|---|---|---|---|---|---|---|
| | Cr | Ni | Co | V | Li | Be | Ga | Mg (%) |
| Tin-bearing tourmaline-bearing gravel in basal conglomerate of Banxi Group | 77.3 | 46.9 | 19.0 | 81.2 | 14.1 | 2.4 | 15.7 | 1.4 |
| Metasedimentary tin-bearing tourmaline rock of Sibu Group | 94.8 | 26.9 | 10.4 | 76.8 | 25.5 | 4.3 | 30.5 | 1.3 |
| Tin ore of sedimeno-metamorphic quartz-tourmaline type of Sibu Group | 253.4 | 27.9 | 14.3 | 103.5 | 25.5 | 3.0 | 24.0 | 0.8 |
| Quartz-tourmaline-bearing rocks in Xuefengian granite | 14.8 | 3.3 | 7.5 | 14.7 | 116.0 | 5.7 | 117.0 | 0.52 |

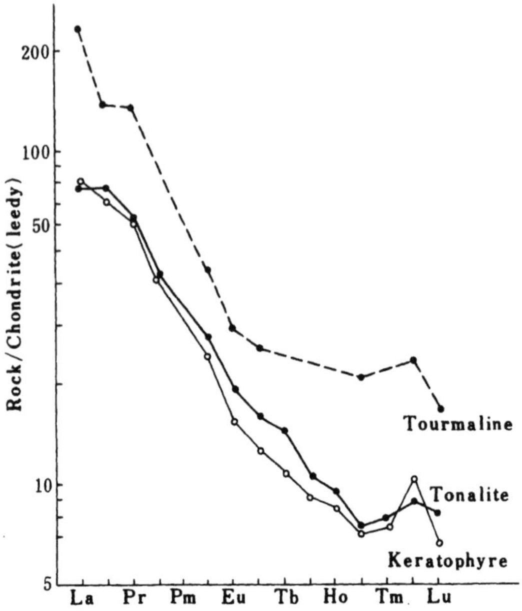


**Fig. 9.** Component pattern of rare earth elements in Sibuian keratophyre and tonalite, and tourmaline from tourmaline-bearing gravel in the basal conglomerates of the Banxi Group

## 2. Xuefengian Granitic Intrusions in Relation to the Formation of Tin Deposits

The role played by the Xuefengian granites in the formation of tin deposits in north Guangxi is two-fold. On the one hand, they leached tin out of such minerals as biotite contained in the rock to form small greisen-type tin deposits at or near the endocontact zone through pneumatogenic processes. On the other hand, more importantly postmagmatic mineralizing hydrothermal fluids removed tin from the Sibuian basic-ultrabasic complexes and volcano-sedimentary rocks, and transported it into fractures within the exocontact zones of granitic bodies to form composite hydrothermal tin deposits.

The postmagmatic reworking and incorporation processes of ore formation are obvious in the Baotan area. Although most of the deposits occur in the Sibu basic-ultrabasic complexes, they are closely distributed around the Xuefengian Pingying granite bodies, and genetically related to silicification, tourmalinization and chloritization, indicating that the postmagmatic hydrothermal processes are essential to ore formation. However, this does not mean that all the ore-forming materials were brought about by granitic magmas, for in general the tin mineralization caused by postmagmatic hydrothermal processes is weak, except where superimposed on tin-bearing basic-ultrabasic complex. For example, the Chongdongwan Pb-Zn deposit, occurring in a NNE fracture zone at the exocontact of the Pingying granite (the model age of Pb of the ores is 773 Ma, nearly identical with the Rb-Sr isochronic age 836 Ma), also shows silicification and chloritization produced by postmagmatic hydrothermal fluids of granitic magmas. Its tin content is only 0.1%, presumably inherited from the Sibu metasedimentary country rocks. Therefore, the tin in the Baotan deposit came chiefly from the basic-ultrabasic complexes through mobilization by granitic activities; only part of the tin came from the granites. This conclusion can also be drawn from the diagram constructed by plotting the log (Li+Be+Ga) as a function of log (Cr+Ni+Co+V+Sc) (Fig. 10) from analyses of tin ores which have different country rocks, including basic-ultrabasic complexes, intermediate-basic volcanic rocks and meta-sedimentary rocks of this region. No matter in what kind of country rocks the tin ores may occur, the contents of elements characteristic of basic and ultrabasic rocks, such as Cr, Ni, Co, V and Sc, are richer. There is a positive correlation (r= 0.91) between the data for tin ores of quartz-tourmaline rocks and quartz-chlorite types in the Baotan ore field and those for the Xuefengian granites. There also is a negative correlation between some of the ores and the Sibu basic-ultrabasic complexes, thus indicating that the postmagmatic hydrothermal fluids of the granitic magma played a part in mobilizing the tin originally dispersed in the basic and ultrabasic rocks. There is a negative correlation (r = –0.83) between biotite-garnet and quartz-tourmaline-bearing tin ores in the Jiumao tin field and the Sibu basic-ultrabasic complexes, showing that the reworking and incorporation processes related to the post-magmatic hydrothermal fluids are not conspicuous, and that ore-forming material mainly came from tin-bearing basic-ultrabasic complexes and volcano-sedimentary rocks by means of activation through regional metamorphic hydrothermal fluids.

## 3. Caledonian Regional Metamorphism and Formation and Transformation of Tin Deposits

Caledonian movements, taking place at the end of the Silurian, had an important effect on the pre-Silurian tin deposit in this region. It not only deformed, reactivated, transported and re-enriched the deposits to form new ore-bodies, but also caused the tin dispersed in the source rocks to be concentrated. This metamorphic mineralization is quite apparent in the Jiumao tin field, where the grade of regional metamorphism was higher. The evidence for this is as follows:

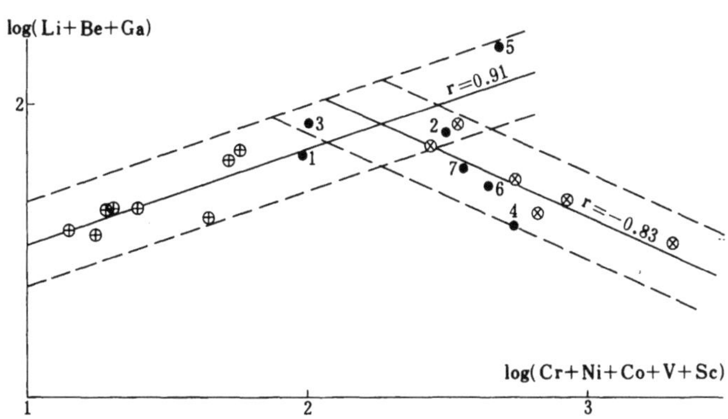


**Fig. 10.** Diagram of log (Li+Be+Ga) versus log (Cr+Ni+Co+V+Sc) showing various types of tin ores in relation to basic-ultrabasic complexes and granites in north Guangxi

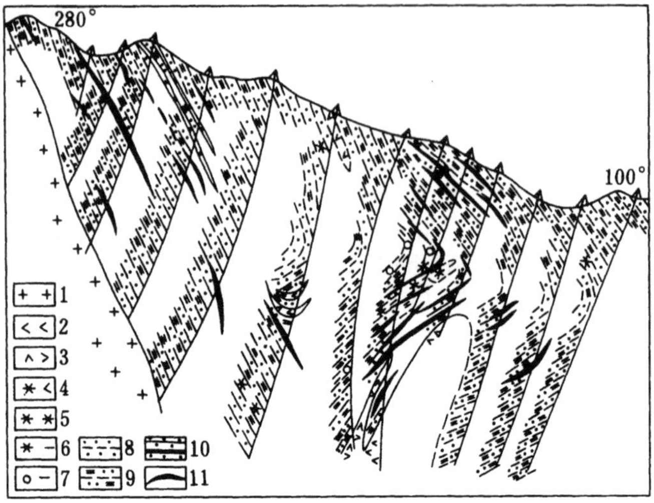


**Fig. 11.** Geological section of the Jiumao tin deposit, Rongshui County, Guangxi

(1) The stratiform tin orebodies are conformable to the schistosity of the country rocks and the general trend of Caledonian structures; the orebodies are folded along with the country rocks (Fig. 11).

(2) Intensity of mineralization entirely depends on the degree of regional metamorphism. Orebodies occurring in rocks of epidote amphibolite facies are greater in size and richer in grade, with some recrystallized cassiterite larger than in greenschist facies rocks.

(3) Intensity of mineralization depends also on the country rocks or the tin content of the source rocks. For example, orebodies from No. 53 ore belt of the Jiumao tin deposits occur in the contact zone of ultrabasic rocks. It is envisaged that these ores, large in size and high in grade, were formed simply because of the higher tin content (87 ppm on average) in these ultrabasic country rocks and the influence of the associated metamorphic-hydrothermal activities. This hypothesis is supported by the fact that the cassiterite and other minerals in these orebodies are richer in elements characteristic of ultrabasic rocks, such as Cr and Ni, which partly show a positive correlation with Sn (Fig. 12). Moreover, the fluid inclusions contain such metamorphic minerals as greenlandite, albite and calcite, indicating genesis by metamorphic-hydrothermal activity (Fig. 13).

(4) In the Jiumao tin deposit, the porphyroblasts of biotite and greenlandite are richer in Sn, Cr, and Ni. Scanning electron micrographs show that the biotite contains cassiterite inclusions (Fig. 14).

(5) According to the model age of galena (391—400 Ma) and the K-Ar age of biotite (434 Ma) of the Jiumao tin deposit, which are basically consistent with the metamorphic age of the Xuefengian granites (322—448 Ma), it is believed that the mineralization is related to metamorphism and took place during the Caledonian orogeny.

The process of tin mineralization in north Guangxi halted at that time, as indicated by cassiterite in the Devonian basal conglomerates.

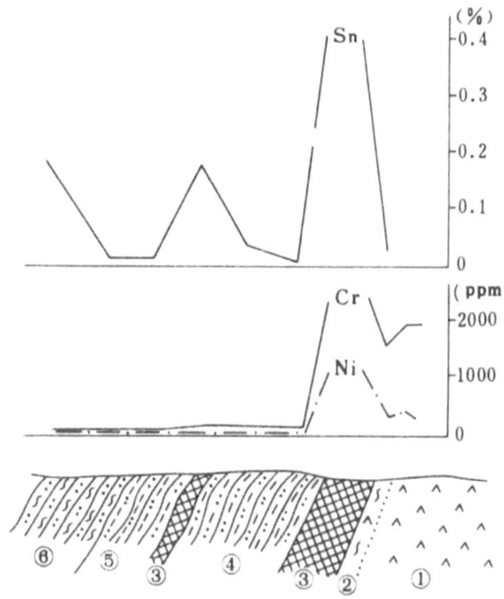


**Fig. 12.** Peak values of Sn, Cr, and Ni in the ore from 453 level, pit PD 58, Jiumao tin deposits

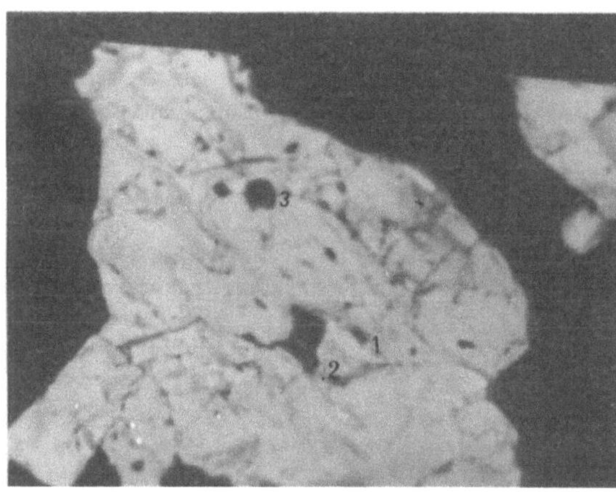


**Fig. 13.** Inclusions of pyrope (1), calcite (2) and albite (3) in cassiterite from the Jiumao ore field

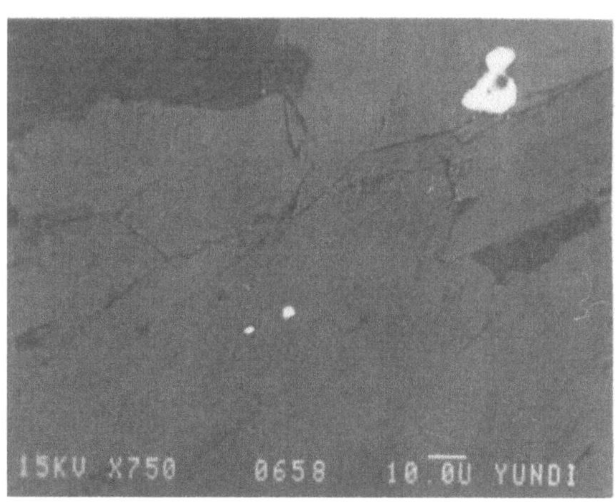


**Fig. 14.** Cassiterite inclusion (white) in biotite porphyroblast from the Jiumao tin deposit

## III. Discussion on Formation of the Deposit

It is clear that the tin deposits in north Guangxi are multi-staged, derived from various sources and accordingly have a composite origin. The process of mineralization may be related to three tectonic events, namely Sibuian, Xuefengian and Caledonian, with the latter two considered most important. Ore-forming materials were contributed mainly by the mantle-derived basic-ultrabasic complexes, and secondarily by anatexis. According to the style of mineralization and the ore components, the tin deposits in north Guangxi can genetically be subdivided into:

1. Repeatedly transformed and incorporated tin deposits formed by granitic hydrothermal activities, such as the Yidong-Wudi deposits.
2. Transformed tin deposits formed during metamorphism, such as the Jiumao deposit.
3. Metasedimentary tin deposits, such as Xiaodongjiang deposit.
4. Greisen type Sn-Cu deposits, such as the Wuyongling deposit.

## References

Cong Bolin, 1979. *Magmatic Activies and Igneous Rock Assemblage,* Geological Publishing House (in Chinese).

Mitchell, A.H.G., 1979. Rift-, Subduction- and Collision-Related Tin Belts. *Geol. Soc. Malaysia Bulletin* 11, 81—102.

Smirnov, V.I., 1981. *Ore Sources for Indogenetic Deposits,* Chinese translation version, Geological Publishing House.

Xu Keqin, Sun Nai, Wang Dezi, Hu Shouxi, Liu Yingchun and Ji Shouyuan, 1982. On the origin and metallogeny of the granites in South China. Proceedings of Symposium on Geology of Granites and Their Metallogenetic Relations (Abstracts), 1—3, Nanjing, China.

## 6.5 China: Gejiu and Other Yunnan Deposits

# 6.5.1 Integrated Geophysical and Geochemical Indicators of the Gejiu Tin Mine and Its Neighbouring Areas

Cao Xianguang[1]


## Abstract

This paper deals briefly with the integrated geophysical and geochemical indicators of the metallogenic belt, ore districts, ore fields, deposits, and orebodies of Gejiu and neighbouring mineralized areas.

The metallogenic belt lies at the second order derivative negative anomaly of the regional Bouguer gravity field or at the gradient zone of alternating positive and negative anomalies, and is accompanied by geochemical anomalies.

The mining districts are all controlled by granites, folded fractures and certain lithologies. Because the granites are characterized by a lower density than the wall rocks, and by weak magnetism, the outcropping or concealed granite masses show low gravity and low magnetic anomalies, associated with geochemical anomalies.

The Gejiu tin ores are commonly distributed around the protuberances of granite masses concealed 200 m to over 1,000 m beneath the surface. Because the granites are over ten times lower than wall rocks in resistance, they can be delineated by the integrated indicators of anomalies for low-resistance bodies defined by electrical sounding corresponding with primary geochemical anomalies.

The integrated geophysical and geochemical indicators for tin fields neighbouring Gejiu consist of magnetic anomalies of certain size, A.C. dual frequency induced polarization anomalies, secondary and primary geochemical anomalies and the existence of ore-bearing layers at corresponding locations.

The orebodies are delimited by primary anomalies and composite profiling anomalies; orebodies between workings are delineated by the radiowave method.

Integrated geophysical and geochemical indicators over cassiterite-sulphide-type orebodies near Gejiu are quite similar to those over the ore fields. Elements such as Sn and W are present over the Sn-W quartz vein type orebodies with groups of quartz veins at corresponding sites.


## Introduction

This paper briefly describes integrated geophysical and geochemical indicators over the metallogenic belt, ore districts, ore fields and orebodies in Gejiu and its

[1] Geophysical Party of Southwest China Geological Exploration Corp., CNNC, Chenggong, Yunnan

neighbouring areas, and gives practical examples illustrating indirect geophysical and direct geochemical prospecting applied at various stages from prognosis of the metallogenic belt to exploration.

## Metallogenic Belt

The Gejiu and neighbouring tin deposits are situated on the same tin-polymetallic ore belt in southeast Yunnan (Fig. 1), which lies at the second order derivative negative anomaly of the regional Bouguer gravity field or at the gradient zone of alternating positive and negative anomalies, suggesting a sedimentary depression area; secondary lead anomalies are also observed.

## Tin Mining Districts

The Gejiu tin-polymetallic ore deposits are mainly controlled by Yenshanian biotite granites, Middle Triassic carbonate rocks and a fold-fracture zone (Fig. 2), the objects for indirect geophysical prospecting. Because the density of granite is 0.15 g $cm^{-3}$ lower than that of the wall rocks, negative gravity anomalies therefore exist over granites or concealed granites (Figs. 2 and 3). The gravity anomalies and geochemical anomalies are coexistent, approximately concordant with the mining district. The upper boundary of the concealed granite mass, defined by the gravity method, has been partially proved by drilling (Fig. 4), and the geophysical and geochemical conceptual model has thus been deduced (Figs. 5 and 6).

The tin-polymetallic ore deposits in the neighbouring areas of Gejiu are distributed chiefly around Yenshanian two-mica granite masses, controlled by Cambrian carbonate rocks intercalated with siliceous rocks and fold-fracture zones. As the gra-

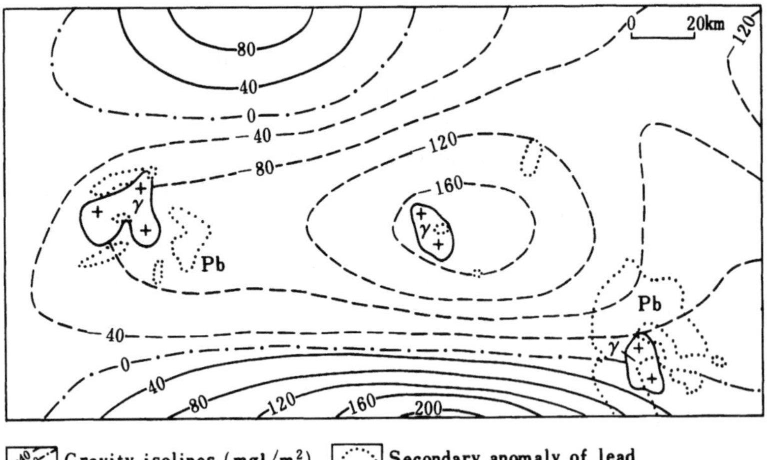


**Fig. 1.** Second order derivative of regional gravity Bouguer field and secondary lead anomalies in the Gejiu tin district and its neighbouring areas

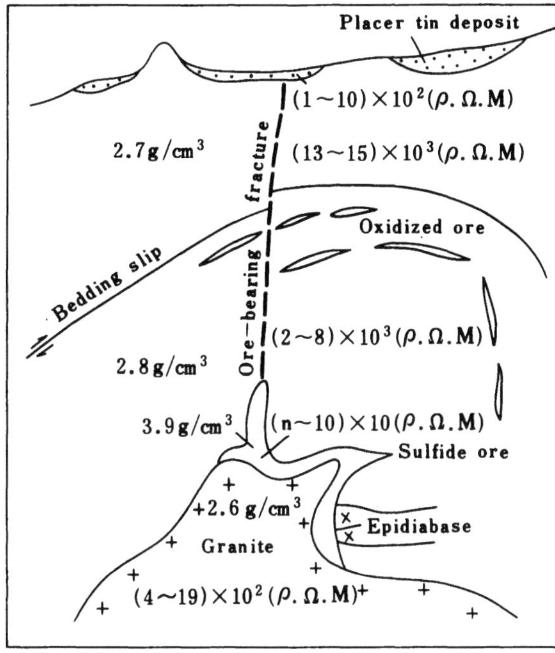


**Fig. 2.** Metallogenic model, density model (g cm$^{-3}$) and resistance model (ς, Ω, M) of the eastern mining district of Gejiu

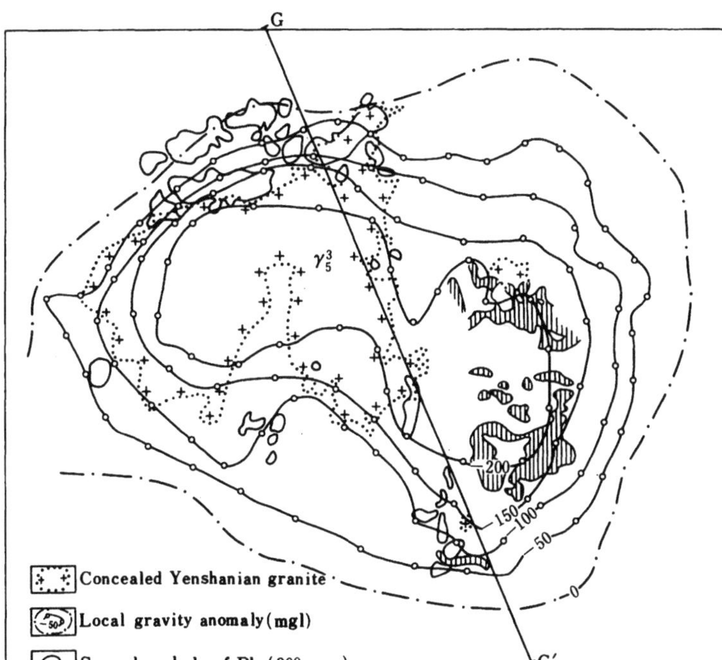


**Fig. 3.** Local gravity anomalies and geochemical anomalies of Gejiu

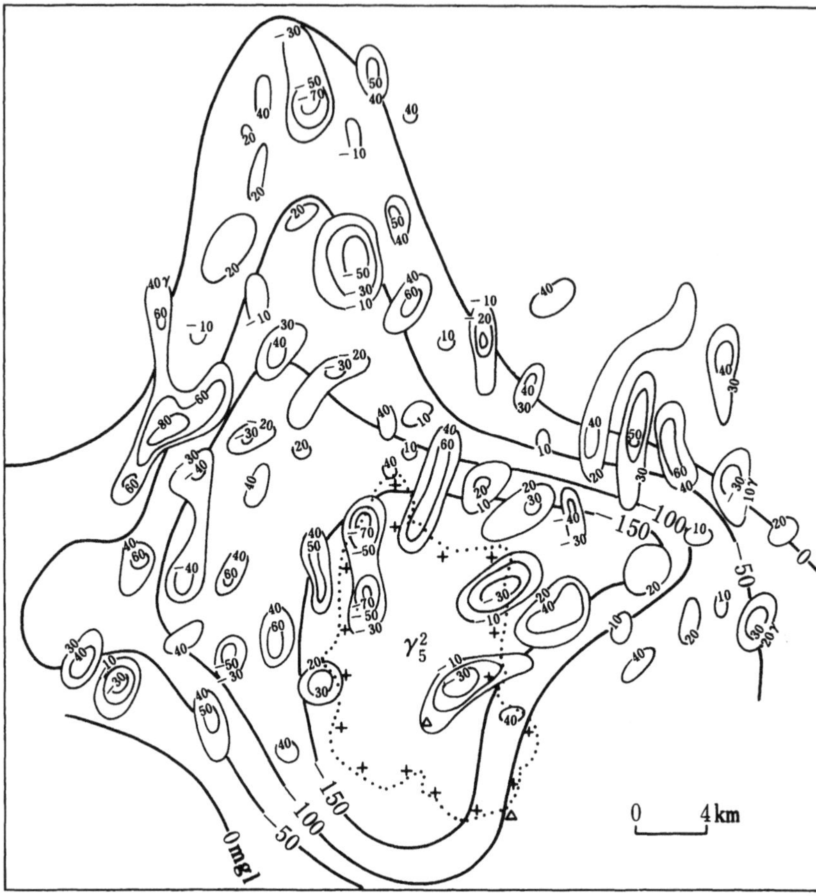


**Fig. 4.** Aeromagmetic △Z⊥ local gravity anomalies in the neighbouring areas of Gejiu

nites are about 0.25 g $cm^{-3}$ lower than the wall rocks in density, and have weak magnetism, there exist over them negative gravity anomalies as well as negative or no anomalies surrounded by circular positive aeromagnetic △Z⊥ anomalies (Fig. 4) which, combined with secondary lead anomalies (Fig. 5), show the outline of the mining district. The exploration geophysical and geochemical model is similar to that of Gejiu.

## Tin Ore Fields

Tin-polymetallic ore fields frequently occur around the granite masses concealed 200–1,000 m below the surface. From the ore-localizing geological conditions such as wall rocks and geophysical characteristics, these ore fields are grouped into two megatypes: one is of cassiterite-sulphide deposits occurring at the top or downwarping part of the granite, the other is of oxidized deposits existing in fractures and interforma-

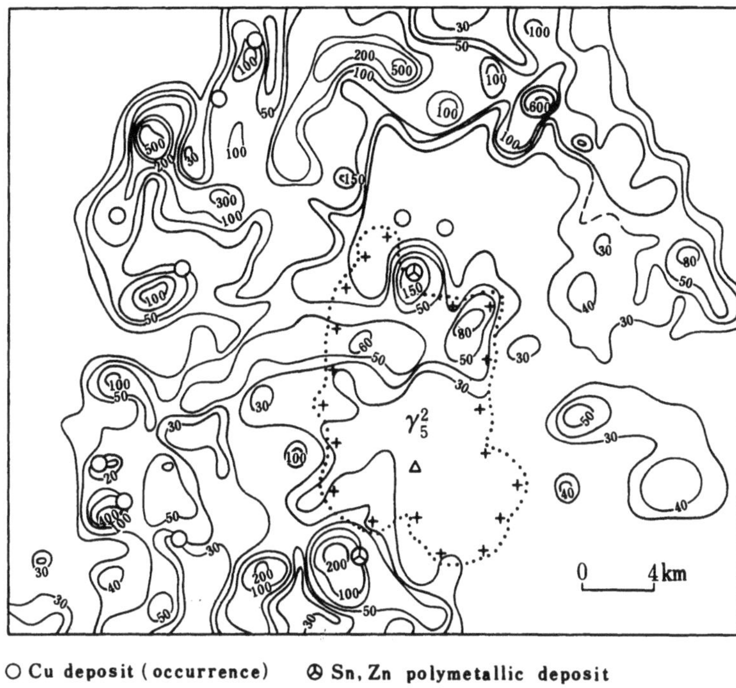


**Fig. 5.** Secondary lead anomalies in the neighbouring areas of Gejiu

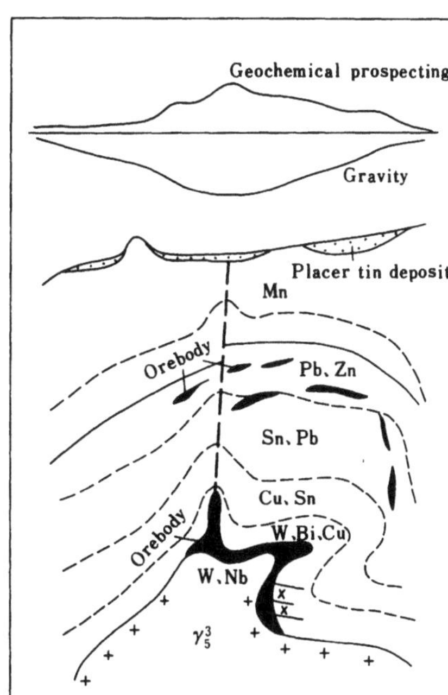


**Fig. 6.** Geophysical and geochemical conceptual model of the eastern mining district of Gejiu

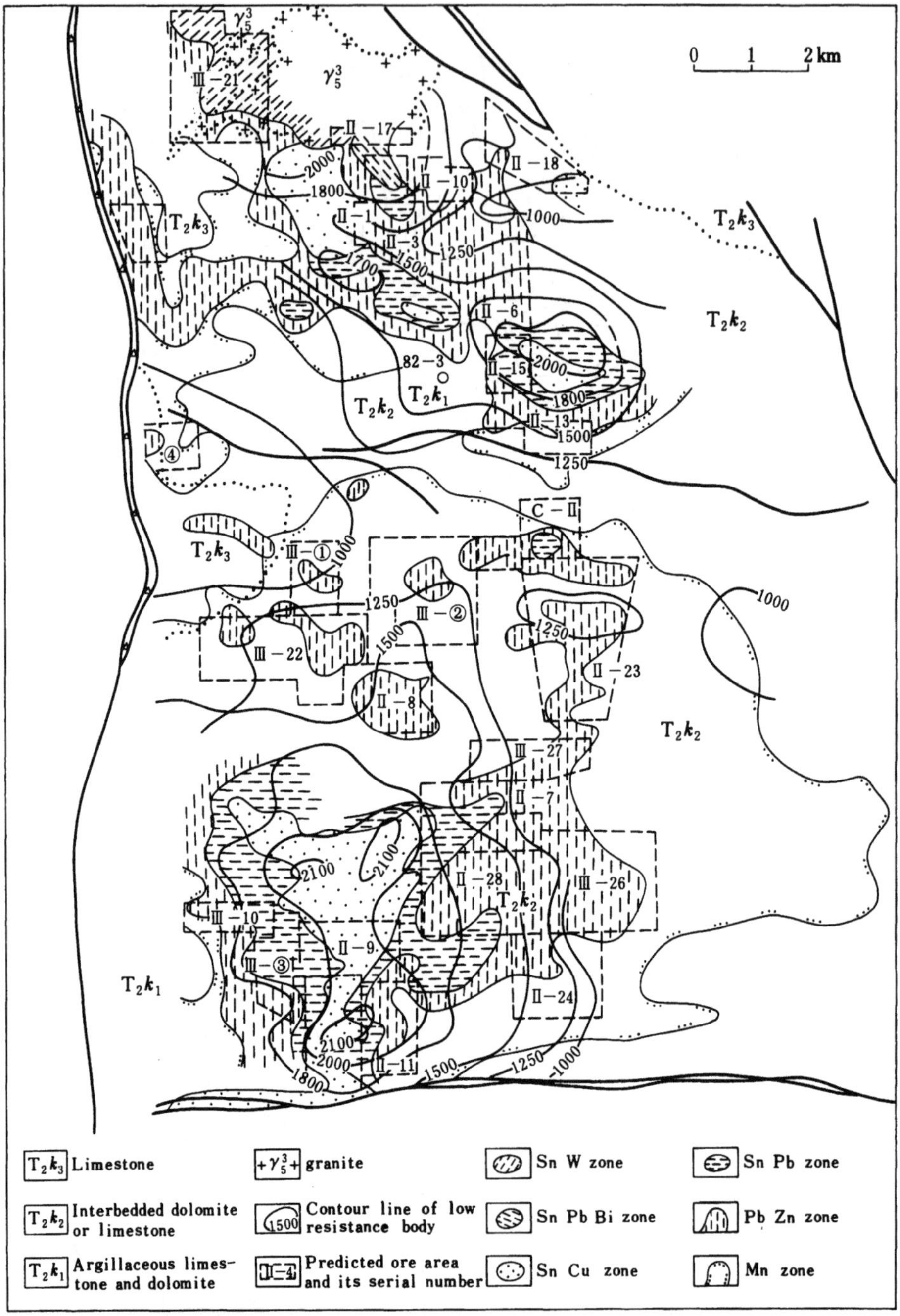


**Fig. 7.** Schematic map showing low resistance body isolines, zonation of primary anomalies and ore prediction in the eastern mining district of Gejiu

tional exfoliations at the exocontact zone. Although ore deposits differ to some extent from the wall rocks in such physical properties such as electricity, magnetism, and density, their great buried depth and strong topographic dissection hinder the proper performance of the magnetic method, self-potential method, and gravity survey in direct ore prospecting. Nevertheless, when the geophysical method is used to locate the protrusive part of the concealed granite and the fractures as a means of indirect ore prospecting, and the geochemical method is used directly for ore prospecting, relatively good results are obtained.

Because the resistance of the granites is over ten times lower than that of wall rocks (Fig. 2), the low resistance relief anomalies delineated by electrical sounding are in conformity with the primary geochemical indicators. Figure 7 represents the result of this investigation, and quite a lot of prospective areas have been proved by drilling to be ore-bearing. It should be pointed out that a primary anomaly is an indicator of vital importance.

Vertically from the endocontact zone to the surface, six zones of metallogenic elements may be distinguished, they are in upward order W, Be, Nb; Cu, W, Bi; Sn, Cu; Sn, Pb; Pb, Zn; Mn (Fig. 6). These zones are not of identical thickness, usually the first zone occurs near the endocontact zone, and other zones are 100—300 m in thickness. The existence or nonexistence of primary anomalies has some identification significance:

(1) *To judge whether the low resistance body anomalies delineated by electrical sounding are caused by granite protrusion or by other factors, such as basic rocks, argillaceous limestone, fracture zones, and lateral influence.* For example, a "protuberance of concealed granite", inferred by anomaly in the electrical sounding, was later denied because of the absence of primary anomalies, and this negation was substantiated by a drill hole of 735 m in depth, which only revealed argillaceous limestone.

(2) *To judge whether there are orebodies in the protrusive part of the granite.* A former ore prospect area predicted by geomathematical methods was subsequently denied due to the nonexistence of geochemical anomalies; later, six drill holes to a depth of over 800 m disclosed only granite, and no orebody was observed. The failure of the geomathematical methods was caused by the improper selection of variables, and this also indicates the importance of using integrated methods in prospecting for concealed ores.

(3) *A comparison of the depth of granite indicated by vertical zoning of element combinations with that inferred by electrical sounding helps to design the exploration engineering.* This has been demonstrated by many practical examples.

Based on the nature of host rocks and geophysical characteristics of ore deposits, the ore fields neighbouring Gejiu may be grouped into two major types: (1) the cassiterite-sulphide skarn type deposits occurring in the exocontact zone of Cambrian marble intercalated with schist, and (2) cassiterite quartz vein type deposits present in granites. The former type makes up the main deposits in these areas with Sn, Zn and other metals bearing pyrrhotite and marmatite as the predominating minerals. The integrated geophysical and geochemical indicators for this type of deposit are therefore made up of considerably-sized and regularly-shaped magnetic anomalies, A.C. dual-frequency IP Fs anomalies, and secondary and primary geochemical anomalies combined with corresponding ore-bearing horizons, as shown in Figure 8. Of these indicators, the Fs profile shows relatively satisfactorily outlines of the orebodies.

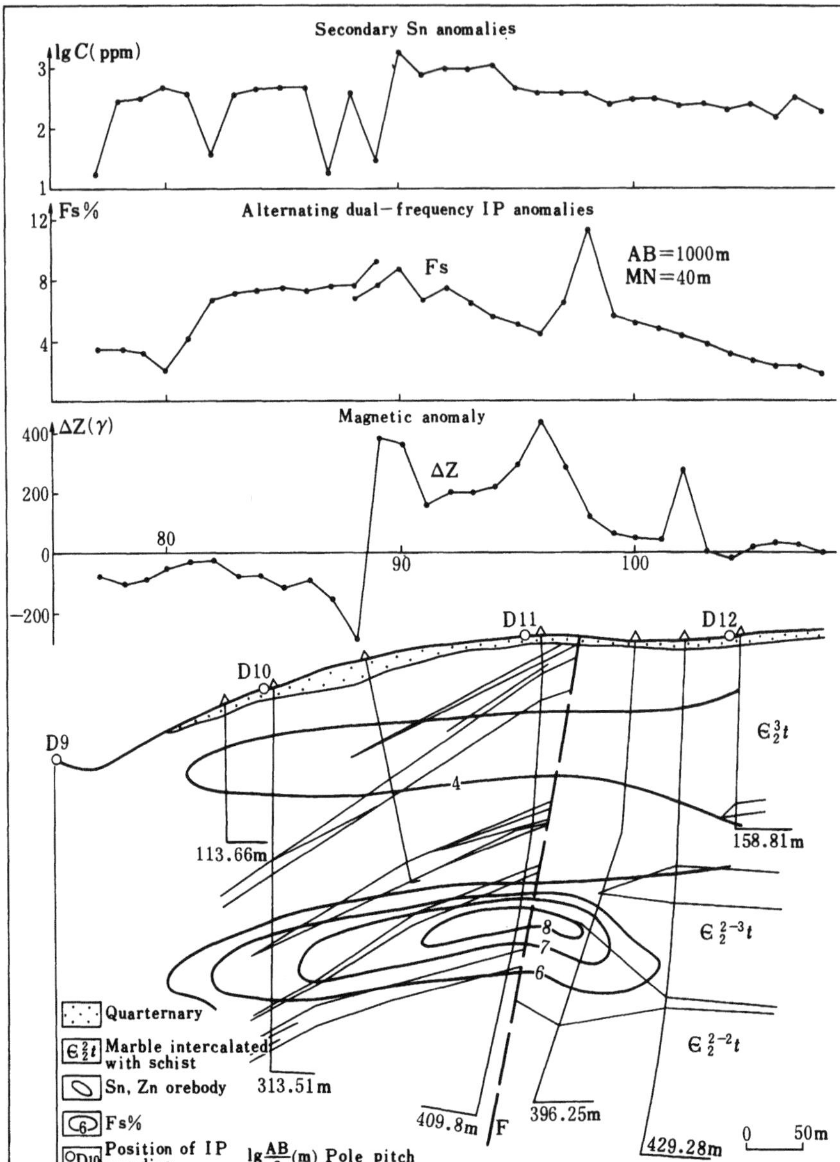


**Fig. 8.** Integrated geological, geophysical and geochemical profile for neighbouring areas of Gejiu

## Orebodies

The primary anomaly is also an important indicator of the orebody. Table 1 indicates six indices of primary anomaly for distinguishing orebodies, mineralizations and ore types, i.e., element associations, near-ore diagnostic elements, element contents, statistical data, characteristics of anomalies and geological settings of anomalies. In

**Table 1.** Six indices for distinguishing ore types, orebodies and mineralization

| Index for ore prospecting / Ore type | Major Element association | | Near-ore diagnostic element | Element content | | Statistical data | | Characteristics of anomalies | Geological setting of anomalies |
|---|---|---|---|---|---|---|---|---|---|
| | | | | Common anomalous value (ppm) | Near-ore anomalous value (ppm) | Cu/Pb | Correlation coefficient | | |
| SnCu type | Sn.Cu<br>Bi.As | Pb.Mn<br>Mo.Ag | Bi | Cu 100–300<br>Sn 300–500 | Cu >1000<br>Sn 800–1500 | >0.5 | $\gamma$SnCu=0.68<br>$\gamma$CuBi=0.62 | Regular halo, obvious centre, noticeable size, high probability of Bi and As | Favourable ore-forming structures, intense mineralization |
| SnPb type | Sn.Pb<br>Cu.Ag<br>Cd. | In.Mo<br>As.Bi<br>Sb.Mn | In<br>Cd | Sn 50–100<br>Pb 300–500 | Sn >1000<br>Pb >2000 | 0.1–0.5 | $\gamma$SnPb=0.82<br>$\gamma$SnZn=0.70<br>$\gamma$CuPb=0.79<br>$\gamma$AgMn=0.83 | Regular halo, obvious centre, noticeable size, with Cd and Ag frequently seen | Favourable ore-forming structures, obvious mineralization |
| PbZn type | Pb.Zn<br>Cu.Sn<br>Cd.Ag | In.Sb<br>Mn.As | Cd<br>Sb | Pb 300–500 | Pb >1000 | <0.1 | $\gamma$PbZn=0.90<br>$\gamma$PbAg=0.80<br>$\gamma$ZnAg=0.97<br>$\gamma$AgCd=0.90 | Regular halo, noticeable size, relatively well-developed Pb, Cd and Ag | Well-developed fractures, obvious mineralization |
| Stockwork SnCu type | Sn.Cu<br>Bi.Be | Pb.Mn | Be | Sn 200–1000 | Sn 1000 | | $\gamma$AgBe=0.97<br>$\gamma$Cu-Be=0.87<br>$\gamma$CuAg=0.87<br>$\gamma$PbAg=0.70 | Erratic halo, high probability of Be | Well-developed tourmaline veinlets |

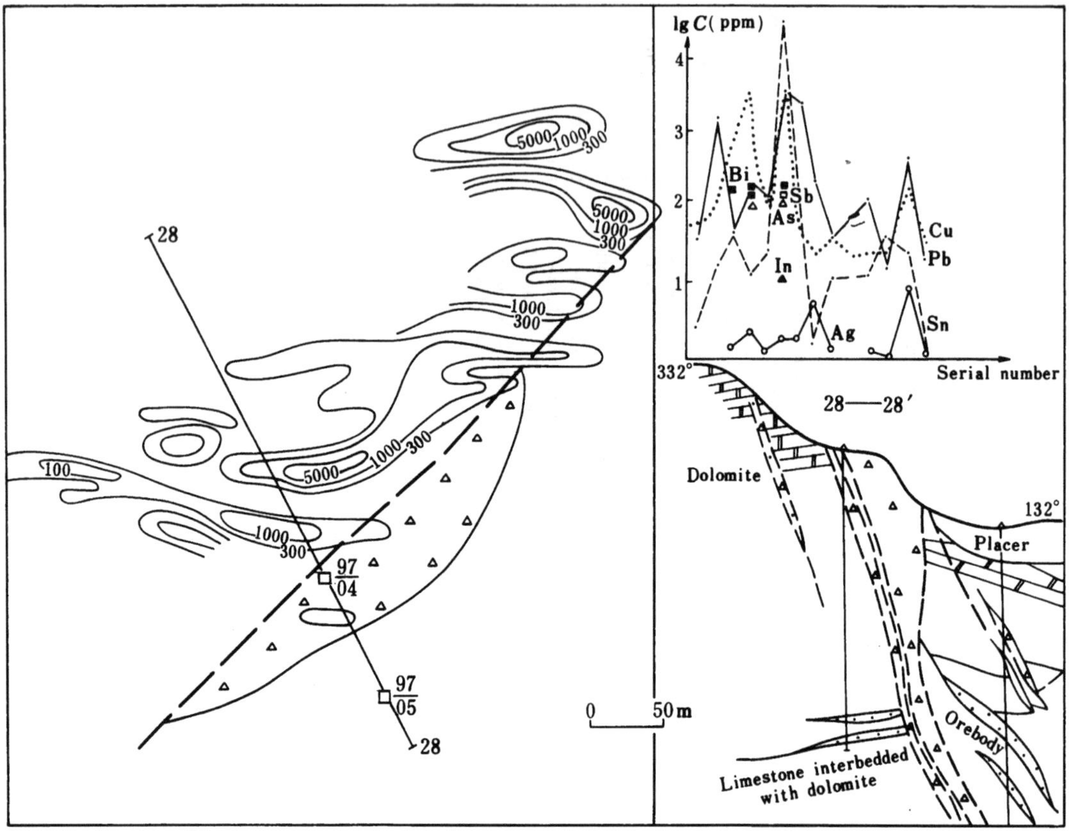


**Fig. 9.** Integrated diagram showing test and verification of primary anomalies by drilling

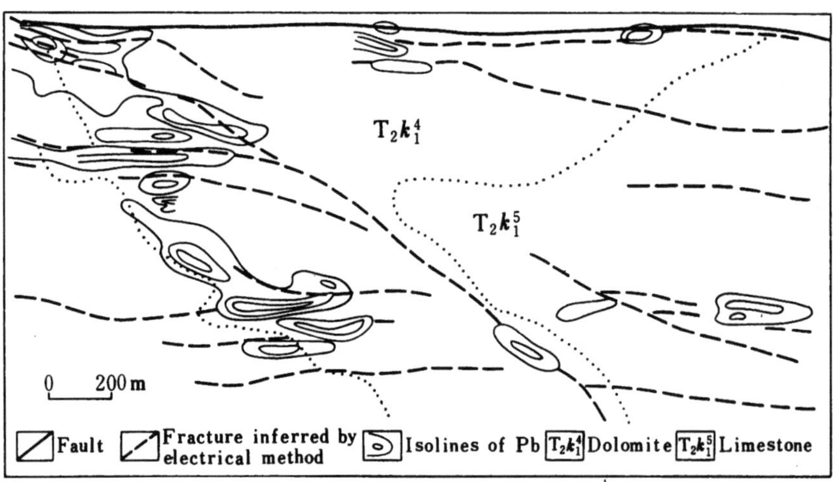


**Fig. 10.** Integrated plan of primary anomalies, composite profiling method and geological setting in Gejiu

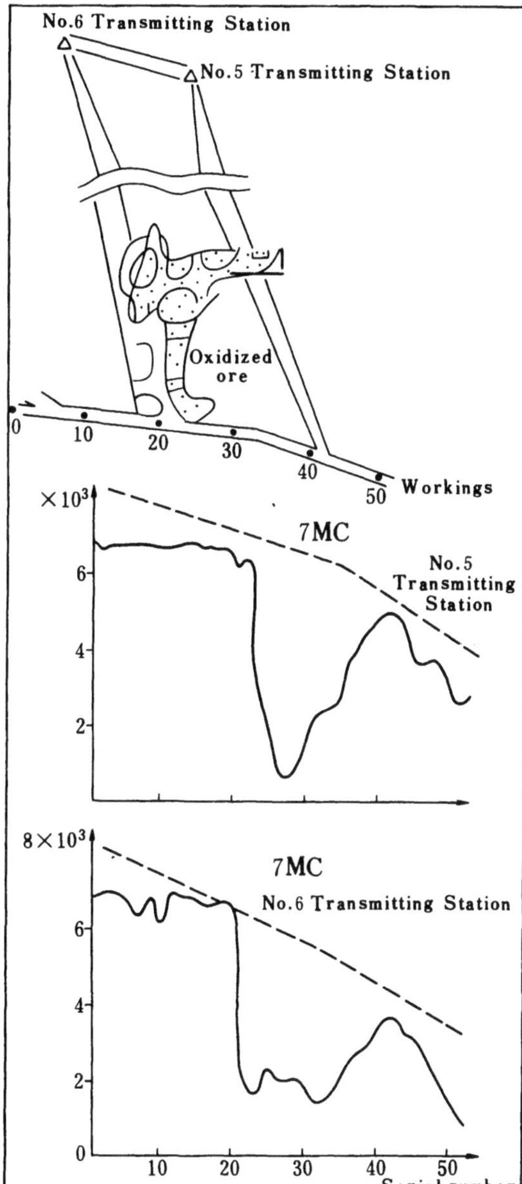


**Fig. 11.** Anomalies of radiowave method in workings of the Gejiu ore district

an area which was formerly regarded as nonproductive but which satisfies these six indices, some Sn-Pb type orebodies have been discovered by drilling (Fig. 9). Therefore, primary anomalies and orebearing fractures delineated by second-degree marks of composite profiling methods – F curve of ς s ratios ($F_A = \varsigma^A_s / \varsigma^B_s$, $F_B = \varsigma^B_s / \varsigma^A_s$) are integrated geophysical and geochemical indicators in the search for orebodies. In a segment of the area shown by Figure 10, geophysical and geochemical anomalies indicate an extension of the ore belt for 2.5 km in a southeast direction.

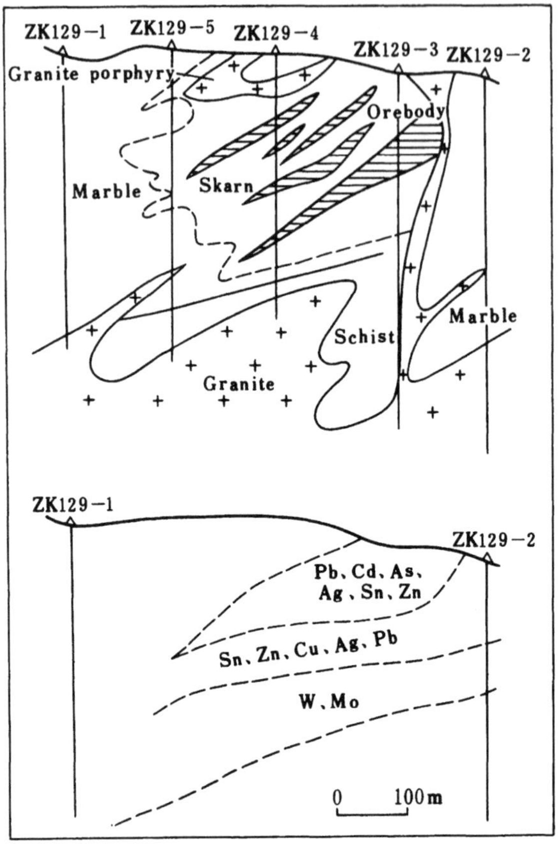


**Fig. 12.** Diagrammatic geological section showing vertical zonation of elements in neighbouring areas of Gejiu

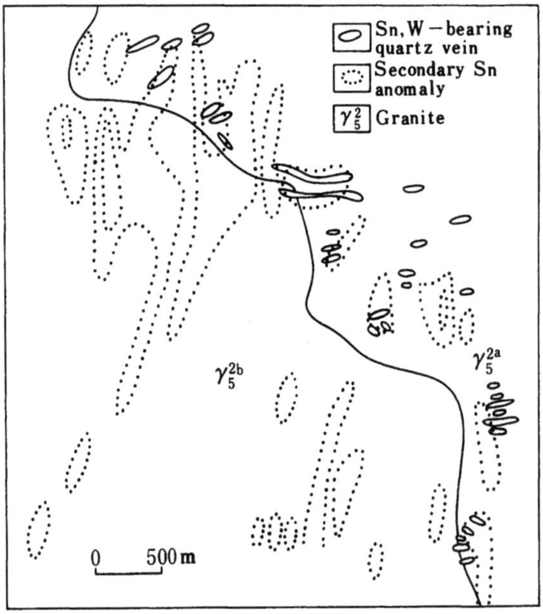


**Fig. 13.** Secondary Sn anomalies from Sn-W-bearing quartz veins in the neighbouring areas of Gejiu

Orebodies lying between workings are delineated by means of radiowave methods performed in the workings. As shown in Figure 11, there exists an obvious radiowave anomaly over the orebody.

The geophysical and geochemical indicators for the major types of orebody-cassiterite sulphide skarn type orebodies in neighbouring areas of Gejiu are quite similar to those in the ore field (Fig. 12). The zonation of the primary anomalies is: Pb, Cd, As, Ag, Be, Sn and Zn over the orebody; Sn, Zn, Cu, Ag, Be and Pb inside the orebody; and W and Mo beneath the orebody.

There also exist Sn and W anomalies over the Sn-W quartz vein type orebodies, accompanied by groups of quartz veins in corresponding positions (Fig. 13).

Quite satisfactory geological and economic results have been gained and ore prospecting and exploration have been accelerated by applying these integrated geophysical and geochemical indicators in predicting metallogenic belts, ore fields, ore districts and orebodies, as verified by large quantities of engineering work.

*Acknowledgements.* The present paper represents the investigation results by exploration geophysical and geochemical workers in our party, and the data of gravity magnetism for metallogenic belts and ore districts have been acquired by the joint efforts of the Exploration Geophysics Section of the Research Institute of Geology for Mineral Resources, CNNC, and our unit. Some research results by Messrs. Zhan Daoneng and Li Xuesheng are also quoted. The author wishes to express his sincere thanks to all the workers mentioned.

# 6.5.2 Origin and Metallization of Gejiu Granites

Yao Yinyan and Wu Mingchao[1]

**Abstract**

The Gejiu tin deposit is an important base area of tin production in China. The polymetallic tin orebodies are closely associated with granites both in space and time. This paper describes the petrology and petrochemistry of the granites, and gives data on Sr isotopes, REE patterns, experimental study of granitic rocks and trace element contents. The origin of the Gejiu tin-bearing granites and their mineralization are also discussed.

**Introduction**

The Gejiu tin mining area, located on the western extreme of the Nanling latitudinal tectonic belt of South China, is abundant in tin and other polymetallic resources. It has become one of the major tin producers in China.

The rocks of this area are mainly thick clastics and carbonates deposited in neritic and/or lagoonal environments during the Triassic. The major host rocks of the tin deposits are medium to thick bedded Middle-Triassic carbonates. Magmatism was strongly developed in Mesozoic time. There is a variety of magmatic rocks, ranging from basic, acidic, to alkalic. The Triassic carbonate rocks have been intruded by this wide range of magmatic rocks, and the surrounding rocks have been extensively altered to some degree. In addition to intrusions; there are also some volcanic rocks. The complexly multi-stage geological processes, especially the extensive development of granites, have resulted in the concentration of non-ferrous mineral resources in this area.

The widely distributed granitoids are closely related to polymetallic tin deposits both in space and time. Many geologists including Yan Yibin, Mao Yanshi, Wang Zhifen et al., have carried out much research work on the geology of this region, especially on the granites.

[1] Research Institute of Geology for Mineral Resources, China National Non-ferrous Metals Industry Corporation, Guilin, Guangxi

1. Petrology and Petrochemistry of the Granitoids

Granitoids are the major magmatic rocks in this area. There are eight granitic bodies forming cusps or stocks of small batholiths (Fig. 1).

They can be subdivided into porphyritic and equigranular granites (equigranular medium-to-coarse and fine-grained granites).

*Porphyritic Granite:* It forms the major portion of the Longchahe rock mass in the western part of this area, with its outcrop occupying about 200 km$^2$. In the eastern part of this area, there is a sparse distribution of porphyritic granite, outcropping only in a small area at Beipaotai in the Malage region. This type of rock is characterized by a pseudoporphyritic texture. The megacrysts are of K-feldspar, plagioclase and quartz. The K-feldspar is of two varieties. One is of euhedral to subhedral thick tabular megacrysts, which show greyish white and yellowish pink colours. The grain-size is generally about 3—5 cm; some may be more than 10 cm. The rock content of such megacrysts is about 20%. This kind of K-feldspar occurs irregularly as large micro-

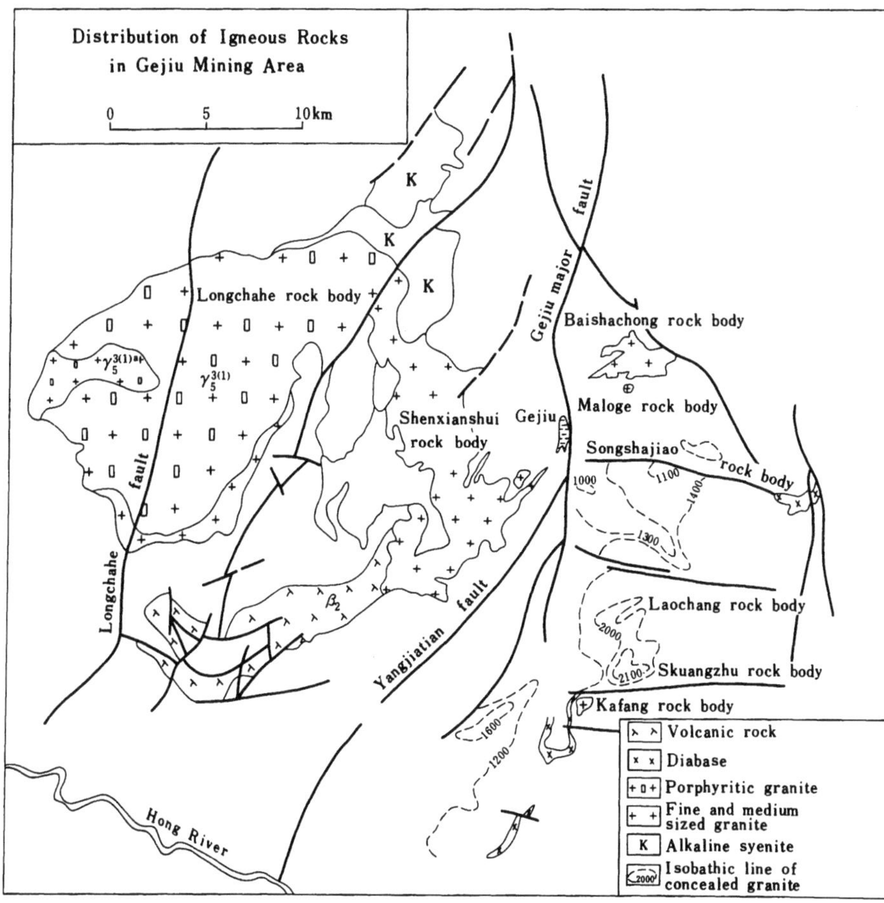


**Fig. 1.** Distribution of igneous rocks in Gejiu mining area

clines, sometimes as pegmatitic veins or patches. The triclinicity (△) and degree of order (S) are 0.8—0.9 and 0.95—1.0 respectively, and -2v is 82—85°. The coarse crystals contain much fine-grained plagioclase, quartz and biotite, indicating that they are porphyroblasts formed at later stage. The second kind of K-feldspar is characterized by small porphyritic crystals, with a grain-size mainly about 0.3—1 cm. The crystal is tabular, and Carlsbad twins and irregular perthitic textures are quite common. Its triclinicity (△), degree of order (S) and -2V are smaller than those of the first type, being characteristic of orthoclase-microperthite. The groundmass is composed of plagioclase, K-feldspar, quartz and biotite, with their contents varying in different rock bodies. Usually, the biotite content in the western area is higher than that in the east, and the same is true for anorthite content of the plagioclase. Accessory minerals, such as sphene, apatite and allanite occur commonly in the western area, whereas zircon is rarer. The accessory minerals of the Masong granites in the eastern area are mainly apatite, zircon and magnetite, while allanite is found sporadically. Petrochemically, the granites contain $SiO_2$ 67.66—72.45%, $TiO_2$ 0.26—0.46%, and high FeO, CaO and $P_2O_5$ contents. The variation in chemistry of porphyritic granites both in the eastern and western areas is consistent with the variation in the mineralogy. Granites in the western area contain lower $SiO_2$ and higher $K_2O$ and $TiO_2$ than those in the eastern area.

Another distinctive feature of this kind of granites is the ubiquitous existence of dark enclaves in the rock bodies, which are especially common in the Longchahe granite, but less common in the Malage and Songshujiao granites in the eastern area. These xenoliths are different both in size and shape and are distributed irregularly. Their sizes vary usually from ten to tens of cm, sometimes even up to several m in diameter. Most of the xenoliths have clear-cut boundaries with the porphyritic granites, with only a few having gradational boundaries. A preliminary study of inclusions from three granites shows that they can be divided into three types, but they may be of only two origins. One is by incorporation of country rocks by granitic magmas on its upward ascent. This kind of xenolith is found in the southern part of Longchahe granite. Lithologically, they are of metamorphosed intermediate-basic volcanic rocks and intermediate-acidic pyroclastic rocks. The second seems to be of comagmatic inclusions, which can be found both in the eastern and western parts. Lithologically, this kind of inclusion can also be divided into two types. One is represented by hybrid-like granodiorite or monzonite, which is mostly found in the Songshujiao granite, and also in the Malage and the Longchahe granite. The other one is of biotite granite, which is more common in the Malage granites. The major rock-forming and accessory minerals, the content of trace elements, and the chemical composition of the cognate inclusions are consistent with those of the host granite, but the amount of dark-coloured minerals and sphene in these dark inclusions is slightly higher and thus they are more basic and have a more compact texture.

*Equigranular Granites:* These rocks are distributed sporadically in this area. Most of them are concealed rock bodies which mainly consist of quartz, K-feldspar, plagioclase and biotite. The content of quartz is commonly higher (30—40%), and the biotite is 2—10%. The plagioclase is mainly oligoclase (20—30%). The triclinicity and degree of order for K-feldspar are both less than 0.5, and therefore, it is orthoclase-microperthite (30—40%).

Zircon is the predominant accessory mineral. The quantity of pneumato-hydrothermal minerals is obviously increased. Fluorite, tourmaline, topaz and muscovite are commonly found. The granites are strongly altered by greisenization, albitization, K-feldspathization, muscovitization (sericitization) and skarnization. Chemical analyses of this kind of rocks exhibit high silica (72—74%) and alkalis ($K_2O$+$Na_2O$ >8.0%); their K/Na ratio is 1.37, and $Al_2O_3$/($K_2O$+$Na_2O$+CaO) ratio varies from 1.3 to 1.4. Therefore, they are peraluminous. The $Fe^{+3}/Fe^{+2}$ ratio is less than 0.4, which indicates a greater depth of rock emplacement, with lower oxidation potential.

## 2. Discussion of Origin of the Granitoids

It is clear that, lithologically and chemically, there are differences between the two granite types. On the whole, the porphyritic granites tend to be more basic in chemical composition. In addition, trace elements show an inherited evolution trend from metabasic porphyritic granite to slightly acidic equigranular granite.

The origin of the granites are evidenced by the following facts:

1) Most dark inclusions in the porphyritic granite are comagmatic and relatively more basic, indicating anatexis of deep-seated source rocks.
2) The diagram of Ca, Na and K (Fig. 2), constructed by plotting more than 40 analyses of collected samples, shows that most of granites are concentrated in the field of magmatic origin and have alkali- and K-rich affinities. Only a few samples fall at the margin of the field of magmatic granites. These represent mostly altered granites.
3) Samples of 14 granites have provided Sr-isotopic initial ratios ranging 0.710—0.715, suggesting an origin by anatexis of continental crust.
4) The REE patterns of the granites are broadly similar. The patterns for almost all rock bodies show steeply inclined curves higher at the left side and lower at the right, with distinctive Eu-depletion (Fig. 3), except the Laocheng granite which shows a smooth curve.

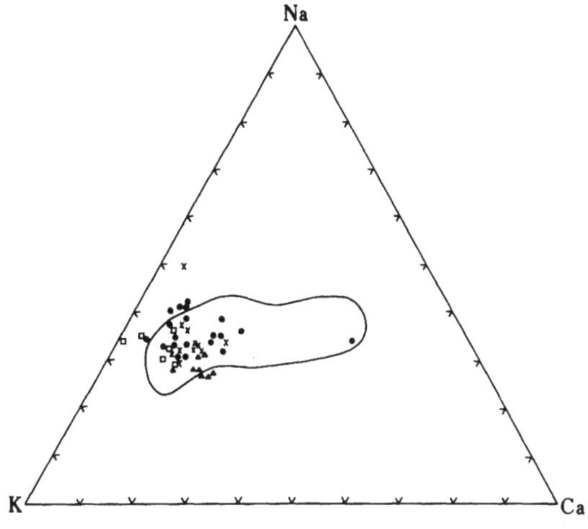


**Fig. 2.** Triangular diagram showing Ca-Na-K relationship of magmatic and replacement granite

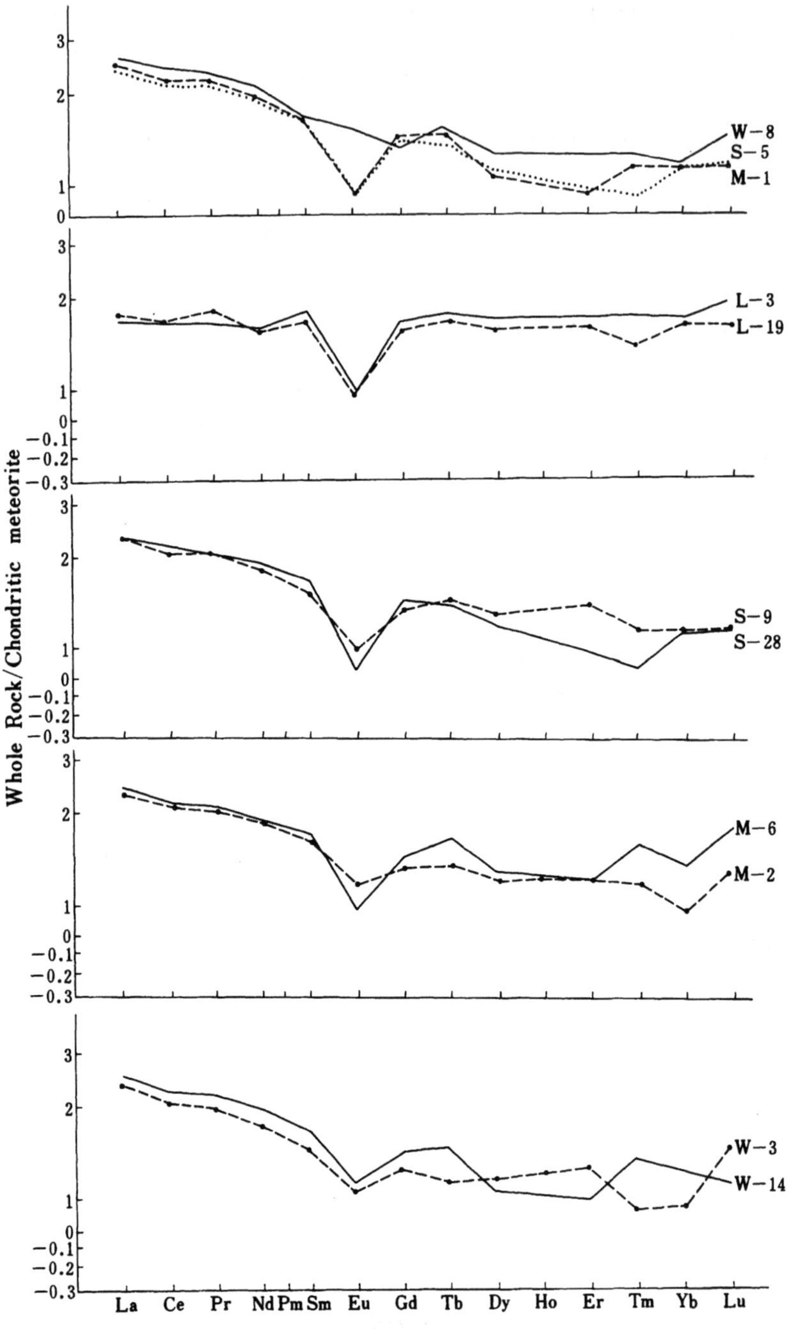


**Fig. 3.** Chondrite-normalized pattern of granites in the Gejiu area

5) According to the experimental data of Zheng Jiliang et al. (1982) at 5,000 kg/cm pressure (reasonably extrapolated on the basis of 3,000 kg/cm measured), the initial melting temperature is 650°C± for the Longchahe granite, 640°C± for the Masong granite, and 615°C± for the Laochang granite.

It is concluded that the granites in the Gejiu area were formed through crustal anatexis and rose up to their present emplacement levels.

## 3. Mineralization of the Granite

Mineralization of Gejiu polymetallic tin deposits is closely associated with certain types of granitic bodies. The known tin deposits almost all occur at the top of small stocks or at the cusps of concealed granites or within their surrounding rocks. Around the cusps of granites, an apparent metallic zoning has developed in the sequence of Cu, Cn-Sn, Sn-Pb-Zn, and Pb-Zn, outward from the centre of the cusp. This phenomenon can be seen in every ore deposit or orebody.

Although porphyritic granites and equigranular granites have the same origin, their mineralizations are quite different because of progressive evolution and differentiation of the magmas. According to the whole-rock Rb-Sr determinations obtained by the Isotope Department of our Institute, the isochron ages of the main Longchahe porphyritic biotite granite, the Beipaotai porphyritic biotite granite in Malage area, the Shenxianshui equigranular granite and the Laochang altered equigranular granite are 147 ± 3 Ma, 90.4 ± 6.3 Ma, 84.4 ± 1.3 Ma, and 81.0 ± 4.9 Ma, respectively, suggesting that the emplacement and consolidation of the Gejiu granites occurred at the middle to late stages of Yenshanian orogeny, and the emplacement of porphyritic granites must have been earlier than that of equigranular granites.

Table 1 shows the evolution process of porphyritic to equigranular granites:

1) Acidity of the rock-forming elements increase with the decrease in Ca, Mg and Fe.
2) Rock-forming K-feldspar changes from large-sized microcline and ortho-microperthite to ortho-microperthite, accompanied by the variation of plagioclase from oligoclase to albite-oligoclase.
3) Sn content varies from 8.6 ppm to 13.8 ppm (up to 20 ppm in the Laochang).
4) F content varies from 1957 ppm to 2742 ppm.
5) The ratios of Mg/Ti, Mg/Li, K/Rb, Ti/Ta, Zr/Sn obviously decrease, and conversely, the Rb/Sr ratio increases.
6) REE and Eu/Eu* ratio decreases with the variation of LREE into HREE.
7) Dark inclusions are well developed within the porphyritic granites; however, they are absent in equigranular granites.
8) The porphyritic granites exhibit less-developed K-feldspathization, silicification and sericitization, whereas equigranular granites show strong greisenization, albitization, K-feldspathization, muscovitization and tourmalinization.
9) Differences in mineralization: The Longchahe granites and their country rocks contain only uneconomic mineralization, no deposits have yet been found within them; the Malage and Songshujiao granites contain large to medium sized polymetallic tin deposits; in addition to extremely large polymetallic tin deposits, the Laochang granites possess also large-sized cassiterite-tourmaline vein-type deposits.

**Table 1.** Contrasting characteristics of the two kinds of granite

| | Porphyritic Granite | | Equigranular Granite | | Evolutional trend |
|---|---|---|---|---|---|
| Rock-forming elements | $SiO_2$ 67.66—72.45%, $TiO_2$ 0.26—0.46% $Fe_2O_3$+FeO 3.52—2.49%, CaO 2.41—1.54 MgO 0.87—0.55% | | $SiO_2$ 73.71—74.42, $TiO_2$< 0.01—0.093%, $Fe_2O_3$ 1.79—1.22. CaO 1.43—1.33%, MgO 0.30—0.08% | | Increase in acidity and decrease in Ca, Mg, Fe |
| Rock-forming minerals | K-feldspar 0.33—1.05, ST 0.3—1.05 (–) 2v 55°—85° Belonging to the large-sized microcline and ortho-microperthite. Plagioclase is oligoclase. | | K-feldspar 0.28—0.5 ST 0.31—0.55 (–) 2v 57—66° Belonging to ortho-microperthite. Plagioclase is albite-oligoclase. | | |
| Sn content | 8.6 ppm (Longcha River 4 ppm) | | 13.8 ppm (up to 20 ppm, Laochang) | | Rich in Sn |
| F content | 1957 ppm | | 2742 ppm | | Rich in F |
| | Variation range | Average | Variation range | Average | |
| Mg/Ti | 1) 1.6 — 2.5 | 2.3 | 0.77 — 0.002 | 0.074 | Mg/Ti |
| | 2) 2.2 — 2.7 | 2.5 | | | |
| | 3) 2.11 — 2.8 | 2.4 | | | |
| Mg/Li | 1) 40.7 — 96.8 | 9.08 | 9.98 — 53.8 | 31.9 | Mg/Li |
| | 2) 41.2 — 65.6 | 53.8 | | | |
| | 3) 60.6 — 73.1 | 68.6 | | | |
| K/Rb | 1) 74.7 — 116.22 | 92.9 | 35.8 — 52.9 | 41.5 | K/Rb |
| | 2) 130.6 — 77.6 | 96.4 | | | |
| | 3) 88.8 — 72.3 | 81.4 | | | |
| Ti/Ta | 1) 432.98 —1243.7 | 838.3 | 22.6 —574.7 | 315.8 | Ti/Ta |
| | 2) 690.9 — 368.5 | 439.7 | | | |
| | 3) 301.5 — 194.4 | 272.9 | | | |
| Zr/Sn | 1) 60 | 60 | 20 — 3.33 | 5.15 | Zr/Sn obviously decrease |
| | 2) 60 — 13.89 | 22.27 | | | |
| | 3) 50 — 13.75 | 26.41 | | | |
| Rb/Sr | 1) 0.259— 0.966 | 0.513 | 2.98 — 63.88 | 24.557 | Rb/Sr obviously increases |
| | 2) 0.849— 4.44 | 2.532 | | | |
| Characters | 1) 339.5 | | 245.09 | | REE content changes from rich LREE to rich HREE; |
| | 2) 319.89 | | | | |
| | 3) 329.91 | | | | |

| | | | |
|---|---|---|---|
| Σ Ce/Σ Y | 1) 7.69<br>2) 5.65<br>3) 5.41 | 0.735 | Eu/Eu* changes from higher to lower |
| Eu/Eu* | 1) 0.36<br>2) 0.45<br>3) 0.195 | 0.165 | |
| Dark inclusions | Well developed | Lacking | |
| Degree of alteration | Weak K-feldspathization, silicification, sericitizaton | Strong greisenization, albitization, K-feldspathization, muscovitization, tourmalinization | |
| Characteristics of mineralization | Only uneconomic tin mineralization exists in the vicinity of Longchahe body; large to medium sized polymetallic tin deposits are observed in Malage and Songshujiao rock bodies. | Gigantic polymetallic tin deposits, accompanied by large cassiterite, are present in Laochang body. | |

* 1) Longchahe body; 2) Malage body; 3) Songshujiao body; $Eu^* = \frac{Sm + Gd}{2}$.

To summarize, the favourable rock bodies for mineralization are the small stock-like bodies or cusps of large concealed granitic bodies formed at the late stage of magmatic evolution. The mineralized equigranular biotite granites are silica-saturated, low in titanium, poor in magnesium, calcium and iron, and rich in tin and fluorine. They have also been subjected to strong post-magmatic alteration, giving rise to albitization, K-feldspathization, greisenization, tourmalinization, muscovitization and skarnization which serve as important indicators in search for this kind of ore deposit.

## References

Department of Geology, Nanjing University, 1981. *On Granitoids of Different Geological Ages and Their Metallogenetic* Relations. Science Press.

Hine, R.H., Williams, I.S., Chappell, B.W., and White, A.J.R., 1978. Geochemical constrasts between I- and S-type granitoids of the Kosciusko Batholith. *Journal of the geol. Society of Australia.* Vol. 25, 219—234.

# 6.5.3 The History of Exploration over the Past Thirty Years in the Gejiu Tin Deposit, Yunnan

PENG CHENGDIAN and CHENG SHUXI[1]


## Abstract

The Gejiu tin deposit is world-famous for its long history of investigation and mining. However, extensive exploration had not been undertaken until the 1950's. In the past thirty years, exploration for tin ore has experienced five stages of development: for placer, for interstratified, for metasomatic, for polymetallic, and for deposits occurring in the concave contact zones below the protrusive granite. Reserves of several hundred thousand to a million tons of tin and other metal ores have been proved at each stage, and experience and some scientific results have been obtained.


## Geological Outline of the Gejiu Tin Deposit

The Gejiu tin deposit is located in the southwestern end of the South China post-Caledonian folded zone bounded by table cratons on the east, south and west. It is an area of protracted depression with three thousand metres of Middle Triassic carbonate rocks (Gejiu Formation) favourable for the formation of ore deposits. The Indosinian and the Yenshanian movements brought about intensive folding and faulting associated with large-scale magmatic activities (acid, basic and alkaline), followed by various mineralizations which form the great tin-polymetallic deposits (Fig. 1).

A N-trending fault (Gejiu fault) divides the ore district into the eastern and the western parts. About ten gigantic, large, and medium-sized ore deposits have been found, of which the Laochang deposit is known for its great size, variety of types and concentration of mineralization. These deposits are mainly controlled by late Yenshanian biotite granite (Fig. 1). The granite bodies tend to occur at a depth of about 200–1000 m with an area of distribution of about 100 $km^2$. All the granites are highly acidic, rich in alkalis, poor in CaO, MgO, and FeO, and contain abundant volatiles. The content of Sn, W, Be, Li, Rb, Nb, Ta, F in these granites is several times that in normal granite. The granite bodies have played a decisive role in the formation of the polymetallic deposit in terms of ore-formation, temporal and spatial distribution, material and energy sources of cassiterite-sulphide deposits. Their control functions are as follows: The deposits tend to occur around protrusions of granite stocks; there is distinct metallic zoning from the granite outward; ore deposits bear geochemical

[1] No. 308 Geological Party, Southwest China Geological Exploration Company, Gejiu, Yunnan

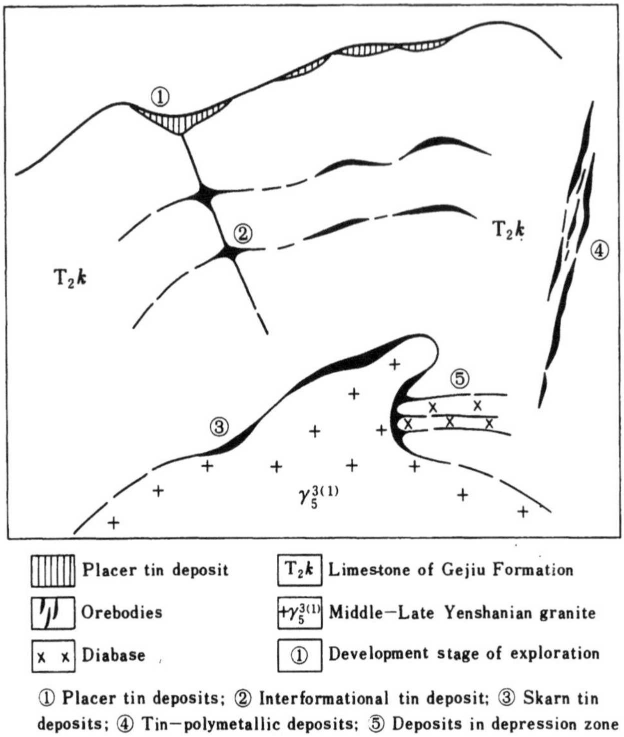


**Fig. 1.** Schematic diagram showing five stage development targets in prospecting and exploration

properties related to granite; the change in tin enrichment and mineralization tend to match the change in magmatic intrusions and evolution as well as in rock constituents and textures. In addition, the size, occurrence, and depth of emplacement and extent of erosion of the granite also have an effect on mineralization.

*Structural Control:* As shown in Fig. 2, several large mineralized areas are controlled by NNE-trending anticlinoria, and N-S and E-W trending fault systems. Second-order fault and fold systems of Ne- and NW-trends control the distribution of the ore deposits.

*Stratigraphic Control:* About 90% of the reserves of Sn and Cu are concentrated in the limestone intercalated with dolomite of the Lower Gejiu Formation. Recent studies indicate that the carbonate rocks from the Gejiu Formation are rich in algae and contain salt and evaporite, which provide abundant pore space for the precipitation and replacement of metallic sulphides as well as some ore-forming materials and media.

The widespread volcanic and subvolcanic rocks, such as the metadiabase sill in the Lower Gejiu Formation, provide new source supply and conditions for the formation of polymetallic deposits.

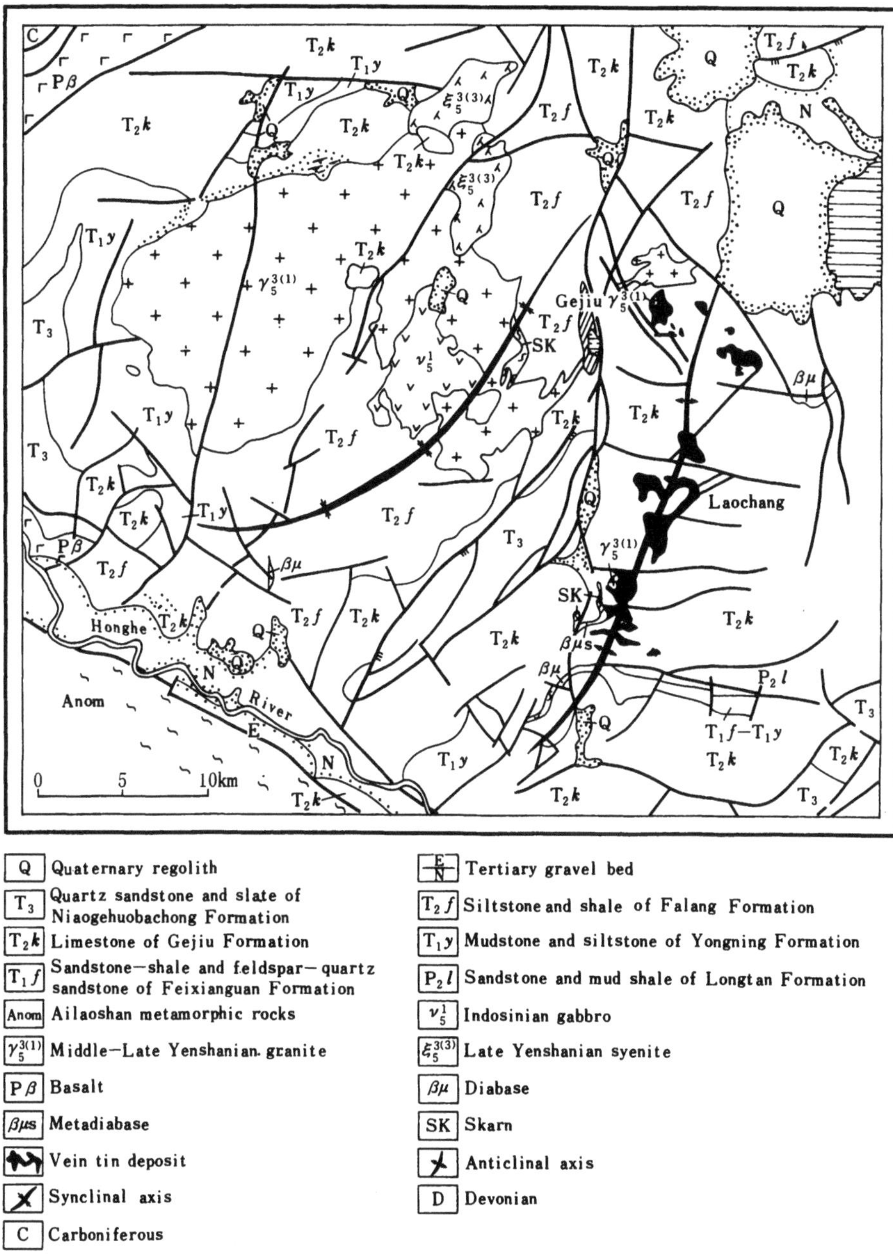


**Fig. 2.** Geological map of the Gejiu mining district

*Phases of Mineralization*

1. Silicate phase, during which skarns were formed;
2. Oxide phase, when tourmaline-veinlet tin deposits were formed;
3. Sulphide phase, the most important phase in which cassiterite-sulphide-polymetallic deposits were formed; and
4. Carbonate phase, in which cassiterite-bearing dolomite deposits were formed. Tin minerals are cassiterite, stannite, franckeite, wood-tin, nordenskioldite, schoenfliesite and varlamoffite.

*Types of ore deposits:* Five main types of tin deposit are distinguished in the district. They are:

1. Placers;
2. Cassiterite-sulphide deposits, divisible into interstratified oxide deposits and contact-zone sulphide deposits;
3. Tourmaline-bearing veinlet-type deposits;
4. Dolomite-type disseminated cassiterite deposits; and
5. Greisen-type cassiterite deposits.

In addition there are Cu, and Au-bearing metadiabase deposits, skarn-type scheelite deposits, and quartz-vein wolframite deposits.

Most of the placer deposits are of eluvial origin, composed of talus accumulations in karst caves and gullies. The primary tin deposits are of hydrothermal origin.

Further understanding was obtained into the salient features of the five big ore fields in the region and tens of large- and medium-sized ore deposits in this district. The types of tin deposits are: placers, cassiterite-sulphide deposits, tourmaline-bearing veinlet-type deposits, tin-bearing dolomite-type and cassiterite-bearing greisen-type deposits. There are more than 20 useful components. In contrast to the large size of ore deposits, the size of orebodies is quite small, as exemplified by the fact that a single deposit usually consists of tens to hundreds of orebodies with complicated shapes. The late Yenshanian biotite granites provide abundant source and energy for mineralization. The ore-containing strata are the limestone and dolomite of the Middle Triassic Gejiu Formation. The deposits are mainly controlled by anticlines and fractures. At the same time we started to make evaluation and investigation into the whole ore cluster or mineralization zones instead of searching for individual orebodies. As the orebodies tend to occur in linear or in pipe-like shape, we developed a method of intercept drilling which proved to be effective and economic.

Stages of the development of exploration in the Gejiu tin deposit are as follows:

1. In the 1950's many young geologists came to the Gejiu mining district. By learning the relevant theories and exploration techniques and experience obtained both in China and foreign countries, they carried out systematic field investigations in a short period of time with emphasis on the exploration for the widespread tin placer deposits in this district. Hundreds of thousand of tons of tin ore reserves were then proved (Fig. 1).

2. To search for primary ore deposits, integrated research and modelling were undertaken in the late 1950's and 1960's, with emphasis laid on the study of the metal-

logenic regularity and mineralization prognosis. Through experience we have progressed from simple metallogenic modelling combinations of geological data, geophysics and geochemistry, which led to the discovery of a number of hypabyssal and intermediate interstratified deposits such as secondary oxide ores of hydrothermal tin deposits and intermediate and deep-seated contact deposits such as skarn-type Sn- and Cu-deposits at the granite contacts. We then diverted our attention to searching for polymetallic deposits, and discovered a series of lead, zinc, copper, tungsten, bismuth and beryllium deposits. The proved reserves of the district were doubled and redoubled several times.

3. After these exploration activities we had to search for new deposits at greater depths, the concealed ores and the surrounding areas of old deposits. By summarizing the experience in the exploration of ore deposits occurring in the lower concavities in the granite cusps, similar Sn and Cu deposits were found in several places. By analysis of about ten tin deposits and hundreds of various orebodies, three metallogenic models have been established in terms of genesis, controlling factors, components, occurrence and distribution regularities (Fig. 3):

*Contact type,* which is dominated by skarn-type sulphide Cu and Sn deposits with two sub-types occurring at the top of the stock, and in the concavity below the granite tongue respectively.

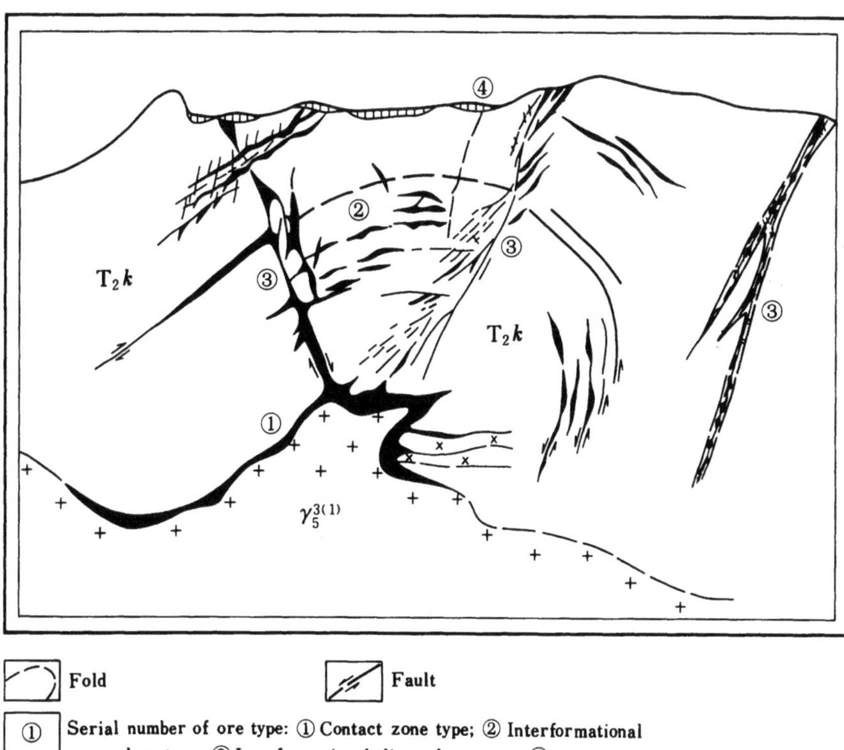


**Fig. 3.** Model for tin deposits in the Gejiu mining district

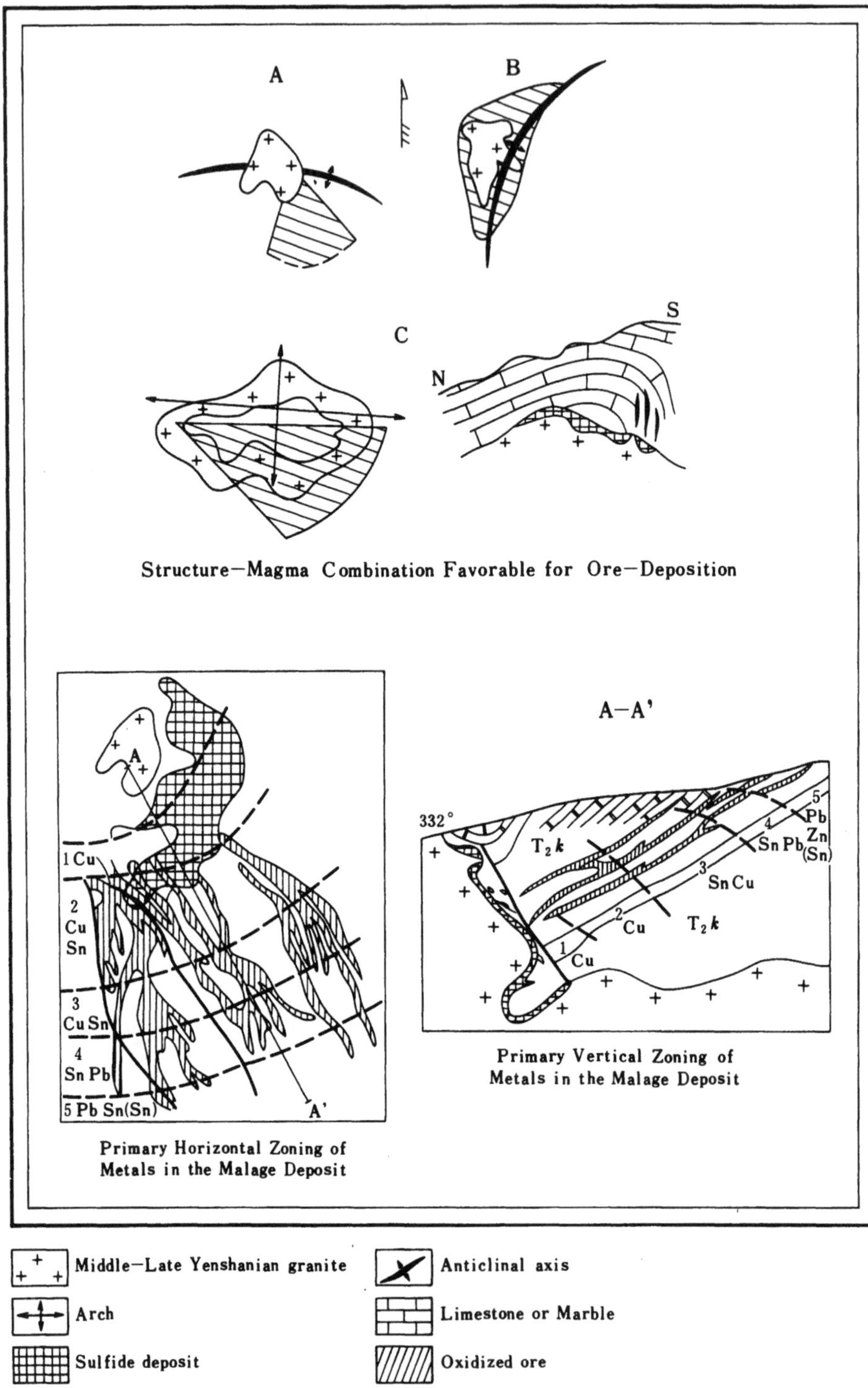


**Fig. 4.** Schematic diagram showing combination patterns of magmatic rocks and primary zonation of metals

*Conformable interstratified type,* which is dominated by oxide deposits distributed along the protrusion of granite controlled by interstratified conformable structures, such as interbed slips and fracture zones. It is characterized by complicated configuration and linear extention of orebodies.

*Unconformable interstratified type,* which is mostly controlled by unconformable structures, such as steep fracture zones, occurring as lode veins and stockworks, and is characterized essentially by lead and zinc mineralizations without clear relations with the granite.

## The Distribution Regularities

The most favourable tectonic setting for the occurrence of the cluster of ore deposits and orebodies is the structural pattern as expressed by an arch structure above them and a projected granite stock below.

There is a distinct zoning around the protrusion of the granite. In an outward order they are Cu (W) zone, Sn zone and Pb zone (Figs. 4 and 5).

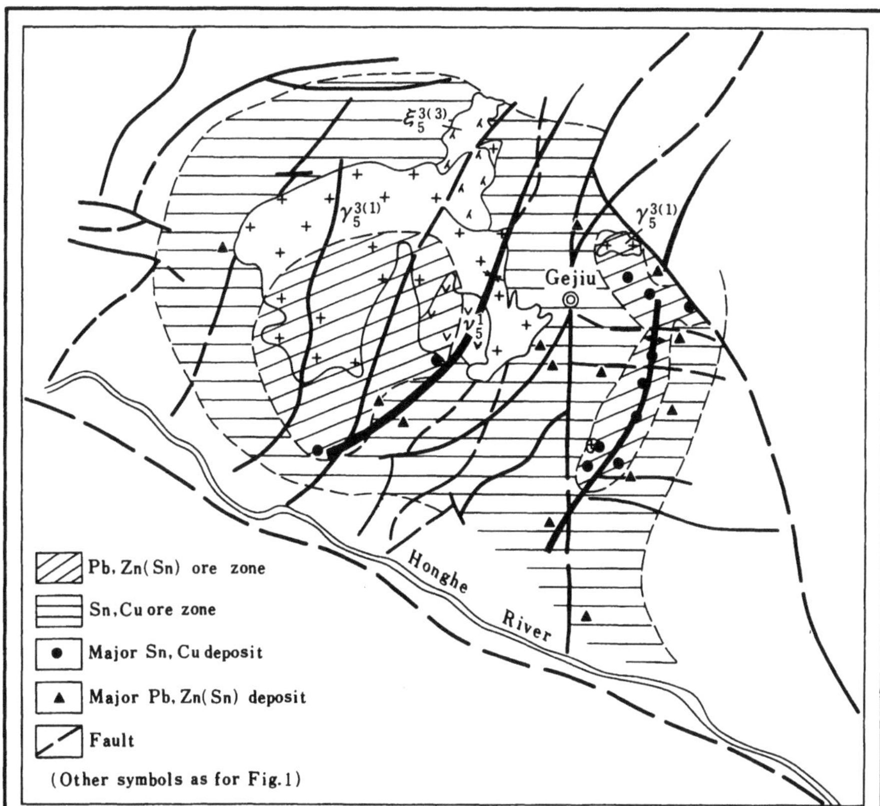


**Fig. 5.** Schematic map showing primary zonation of metals in the Gejiu tin mining district

The ore deposits tend to occur in a regular pattern in the vertical direction: the placer deposits in the top part, interstratified deposits in the middle part, and the contact-zone deposits in the lower part. The discovery of one or two types of these mineralizations may lead to the discovery of the other in the corresponding position. Being controlled by faults, interbeds and interformation structures, the oxide ore deposits often occur in clusters or in zones. The sulphide ore deposits in the contact zone tend to occur at the junction of the faults and dykes with the granite. The lead, zinc and tin deposits are mainly concentrated in the E- and NE-trending fracture zones.

The fluorine content is the major criterion in evaluating the tin-bearing granite. The presence of diabase is an important prerequisite for the formation of tin and copper deposits in the concavities of the projected part of the granite.

The close combination of geological, geophysical, and geochemical field investigations yield good results. By using electric sounding we identified the concealed protrusion of granite, and with the aid of primary haloes we defined the mineralization zone. We have also developed effective methods and means to find the concealed deposits occurring in the depression.

With the above regularity pattern and methods, we have discovered a large tin- and copper deposit at a depth of 1,000 m in the Changfengshan area, which represents a good example in the search for deep-seated concealed ores (Fig. 2).

4. Recently we have entered into a second round of prospecting and exploration in the Gejiu district which requires higher-level research, more information on mineralization so as to find new types, and still deeper concealed deposits with new kinds of minerals. We have undertaken research on many new subjects related to statistical geology, isotope geology and diagenesis, and achieved scientific result in many areas, such as the discovery of nordenskloldine and schoemfliesite as well as tin-bearing skarns and tin porphyry deposits. Through statistics and integrated prediction, interstratified rich tin and lead deposits have been found at a depth of 600–700 m. Attention has been paid to the advantage of old mining areas which possess abundant data related to ore deposits, detailed studies of metallogenic regularity and conditions, and higher economic effectiveness and good accessibility to the areas of anomalies. At the same time, prospecting for gold and silver has been started in surrounding areas of polymetallic tin deposits to make the mine district more valuable.

# 6.5.4 The Relationship Between the Fluid Inclusions in Minerals from Magmatic Rocks and the Mineralization of the Tin Polymetallic Deposit in Gejiu, Yunnan, China

WANG ZHIFEN and ZHU QIJIN[1]


## Abstract

Following the study of inclusions in tin deposits of Gejiu, an investigation was made into the shape, salinity, composition as well as formation temperature and pressure of the inclusions in various magmatic rocks, which reveals the existence of obvious similarity and inheritance between inclusions in different deposits and those in various granite rocks. The boiling inclusions in granite are closely related to tin mineralization. Both pneumato-hydrothermal deposits and cassiterite sulphide deposits are derived from the Laoka granite body, subjected mainly to gas-phase boiling, while only the cassiterite sulphide deposits are produced from the Masong rock body which underwent liquid-phase boiling. As for the rock bodies poorly related to tin mineralization, boiling is observed merely in their vicinities or on both sides of fractures. This characteristic can serve as one of the indicators for ore prospecting.


## Introduction

The tin-polymetallic deposits of the Gejiu Mine in Yunnan Province, China, are genetically related to the middle-late Yenshanian granites in both time and space. On the basis of studies on the fluid inclusions in the minerals from the ore deposits, the authors focused their researches on the inclusions in minerals from magmatic rocks, with emphasis on their morphology, salinity, composition, and forming temperature and pressure, for the better understanding of the genesis of the ore deposits and physical-chemical conditions of the rock formations and mineralization. These fluid inclusions are believed to be one of the indicators for ore-exploration.

The igneous rocks in Gejiu are a basic-acidic-alkalic complex predominated by granite (Fig. 1). The granite shows a distinct regularity of temporal and spatial evolution in its chemical composition, mineral content, accessory mineral features, texture, structure, and deuteric alteration. The late stage granite, which is more acidic and alkalic, poor in CaO, FeO and MgO and rich in volatile, is believed most favourable for tin mineralization.

[1] No. 308 Geological Party of Southwest China Geological Exploration Corp., CNNC, Gejiu, Yunnan

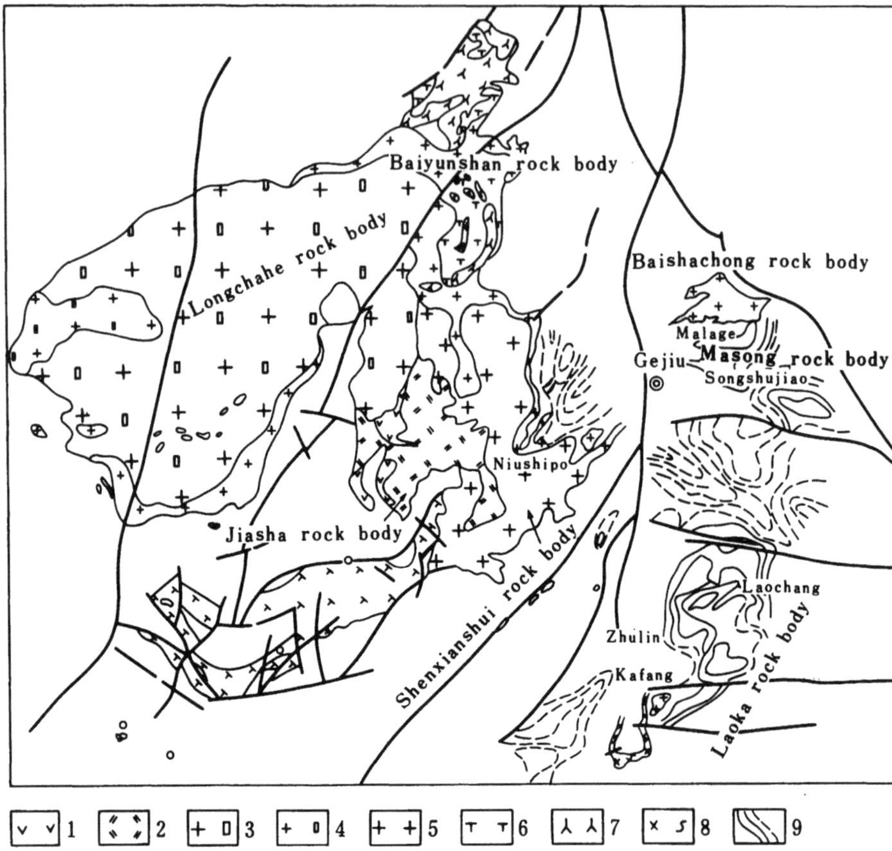


**Fig. 1.** Geological map showing the distribution of igneous rocks in Gejiu, Yunnan Province. 1-Gabbro; 2-Monzonite; 3-Porphyritic biotite granite; 4-Fine-grained porphyritic biotite granite; 5-Granular biotite granite; 6-Syenite; 7-Nepheline syenite; 8-Metadiabase; 9-Contour-lines of granite

## I. The Classification and Occurrence Features of the Inclusions

325 samples from nine rock bodies have been determined for the homogenization temperatures of the inclusions, using Leitz 1350 and 350 type heating stages. In accordance with their physical properties, the inclusions are classified into six types and nine subtypes (Table 1):

(1) Liquid inclusions: This type of inclusion, mostly secondary, contains only liquid and they are characterized by small size (less than 2 $\mu$m), sub-rounded shapes, and light red colour.

(2) Fluid inclusions: The gas/liquid ratios of this type vary from 10% to 60%, according to which the inclusions are divided into two subtypes, one with gas/liquid ratio of 10—30% and the other 30—60%. They are light red in colour, irregular, sub-rounded, square, oval, and negative crystalloid in shape, and generally 1—10 $\mu$m in size, with some bigger ones ranging from 10 to 20 $\mu$m and a few individuals reaching

**Table 1.** The characteristics of inclusions in the minerals from igneous rocks of Gejiu

| | Liquid inclusion | Gas-liquid inclusion | | Gas inclusion | $CO_2$-bearing inclusion | | Daugh-mineral inclusion | Melt inclusion | |
|---|---|---|---|---|---|---|---|---|---|
| | | 10–30% | 30–60% | | bi-phase | tri-phase | | Crystal inclusion | Solid inclusion |
| Jiasha rock body | | +++ | | + | | | | + | + |
| Longchahe rock body | +++ | +++ | + | + | | + | + | + | + |
| Shanxianshui rock body | +++ | +++ | + | | | + | + | | + |
| Baishachong rock body | +++ | ++ | + | | | + | + | | + |
| Daimoshan rock body | | ++ | + | + | | + | ++ | | |
| Malage rock body | | +++ | + | + | | ++ | ++ | + | + |
| Songshujiao rock body | | +++ | ++ | ++ | + | ++ | +++ | + | + |
| Gaofongshan rock body | | ++ | + | | | ++ | ++ | | |
| Laochang rock body | | ++ | ++ | +++ | + | ++ | + | + | + |
| Zhulin rock body | | ++ | ++ | ++ | +++ | + | + | + | + |
| Kafang rock body | | ++ | + | ++ | | + | + | + | |
| Baiyunshan rock body | | ++ | + | ++ | | | | | |
| Tabai volcanic rock | | ++ | | | | + | | | |

+++ high content, ++ moderate content, + low content.

30–40 $\mu m$. This type is made up of not only primary but also secondary inclusions. The low ratio subtype contains more inclusions than the high ratio one.

(3) Gas inclusions: This type has gas/liquid ratios >60%. They are generally 5–10 $\mu m$, black, and sub-round, negative crystalloid and irregular with very few as irregular fluid inclusions.

(4) $CO_2$-bearing inclusions: They are to two subtypes: the bi-phase $CO_2$-bearing inclusions consisting of both liquid and gaseous $CO_2$, and tri-phase $CO_2$-bearing inclusions made up of gaseous, liquid and fluid $CO_2$. The former subtype inclusions, probably formed under low temperature, have homogenization temperatures ranging from 17 to 27°C. These inclusions, sub-rounded or irregular, have gas/liquid ratios varying from 10% to 30% with some reaching 70–80%. The latter subtype inclusions, black, around 10 $\mu m$, and sub-rounded and irregular, have homogenization temperatures of 200°C–380°C, with some down to 26–30°C.

(5) Saline daughter mineral-bearing inclusions: This type is composed of inclusions of two phases made up of liquid and daughter minerals, three phases of gas, liquid, and daughter minerals; and four phases of gas, liquid, $CO_2$ and daughter min-

erals. The inclusions contain 1—4 kinds of saline daughter minerals which are difficult to identify, excepting NaCl and KCl. Inclusions of this type are chiefly irregular in shape with sizes varying from 5 to 10 $\mu$m, some reaching 15—20 $\mu$m.

(6) Melting inclusions: They may be further classified into two subtypes: the crystalline melting inclusions and glassy melting inclusions. The former subtype chiefly consists of apatite crystals, generally about 10—20 $\mu$m in size with some bigger ones up to 40 $\mu$m, which have either bubbles or fluid. The later subtype inclusions are irregular or subrounded, light-green, and <10 $\mu$m and contain round or irregular black bubbles edged by glassy matter and consolidated magma.

## II. Main Features of the Inclusions in Minerals from Various Igneous Rocks

1. *Jiasha gabbro-monzonite body:* No inclusion has ever been recognized in the basic gabbro, while some glassy melting inclusions and fluid inclusions as well as a few gas inclusions have been found in augite, apatite and quartz from quartz diorite hybridized with granite (Table 2).

2. *Longchahe, Shenxianshui, and Baishachong Rock Bodies:* They are respectively composed of porphyritic biotite granite and medium-grained equigranular granite, containing inclusions of similar features. The inclusions, mainly of liquid phase and fluid phase with low gas/liquid ratio, smaller than 5 $\mu$m in size and in subrounded or irregular shapes, generally occur in quartz; while fluid and gas inclusions, tri-phase $CO_2$-bearing inclusions and saline daughter mineral bearing inclusions of multiple phases are restricted to the inner alteration zone of the intrusion front and to the vicinities of mineralized areas and fracture zones. The inclusions with homogenization temperature of 160—220°C are secondary. The inclusions of 220—360°C and 300—420°C are closely related to the mineralization while those of 460—700°C have nothing to do with mineralization. The high salinity (>23.3%, calculated as NaCl) indicates that the solution is oversaturated.

3. *Malage and Songshujiao Rock Bodies:* They are chiefly made up of porphyritic biotite granite characterized by higher acidity, stronger deuteric alteration, high volatile content, and wide distribution of inclusions of multiple phases. The inclusions reveal a strong deuteric pneumatolito-hydrothermal process during the magmatic stage of the rock bodies. The homogenization temperatures are 230—360°C for fluid inclusions and 360—440°C for gas inclusions, being closely related to the mineralization. The rock bodies contain quite a few saline daughter mineral-bearing inclusions, up to 50% in some parts of the Songshujiao body. Inclusions of high salinity and $CO_2$ and with different gas/liquid ratios are better developed in the upper part than in the lower part, richer at the 1995 m than at the 1820 m level. The inclusions at the 1360 m level are different from those in the upper part, characterized by less development, small size and quantity, excepting high salinity.

4. *Laochang, Zhulin, and Kafang Rock Bodies:* Chiefly consisting of light coloured medium- and fine-grained granite, they are characterized by high acidity and alkalinity, strong deuteric alteration, and abundant volatiles F, Cl and B. The rock bodies

**Table 2.** The thermodynamic status of inclusions in minerals from igneous rock of Gejiu

| Rock body | Ht* of liquid-gas inclusion and gas inclusion | | | | Ht of $CO_2$-bearing inclusion | | Inclusion of solidified magma | | Chrystalline apatite inclusion |
|---|---|---|---|---|---|---|---|---|---|
| | I | II | III | IV | partially Ht | Ht | Homogeni-zation method | Quench method | |
| Jiasha rock body | | 200–220°C | | > 600°C | | | | | |
| Longchahe | 160–220°C | 240–280°C | 300–350°C | > 700°C | 30°C | 200°C | | | > 800°C |
| Shenxianshui rock body | 100, 120°C<br>160–220°C | 200–340°C | 340–420°C | 480°C | 28°C | 360°C | | 1140°C | |
| Baishachong rock body | 160–220°C | 280–360°C | | | | | | | |
| Daimoshan rock body | 120–170°C<br>200–220°C | | 400–460°C | | | | | | |
| Malage rock body | 160–220°C | 220–360°C | 360–440°C | 700–730°C<br>580°C | 17–29°C | 360°C | 850°C, 1000°C | | > 800°C |
| Songshujiao rock body | 100–140°C<br>180–200°C | 240–400°C | 360–400°C | 680°C | 28°C, 29°C | 280°C<br>300–320°C | > 800, 1160°C | > 1000°C | > 700°C |
| Gaofongshan rock body | 140–180°C | 300–320°C | 400°C | 550°C | | | | | |
| Laochang rock body | 140, 150°C<br>180–200°C | 240–400°C | 400–460°C<br>500°C | > 550°C | 26°C, 27°C | 340°C, 380°C<br>440°C | | 1140, 1170°C | > 800°C |
| Zhulin rock body | 100–120°C<br>160–220°C | 240–340°C | 360–400°C | 460–550°C | 17.20.23°C<br>26.27.28°C | 300°C | 1140°C | | |
| Kafang rock body | 110°C<br>160–200°C | 240–420°C | 340°C | > 500°C | 26°C | 300°C | 900°C | | |
| Baiyunshan Nepheline syenite | 170–220°C | 250–300°C | 350°C | 780°C<br>800–860°C | | | | | |
| Tabai volcanic rock | 140, 160°C | 200, 210°C | 340°C | > 640°C | | | | | |

* Ht – Homogenization temperature.

contain inclusions dominated by gas phase, suggesting stronger pneumatolito-hydrothermal processes. Bi-phase $CO_2$-bearing inclusions occur in great abundance, even up to 80% in the Zhulin body. Tri-phase $CO_2$-bearing inclusions and saline daughter mineral bearing inclusions become more abundant towards the ore-bodies, but the saline daughter mineral bearing inclusions are less than in the Malage and Songshujiao bodies and their salinity is lower. The homogenization temperatures of the inclusions range 240—420°C and 330—440°C, genetically related to the pneumoto-hydrothermal deposits.

5. *Damoshan Aplitic Granite and Gaofengshan Vein Rocks:* They contain various fluid, gas, $CO_2$- and saline daughter mineral-bearing inclusions, the salinity and complexity of which are higher than that of the main intrusives.

6. *Baiyunshan Nepheline Syenite and Alkali-Syenite:* Inclusions with low gas/liquid ratios have been recognized in aegirine, nepheline, fluorite, and sodalite. The gas inclusions occurring along the crystal structures of nephelines show facies change when heated to 780—800°C, with some remaining unchanged, suggesting the forming temperature of sodalite.

7. *Tabai Volcanic Rocks:* They are predominated by intermediate-acid volcanic tuffs and intermediate-basic lavas, containing melting inclusions and fluid inclusions in quartz.

## III. The Relationships Between the Inclusions in the Minerals of Granite and Mineralization

1. *Crystallization Temperature of Granite Determined with Melting Inclusions:* No phase took place when crystalline apatite inclusions were heated to 800—860°C using the Leitz 1350 heating stage. The homogenization temperature of glassy melting inclusions, measured by homogenization and quench methods, is 850—1170°C, higher than the initial melting temperature of granite obtained from the high temperature and high pressure experiments made by the Research Institute of Geology for Mineral Resources, China National Nonferrous Metals Industry Cooperation (Table 2). In view of the high fugacity of water and great abundance of volatiles, this temperature (850—1170°C) seems too high. The forming temperatures of fluid inclusions may be divided into four groups: (1) 160—220°C (some down to 100°C±): Falling into this group are post-mineralization secondary inclusions characterized by various sizes ranging to less than 10—30 $\mu$m, low gas/liquid ratio, irregular and subrounded shapes, pink colour, and linear arrangement along fissures; (2) 200—420°C: This group is represented by fluid inclusions with various gas/liquid ratios, which are of 5—10 $\mu$m size and irregular, subrounded, and negative crystalloid in shape and closely related to the hydrothermal mineralization. The fluid inclusions in the fluorine-associated sulphides have similar features, the homogenization temperature of which ranges 290—386°C; (3) 300—440°C: This group is predominated by subrounded and negative crystalloid, black, and 10 $\mu$m± sized gas inclusions which are related to the pneumatolitic mineralization. The similar inclusions from tourmaline veinlet deposits,

the forming temperatures of which have been measured as 358—248°C, are of gaseous phase in the early stage and fluid facies in the late stage. The cassiterites of the Gejiu deposits contain rare inclusions for which the homogenization temperature has been measured by the decrepitation method as 275—390°C. The homogenization temperatures of the inclusions in the cassiterites from the Damochang deposits are 360°C (liquid phase) and 400°C (gaseous phase); and (4) 500—700°C and >700°C: This group is represented by inclusions showing a scattered distribution. This temperature has not much relation with mineralization, but seems to be the forming temperature of the residual granite magma. The inclusions in the ore deposits are similar to and have an inheriting relation with those in granite, suggesting that the deuteric solution (or fluid) was the product of differentiation during crystallization of the magma.

2. *Salinity of Inclusions (Table 3):* The salinity of inclusions determined by the method of freezing with liquid nitrogen indicates that solutions inside the inclusions are mostly oversaturated with NaCl. The inclusions contain 1—4 kinds of saline daughter minerals such as NaCl. The dissolution temperatures of daughter minerals have been measured using the heating method at 100°C for KCl and 300—520°C for NaCl and the total salinity and relative contents of individual components have been obtained from phase diagrams of the relevant salt system, up to 77% for KCl+NaCl, indicating that the deuteric solution was highly saline. It is believed that the salinity of the inclusions in porphyritic granite is higher than that of the inclusions in granular granite, e.g. the salinity of inclusions in the Malage and Songshujiao rock bodies is higher than that of the inclusions in the Shenxianshui, Baishachong, Laochang, and Kafang rock bodies. Of the same type of granite, the late evolved bodies contain inclusions with salinity higher than the early evolved ones, e.g. the salinity of the Malage body is higher than that of the Longchahe body, and Damoshan body higher than the Baishachong body, and Songshujiao body is the highest. The late veins have higher salinity than the main rock bodies, e.g. the medium- and fine-grained granite veins in the Gaofengshan, Baishiyan, and Malage bodies all have higher salinity than the main bodies. The salinity becomes higher near the mineralized sections in a rock body. The skarn minerals also contain many saline daughter minerals bearing inclusions.

The $CO_2$-bearing inclusions are governed almost by the same regularity as the saline daughter mineral inclusions. The rock-forming pressure determined with $CO_2$ suggests that the intrusion depth of the rock body is approximately 2.5 km, almost the same as the depth estimated from the mineral phases, but less than that calculated from the thickness of the overlying rocks.

3. *Composition of the Inclusion Solution in Quartz:* The composition of the inclusion solution in quartz has been analyzed showing that the solution is weakly acid with pH values ranging from 5.95—6.55, favourable to the migration of Sn, Cu, Pb, and Zn etc. The low Eh (0.380—0.390) shows that the deposits formed in a reducing environment are mostly of sulphide type. The $CO_2$, Na, Cl, and K contained in the inclusion solutions become richer in the mineralization zone, for which the mineralization might be responsible (Table 4).

For better understanding the origins of the ore-forming materials, the deuterium and carbon contents of fluid inclusions in quartz and the hydrogen content of hyd-

**Table 3.** The salinity of inclusions in the minerals from the igneous rocks of Gejiu

| | Salinity of inclusion by nitrogen freezing | | Dissolution temperature of daughter mineral inclusions | | | | | | |
|---|---|---|---|---|---|---|---|---|---|
| | | | KCl | | NaCl | | Others | | Total salinity |
| | Freezing point | Concentration | Dissolution temperature | Salinity | Dissolution temperature | Salinity | Dissolution temperature | Salinity | |
| Longchahe rock body | -19.5 to 38°C | 21.18%-over saturated | | | | | | | |
| Shenxianshui rock body | -37.3 to 38.5°C | over saturated | | | | | | | |
| Daimoshan rock body | | | 110°C | 33% | 350—480°C | 41—55% | > 710°C | | |
| Malage rock body | -26.4°C | over saturated | 100—110°C | 32—34% | 360—500°C | 42—>60% | | | 72% |
| Songshujiao rock body | -17°C to 49°C | 19.7%-over saturated | 100—260°C | 32% | 240—520°C | 32—>60% | | | 77% |
| Gaofongshan rock body | -18.4 to -36°C | | | | 340—450°C | 40—53% | | | |
| Laochang rock body | | 21.2%-over saturated | | | 380—430°C | 42—50% | | | |
| Zhulin rock body | 100—270°C | 32% | 400—520°C | 44—>60% | | | | | |
| Kafang rock body | -32 to -34.6°C | over saturated | | | 400—520°C | 44—>60% | | | |

**Table 4.** The chemical composition of inclusions in quartz from the granite of Gejiu

| Rock body | Sample No. | Gas phase composition ml/100 g | | | | | pH | Conductivity $\mu\gamma$/Cm | Eh | Liquid phase composition (ppm) | | | | | | | | |
|---|---|---|---|---|---|---|---|---|---|---|---|---|---|---|---|---|---|---|
| | | $CO_2$ | $H_2$ | $N_2$ | CO | $CH_4$ | | | | $SiO_2$ | CaO | MgO | $K_2O$ | $Na_2O$ | $Fe_2O_3$ | $F^-$ | $Cl^-$ | $HCO_3$ |
| Longchahe rock body | 80-F-8 | 3.3 | 0.88 | 2.65 | 0.57 | 0.34 | 6.10 | 0.05$^2$ $10^2$ | 0.389 | 1.2 | 0.31 | 0.04 | 0.56 | 0.51 | 0.09 | 0.17 | 2.12 | 10.05 |
| Longchahe rock body | 80-F-11 | 3.47 | 0.85 | 1.76 | 0.52 | 0.41 | 6.26 | 0.038 $10^2$ | 0.387 | 1.13 | 0.06 | 0.05 | 2.01 | 0.23 | 0.11 | 0.11 | 4.95 | 5.02 |
| Longchahe rock body | 80-F-12 | 4.95 | 0.88 | 1.42 | 0.61 | 0.40 | 6.22 | 0.04 $10^2$ | 0.385 | 1.37 | 0.13 | 0.07 | 0.60 | 0.22 | 0.13 | 0.11 | 4.24 | 6.69 |
| Longchahe rock body | 80-F-18 | 4.38 | 0.60 | 2.12 | 0.46 | 0.26 | 6.16 | 0.042 $10^2$ | 0.383 | 1.06 | 0.22 | 0.06 | 0.53 | 0.27 | 0.13 | 0.1 | 5.65 | 13.39 |
| Shenxianshui rock body | 80-F-62 | 4.80 | 0.67 | 1.36 | 0.44 | 0.33 | 6.20 | 0.05 $10^2$ | 0.384 | 0.53 | 0.22 | 0.07 | 0.60 | 0.23 | 0.13 | 0.11 | 3.53 | 10.05 |
| Shenxianshi rock body | 80-F-56 | 4.36 | 0.68 | 1.31 | 0.46 | 0.27 | 6.24 | 0.04 $10^2$ | 0.391 | 0.76 | 0.22 | 0.02 | 0.48 | 0.58 | 0.09 | 0.012 | 3.53 | 6.69 |
| Shenxianshi rock body | 80-F-43 | 4.37 | 0.85 | 1.39 | 0.64 | 0.51 | 0.05 | 0.391 $10^2$ | 0.53 | 0.18 | 0.05 | 0.96 | 0.61 | 0.09 | 0.10 | 4.95 | 6.69 | |
| Baishachong rock body | 80-F-313 | 3.58 | 0.73 | 1.29 | 0.49 | 0.31 | 6.55 | 0.13 $10^2$ | 0.395 | 1.14 | 0.12 | 0.02 | 1.43 | 0.19 | 0.13 | 0.026 | 7.07 | 3.35 |
| Songshujiao rock body | 80-F-171 | 9.42 | 0.84 | 2.05 | 0.66 | 0.57 | 5.96 | 0.14 $10^2$ | 0.388 | 0.76 | 0.63 | 0.03 | 1.45 | 1.53 | 0.18 | 0.10 | 7.06 | 10.05 |
| Songshujiao rock body | 80-F-151 | 14.60 | 0.82 | 2.92 | 0.90 | 0.90 | 6.22 | 0.10 $10^2$ | 0.380 | 0.85 | 0.22 | 0.06 | 0.97 | 1.45 | 0.27 | 0.14 | 8.47 | 10.05 |
| Niushipo granite porphyry | 80-F-63 | 7.99 | 0.86 | 1.27 | 0.64 | 0.67 | 6.12 | 0.09 $10^2$ | 0.386 | 0.60 | 0.31 | 0.05 | 0.60 | 1.16 | 0.18 | 0.07 | 4.95 | 10.05 |
| Niushipo pegmatite beryl | 80-F-61 | 5.46 | 0.88 | 2.10 | 1.45 | 0.99 | 6.29 | 0.075 $10^2$ | 0.385 | 1.20 | 0.22 | 0.04 | 2.05 | 1.03 | 0.15 | 0.36 | 4.95 | 6.69 |

roxyl in biotite found in the Gejiu granite have been determined by the Department of Geology, Beijing University. The isotopic compositions of D and O are: δ D –58.5 to –99‰, averaging –83‰ and $\delta^{18}O$ 9.3 to 9.9‰, averaging 9.7‰, generally agreeing with the isotope range of "magmatic water" defined by Taylor which reached equilibrium in the same rock body under rock forming temperature. The δ D values of the fluid inclusions in cassiterites are respectively –86‰ at the oxide stage and –112‰ at the sulphide stage. The $\delta^{18}O$ value of ore-forming fluid calculated with the fractionation equation is 6.0‰ for the oxide stage, 4.9‰ for the sulphide stage, and – 5.3‰ for the silicate stage. Isotopic determinations on carbon have also been made. The $\delta^{13}C$ values of the inclusions in the Longchahe rock body range from –50 to – 28‰ and the $\delta^{13}C$ values obtained from the inclusions in the calcites in hydrothermal veins are – 7.09‰ for the carbonate stage and –0.36 to –0.10‰ for the sulphide stage, revealing that there exist two different kinds of carbon, the deep-sourced carbon whose range of isotope content is almost the same as that of primary carbon and the organic carbon, indicating that organic carbon from the sedimentary country rocks was involved in the mineralization. Crystalline and glassy melting inclusions have been recognized in different types of granite in the Gejiu deposits. Thus it can be believed that the Gejiu granite was the outcome of a melting process instead of the product of metasomatic granitization. On the basis of the values $Sr^{87}/Sr^{86}=0.7095-0.7208$ for the apatite in the granite and $Sr^{87}/Sr^{86}=0.7102-0.7142$ and $\delta^{18}O>10$‰ for the whole rock and considering the isotopes δ D, $\delta^{15}O$ and $\delta^{13}C$ in the rock-forming fluids, the Gejiu granite is believed to be anatectic of crustal sources type. The similarities and inheritance of forms, $CO_2$ content, salinity, and composition between the inclusions in magmatic rocks and in ore deposits and the δ D, $\delta^{18}C$ and $\delta^{13}C$ compositions in the fluid inclusions in granite are also the evidence showing that the mineralization resulted from "magmatic water" with a small amount of atmospheric water added which increased as the carbonatization took place.

The liquid and gas inclusions with high $CO_2$ content and salinity and various gas/liquid ratios, which were developed simultaneously and occurred rhythmically, were generated from the boiling of the ore fluid. The boiling of the ore fluid was also responsible for the separation of acid from alkali, gas from liquid, and salt from water, which promoted the precipitation of minerals. It is quite obvious, therefore, that the boiling and pressure-reducing boiling of ore fluid are closely related to the mineralization of the Gejiu tin deposits. The evidence of boiling of ore fluid has been extensively recognized in the Laochang, Kafang, Malage, and Songshujiao rock bodies. Large pneumatolito-hydrothermal tourmaline veinlet deposits are associated with cassiterite sulphide deposits in the Laochang and Kafang rock bodies which were subjected principally to pneumatic boiling, producing large amounts of gas inclusions associated with some other kinds of inclusions. The Malage and Songshujiao rock bodies characterized by pneumatolito-hydrothermal sulphide mineralization mainly underwent liquid boiling which became stronger in the better mineralized zones. In the poorly mineralized rock bodies, the evidence of boiling of ore fluid can only be found in the vicinities of mineralized zones (Fig. 2). These facts may be taken as indicators for ore exploration.

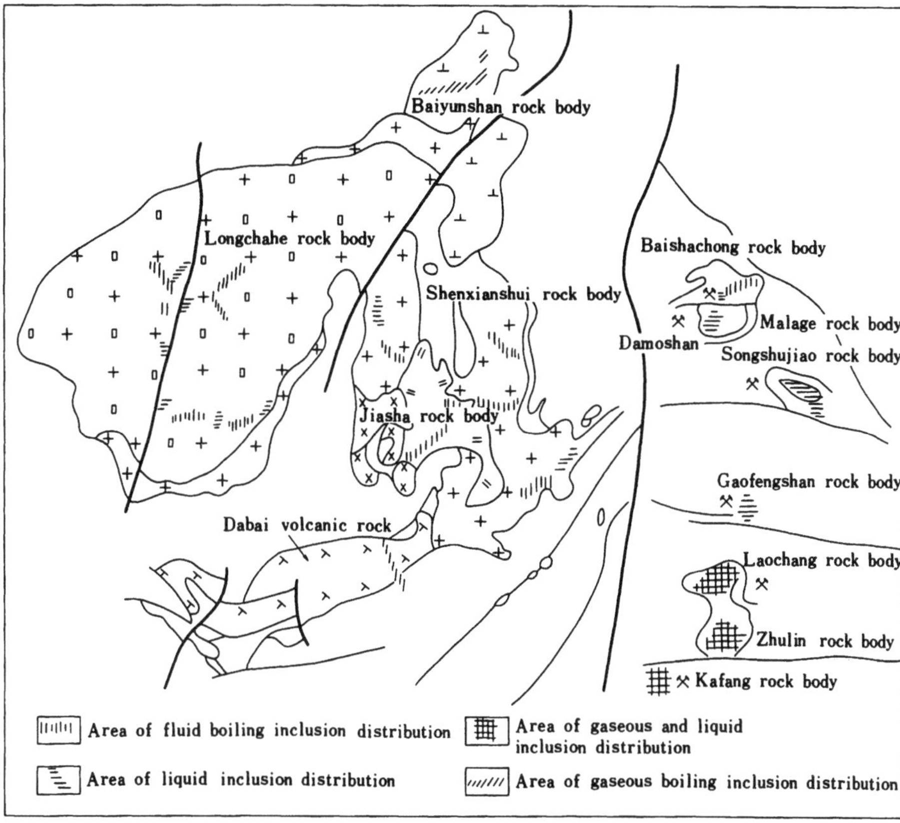


**Fig. 2.** A Sketch Showing the Relationship between the Inclusions and Mineralization in the Gejiu Granite

*Acknowledgements.* Sincere gratitude is given to Li Zhaolin, Associate Professor of the Department of Geology, Nanjing University, under whose directions this project was carried out. Wu Quizhi, Jiang Haosheng and Zhao Meifang, lecturers of the University who took part in the experimental work, are also acknowledged.

## References


Chen Yinhan, 1981. *Geochemistry of Inclusions in Minerals.* Hebei College of Geology (in Chinese).

Huang Fusheng, Mu Zhiguo, Chen Cheng and Wang Zhifen, 1983. The Study of isotopic composition of oxygen, hydrogen and carbon in granites of tin deposits, Gejiu. *Acta Petrologica Mineralogica Et Analytica,* Vol. 2, No. 4, 241—247 (in Chinese).

Lu Huanzhang, Wang Zhifen et al., 1981. Study of peculiarities of gas-liquid inclusions and their formation temperature at Gejiu tin deposits, Yunnan Province. *Proceeding of Symposium of National Meeting on Experiment Studies of the Inclusions in Minerals and the Genesis of Rocks and Minerals.* Science Press, Vol. 1, 15—20 (in Chinese).

Wang Zhifen, Yin Chengyu, 1982. Evolutionary rule and minerogenetic relation of the tin granite in Gejiu. *Proceedings of International Symposium on Geology of Granites and Their Metallogenetic relation,* 55—56 (Abstracts in English).

## 6.6 China: Porphyry and Ignimbrite Tin Deposits

# 6.6.1 A New Type of Tin Deposit – The Yinyan Porphyry Tin Deposit in China

GUAN XUNFAN, SHOU YONGQIN, XIAO JINGHUA, LIANG SHUZHAO, and LI JINMAO[1]

**Abstract**

The Yinyan tin-bearing granites are composite. They yield a K-Ar age of about 92 Ma, therefore they appear to be late Yenshanian. They are oversatured, peralkaline and rich in fluorine, tin, molybdenum and bismuth. They have a REE pattern which is similar to the S-type granites of south China. Explosive volcanic phenomena can generally be seen at the top of granites.

Orebodies occur at the upper part of these granites, which have the shape of an inverted cup. Ore minerals are mainly major cassiterite and subordinate wolframite, bismuthinite and molybdenite. Ores are characterized by a hypidiomorphic texture and veinlet-impregnation structure. Four alteration zones can be recognized: (1) chlorite hornfels zone; (2) topaz-greisen zone, which forms the main portion of mineralization; (3) the incorporated potassium-silicate zone; and (4) the potash-silicate zone. Fluid inclusion study has demonstrated that the temperature at the time of mineralization varied from 160 to 360°C.

## I. Geological Setting

The deposit is situated on the axial part of the arcuate Datianding structural zone which is developed along the Yunkai Caledonian uplift, lying between Guangdong and Guangxi west of the Wuchuan-Sihui deep fault, and striking NE-SW. The foreland has a convex shape towards the southwest. The country rocks consist of Pre-Devonian mica-quartz schists, gneisses, migmatites, etc. Within the deposit proper, there is an anticline complicated by sets of faults.

## II. Porphyry

The host rocks of this tin deposit are of complex granite-porphyry and quartz-porphyry intruded into an anticlinal axis. The porphyries are not exposed, with only several veins outcropping. Downward, they converge into one granite-porphyry body,

[1] 704 Geological Party, The Bureau of Geology and Mineral Resources of Guangdong Province, Zhanjiang, Guangdong

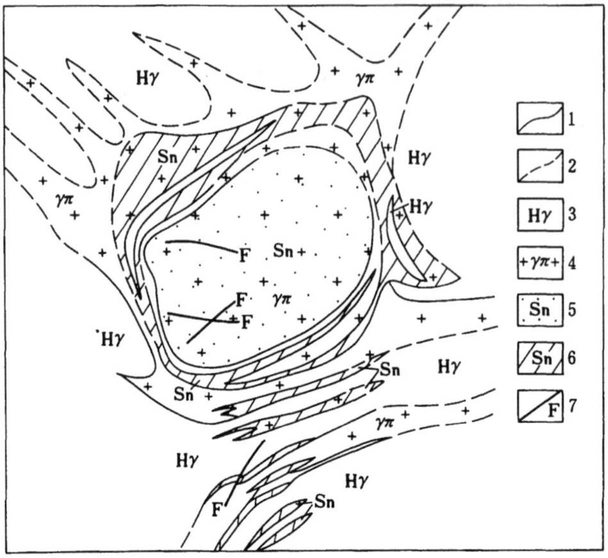


**Fig. 1.** The Yinyan tin-bearing porphyry at the 1057 level (scale 1:5,000)

and at the depth of 200 m it occupies an area of 0.06 $km^2$, in plane assuming an elliptic shape (Fig. 1) and pipe-like in section, with a steeply inward inclined contact zone.

Two samples from the slightly altered granite-porphyry rocks give a whole-rock K-Ar age of 80 and 83 Ma. The biotite from a pegmatite vein penetrating the granite-porphyries has provided a K-Ar age of 83 Ma. Another sample from a fresh pegmatite yielded a whole-rock K-Ar age of 92 Ma. It may therefore be inferred that the granite-porphyries were emplaced earlier than 92 Ma and they probably are late Yenshanian. The alteration and mineralization might have occurred during the period 80 to 83 Ma, and activities of mineralized hydrothermal fluids and formation of the genetically related pegmatite may have taken place almost at the same time.

The chemical composition of granite-porphyries is as follows: $SiO_2$, 75.25%, $TiO_2$ 0.02%, $Al_2O_3$ 12.34%, $Fe_2O_3$ 0.42%, FeO 1.54%, MnO 0.04%, MgO 0.30%, CaO 0.63%, $Na_2O$ 2.22%, $K_2O$ 4.96%, $SnO_2$ 0.02%, $WO_3$ 0.01%, $P_2O_5$ 0.02%, F 0.69%, and loss on ignition 0.98%. It can be seen that they are highly silica-saturated, alkali-rich (with $K_2O>Na_2O$), halogen-rich and poor in $R^{+2}$ (FeO+MnO+MgO+CaO=2.51). The granite-porphyry and quartz porphyry are characterized by high contents of fluorine (0.69–4.96%) and trace elements, such as Sn, W, Bi, Cu, Pb, Zn and Ag (ten to hundred times more than their clark values), as well as Li and Y.

They have a typical porphyritic texture (Fig. 2). Phenocrysts are of quartz, potash-feldspar, plagioclase and biotite with the matrix of fine to micro-grained subhedral potassium-feldspar and quartz.

Explosive-volcanic features are evident at the top of porphyries, where breccia and fissure-filling stockworks are quite common.

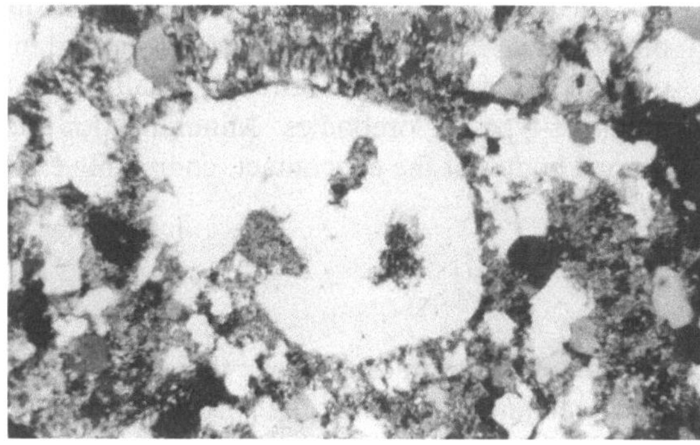

**Fig. 2.** Blastoporphyritic texture of the altered granite-porphyry, with a phenocryst of quartz shown in the centre. The matrix consists of quartz, topaz, etc. crossed polars 32x

## III. Orebodies

Four types of orebodies have been distinguished:

1) Inverted cup-like tin orebody occurring at the top of the porphyry. The porphyry is pervasively mineralized, lying at a depth of 200 m, and bifurcates into several tin-bearing veins extending up to the surface. Being in the form of an inverted cup, the marginal part of the orebody extends clearly deeper than the central part. Its mineralization was probably controlled by multiple alterations and strictly confined to fracture zones. These ores are predominated by subhedral textures and characterized by veinlet-dissemination (Fig. 3). Ore formation includes cassiterite-topaz, greisenized aplite and cassiterite-chlorite greisenized aplite, with their total thickness reaching 200 m or more.
2) A W-Mo mineralized core occurs at the centre of porphyry and is surrounded by the inverted cup-like tin orebody. The alternating molybdenite and wolframite mineralizations appear in the shape of sheets within this type of orebody. The pattern of ores is also characterized by a stockwork structure. The ore formation is a superimposed association of greisenized aplite and potassium silicates containing molybdenite and wolframite.

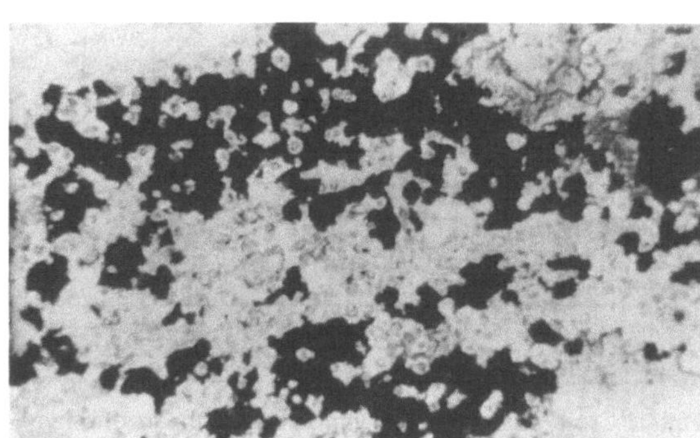

**Fig. 3.** Stock-work tin ore showing veinlet-dissemination texture. The dark colour is cassiterite and light coloured minerals are topaz and quartz, linearly-polarized light 32x

3) The near-surface tin orebody occurring in the quartz-porphyry dykes. This type of orebodies is seen as veins and lenses localized in some parts of the dykes. The size of this kind of orebody is generally small.
4) Sulphide-type tin orebodies. Mineralization occurs as lenticular and lotus-root shaped bodies at the exocontact, commonly exhibiting pinch and swell structure.

## IV. Ores

From top downward, it can be seen that, near the surface, there is cassiterite-sulphide type of orebodies; in the middle part it has changed to cassiterite greisenized aplite; while in the deeper part, there are molybdenite- and wolframite-bearing silicates and greisenized aplite (Fig. 4). The metallic minerals are cassiterite, wolframite, molybdenite, bismuthine, chalcopyrite, pyrite, specularite and a small amount of galena, sphalerite, hematite, magnetite, xenotime, etc., whereas the main nonmetallic minerals are quartz, sericite, hydromuscovite, hydromica, chlorite, and topaz, the subordinate are muscovite, fluorite, biotite, potassium feldspar, garnet, zircon, etc.

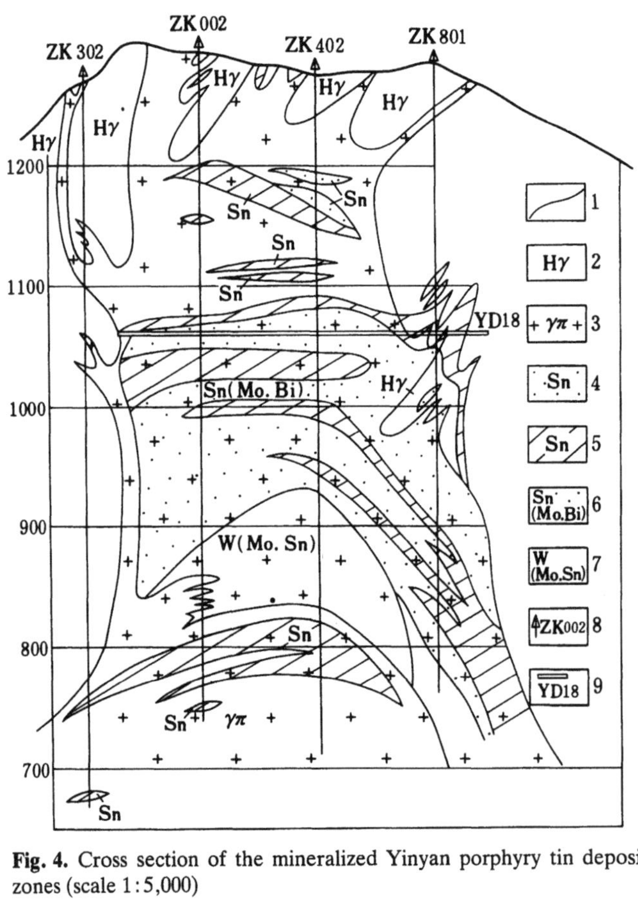


**Fig. 4.** Cross section of the mineralized Yinyan porphyry tin deposit showing the mineralization zones (scale 1:5,000)

## V. Alteration Zoning

Four alteration zones may be discerned (Fig. 5):

1) Chlorite hornfels zone: This zone occurs in country rocks near the porphyries. The alteration minerals are mainly chlorite, topaz, quartz, sericite, hydromuscovite, etc. Thermometamorphic minerals are biotite, cordierite, and andalusite, quartz, feldspar, garnet, etc. There is only weak mineralization of cassiterite developed within this zone, though in some places it may attain commercial grades.
2) Topaz-greisenized aplite zone: It is distributed in the middle-upper part of the porphyries. The altered minerals are mainly quartz, sericite, hydromuscovite, hydromica, chlorite, topaz, cassiterite, fluorite, muscovite, kaolinite, dickite, etc. It is the zone in which the major tin mineralization is developed.
3) Superimposed zone of potassium-silicate and greisenized aplite: The greisenization is overprinted on the potassium-silicate, giving rise to joint development of molybdenite, wolframite and cassiterite in this very zone.
4) Potash-silicate zone: This zone is located in the lower part of the porphyries. The major altered minerals are biotite and potash feldspar, with some quartz, hydromuscovite and chlorite added. Mineralization is very subdued in this zone.

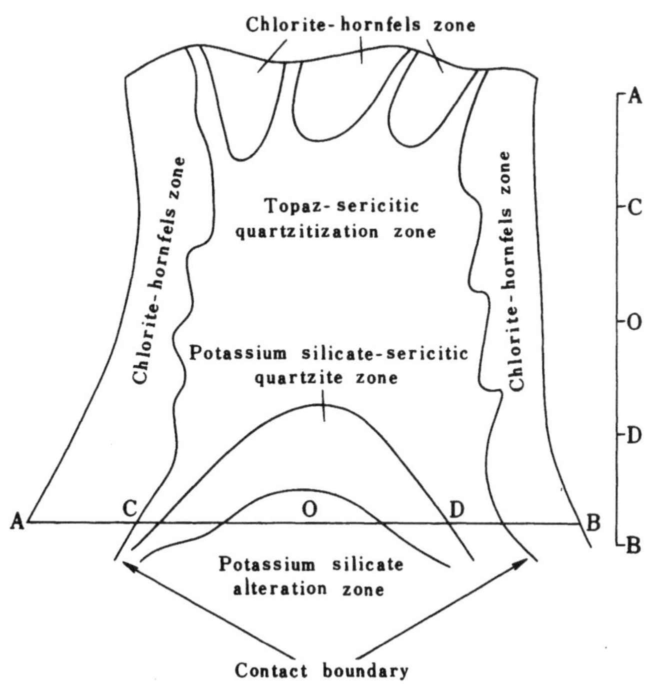


**Fig. 5.** Alteration model of the Yinyan porphyry tin deposit

## VI. Fluid Inclusions

Gas inclusions, liquid inclusions and multiphase inclusions of high salinity coexist closely (Plates 3, 4) (Figs. 6 and 7). The gas inclusions in phenocryst and matrix minerals of grano-porphyries have homogenization temperatures ranging from 300°C to 600°C; the homogenization temperature for multiphase inclusions ranges from 200°C to 570°C; the temperature for liquid inclusions ranges from 100°C to 520°C. A variety of gas inclusions from various quartz veins provided homogenization temperatures ranging from 320°C to 510°C; the same temperature for multiphase inclusions ranges from 200°C to 370°C; while for liquid inclusions from 140°C to 430°C. Therefore, the conclusion can probably be drawn that hydrothermal alteration and ore precipitation came into play under a boiling conditions when the pressure of the hydrothermal fluids was being decreased. Hydrothermal activity began at the late stage of magmatism and did not finish until middle-to-low temperature. The existence of a large amount of gas inclusions shows that both emplacement of the grano-porphyries and hydrothermal activity occurred at a very high level. The existence of a large number of multiphase inclusions of high salinity shows that the hydrothermal fluids probably contain a great quantity of HF and HCl, and resulted in resorbing of feldspar and quartz in granitic rocks. During the formation of the greisenized aplites a large amount of $Na^+$ and $K^+$ might have been introduced into these fluids, resulting in the formation of a large amount of daughter minerals of NaCl and KCl.

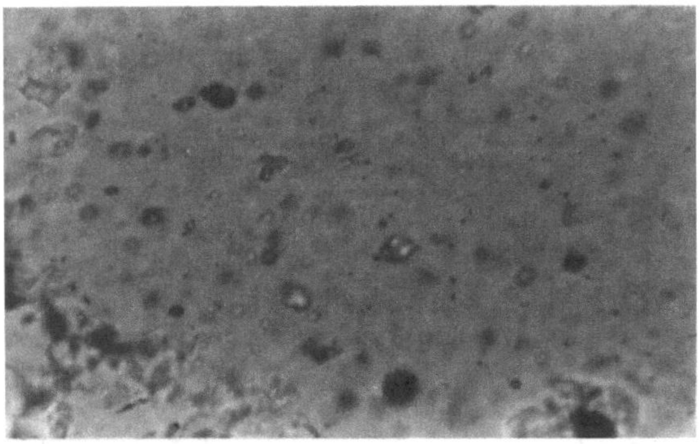

**Fig. 6.** Various inclusions in topaz greisenized aplite, linearly polarized light 320x

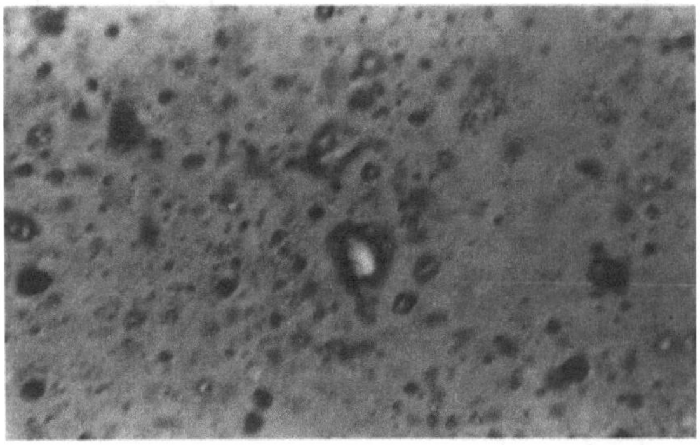

**Fig. 7.** Various inclusions in quartz genetically related to silicification, linearly polarized light 320x

## VII. Mechanism of Ore Formation

The Yinyan ore-bearing porphyry seems to be a product of water-rich magmas. It is speculated that when these magmas moved up to a high or extremely high crustal level, they still maintained a connection with their magma chamber for quite a long time. With the withdrawing of the deep-seated magmas, a great quantity of hydrothermal fluids would be continuously separated and rise up to the top of the porphyries, and at the same time the top part of the grano-porphyries was cooling and would already have a crystallized outer carapace. Enough mechanical energy would be concentrated at the top of grano-porphyries to lead to explosion of the encrusted shell, resulting in the formation of a system of fissures and explosive breccia. The fissure system would create a suitable environment for the mineralizing hydrothermal fluids to circulate in, and also to precipitate under favourable conditions. It is clear that the evolution of hydrothermal fluids and change in physical and chemical gradients would cause the formation of special zoning of alteration and mineralization around the porphyry. A study of the rare-earth emements shows that the REE (La-Lu+Y) values vary from 117 to 497 ppm; LREE (La-Eu) from 24 to 156 ppm; HREE (Gd-Lu+Y) from 92 to 350 ppm; HREE approximately from 1.5 to 2.5; and Eu from zero to 0.115 (Table 1). The diagram showing the pattern of rare-earths is similar to that of anatectic S-type crustal granites in south China (Fig. 8). It is evident, therefore, that the source of the tin-bearing Yinyan porphyries is quite different from that of the copper-bearing porphyries. The latter are assumed to be derived from deeper basaltic sources, either from the lower crust or the upper mantle, whereas the tin porphyries must have been produced from the sialic crust. Therefore, the Yinyan porphyries represent a new type of tin-bearing deposit in China. In many aspects, they may be compared with the Bolivian tin porphyry deposits.

**Table 1.** Data on REE elements for the Yinyan granoporphyries (ppm)

| Element \ No. of samples | Zk 302−147 | Zk 302−584 | Zk 801−276.5 |
|---|---|---|---|
| La | 25.05 | 29.31 | 24.94 |
| Ce | 62.44 | 74.13 | 66.33 |
| Pr | 9.25 | 9.77 | 8.86 |
| Nd | 35.68 | 31.94 | 28.79 |
| Sm | 14.39 | 11.60 | 10.05 |
| Eu | 0.14 | 0.02 | 0.05 |
| Gd | 20.17 | 13.02 | 10.48 |
| Tb | 4.55 | 3.29 | 2.68 |
| Dy | 31.86 | 22.68 | 17.84 |
| Ho | 7.01 | 4.95 | 3.85 |
| Er | 20.75 | 15.55 | 12.37 |
| Tm | 3.83 | 3.16 | 2.54 |
| Yb | 26.05 | 23.90 | 19.05 |
| Lu | 4.18 | 3.81 | 3.15 |
| Y | 231.70 | 123.8 | 89.97 |

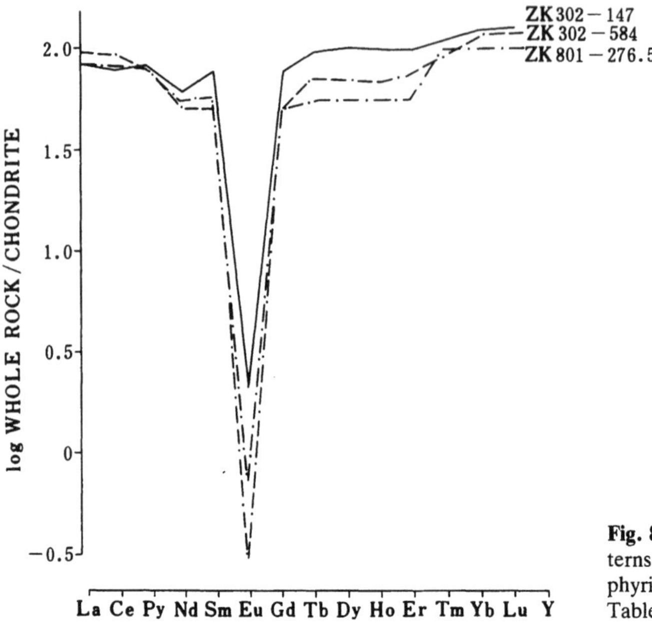


**Fig. 8.** REE distribution patterns of the Yinyan granite porphyries (for analytical data see Table 1)

*Acknowledgements.* The authors are grateful to our colleagues, Messrs. Rui Zongyao and Zhang Hongtao from the Institute of Mineral Deposits, Chinese Academy of Geological Sciences, for checking and revising this paper.

## References

Guiyang Institute of Geochemistry, Academia Sinica, 1979. *Geochemistry of Granite in South China.* Science Press (in Chinese).

Guo Wnkuei, 1982. On Granitoids Relevant to Metallogeny, *Regional Geology of China,* No. 2, 15–30 (in Chinese).

Li Xinqing, Rui Zongyao and Cheng Laixian, 1981. Fluid Inclusions and Mineralization of the Yulong Porphyry Copper (Molybdenum) deposit. *Acta Geologica Sinica,* Vol. 55, No. 3, 216–231 (in Chinese).

Rui Zongao, Huang Chongke, Xu Jue and Qi Guoming, 1982. Discrimination Between Ore-bearing Porphyry and Barren Porphyry from Fluid Inclusions in the Yu Long Porphyry Copper (Molybdenum) Deposit Belt of the Eastern Tibet Autonomous Region, *Bulletin of the Insitute of Mineral Deposits,* Chinese Academy of Geological Sciences, No. 2, 159–174 (in Chinese).

Zhu Xun, Huang Chongke, Rui Zongyao, Zhou Yaohua, Zhu Xianjia, Hu Congsheng and Mei Zhankui, 1983. *The geology of the Dexing Porphyry Copper Ore Field,* Geological Publishing House (in Chinese).

# 6.6.2 Geological Characteristics of the Ignimbrite-Related Xiling Tin Deposit in Guangdong Province

Lin Guiqing[1]


## Abstract

The Xiling tin deposit occurs exclusively in the upper Jurassic tin-rich (average Sn=86 ppm) rhyolitic ignimbrite. Orebodies are localized essentially at the strongly jointed bottom of the ignimbrite facies with medium-welded tuff texture, keeping a certain distance from the unconformity interface. This deposit is distinguished for its high economic value.

The tin orebodies consist of irregularly interwoven cassiterite veinlets, cassiterite-(feldspar)-quartz and cassiterite-calcite veinlets, occurring in pockets and stockworks. The host volcanic rocks are characterized by distinctive pyrophyllitization, sercitization, and silicification. Both mineralization and alteration were controlled by primary joints. This deposit, genetically related to ignimbrite, is thought to be of volcanic-subvolcanic hydrothermal type.

The Xiling tin deposit, located about 290 km east of Guangzhou, lies in the Circum-Pacific tin belt, and the southern end of the Zhe(Zhejiang)-Min(Fujian)-Yue(Guangdong) Mesozoic volcanic arc. The volcanic activity in the western part of the deposit is characterized mainly by quiet eruption, while that in the eastern part by strong eruption. The deposit occurs along the southeast margin of the Jianfengdong Mesozoic volcanic basin, a typical volcanic eruptive centre, which is situated at the junction of the Zhaoqing-Huidong latitudinal tectonic zone with the Chaoon-Pogou NE- and the Guazhai NW-trending faults (Fig. 1).

The Xiling tin deposit is confined to continental volcanic rocks and is of medium size and high grade. In addition, this deposit is distinguished for its big cassiterite crystals, good separability of the ore and its high recovery ratio, and its ore is easily dressed and mined.


## The Geological Features of the Deposit

The Xiling tin deposit occurs in rhyolitic pebble-bearing crystal-lithic ignimbrite of the upper Jurassic Guachiping volcanic series (Fig. 2) resting unconformably on the lower Jurassic slate and sandstone. No thermal contact metamorphism has been observed in the underlying rocks. Locally some fractures are seen on the unconform-

[1] Institute of Guangdong Geologic Exploration Corporation, CCN, Guangzhou, Guangdong

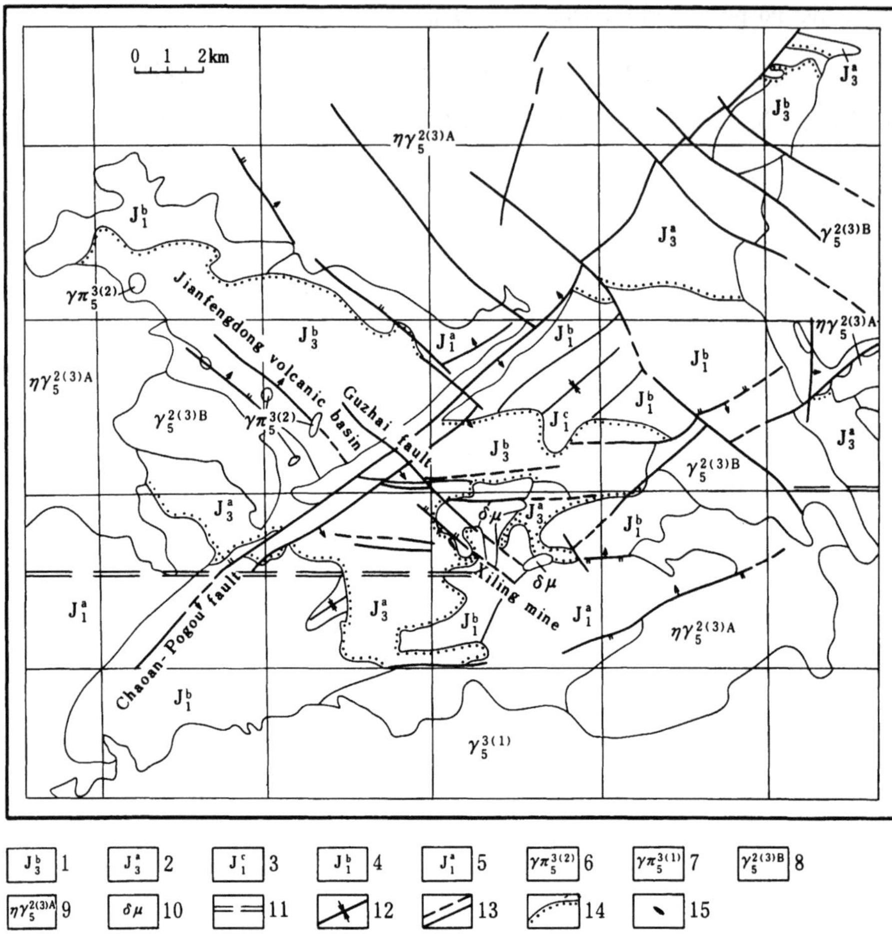


**Fig. 1.** Regional geological map of the Xiling tin deposit. (modified after the Institute of Geology annd Mineral Resources of Guangdong Province)

able contact with no apparent displacement, and they are not related to mineralization: There are two principal post-ore faults $F_1$ and $F_2$ in the district, which are all gravity faults striking N10°–40°E and dipping to southeast, and have no destructive influence on the orebodies.

The ignimbrite composed of variously welded tuffs is grey and greyish green. It contains many crystal, lithic and vitric clasts, plastic vitric fragments and plastic rock fragments. Opacitization has taken place along the cleavages of the biotite crystal fragments, which usually appear flattened and orientated. The lithic fragments are sharply angular, semi-round or round. The clasts are complex, consisting predominantly of slate and siltstone, with lesser sandstone, quartz porphyry, rhyolite, lamprophyre and aplite. They are unsorted and mixed together. The plasti-clasts appear broken and flame-like, and some of them are flattened. Welded tuff texture, double

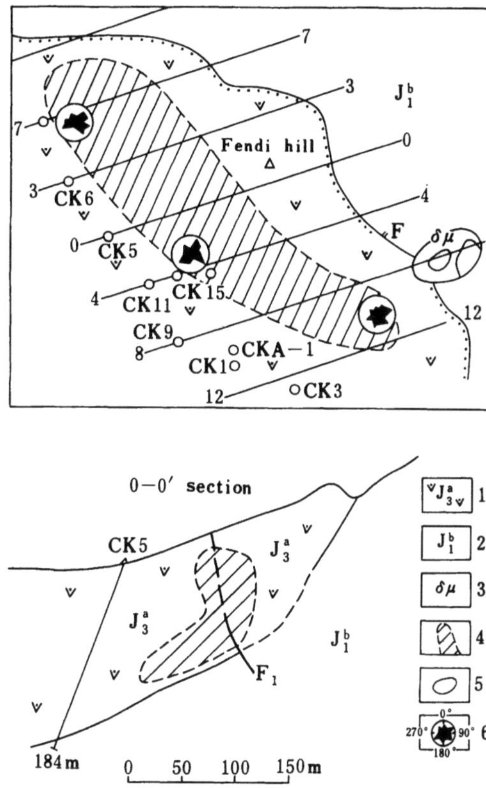


**Fig. 2.** Geological map of the Xiling tin mine (The lower diagram is the geological section along the o-o line). (modified after No. 911 Geological Exploration Party of Guangdong Province, 1958)

devitrification (Fig. 3) and pseudorhyolitic structure (Fig. 4) are often seen in the ignimbrite under the polarizing microscope.

The abundance of major element oxides and selected trace elements in the ignimbrite is listed in Table 1. On the basis of the data in Table 1, the ignimbrite can be described as: 1) It is aluminium-oversaturated. 2) The calculated norm places the ignimbrite in the rhyolitoid field of Streckeisen (1978) (Fig. 5). 3) The Peacock calc-alkali index is CA=76, and the Rittman index $\delta$ <1.8, indicating that the rock is representative of the Pacific-type calc-alkali volcanic series. 4) The differentiation index, DI=70−80, shows a relatively high degree of fractional crystallization. 5) The solidification index, SI=5.9−8.7, is quite different from mantle source values. 6) The Mg/Ti, Mg/Li, K/Rb and $TiO_2$/Ta ratios characteristic of the ignimbrite approximate to the geochemical indicator of stanniferous granitoids in various tin-bearing regions of the world, as suggested by Tischendorf (1977). The primary tin content of the volcanics is 86 ppm, which is 29 times as much as the common acidic igneous rocks, indicating that the magmatic source was likely to be rich in tin.

The volcanics occurring in the Xiling area may be subdivided into three facies which are given in the following ascending order:

1) *Ash-flow tuffaceous facies:* In general it is less than one meter in thickness, non-sorted, showing tuffaceous texture and no bedding. Primary joints are poorly developed, and no mineralization has been discerned in this facies.

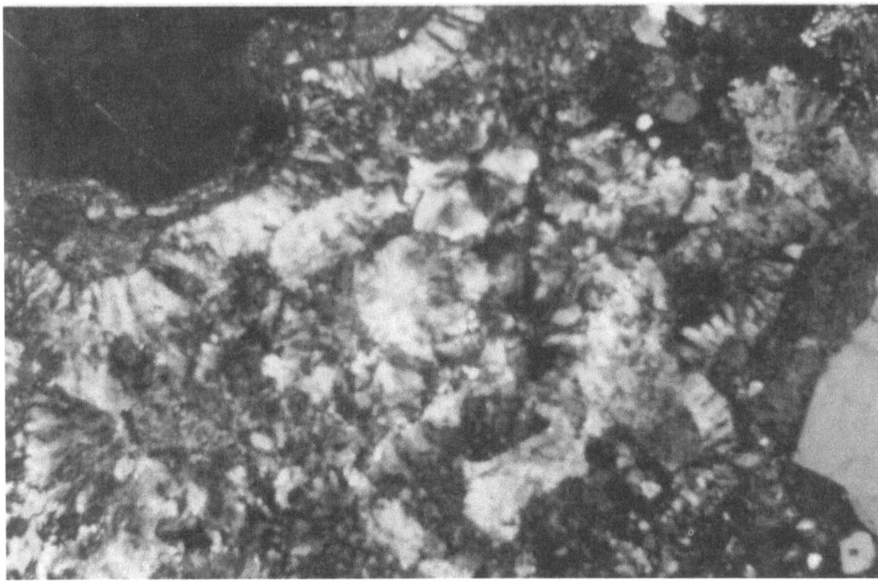

**Fig. 3.** Photomicrograph of double devitrification texture of ignimbrite with the rim exhibiting comb-structure and interior showing mosaic texture (linearly polarized light)

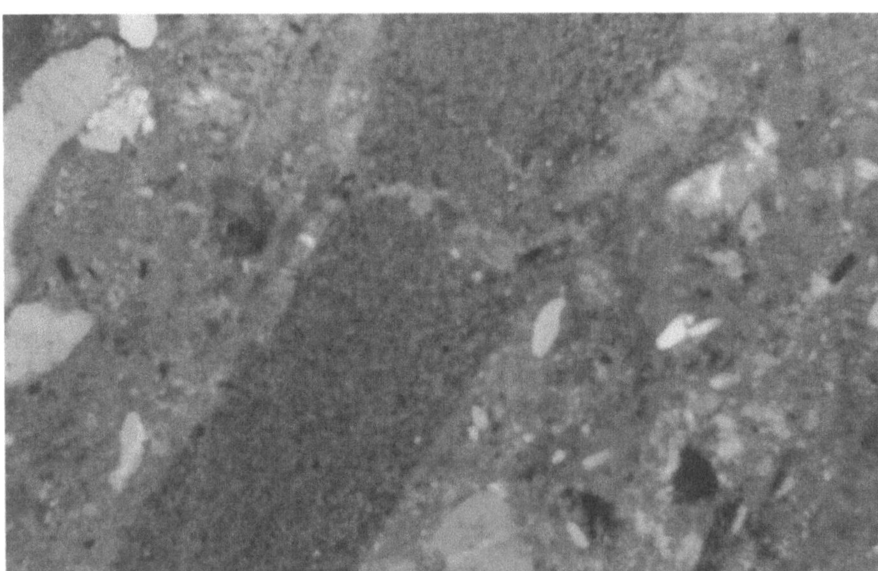

**Fig. 4.** Photomicrograph of pseudorhyolitic structure of ignimbrite with plastic lithic (dark green) and plastic vitric (yellow) clasts flattened and orientated (linearly polarized light)

**Table 1.** Abundance of major element oxides and selected trace elements in ignimbrite of the Xiling Tin Deposit

| Sample No. | weight % | | | | | | | | | | | | | | | | | $Al_2O_3$ / ($K_2O+Na_2O+CaO$) |
|---|---|---|---|---|---|---|---|---|---|---|---|---|---|---|---|---|---|---|
| | $SiO_2$ | $TiO_2$ | $Al_2O_3$ | $Fe_2O_3$ | FeO | MnO | MgO | CaO | $Na_2O$ | $K_2O$ | $P_2O_5$ | $H_2O$ | ignition loss | Total | δ* | DI* | SI* | |
| 1. | 72.64 | 0.25 | 14.04 | 0.48 | 2.06 | 0.08 | 0.52 | 2.98 | 2.51 | 3.12 | 0.07 | 1.12 | 1.06 | 100.93 | 1.07 | 78 | 5.9 | 1.63 |
| 2. | 72.08 | 0.24 | 14.19 | 0.51 | 2.04 | 0.12 | 0.59 | 3.42 | 2.22 | 3.68 | 0.07 | 1.12 | 0.93 | 101.21 | 1.20 | 76 | 6.5 | 1.52 |
| 3. | 71.52 | 0.15 | 13.97 | 0.47 | 2.60 | 0.07 | 0.75 | 2.50 | 2.86 | 3.68 | 0.07 | 1.38 | 1.05 | 101.07 | 1.50 | 79 | 7.2 | 1.54 |
| 4. | 71.30 | 0.33 | 14.03 | 0.55 | 2.38 | 0.05 | 0.83 | 2.66 | 1.77 | 3.88 | 0.04 | | 1.93 | 99.75 | 0.69 | 77 | 8.7 | 1.69 |
| 5. | 71.29 | 0.33 | 13.80 | 0.61 | 2.78 | 0.06 | 0.74 | 1.86 | 2.62 | 4.36 | 0.04 | | 1.23 | 99.73 | 1.72 | 80 | 6.6 | 1.56 |

| Sample No. | parts per million | | | | | | | | | | | | | | |
|---|---|---|---|---|---|---|---|---|---|---|---|---|---|---|---|
| | Sn | $WO_3$ | Bi | Mo | F | B | Li | Rb | Nb | Ta | TRO | Mg/Ti | Mg/Li | K/Rb | $TiO_2$/Ta |
| 1. | 280 | 80 | 20 | 20 | 50 | <10 | 29 | 201 | 10 | 8 | 360 | 0.48 | 107 | 117 | 187.5 |
| 2. | 90 | 60 | 20 | 20 | 30 | <10 | 29 | 219 | 7 | 8 | 390 | 0.39 | 124 | 139 | 175 |
| 3. | 20 | 80 | 30 | 30 | 40 | <10 | 25 | 255 | 17 | 7 | 400 | 0.20 | 180 | 129 | 128 |

Note: *δ-Rittman index; DI-Differentiation index; SI-Solidification index.
No.1, No. 2, No. 3 are from Guangdong 931 Party (1979); No. 4, No. 5 are from Institute of Bureau of Geology and Mineral Resources of Guangdong Province (1982).

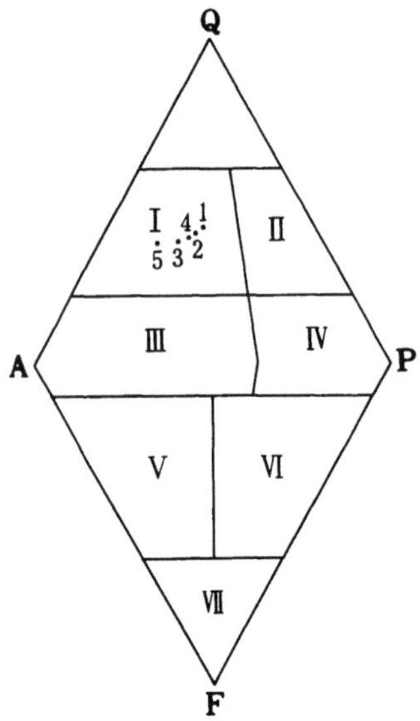


**Fig. 5.** The classification of volcanic rocks (generalized group names)

2) *Weakly welded tuffaceous facies:* The thickness varies from several to tens of meters; tuffaceous texture is substantially preserved; the vitric clasts are convex and polygonal in form, and are not apparently deformed. Radiated and sheaf-like devitrification structure is pronounced in this facies. Primary joints are not well developed and mineralization is weak.

3) *Ignimbrite facies:* Tuffaceous texture is practically not discerned; most of the vitric clasts are flattened and bent around rigid fragments, exhibiting pseudorhyolitic structure. The density of the ignimbrite is 2.5. Primary joints are well developed, especially at the bottom of this unit. This was presumably due to the reducing of the ignimbrite volume by one third to one half – the density varies from 1.3 to 2.5 (Wang and Zhou, 1982) in the welding process, in contrast with the ash-flow tuffaceous facies and weakly welded tuffaceous facies, the volume of which did not change greatly, leading to non-equilibrium in the shrinkage of different petrographic units. As a result, the joint system was created to trap mineralizing fluids, so that a strong mineralization is observed in this facies.

There is a gradational contact between these facies. Orebodies occur at the bottom of the jointed ignimbrite facies, not at the interface of the unconformity, but keeping always a certain distance from it, as shown in Figure 6.

The orebodies consist of quantities of irregularly interwoven cassiterite veinlets, cassiterite-quartz and cassiterite-calcite veinlets (Fig. 7), and the cassiterite veinlets are predominant. The mineralization is recognized to be multi-staged; the cassiterite-quartz veinlets were the earliest, the cassiterite veinlets later, and the cassiterite-bearing and -barren calcite veinlets the latest. As shown in Fig. 8, the ore-forming fluids filled the primary joints and replaced the ignimbrite to form veinlets with ore shoots at the confluence of the intersecting joints.

Ore constituents are rather simple with cassiterite as the only metallic mineral and pyrophyllite, sericite, quartz, calcite and feldspar as gangue. The cassiterites are dark

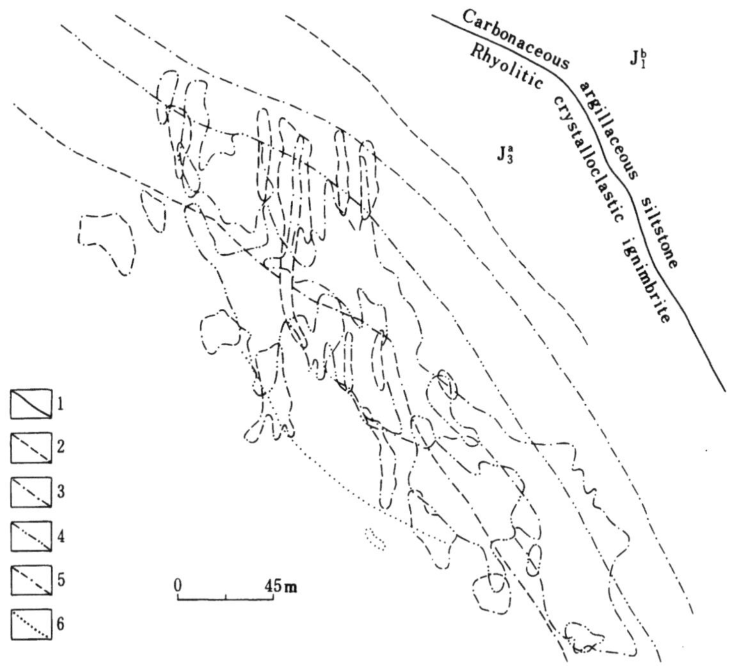


**Fig. 6.** The distribution of worked-out areas at the main levels in the Xiling tin mine, showing the ocurrence of orebodies keeping a certain distance from the unconformity interface

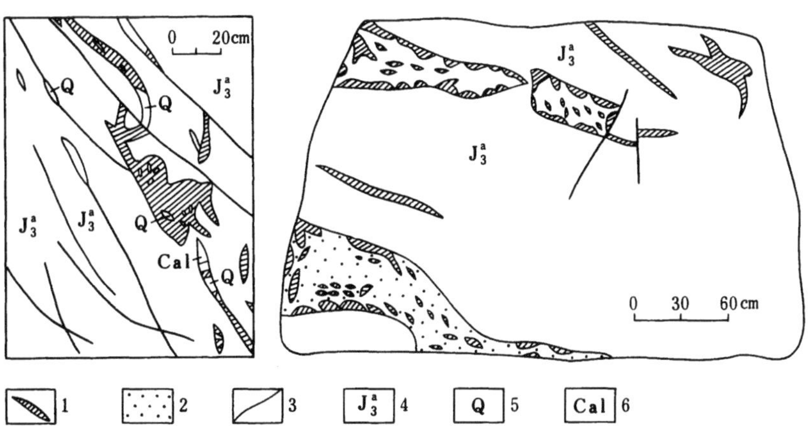


**Fig. 7.** Orebodies consisting of irregularly interwoven cassiterite, cassiterite-quartz and cassiterite-calcite veinlets

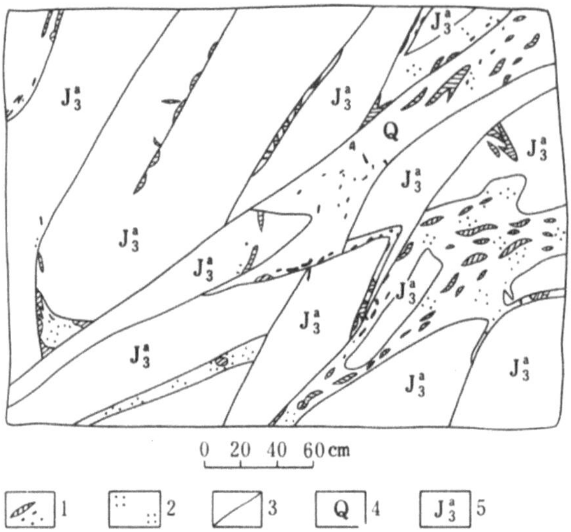


**Fig. 8.** Primary joints in the ignimbrite filled with ore-forming fluids

brown non-pleochroic, exhibiting a yellowish brown to reddish brown colour and zoned structure under the polarizing microscope (Fig. 9). The cassiterite crystals form elongated tetragonal prisms (Fig. 10) or are needle-like and rather big in size, between 0.2 and 5 mm, occurring usually as irregular drusy aggregates, sometimes with penetration twins. The trace element analysis (Table 2) reveals that the cassiterite is poor in Nb and Ta. The cell parameter data for the cassiterite are given in Table 3. These features indicate that the cassiterite is a product of epithermal mineralization.

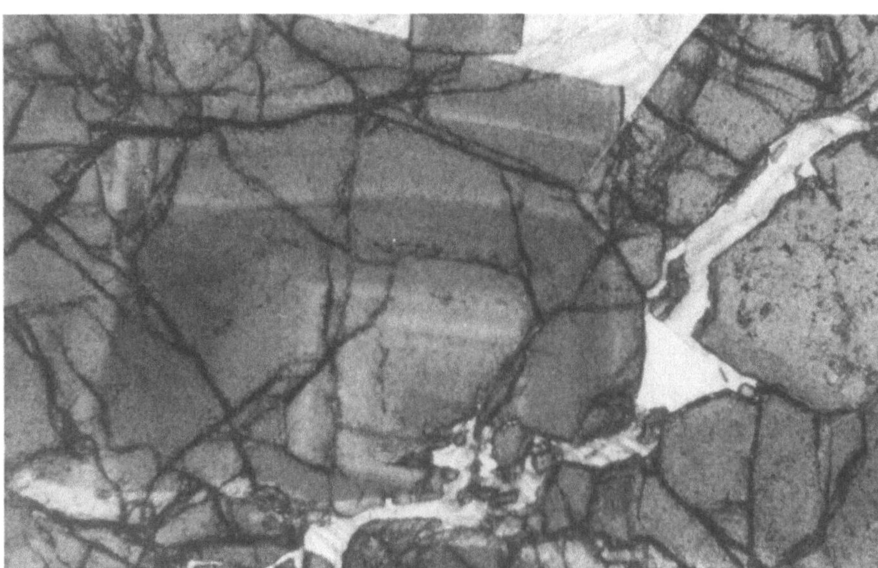

**Fig. 9.** Photomicrograph of a cassiterite crystal showing zoned structure (crossed polars)

**Fig. 10.** Photomicrograph of a cassiterite crystal appearing as an elongated tetragonal prism (crossed polars)

**Table 2.** Trace element analysis of cassiterite in wt%

| $Nb_2O_5$ | $Ta_2O_5$ | In | Ti | Total Fe | Mn | Sc | Zr |
|---|---|---|---|---|---|---|---|
| 0.0032 | 0.00 | 0.0107 | 0.0403 | 0.1680 | 0.0318 | 0.0030 | 0.0037 |

**Table 3.** Geological features of the Xiling tin deposit in comparison with typical porphyry tin deposits in China and Bolivia

| Item | Xiling tin deposit | Yinyan porphyry tin deposit | The porphyry tin deposits in Bolivia |
|---|---|---|---|
| Size of deposit | Medium | Large | Large |
| Depth extent | 120 m | Hundreds of meters | Hundreds of meters |
| Main-ore-controlling factor | Volcanic facies | Porphyry body | Subvolcanic stock |
| Main alteration-controlling factor | Primary joints in volcanics | Porphyry body | Subvolcanic stock |
| Hydrothermal breccias | Absent | Present | Present |
| Ore composition | Simple | Complex | Complex |
| Content of volatile-rich minerals | Low | High | High |
| Average grade of ore and grade variation | 1% and great variation in grade | 0.46% and little variation in grade | 0.5% and little variation in grade |
| Grain size of cassiterite | 0.2–5 mm | 0.02–0.2 mm | Very small |
| Cell parameter of cassiterite in Å | a.=4.7347<br>c.=3.1860 | a.=4.7376<br>c.=3.1889 | a.=4.738–4.741<br>c.=3.186–3.190 |
| Inclusions trapped into boiling fluid | Absent | Present | Present |
| Salinity of inclusions | Low | High | High |

After Wu Zhiliang, Liang Shuzhao, R.G. Taylor, R.H. Sillitoe, C. Halls et al.

Late stage pyrophyllitization, silicification, sericitization and carbonatization are predominant. Among them the pyrophyllitization and silicification were closely connected with mineralization. All alterations were controlled by the primary joints.

## Genetic Type of the Deposit

There are two different views on the genetic type of the Xiling tin deposit, some people hold that it is of porphyry type and others assign it to a volcanogenic hydrothermal origin.

It is obvious that the Xiling tin deposit greatly differs from the typical porphyry tin deposits both of China and foreign countries in geological features and ore-forming factors (Table 3). It shows the following characteristics, which indicate late phase volcanic hydrothermal mineralization.

1) The tin deposit exclusively occurs in volcanics and the ore localization is controlled by the ignimbrite facies.
2) Morphologically, the orebodies appear in the form of pockets and stockworks; they are of small size, pinching out rapidly towards depth, and are rich in tenor, particularly those occurring in unusual pockets.
3) The characteristic alteration of the wall rocks shows that the ore-forming fluids were acidic and genetically related to the volcanics. And among all alterations, pyrophyllitization, as a typical volcanic hydrothermal activity, is closely connected with mineralization.
4) Volatile-rich minerals are rarely seen in the ore.
5) There are few, if any, inclusions in the cassiterite; the homogenization temperature (Th) of quartz inclusions ranges from 171°C to 241°C. The characteristics of the cassiterite suggest that the Xiling deposit was likely to have formed in a relatively stable geochemical environment at a late stage in the formation of the ignimbrite when volcanism basically ended.
6) The primary tin content of the volcanics of the Xiling deposit is 86 ppm, as contrasted to 33 ppm in the immediate country rocks, which suggests that tin, as an ore-forming element, had been partly removed from the country rock into the ore-forming fluids in the process of ore-formation, so that the tin concentration in the ore-forming fluids increased.

Based on these characteristics of the Xiling tin deposit, the author considers it to be of volcanic-subvolcanic hydrothermal type, rather than of porphyry type.

## Metallogenetic Model

From the geological features and considering 1) the poor content of sulphide indicating a low $H_2S$ environment in which mineralization took place, and 2) cassiterite-bearing or -barren calcite veinlets and their country rocks being generally carbonatized, which shows a rich content of $HCO_3$ in ore-forming fluids, the author proposes a metallogenetic model for the Xiling tin deposit (Fig. 11). The metallogenetic mechanism is assumed as follows: the tin-rich rock fluid secreted in the process of

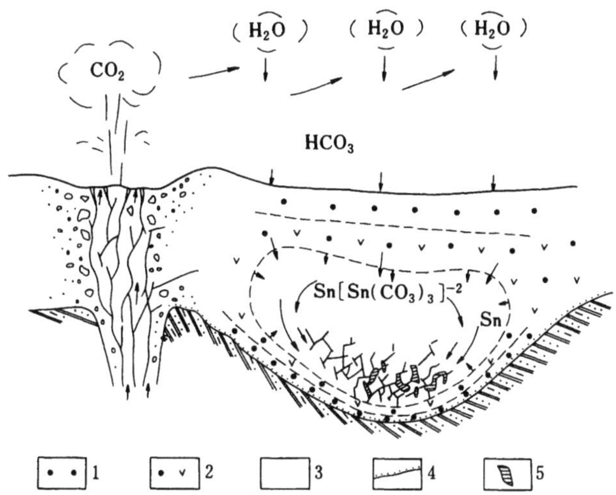


**Fig. 11.** Metallogenetic model

ignimbrite welding, and partial remnant volcanic hydrothermal fluids were mixed with the acidic surface water emanated by volcanism through circulation in the joint system so as to produce $HCO_3$-rich fluid and further extract tin from the adjacent volcanic rock, i.e. in superhypabyssal environment. The primary tin from the volcanics tended to be combined with $HCO_3$ or $CO_2$ and $O_2$ from the hydrothermal fluid to form a stannic carbonate complex (see reaction equation in Fig. 11), which moved along the joints and, with the decrease of temperature and pressure of the hydrothermal fluid, would form cassiterite or tin-bearing calcite veinlets when combined with calcium oxide from the country rock (see reaction equation Fig. 11). In a rough estimate, about 500 t of metallic tin could have been leached only from the host volcanic rocks, thus producing this rich deposit.

## Concluding Remarks

Judging from its economic aspect, the Xiling deposit is highly commercial, ranking with the best known tin deposits. In respect of the geology, it occurs in a distinctive environment and possesses a unique metallogenetic mechanism. It should be noted that this is not the only deposit of this type in Guangdong. In the light of available data, another two small tin deposits, quite similar to Xiling, have been found on the east margin of the same volcanic basin. It is believed that a further investigation of the ancient volcanic edifice, ancient geomorphic features and petrographic facies would open up prospects for finding more deposits of this type in Guangdong Province.

## References


Boissavy-Vinau, M. and Roger, G., 1980. The $TiO_2$/Ta ratio as an indicator of the degree of differentiation of tin granites. *Mineral. Deposita,* Vol. 15, 231–236.

Ehlers, E.G., and Blatt, H., 1982. Petrology. Copyright by W.H. Freeman and Company, U.S.A.

Grant, J.N., Halls, C., Sheppard, S.M.F., and Avila, W., 1980. Evolution of the porphyry tin deposits of Bolivia. *Mining Geology,* Special Issue No. 8.

Guo Lingzhi, Shi Yangshen and Ma Ruishi, 1980. *Tectonic framework and earth's crustal evolution in south China.* Geological Publishing House (in Chinese).

Sillitoe, R.H., Halls, C., and Grant, J.N., 1975. Porphyry tin deposits in Bolivia. *Econ. Geol.* Vol. 70, No. 5, 913–927.

Sillitoe, R.H., Halls, C., and Grant, J.N., 1976. Porphyry tin deposits in Bolivia - A reply. *Econ. Geol.,* Vol. 71, No. 6, 1065–1067.

Streckeisen, A., 1978. Classification and nomenclature of volcanic rocks, lamprophyres, carbonatites and melitic rocks. *N. Jb. für Mineralogie Abh.,* 134, 1–14.

Taylor, R.G., 1976. Porphyry tin deposits in Bolivia. *Econ. Geol.* Vol. 71, No. 6, 1064–1065.

Taylor, R.G., 1979. *Geology of tin deposits.* Elsevier Scientific Publishing Company.

Tischendorf, G., 1977. Geochemical and petrographic characteristics of silicic magmatic rocks associated with rare-earth mineralization. In: *Metallization Associated with Acid Magmatism.* Eds. Stemprok, M., Burnol, L., and Tischendorf, G. International Geol. Correlation Programme Vol. 2, 41–96.

Wang Dezi and Zhou Xinmin, 1982. *Petrology of volcanic rocks.* Geological Publishing House (in Chinese).

Zhang Shuye and Liu Ruxi, 1982. *Atlas of textures in volcanic rocks.* Geological Publishing House (in Chinese).

# 6.6.3 Volcanic Activity in Xiling Mine Area, Guangdong Province, and Its Genetic Relationship with Tin and Polymetallic Sulphide Deposits

Yu Zhunggi, Wang Shenyu, and Liao Guoxin[1]


## Abstract

The present paper deals with the temporal and spatial relationships between the tin and polymetallic sulphide deposits in the Xiling Mine area and the late Jurassic volcanic rocks in eastern Guangdong of the south-eastern coastal region of China. Judging from the geological characteristics of these deposits, the chemical composition of rocks and minerals, the inter-relations of various useful components, the rare-earth element distribution patterns, and sulphur and oxygen isotopes, it is evident that the tin and polymetal mineralizations at Xiling have a close genetic relationship with the volcanic activity of the area.


## Introduction

Along the coastal terrain of southeast China from the provinces of Zhejiang and Fujian to the eastern part of Guangdong, Mesozoic volcanic rocks are extensively distributed, where the Xiling mine area lies on the southeastern flank of the Lianhua mountain range in east Guangdong. In respect of regional tectonics, the area is situated in a composite position between the southern end of the Caho'an-Pogou Neochathaysian tectonic belt trending NE and the W-E Zhauqing-Huilai structural belt (Fig. 1).

The outcrops in the mine area are of Jurassic volcanic rocks and Lower Jurassic sedimentary rocks. The latter form a thick formation of sandy shale in alternating marine and continental facies, the known thickness of which is over 1,135 m. The former is an eruptive rhyolite-dacite formation of continental facies; mainly of lava intercalated with shale and volcanoclastic rocks. The intermediate acidic lava is characterized by rich salic and high K-Na but poor Ca-Fe content. In addition, such volcanic rocks commonly have microcontents of Y, Yb, B, Li, Sc, Sn, Be, Mo, Cu, Pb, Zn, etc., generally higher than the Clark values. The common accessory minerals include ilmenite, monazite, zircon, garnet, tourmaline, etc.

Tectonically, the mine area is situated at the juncture of three sets of geological structures trending W-E, NE, and NW. The western part of the mine area is penetrated

[1] Institute of Geology and Minerals Resources, Bureau of Geology and Mineral Resources of Guangdong Province, Guangzhou

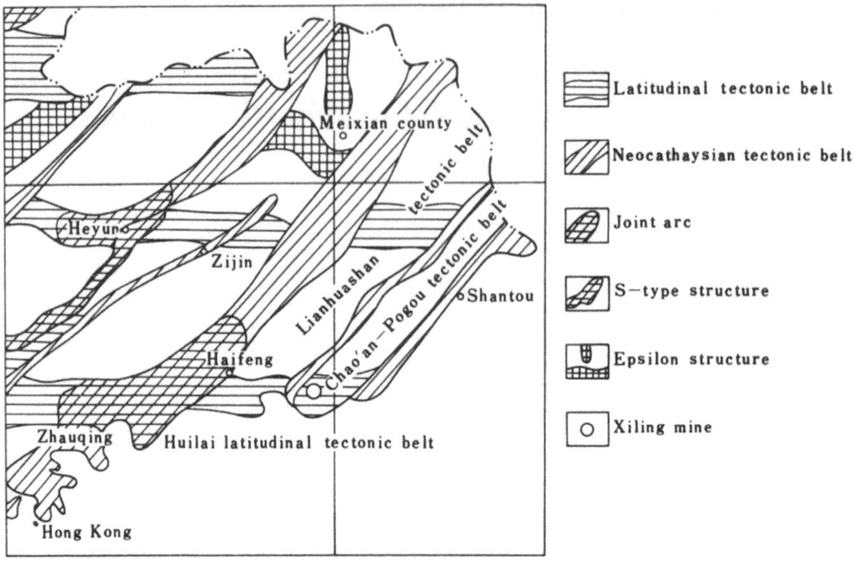


**Fig. 1.** Location of the Xiling Mine

by compresso-shear fractures of the latitudinal Zhauqing-Huilai tectonic belt; the eastern part is distributed with NE trending anticlines and synclines in association with the Cau'an-Pogou Neocathaysian fracture system, and in the middle part tensotorsional fractures trending NW are developed, accompanied by a faulted Jianfengdong volcanic basin. The latter two might be the minor structures of the Neocathaysian or the products of the interaction between the latitudinal and the Neocathaysian structures.

Quartz-diorite-porphyrite veins occur along the fracture zones, and pipe-like secondary dacite-porphyrite and dacite-rhyolitic tuffaceous lava bodies occur at the junctures of the fractures. Pyritization is very common in the quartz-diorite-porphyrite. Tin deposits are mainly related to tuffaceous lava, and polymetallic sulphide deposits are genetically connected with the secondary dacite-porphyrite (Fig. 2).

The case histories of the Fengdishan tin deposit and Sauzhoudi polymetallic sulphide deposit are described in the following paragraphs to illustrate the genetic relationship between the volcanic activity in the Xiling area and the tin and polymetallic sulphide deposits.

## I. Fengdishan Tin Deposit

The Fengdishan tin deposit is located in the southern part of the area, on the southeastern margin of the NW oriented Jianfengdong faulted volcanic basin. At the intersection of the two sets of the NW and W-E trending fractures occurs dacite-rhyolitic tuffaceous lava (hereafter called the tuffaceous lava body), which is of elliptic shape, 300 m long and 80—100 m along the short axis, converging downward into a long

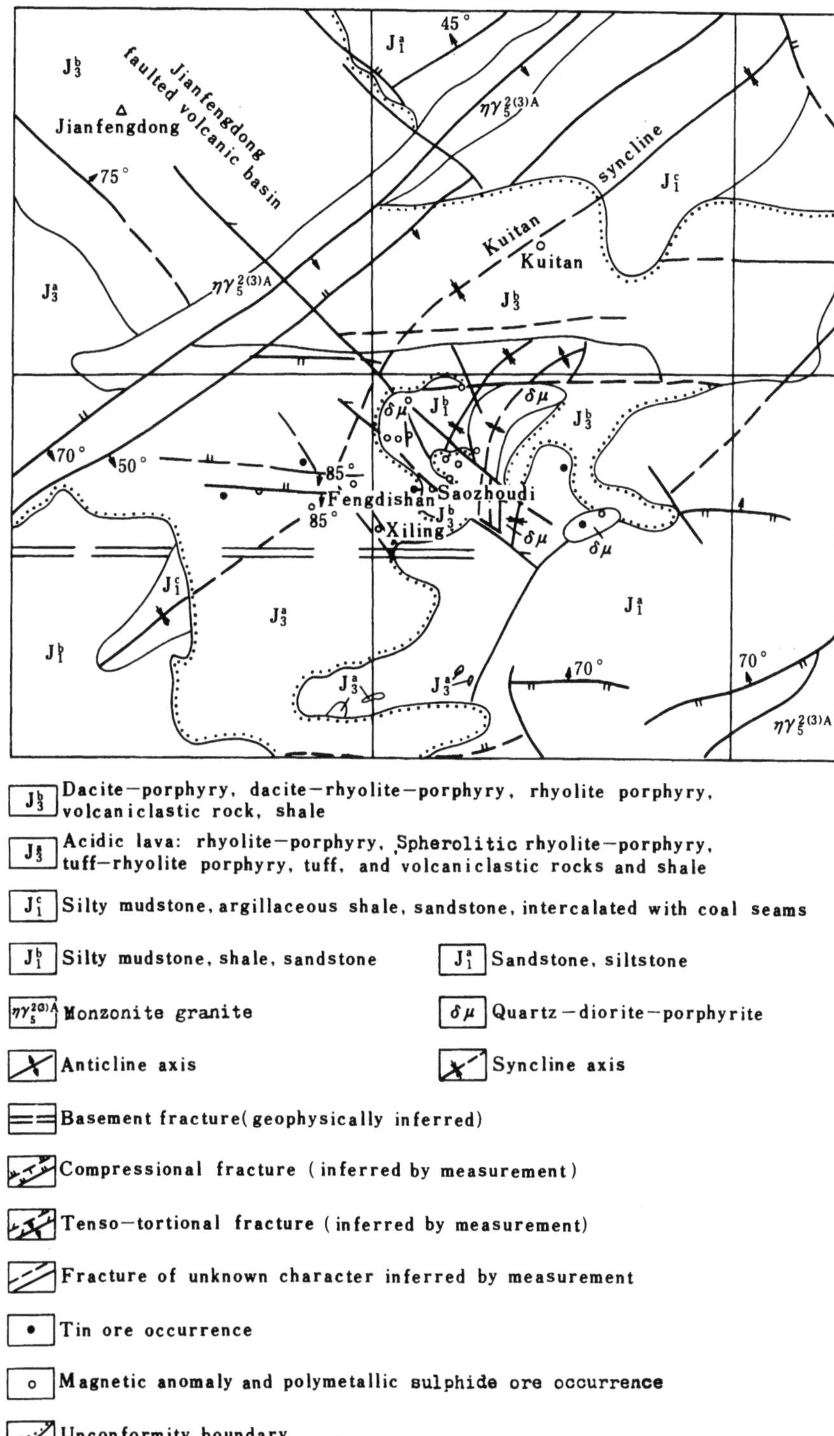


**Fig. 2.** Geological map of the Xiling mine area

stripe, and revealed to be funnel-shaped. The tuffaceous lava body disappears at depth because it was cut by faults after its mineralization. The rock body is in fault contact on the northeastern side with carbon-bearing argillaceous siltstone and in transitional relation on the southwestern side with dacite-rhyolitic breccia-bearing ignimbrite (ignimbrite hereafter), which plunges southeast, striking northwest, and dips southwest, with its dip angle steep in the upper part (50—60°) and relatively smooth in the lower part.

The tuffaceous lava body and the ignimbrite are similar in mineral association, lithochemical composition, and rare earth element content and distribution, but different in lithological structure and content of volcanic clastics.

The mineral constituents of both consist of plagioclase (An 10±), potash feldspar, quartz and volcano-clastic materials, with a felsic matrix. The average chemical compositions of the rocks are listed in Table 1.

It is shown in Table 1 that the alkali content decreases either in tuffaceous lava or in ignimbrite, and the Na content decreases even more sharply. This is because the alteration process caused by volcanic heat has induced feldspar particularly plagioclase to be metasomatized into aggregates of sericite or sericite-quartz. From the composition of the rocks as a whole, they have originated from acidic magma. In addition, the tuffaceous lava body has an iron content more than ferrous with an oxidation coefficient as high as 0.60, which indicates that oxygen from the atmosphere was mixed in during the formation process, and consequently the lithological character seems to be that in between hypabyssal and eruptive facies. The contents of rare earth elements in the tuffaceous lava body and ignimbrite are listed in Table 2. The REE distribution pattern is given in Figure 3.

The tuffaceous lava body commonly has a sand-like crystalloclastic-porphyritic structure with a felsitic lava-cemented groundmass. Whereas the ignimbrite is mainly of clastic, crystalloclastic, vitroclastic, or palimpsest volcanic-ash structure, it is cemented by volcanic ash in addition to felsic minerals. The latter usually come from the late recrystallization of the volcanic ash.

In the tuffaceous lava body the content of volcanic clastics is normally less than 50%, and argillaceous siltstone breccia of the basement rock can be seen on the periphery. However, in the ignimbrite the content of volcanic clastics may reach as high as 50%, the maximum being 80%. Breccias of the basement rock greater than 1 cm in diameter are observed in large amount and fine-grained clastics and vitrosubstances make up about 18%.

The content of porphyritic crystals in complete automorphous or semiautomorphous forms are more than that of crystalloclastics in the tuffaceous lava body, while in

**Table 1.**

| Rock | Number of samples | Chemical analysis (in percentage) | | | | | | | | | | | |
|---|---|---|---|---|---|---|---|---|---|---|---|---|---|
| | | $SiO_2$ | $TiO_2$ | $Al_2O_3$ | $Fe_2O_3$ | FeO | MnO | MgO | CaO | $Na_2O$ | $K_2O$ | $P_2O_5$ | Ignition |
| Tuffaceous lava | 7 | 72.15 | 0.35 | 14.64 | 2.07 | 1.40 | 0.04 | 0.67 | 1.21 | 0.83 | 4.18 | 0.03 | 2.41 |
| Ignimbrite | 6 | 72.11 | 0.34 | 14.06 | 1.45 | 1.84 | 0.05 | 0.64 | 1.36 | 1.81 | 4.35 | 0.04 | 1.81 |

**Table 2.**

| Rock | Content of REE (ppm) | | | | | | | | | | | | | | | Light REE: Heavy REE | | | | |
|---|---|---|---|---|---|---|---|---|---|---|---|---|---|---|---|---|---|---|---|---|
| | La | Ce | Pr | Nd | Sm | Eu | Gd | Tb | Dy | Ho | Er | Tm | Yb | Lu | Y | Total REE content TRR (ppm) | Total Ce content Ce (ppm) | Total Y content Y (ppm) | Ce/Y | Eu |
| Tuffaceous lava | 19.76 | 74 | 8.94 | 28.54 | 8.65 | 1.09 | 6.36 | 1.33 | 8.18 | 1.23 | 3.11 | 0.44 | 3.35 | 0.30 | 20.22 | 185.50 | 140.98 | 44.52 | 3.17 | 0.47 |
| Ignimbrite | 21.22 | 78.47 | 9.51 | 29.57 | 9.07 | 1.09 | 7.03 | 1.25 | 7.74 | 1.21 | 3.35 | 0.45 | 3.22 | 0.33 | 21.51 | 195.02 | 148.93 | 46.09 | 3.23 | 0.44 |

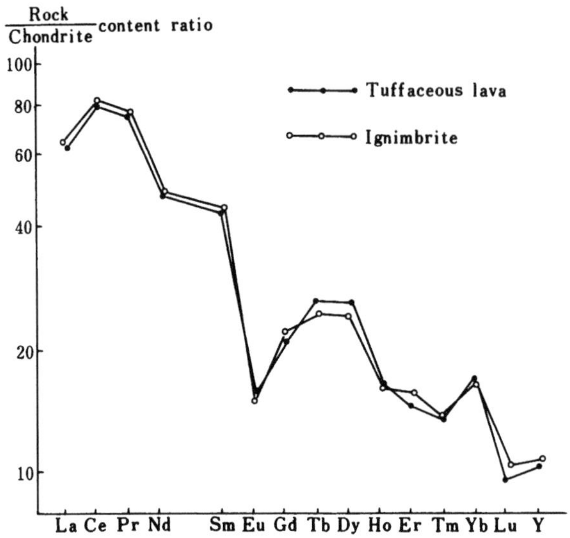


**Fig. 3.** Distribution pattern of rare earth elements

the ignimbrite crystalloclastics in sharp angular and resorption-embayed shapes or terraced fractures are more abundant than porphyritic crystals, among them the quartz crystalloclastics often have cracks caused by sudden cooling of the eruption.

The volcanoclastic substances gradually decrease from the centre toward the outer part of the tuffaceous lava body, but the crystallization grows intense with the increasing content of feldspar and quartz porphyritic crystals, and there generally occur pyritization, silicification, sericitization, chloritization, pyrophyllitization, etc. The rock is mainly silicified and pyritized in the upper part and becomes more and more serizitized, chloritized and pyrophyllitized downward, turning into a smooth texture of bluish grey colour, which feels lubricated. All these phenomena indicate that the tuffaceous lava is probably a variety of acidic lava of a volcanic-neck, a consanguineous product of the same volcanic mechanism as the eruptive ignimbrite.

From the semiquantitative spectrometric analysis of 162 rock samples from four profiles, the tin content averages 152 ppm for the tuffaceous lava body, 33 ppm for the ignimbrite, and 32 ppm for the carbon-rich argillaceous siltstone of Lower Jurassic age. It is obvious that the first has a tin content five times that of the latter two, increasing from upper to lower.

The Fengdishan tin deposit is developed in ignimbrite on the southwestern side of the tuffaceous lava body, in semi-ringed arc distribution closely round the latter from upper toward lower (Figs. 4, 5, 6, and 7).

The tin deposit is constituted of dense veinlets in parallel arrangement interwoven with plume veins, forming a special and complex stockwork type of commercial orebody. The occurrence of the orebody is in concordance with that of the tuffaceous lava body, dipping toward the southeast. The fine veins in the NW part of the orebody strike nearly N-S, dipping steeply east; in the southeastern part they strike NW, dipping moderately NE; in the middle part they strike nearly W-E, dipping steeply north. The ore veins are generally short, 5—20 cm in width. The plume ore veins dominate along the veinlets or in between the veins, occurring in intricate directions. Individual

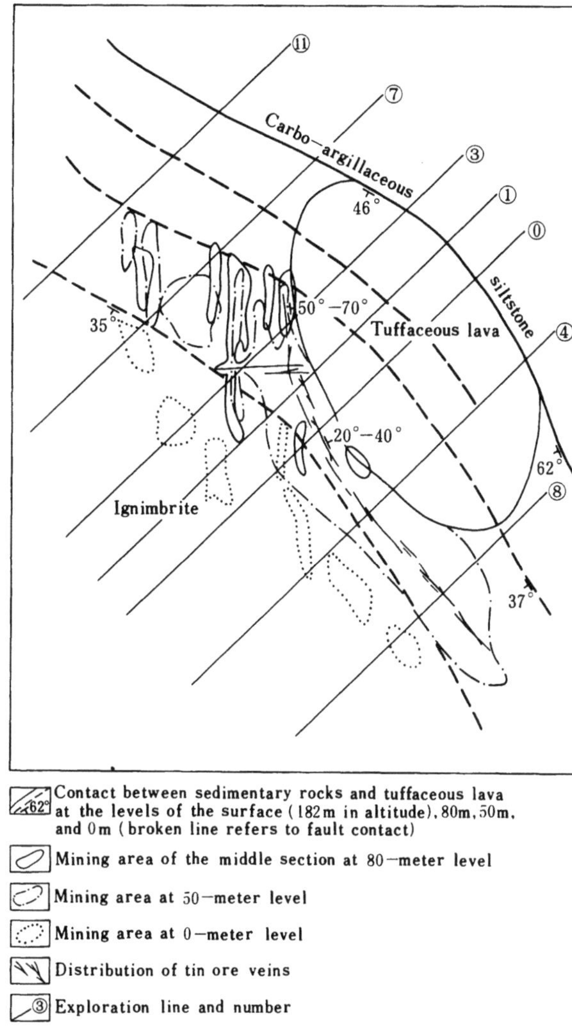


**Fig. 4.** Plan sketch showing the middle mining area at 80, 50, and 0 m levels, Fengdishan tin deposit

single veins are not arranged en echelon, inclining towards the tuffaceous lava body and distributed in a semi-ringed arc pattern round the rock body but not entering it.

The Fengdishan tin deposit is mainly of fissure-filling type; metasomatism is not important. The ores can be distinguished as three types: cassiterite-quartz; cassiterite-quartz-feldspar; and cassiterite-quartz-calcite; with the first being the most important. The associated metallic minerals are mainly cassiterite, with minor contents of pyrite, sphalerite, galena, magnetic pyrite, chalcopyrite, arsenopyrite, etc.; the non-metallic minerals include alteration products such as quartz, feldspar, calcite, sericite, pyrophyllite, chlorite, siliceous limestone, etc. and minor contents of pneumatolytic high-temperature minerals such as topaz and fluorite. The tenor of tin in the ore is generally as high as about 1%, but is distributed inhomogeneously. The rich ore often occurs in "massive", "length", "package", or "nest" form. The enriched cassiterite

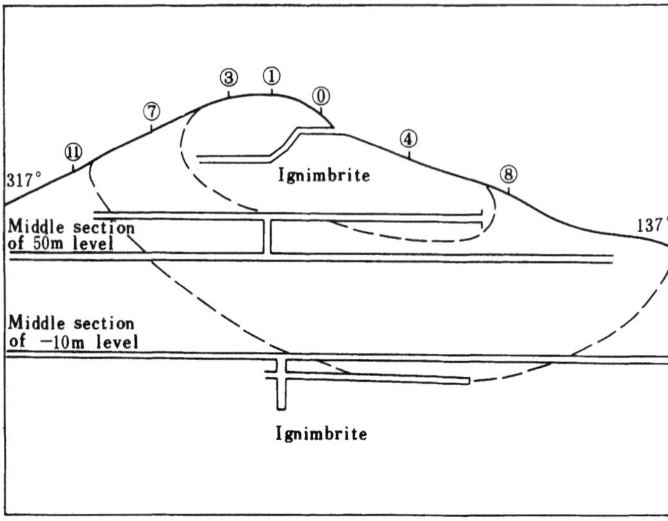


**Fig. 5.** Profile sketch of mined-out area of the Fengdishan tin deposit

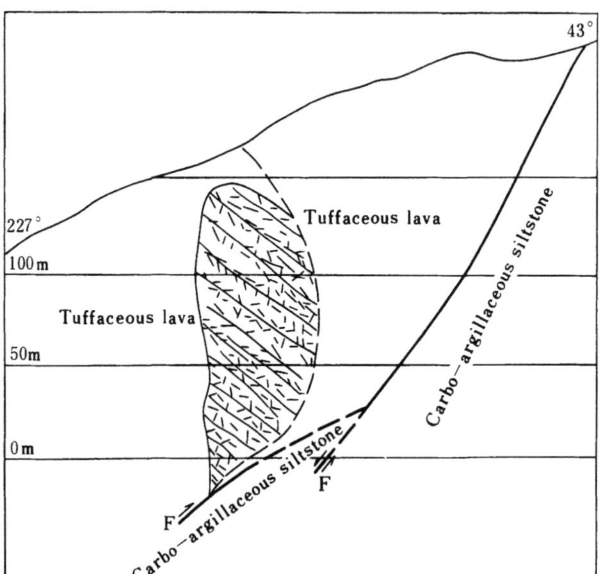


**Fig. 6.** Profile of exploration line No. 1

sometimes appears as massive ore. The occurrence of rich ore is controlled by fractures, generally in the section where fissures are densely distributed or at the intersecting parts of differently oriented fissures. Under such condition of concentrated ore veins, the mining is often extended throughout the entire ore zone.

The crystal form of the cassiterite is usually prismatic, prismatic with single terminations, granular, or needle-like. The faces of the prisms or pyramids are well

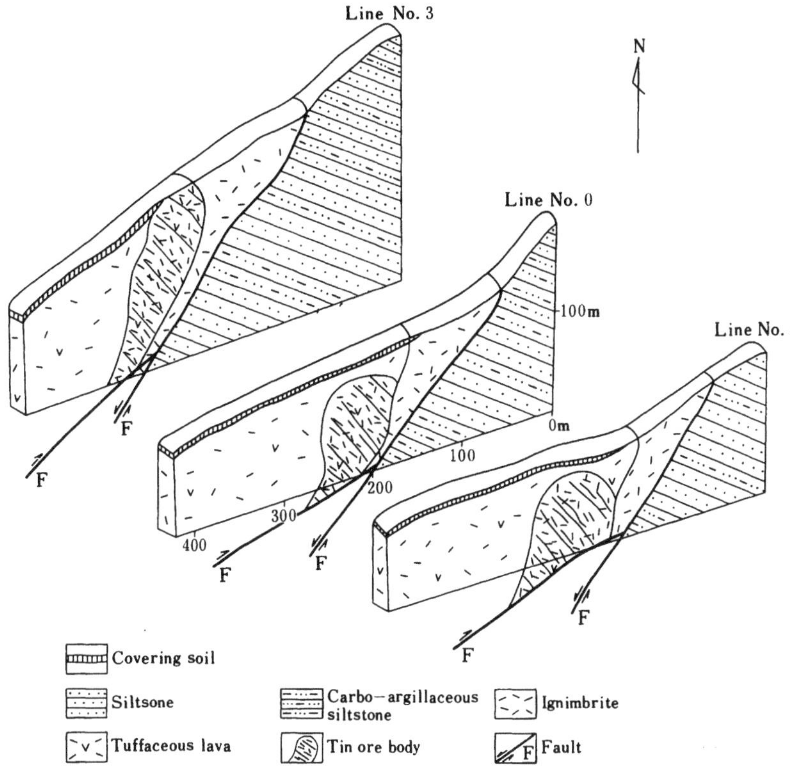


**Fig. 7.** Block diagram of the Fengdishan tin deposit

developed. The grain size is generally 0.2—5 mm. The length-width ratio is 3:1 to 5. Specific gravity is 6.77; colour varies from dark to light brown. The forming temperature (decrepitation temperature) is 380 to 430°C. The chemical composition is $SnO_2$ 98.28%, total Fe 0.87%, $WO_3$ 0.53%, $TiO_2$ 0.27%, $Nb_2O_5$ 0.038%, $Ta_2O_5$ 0.002%.

The alteration of the country rock of the deposit is represented by silicification, sericitization, potash-feldspathization, prehnitization and carbonatization. The first two are closely related to tin ore veins, usually developed in the fissure-concentrated segments or on either side of an ore vein. The tin content is not high, normally 50—100 ppm, approximately corresponding to the mineralization intensity of tuffaceous lava. The prehnitization and carbonatization are in the form of penetrating veinlets, which are the late-stage product of ore fluid mobilization. The potash-feldspathization occurs mainly at deeper levels; where it is most intense the tin mineralization is weak.

The oxygen isotope $\delta^{18}O$ of quartz in the ore veins is +7‰, in accord with the +6 — +9‰ value of primary magmatic water.

In general, the Fengdishan tin deposit is featured by the characteristics of high-temperature hydrothermal deposits: with coarse-grained cassiterite and simple associated minerals, of high tenor and of shallow burial. It is appreciated by mining departments because of its readiness for exploitation and ore dressing.

## II. Saozhoudi Polymetallic Sulphide Deposit

The Saozhoudi polymetallic sulphide deposit is located about 60 m southeast of the Fengdishan tin deposit. The pipe-like orebody in the dacite porphyry, at the intersecting part of the two sets of fractures trending NW and N-S, occurs within the Lower Jurassic intensely silicified carbo-argillaceous siltstone. The orebody and the secondary dacite-porphyry are both of irregular round shape in plan section, with the former enclosed by the latter. The orebody occupies nearly 4/5 of the rock mass. The secondary dacite-porphyry is only 3 to 5 m thick round the orebody, which is very typical (Figs. 8, 9). The pipe-like occurrences of the rock and orebody are controlled by the two sets of fractures, dipping generally towards the SEE, with a dip angle of 70—75°, in distinct penetrating relationship with the country rock of carbo-argillaceous siltstone.

The secondary dacite-porphyry body is known to extend to 124 m depth. It is greyish green with relatively intense silicification and metallic sulphide mineralization. The contained porphyritic crystals are mainly sericitized plagioclase (25—30%), biotite (5—10%), quartz (23—25%), and a minor content of potash feldspar. The groundmass is mainly of quartz (23—25%), and sericite and clay minerals altered from feldspar (24%). Minor minerals include apatite, titanite, rutile, zircon, etc. The lithochemistry is listed in Table 3. Rare earth elements contained in the rock are shown in Table 4, and Figure 10 gives the distribution pattern of the REE.

The orebody is known to extend to a depth of 124 m. On the surface there are two gossans of different size distributed in an area larger than that of the primary deposit. The oxidization zone is about 10 m deep. The plan section of the orebody may have an area of several hundred square meters. The ore texture presents a dense massive or band shape. The minerals of highest content in the orebody are magnetic pyrite and

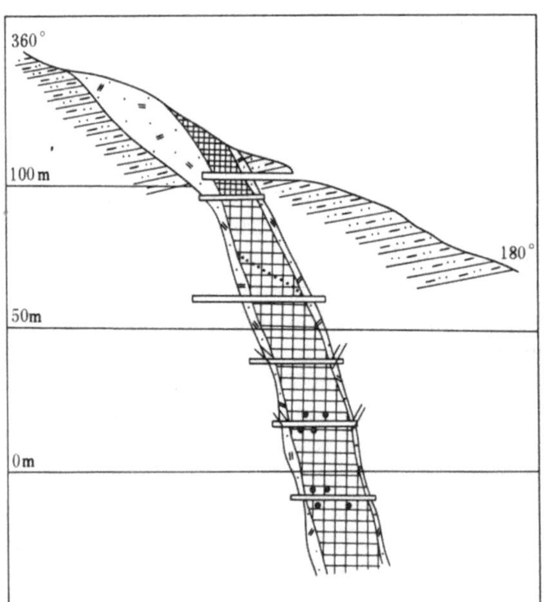


**Fig. 8.** Section through the Saozhoudi pipe-like orebody. Legend as in Fig.9

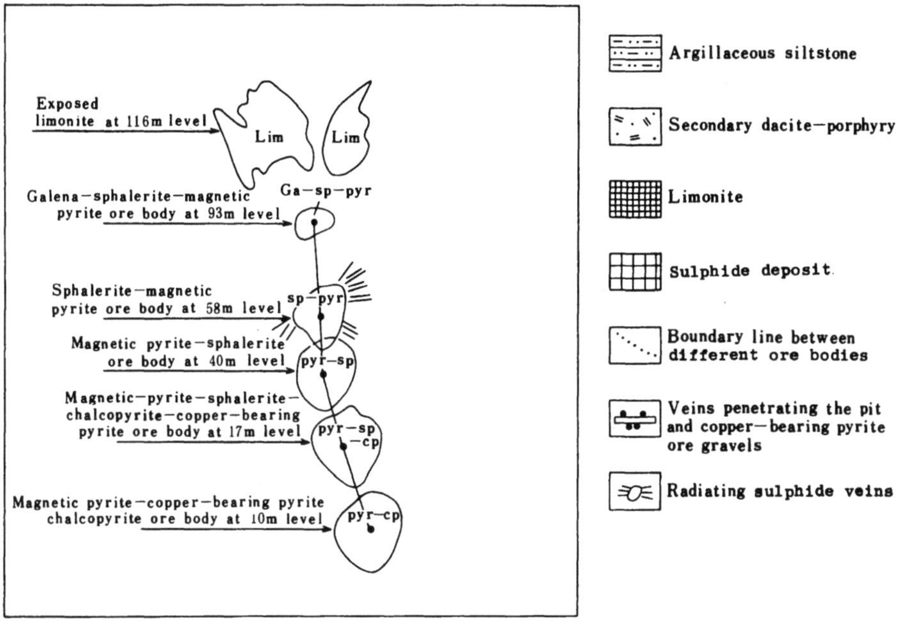


**Fig. 9.** Plan of the pipe-like orebody showing mineralogy and shape at different levels

marmatite, next come pyrite and galena, and minor ones include chalcopyrite, arsenic, etc. The mineral zoning of the ores is very distinct in a vertical direction: Galena and sphalerite appear at the topmost part of the orebody, with the galena gradually decreasing downwards while the sphalerite increases; still downwards it will be transitional toward magnetic pyrite (Fig. 9). In the ores from depth, coarse grains of copper-bearing pyrite are abundant (Fig. 8), the larger ones occurring as egg shapes and the smaller ones as grain size, all unevenly distributed in the magnetic pyrite ores. Such coarse ore grains increase towards the lower part, particularly near the orebody. According to the preliminary statistics, 76–80 grains can be found within an area of one square meter, the maximum diameter being 8 cm, generally 3–5 cm. The chalcopyrite veinlets also increase slightly towards depth. The variation of such ore associations in vertical direction indicates that the minerogenetic activity is multistaged.

Because the galena and sphalerite occur in the top part of the orebody and the pyrite and magnetic pyrite are mainly distributed in the lower part, and the mineralization temperature (decrepitation temperature) of the sphalerite (335–350°C) is much higher than that of pyrite (215–235°C) and magnetic pyrite (128–160°C), and

**Table 3.** Chemistry of the secondary dacite porphyry

| Rock | Chemical analysis (%) | | | | | | | | | | | |
|---|---|---|---|---|---|---|---|---|---|---|---|---|
| | $SiO_2$ | $TiO_2$ | $Al_2O_3$ | $Fe_2O_3$+FeO | MnO | MgO | CaO | $Na_2O$ | $K_2O$ | $P_2O_5$ | $H_2O^+$ | $H_2O^-$ |
| Secondary dacite-porphyry | 63.55 | 0.55 | 14.69 | 10.51 | 0.03 | 0.35 | 0.80 | 0.29 | 4.06 | 0.14 | 1.77 | 0.29 |

**Table 4.** Rare earth content of the secondary dacite porphyry

| Rock | REE content (ppm) | | | | | | | | | | | | | | Total REE (TR) (ppm) | Light RE (total Ce) Ce (ppm) | Heavy RE (total Y) Y (ppm) | Ce/Y | Eu |
|---|---|---|---|---|---|---|---|---|---|---|---|---|---|---|---|---|---|---|---|
| | La | Ce | Pr | Nd | Sm | Eu | Gd | Tb | Dy | Ho | Er | Tm | Yb | Lu | | | | | |
| Secondary Dacite-porphyry | 15.30 | 38.88 | 6.14 | 19.61 | 4.73 | 1.12 | 4.40 | 0.78 | 3.74 | 0.93 | 2.30 | 0.46 | 2.15 | 0.48 | 119.35 | 85.78 | 33.57 | 2.56 | 0.79 |

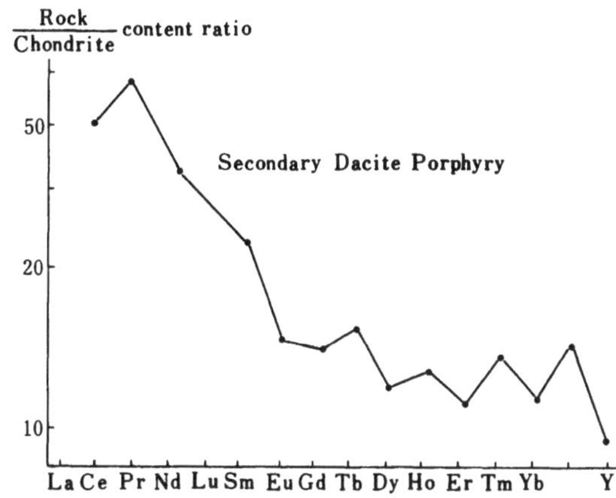


**Fig. 10.** Chondrite-normalized REE pattern

the sphalerite is enclosed by the magnetic pyrite or replaced in drips, it can be concluded that sphalerite and galena were formed much earlier, whereas pyrite and magnetic pyrite later. The chalcopyrite associated with quartz, distributed in veinlet dissemination in the magnetic pyrite or in the sphalerite, were formed the latest whereas copper-bearing pyrite in granular distribution in the magnetic pyrite the earliest, which may have been derived from depth. The diagenetic sequence of the mineral appears approximately: copper-bearing pyrite → ferro-sphalerite → galena → pyrite → magnetic pyrite → chalcopyrite. Moreover, silver and minor gold are also contained in the ores. However, their occurrence is not very clear, except silver, which partially has an interdependent relationship with galena.

Fissure-filling veins and disseminated mineralization of metallic sulphide are developed in the secondary dacite porphyry round the pipe-like orebody and its country rock. The disseminated mineralization is relatively strong in the secondary dacite porphyry and rather weak in the sedimentary rocks. The distribution area of the fissure-filling veins has not been delineated and their mineral association is certainly similar to that of the pipe-like orebody, basically of sphalerite, galena, pyrite, and magnetic pyrite. From the exposures in the pits, there are three sets of fissure-filling veins trending NE, NNE and WNW, dipping towards 140°, 100°, and 190—210° with dip angles of 60°, 30°, and 60° respectively. The NE and NNE sets occur NE and SWW of the orebody respectively, while the WNW set appears SE of the orebody. They are distributed in a radiating pattern. The fissure veins are denser in the eastern part, 10—20 per metre, than in the western part, 3—5 per metre. The veins are straight and regular, generally within the range between 1 and 5 cm. Such mineralized country rock may contain lead and zinc up to the standard of poor grade ore.

A magnetic anomaly of circular shape, which is induced by the magnetic pyrite, can be detected over the ground surface of the Sauzhoudi polymetallic sulphide deposit. The range of the anomaly roughly reflects the distribution area of the pipe-like orebody and its surrounding fissure veins. The centre of the anomaly coincides with that of the orebody. The determination of 24 samples of sulphide minerals from the pipe-like deposit indicates a very stable sulphur-isotope composition in the ores, with distinct "tower effect". The value of $\delta^{34}S$ is +4.01 — +6.15‰, averaging +4.93‰, within which magnetic pyrite has a value +4.93‰, pyrite +4.54‰, and sphalerite +5.73‰. The slight variation shows that the ore-forming substances of the deposit are of similar mantle source.

## III. Conclusions

1. The Fengdishan tin deposit has a close spatial relationship with the tuffaceous lava rock body. The rock body occurs in an oval shape at the surface, narrowing downwards into a funnel shape. The rock is chemically acidic, with higher ferric content than ferrous. The oxidation coefficient is 0.6, which indicates the admixture of atmospheric oxygen during the rock formation. Lithologically the rock should fall between epizonal and eruptive facies. The site of the tuffaceous lava rock body was probably a crater or a volcanic neck.

The Fengdishan tin deposit occurs as a belt shape on the southeastern side of the tuffaceous lava body and as an arc shape round the rock body. The densely concentrated cassiterite-quartz veinlets in the ore belt are in en echelon arrangement, and the plume veins are developed on both sides of the ore belt. A fine vein belt of ore deposit is thus formed inclining towards the southeast. The cassiterite-quartz veins occur in the ignimbrite, dipping towards the tuffaceous lava rock body but not extending into it.

These features demonstrate that the Fengdishan tin deposit should be a veinlet ore belt occurring along the outer part of the crater, and perhaps structurally related to the crater.

2. The tuffaceous lava rock body is in transitional relationship with the ignimbrite. The two rocks have the same chemical composition and appear quite similar in REE distribution pattern. Along the periphery of the lava body there often exist clastics of the basement rock, and the volcanic clastic substances gradually decrease towards the inner part of the rock body while the porphyritic crystals of feldspar and quartz increase and the crystallization becomes intense. This verifies that the tuffaceous lava might be an acidic lava type of volcanic-neck rock, which together with the eruptive ignimbrite are the comagmatic products of the same volcanic mechanism.

3. The tuffaceous lava exhibit pronounced alteration caused by volcanic gas. The marked decrease of alkalic composition, particularly the decrease of sodium, is caused by the metasomatism of feldspar, especially the replacement of plagioclase by sericite or sericite-quartz association during the process of volcanic gas heat alteration, which includes pyritization, silicification, seritization, chloritization and pyroxenization. In the upper part silicification and pyritization are dominant, whereas towards the lower part seritization, chloritization and pyroxenization become intense.

4. The tuffaceous lava rock body has an average tin content of 152 ppm, five times that of the Lower Jurassic carbo-argillaceous siltstone and that of the ignimbrite. The mineralization intensity of the rock body generally corresponds to that of the altered country rock of the orebody. It tends to increase from upper to lower. This indicates that the tuffaceous lava rock body is also closely related to the tin deposit at the time of mineralization.

5. The $\delta^{18}O$ value of the quartz in the Fengdishan tin ore is +7‰, which indicates the existence of magmatic water. Meanwhile, considering the REE distribution pattern of both the tuffaceous lava rock body and the ignimbrite, it seems quite clear that the Fengdishan tin deposit and related diagenetic and metallogenetic materials in the volcanic rocks mainly originated from the upper mantle, with an admixing of crustal substances during the upwelling. Genetically the ore deposit should belong to the hydrothermal type occurring at the outer side of the crater in the post volcanic stage.

6. The formation of the Fengdishan tuffaceous lava rock body is dated at 170 Ma by uranium-lead age determination on zircon from the rock body (Middle Jurassic).

7. The fact that the Sauzhoudi polymetallic sulphide deposit exists in pipe-like occurrence within the pipe-like body of the secondary dacite porphyry shows that the ore deposit is closely related to the secondary dacite porphyry. The sulphur isotope data of the deposit and the REE distribution pattern of the secondary dacite porphyry also indicate that the deposit is composed of diagenetic and metallogenetic materials from the upper mantle and should be grouped under the secondary volcanic rock type. Zircon U-Pb dating shows the age of the secondary dacite porphyry should be 161 Ma, later than the formation of the tuffaceous lava rock body.

*Acknowledgements.* The authors wish to express their thanks to Mr. Chen Tingxian from the Guangdong Bureau of Geology and Mineral Resources, whose materials were referred to while writing the paper. Thanks are also due to the Regional Geological Party of the Guangdong Bureau of Geology and Mineral Resources, 940 and 931 exploration parties of the Guangdong Metallurgy and Geological Exploration Company, and the Xiling Tin Mine Administration of the Shantou Metallurgy Bureau, Guangdong, who have readily contributed their information and data.

## 6.7 China: Guangdong Deposits

# 6.7.1 Geological and Metallogenic Characteristics of Tin Deposits in the Middle Segment of the Lianhuashan Fracture Zone of Guangdong Province

YU JINENG and YAN GONGSHENG[1]

**Abstract**

The paper deals with the geological and metallogenic characteristics of a group of tin deposits which are controlled essentially by metamorphism of the Lantang Group of Late Triassic-Early Jurassic sandstone and shale along the Lianhuashan fracture zone. The characteristics of spatial distribution and mineral association of the tin deposits suggest that their formation is closely related to fracturing and metamorphism in the fracture zone. The syngenetic sandstone ellipsoids in some ore deposits are rich in metallic elements, such as Sn, Pb, Zn and Cu, while the oolites are rich in Sn. This indicates that both the ore-forming and the rock-forming elements must have been deposited from the sea water. The oxygen and sulphur isotopes tend to show that the major ore-forming elements and fluids were not of a mantle source.

## 1. Geological Setting

The study area lies in the southwest segment of the NE-SW-trending Haifeng-Lishui fracture zone and the well-known parallel Lianhuashan fracture zone. The exposed geology is relatively simple, consisting mainly of sandstone and shale of the Late Triassic-Lower Jurassic Lantang Group, acid volcanics of the Upper Jurassic Gaojiping Group and the third and fourth phases of Yenshanian granites (Fig. 1).

Two subgroups of the Lantang Group are exposed. Subgroup a is of thick-layered conglomerate, gravel-bearing coarse-grained sandstone and fine-grained quartz sandstone intercalated with thin-layered carbonaceous and argillaceous shale in the lower part, grading upward into fine-grained sandstone, black siltstone and shale. Subgroup b has a thin-layered gravel-bearing sandstone at the bottom, grading upward into alternating fine-grained sandstone, shale and sandy-shale, then becomming sandstone again, forming two obvious rhythms. Under the influence of dynamic metamorphism, a suite of low-medium grade metamorphosed rocks was formed, characterized by both rupture metamorphism (Fig. 2) and regional dynamothermal metamorphism (Fig. 3); the dynamic metamorphism finds expression in the crushing

[1] No. 756 Geological Party, Bureau of Geology and Mineral Resources of Guangdong Province, Huidong, Guangdong

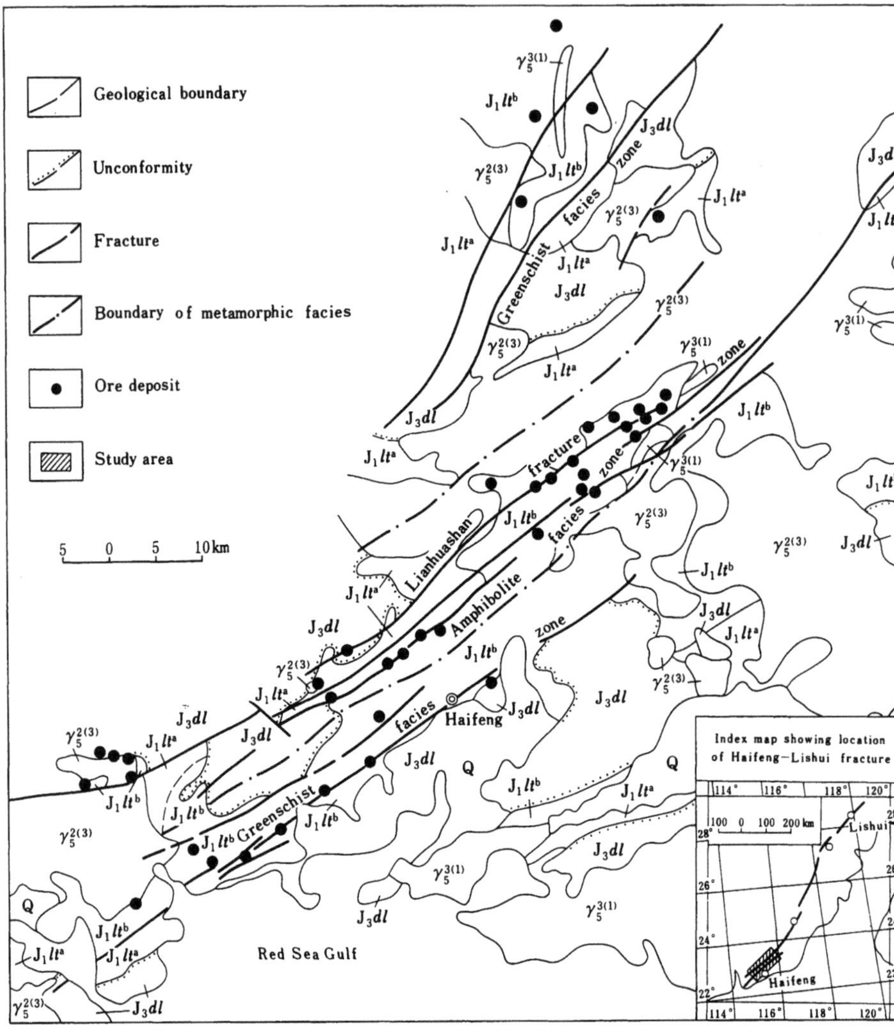


**Fig. 1.** Geological map of the middle segment of the Lianhuashan fracture zone in Guangdong

and elongation of the mineral grains and the orientational arrangement of schistose minerals, while the dynamothermal metamorphism results in the increase of dark minerals, crumpling and orientational arrangement of the mineral grains.

From the assemblages of the major metamorphic minerals, two metamorphic facies zones can be distinguished – greenschist facies and amphibolite facies zone (Fig. 1).

## 2. Geological Characteristics of the Ore Deposits

The primary tin ore deposits may be grouped into four types on the basis of the association of ore minerals (Table 1).

**Fig. 2.** Crushing and partial dislocation of quartz

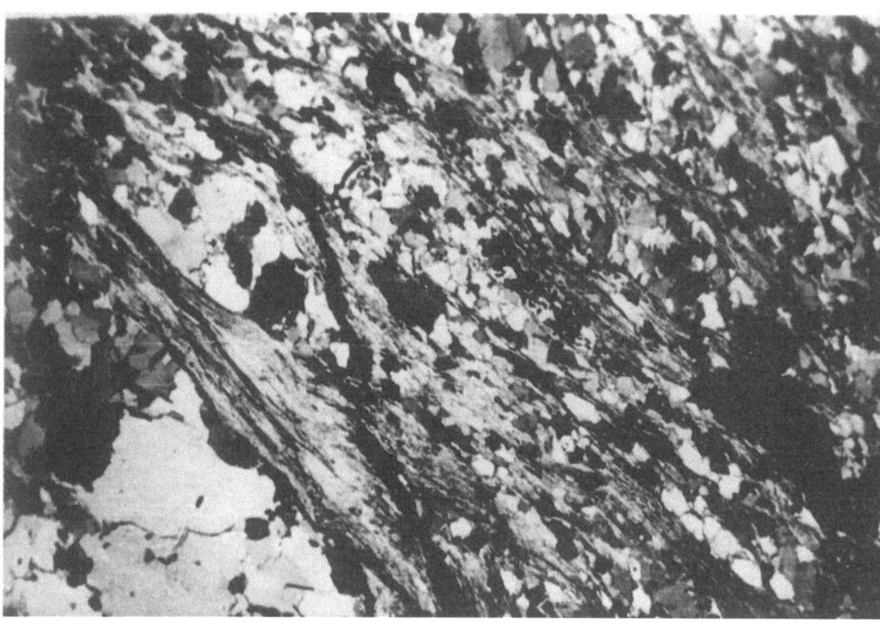

**Fig. 3.** Oriented arrangement of mica, and quartz in the form of lenses after being elongated

**Table 1.** Characteristics of different types of ore deposit

| Metamorphic facies (zone) | Type of ore deposit | | Mode of occurrence and shape of orebody | Mineral assemblage | Texture and structure of ore | Characteristics of cassiterite | Wall rock alterations | Grade | Size of ore deposit | Example |
|---|---|---|---|---|---|---|---|---|---|---|
| Greenschist facies (outer zone) | Cassiterite sulphide type | | Orebodies occur in interformational fracture zone in sandstone-shale of subgroup b of Lantang Group, assuming mostly stratiform, stratoid or lenticular shapes, with some also in vein-like form. They are mainly in conformity with oblique crossing also observed. | Cassiterite, pyrite, galena, sphalerite, pyrrhotite, chalcopyrite, molybdenite, arsenopyrite, quartz, chlorite, sericite, tourmaline, fluorite, etc. | Porphyritic crushing, blastopsammitic, enclosing, porphyroblastic or fibroblastic, skeletal, emulsive, metacolloidal and relict textures as well as oolitic and basement cementation textures; brecciated, banded and dissemination structure. | Brown, brownish-yellow or cream-coloured, rarely oolitic. Cassiterite replacing muscovite and cassiterite enclosing tourmaline are sometimes observed. | Mainly seritization, pyritization, tourmalinization; less chloritization, silicification and greisenization. | 0.2—0.75 | Middle-sized | Changpu, Tiezhang |
| Amphibolite facies (inner zone) | Cassiterite-garnet-chlorite (muscovite) type | Subtype I | Orebodies occur in interformational fracture zones and fissure zones of subgroup b and subgroup a of Lantang Group, mostly in the stratoid, lenticular and vein-like forms. Major orebodies are approximately conformable with the strata. | Cassiterite, pyrite, garnet, chlorite, muscovite (sericite), quartz, biotite, arsenopyrite, fluorite, etc. | Blastopsammitic, enclosing, wormlike, fibroblastic, porphyritic crushing and relict textures; dissemination, banded, massive and miarolitic structures. | Brownish black or dark grey; mainly granular aggregates, poorly euhedral; replacement of muscovite by cassiterite is observed. | Chiefly chloritization, garnetization, sericitication, and to the less amount, silicification and pyritization. | 1.75 | Middle-sized | Yinping |

| | | | | | | | | | | |
|---|---|---|---|---|---|---|---|---|---|---|
| Predominantly biotitization, muscovitization, garnetization and secondarily sodium alteration. | | Sub-type II | Orebodies occur in endo- and exocontact zone between quartz porphyry and strata of Lantang Group, assuming lenticular, veinlike and irregular forms. | Cassiterite, pyrite, garnet, chlorite, muscovite (sericite), biotite, stannite, cordierite, staurolite, etc. | Banded, dissemination and gneissic structures | Dark-grey, grey-light grey, and less commonly light brown; mainly granular, poorly euhedral, cassiterite replacing muscovite is seen. | | 0.2—0.48 | Small-sized | Tashan |
| | Cassiterite-wolframite type | | Orebodies occur at endo- and exocontact zone of small granite stock on the flank of fractures in strata of Lantang Group, mainly as veinlet and dissemination forms. | Cassiterite, wolframite, pyrite, chalcopyrite, quartz, muscovite, chlorite, tourmaline, fluorite, etc. | | | Mainly greisenization, secondarily tourmalinization and pyritization. | 0.03—0.41 | Ore occurrences | Puzhidong |

Cassiterite-sulphide type and cassiterite-garnet-chlorite (muscovite) type are the two major commercial types of tin deposit in this area and are relatively large-sized. The former are predominantly distributed on two flanks of the Lianhuashan fracture zone and occur in the greenschist facies zone, while the latter are mainly seen in the middle part of this fracture zone within the amphibolite facies zone. Cassiterite-wolframite type and cassiterite-quartz type deposits occur less frequently in this area. They are of small size, low grade and rather erratic in shape.

The major orebodies of cassiterite-sulphide type have stratiform, stratoid or lenticular forms and occur in conformity with the strata (Fig. 4). The less important orebodies are of complex shapes, as various forms of veins which cut through the strata. In the arenaceous country rocks of these deposits, "ellipsoids of syngenetic sandstone" over ten cm to tens of cm in length and rich in metallic elements such as Sn, Pb, Zn, Ag and Cu are sometimes observed (Fig. 5). Apart from containing abundant metallogenic elements, these ellipsoids show no difference from the country rocks. In carbonaceous mudstone, there exist sporadic stanniferous oolites (Fig. 6) 0.005—0.05 mm in diameter. Carbonaceous and argillaceous cores can be seen in the larger oolites.

Orebodies of cassiterite-garnet-chlorite (muscovite) type mostly occur in stratiform, stratoid or striped forms and have the same attitude as the strata. They grade into and show indistinct boundaries with the country rocks. Cassiterite assumes anhedral minute-to fine-grained shapes, and is frequently contained as inclusions in such minerals as greenlandite, biotite and albite, making up metamosaic sieve and helicitic textures (Fig. 7); in some places, they occur alternately with andalusite, quartz, mica or chlorite, forming striped structure (Fig. 8).

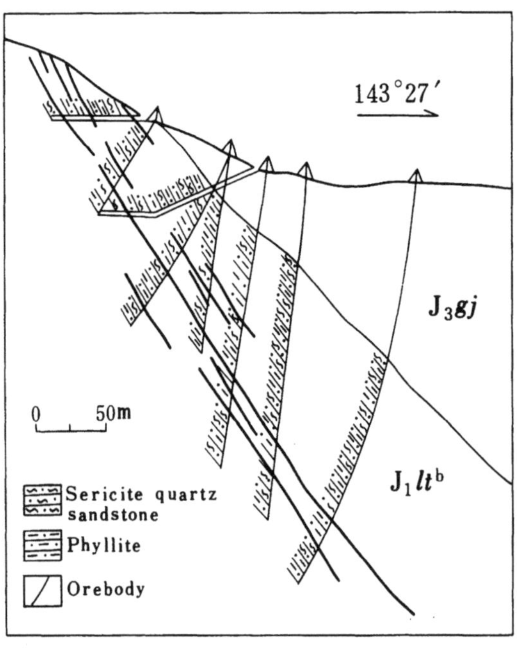


**Fig. 4.** Section showing conformable occurrence of orebodies in stratiform or stratoid forms (based on line No. 4 of Changpu)

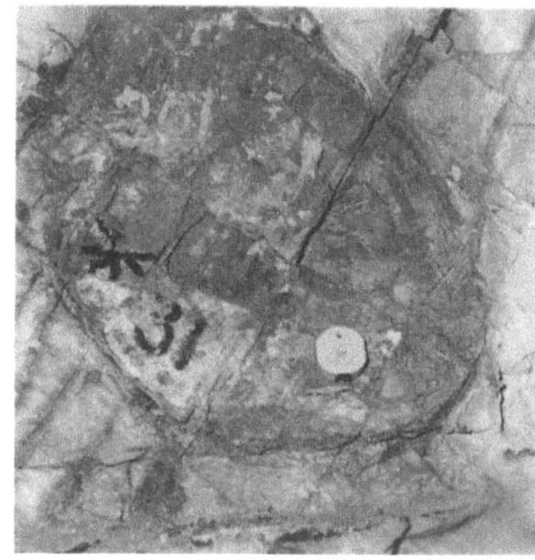

**Fig. 5.** Sn-rich feldspar-quartz sandstone ellipsoid enclosed in feldspar-quartz sandstone

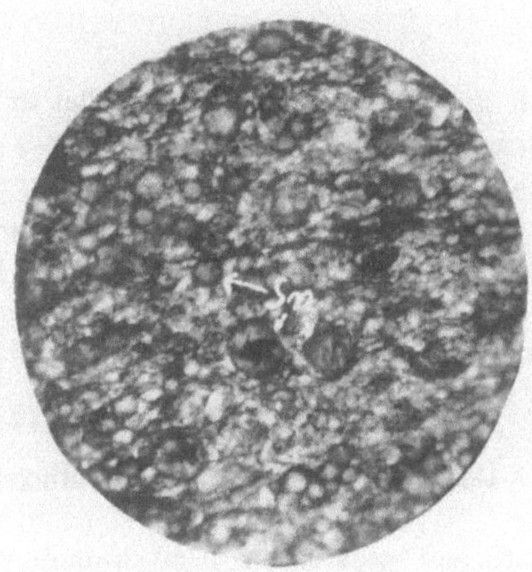

**Fig. 6.** Sn-rich oolites (crossed polars, 150x)

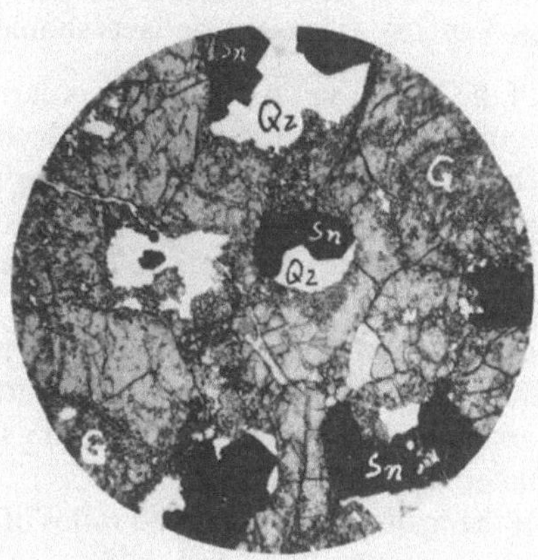


**Fig. 7.** Cassiterite (Sn) enclosed in garnet (G), making a poikiloblastic texture (linearly polarized light, 13x)

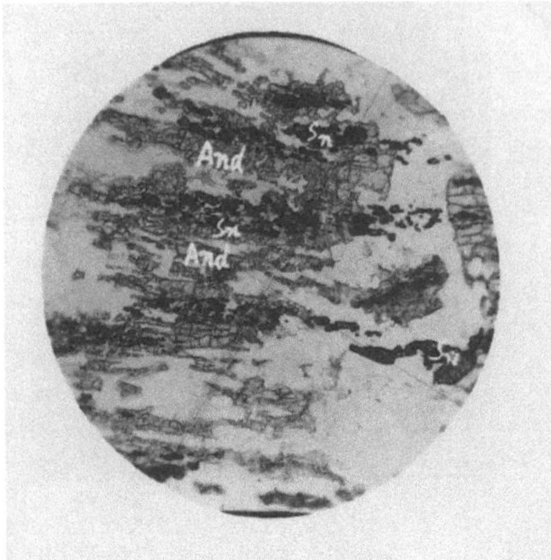


**Fig. 8.** Cassiterite alternating with andalusite (And), forming striped distribution (linearly polarized light, 30x)

It may be seen from Table 1 that the four types of deposit are mostly spatially distributed in the Late Triassic-Upper Jurassic Lantang Group and are closely associated with the dynamic metamorphic zone (Fig. 1). In the greenschist facies zone, the ore types are relatively simple, dominated by cassiterite-sulphide type deposits, while in the amphibolite facies zone, all four ore types are developed.

All this indicates that there exists an obvious correlation between orebodies and strata, ores and country rocks, as well as ore type and grade of metamorphism.

## 3. A Discussion on Sources of Metallogenic Material

Insufficient research has been done on sources of metallogenic materials in this region. The author's preliminary study suggests that tin was derived mainly from the strata of the Late Triassic-Lower Jurassic Lantang Group, especially from subgroup b. In addition to direct evidence of the precipitation of cassiterite (or Sn) from sea water (Figs. 5 and 6), the following facts should be noted:

1) Of the known 46 tin deposits or occurrences in this region, 35 occur in the Lantang Group with 29 in its subgroup, which account for 76% of the total ore deposits. The major commercial deposits are without exception present in strata of this group.
2) Analyses of rock samples collected from various rock bodies and strata and soil samples show from their distribution pattern that the Lantang Group, especially its subgroup b, possesses the highest tin content (Tables 2–4).
3) Trace element compositions of cassiterite from strata of the Lantang Group are quite similar to those of cassiterite from the ore deposits, suggesting the same material source (Table 5).
4) In the major deposits, Co/Ni ratios of pyrite associated with cassiterite are mostly <1 (generally 0.5), and $\delta^{34}S$‰ of sulphides are without exception negative (Table 6).

**Table 2.** Average regional tin content of strata and rocks in coastal areas of eastern Guangdong

| Age | Name of strata or rocks | Tin content (ppm) | Remarks |
|---|---|---|---|
| Late Triassic-Lower Jurassic | Sandy shale of Lantang Group | 21–23 | Data from 1:200,000 regional survey |
| Mesozoic Yenshanian period | Granites | 8–10 | |
| Upper Jurassic | Volcanics of Gaojiping Group | 10 | |

**Table 3.** Data of geochemical soil survey in areas of various strata and rocks

| Distribution area of strata or rocks | Area under investigation ($km^2$) | Average content of tin (ppm) |
|---|---|---|
| Subgroup b of Lantang Group | 220 | 21.2 |
| Subgroup a of Lantang Group | 25 | 15.3 |
| Granites | 76 | 7.5 |
| Volcanics of Gaojiping Group | 60 | 4 |

**Table 4.** Comparison of tin content of various rock types

| Strata and rocks | | Area | Sample number | Tin content (ppm) | Average value (ppm) | Remarks |
|---|---|---|---|---|---|---|
| Lantang Group | Sandstone | Changpu-Yinping | 6 | 890 | | |
| | Sandstone | | 2 | 250 | | |
| | Sandstone | | 4 | 230 | | |
| | Sandstone | | 18 | 230 | 284.3 | |
| | Carbonaceous shale | | 29 | 230 | | |
| | Spotted slate | | 8 | 260 | | |
| | Schist | Tashan | 42 | 270 | | |
| Quartz porphyries | | Changpu-Yinping | 2 | 280 | 191.92 | |
| | | | 25 | 300 | | |
| | | Tashan | 290 | 182 | | |
| | | Buge | 384 | 170 | 77 | Data from Geological Survey Party |
| | | Gaotan | | 10 | | |
| | | Hetian | | 50 | | |
| Gaojiping Group | Tuff | Chishi | 1 | 100 | 100 | Spectral analysis |
| | Crystal tuffaceous lava | Changpu | 1 | 100 | | |
| | Volcanic breccia | Dadonmen | 1 | 100 | | |
| | Lava | | 1 | 100 | | |
| | | | 1 | 100 | | |

**Table 5.** Comparison of trace element compositions of cassiterite

| Element % / Sample | $SiO_2$ | $Al_2O_3$ | MgO | $Fe_2O_3$ | $MnO_2$ | $TiO_2$ | Be | Zr | Sn | W | Nb | Cu | Zn | Y | Yb | Sampling site |
|---|---|---|---|---|---|---|---|---|---|---|---|---|---|---|---|---|
| RZ 11 | 3–5 | 0.3 | ⩽0.001 | ⩽0.1 | <0.001 | 0.1 | x | ⩽0.01 | >10 | 0.02 | 0.03 | <0.0001 | <0.005 | 0.003 | ⩽0.001 | Strata of |
| RZ 52 | 3–5 | 0.2 | 0.003 | ⩽0.1 | 0.01 | 0.1 | 0.001 | 0.03 | >10 | | x | 0.0003 | <0.005 | x | 0.001 | of Lantang Group |
| RZ 42 | 5–10 | 0.3 | 0.003 | ⩽0.1 | <0.001 | 0.1 | 0.003 | ⩽0.01 | >10 | 0.03 | x | 0.0003 | x | x | x | Cassiterite-sulphide-type deposit |
| RZ 51 | 3–5 | 0.2 | 0.003 | ⩽0.1 | <0.001 | 0.1 | 0.001 | 0.01 | >10 | 0.03 | x | 0.0003 | <0.005 | x | x | Cassiterite-garnet chlorite (muscovite) type deposits |

Note: 1. W content of RZ 52 sample could not be determined due to insufficient quantity of sample. 2. Contents of other elements are below detection.

**Table 6.** Sulphur isotopic composition

| Ore district | Sample number | $^{34}$S ‰ | | | | Remarks |
|---|---|---|---|---|---|---|
| | | Pyrite | Pyrrhotite | Sphalerite | Galena | |
| Changpu | G102 | −1.36 | | | | Data from |
| | G 103 | −3.36 | | | | Institute |
| | G 105 | −2.65 | | − 3.70 | | of Geological |
| | G 106 | −1.92 | −3.57 | − 3.41 | | Sciences, |
| | G 107 | −2.58 | | −16.33 | | Bureau of |
| | −1 | −2.12 | −2.46 | − 3.74 | | Geology and |
| | 007 | −2.41 | −6.93 | − 3.04 | | Mineral |
| | IIIT 11-1 | −2.74 | | | | Resources |
| | IIIT 11-2 | −3.03 | | | | of Guangdong |
| | IIIT 11-3 | −2.91 | | | | Province |
| | IIIT 11-4 | | | − 2.87 | | Average value of |
| | IIIT 64 | −2.46 | | | | $\delta^{34}$S: −3.58%. |
| Jishuimen | KC 1 | −2.55 | −2.05 | − 1.30 | − 5.96 | |
| | KC 2 | | −2.39 | − 3.56 | − 7.57 | |
| | KC 3 | −1.26 | | | | |

**Table 7.** Oxygen isotopic composition of cassiterite and its associated minerals and $\delta^{18}_{H_2O}O$ values of mineralized fluids

| Ore district | Sample number | Mineral | $\delta^{18}$O ‰ | T(°C) | $\delta^{18}_{H_2O}O$ | Remarks |
|---|---|---|---|---|---|---|
| Changpu | Chang-1 | | +9.2 | 143 | −6.9 | Data of |
| | | | | 327 | +3.2 | Chang 11–1 |
| | Chang-2 | | +8.5 | 143 | −7.6 | and Chang |
| | | | | 327 | +2.5 | 11–2 are |
| | Chang-1 | | −1.9 | 143 | −9.3 | from Yichang |
| | | | | 327 | −0.3 | Institute |
| | RZ 5 | | +2.1 | 143 | −5.3 | of Geology |
| | | | | 327 | +3.7 | and Mineral |
| | RZ 4 | | +9.6 | 143 | −6.5 | Resources |
| | | | | 327 | +3.6 | |
| Tashan | Ta-1 | | +6.7 | 270 | −1.36 | |
| | | | | 300 | −0.19 | |
| | Ta-2 | | +4.8 | 270 | −3.16 | |
| | | | | 300 | −2.09 | |
| | Ta-1 | | +1.3 | 270 | +1 | |
| | | | | 300 | +2.1 | |
| Yinping | RZ 21 | | −2.5 | | | |
| | RZ 51 | | −0.7 | | | |
| Niutou-shan | RZ 12 | | +7.5 | | | |
| | RZ 12 | | +3.3 | | | |
| | RZ 13 | | +4.2 | | | |
| Shan-jiaowo | Jiao 2 | | +5.1 | | | |
| | Jiao 2 | | +7.1 | | | |
| | Jiao 3 | | +5.1 | | | |

Note: Calculation formula for $\delta^{18}_{H_2O}O$: For quartz: $1000d=3.38x10^6T^{-2}-3.40$ (Clayton, 1972); for wolframite and cassiterite: $1000d=3x10^6T^{-2}-9.9$ (Landis, 1974) (T stands for absolute temperature).

5) Oxygen isotopic compositions of cassiterite and its associated quartz and wolframite imply that the metallogenic elements are mainly of crustal sources, and the metallogenic media are predominantly mixed fluids of meteoric water, syngenetic water and metamorphic water (Table 7).
6) Cassiterite makes sieve textures with such metacrystal minerals as greenlandite, indicating its low temperature metasomatic origin; and minute-grained cassiterite is in striped distribution, showing the characteristics of original precipitation (Figs. 7 and 8).

## 4. Some Ideas on Metallogenic Mechanism

Though characterized by somewhat different metallogenic evolutional processes, the tin deposits and occurrences of this region are all related for their metallogenic source to sandstone and shale of the Late Triassic-Early Jurassic Lantang Group, especially to its subgroup b. Their metallogenic activities are evidently controlled by the Lianhuashan fracture and its associated metamorphism. Their metallogenic mechanism may have been as follows:

In Late Triassic-Early Jurassic, this region was a stable neritic facies environment (Fig. 9) and the old landsurfaces such as mountainous areas and peninsulas (e.g. Fogang peninsula and Jiulianshan mountainous area) were weathered and denudated, and the related materials, including tin probably in the form of chlorine-tin complexes, were transported into the sea, and were precipitated through chemical

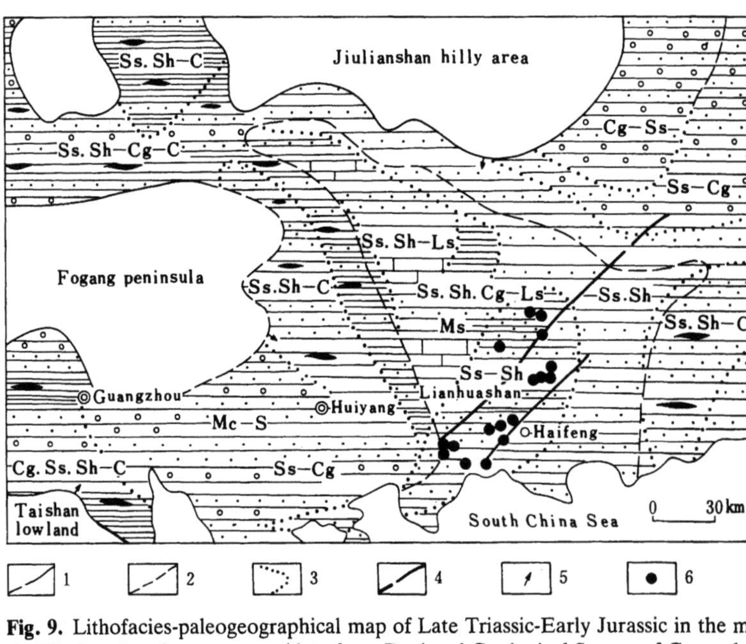


**Fig. 9.** Lithofacies-paleogeographical map of Late Triassic-Early Jurassic in the middle segment of the Lianhuashan fracture zone (data from Regional Geological Survey of Guangdong Province)

reaction or absorption by organic materials and clay minerals in a neutral to slightly alkaline and a suitable oxygen fugacity environment.

During Late Mesozoic time, with the intense thermal event in the southeast coastal area, the Lianhuashan fracture also underwent strong thermodynamic metamorphism, which led to a series of dynamothermal metamorphism of the stanniferous strata of the Lantang Group, such as ubiquitous compression, recrystallization and partial melting (Fig. 10). Under the influence of dynamothermal metamorphism, tin in the strata of the Lantang Group partly formed minute-grained cassiterite through in-situ recrystallization, and partly went into solutions or magmas, which migrated along certain channelways by means of diffusion, filtering or reboiling. With the variation in physical-chemical conditions, metallogenic elements in solutions or magmas were gradually concentrated, forming different types of deposit, in diverse environments.

## 5. Conclusion

The major orebodies in this region are mostly stratiform, stratoid or lenticular, while the minor ones are veinlike, chambered and irregular, and have the characteristics of common hydrothermal deposits. Oxygen and sulphur isotopic compositions imply that the major metallogenic elements and ore-bearing fluids are not of mantle deriva-

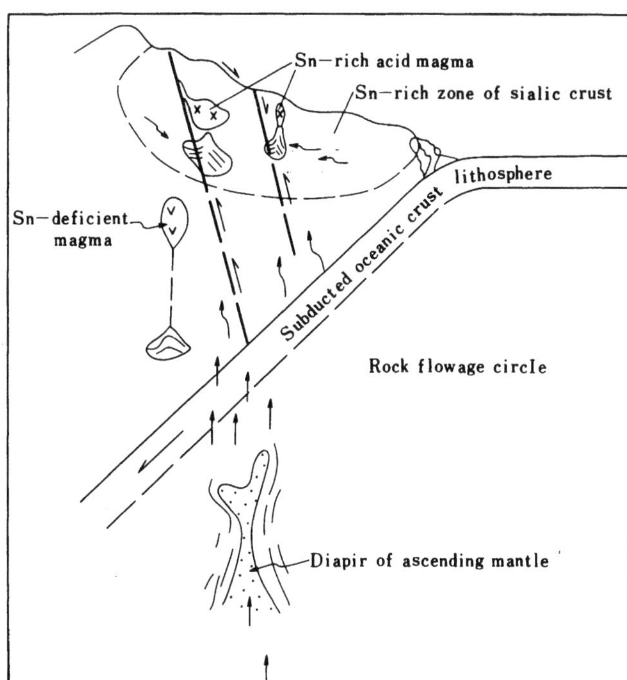


**Fig. 10.** Schematic diagram showing formation mechanism of tin-rich acid magma

tion, thus demonstrating that the formation of tin deposits in this region must have been controlled mainly by strata of the Lantang Group, Lianhuashan fracture and their metamorphism. Therefore, in consideration of metallogenic processes and major sources of ore-forming materials, these deposits should be assigned to the stratabound type.

*Acknowledgements.* In the second-stage research work, Engineer Yan Gongsheng took part in the work and provided part of the microscopic data; Associate Chief Engineer Li Jiucheng went over the preliminary manuscript; and Engineer Zhang Yanlong made some comments. The author should like to express his sincere thanks to all of them for their contributions.

## Reference

Man Fasheng, Bai Yuzhen, Ni Shoubin and Li Tong, 1983. Preliminary isotope studies of the Lianhuashan tungsten ore deposit. *Mineral Deposits,* Vol. 2, No. 4, 35–42 (in Chinese).

## 6.8 Indonesia

# 6.8.1 Geochemistry and Tin Mineralization in Northern Sumatra, Indonesia

S. Johari[1]


## Abstract

In 1975 a major five-year geochemical project was started in Sumatra north of the equator. This project, undertaken jointly by the British Geological Survey and the Directorate of Mineral Resources (Republic of Indonesia), covered an area of 190,000 km$^2$, or about 10% of the total land surface of Indonesia, and combined systematic geological mapping with regional geochemical prospecting.

The survey and follow-up mineral exploration studies have revealed tin-bearing regions of primary exploration interest. One result has been the recognition that tin is much more widely distributed than was known previously. In particular, stream sediment geochemistry shows that some parts of the eastern Barisan are anomalous in tin. Regional geological correlations indicate that this area has a similar geological history to that of South Thailand and Peninsular Malaysia.

Preliminary follow-up work in two areas, Mabundar-Kutacane and Hatapang-Rantau Prapat, has confirmed the presence of tin mineralization as cassiterite. Both areas are underlain by sediments of Carbo-Permian age intruded by Late Cretaceous granites. The Hatapang pluton consists mainly of coarse-grained porphyritic, medium, equigranular biotite granite. According to major and trace element contents, these granites are of 'S' or ilmenite-type and have the character of specialized granites, which are usually associated with Sn/W deposits. They resemble granites in the Phuket tin mineralization zone of Thailand.


## Introduction

This work is part of a larger five-year programme of systematic geological mapping and regional geochemical prospecting in Sumatra north of the equator undertaken jointly by the Directorate of Mineral Resources (Republic of Indonesia) and the Institute of Geological Sciences, now the British Geological Survey (Page et al., 1979a, b). The project started in 1975 and forms part of the British Government's technical cooperation programme with developing countries.

[1] Southeast Asia Tin Research and Development Centre, Tiger Lane, Ipoh, Malaysia
Present address: Procurement Officer (Consultant Matters), Geological and Mineral Survey, Project (ADB LN NR. 641-INO), Jalan Diponegoro 57, Bandung 40122, Indonesia

## Climate, Vegetation and Terrain

Sumatra is a mountainous area largely covered by tropical rainforest. Its climate is humid tropical. Precipitation is 4 to 5 m annually, producing a high drainage density. Seventy percent of its surface is covered by rainforest and approximately 15% is under productive agriculture. Relief is very rugged with several mountains greater than 3,000 m. At the other extreme there are large tracts of low lying swamp.

## Outline of the Geology of N. Sumatra

The stratigraphic nomenclature used here was developed as a result of the five year North Sumatra Project and is summarised from a much fuller account (Cameron et al., 1980). Figure 1 is a simplified geological map.

The Upper Permian-Triassic Peusangan Group, which overlies the Tapanuli Group, is made of the Silungkang Formation, composed of semi-continental basic volcanic rock and *Fusulina* limestone, and the Kuala Formation, composed of Triassic limestone, radiolarion chert, wacke sandstone and siltstone. The Kuala Formation unconformably overlies the Tapanuli Group and the Silungkang Formation.

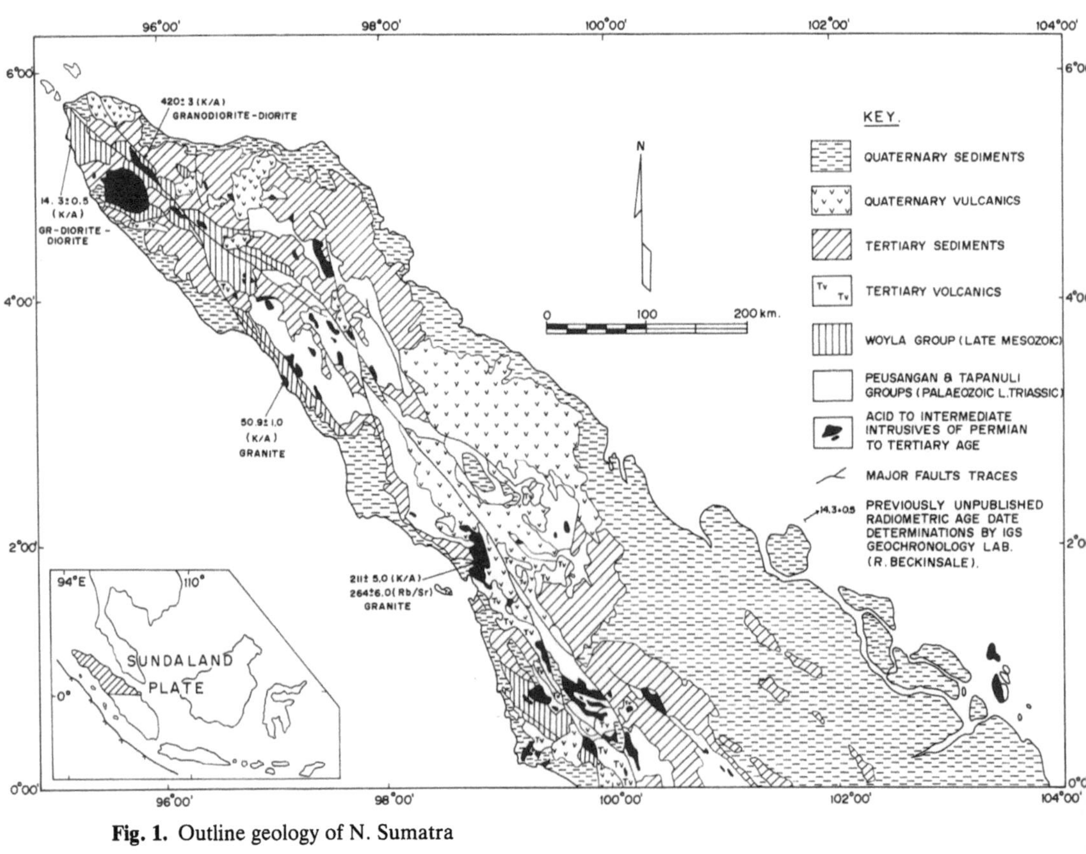


**Fig. 1.** Outline geology of N. Sumatra

The Silungkang Formation is distributed along the Takengon tectonic line in Central Aceh as well as in the Muara Sipongi region and is correlative with the Sinlungkang Formation south of the equator (Silitonga, Kastowo; 1975). The Kuala Formation outcrops on the slopes of the Barisan Mountains from Rantauprapat to Prapat.

During the Cenozoic era there was periodic volcanic activity, during which most of the sedimentary basins were formed. Among these basins are the Central Sumatra Basin, the North Sumatra Basin, the North-west Aceh Basin and the West Sumatra Basin. Volcanism accompanied uplift and the most dramatic example was the Late Pleistocene eruption of the Toba Tuffs from rift released fissures in the roof of an elongated dome centering on the present Lake Toba (Aldiss et al., 1980). The ash-flow tuffs blanketed a large area of Central Northern Sumatra and still remain over wide areas, in some places filling pre-eruption river valleys. Following the eruption, the roof of the magma chamber collapsed, giving rise to the fault-bounded depression now the site of Lake Toba. The remainder of the dome still has a topographic expression and exposes pre-Tertiary units including granites well to the east of those of the main geanticline. At the present time most coasts are prograding and the volcanoes are characterised only by fumarolic activity.

Plutonic activity involved the intrusion of granitoid rock during the Paleozoic, Mesozoic and Tertiary. Mineralization associated with granite emplacement included copper, lead and zinc replacement deposits associated with Jurassic granites, tin and tungsten with Cretaceous granites, and porphyry copper deposits with Tertiary intrusions.

## Sampling Media and Method

### Regional Geochemistry

Drainage sediments were collected at a density one per ten km$^2$. They were wet-sieved in the field to obtain the -1 mm size fraction. From these field samples to -150 to +100 micron fraction was used for the analysis of seventeen major, minor and trace elements by atomic absorption spectrophotometry and colorimetric techniques in the project's analytical laboratory (Page et al., 1979a). Data processing, cleaning and interpretation were carried out on an in-house Hewlett Packard 9820A and an IBM 370.

The major and minor elements in three samples of granite were determined by XRF, electron probe and Beta-probe in the British Geological Survey Central Chemical Laboratories, London (Clarke and Beddoe-Stevens, 1982).

### Detailed Geochemistry

Drainage sediments were collected at 200—400 m intervals along rivers and side streams. They were sieved in the field to -200 microns. Panned concentrates were obtained from approximately 8 kg of sediment screened through a 3 mm sieve before panning. Care was taken not to overpan so that only the quartz, feldspar and mica etc. were removed, the black heavier mineral concentrate remaining.

Stream sediments and panned concentrate samples were analysed for Cu, Pb, Zn, Co, Ni, Ag, Sn, W and Mo in the geochemical laboratories of the Directorate of Mineral Resources in Bandung. This paper reports results for Sn determined by the colorimetric method (gallein) after ammonium iodide fusion. Panned concentrates from Mabundar and Rantauprapat areas were examined by binocular microscope and a simple laboratory test made for cassiterite (Hosking, 1974).

## Regional Distribution of Tin

Analysis of stream sediments detected Sn in 1382 samples out of a total of 9000. The lower limit of detection was 10 ppm. Figure 8 summarises raw data and shows values of Sn on a logarithmic scale. Values above 500 ppm Sn are not shown.

Figure 2 shows localities of twelve Sn anomalies with clusters of 2 to 10 values above 200 ppm. It should be realised that this approach is designed only to demonstrate that Sn anomalies are quite widespread, it does not represent the level of statistical analysis routinely applied (c. f. Page and Young, 1981). Five of the twelve anomalous clusters can be related to nearby intrusive bodies (Sikoleh, Mabundar, Ketambe, Rantauprapat, Tanjung Medan), 5 are in areas of Pleistocene alluvium at a considerable distance from the nearest intrusive (Kecamatan Nangka Payung, Tan-

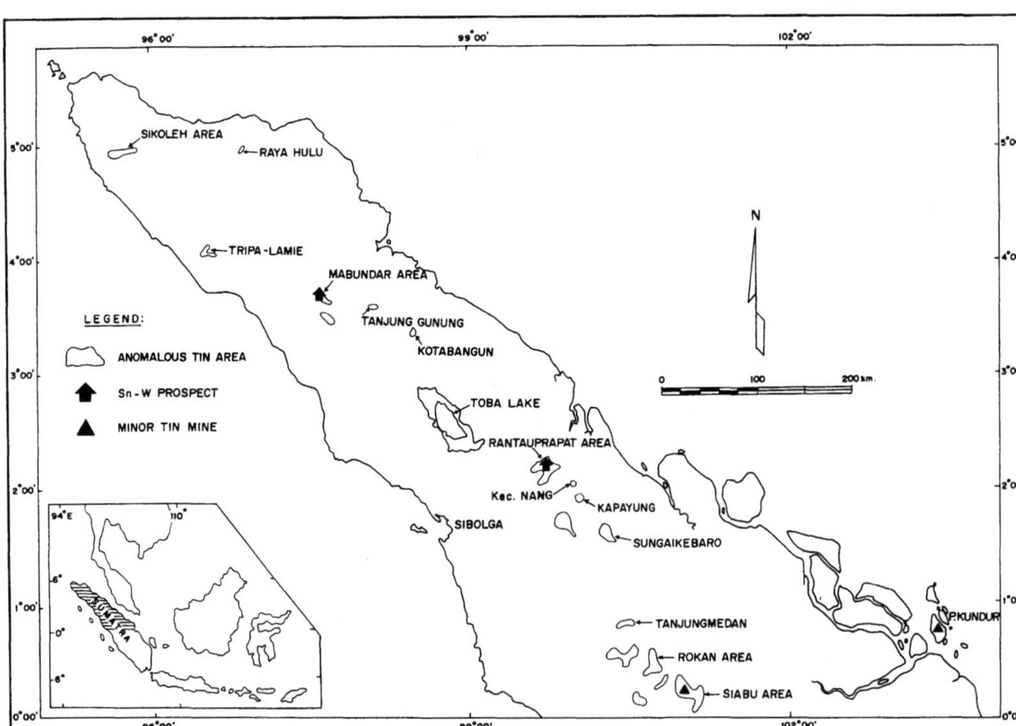


**Fig. 2.** Locality map of tin anomalies of Northern Sumatra

jung Gunung, Siabu, S. Kebaro and Tripe-Lamie) and two are in areas of Quaternary volcanics (Kotabangun and Tanjung Gunung).

Strong Sn anomalies are seen to be associated with known minor Sn mines and prospects in the Siabu-Rokan zone. This correlation between known Sn occurrence (Bemmelen, 1949) and geochemical anomalies encouraged us to believe that some, at least, of the other anomalies might also be associated with significant Sn occurrences. Subsequent field work has demonstrated that this is true for the Mabundar and Rantauprapat anomalies.

## Distribution of Tin at Mabundar

### Location and Access

The area of interest for tin mineralization occurs west of Lawe Alas near Mabundar village, some 40 km NW of Kutacane, at latitude 3° 43' N and longitude 97° 37' E. This area is within a wild life reserve which could seriously affect the commercial exploitation of any mineralization found.

### Geology

The geology of the area is included in the report on the Medan Quadrangle (Cameron et al., 1980). During the follow-up studies a geological sketch map was prepared (Fig. 3). Its major feature is a large fault zone (a strand of the Sumatra Fault System) which follows the course of the Lawe Alas. The area west of this fault is mostly underlain by granitic material. Two major intrusive types occur. The older microdiorite is sheared and altered but occasionally contains large feldspars and disseminated pyrite. The younger intrusive body is a fresh biotite granite which frequently contains large feldspars and heavy pyrite disseminations. It also commonly contains xenoliths of microdiorite which range in size from 2 x 2 cm to 15 x 25 cm. The xenoliths are thought to be derived from the older complex. Aplitic rocks cross-cut the two granite bodies.

The associated sedimentary rocks include meta-sandstone, carbonaceous meta-siltstone and meta-limestone, all assigned to the Tapanuli Group. They have suffered low grade metamorphism. Some andesitic rocks, probably dykes, also occur.

### Tin Geochemistry

Stream sediment and panned concentrate samples were collected at forty six localities in the Mabundar area. The stream sediment and panned concentrate results shown in Figsures 4 and 5 indicate that tin values of up to 1200 ppm and 18000 ppm occur in the sediments and panned concentrates, respectively. This confirmed the results of the regional stream sediment survey.

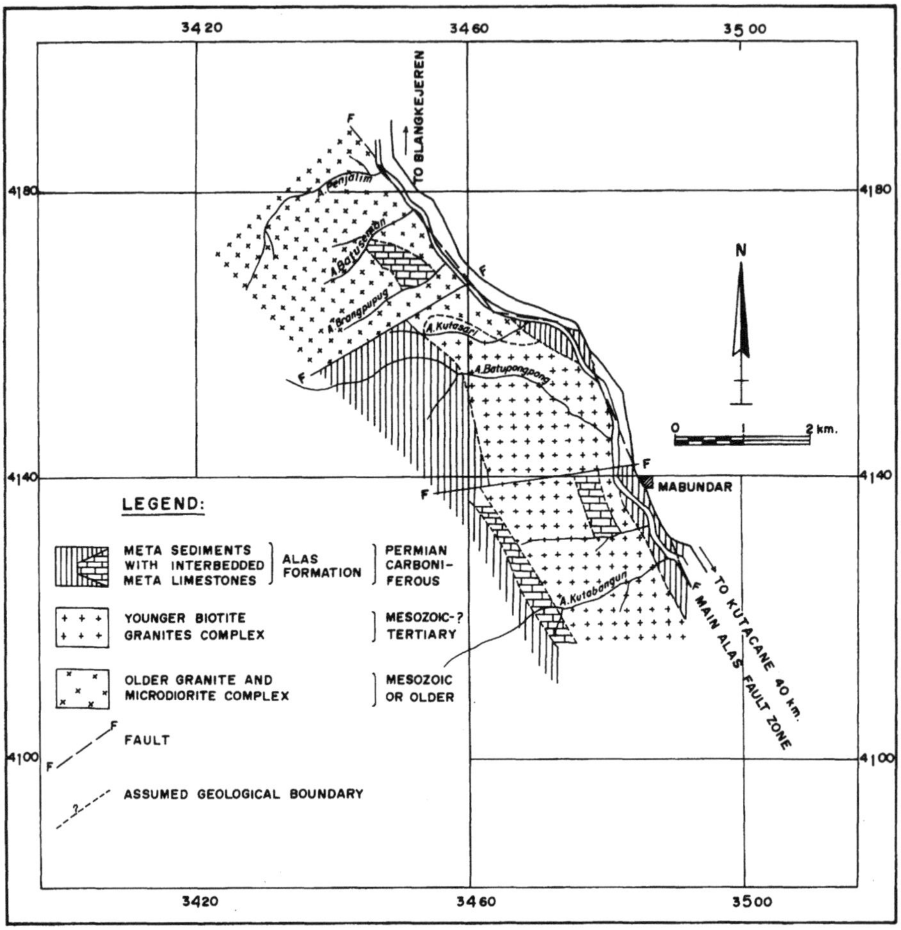


**Fig. 3.** Geological sketch map of the Mabundar area, Kutacane

## Distribution of Tin at Hatapang

### Location and Access

The area lies between longitude 99° 30' and 100° 15' E and between latitude 2° 00' and 2° 30' N in the province of North Sumatra and includes the town of Rantauprapat. There is good access with the Trans Sumatra Highway crossing the area from north to south-east. The journey from Medan to Rantauprapat takes about 8–9 hours by road. The Hatapang survey site is situated approximately 2° 10' N and 99° 37' E.

### Geology

The Hatapang granite lies in rugged forested country 25 km W of Rantauprapat and has an area of approximately 60 km$^2$ (Fig. 6). The country rocks are low grade region-

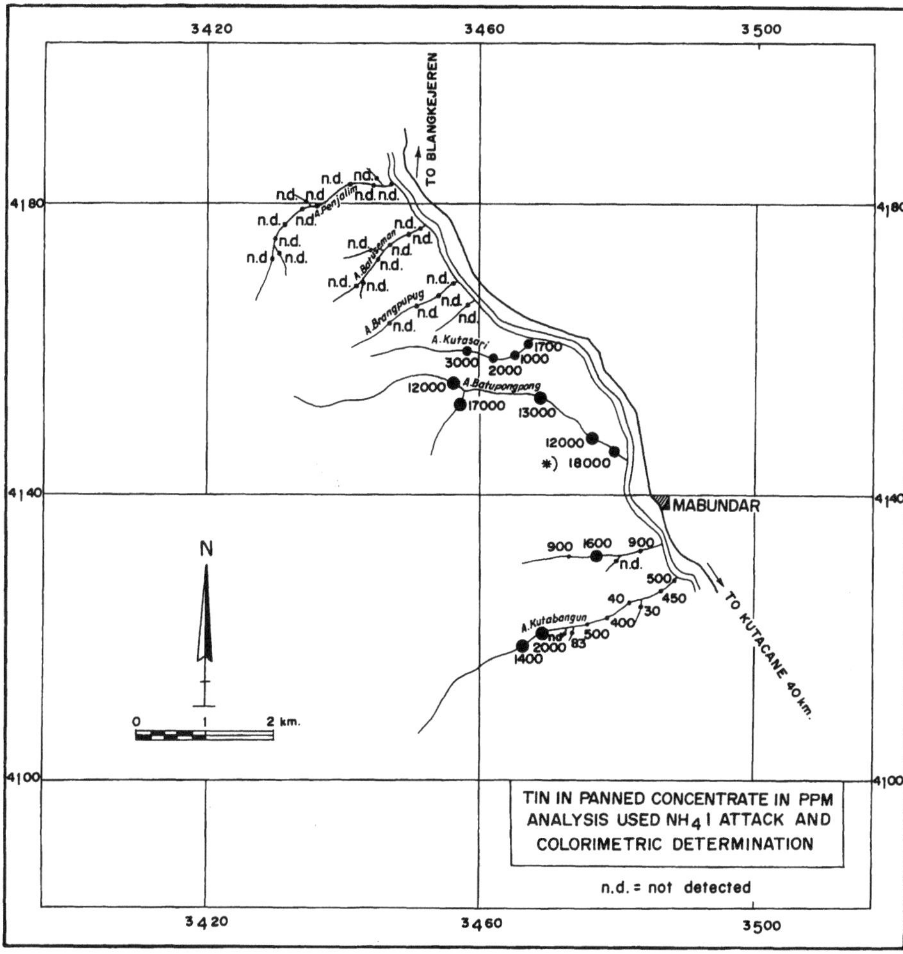


**Fig. 4.** Tin in stream sediments, Mabundar area, Kutacane

ally metamorphosed and tightly folded argillites of the Tapanuli Group (Clarke et al., 1980).

To the east and separated by a tectonic break, the Triassic Kualu Formation crops out. This formation is much less thightly folded and shows no obvious metamorphism. Further east, fluviatile sandstones and conglomerates of Lower Miocene age unconformably overlie the Triassic rocks and dip gently NE and E passing into more marine lithologies characteristic of the back arc basinal area of Sumatra. These in turn are unconformably overlain by Quaternary deposits of both alluvial and volcanic (resulting from the Toba eruption) origin (Clarke et al., 1981).

Marked thermal metamorphic effects are seen near granite contacts resulting in the development of pelitic hornfels and locally of mica schist. The main phase granite is porphyritic and carries biotite and minor muscovite.

The absence of hornblende and presence of biotite indicate that it is probably an S-type granite in the sense of Chappell and White (1974). Microgranite dykes and pegmatites carrying muscovite and tourmaline cut both granite and country rocks near

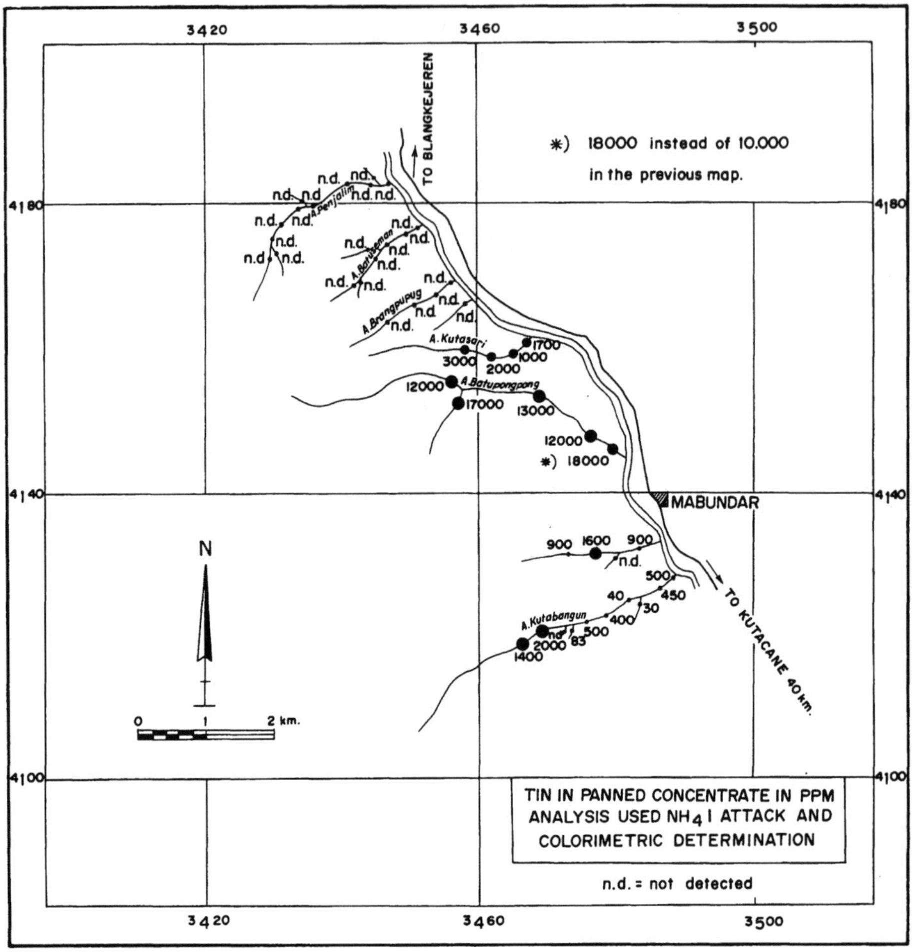


**Fig. 5.** Tin in panned concentrates, Mabundar area, Kutacane

the contacts. Frequently quartz veins cut the surrounding hornfelses, some carrying iron sulphides. Small scale blue-green fluorite-pyrite mineralisation in fractured hornfels and greisenisation of the country rock adjacent to quartz-tourmaline veins also occur.

Three samples of the Hatapang granites were dated by the K : Ar method. The porphyritic biotite granite gave 78.8 Ma and 76.2 Ma and aplite 65.3 Ma, that is late Cretaceous.

## Stream Sediment Geochemistry

The regional stream sediment dispersion pattern for tin in the Rantauprapat area is shown in Figure 7. The total number of samples analysed was 167, giving an average density of 6.7 samples per 100 km$^2$. This figure excludes the low swampy area to the

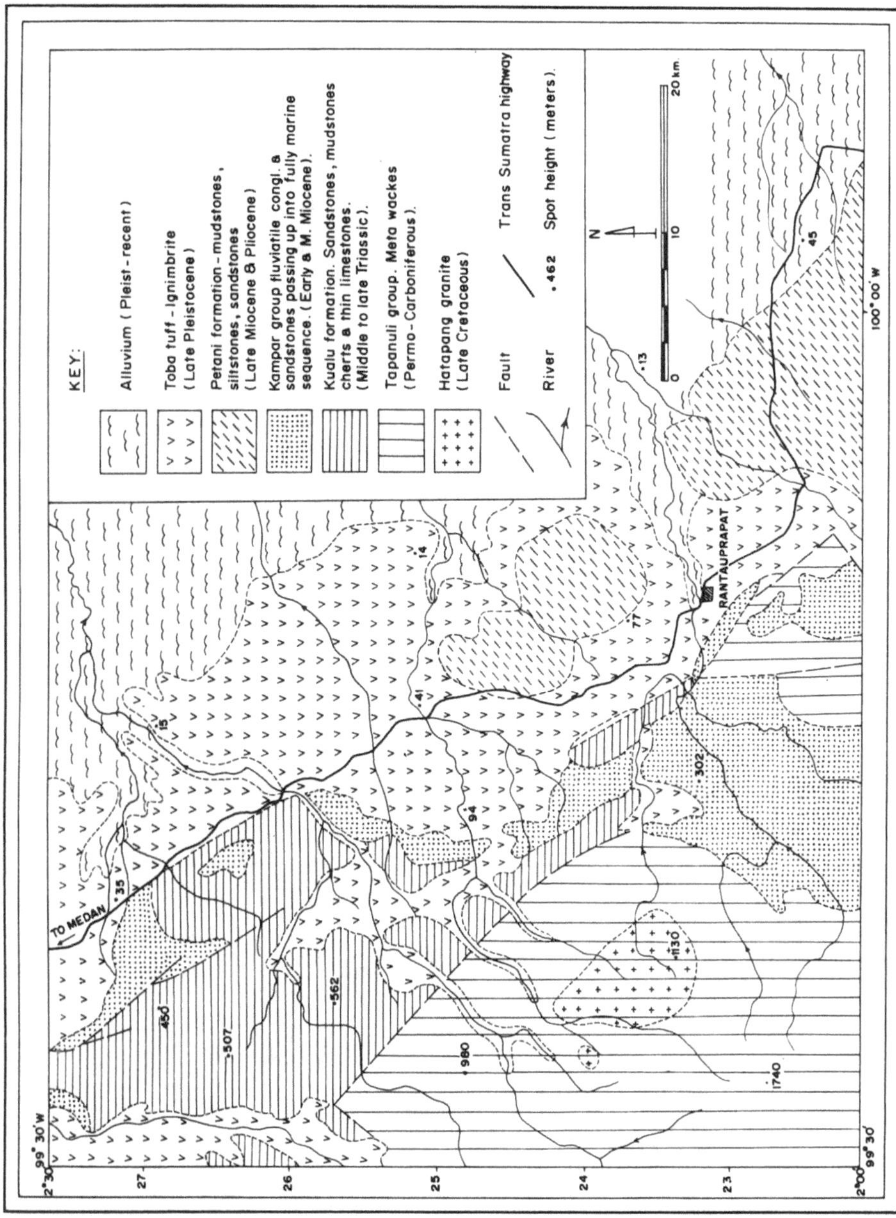


**Fig. 6.** Geological map of the Rantauprapat area

east which was not sampled. Two anomalous populations and one background Sn population are recognised. The background population represents the regional level of response. The anomalous populations are defined as '1st order' (for high values greater than 10 times average background) and '2nd order' (for values between 3 and 10 times average background). The strong '1st order' anomalies centre directly over or occur in close proximity to the known exposures of the Hatapang Granite. These

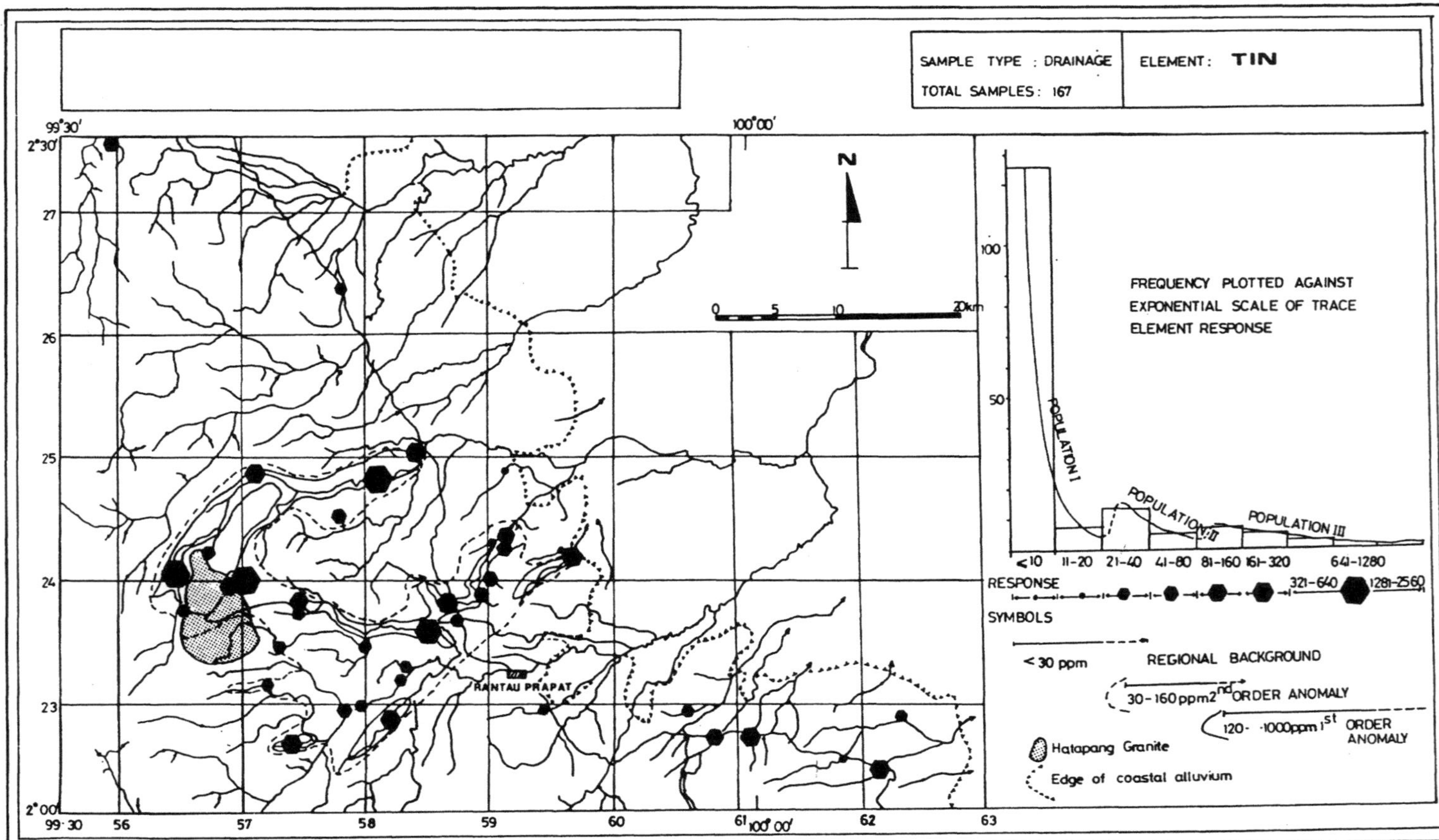


**Fig. 7.** Preliminary geochemical result for the Rantauprapat area

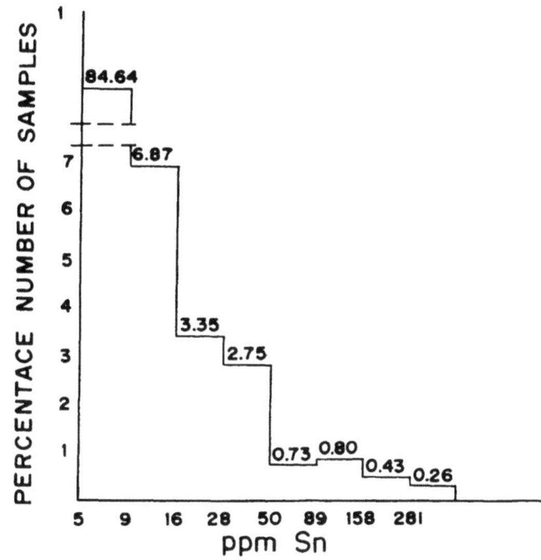


**Fig. 8.** Histogram showing the tin distribution pattern in stream sediments, in Rantauprapat area

anomalies also extend for a considerable distance along the main rivers draining eastwards from the granite. With the exception of a small area to the east of Rantauprapat, values for tin over the rest of the area never exceed background levels of response (Fig. 8).

In addition to Sn, samples collected in the Rantauprapat area were analysed for Cu, Pb, Zn, Co, Ni, Mn, Ag, Li, K, Cr, Sb, As, Mo, V and W. Of these elements only As, Li, Mo and W showed any anomalous response that could be related directly to the Hatapang Granite.

Following inspection of the stream sediment results a detailed examination of rocks samples collected during the reconnaissance mapping was made. This revealed that cassiterite (J. Bowles pers. comm.) is present in late stage granite and quartz veins of several cm width, thus confirming that the granite is the primary source of the widespread stream sediment anomalies. However, the anomalous dispersion patterns particularly in the SE part of Figure 7 cannot all be related to the known outcrop of granite. Some appear to indicate a source in the area underlain by Tertiary sediments. As the oldest of these are essentially fluviatile deposits it appears likely that reworking of fossil placers may have taken place. Similarly, reworking of Pleistocene alluvium is also occurring and this may account for the spread of tin results east of Rantauprapat.

## Mineralization

From the field evidence and petrographic studies, it was found that tin mineralization is associated with fine-grained granite and quartz veins in the Hatapang River basin and the lower reaches of the Batu Jongjong River in the north-eastern section of the Hatapang granite stock. Along the Batu Jongjong survey route, quartz veins and dykes of fine-grained two mica granite become numerous some 200 m from the hornfels side of the Hatapang granite contact. Cassiterite was found within the quartz veins and tourmaline was universal. Eight quartz veins ranging from 2 to 20 cm in width were found within a space of 2 m along a tributary of the Hatapang River. A value of 0.83% Sn was obtained from a vein 10 cm wide and 0.06% from one 15 cm in width.

Quartz veins which strike 60° are also numerous in the Mabat River Basin in the marginal area between the Hatapang Formation and the Hatapang granite. These quartz veins show hematite stains and are accompanied by tourmaline. Fine-grained two mica granite and weak argillization zones have been identified in this area. The maximum grades obtained in a chemical analysis of the quartz veins were 6300 ppm of Sn and 410 ppm of W.

## Rock Geochemistry

Analyses of granites at Hatapang are given in Tables 1 and 2 (after Clarke and Beddoe-Stevens, 1982).

From the above data the values for Hatapang samples using parameters 1, 2, 3, 5 and 6 of Table 3 agree with those of 'S' type granites. Parameter 4 is used by Ishihara et al. (1979) to distinguish ilmenite from magnetite-series granite in Malaysia. 'Specialised' 'S' and ilmenite-series granites are favourable hosts for Sn and W vein/greisen mineralization in many parts of the world. Considering its trace element content, the Hatapang Granite (Table 4) has the character of a 'specialized' granite in the sense of Tischendorf (1977). 'Specialized' granites are associated with Sn/W deposits.

**Table 1.** Chemical Analyses of Hatapang Granite (after Clarke and Beddoe-Stevens, 1982) A. Major Elements (%)

| Element | Sample Field Number | | | Average |
|---|---|---|---|---|
| | 4009 | 4059 | 4061 | |
| + $SiO_2$ | 77.5 | 75.2 | 72.9 | 75.2 |
| + $TiO_2$ | 0.08 | 0.13 | 0.19 | 0.13 |
| + $Al_2O_3$ | 12.7 | 13.8 | 14.8 | 13.77 |
| + $Fe_2O_3$ | 0.53 | 0.44 | 0.51 | 0.49 |
| + FeO | 0.94 | 1.27 | 1.21 | 1.14 |
| + MnO | 0.03 | 0.06 | 0.04 | 0.04 |
| + MgO | 0.2 | 0.3 | 0.4 | 0.3 |
| + CaO | 0.9 | 0.3 | 1.1 | 0.77 |
| + $Na_2O$ | 3.1 | 2.9 | 3.0 | 3.0 |
| + $K_2O$ | 4.6 | 5.5 | 5.8 | 5.3 |
| + $P_2O_5$ | 0.06 | 0.04 | 0.10 | 0.07 |
| * S | 0.13 | 0.13 | 0.10 | 0.12 |
| F | 0.37 | 0.46 | 0.27 | 0.37 |
| $H_2O$ | 0.69 | 0.68 | 0.73 | 0.7 |
| Total | 101.83 | 101.11 | 101.15 | 101.4 |

* XRF – D.J. Bland. + Betaprobe – A.E. Davies

**Table 2.** Chemical Analyses of Granites (after Clarke and Beddoe-Stevens, 1982)
B. Minor Elements (ppm)

| Element | Sample Field Number | | | Average |
|---|---|---|---|---|
| | 4009 | 4059 | 4061 | |
| * CL | 5 | 80 | 10 | 32 |
| o Li | 192 | 506 | 168 | 289 |
| o Be | 5 | 17 | 10 | 11 |
| o B | 254 | 9 | 63 | 109 |
| o Cu | 1 | 4 | 1 | 2 |
| o Zn | 88 | 53 | 41 | 62 |
| o Ga | 17 | 22 | 22 | 20 |
| * Rb | 650 | 1110 | 580 | 780 |
| * Sr | 60 | 65 | 75 | 67 |
| * Y | 70 | 15 | 20 | 35 |
| * Zr | 115 | 160 | 150 | 142 |
| * Nb | 35 | 30 | 60 | 42 |
| o Mo | 3 | 3 | 3 | 3 |
| o Ag | 1 | 1 | 1 | 1 |
| o Cd | 1 | 1 | 1 | 1 |
| * Sn | 85 | 130 | 60 | 92 |
| o Ba | 74 | 116 | 184 | 125 |
| o La | 30 | 45 | 60 | 45 |
| * W | 35 | 10 | 12 | 19 |
| * Pb | 75 | 75 | 80 | 77 |
| * Bi | 10 | 10 | 10 | 10 |
| * Th | 70 | 75 | 75 | 73 |
| * U | 15 | 10 | 15 | 13 |
| * F | 3700 | 4600 | 2700 | 3700 |

* XRF – D.J. Bland. o Optical Emission – K.A. Holnes et al.

**Table 3.** Hatapang Granite chemical characteristics compared with granites elsewhere in the world (after Clarke and Beddoe-Stevens, 1982)

| Element | Range in 'normal' granite | Sample 4009 | Field 4059 | Number 4061 |
|---|---|---|---|---|
| 1. Mol $\frac{Al_2O_3}{Na_2O + K_2O + CaO}$ | = >1.1 ) | 1.48 | 1.59 | 1.49 |
| 2. $Na_2O$ content in rocks approx. 5.0% $K_2O$ | = <3.2 ) 'S' type of Chapell and White (1974) | 3.1 | 2.9 | 3.0 |
| 3. % CIPW normative corundum | = >1% ) | 1.13 | 2.63 | 1.83 |
| 4. $\frac{FeO}{Fe_2O_3}$ | Ilmenite series ≥ 1.0 (Ishihara et al., 1979) | 1.77 | 2.88 | 2.37 |
| 5. $\frac{K}{K + Na}$ | < 0.5 ('S' Type in Fig. 7, Takahashi et al., 1980) | — | 0.66 | — |
| 6. $\frac{Fe^{+++}}{Fe^{+++} + Fe^{++}}$ | Centre of 'S' type field, Fig. 7, Takahashi et al., 1980 | — | 0.16 | — |

**Table 4.** Trace element contents of specialised and normal granites as proposed by Tischendorf (1977) compared with average of 3 Hatapang samples (after Clarke and Beddoe-Stevens, 1982)

| Element | Range in 'normal' granite | Proposed average content in specialised granite | Hatapang average |
|---|---|---|---|
| Sn | 2 – 8 | 30 ± 15 | 92 |
| Li | 30 – 150 | 200 ± 100 | 299 |
| Pb | 130 – 270 | 500 ± 200 | 780 |
| W | 1 – 2.7 | 7 ± 3 | 19 |
| Mo | 2.5 | 4 ± 2 | 3 |
| Be | 3 –8 | 13 ± 6 | 11 |
| F | 250 – 1500 | 3700 ± 1500 | 3700 |

## Discussion

Previous to the survey described here, there was very little indication of tin in mainland Northern Sumatra other than in the Rokan Siabu Zone. However, widespread occurrences of anomalous tin concentration in stream sediments (Figure 2) and the confirmation of cassiterite/wolframite occurrences in the Mabundar and Rantauprapat area, respectively, indicates that the SE Asian tin/tungsten metallogenic province extends across much of N Sumatra. Further supporting evidence is that the Late Palaeozoic and Early Mesozoic stratigraphic sequences of S Thailand, NW Malaysia and N Sumatra are similar and have had a similar history of granite intrusion during the pre-Tertiary (Cameron et al., 1980).

The Hatapang granite is similar to the Phuket tin-bearing granite at the westernmost margin of the Thailand-Malaysia-Indonesia tin mineralization zone; these are porphyritic granite of Late Cretaceous age (Mitchell, 1979; P. Nutalaya, 1979; Ishihara, 1979). This suggests that the Hatapang biotite granite is possibly a southern extension of the tin-bearing granite zone of the Phuket-Phangnga-Krabi regions of southern Thailand. This relationship is illustrated in Figure 9 (after Mitchell, 1979).

## Conclusions

1) Stream sediment anomalies for tin are widespread in Sumatra and many can be related to granite intrusions; others are underlain by Quaternary volcanics and sediments.
2) Limited follow-up work in two areas has shown that high values of tin as cassiterite occur in panned concentrates at Mabundar.
3) The relatively easy access and widespread occurrence of Pleistocene alluvial sediments makes the area around and east of the Hatapang granite an attractive exploration target for Sn placers.
4) The K-Ar age of the Hatapang granite is 78.8 Ma and 76.2 Ma (i.e. Late Cretaceous).
5) The Hatapang granites are of 'S' or ilmenite type and have the character of specialised granites, which are usually associated with Sn/W deposits.
6) According to points 4 and 5 above, this suggests that the Hatapang biotite granite is a southern extension of the tin-bearing granite zone of the Phuket-Phangnga-Krabi regions of southern Thailand.

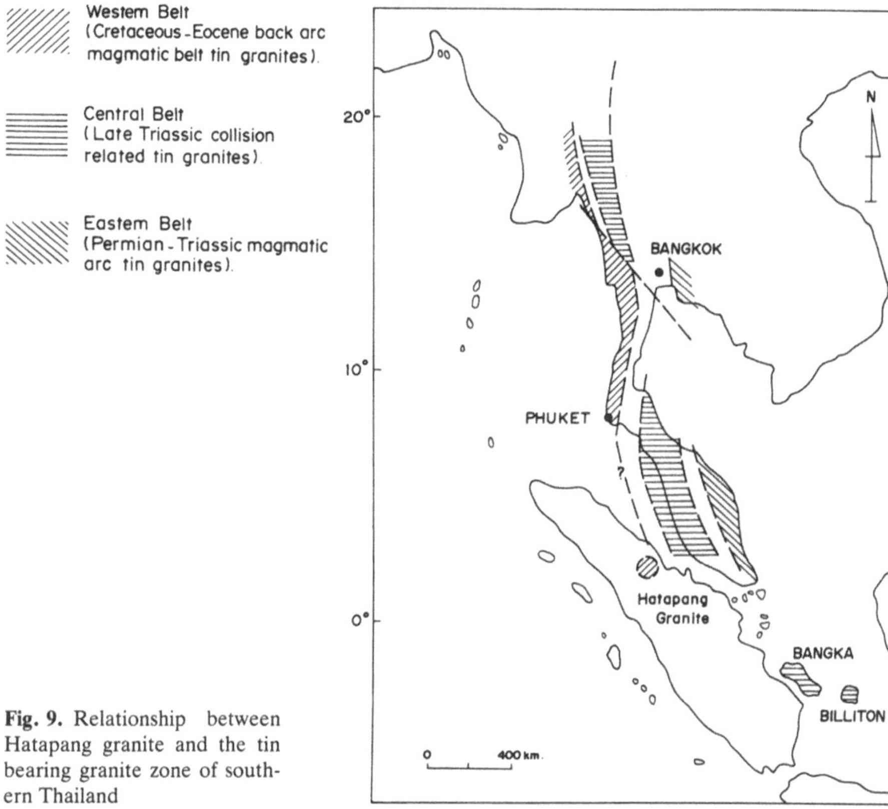


**Fig. 9.** Relationship between Hatapang granite and the tin bearing granite zone of southern Thailand

*Acknowledgements.* The author would like to thank Ir Salman Padmanagara, Director of the Directorate of Mineral Resources, Indonesia for permission to publish the paper. I wish to thank Dr. Abdullah Hasbi b Hj Hassan, Director of SEATRAD Centre for providing facilities for the preparation of this paper and the opportunity to participate in the conference.

Dr. W.K. Fletcher, UN Team Leader, SEATRAD Centre, offered constructive comments and the United Nation Development Programme provided financial support for participation in this conference. This financial source is gratefully acknowledged. Miss Low Yoke Yin typed the manuscript and Miss Suraiya bt Hashim drew all the maps. I thank them very sincerely for their care and diligence.

## References

Aldiss, D.T., Whandoyo, R., Ghazali, S.A. and Kusyono (in press). Geology of the Pangururan and part of the Sinabang quadrangles, Northern Sumatra. Geological Research and Development Centre, Bandung.

Beckinsale, R.D., 1979. Granite magmatism in the tin belt of Southeast Asia. *Origin of granite batholiths-geochemical evidence* (Atherton, M.P. and Tarney, J., eds.). Shiva Publishing Ltd. U.K., 34–44.

Bemmelen, R.W. Van, 1949. *The Geology of Indonesia.* Government Printing office, The Hague.

Cameron, N.R., Clarke, M.C.G., Aldiss, D.A., Aspden, J. and Djunuddin, 1980. The Geological Evolution of Northern Sumatra. *Proceedings of the Ninth Indonesian Petroleum Association Convention,* Jakarta, Indonesia, 149–187.

Cameron, N.R., and 12 others (in press). Geology of the Medan Quadrangle, North Sumatra. Geological Research and Development Centre, Bandung.

Chappell, B.W. and White, A.J.R., 1974. Two Contrasting Granite Types, Pacific Geology, 8, 173–174.

Clarke, M.C.G. and Beddoe-Stevens, B., 1982. Interim Report on recent analytical work on mineralized samples of the Hatapang Granite, Northern Sumatra (unpublished report).

Clarke, M.C.G., Ghazali, S.A., Harahap, H., Kusyono and Stephenson, B. (in press). Geology of the Tanjung Balai Quadrangle, North Sumatra. Geological Research and Development Centre, Bandung.

Clarke, M.C.G. and Kusyono (in press). Tungsten mineralization and the Hatapang granite, Northern Sumatra. Proc. of D.M.R. Colloqium, Bandung 1980.

Hosking, K.F.G., 1974. Practical aspects of the identification of cassiterite ($SnO_2$) by the tinning test. *Bull. Geol. Soc. Malaysia,* 7, 17–21.

Hutchison, C.S. and Taylor, D., 1978. Metallogenesis in S.E. Asia. Jl. Geol. Soc. Lond. V 135, 407–428.

Ishihara, S., Sawata, H., Arpornsuwan, S., Busaracome, P. and Bungbrakearti, N., 1979. The magnetite-series and ilmenite-series granitoids and their bearing on tin mineralisation, particularly of the Malay Peninsula region. *Geol. Soc. Malaysia, Bulletin,* 11, 103–110.

Johari, S., Clarke, M.C.G., Djunuddin, A. and Stephenson, B., 1982. New evidence of tin and tungsten mineralization in Northern Sumatra. *Proc. Fourth Reg. Conf. on the Geology of South East Asia,* Manila, Philippines, Geol. Soc. Philippines, 497–508.

Katili, J.A., 1968. Permian volcanism and its relation to the tectonic development of Sumatra. *Bull. Ntl. Inst. Geol. Mining,* Bandung 1, 3–13.

Mitchell, A.H.G., 1979. Rift subduction and collision - Related Tin Belts. *Geol. Soc. Malaysia, Bulletin,* 11, 81–102.

Page, B.G.N. and Young, R.D., 1981. Anomalous geochemical patterns from Northern Sumatra: their assessment in terms of mineral exploration and regional geology. *J. Geoch. Expl.,* 15.

Page, B.G.N., Bennet, J.D., Cameron, N.R., Bridge, D.M., Jeffrey, D.K., Keats, W. and Thaib, J., 1979a. Regional geochemistry, geological reconnaissance mapping and mineral exploration in Northern Sumatra, Indonesia. Proc. of 11th Comm. Mining and Met. Congr. Hongkong, 1978. Ed. Jones, M.J., Instit. Mining and Met. London, 455–462.

Page, B.G.N. and six others, 1979b. A review of the main structural and Magmatic features of Northern Sumatra. *J. Geol. Soc. Lond.* 136(5), 569–579.

Tischendorf, G., 1977. Geochemical and petrographic characteristics of silicic magmatic rocks associated with acid magmatism 2", *Geol. Surv. Prague, Czechoslowakia,* 41–46.

Takahashi, M., Aramaki, S. and Ishihara, S., 1980. Magnetite-series/Ilmenite-series vs I-type/S-type granitoids. *Mining Geology Special Issue* (Japan), No. 8, 13–28.

# 6.8.2 Application of Geophysical Methods to Investigate the Extention of Primary Tin Deposits in the Pemali Open Pit Mine, Bangka, Indonesia

EMPON RUSWANDI[1]


## Abstract

A study on the application of geophysical methods to investigate primary tin deposits at Pemali has been conducted since early 1970 by the Geological Survey of Indonesia. In 1980, an integrated gravity, magnetic, resistivity sounding and induced polarization survey was conducted along profiles across the tin-bearing zone.

Magnetic and resistivity sounding methods fail to show anomalies relating to the tin-bearing zone. Metal pipes and high voltage power lines in the mine area produce noise which masks the low response anomalies. Furthermore, the complicated subsurface structure produces unreliable resistivity sounding results.

Bouguer gravity anomalies show a low value ranging between 2 and 3 mgals stretching from south to north bending toward the northeast. This low gravity anomaly coincides with the alteration zone, along the contact between granite and metasediment.

Induced polarization shows relatively low apparent resistivity with a high percent frequency effect (PFE) along the alteration zone. The apparent resistivity anomalies ranging between 10 and 500 ohm-m occur in the south, whereas in the north the anomalies range between 100 and 1,100 ohm-m. These anomalies coincide with a PFE of (1–3) % and (2–5) % respectively.


## Introduction

Pemali mine in Bangka, composed of relatively flat terrain with some nearby granite hills, is situated about 75 km north of Pangkalpinang, the capital of Bangka (Fig. 1).

The mine lies in the contact zone between granite in the east and metasediments in the west, 1,000 m long (N-S) and 500 m wide (E-W). Primary tin deposits in the Pemali area have been mined by open pit methods since the early forties. Up to 1982, mining activity has reached 30 m below sea level, or about 75 m below the ground surface, and produced about 6.8323 tons of tin ore (Mulyadi, 1983).

Up until now, exploration by drilling, still being intensively carried out, has been the only source of geological information. For a period of 43 years, from 1940 to 1983,

[1] Directorate of Mineral Resources, Bandung, Indonesia

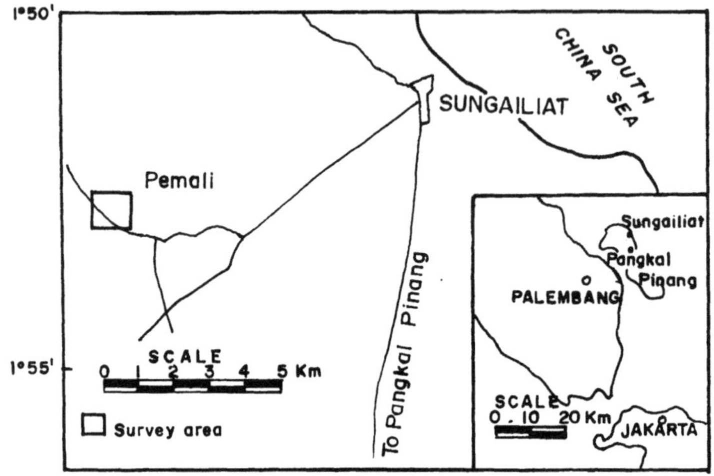


**Fig. 1.** Index Map of the Survey Area, Bangka

drilling had reached a tremendous cumulative depth of 22,763 m within a relatively small area. Such a programme has been enormously expensive, representing a major problem for this and other mines throughout Southeast Asia.

Because of this problem, the Southeast Asia Tin Research and Development Centre (SEATRAD Centre) initiated a geophysical survey in the Pemali area. The programme was carried out by the Directorate of Mineral Resources of Indonesia in collaboration with McPhar Geoservices (Philippines) Inc. in 1980.

## Geophysical Survey

The geophysical survey consisted of gravity, magnetic, induced polarization (IP) and resistivity sounding methods along profiles crossing the tin producing zone. The survey lines, 1,000 m in length trending east-west, were laid out every 50 m. The geophysical survey was done on 24 lines, covering an area of 1.2 km$^2$. Gravity and magnetic observations were done at stations of 10 m intervals along each profile. The induced polarization method was conducted by setting up a dipole-dipole arrangement for every 50 m dipole length. The resistivity soundings were needed as controls and thus only done at 21 selected stations 250 m apart.

Alteration processes, which followed the granite intrusion into the metasediments, reduced the rock densities 20 to 25 %. The alteration has also changed the electric properties through the formation of clay minerals. These changes have rendered the rocks susceptible to gravity and electrical methods. Cassiterite and sulphide minerals in the alteration zone cause an enhancement of the induced polarization response.

## Geological Setting

The rocks observed in the mine area are classified into three groups, from old to young respectively: metasediments, granite and alluvial deposits.

The Permian to U. Triassic metasediments (Priem et al., 1975; Cobbing and Mallick, 1984) consist mainly of slate and schist with quartzite lenses.

The metasediments generally strike NW-SE, dipping steeply northeast. Granite intrusion took place around the Triassic/Jurassic boundary (Cobbing and Mallick, 1984) through the bedding plane fractures of the strata. The contact zones are thin and narrow, dipping east at 70°.

Four types of granite are noted in the mine (Schwartz and Suyono, 1984):

1) *Medium grained foliated granite,* observed along the tin-producing zone (some workers name it the Pemali granite). At many places the granite is altered to greisen and is very rich in tin ore (Fig. 2).
2) *Feldspar-megacrystic medium grained granite,* known as the Klabat granite, is observed in the southeast corner of the mine.
3) *Fine grained equigranular granite,* observed in very limited places east of the Pemali granite.
4) *Porphyritic (quartz) fine-grained granite* is also noted, see Fig. 2. These granites belong to the variable assemblage of coarse, medium and fine grained alkali-feldspar-megacrystic granite of the Tanjung Raya Complex (Cobbing and Mallick, 1984).

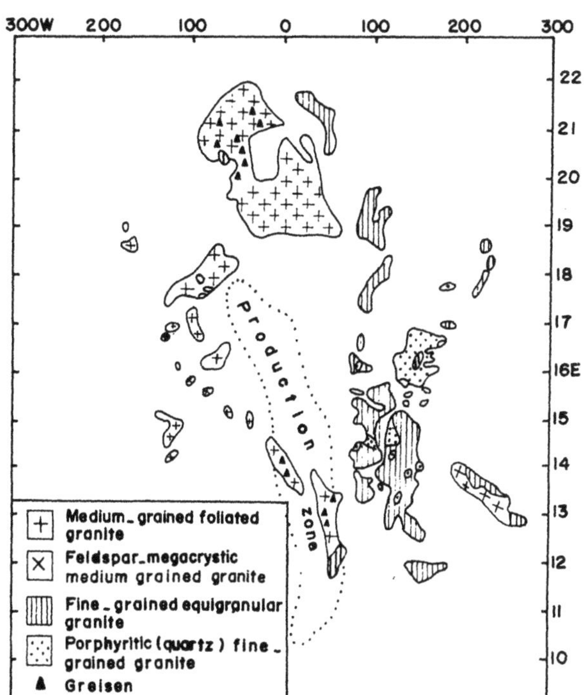


**Fig. 2.** Geological map of the Pemali mine (after Schwartz and Suyono, 1984). The Fig. 10 to 22 refer to the geophysical lines, and 100 to 300 are distances in m from the base line (O)

The metasediments and granite experienced alteration during a late magmatic process, characterized by the alteration of feldspar into sericite and clay minerals, along with the formation of veins that contain quartz, sericite, tourmaline, topaz and fluorite (Setiawan et al., 1984).

The alluvial deposits generally are very thin and are observed at the southern part of the map. Some places are also covered by artificial deposits. The alluvial deposits consist mainly of clay and sand.

## Tin Mineralization

Metasomatic and hydrothermal processes, which followed the granite intrusion into the metasediments, resulted in the mineralization of cassiterite and its associated minerals in the contact zone.

The ore bodies can be found either in the granite or in the metasediments as stockworks along the contact zone. Greisenization has led to the formation of pockets and lenses of ore bodies within the granite. So far, the greisen is the primary host of cassiterite ore in the Pemali mine. The richest ore production zone is associated with the greisen.

Mulyadi (1983) stated that the tin enrichment is very much affected by fault structures along the contact zone. He observed that cassiterite in the metasediments is always associated with fault breccia, and particularly the fault close to the contact zone shows a higher grade of cassiterite when it contains fault breccia.

Commonly associated minerals in the ore are sulphide, tourmaline and topaz. Noer (personal comm.) has noted from core samples that marcasite is common in the area.

## Geophysical Anomalies

1. Gravity

Figure 3 shows a simple Bouguer anomaly map of the Pemali area standardized against the gravity national network at Kemayoran Airport in Jakarta and is calculated using a density of 2.67 ton-m$^{-3}$.

On the west side, a high gravity anomaly is noted over the metasediments and is part of the high gravity anomaly existing west of the map (Oentoeng, 1972). The coincidence of this high gravity with the metasediments suggests an association of high density metasediments with the regional tectonic trend of Bangka.

It is very difficult, however, to find samples of fresh metasediments for direct measurement, but there is no evidence of the existence of intrusive bodies around the anomaly. In this region, the anomaly shows a NW-SE contour lineation parallel to the main structural trend of Bangka. In the actively mined area, a low gravity anomaly with a northerly trend is observed (Fig. 4). The anomaly inclines northeast towards that corner of the map.

Field evidence shows that this low anomaly is mainly related to the altered rocks in the area. Here, as previously mentioned, late magmatic processes have caused the alteration of feldspar into sericite and clay minerals, well known for their low density.

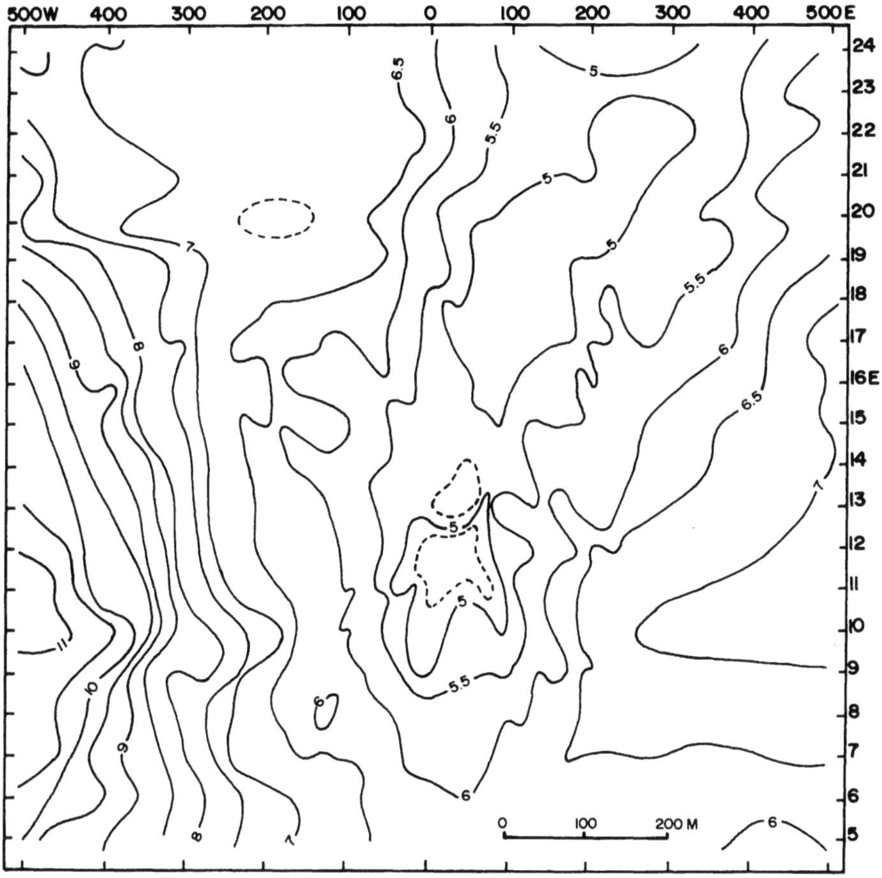


**Fig. 3.** Simple Bouguer gravity anomaly map of the Pemali Area. The contour interval is 0.5 mgal

Within this low gravity zone, a relatively positive anomaly with very low amplitude is observed just over the Pemali granite in the mine. The pattern of the anomaly shows that the granite body is relatively thin, forming a dyke which dips steeply eastward at about 70°. This Pemali granite consists of fine to medium grained minerals altered to greisen.

In the southeast corner of the map, the anomaly is very weak and almost flat, of 6 to 7 mgals magnitude. This anomaly typically coincides with the Klabat medium to coarse grained granite. The anomaly pattern suggests that the granite extends southeastward to the southern tip of the mine. A similar gravity anomaly in the northwest of the map may be related to the granite underneath. Some granites have been discovered east of the anomaly at the northern end of the mine. Smaller gravity anomalies that differ from the earlier mentioned anomalies are interpreted as related to fault structures, and are grouped into two general strikes. The first one is correlated with normal faults; the second with wrench faults. These anomalies suggest that the second group of faults most likely occurred later.

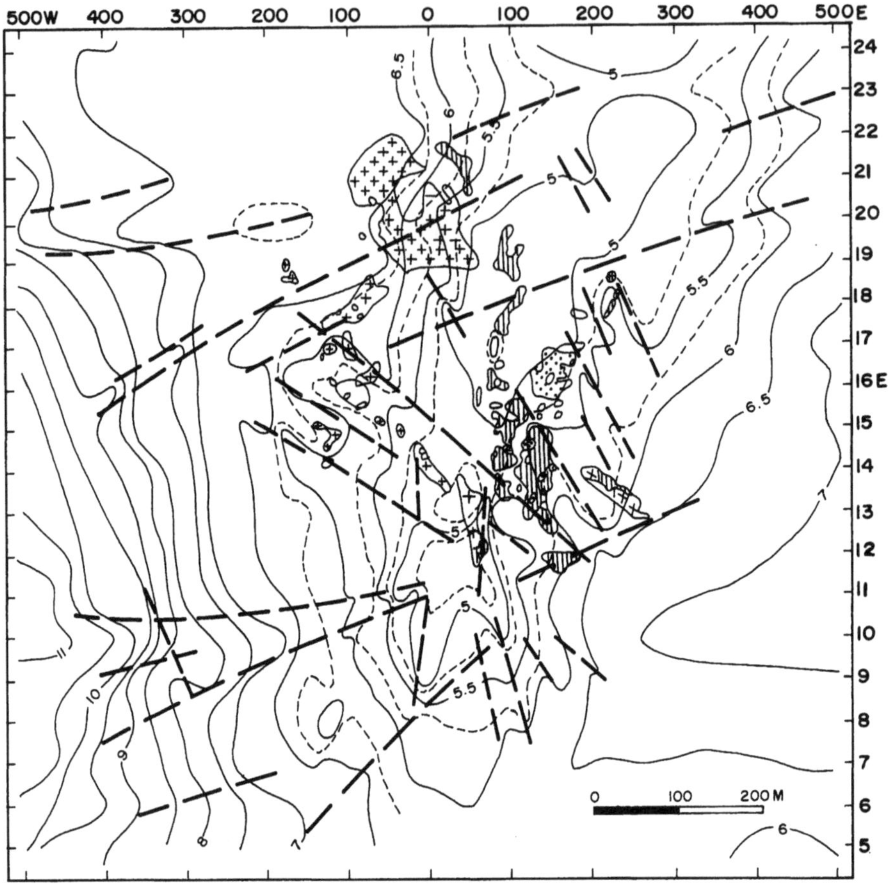


**Fig. 4.** Geology and Bouguer gravity map of the Pemali Area. Faults interpreted from the gravity are shown as dashed lines. For legend see Figures 2 and 3

Some north-south trending faults are also apparent, particularly one that can be seen at the tin-producing zone.

The total magnetic anomaly shows an irregular pattern as predicted. Man-made objects such as metal pipes, electric power poles, and high voltage power lines can produce a magnetic anomaly as high as 1,500 gammas, while the magnetic anomaly associated with granite ranges only between 20 and 100 gammas (Oentoeng, 1972). Thus, it is clear that such high noise seriously mantles the low magnetic signal produced by naturally occurring granite.

The filtering process applied to the observed data did not agree with the prediction since the process reduced the size of anomaly by about 15%. The filtered data show a clear anomaly in the 400 by 400 m area, and it indicates a relatively noise-free anomaly with magnetic poles at lines 14, and 17–18. An "upward continuation" for h=2 shows a much clearer anomaly (Fig. 5). The magnetic pole at line 14 is clearly visible and is associated with the occurrence of fine grained granite. Meanwhile, the magnetic pole at line 17–18 is not very clear, but it may be related to granite in the area.

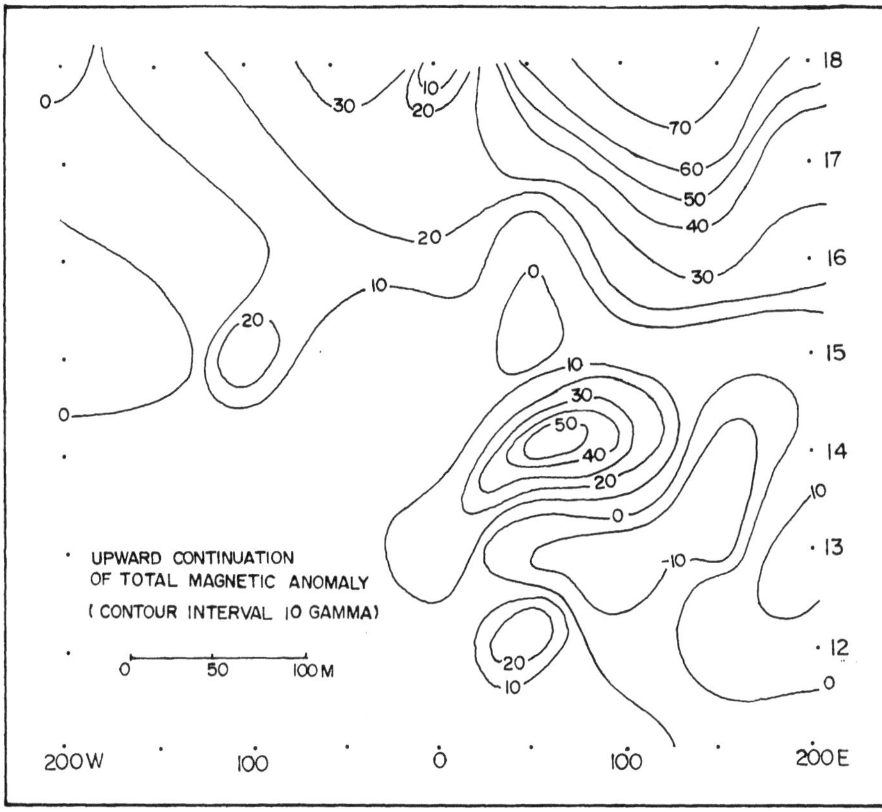


**Fig. 5.** Upward continuation of total magnetic anomaly for h=2

On the whole, the magnetic method does not seem to work effectively in a place like Pemali.

## 2. Induced Polarization

We must first discuss three parameters: resistivity, percent frequency effect (PFE) and metal factor (MF). The resistivity anomaly, related to the cassiterite zone in the low gravity zone, shows variable values between 10 ohm-m and 1,100 ohm-m. The background resistivity anomaly ranges between 200 and 700 ohm-m for metasediments and between 200 and 1,800 ohm-m for granite.

In the south, the resistivity anomaly related to the alteration zone shows low values of 10 to 500 ohm-m, while in the north, the anomaly shows a higher value of 75 to 1,144 ohm-m. These results may be related to the mineralization in those mentioned areas (Table 1).

One of the most striking features of the anomaly shown by resistivity profiles is the mineralization observed at a contact zone between low and high resistivity contours, showing that the mineralization took place in the contact zone between granite and metasediments (Figs. 6 and 7).

**Table 1.** List of IP parameters correlated to the drilling log

| Line No. | App. resistivity (ohm-m) | PFE (%) | MF (mhos) | Remark |
|---|---|---|---|---|
| 8 (S) | 200 | 1 – 3 | 20 – 200 | Contact zone |
| 9 | 10 – 500 | 1 – 3 | 10 – 300 | Contact zone |
| 10 | 100 – 300 | 2 | 20 – 50 | Contact zone |
| 11 | 100 | 3 | 30 | Greisen |
| 12 | 30 – 300 | 2 – 4 | 20 – 50 | Greisen |
| | 20 – 80 | 1.6 | | Contact zone |
| 13 | 200 | 3 | | Greisen |
| 14 | 114 – 1000 | 4 – 5 | 10 – 30 | Greisen |
| 15 | 100 – 500 | 2 – 5 | 10 – 20 | Greisen |
| 16 | 108 – 1144 | 3.5 – 5 | 10 – 20 | Greisen |
| 17 | 1000 | 5 – 6 | 10 – 20 | Greisen ? |
| 18 | 1000 | 5 – 6 | 10 – 20 | Greisen |
| 19 (N) | 500 | 2.5 | 10 – 30 | Greisen |
| | 700 | 5 | | Contact zone ? |

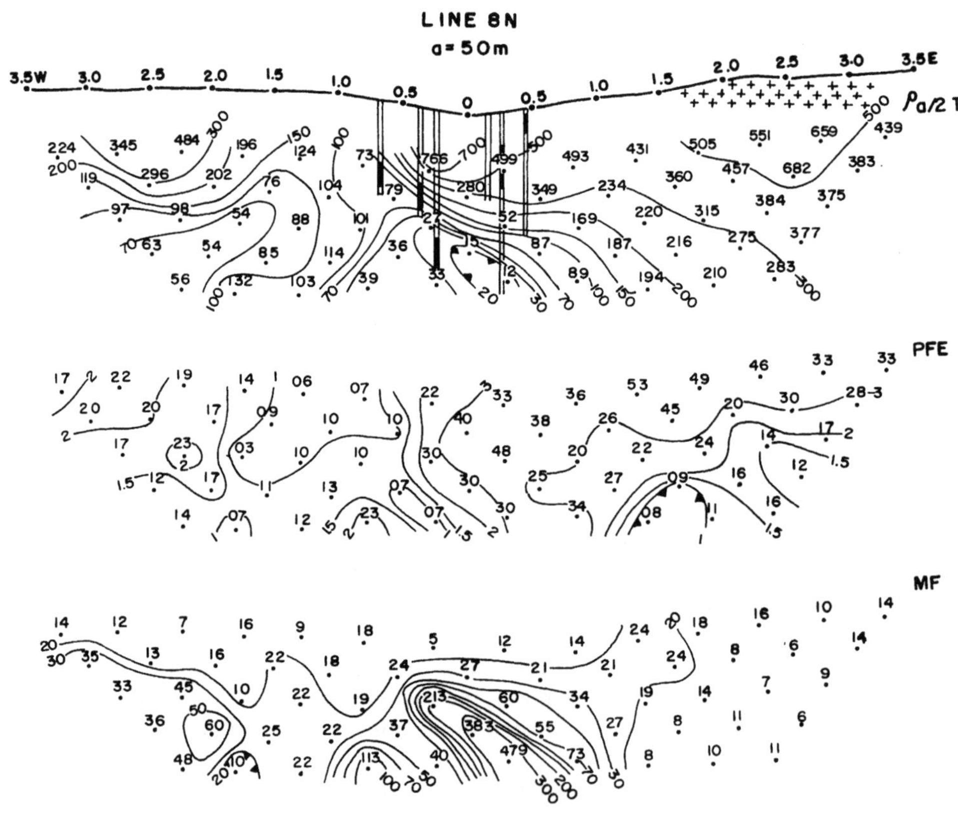


**Fig. 6.** Profile showing IP pseudo section and drilling log. Blackened drilling log section indicates tin content prospect

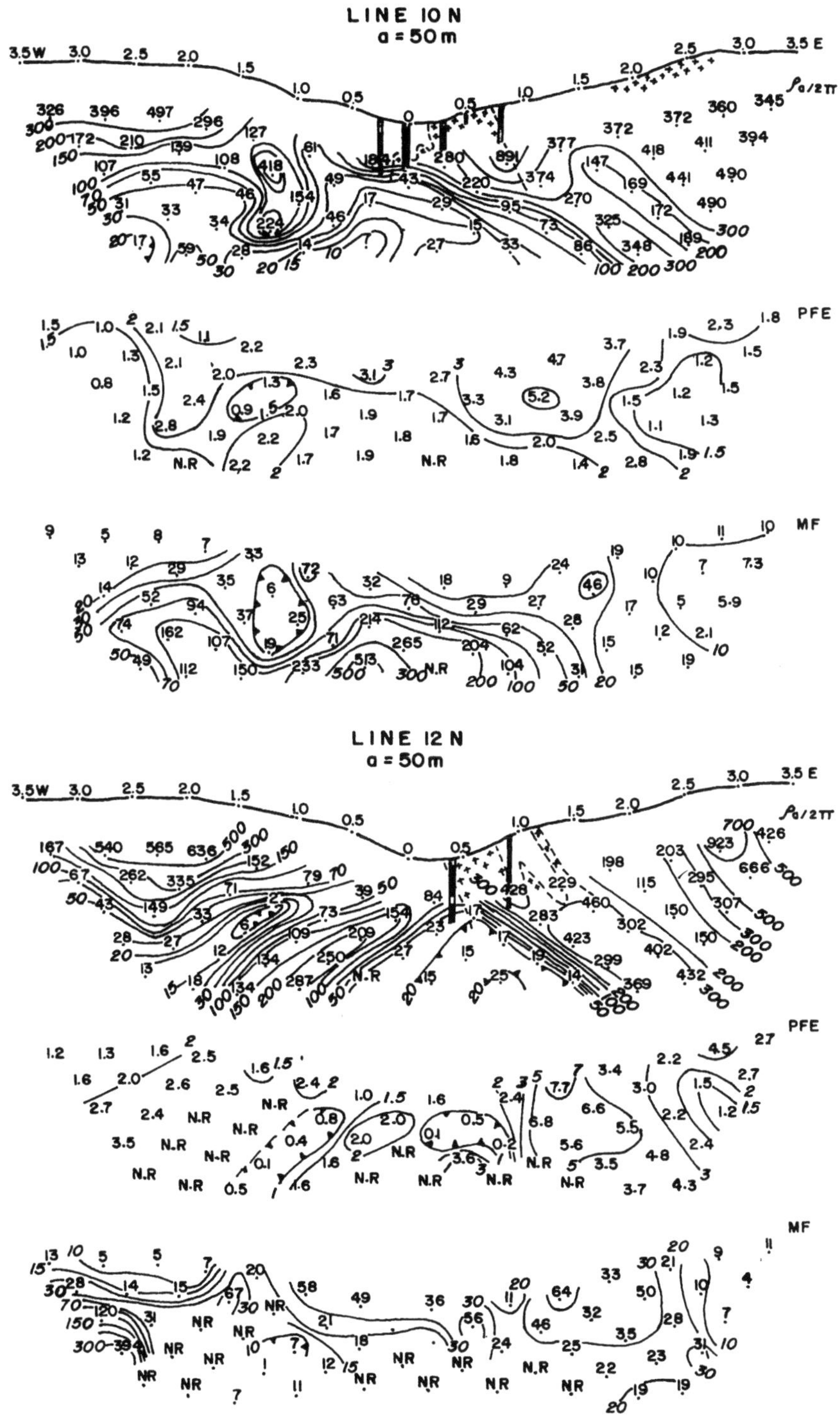


**Fig. 7.** IP pseudo sections for lines 10 and 12

The PFE values of the profiles in general show moderate to high anomalies ranging between 1% and 6%, as expected.

Cassiterite is believed to be one of the oxide minerals which is an electric conductor and can be expected to give IP effects (Madden and Cantwell, 1967). Thus cassiterite is a valid IP target.

The tendency for two different resistivity ranges, as mentioned above, is also observed in the PFE anomalies. In the south, the PFE anomaly ranges between 1% and 3%; in the north between 1 and 6%. The transition area is between line 12 and line 13.

The high PFE anomaly in the north is associated with sulphide mineral in the cassiterite ore which doubles the PFE anomaly. The IP effect will also occur in the presence of a small quantity of clay minerals/particles (membrane) in pores, called "membrane polarization".

A large quantity of clay minerals existing in the contact zone (about 30% by volume) will result in less polarization than if a smaller quantity of clay minerals were present (about 10%) (Marshall and Madden, 1959). In the presence of large quantities of clay minerals, the IP effect will not occur since the membranes will be bypassed by purely resistive paths (Ward and Fraser, 1967). We believe that the clay mineral content in the alteration zone could be as high as 30%, therefore the occurrence of relatively high PFE values here is considered to be associated with the presence of cassiterite. Drilling data in the area confirms this hypothesis.

The metal factor (some workers call it the metal conduction factor) is another way of expressing the IP anomaly. It is done by normalization of PFE to its resistivity and is a measure of the amount of conduction that involves polarization blockage.

The metal factor, which is considered to be associated with the cassiterite, ranges between 20 and 50 mhos (Table 1). A higher value of metal factor seems due to low resistivity caused by clay minerals.

Some twenty-one resistivity soundings were conducted at selected points, and five of them were done on line 5 of the IP survey. The interpreted curves and IP pseudo sections are drawn together as shown in Fig. 8.

In general, the figure shows that there are three different layers: fresh layers on top and bottom, and an altered layer in between.

The top layer with resistivity ranging between 976 and 2079 ohm-m is interpreted to be granite (S7, S9, S4 and S15); S14 with a resistivity of 1357 ohm-m is interpreted to be metasediments.

The altered rock, with its resistivity ranging between 57 and 409 ohm-m, thickens on both sides. Meanwhile, the bottom layer, with its resistivity of 50 to 1364 ohm-m and interpreted as granite, shows a shallow section around 135 meters between S15 and S4. The situation contradicts the IP data here for the IP anomaly shows that this area is still of low resistivity. This phenomenon repeats at S9.

It is apparent that the resistivity sounding tends to show a more shallow depth of rock compared to the IP.

Heterogeneity of the rocks or a layering system causes errors in the sounding method and fails to render reliable information.

The sounding method has not properly worked in areas similar to Pemali.

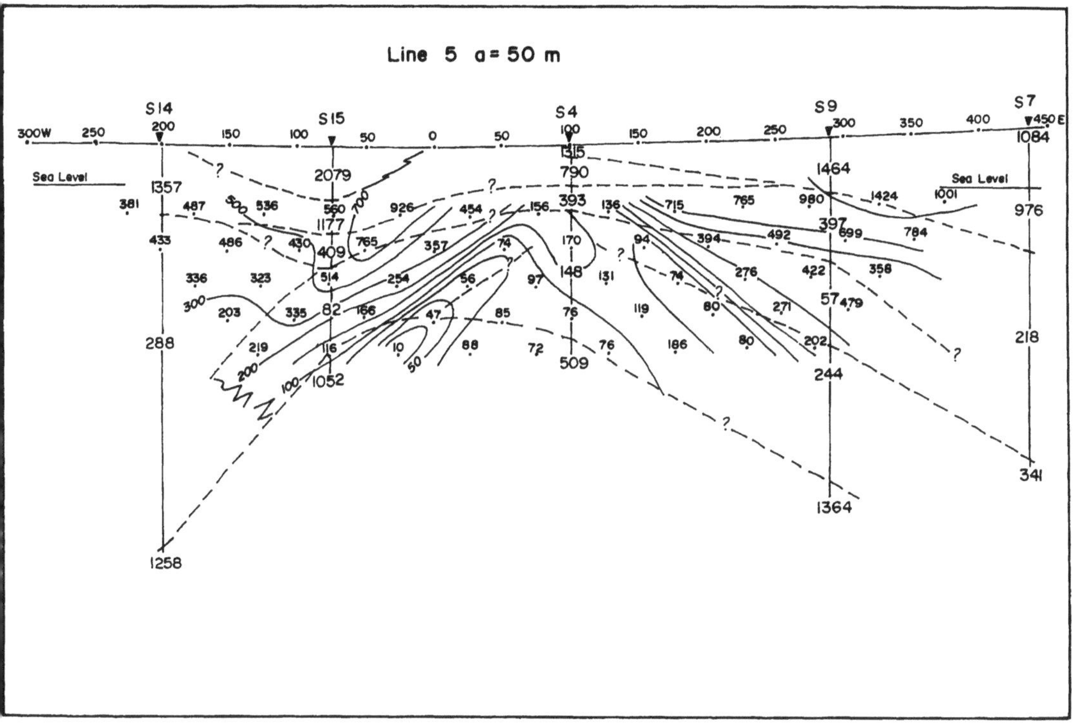


**Fig. 8.** The IP pseudo section of line 5 with the resistivity sounding profile along the line. Sounding numbers are shown as S4 etc. Larger size numbers refer to the resistivity values from the resistivity sounding data

## Discussion

The IP anomaly is superimposed over the Bouguer gravity anomaly in Fig. 9.

Here, the occurrence of the IP anomaly is clearly seen over the low gravity anomaly of the alteration zone.

The figure also shows that the suspected mineralization zone coincides with a moderate slope of the gravity anomaly. This may be related to the fact that the cassiterite is always located at the contact between granite and metasediments, both of which have been altered.

The alteration zone in the magnetic anomaly, in general is shown by a weak anomaly less than 20 gammas (Fig. 5). The alteration process has reduced the magnetic susceptibility of the rocks, so that the contrast between granite and metasediments is very small. The magnetic poles at lines 14 and 17–18 that coincide with the occurrence of fine-grained granite are merely due to that granite.

This suggests that the fine-grained granite is likely to be more susceptible to the alteration process. On the other hand, the difference between granites is not clearly defined by the gravity anomaly.

A weak but positive gravity anomaly that coincides with the Pemali granite is well outlined by the IP anomaly pattern. In many cases, the definite IP anomaly directly

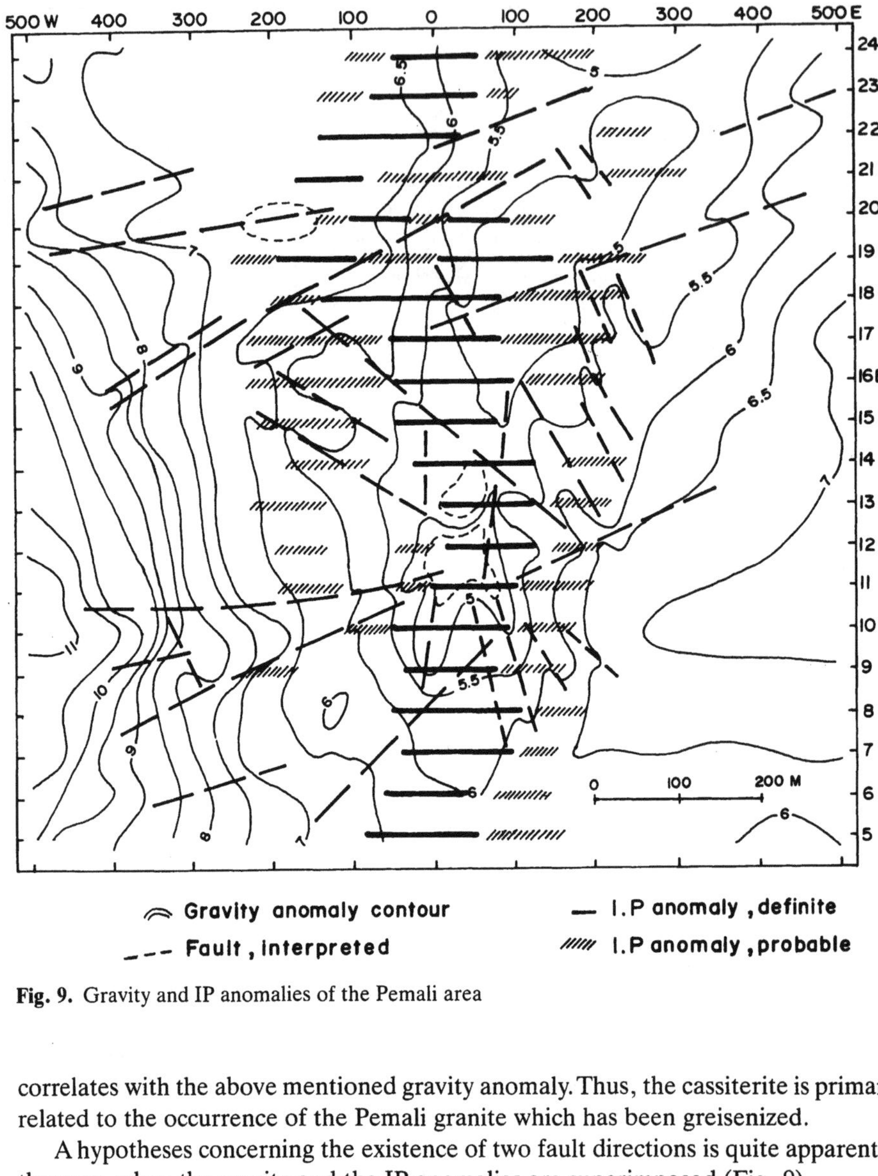


**Fig. 9.** Gravity and IP anomalies of the Pemali area

correlates with the above mentioned gravity anomaly. Thus, the cassiterite is primarily related to the occurrence of the Pemali granite which has been greisenized.

A hypotheses concerning the existence of two fault directions is quite apparent on the map, when the gravity and the IP anomalies are superimposed (Fig. 9).

The northwest and north-northwest faults are more likely pre-mineralization structures that are related to the cassiterite enrichment (Fig. 10). Meanwhile, the northeast faults are clearly post-mineralization. Lateral displacement of rocks is quite distinct. The patterns of the gravity and the IP anomalies support this idea.

Fig. 10 also displays that the first mentioned faults are mostly observed in the northern part of the area. The IP anomaly, either the resistivity or the PFE values, observed in this region is noted to be of higher grade. The coincidence suggests that the high IP anomaly may be related to the occurrence of sulphide minerals in the areas affected by the granite intrusion. Hamzah and Rosadi (1979) noted a high PFE

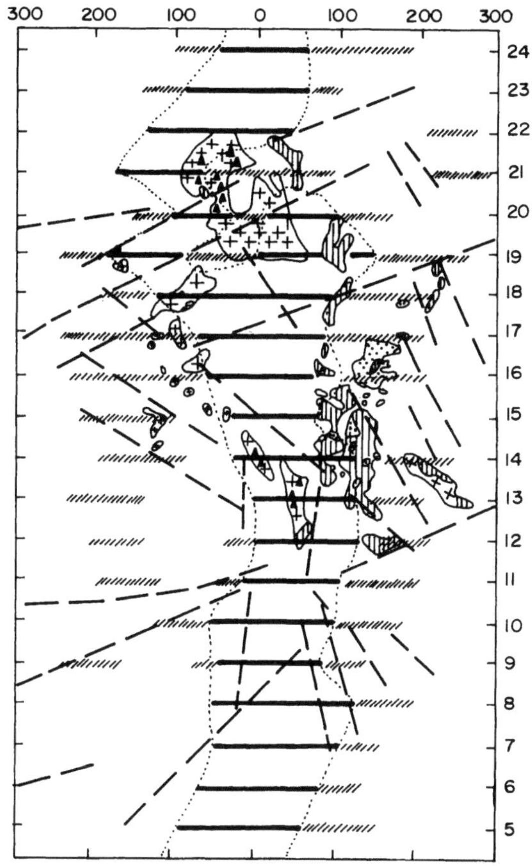


**Fig. 10.** Geophysical anomalies of the Pemali area. See previous figures for legend

anomaly of 5% to 8% during the measurement of IP along the greisen, which also contains sulphide minerals.

## Summary

The alteration zone at Pemali is well-defined by a low gravity anomaly of 2 to 3 mgals amplitude, stretching from south to north and bending northeast. In the alteration zone, the existing weak but positive anomaly is believed due to the occurrence of cassiterite-bearing granite. Flat gravity anomalies at the southeast and the northwest corners of the map are due to granite bodies. Meanwhile, in the west a steep gradient is related to the high density metasediments, associated with the regional tectonic pattern of Bangka.

Finally, the two fault directions are shown by the gravity data. The existence of the Pemali granite in the low gravity zone is clarified by the apparent low resistivity and the high percent frequency effect of the IP. In general, the IP anomaly shows a north-south direction close to the base line, and the definite IP anomaly coincides with the tin-bearing granite.

Therefore, a possible extension of the mine, based on the gravity and induced polarization anomalies, is expected in a north-south direction. In the south, the IP anomaly indicates a deviation towards the south-southeast. The occurrence of granite in the area is clearly shown by the gravity anomaly.

A combined geophysical survey of gravity, induced polarization and magnetic methods is a good exploration method for primary tin mineralization.

*Acknowledgements.* I am grateful to Mr. Salman Padmanagara, Director of Mineral Resources of Indonesia, for his guidance and support. I am also indebted to all the people of the Geophysical Exploration Division for their continuous help during the preparation of the paper. Special thanks are due to Miss Sarah R. Whitmore for her assistance in correcting this paper.

## References

Cobbing, E.J. and Mallick, D.I.J., 1984. South East Asia granite Project. Preliminary Report-Indonesia. Report No. 1984/2. British Geol. Survey, Overseas Division.

Hamzah, E. and Rosadi, Yos, 1979. Induced polarization survey for primary and secondary tin deposits in Pemali, Bangka (in Indonesian). Directorate of Mineral Resources of Indonesia.

Madden, T.R. and Cantwell, T., 1967. Induced polarization. A Review. *Mining Geophysics* vol. II, part D, 373–400.

Marshall, D.J. and Madden, T.R., 1959. Induced polarization, a study of its causes. *Geophysics,* vol. 24 (4), 790–816.

Mulyadi, H., 1983. Primary tin deposits in Pemali area (in Indonesian). PT. TIMAH, Indonesia (unpublished).

Oentoeng, M., 1972. Gravity and magnetic survey for tin primary deposits, Pemali, Bangka Indonesia (in Indonesian). Geological Survey of Indonesia.

Priem, H.N.A., Boelrijk, N.A.I.M., Bon, E.H., and others, 1975. Isotope geochronology in the Indonesian tin belt. *Geol. Mijnbouw,* 54, 61–70.

Schwartz, M. and Suryono, 1984. Pemali mine, Bangka, primary tin deposits (unpubl.). Directorate of Mineral Resources, Indonesia.

Setiawan, R., Kridhoharto, P., and Sunarya, N., 1980: A combined geophysical exploration for primary tin deposits in Pemali, Bangka Indonesia. Directorate of Mineral Resources.

Van Bemmelen, R.W., 1949. *The geology of Indonesia* vol. IA. Martinus Nijhoff, The Hague, 315–318.

Ward, S.H., and Fraser, D.C., 1967. Conduction of electricity in rocks. *Mining Geophysics,* vol. II, Part B, 197–223.

# 6.8.3 Granitoids of Sumatra and the Tin Islands

WIKARNO, D.A.D. SUYATNA, and S. SUKARDI[1]

## Introduction

The granitoid study by the Geological Research and Development Centre (GRDC) started in 1980. The investigation began in the Lampung area of the southernmost part of Sumatra and gradually moved northward to the Bengkulu, Jambi, and Tapanuli areas. In 1983 the investigation was carried out in collaboration with Dr. E.J. Cobbing and Dr. D.I.J. Mallick, geologists of the British Geological Survey (BGS). Since 1982, the Overseas Division of the BGS has carried out a South East Asian Granite Project which covers Thailand, Malaysia and Indonesia. Together with GRDC staff, they undertook six weeks of field work on the tin island granitoids and spent two weeks in West Sumatra and the South Tapanuli granites.

A study of the physical properties of granitic rocks was also carried out by scientists from Thailand, Malaysia and Indonesia, co-ordinated by the Regional Mineral Resources Development Centre (R.M.R.D.C.). Coincidently both projects were carried out in the same region, especially in the Southeast Asian Tin Belt, which is the world's foremost tin producing province.

Tin is closely related to granite, but to which specific types is not yet clearly defined. Therefore many scientists have attempted to study granitoid rocks in order to clarify the relationship. Among others, Chappell and White (1974) with their I-type and S-type granites, and Ishihara (1977) with the Ilmenite-series and Magnetite-series, made significant advances. However, problems still remain as to the association and location of tin mineralization in granites. Each classification given above has its own advantages and disadvantages. Nevertheless both will help to bring about an understanding of tin occurrence. Mineralization of such metals as copper, lead, zinc and gold is normally related to the I-type granitoids of Chappell and White (1974) and the Magnetite-Series of Ishihara (1977). The S-type and Ilmenite-Series granitoids, on the other hand, seem to have an affinity for tin and tungsten mineralization.

It is important to mention here that not all Ilmenite-series granitoids are of S-type granite, and some may be I-type.

The I and S-type classification is chemically based, whereas the Ilmenite and Magnetite series classification is descriptive, in which granites are classified solely on their opaque mineral content. Granite containing less than 0.1% by volume of magnetite is classified als Ilmenite series. In terms of magnetic susceptibility, the read-

[1] Geological Research and Development Centre, Jalan Diponegoro 57, Bandung, Indonesia

ing should less than 40 x $10^{-3}$ SI units (Ishihara et al., 1979). The magnetic susceptibility may be measured directly in the field by a Kappameter. In order to distinguish the I-type and S-type, careful mineralogical identification and chemical analysis are necessary. The I-type granite is considered to be generated from igneous source material, while the S-type granite derives from sedimentary material (White and Chappell, 1977). According to Chappell and White (1974), the I-type contains relatively high $Na_2O$ and CaO and displays a broad spectrum of mineralogical composition ranging from felsic to mafic varieties and shows a regular inter-element variation within a pluton. On the other hand, the S-type granite contains relatively low $Na_2O$ and CaO and has a restricted mineralogical composition (Takahashi et al., 1980).

## Granitoid Distribution

### Sumatra

Granitoids are distributed along the axis of the island following the great Fault Zone of Sumatra. They consist of independent plutons and batholiths of small and large size up to several hundred km length. Most of these granite bodies are elongated in a NW-SE direction parallel to the length of the island. Many of them have sharp boundaries, coinciding with the fault zone. This feature is very clearly shown in central Sumatra in the Padang Highland granite Complex (Fig. 1). From north to south the granite distribution is as follows:

In the Tapanuli area, are the Sibolga, Padang, Sidempuan, Penyabungan, Muara Sipongi, and Pangarayan plutons.

The Padang Highland granite complex consists of the Anai, Lassi, Sulit Air, Mandiangin, and Tanjung Gadang plutons. The latter is the largest in this complex.

In south Sumatra the granite bodies are the Talang Gadang, Bukit Payung, Bukit Garba, Bandar, Tanjung Gadang, and Tanjung Karang plutons.

In the Jambi and south Sumatra area, there are plutons which are situated east of Sumatra Fault Zone. They are the Bukit Dua Belas, Bukit Tiga Puluh and the Bukit Batu plutons. These plutons need to be studied carefully because they occur in different geological settings.

According to Katili (1962), the granite emplacement in Sumatra is syntectonic with folding phases in Early Cretaceous or Late Jurassic time. Radiometric age determinations on granites from various plutons in Sumatra have yielded different ages ranging from Carboniferous to Cretaceous; many of them are Triassic (Table 1).

In the Tapanuli area, the largest granite body is the Sibolga pluton. Its Rb : Sr age determination are mostly Triassic; but there is no information on the sample localities. These radiometric ages indicate that there are old and young granites in north Sumatra, and even Tertiary granites also occur.

In the Padang Highland, the Tanjung Gadang pluton is the largest. The radiometric age of granite plutons of this area are Triassic to Cretaceous, but older ages of Permian and Carboniferous are also found. From south Sumatra, a number of granite ages have been determined, and the results are mostly of Cretaceous age.

On the basis of radiometric age dating results (Table 1), it may be summarized that Sumatra contains Paleozoic, Mesozoic and Tertiary granites.

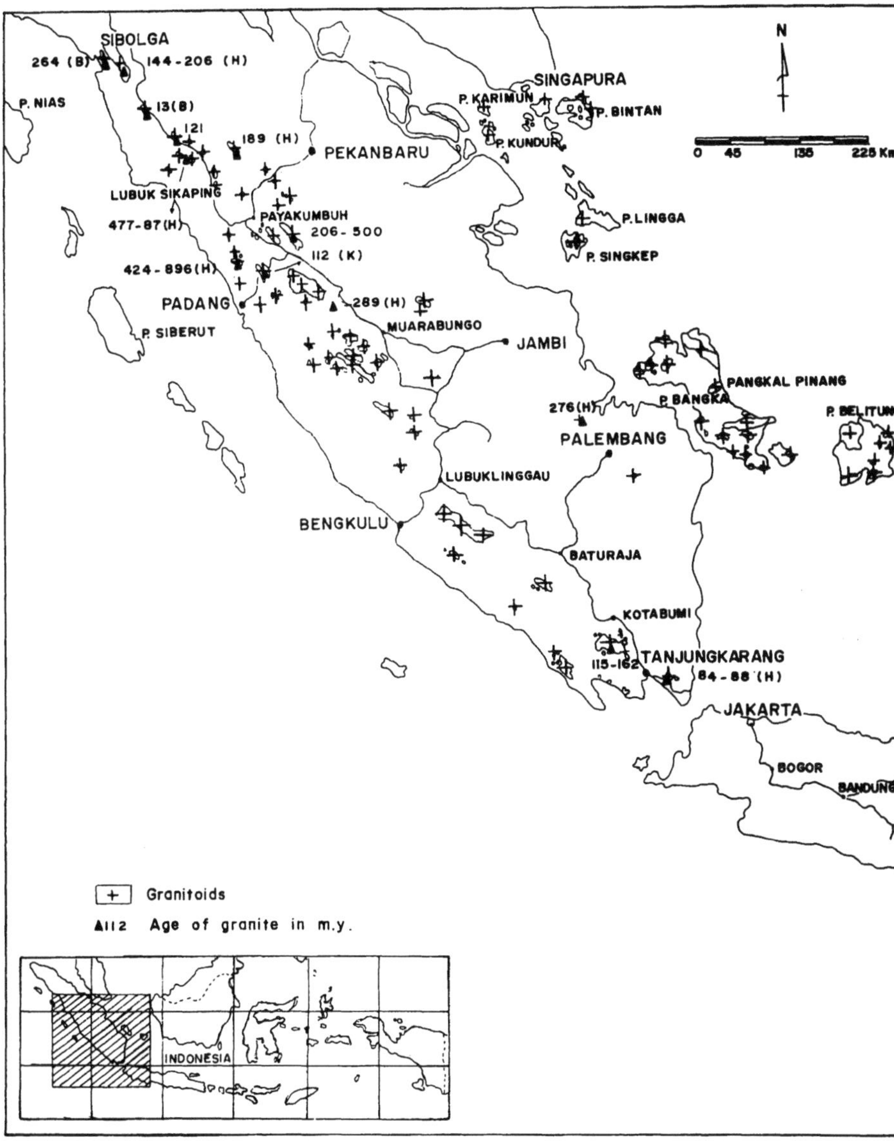


**Fig. 1.** Granitoids distribution and sampling locations on Sumatra

## Bangka

Granitoid bodies occupy about one third of the total area of the island and about 30% of the coast line is of granite outcrops. More than 14 granite plutons are encountered (Fig. 2). According to Katili (1967), there are three conspicuous trends of granite bodies on this island. Those are the S.E., E.W. and N.E. trends. The first two are parallel to the fold axes of the island. The S.E. trend is demonstrated by the southern boundary of the Klabat batholith. The Babuluh batholith, which is situated in the

**Table 1.** Radiometric age dating results of Sumatran granites

| Granite Pluton | Rock | Mineral type | Method analyzed | Age in Ma. |
|---|---|---|---|---|
| *South Tapanuli* | | | | |
| Sibolga | granite | biotite | K:Ar | 257 ± 24 |
| | granite | | Rb:Sr | 264 ± 6 |
| Tarutung | granite | hornblende | Rb:Sr | 206.1 ± 2.5 |
| Tarutung | granite | biotite | Rb:Sr | 218.8 ± 3.9 |
| Sibolga | granite | biotite | Rb:Sr | 217.4 ± 4.4 |
| Sibolga | granite | biotite | Rb:Sr | 211.5 ± 2.6 |
| *North Sumatra* | granite | biotite | Rb:Sr. | 147 ± 2.4 |
| *West Sumatra* | | | | |
| Tanjung Gadang | granite | Kf-hbl. | Rb:Sr | 110 ± 155 |
| Sisawah | granite | Kf-hbl. | Rb:Sr | 216 |
| Sisawah | granite | plagioclase | Rb:Sr | 214 |
| Sisawah | granite | biotite | Rb:Sr | 215 |
| Sisawah | granite | muscovite | Rb:Sr | 294 |
| Lassi Granite | granite | whole rock | Rb:Sr | 112 ± 24 |
| Panti Granite | granite | whole rock | K:Ar | 427 |
| Sumatra Barat | granite | biotite | K:Ar | 20.6 ± 25 |
| Sumatra Barat | granite | hornblende | K:Ar | 206 ± 3.4 |
| Sumatra Barat | granite | biotite | K:Ar | 215 ± 3 |
| Sumatra Barat | granite | biotite | Rb:Sr | 135 ± 55 |
| Sumatra Barat | granite | biotite | K:Ar | 196.7 ± 24 |
| Sumatra Barat | granite | muscovite | K:Ar | 287 ± 3.5 |
| Sumatra Barat | granite | muscovite | Rb:Sr | 256 ± 6 |
| Sumatra Barat | granite | alk.-feldspar | Rb:Sr | 145 ± 4 |
| *South Sumatra* | | | | |
| South Sumatra | granite | biotite | K:Ar | 81.3 ± 1.3 |
| Peg. Balik Batang | granodiorite | zircon | FT | 53.53 |
| Bangunrejo Lp. | granite | zircon | FT | 120.9 ± 8.75 |
| Wai Welirang | granite | zircon | FT | 115 – 116 |
| Jati Baru | granite | zircon | FT | 48.37 |

eastern part of Bangka, is made up of the Kulur and Pading plutons. These granite bodies which stretch in an E-W direction are also parallel to the fold axes of the district. The N.E.-trending granite bodies are the Mangkol pluton in central eastern Bangka, the Kelapa pluton, the Panagas pluton and the Menumbing pluton in the west.

## Belitung

As in Bangka, large granite plutons are distributed on the outer parts of Belitung island (Fig. 3). The Tanjung Pandan pluton in the north western corner is the largest body. The boundary of the granite bodies also indicates that the trend is parallel to the fold axes (Katili 1967).

**Fig. 2.** Granitoid distribution and sampling locations on Bangka

## Other Islands

Almost the entire islands of Karimun and Kundur are made up of granite (Fig. 4). On Batam granite occurs only at the north eastern part, and on Bintan, granite occurs in the north and eastern part of the island, outcropping over about 50% of the island. On Lingga, granite crops out only in the southern part of the island, and on Singkep, two granite bodies are found in the N.E. and south central. On Selayar, which is situated between Lingga and Singkep, almost the entire island is made up of granite.

**Fig. 3.** Granitoid distribution and sampling locations on Belitung

The age of granitoids of the tin belt still has to be examined carfully because different opinions exist. Some structural geologists believe that the tin granitoids are Upper Jurassic, synchronous with the mountain building phase (Katili, 1967). Radiometric age determinations on granites from Bangka and Belitung range from Triassic to Cretaceous (Priem et al., 1975). This means that there are older and younger granites (Table 2).

**Table 2.** Radiometric ages of granites of Bangka and Belitung

| Area | Rock name | Analysis mineral | Method, Age in m.y. | |
|---|---|---|---|---|
| Kelapa Pluton Bangka | granodiorite | whole rock | K-Ar | 213 ± 6 |
| Muntok Bangka | granite | whole rock | K-Ar | 211 ± 10 |
| Belitung Belitung | | | | |
| Bebulu Belitung | biotite granite | biotite | K-Ar | 213 ± 10 |
| Bebulu Belitung | biotite granite | hornblende | K-Ar | 224 ± 20 |
| Bebulu Belitung | biotite granite | biotite | K-Ar | 212 ± 10 |
| Bebulu Belitung | biotite granite | hornblende | K-Ar | 213 ± 9 |
| Burung Mandi Complex | hbl. diorite | hornblende | K-Ar | 120 ± 4 |

## Granitoid Petrology

### Sumatra

The granitoids of Sumatra are mainly composed of alk.-feldspar, plagioclase, and quartz. The mafic minerals may consist of biotite, muscovite or hornblende. Pyroxene is scarce. The accessory minerals are magnetite, ilmenite, sphene, zircon, and apatite. Cassiterite and tourmaline are also found among the accessory minerals in some granites. Hornblende and sphene are the diagnostic minerals for the magnetite-series granitoids and the amount of biotite is usually less than 15 % by volume.

The alk.-feldspar varies from colourless, grey, to pink; quartz is light grey to white, and biotite is greenish. The presence of hornblende, sphene, apatite, monazite and the amount of biotite and magnetite present may indicate the type of granite. Based on the mineral constituents, the granitoids of Sumatra are of biotite granite, granodiorite, granite, quartz monzonite, tonalite (quartz diorite), biotite-hornblende granite, muscovite granite, and hornblende granite (Table 3). The most common rock types of the granitoid bodies are biotite granite, granodiorite and normal granite.

The degree of crystallinity varies from fine equigranular, medium to very coarse grained, and from weak porphyritic to strongly porphyritic containing megacrysts. The megacrysts are of alk.-feldspar and plagioclase. Biotite and hornblende very

**Table 3.** The granite types encountered in the granite plutons of Sumatra

| No. | Rock Name | Number of occurrences | | | | |
|---|---|---|---|---|---|---|
| | | North Sumatra | West Sumatra | Jambi | Bengkulu | Total |
| 1. | Granite | — | 6 | 4 | – | 10 |
| 2. | Biotite granite | 11 | 1 | 9 | 1 | 22 |
| 3. | Muscovite granite | 5 | 1 | 1 | – | 7 |
| 4. | Biotite-hornblende granite | 1 | – | 3 | – | 4 |
| 5. | Hornblende granite | — | – | 3 | 3 | 6 |
| 6. | Quartz monzonite | 8 | – | – | | 8 |
| 7. | Granodiorite | 12 | 1 | – | | 13 |
| 8. | Tonalite | 3 | 3 | 2 | | 8 |
| 9. | Pegmatite granite | 2 | – | – | | 2 |

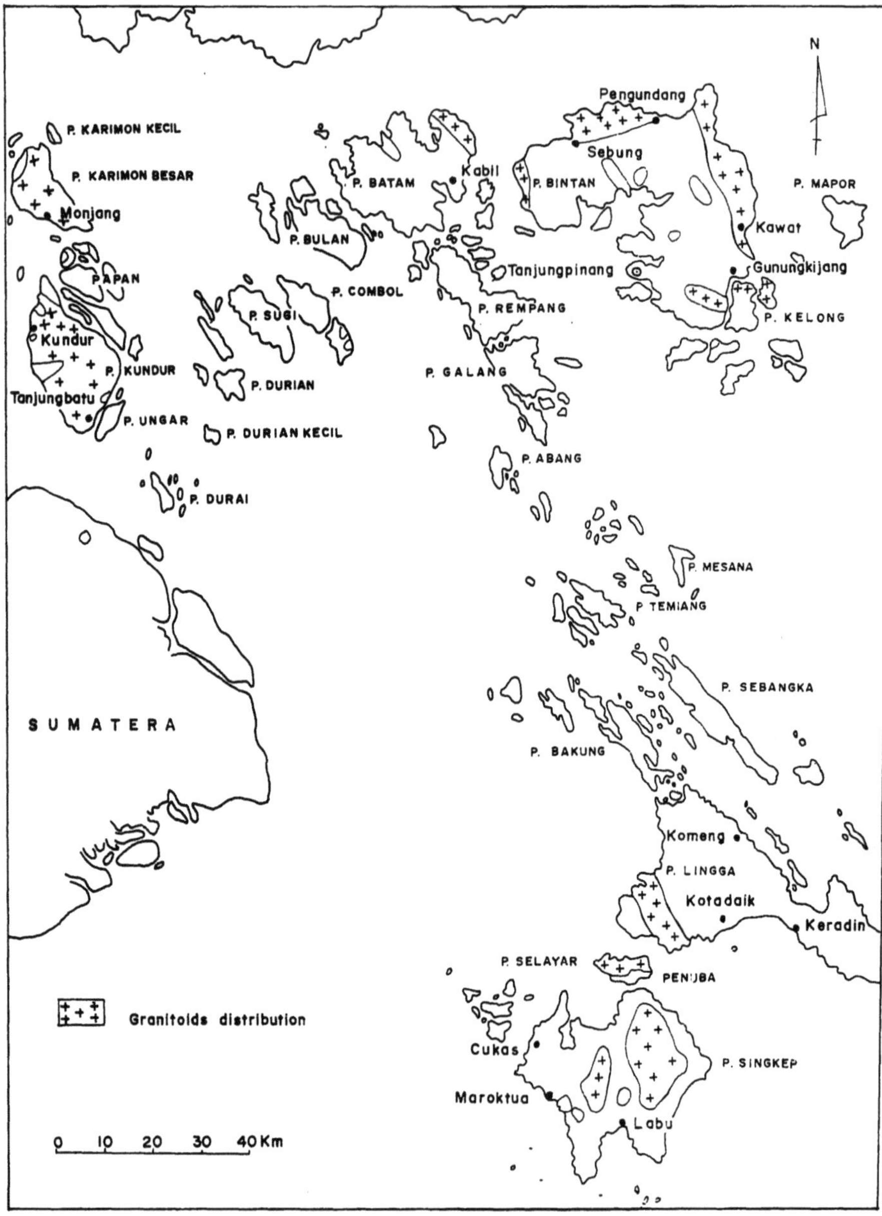


**Fig. 4.** Granitoid distribution on Riau Archipelago

often occur as clots or clusters. Xenoliths of more basic rock are found in some granite bodies such as the Sulit Air pluton.

Intergrowths between alk.-feldspar and plagioclase, quartz and plagioclase, and alk.-feldspar are common. Many plutons, especially those which have been affected by tectonic deformation, show foliated, gneissic or brecciated textures.

### The Tin Islands

Throughout the tin islands the granitoids are characterized by a great textural variety even within individual plutons. The range is from medium grained equigranular to a strongly porphyritic megacrystic texture. The mineralogical composition on the other hand in general is similar from one pluton to another. The main constituent are alk.-feldspar, quartz, plagioclase, biotite and hornblende; and muscovite may also be present. The frequently occurring accessory minerals are magnetite, ilmenite, zircon, apatite, sphene, fluorite, cassiterite, monazite and sulphide minerals such as pyrite, pyrrhotite and galena. The alk.-feldspar is white to light grey, and occurs as phenocrysts or in the groundmass. Under the microscope, inclusions of biotite or oxide minerals are common in the megacrysts of alk.-feldspar. At several locations such as in the Belinyu and Toboali plutons in Bangka, at Bukit Tumang and north of Tanjung Pandang on Singkep, the alk.-feldspar crystals are arranged in sub parallel orientation in a randomly arranged matrix. This alignment is not related to metamorphism, and indicates that the alk.-feldspar crystallised before the magma was solidified. It also means that the temperature of the magma was not too high and cooled relatively rapidly. Colourless to greyish-white plagioclase is found as phenocrysts and in the groundmass. It is of oligoclase, andesine or labradorite, and commonly contains inclusions of Fe-Ti oxides and biotite.

Biotite is the most common mafic mineral constituent of the tin island granites. Under the microscope it is brown and occurs as single crystals or as clusters. The colour of biotite from Sumatra is generally green and biotite from the tin island granites is brown. This is one among other differences recognized between Sumatran and tin island granites. Hornblende occurs only at several locations in the Kelapa and Toboali plutons in Bangka, south Belitung, north Bintan and east Batam. The volume percent of hornblende is less than 5.

If a rock contains tourmaline topaz and fluorite among the accessory minerals, this may be an indication of primary tin mineralization.

The granitoids of the tin islands may be classified as biotite granite, biotite-hornblende granite, granite, muscovite granite, quartz monzonite, granodiorite, bitotite-hornblende granodiorite and tonalite (Table 4).

## The I/S Type and Magnetite/Ilmenite Series Granitoids

A rough classification into I/S type and Magnetite/Ilmenite series granitoids has been prepared according to the mineralogical composition, magnetite content, and magnetic susceptibility value. This is a temporary classification and may be changed after a detailed examination of the respective plutons.

**Table 4.** The granite types encountered in the granite plutons of the tin islands

| No. | Rock name | Number of occurrences | | | | | | | | | |
|---|---|---|---|---|---|---|---|---|---|---|---|
| | | Bangka | Belitung | Batam | Bintan | Karimun | Kundu | Singkep | Lingga | Selayar | Total |
| 1. | Granite | 3 | 1 | – | – | – | – | – | – | – | 4 |
| 2. | Biotite granite | 8 | 5 | 1 | 9 | 1 | 1 | 1 | 1 | – | 27 |
| 3. | Muscovite granite | 1 | – | – | – | – | 2 | – | – | – | 3 |
| 4. | Biotite-hornblende granite | 4 | 1 | 1 | 6 | – | – | 1 | – | 1 | 14 |
| 5. | Quartz monzonite | – | 1 | – | – | – | – | – | – | – | 1 |
| 6. | Granodiorite | 1 | – | – | – | – | – | – | – | – | 1 |
| 7. | Biotite-hornblende granodiorite | – | 1 | – | – | – | – | – | – | – | 1 |
| 8. | Tonalite | – | 1 | – | – | – | – | – | – | – | 1 |

In Sumatra, the distribution of both I/S type and Ilmenite/Magnetite series granitoids is mixed. Figure5 shows the distribution of Ilmenite series/S-type granite, Ilmenite series/I-type, and Magnetite series/I-types. Many of the granite bodies are classified as I-type and Magnetite series (Table 5). These granites generally have affinities to base metal mineralization. There are only two granites that seem to be of the Ilmenite series and S-type granite. Those are of the Panyabungan pluton and a small granite occurrence on the N.E. part of Lake Singkarak. The Ilmenite series and S-type granites are generally related to tin mineralization. Therefore a detailed study of the Penyabungan granite pluton need to be conducted. Other granite plutons with indications of Ilmenite series also need to be studied further.

**Table 5.** Classification of Sumatran granitoids into Ilmenite/Magnetite series, and I/S type and their relation to tin deposits

| No. | Granite Pluton | Rock name | Series | | Type | | Tin-Association |
|---|---|---|---|---|---|---|---|
| | | | Ilmenite | Manetite | I | S | |
| 1. | Tanjung Gadang | biotite-hornblende granite | – | v | v | – | – |
| 2. | Lassi | tonalite/foliated granite | – | v | v | – | – |
| 3. | Lolo | tonalite | – | v | v | – | – |
| 4. | Muara Gadang | granite | – | v | v | – | – |
| 5. | Sulit Air | monzodiorite | – | v | v | – | – |
| 6. | Singkarak | foliated muscovite granite | v | – | – | v | – |
| 7. | Anai Pluton | granite | v | – | v | – | – |
| 8. | Salibutan | granite (weathered) | v | – | – | ? | – |
| 9. | Matur | granite (weathered) | – | – | – | – | – |
| 10. | Sibolga | granite | v | – | v | – | – |
| 11. | Penyabungan | granite | v | – | – | v | – |

(Classification of granites after Cobbing and Malick, 1984)

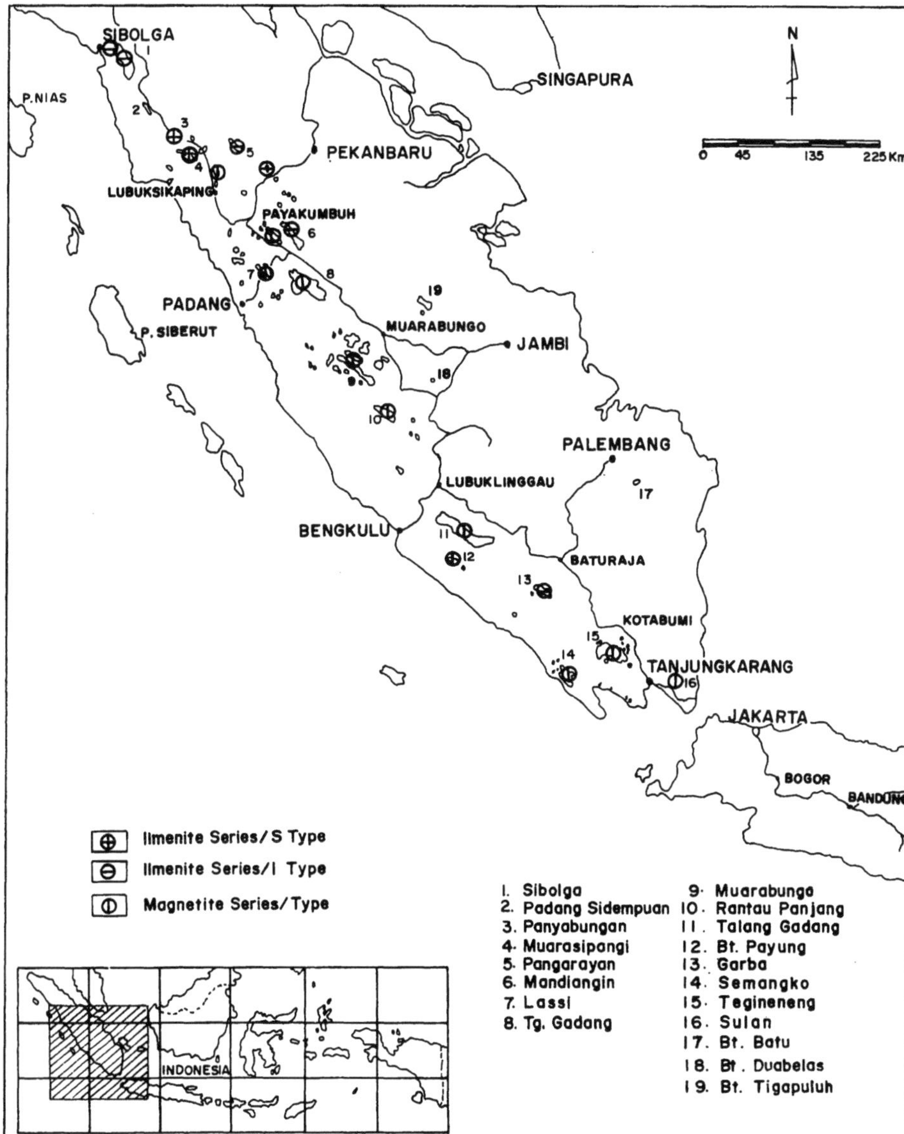


**Fig. 5.** Distribution of Ilmenite series/S-type, Ilmenite series/I-type and Magnetite series/I-type of Sumatra granitoids

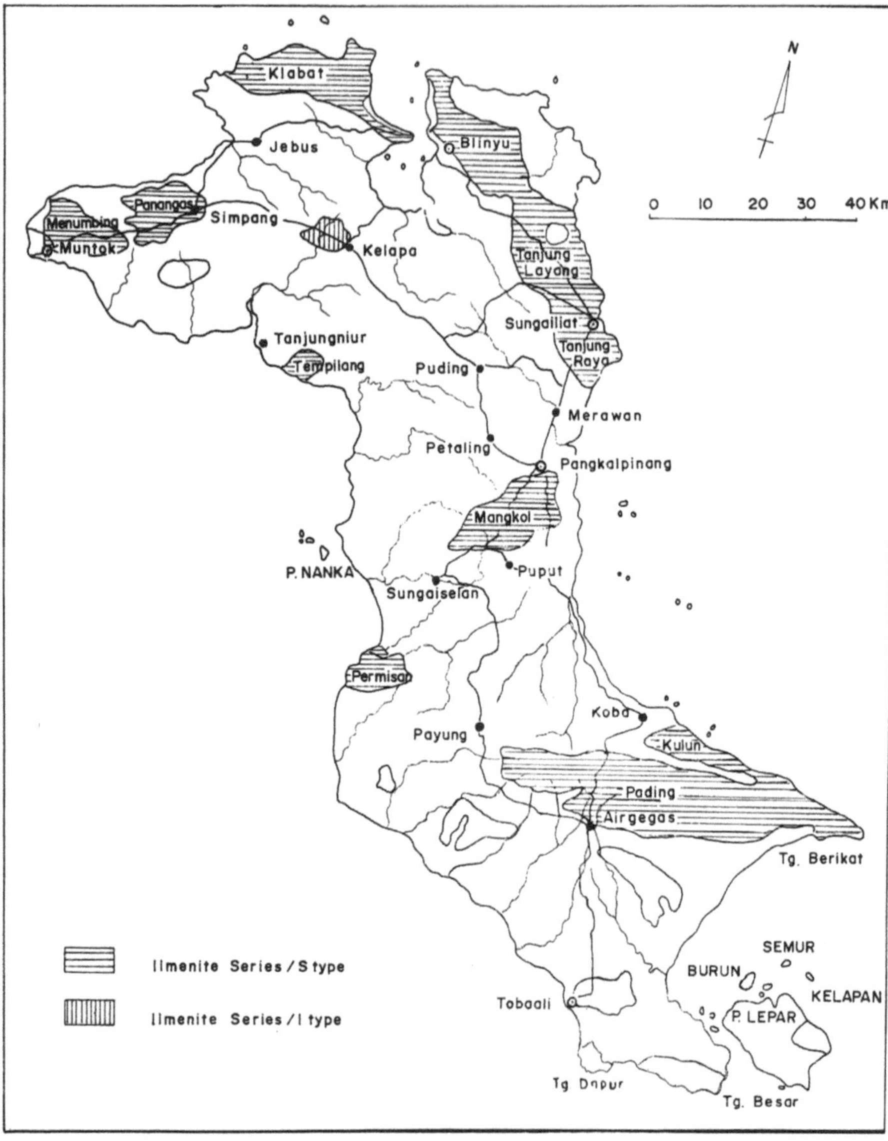


**Fig. 6.** Distribution of Ilmenite series/S-type, Ilmenite series/I-type and Magnetite series/I-type of Bangka granitoids

**Table 6.** Classification of Bangka granitoids into Ilmenite/Magnetite series, I/S type and their tin deposit association

| No. | Granite Pluton | Rock name | Series | | Type | | Tin Association |
|---|---|---|---|---|---|---|---|
| | | | Ilmenite | Magnetite | I | S | |
| 1. | Tanjung Raya | Megacryst granite | v | – | – | – | Alluvial tin |
| 2. | Tanjung Layang | Biotite granite | v | – | – | v | – |
| 3. | Belinyu | granite | v | – | – | v | Extensive alluvial tin |
| 4. | Panangas | granite | v | – | – | v | Extensive alluvial tin |
| 5. | Bt. Bias | granite | v | – | – | v | Alluvial tin |
| 6. | Tanjung Batu | granite | v | – | – | v | – |
| 7. | Menumbing | granite | v | – | – | v | Extensive alluvial tin |
| 8. | Tempilang | granite | v | – | – | v | Alluvial tin |
| 9. | Kelapa | Biotite horn-blende granite | v | – | v | – | – |
| 10. | Mangkol | granite | v | – | – | v | Aluvial tin |
| 11. | Permisan | Biotite granite | v | – | – | ? | Traces of alluvial tin |
| 12. | Toboali | Biotite-horn-blende granite | v | – | ? | – | Extensive alluvial tin, origin not known |
| | | granite | v | – | – | v | |
| 13. | Nama | granite | v | – | – | ? | |
| 14. | Kulur | Hornblende granite Granodiorite | v | – | v | – | |
| | | Gabbro | – | v | v | | |
| 15. | Pading | granite | v | – | – | v | Alluvial tin found |

(Ilmenite/Magnetite series and I/S type classification after Cobbing and Mallick, 1984).

The granite plutons of Bangka and Belitung are classified as Ilmenite series and S-type (Table 6 and Figure 6). Only a few plutons: the Kelapa pluton, and a part of the Toboali pluton of Bangka are classified as Ilmenite series and I-type. Alluvial tin fields are scattered on many granite plutons on Bangka (Table 6). Those are on the Tanjung Raya, Belinyu, Batu Bias, Menumbing, Tempilang, Mangkol, and Toboali granites.

**Table 7.** Classification of granitoids in Belitung into Ilmenite/Magnetite series, I/S type and their tin deposit association

| No. | Granite Pluton | Rock name | Series | | Type | | Tin Association |
|---|---|---|---|---|---|---|---|
| | | | Ilmenite | Magnetite | I | S | |
| 1. | Gunung Mang | Granodiorite | v | – | v | – | – |
| | | Diorite | v | – | v | – | – |
| 2. | Gunung Legau | Granodiorite | v | – | v | – | |
| 3. | Bt. Besi | Leucogranite | v | – | – | v | Alluvial tin in progress |
| 4. | Tanjung Pandan | | | | | | Tin field |
| | Tungkusan | Biotite granite | v | – | – | v | scattered over |
| | Sijuk | granite | v | – | – | v | the outcrop and |
| | Kelayang | granite | v | – | – | v | off shore |
| 5. | Kelumpang | Hornblende granite | v | – | v | – | – |
| 6. | Parang Buloh | Hornblende granite | v | – | v | – | – |
| 7. | Lilangan weathered | granite | – | – | – | – | Alluvial tin working |
| 8. | Buntar weathered | granite | – | – | – | – | Cut by quartz tourmaline cassiterite veins |

(Ilmenite/Magnetite series and I/S type classification after Cobbing and Mallick, 1984).

On Belitung granitoids of the Ilmenite series and S-type are found at Batu Besi, Tungkusan, and Lilangan (Table 7 and Figure 7). The other granite plutons are classified as Ilmenite series and I-type.

All the alluvial tin fields in Belitung occur in the areas surrounding Ilmenite series and S-type granite, except for the extensive alluvial tin field of the Gunung Mang pluton. In this area there are quartz-tourmaline veins within metasediments. Another locality of quartz-tourmaline-cassiterite veins is found at the Buntar granite.

On Karimun and Kundur, all the granite plutons are of Ilmenite-series/S-type (Figure 8). Large scale tin workings are in operation onshore and offshore in the N.W. corner of Kundur island. The Dabo pluton of Singkep is of Ilmenite series/S-type granite. Associated with it extensive alluvial tin deposits are being worked (Table 8).

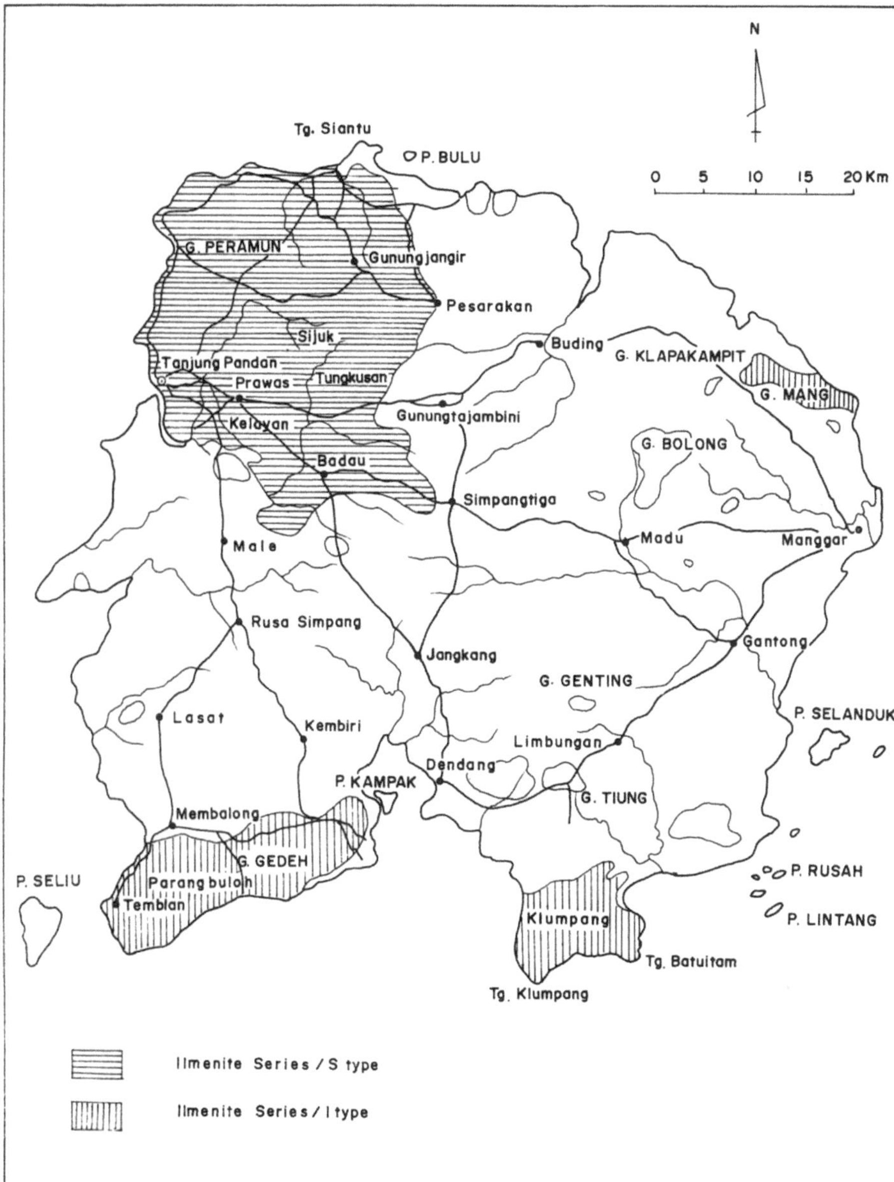


**Fig. 7.** Distribution of Ilmenite series/S-type and Ilmenite series/I-type of Belitung granitoids

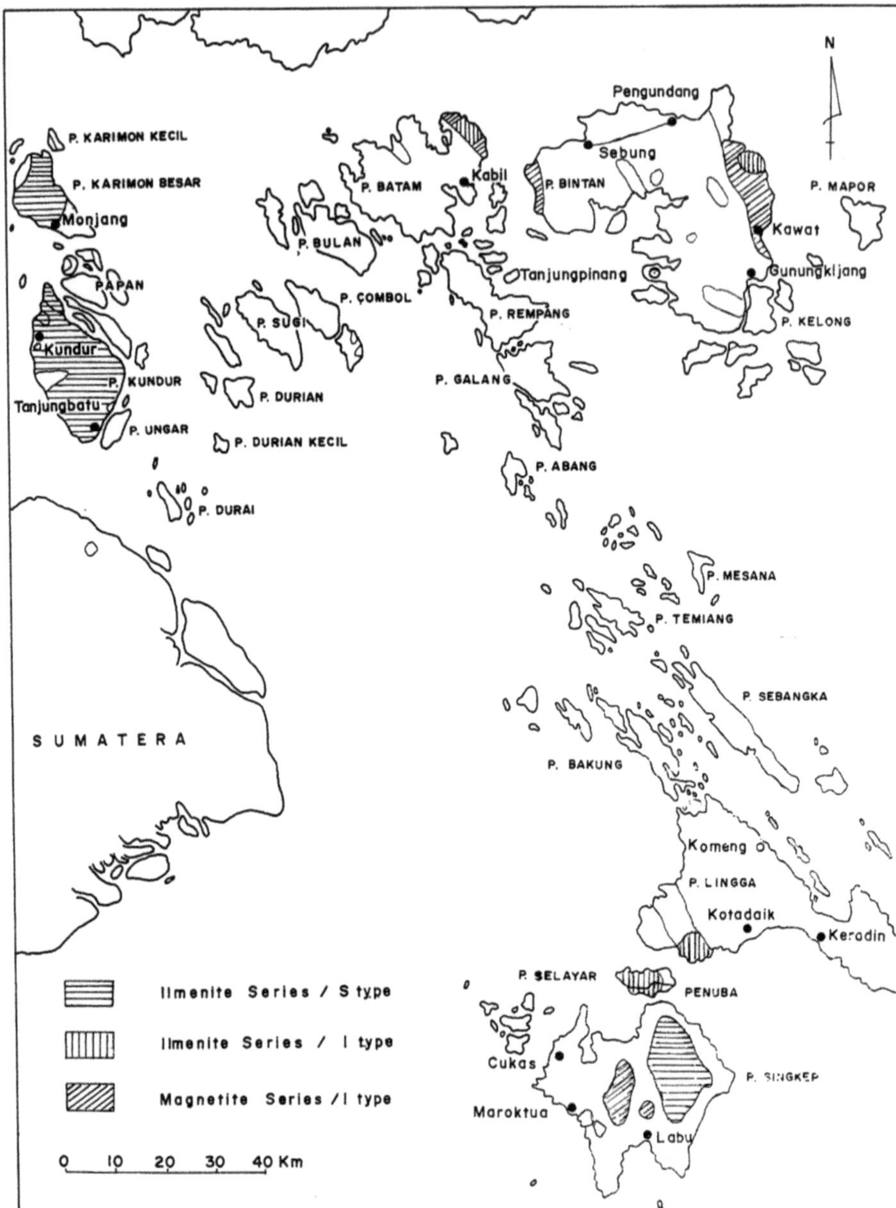


**Fig. 8.** Distribution of Ilmenite series/S-type and Ilmenite series/I-type of Riau Archipelago granitoids

**Table 8.** Classification of granitoids of Batam, Bintan, Karimun, Kundur, Singkep, Lingga and Selayar and their tin associations

| No. | Granite Pluton | Rock name | Series | | Type | | Tin Association |
|---|---|---|---|---|---|---|---|
| | | | Ilmenite | Magnetite | I | S | |
| a. | *Batam* | | | | | | |
| 1. | North east corner variable granite | Coarse gr. granite | – | v | v | – | – |
| | | Medium gr. granite | v | – | v | – | – |
| | | Coarse pink granite | v | – | v | – | – |
| | | Plagioclase porphyry | – | v | v | – | – |
| b. | *Bintan* | | | | | | |
| 1. | Loban Laut | Quartz, Plag. Hbl.-porphyry | – | v | v | – | – |
| 2. | Lagoi | granite | – | v | v | – | – |
| 3. | Ankulai | granite | – | v | v | – | – |
| 4. | East Bintan | | | | | | |
| | Kanwal | Hbl.-biotite granite | – | v | v | – | – |
| | Hantu | Hbl.-biotite granodiorite | – | v | v | – | – |
| | Belangkap | Biotite granite | v | – | v | – | – |
| | West Kanwal | Biotite granite | – | v | v | – | – |
| | Buton | Biotite granite | v | – | v | – | – |
| | Lena | Biotite granite | v | – | v | – | – |
| | Bakau | Biotite granite | v | – | v | – | – |
| | Pangkil Besar | granite | – | v | v | – | – |
| c. | *Karimun* | | | | | | |
| | | Biotite-Musc. granite<br>Biotite granite | v | – | – | v | – |
| d. | *Kundur* | | | | | | |
| 1. | Karang Sandung | Biotite-Musc. granite<br>Biotite granite | v | – | – | v | Large scale tin workings on shore and off shore (NW Corner) |
| 2. | Batu Delapan | Biotite-Musc. granite (foliated) | v | – | – | v | |
| e. | *Singkep* | | | | | | |
| 1. | Dabo | Biotite granite | v | – | – | v | Extensive alluvial tin at Tumang, rich in quartz-cassiterite veins |
| | | | v | – | – | | |
| 2. | Batu P. | Biotite granite<br>Biotite-Hbl. granite | – | v | v | – | – |
| 3. | S.W | granite | v | – | – | v | – |
| f. | *Lingga* | granite | v | – | v | – | – |
| g. | *Selayar* | Biotite-Hbl. granite | v | – | v | – | – |

## Concluding Remarks

1) The degree of crystallinity and textural features of the tin island granite is quite variable between and within individual plutons. However the mineralogical composition of these granites is relatively similar. The biotite colour of the tin island granites is mostly brown, whereas those from Sumatra are generally green.
2) Detailed chemical studies especially of the Ilmenite series and S-type granite have to be continued.
3) Radiometric age determinations yield a wide range; in order to have a better understanding, a careful re-evaluation of the data is needed.
4) Most of the granite plutons in Sumatra are classified as Magnetite series and I-type, except for the Penyabungan and Singkarak granites. The granite plutons of the Tin Island Belt, on the other hand, are classified as Ilmenite series and S-type.
5) Most of the tin fields are associated with Ilmenite series and S-type granite; in other words tin has strong affinities for Ilmenite series/S-type granites.

*Acknowledgements.* The authors wish to thank all the members of the granite study project of GRDC and both Dr. Cobbing and Dr. Mallick for their field and laboratory data. Great appreciation is also directed to the Director of GRDC and to the representative of UNESCO in Indonesia. Many thanks also are due to Mr. Djoko Karsono, officer to the Regional Mineral Resources Development Centre, ESCAP, and to officers of the Geology Division of GRDC. Deep appreciation is especially directed to Professor Hutchison for his revision of the manuscript.

## References

Chappell, B.W. and White, A.J.R., 1974. Two contrasting granite types. *Pacific Geology,* 8, 173–174.

Cobbing, E.J., Mallick, D.I.J., 1984. South East Asia Granite Project Preliminary Report – Indonesia, *Report No. 184/2, Overseas Division Institute of Geological Sciences.*

Hehuwat, F., 1975. Isotopic age determination in Indonesia, The State of The Art, *CCOP Proc. 1976,* Bangkok.

Hutchison, C.S., 1978. South east Asian Tin Granitoids of contrasting tectonic setting, *J. Phys. Earth, 26, Suppl:,* 221–232.

Ishihara, S., 1977. The magnetite series and ilmenite-series granitic rocks. *Mining Geology,* 27, 293–305.

Ishihara, S. et al., 1979. The magnetic-series and Ilmenite series granitoids and their bearing on tin mineralization, particularly of the Malay Peninsula region, *Geol. Soc. Malaysia, Bull.* 11 Des., 1979, 103–110.

Ishihara, S. et al., 1980. Granites and Sn-W Deposits of Peninsula Thailand, *Mining Geology Special Issue,* No. 8, 223–241.

Kardana, H., 1983. Laporan Penelitian Lapangan dan Petrologi batuan granitis daerah Lembar peta Lubuk Sikaping dan Pakanbaru, Sumatra Barat-Riau Daratan, unpublished report GRDC.

Kastowo, Gerhard W. Leo, 1973. Geologic map of the Padang Quadrangle, Sumatra, *Geol. Surv. Indonesia.*

Katili, J.A., 1962. On the age of the granitic rocks in relation to the structural features of Sumatra, Geotechtonics of Indonesia. A modern view 1980. Reprinted from Crust of the Pacific Basin, *Geophysical Monograph* No. 6, 1962.

Katili, J.A., 1967. Structure and age of the Indonesian tin belt with special reference to Bangka, Geotectonic of Indonesia, A modern view, 1980, Reprinted from Tectonophysics, 4 (4–6), (1967), 403–4018.

Katili, J.A., Hehuwat, F., 1967. On the occurrence of Large Trunscurrent Faults in Sumatra, Indonesia, Geotectonics of Indonesia, A modern view, 1980, Reprinted from Journal of Geosciences, Osaka City University, Vol. 10, Art 1-1., 1967.

Pitcher, W.S., 1983. Granite Type and Tectonic Environment. Chapter 1—3, Department of Geology Liverpool 693 BX. UK.

Rosidi, H.M.D., Tjokrosapoetro, Pendowo, B., 1976. Geologic Map of the Painan and north eastern part of the Muara siberut Quadrangles, Sumatra, *Geol. Surv. Indonesia.*

Suminto, 1982. Laporan Penelitian batuan granitik di wilayah Propinsi Jambi dan Bengkulu, unpublished Report, GRDC.

Suyatna, D.A.D., 1982. Laporan Pendahuluan mengenai batuan granitik dari daerah Sumatra Barat, unpublished Report, GRDC.

Suyatna, D.A.D., 1982. Laporan Sementara Hasil Penelitian Batuan Granitis dari Kepulauan Riau, unpublished Report, GRDC.

Suyitno, S., 1984. Exploration for offshore tin placer in Indonesia. P.T. Tambang Timah (Pesero).

Takahaṣhi, M., Aramaki, S. and Ishihara, S., 1980. Magnetite-Series/Ilmenite Series vs. I-type/S-type granitoids, Granitic Magmatism and related mineralization, *Mining Geology Special Issue,* No. 8, 223—241.

White, A.J.R. and Chappell, B.W., 1977. Ultrametamorphism and granitoid genesis. *Tectonophysics,* 43, 7—22.

Wikarno, 1975. Status of Fission Track Dating in the Geological Survey of Indonesia, CCOP Proc. 1976.

Wikarno, 1980. Hasil Pentarikhan Radiometri terhadap contoh dari pulau-pulau di Indonesia, unpublished report.

## 6.9 Malaysia

# 6.9.1 Primary Tin Mineralization in Malaysia: Aspects of Geological Setting and Exploration Strategy

Chu Ling Heng, Fateh Chand, and D. Santokh Singh[1]


## Abstract

Primary tin mineralization in Peninsular Malaysia can be grouped into four broad classes: (i) pneumatolytic-hydrothermal lodes, veins, stockworks and stringers, (ii) pyrometasomatic skarns which can be subdivided into stanniferous and stanniferous-iron varieties, (iii) tin-bearing pegmatites and aplites, and (iv) stanniferous polymetallic sulphide bodies.

Tin mineralization is largely disposed within the Western and Eastern Tin Belts where it shows a close spatial relationship with acid intrusives and their immediate surroundings. Hydrothermal veining is ubiquitous. Stanniferous skarns, pegmatites and aplites are present mainly in the Western Belt, whereas the Sn-Fe skarns are confined to the Eastern Belt. A few stanniferous sulphide bodies occur in the Western Belt. One unusual deposit, the Manson's Lode, is found at the northern extremity of the Central Belt.

These primary deposits have been variously worked for their tin content. At present, the large Cornish-type lodes at Sungai Lembing[2], Pahang, and some open-cast "porphyry-type" tin deposits within the Kinta Valley and Klian Intan, Perak, continue to contribute to the country's tin production.

Virtually all primary tin mineralization has been uncovered through the mining of secondary tin deposits. Despite the success of this indirect means, the importance of utilizing geological principles in primary tin exploration programmes has been increasingly recognised since the early 1900s. Exploration techniques have included airborne geophysics in an attempt to delineate granite-sediment contacts, and the applications of geochemistry to stream sediments, heavy mineral concentrates, and soil surveys. Soil sampling has been successfully used in the Sungai Lembing area.

Based on the features of Malaysian Sn-Fe skarns, which are similar to the important Sn-Fe-S carbonate replacement-type skarns of Australia, it is imperative that Malaysian examples be thoroughly investigated. This necessitates an evaluation of the known pyrometasomatic iron deposits, and careful studies of the calcareous environments for feeder veins and shatter zones in the vicinity of high-level granitoids. The available geological, geochemical and geophysical data should prove invaluable for this purpose.


[1] Geological Survey, P.O. Box 1015, Ipoh, Perak, Malaysia

[2] Note: The Sungai Lembing Mine has not been operating since February 1987

## Introduction

Malaysia, the world's leading producer of tin, continues to be in the forefront whenever discussions on this important commodity arise. It is hardly surprising, therefore, that literature on tin mineralization in Malaysia is varied and voluminous. Despite this, certain concepts, such as the validity of the Western and Eastern Tin Belts, first recognised by Scrivenor (1928), and the undisputed spatial relationship of tin mineralization to granitoids, continue to influence thinking on the geological setting of the Malaysian deposits. Research workers have studied in detail the mineralogical and chemical characteristics of various tin species, and have propounded exploratory techniques for the search for the most important of them all – cassiterite.

This summarizes the present knowledge of some aspects of primary tin mineralization, with particular reference to the geology of the various types. Whilst the published literature has been freely used, unpublished information, mainly through the work of the Geological Survey, and the views of the mining fraternity, have also been incorporated.

## Primary Tin Mineralization

### Geological Setting

Tin deposits in Peninsular Malaysia are spatially and genetically related to granitoids. The most significant deposits fringe the granitoids of the Western and Eastern Tin Belts (Figs. 1 and 2). The former, which includes the Main Range granitoid batholith, is by far the most important in terms of the number of tin mines. Workable tin deposits are less abundant within the Eastern Belt. The granitoids of these two stanniferous belts are thought to belong to the ilmenite-series, as opposed to the magnetite-series plutons of the intervening tin-poor Central Belt (Ishihara et al., 1979). Initial studies, jointly undertaken by the British Geological Survey (BGS) and the Geological Survey of Malaysia (GSM), on tin granites of Peninsular Malaysia suggest that the Western Belt represents mainly S-type plutons whereas the Eastern Belt has both I- and S-types.

Primary tin mineralization is commonly associated with the highly-differentiated late-phase acid intrusives. Mineralization is also important within the contact zones and the intruded country rocks. As a corollary, the secondary tin deposits are generally similarly disposed.

Currently, a joint study by Bundesanstalt für Geowissenschaften und Rohstoffe (BGR) and GSM on tin-bearing and tin-barren granitoids is underway. Investigations will cover a variety of aspects, including petrography, whole-rock/mineral chemistry, fluid inclusion studies and isotopic work. Rock alteration haloes of Sn granitoids will feature prominently in the study and it is hoped that meaningful exploration guidelines can be established for Sn/W/Ta/Nb deposits.

At present several types of primary tin mineralization are known. These have been variously classified by Bradford (1957), Hosking (1969), and Rajah and Chand (1973). The last named classification is currently being adopted by the GSM, and forms the basis for the present paper.

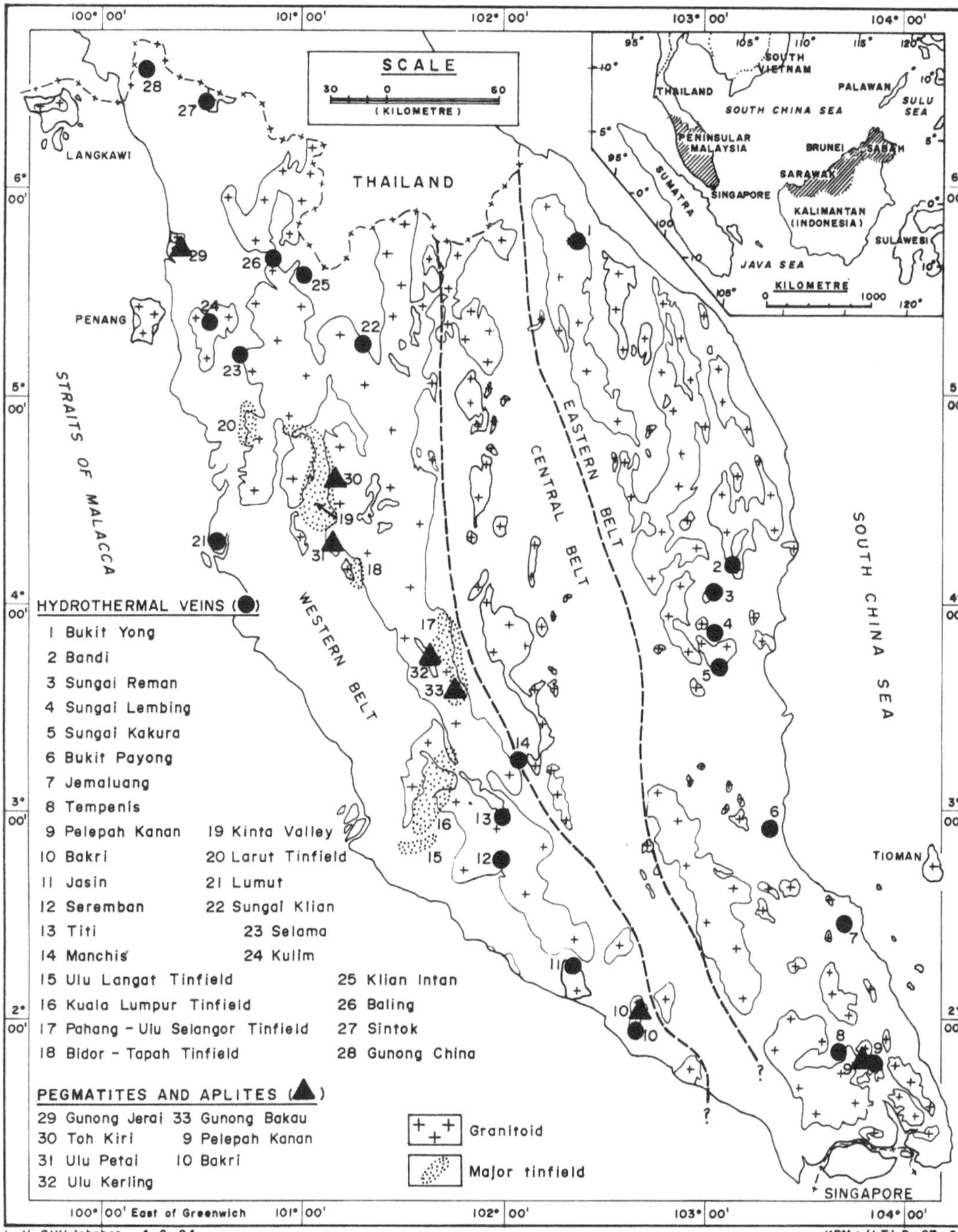


**Fig. 1.** Distribution of significant stanniferous hydrothermal veins, pegmatites and aplites

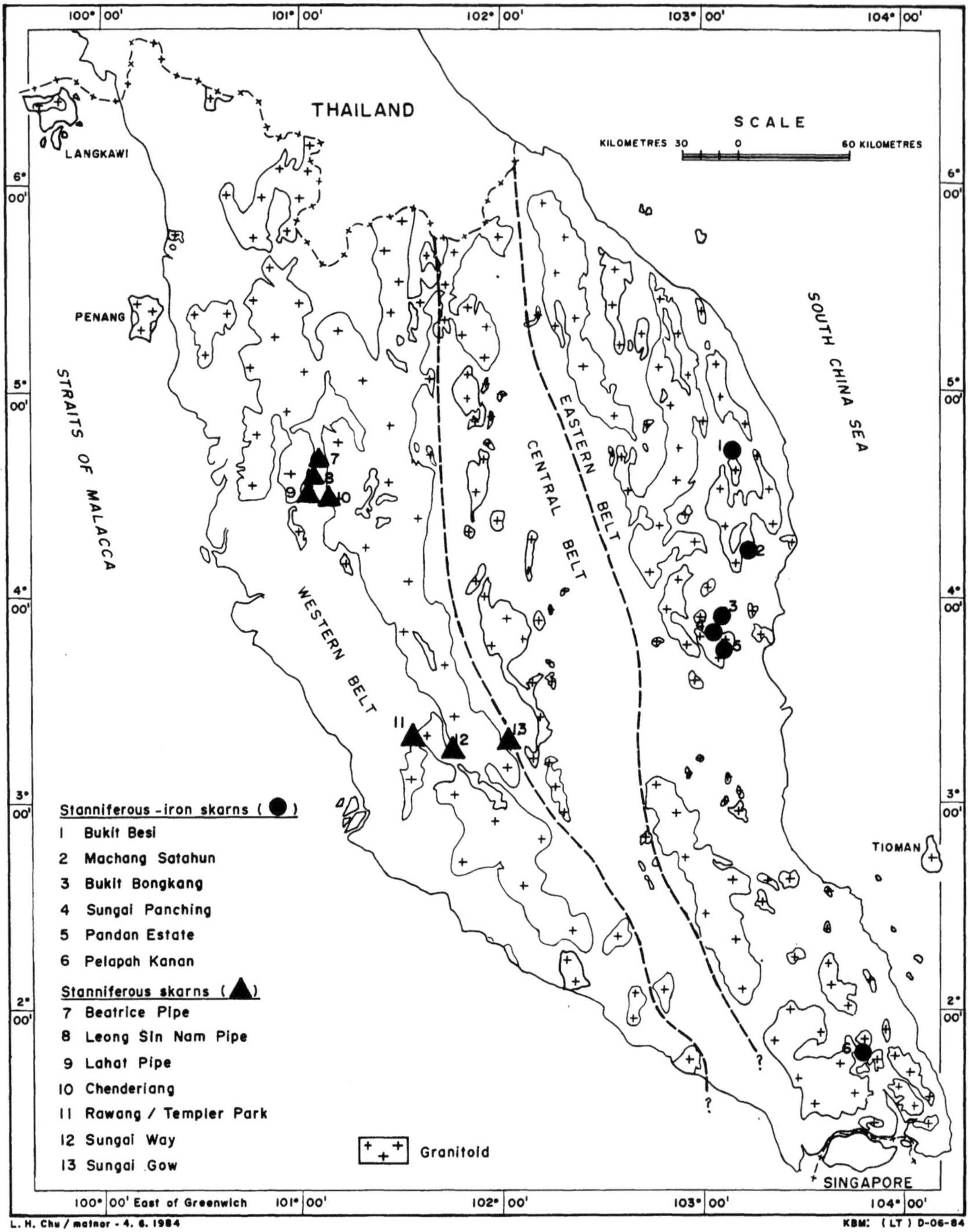


**Fig. 2.** Distribution of tin-bearing skarns

## 1. Pneumatolytic-Hydrothermal Mineralization

This class predominantly constitutes the most significant source of primary tin. It displays a wide range of grade, tonnage, and mineralogical complexity. High- and low-temperature minerals may be present together, and wall rock alteration in the form of greisenization, tourmalinization, hematization, kaolinization and silicification is common.

### *a) Cornish-Type Lodes*

These relatively large and extensive tin-lodes are comparatively rare. The best examples are in the Sungai Lembing area, Pahang, which falls within the Eastern Belt (Fig. 1). These lodes, sometimes termed "Cornish-type", are restricted to the heavily-faulted Early Carboniferous sedimentary cover flanking a deep-seated Late Permian granitoid (Figs. 3 and 4). They appear in a zone approximately 600 m away from the granitic contact (Fitch, 1952; Pun and Singh, 1978), and have been worked since the 1880s by the Pahang Consolidated Company Limited. In general the lodes dip between 70° and vertical. A few however have dips as shallow as 20°.

Ore deposition is localized in faults restricted to argillaceous beds. Cassiterite is the main mineral, followed by pyrite, chalcopyrite, arsenopyrite, stannite, sphalerite, galena, pyrrhotite, quartz and chlorite. The sulphides contain traces of gold. There is no marked primary zoning, although copper is relatively more abundant in the upper levels. The lodes show an intimate cassiterite-chlorite association and display more than one generation of quartz.

Outcropping lodes have suffered oxidation and weathering in the upper levels. Weathering may extend down to 300 m, as in the Willinks section to the west of Willinks fault. The oxidised and weathered parts of the lodes contain cassiterite commonly associated with iron-stained clay, and some chlorite.

The Company presently operates three mines, namely Willinks, Myah and Gagak (Pahang Consolidated Public Ltd., 1978; 1984). The grade presently being worked is 1% Sn. The final mill product contains approximately 63% Sn and 10% Cu. As at July 1983, estimated ore reserves of the Company were 281,846 t, averaging 1.38% Sn. Lately, four additional tin lodes have been uncovered following trenching over newly-delineated geochemical soil anomalies.

### *b) Veins, Stringers, Stockworks*

These are more prolific, being found in many places within both the stanniferous belts (Fig. 1). They are, however, more abundant within the Western Belt. Mineralization is commonly accompanied by greisenization which is expressed mainly by muscovite-tourmaline and quartz-topaz assemblages.

Some of the best documented examples of this type have been described from such renowned tin-fields as the Kinta Valley and the Klian Intan area in Perak, the Kuala Lumpur and Ulu Langat districts, and the Ulu Selangor-Pahang border region occupying part of the rugged Main Range. Roe (1951) and Alexander (1968) have

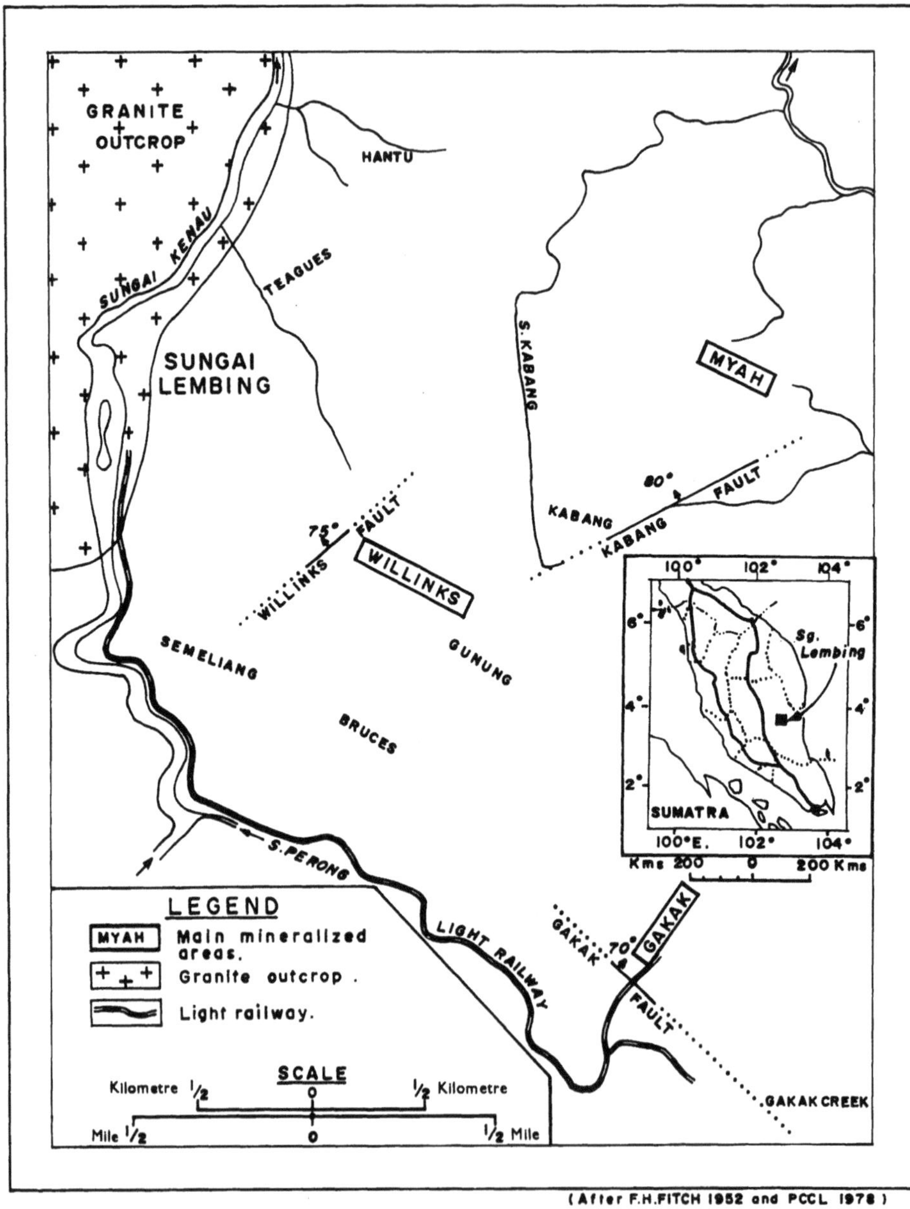


**Fig. 3.** Mineralization at Sungai Lembing (after Fitch, 1952)

given a comprehensive account of this latter locality which had once been the scene of active tin-mining (Fig. 5). Most of the tin veins are small, generally less than 0.5 m in width, but in places veins attain widths of up to 4 m. Tourmaline is a commonly-associated mineral. The most notable features of this locality is the occurrence of quartz-topaz-cassiterite sills at Gunung Bakau, and local torbernite in some of the stanniferous veins.

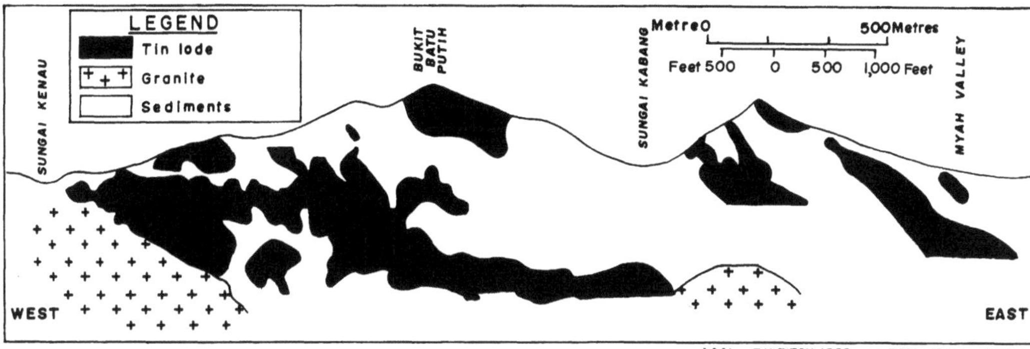


**Fig. 4.** Cross-section of the Sungai Lembing Mine, looking north

The Kinta Valley, which is the foremost placer tin-producing district in the world, shows the greatest development of hydrothermal tin veining in the country. A central zone comprising schist, phyllite and limestone is bordered on the east by the Main Range Granite, and on the west by granitoids of the Kledang Range (Scrivenor, 1928; Ingham and Bradford, 1960; Bradford, 1961; Rajah, 1979). Figure 6 depicts some of the more important occurrences of primary tin mineralization.

Several of these primary tin deposits flanking the Kledang and Main Range Granites are being mined by the open-cast method, of which the Tekka deposit is the best known (Fig. 7). The intra-granitic tin veins at Tekka occur in two vein swarms separated by an elongated schist roof pendant. The northerly-dipping veins also extend into the overlying schist. The veins are mineralogically more complex than previously thought. Described by Jones (1925) as small quartz-cassiterite stockworks in metasediments, these veins are now found to display a strong telescoping of high- and low-temperature minerals (Hosking, 1973; Teh, 1981). Associated with the cassiterite are scheelite, wolframite, stannite, varlamoffite, columbite-tantalite, galena, stibnite, hematite and fluorite. The cassiterite commonly contains exsolved tapiolite. Santokh Sing and Bean (1967) showed that the varlamoffite was formed at the expense of stannite.

In Klian Intan in north Perak (Fig. 1), stanniferous stockworks are being mined by the open-cast method. The quartz-cassiterite veins, emplaced in phyllite, commonly trend east-west, and vary from a few mm to more than 15 cm in width. These stockworks contain tourmaline and pyrite, with minor galena, sphalerite, arsenopyrite, pyromorphite and native bismuth. Hematization is widespread.

These low-grade, large-tonnage deposits are amenable to open-cast mining and may be construed as examples of bulk deposits. Some authors have even used the term 'porphyry-type' from a mining point of view.

Because of the high degree of weathering, some of these veins, stringer and stockwork deposits are being mined by opencast/gravel pump methods. As hardrock depositions they will not be economic at present day prices and market conditions.

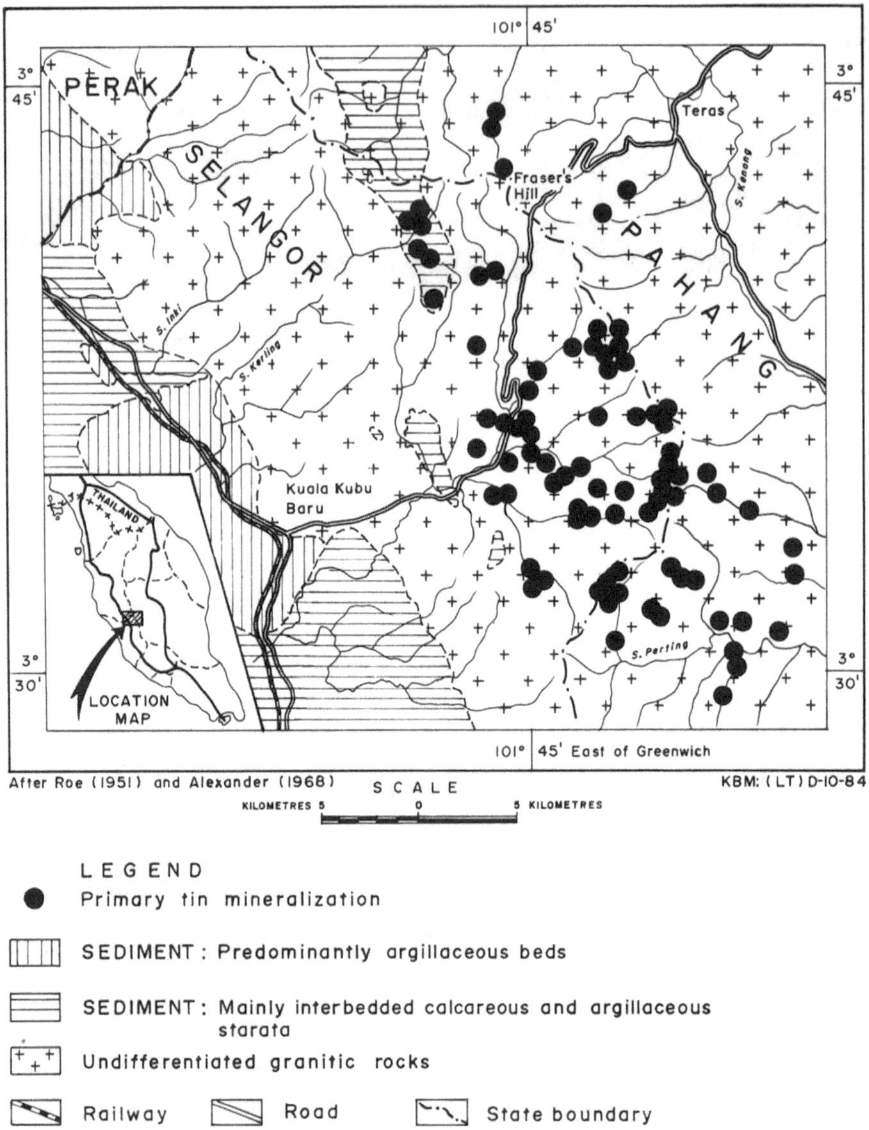


**Fig. 5.** Primary tin mineralization in the Ulu Selangor – Pahang interstate area. After Roe (1951) and Alexander (1968)

## 2. Pyrometasomatic (Skarn) Mineralization

Mineralization is characterised by a suite of calc- and/or magnesian silicates. Hosking (1973) further subdivided this type into stanniferous-iron, and stanniferous groups. This subdivision is also supported through the work of the GSM.

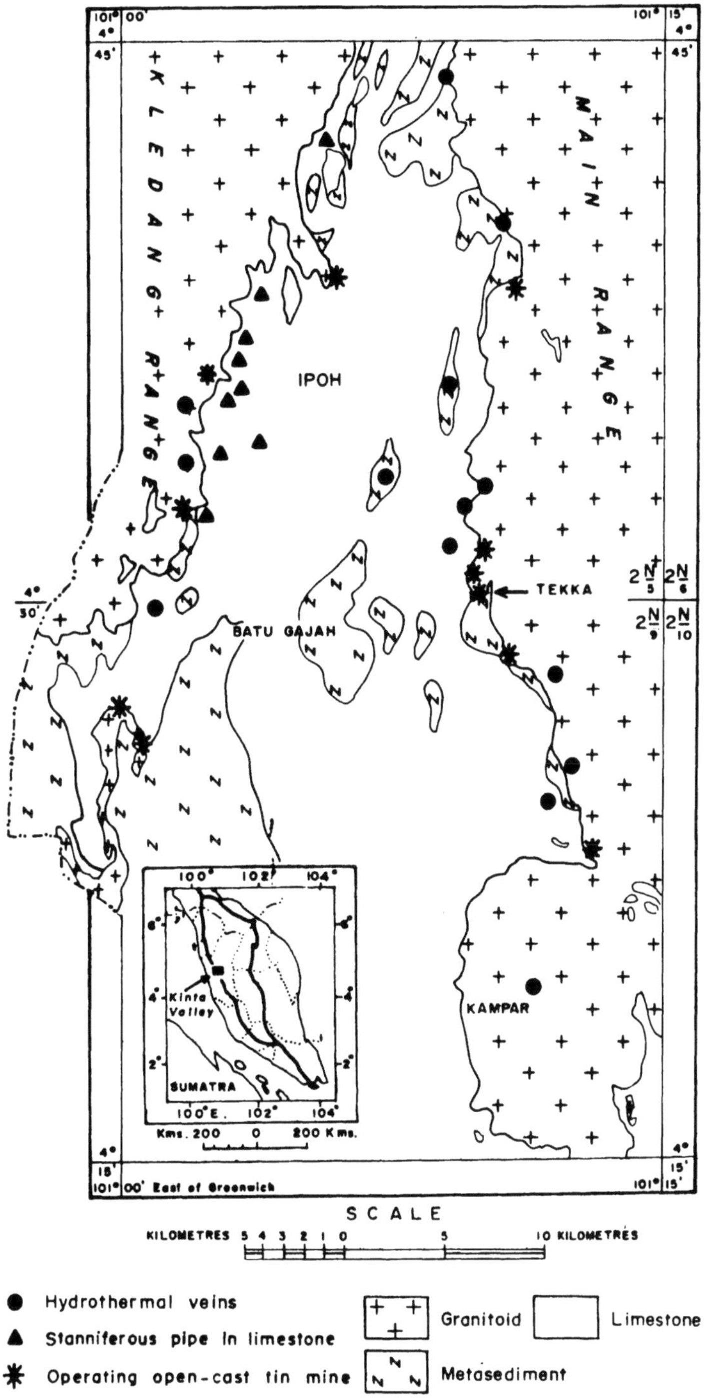


**Fig. 6.** Primary tin mineralization in the Kinta Valley

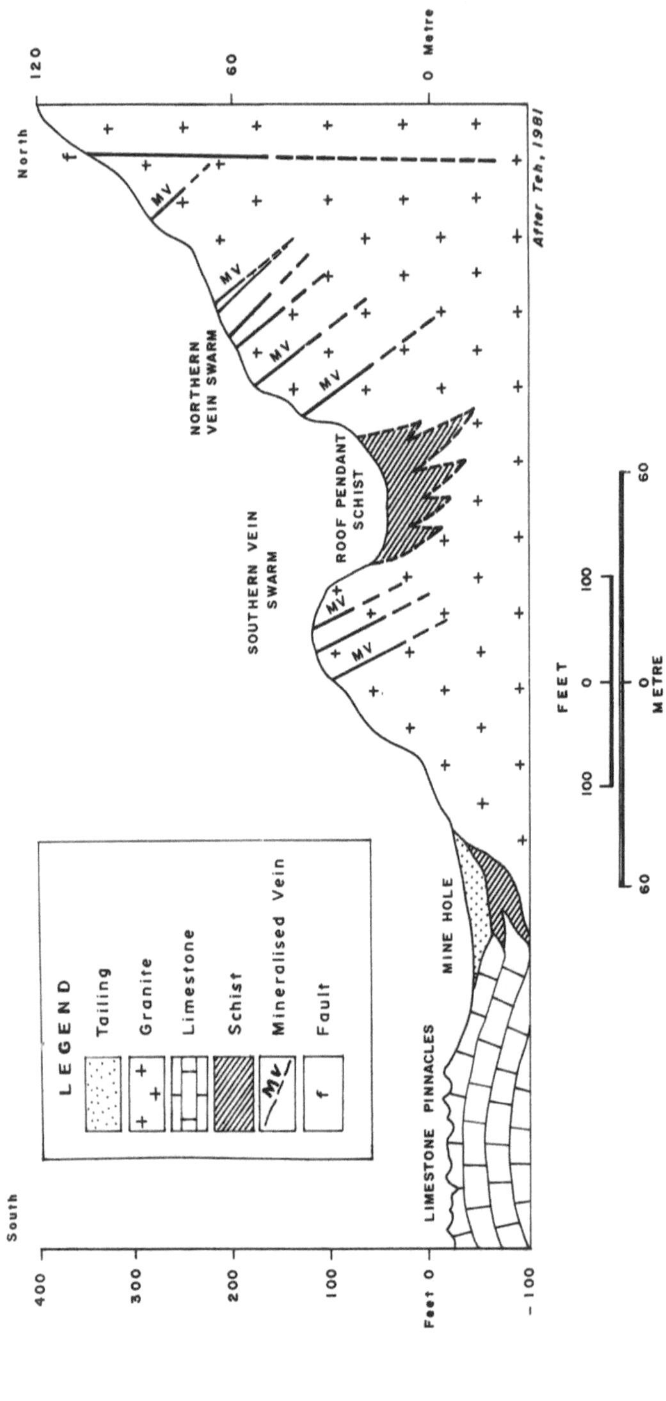


**Fig. 7.** Sectional view of the Tekka Area, Perak. Vertical scale equals horizontal scale. After Teh (1981)

*a) Stanniferous-Iron Skarns*

Stanniferous-iron skarns are restricted to the Eastern Belt (Fig. 2); the better known localities are Pelepah Kanan (Johore) and Bukit Besi (Terengganu). Both deposits have been mined, the former primarily for tin and the latter mainly for iron. Tin-iron mineralization is closely associated with stanniferous calc-silicate skarn zones, and the bulk of the tin mineralization is epigenetic and fracture-controlled.

Fig. 8 shows a diagrammatic section through the Pelepah Kanan mine which has been described by Roe (1941), Burton (1959), Bean (1969), and Khoo (1969). The hill on which the iron deposit occurs is a quartzite-calc-silicate hornfels roof pendant overlying granite porphyry. It is capped by a magnetite-martite-hematite lens connected downwards to a feeder zone comprising magnetite-quartz-cassiterite-fluorite-loellingite-scheelite-sulphides (Yeap, 1982; 1984). In places, rhythmically-banded magnetite-fluorite-cassiterite "wrigglites" of the feeder zone, similar to some of the Australian wrigglite skarns, have been recorded. The main body itself is cut by numerous pegmatite and quartz veins carrying cassiterite, arsenopyrite, pyrite, scheelite, fluorite, chlorite, and loellingite. Up to 1% Sn has been reported from random samples of iron-ore. Cassiterite and magnetite are also in intimate association within the calc-silicate hornfels (Bean, 1969).

Farther north in Terengganu, the Bukit Besi iron mine, considered as a classic pyrometasomatic Sn-Fe deposit, was at one time the largest opencast property in Southeast Asia. The bulk of the ore consists of magnetite and hematite which are replacements in shale and limestone. Towards the west end of the property (Fig. 9), fine-grained cassiterite is intimately associated with iron-ore in two northerly-striking easterly-dipping lenses close to the granite (Bean, 1969). The ore in general averages 0.07% Sn.

Other skarn-type iron deposits in which tin forms an important constituent include those at Machang Satahun (Terengganu), Bukit Bangkong, Pandan Estate, and Sungei Panching (all in Pahang). It is pertinent to note that Hutchinson (1984) regards Sn-Fe skarn mineralization in general to have a volcanic-exhalative origin, a

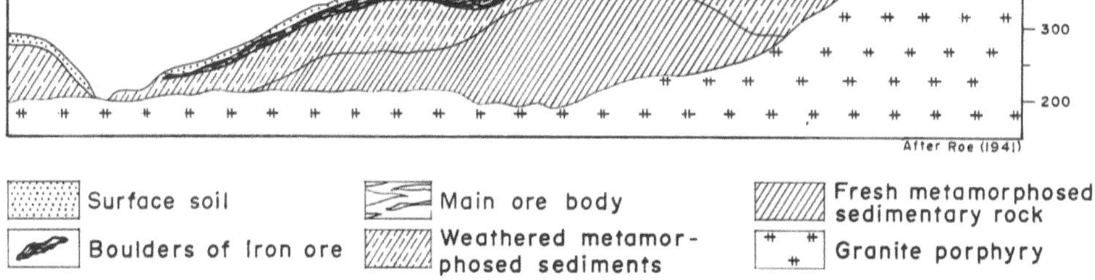


**Fig. 8.** Diagrammatic section across the Pelepah Kanan Deposit. After Roe (1941)

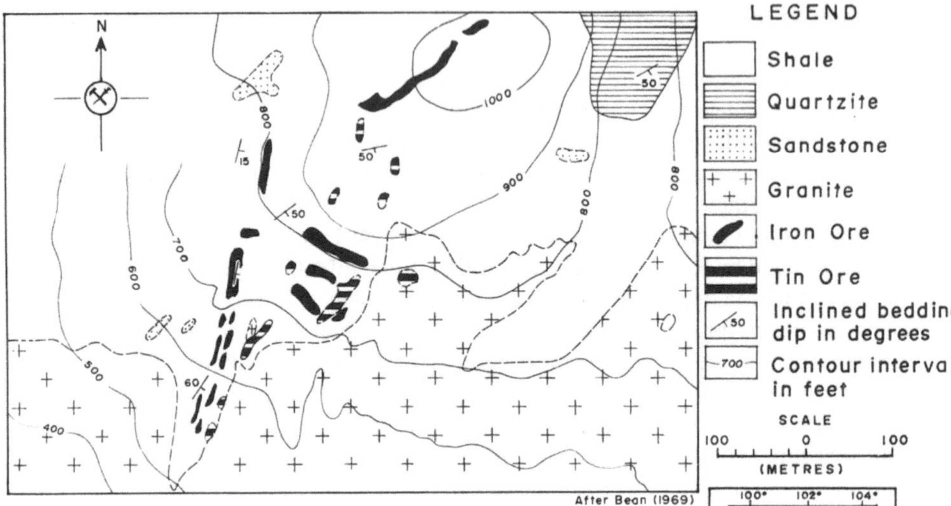


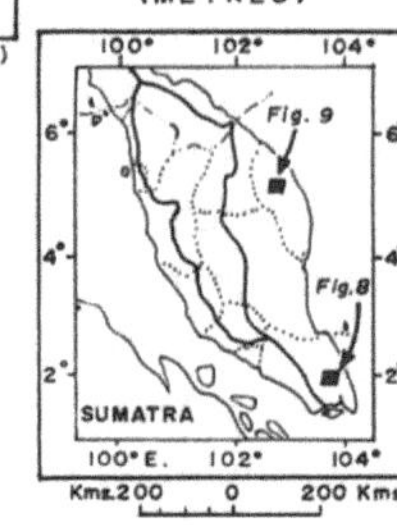


**Fig. 9.** Sri Bangun Area, Bukit Besi Mine. After Bean (1969)

concept which has been debated recently. The writers, however, find a pyrometasomatic origin more consistent with field evidence, in that they are in all cases spatially related to granitic intrusives.

### *b) Stanniferous Skarns*

Stanniferous skarns of this group (Fig. 2) are found mainly within the Western Belt, the most important being localised in the limestone bedrock of the Kinta Valley. Mineralization exists as stanniferous veins and pipe-like bodies in limestone intruded by granite. The best known example is the Beatrice Pipe (Fig. 10) which Taylor (1979) and Kwak (1984) regard as a Sn-S type of skarn development. It was a pipe-like body extending for 150 m horizontally in limestone, and then 90 m vertically before being faulted off by granite (Ingham and Bradford, 1960). The deposit contained cassiterite, stannite, arsenopyrite, pyrite, pyrrhotite, molybdenite, bornite, tremolite, fluorite, and fluoborite. Cassiterite occurred in bands parallel to the border of the pipe, and interstitially to arsenopyrite in the main body.

Another noteworthy example of this type of mineralization is the Leong Sin Nam Pipe near Menglembu. Localized in dolomite, it consisted of cassiterite, arsenopyrite, pyrite, stannite, jamesonite, tennantite, tremolite, and calcite. The nearby Lahat Pipe is yet another sulphide-bearing stanniferous skarn in limestone.

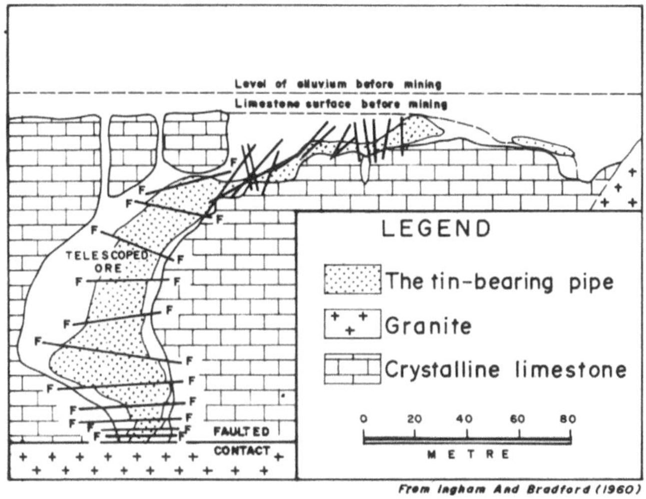


**Fig. 10.** Section of the Beatrice Mine, Selibin, Perak. After Ingham and Bradford (1960)

Not recorded in all of these examples is the mineral malayaite, which was first discovered in Batang Padang, Perak (Alexander and Flinter, 1965), and which seems to be fairly common in several of the smaller stanniferous skarn occurrences. The mineral has been recorded at Rawang, Templer Park, Sungei Way (Selangor), and at Sungei Gow (Pahang).

## 3. Stanniferous Pegmatites and Aplites

These are generally small and unzoned (Fig. 1). A few have been mined for their cassiterite, for example the topaz-aplite at Gunung Bakau, the Toh Kiri aplite in the Kinta Valley, the pegmatite at Ulu Kerling, Selangor, and those at Ulu Petai, Perak. Although Jones (1925) reported pegmatites "with a good deal of tinstone" at Gopeng Consolidated Mine in the Kinta Valley, no mention was made if these were mined. The best known stanniferous pegmatites, however, are near Gunung Jerai, Kedah, and Bakri, Johore. At both localities, columbite-tantalite is associated with cassiterite. Bradford (1972) reported that cassiterite from the Gunung Jerai area contains a higher content of Ta than Nb.

Fig. 11 shows the general geology of the Gunung Jerai area. Metasediments of the Upper Cambrian – Lower Ordovician Jerai Formation are domed by biotite granite which is exposed only along the southern flank of the mountain. Pegmatites, cutting both the granite and the intruded metasediments, abound within the southern half of the area, whereas quartz veins are more numerous towards the north. The quartz veins and pegmatites are both tin-bearing, but columbite appears to be more common in the pegmatite. Weathering of these late-phase minor intrusives has given rise to the economic placer deposits around the townships of Bedong and Semiling. Minerals associated with the cassiterite and columbite-tantalite in the placers include tour-

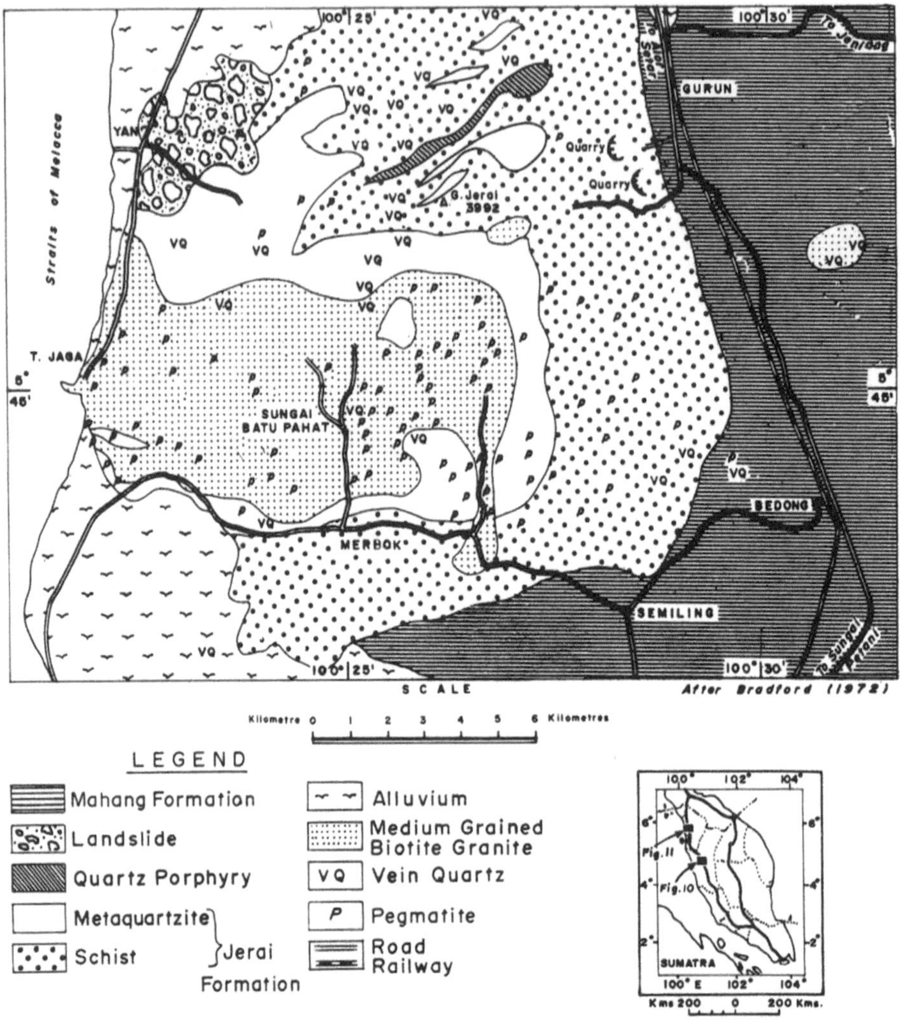


**Fig. 11.** Geological Map of the Gunong Jerai Area, Kedah. After Bradford (1972)

maline, almandine garnet, muscovite, monazite, ferroan gahnite, topaz, betafite and complex Nb-Ta-Ti oxides (Bradford, 1972).

The cassiterite grains have been examined in detail by Santokh Singh and Bean (1967). The characteristic features noted are:

1) they are paramagnetic
2) $Sn^{4+}$ is substituted partly by pentavalent Ta and Nb in the crystal structure, without causing any appreciable effects on the cell parameters of the cassiterite
3) translucent cassiterite grains are strongly pleochroic, from deep red to light orange or green, reflecting the presence of Ta-Nb in the mineral in solid solution
4) Ta also occurs in the form of tapiolite and/or tantalite inclusions in the cassiterite host

5) there are subordinate amounts of magnetite inclusions.

Towards the southern end of the country, in the Bakri area, stanniferous pegmatites and quartz-cassiterite veins cut the sedimentary rocks over which economic tin placers have been developed. Columbite and monazite are important by-products.

4. Stanniferous Polymetallic Sulphide Bodies

These contain a high proportion of sulphide minerals, but unlike the skarns, they lack calc- or magnesian-silicate minerals. Hosking (1973) refers to them as xenothermal bodies because of their mineralogical complexity and strongly telescoped nature. Some examples are known from the Kinta and Kuala Lumpur tinfields, but probably the best-documented is the Manson's Lode in Kelantan (Fig. 12).

The Manson's Lode is an argentiferous-auriferous polymetallic sulphide manto localised in Permian limestone and along limestone-phyllite contacts (MacDonald, 1967; Smith, 1969; Rajah, 1970). Quartz porphyry dykes are found close to the body, and nearby there is also an extensive development of acid to intermediate volcanics. Although commonly regarded as an epigenetic replacement of brecciated limestone, Gan (1980) and Hutchinson and Burton (per. com) have recently suggested that a submarine volcanic-exhalative origin for this sulphide body should not be discounted.

On the surface, Manson's Lode is characterised by limonitic and manganiferous gossan. The primary sulphides are chiefly galena and sphalerite forming the main part of the body which is enveloped by a pyrite-pyrrhotite shell. Minor amounts of cassiterite, stannite, chalcopyrite, arsenopyrite, magnetite, gold, and argentite are present. Cassiterite has formed early in the paragenetic sequence, and is commonly rim-replaced by stannite.

## Age of Granitoids and Tin Mineralization

Although a considerable amount of isotopic data (Rb/Sr, K/Ar and U/Pb) are available for the granitoids of Peninsular Malaysia, recent work suggests that some of the existing age determinations are suspect. This is because of insufficient field control during collection of samples. Yap (per. com.) is of the opinion that widespread intrusive activity occurred in Late Triassic as documented in plutons along the entire length of the Western Belt. A Late Carboniferous/Early Permian intrusive episode has been dated. However this episode has to be substantiated with more work since some younger (Late Triassic-Jurassic) U/Pb zircon ages have been recorded for some of the same intrusives (Liew, 1983).

In the Eastern Belt the major intrusive episode occurred in Late Triassic. However, older granitoids dated as Late Carboniferous/Early Permian are known and Late Permian igneous activity has been inferred from a regional Rb/Sr isochron (Bignell and Snelling, 1977).

Radiometric ages of around 70 Ma (Late Cretaceous) have been recorded from small plutons from the Eastern and Central Belts.

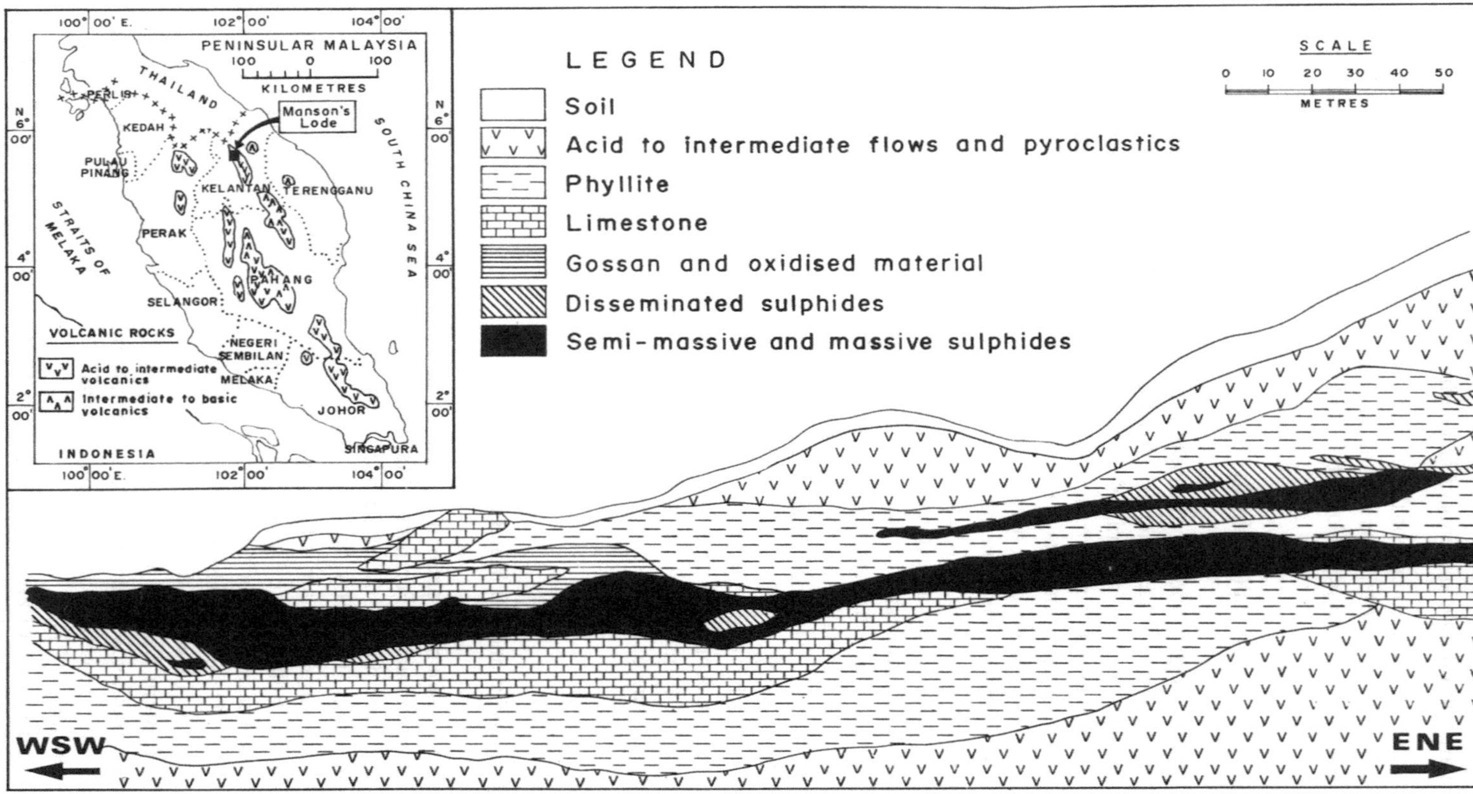


**Fig. 12.** A longitudinal cross section of the Manson's Lode. After Rajah (1970)

The pattern of K/Ar mica ages from the Western Belt shows an almost continuous spread from 220–85 Ma. There is a complete absence of ages older than Late Triassic. This may be due to re-setting of ages by thermal/tectonic events. A marked correlation has been noted between the areal distribution of disturbed K/Ar ages and areas affected by major faulting.

It is generally believed that tin mineralization is slightly younger than the main intrusive activity. Field evidence suggests the mineralization is associated with the late acid intrusives. Recent isotopic dates for pegmatites from the Kuala Lumpur tin field have given a 190 Ma age, about 20 Ma younger than the main host intrusive. However more work is required to be done before a meaningful picture can emerge.

## Exploration for Primary Tin in Peninsular Malaysia

### Exploration Techniques

It is perhaps justifiable to generalize that most primary tin deposits and occurrences have been discovered as a consequence of the mining of secondary tin deposits. Although not all these primary sources are workable, there are some, like the famous Beatrice Pipe in the Kinta Valley, and the stanniferous veins at the Ulu Selangor – Pahang border which were orebodies in their own right. The techniques instrumental in the discovery and subsequent evaluation of the secondary deposits in the first instance are the simple method of preliminary assessment by heavy mineral panning, and Banka drilling during the final phase of evaluation.

Despite the apparent success of this indirect approach in the discovery of primary tin mineralization, geological concepts have been increasingly incorporated into tin exploration programmes since the early 1900s. The spatial relationship of primary tin mineralization to granitoids and their immediate surroundings was one concept considered significant enough to have inspired the GSM to seriously consider delineating granitic bodies on a regional scale over certain favourable areas. Thus between 1956 and 1957, an airborne magnetometer and scintillation counter survey was flown over six selected areas (Fig. 13) primarily for this purpose, although an evaluation of their radiometric potential and the identification of zones of sulphide mineralization were simultaneously attempted (Agocs, 1958). Some 41,000 $km^2$ was covered under this Colombo Plan project. Additional airborne data have recently become available following the completion of a helicopter-flown magnetic and spectrometric survey covering some 31,000 $km^2$ of north-central Peninsular Malaysia (Fig. 13). This airborne survey was conducted in 1980 as part of the Central Belt Project. Such data have resulted in a better knowledge of the definition of igneous contacts and the distribution of sub-outcropping intrusives within a large part of the country.

Since the 1960s, exploration geochemistry has assumed an increasingly vital role in exploration programmes. Geochemical programmes entailing the use of stream sediment and heavy mineral sampling are now undertaken routinely. A detailed account of exploration methods is given by Chand (1981). For tin, heavy mineral sampling is probably a more reliable guide in view of the resistate nature of cassiterite grains whose presence in a sample can be visually checked. If necessary a quick field-tinning test can be made.

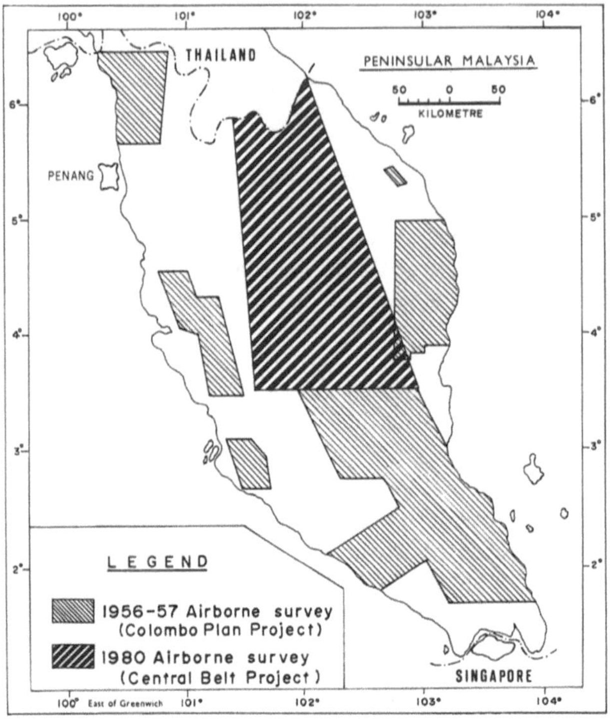


**Fig. 13.** Airborne survey coverage of Peninsular Malaysia

Although several tin anomalies have been delineated on the basis of the Sn content of the heavy mineral concentrates rather than that of the sediments (Chu et al., 1982), it is common for both media to complement each other. In retrospect, stream sediments may prove to be the ideal sampling medium if cassiterite exists in a finely-divided state, as suggested by the findings of a reconnaissance geochemical survey undertaken in south Pahang (Foo, 1970).

In general, stream sediment tin values tend to be erratic. Fletcher et al. (1984) contended that the erratic distribution of Sn in drainage sediments reflects the local enhancement of cassiterite in response to hydraulic conditions, and have suggested the use of Sn/magnetite ratios for improved data treatment. However, the present authors feel that while this data treatment technique shows promise, the method may suffer the disadvantage of actually suppressing the geochemical significance of the anomaly in areas of tin-magnetite mineralization.

Analysis of residual soil in tin exploration is another common geochemical method. Although the -80 mesh fractions are normally utilised, it has been demonstrated that the coarser fractions tend to give a better contrast. At the Pahang Consolidated Company Limited Mines in Sungai Lembing, -20 mesh soil fractions have been shown to produce clearer anomaly peaks (Tooms and Kaewbaidhoon, 1961). At this mine, ridge and spur sampling is practised because of the rugged terrain. Soil samples are collected at 7 to 9 m intervals at depths ranging from 45 cm to 1.7 m, depending on the overburden thickness (Pun, 1981). The upper limit of background fluctuation based on -80 mesh fractions has been established at 50 ppm Sn, and any area showing a higher Sn content is trenched. However, it is necessary to lower this limit to 20 ppm Sn in areas where the lodes are anticipated to be more deeply emplaced, since the resulting anomaly will be weaker (Joginder Singh, per. com.).

The effectiveness of pathfinder elements in soil surveys has been demonstrated in a number of instances. At Sungai Lembing, As is a reliable pathfinder (Pun, 1981). At Tekka, As, W, F, Sn show a positive response over the suboutcropping stanniferous-sulphide veins (Wong, 1975). Subsequent work by Teh (1979) at Tekka not only substantiated this observation, but has also indicated the usefulness of soil conductivity measurements in tracing the veins.

## Exploration Strategy

In Malaysia, production of tin from hydrothermal lodes, veins and stockworks has always surpassed that from other primary deposits. The search for stanniferous hydrothermal deposits will continue to feature prominently in exploration strategies for tin, and exploration decisions will naturally be influenced by the geological setting of the known occurrences. This implies there is a necessity to examine areas of intense faulting, centres with a high concentration of late-phase acid intrusives, greisen zones, and areas likely to be underlain by concealed granitic ridges or cusps. In this regard, the findings of the BGS/GSM regional study on granitoids and that of BGR/GSM research on tin-bearing and tin-barren plutons will certainly help to refine existing, or evolve new exploration models for the search for intra-granitic tin deposits as well as for blind cupola-related ones.

Another important strategy that should perhaps be adopted is to explore for large carbonate-replacement type tin deposits similar to those of Australia. Some of the Australian Sn-Fe-S bodies, like those at Renison Bell, Cleveland, and Moina, are important sources of tin (Taylor, 1979). Their geological setting appears to be analogous with that shown by the Sn-Fe skarns found within the Eastern Belt. An in-depth study of these skarn zones, and a critical review of other replacement-type iron bodies within this belt, is warranted. In this respect, the airborne data should be fully utilised. Similarly, areas of strong carbonate development need to be examined for fracture zones and feeder veins which are related to the emplacement of the relatively high-level plutons within this stanniferous belt.

A re-evaluation of the former lode tin-mining areas may likewise be justifiable, since such areas have been rarely systematically investigated. It may also be worthwhile to assess areas supporting, or are known to have sustained, important eluvial tin deposits, in the hope that their primary sources may be uncovered.

## Conclusions

Primary tin deposits in Malaysia exist in several varieties, ranging from the mineralogically simple ones to those which are complex and telescoped. Since very early times, they have variously contributed to the country's tin production. Today the large "Cornish-type" stanniferous lodes at Sungai Lembing and the several open-cast, low grade deposits within the Western Tin Belt continue to be important sources of primary tin. Probably more deposits of this nature may be found along unexplored granitic contacts and cusps. The available airborne data should be very useful, particularly for detecting concealed granitic cusps and ridges.

Analogies have been drawn between the Sn-Fe skarns of the Eastern Tin Belt and the important carbonate-replacement types of Sn-Fe-S deposits of Australia. Although opinion has lately been expressed that such deposits may have been formed through submarine volcanic-exhalative activity, field evidence suggests their genesis to be more consistent with a contact-pyrometasomatic origin. Regardless of their genesis, the significance of the Australian examples should warrant the formulation of a suitable exploration strategy to critically assess the known stanniferous iron skarns, pyrometasomatic iron deposits, and calcareous environments close to high-level granitoids. The available geological, geochemical and airborne geophysical data are invaluable in this regard.

*Acknowledgements.* The writers thank their colleagues, in particular Mr. Senathi Rajah for reviewing this manuscript, and Mr. Aw Peck Chin for lively discussions. Mr. Redzuan Sumun of the Mines Department is also thanked for his comments. The Management of the Pahang Consolidated Co. Ltd. is gratefully thanked for information readily supplied. Appreciation is accorded to the Cartographic Section of the Geological Survey for drafting the text figures.

This paper is published with the permission of the Ministry of Primary Industries, Malaysia.

## References

Agocs, W.B., 1958. *Report of airborne magnetometer and scintillation counter survey of Kedah, Perak, Selangor, Terengganu, Pahang and Johore, Federation of Malaya.* Geol. Surv. Malaysia Intern. Rep., Ipoh, Malaysia.

Alexander, J.B., 1968. Geology and Mineral Resources of the Bentong Area, Pahang. *Geol. Surv. Malaysia, District Memoir* No. 8.

Alexander, J.B., and Flinter, B.H., 1965. A note on varlamoffite and associated minerals from the Batang Padang district, Perak, Malaysia. *Mineralogical Mag.* 35, 622—627.

Bean, J.H., 1969. The iron-ore deposits of West Malaysia. *Geol. Surv. Malaysia, Econ. Bul.* No. 2.

Bignell, J.D. and Snelling, N.J., 1977. Geochronology of Malayan granites. *Overseas Geol. and Min. Resources,* No. 47.

Bradford, E.F., 1957. The occurrence of tin and tungsten in Malaya. *Ninth Pacific Science Congress,* No. 12, 378—398.

Bradford, E.F., 1972. The geology and mineral resources of the Gunung Jerai Area, Kedah. *Geol. Surv. Malaysia, District Memoir* No. 13.

Burton, C.K., 1959. A note on the geology of the tin and iron-bearing deposits located near Sungei Pelepah Kanan district of Kota Tinggi, Johore. *Geol. Surv. Malaysia Papers* Vol. 3, 1980.

Chand, F. (Compiler), 1981. A manual of Geochemical Exploration Methods. *Geol. Surv. Malaysia,* Spec. Paper 3.

Chu, L.H., Muntanion, H., Sidik, A., Chand, F., and Troup, A., 1982. Regional Geochemistry of South Kelantan. *Geol. Surv. Malaysia, Geochem. Rep.* I.

Fitch, F.H., 1952. The geology and mineral resources of the neighbourhood of Kuantan, Pahang. *Geol. Surv. Malaysia, Memoir* No. 6.

Fletcher, W.K., Dousset, P.E., and Yusoff Ismail, 1984. Behaviour of tin and associated elements in a mountain stream, Bujang Melaka, Malaysia. Abs. of Papers, GEOSEA V. Geol. Soc. Malaysia, p. 9.

Foo, K.Y., 1970. Geological and geochemical reconnaissance of the Bahau – Kuala Rompin area. *Geol. Surv. Malaysia Intern. Rep.*

Gan, L.C., 1980. Manson Lode, a stratabound base metal-silver deposit in North Kelantan, Malaysia. *Ph.D. dissertation, Univ. Leoben, Austria* (unpub.).

Hosking, K.F.G., 1969. Aspects of the geology of the tin fields of southeast Asia. *A second tech. conf. on tin,* Vol. I, Intl. Tin Council, 41–80.

Hosking, K.F.G., 1973. Primary mineral deposits. Gobbett, D.J. and Hutchison, C.S. (eds.) *Geol. of the Malay Peninsula.* Wiley-Interscience, New York, 335–390.

Hutchinson, R.W., 1984. Massive sulfide deposits and their significance. Keynote Pap., GEOSEA V. *Geol. Soc. Malaysia.*

Ingham, F.T., and Bradford, E.F., 1960. The geology and mineral resources of the Kinta Valley, Perak. *Geol. Surv. Malaysia, Memoir* No. 9.

Ishihara, S., Sawata, H., Arpornsuwan, S., Busaracome, P., and Bungbrakearti, N., 1979. The magnetite-series and ilmenite-series granitoids and their bearing on tin mineralization, particularly of the Malay Peninsula region. *Geol. Soc. Malaysia, Bulletin* 11, 103–110.

Jones, W.R., 1925. *Tinfields of the world.* London: Mining Publications Ltd.

Khoo, H.P., 1969. Mineralization at Pelepah Kanan, Kota Tinggi, Johore, West Malaysia. *Unpub. B.Sc. Hon. thesis,* Dept. of Geology, Univ. Malaya.

Kwak, T.A.P., 1984. An overview of Sn-sulphide replacement style deposits in Australia. *Abs. of Papers, GEOSEA V.* Geol. Soc. Malaysia, p. 17.

Liew, T.C., 1983. Petrogenesis of the Peninsular Malaysian granitoid batholiths. *Unpubl. Ph.D. thesis, Australian Nat. Univ., Canberra.*

MacDonald, S., 1967. The geology and mineral resources of north Kelantan and north Trengganu. *Geol. Surv. Malaysia, Memoir* No. 10.

Pahang Consolidated Company Ltd., 1978. *Information on the operations of the Pahang Consolidated Company Ltd.*

Pahang Consolidated Public Ltd. Company, 1984. *Annual Rep. for 1983.*

Pun, V.T., 1981. Some aspects on prospecting of primary tin deposits with special reference to diamond drilling techniques. *Proc. drilling and sampling techniques in tin prospecting.* SEATRAD Centre, Ipoh, Tech. Publ. No. 1, 152–171.

Pun, V.T., and Joginder Singh, 1978. Geology and mineralization of Sungei Lembing tin deposits. *Abs. of Papers, Int. symposium on Geology of tin deposits.,* Geol. Soc. Malaysia, Annex to Newsletter, Vol. 4, March–April, 7–8.

Rajah, S.S., 1970. The geology and sulphide mineralization in the Ulu Sokor area, Kelantan, Malaysia, with special reference to Manson's Lode. *DIC dissertation Imperial College, Univ. London* (unpubl.).

Rajah, S.S., 1979. The Kinta Tinfield, Malaysia. *Geol. Soc. Malaysia, Bulletin* 11, 111–136.

Rajah, S.S., and Fateh Chand, 1973. Brief outline of the mineral resources of Peninsular Malaysia. *Geol. Surv. Malaysia Ann. Rep.,* 82–92.

Roe, F.W., 1941. Report on the deposit of tin-ore and iron-ore at Pelepah Kanan, Kota Tinggi, Johore. *Encl. No. 19 in Geol. Surv. File 61/41* (unpub.).

Roe, F.W., 1951. The geology and mineral resources of the Fraser's Hill area, Selangor, Perak and Pahang, Fed. of Malaya. *Geol. Surv. Malaysia, Memoir* No. 5.

Santokh Singh, D. and Bean, J.H., 1967. Some general aspects of tin minerals in Malaysia. *A tech. conf. on tin,* Vol. 2, Intl. Tin Council, 459–478.

Scrivenor, J.B., 1928. *The geology of Malayan ore-deposits.* London: Macmillan.

Smith, J.M., 1969. Report on mineral investigations in the Ulu Sokor area. *Geol. Surv. Malaysia Intern. Rep.*

Taylor, R.G., 1979. *Geology of tin deposits.* Dev. in econ. geology 11, Elsevier Scientific Pub. Co.

Teh, G.H., 1979. Geochemical studies around the Tekka area, Perak, Peninsular Malaysia. *Geol. Soc. Malaysia, Bulletin* 11, 353–373.

Teh, G.H., 1981. The Tekka tin deposit, Perak, Peninsular Malaysia. *Geol. Soc. Malaysia, Bulletin* 14, 101–118.

Tooms, J.S., and Kaewbaidhoon, S., 1961. Dispersion of tin in soil over mineralization at Sungai Lembing, Malaya. *Trans. IMM,* Vol. 70, part 8, 475–490.

Wong, Y.C., 1975. A pedogeochemical survey on Tekka hill, Perak. *Geol. Surv. Malaysia, Ann. Rep.,* 129–130.

Yeap, E.B., 1982. Reinterpretation of the Fe-Sn mineralization at the Waterfall Mine, Pelepah Kanan, Johore. *Geol. Soc. Malaysia Newsletter* Vol. 8, No. 5, Sep–Oct. 1982, 228–229.

Yeap, E.B., 1984. Geology of some Malaysian Fe-Sn deposits and their significance. *Abs. of Papers, GEOSEA V.* Geol. Soc. Malaysia, p. 39.

## 6.10 Nepal

# 6.10.1 Geology and Exploration for Tin-Mineralization in the Himalayas of Nepal

P.R. Joshi[1]


## Abstract

In the Himalayas of Nepal, out of five known distinct granitic terrains covering 8,500 $km^2$, approximately 700 $km^2$ of favourable area have been explored for tin. Minor scattered tin mineralization has been recorded in association with the Lesser Himalayan cordierite-bearing two-mica granites and granitic rocks of the 'Central Crystalline' Higher Himalaya. The other three granitic terrains, namely: (i) granitic intrusives within the Lesser Himalayan Sequence, (ii) Augen gneiss south of the Main Central Thrust, and (iii) Leuco-granites of the Higher Himalaya, have not so far revealed any tin mineralization. These granitic rocks display a wide age range from Precambrian to Late Oligocene, and intrude mainly medium grade metamorphic formations of Precambrian to Cambrian age.

Within the contact zones of the Lesser Himalayan cordierite-bearing two-mica granites, three types of mineralization with different mineral assemblages viz. (i) Cassiterite-pyrite with subordinate amounts of chalcopyrite, pyrrhotite and arsenopyrite, (ii) Cassiterite-molybdenite-bearing quartz-tourmaline rocks and (iii) Cassiterite-bearing pegmatites, have been found at the northern contact zone of the Dandeldhura granite in silicified and chloritised phyllites of Precambrian age, at the southern contact zone of the Palung granite in biotitic schists (Kalitar Formation) of Precambrian to Lower Cambrian age, and at the southern contact zone of the Ipa and Narayanthan granites in Precambrian to Lower Cambrian biotite schists of the Kulikhani and Kalitar Formations. Cassiterite and wodginite, together with columbite-tantalite as well as pyrochlore, have infrequently been noticed in polychrome tourmaline-, and aquamarine-bearing pegmatites in staurolite-almandine-biotite schists and calc schist in the Sankhuwa Shaba region, eastern Nepal.


## Introduction

In the Kingdom of Nepal attempts to explore for tin were carried out in the past decade. The presence of tin in the heavy concentrates of the Lesser Himalayan granite terrain particularly along the contact zones of the Dandeldhura granite at Melmura and the Palung granite at Mandu Khola, were first reported by Talalov (1972). After

[1] Senior Geologist, Department of Mines and Geology, Lainchaur, Kathmandu, Nepal

carrying out geological investigations around the Palung granite in 1973, around the Ipa and Narayanthan granites in 1974, and around the Dandeldhura granite in 1977 and 1978, Joshi (1977 and 1978, 1980) established the occurrence of primary tin mineralizations in the contact zones of the Lesser Himalayan cordierite-bearing two-mica granites and in the pegmatites occurring in the Central Crystallines of the Higher Himalaya.

In old mining records of Nepal, no mention of tin exploitation has been found, despite the extensive mining activities for base metals.

## Geological Setting of Granitic Rocks and Their Tin Potential

Favourable granitic terrains for primary tin mineralization occur in five distinct tectonic settings in the Himalayas of Nepal (Fig. 1). Among them to-date tin mineralization is reported to be associated with (1) Lesser Himalayan cordierite-bearing two-mica granites in the klippen of the Higher Himalaya, and (2) pegmatites of the "Central Crystallines" of the Higher Himalaya.

### 1. Granitic Intrusives Within the Lower Himalyan Sequence

Uncommon and small bodies of biotite or biotite-muscovite granite and adamellite, presumably formed earlier than the Himalayan orogeny, showing weak foliation or gneissic character, have been reported from the Upper Precambrian to Lower Paleozoic greenschist facies of the Kuncha Formation in Central Nepal (Mitchell, 1979).

Due to incomplete mapping as well as the small size of such granitic bodies, their economic significance for tin is not yet known.

### 2. Augen Gneiss South of the Main Central Thrust

The highly metamorphosed Melung Augen Gneiss is possibly the oldest among the Himalayan granitic rocks, comprising a discontinous zone of gneiss along the southern margin of the Lower Himalayan Zone. It comprises a thrust sheet of inferred Precambrian rocks in Eastern Central Nepal (Hashimoto et al., 1973) and the Ulleri Augen Gneiss in Western Central Nepal, a metamorphic product of felsic volcanism erupted in Middle or Lower Paleozoic time (Le Fort, 1975b). Stöcklin (1980) observed these gneisses with locally thrust contacts against graphitic schist, marble, tremolite schist and quartzite, equivalent to the Benighat slate of the Upper Nuwakot group of Paleozoic age. Other correlatives of the augen gneisses of Nepal, as mentioned by Pecher and Le Fort (1977), are the gneissic quartz porphyry of Ramgarh (Heim and Gansser, 1939), and the Chiplokot augen gneiss in the Kumaon Himalayas. However Mitchell (1981) suggested it could possibly be equivalent to meta-igneous rocks of Birnags (Frank et al., 1977) which have yielded a Rb/Sr whole rock isochron age of 1840 ± 70 Ma.

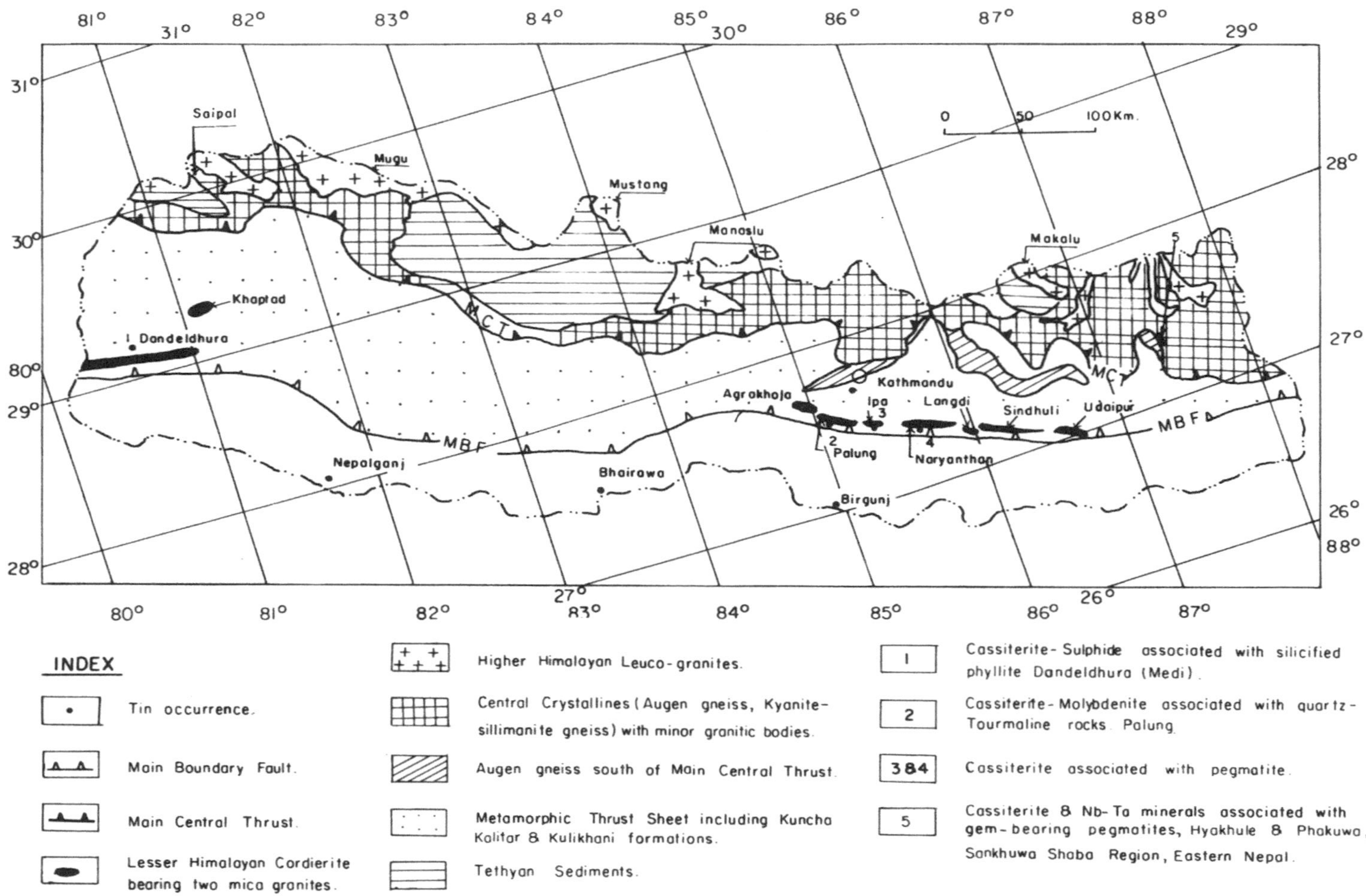


**Fig. 1.** Tin occurrence and general geology of the Himalayas of Nepal

Despite their extensive cover (3,500 $km^2$) south of the Main Central Thrust, a small portion of the Mailung and Ulleri gneisses were searched for tin occurrences but did not show positive results. Nevertheless minor arsenopyrite and Cusulphide have been observed in the upper boundaries of the Ulleri Gneiss (MEP, 1976). Furthermore, the prevailing gneissic character is indicative of deep erosion hence there is less possibility for the preservation of intrusive-related hydrothermal deposits (Mitchell, 1981).

## 3. Granite Within the Klippen of the Higher Himalaya

Massive and weakly foliated to gneissic, mostly inclined slab-like granitic bodies of crystalline thrust sheets or klippen, have long been known in the Lesser Himalaya of Kumaon (Gansser, 1964) and Nepal (Hagen, 1969). Such granitic bodies of the metamorphic terrain occur structurally above the Lower Himalayan sequence as klippen of Higher Himalayan rocks (Andrieux et al., 1977). Le Fort et al. (1980) grouped them into a Lesser Himalayan cordierite-bearing two-mica granite belt which occurs as fifteen independent plutons over a stretch of 1,600 km from Mansera, Pakistan, to Udaipur, Nepal.

Furthermore these granitic bodies, once considered to be the relicts of the Higher Himalayan ones, have given much older ages such as a 486 ± 10 Ma Rb/Sr isochron age with an initial $Sr^{87}/Sr^{86}$ ratio of 0.720 for the Palung granite (Mitchell, 1981) which intrudes the meta-sedimentary rocks of the Late Precambrian to Cambrian Kathmandu Complex. Similar Rb/Sr isochron ages of 512 ± 16 Ma, with an initial $Sr^{87}/Sr^{86}$ ratio of 0.71, for the granite within the Lahul Crystallines (Frank et al., 1977); and 516 ± 16 Ma, with an initial $Sr^{87}/Sr^{86}$ ratio of 0.719, for the Mansera granite (Le Fort et al., 1980), have also been reported.

Among the eight such granitic bodies covering 1,500 $km^2$ in the Lesser Himalaya of Nepal, to-date scattered minor tin occurrences have been detected in the contact zones of Dandeldhura, Palung, Ipa and Narayanthan granitic bodies. The other eastern granitic bodies still remain unexplored.

## 4. Granitic Rocks of the 'Central Crystallines' Higher Himalaya

Migmatites and numerous minor bodies of tourmaline-rich leuco-granites and pegmatites, most probably of Tertiary age, resulting from heating and metamorphism associated with the India-Asia collision as well as movement on the Main Central Thrust, occur locally traversing the schist and gneiss (Heim and Gansser, 1939) of the 'Central Crystalline unit' or 'Tibetan Slab' (Le Fort, 1975a). The gneissic host underlies the Ordovician rocks of the Higher Himalayas without any obvious structural break, suggesting a Precambrian age. Similar granite sills also intrude the lowest part of Tethyan calcareous sedimentary succession of probable Cambrian age.

Many such granites and pegmatites have not been mapped out, except a few in the Sankhuwa Shaba region of Eastern Nepal, where the polychrome tourmaline- and aquamarine-bearing pegmatites of a few metres thickness have indicated the presence of Sn, Ta, Nb and U-minerals.

### 5. Higher Himalayan Leucogranite

A chain of tourmaline-rich two-mica leucogranites, namely Makalu, Manaslu (Late Oligocene, 28 Ma age, Hamet and Allegre, 1976), Mugu, and Mustang granites, having injected into the Phanerozoic succession above the Central Gneiss, are known in the Higher Himalayas (Gansser, 1977). The huge size as well as comparatively northward position of these granite bodies distinguishes them from the granitic rocks of the 'Central Crystallines' or 'Tibetan Slab'. One of the plutons, the Manaslu granite as described by Le Fort (1975a), is more than 10 km thick. This north-dipping slab-like granitic body comprises a distinctly foliated two-mica tourmaline-bearing granite core surrounded by a fringe of tourmaline-rich fine-grained granite.

So far the Higher Himalayan granites of Nepal, except the Manaslu granite, are poorly investigated. The presence of minor disseminated pyrite, chalcopyrite, arsenopyrite, sphalerite, galena, molybdenite, bismuthinite and native bismuth have been reported in the overburden derived from the Manaslu granite (Le Fort, 1975a).

## Tin Mineralization

To-date tin mineralization is known from associations with (i) Granites within the klippen of the Higher Himalaya, and (ii) granitic rocks of the 'Central Crystallines' of the Higher Himalaya (Figure 1).

### 1. Tin Associated with the Granites Within the Klippen of the Higher Himalaya

Three types of epigenetic Sn-mineralization are encountered in the contact zones of the cordierite-bearing two-mica granites of the Lesser Himalaya or granites within the klippen of the Higher Himalaya.

| *Type of Mineralization* | *Locality* |
|---|---|
| 1. Cassiterite-pyrite mineralization with subordinate amounts of chalcopyrite in silicified and chloritised phyllite of Precambrian age. | About 1 km away from the contact of the Dandeldhura granite in the hanging wall, Meddi, Dandeldhura. |
| 2. Cassiterite-molybdenite mineralization in quartz-tourmaline rocks traversing the biotitic schist of the Kalitar Formation of Precambrian to Lower Cambrian age. | Within 100 m from the southern contact of the Palung granite in the foot wall side, Chau Khola. |
| 3. a) Cassiterite in pegmatites traversing the biotitic schist of the Kulekhani Formation of Precambrian age. | Approximately 100 m away from the southern contact of the Ipa granite, Asp Khola. |

| | |
|---|---|
| b) Cassiterite in pegmatite traversing the biotite schist of the Kalitar Formation. | Approximately 150 m away from the southern contact of the Narayanthan granite, Chhyangchhyang Khola and Harbidanda Khola. |

2. Tin Associated with the Granitic Rocks of 'Central Crystallines', Higher Himalaya

The presence of finely disseminated cassiterite and wodginite (Ta, Nb, $Sn)O_2$ together with columbite-tantalite as well as pyrochlore (Na, Ca, U) 1-x (Nb, Ta, Ti)' x $O_6$ (OH, $F)_{1-2}$, or some unnamed mineral (Bi, Ca) (Ta, $Nb)_2O-6$(OH) has been observed infrequently in the polychrome tourmaline- and aquamarine-bearing pegmatites occurring in staurolite-almandine-biotite schist and calc schist of the 'Central Crystallines' in the Sankhuwa Shaba region, Eastern Nepal.

## Description of Mineralization

1. Cassiterite-Pyrite Mineralization

Irregular and minor vein and lens-shaped, usually foliation-controlled, but sometimes of discordant-type, of cassiterite-pyrite mineralization with subordinate chalcopyrite, pyrrhotite and arsenopyrite, formed under hypothermal to mesothermal conditions, are observed in Precambrian chloritised and silicified phyllites approximately 1 km away from the contact of the Dandeldhura granite. Such mineralised veins and lenses of up to 0.1 x 0.5 x 2 $m^3$ extension, with a weighted grade of 0.05 % Sn, are found distributed over an 80 m thick zone in the host rock.

Cassiterite occurs as dark irregular and fractured, sometimes tabular and acicular crystals with dimensions varying from 0.01 x 0.01 to 5 x 5 $mm^2$. They infrequently contain minute inclusions of pyrite and the tantalite-columbite group. The cassiterite is rarely free of pyrite. Quartz, sericite and chlorite are the main gangue minerals.

This mineralization shows a spatial relation to the Dandeldhura granite but the genetic and temporal relationships to the granite are unknown (Fig. 2).

2. Cassiterite-Molybdenite Mineralization

Insignificant amounts of disseminated cassiterite-molybdenite mineralization occur in minor quartz-tourmaline rocks which have a lateral extension of a few metres. Locally a few quartz-tourmaline bodies inject the biotite schist of the Precambrian to Lower Cambrian Kalitar Formation within 100 m of the southern contact of Palung granite at Chau Khola.

3. Cassiterite in Pegmatite

A few pegmatite blocks occur in the overburden over the biotite schists of the Kulekhani and Kalitar Formations, considered to be derived from the pegmatitic offs-

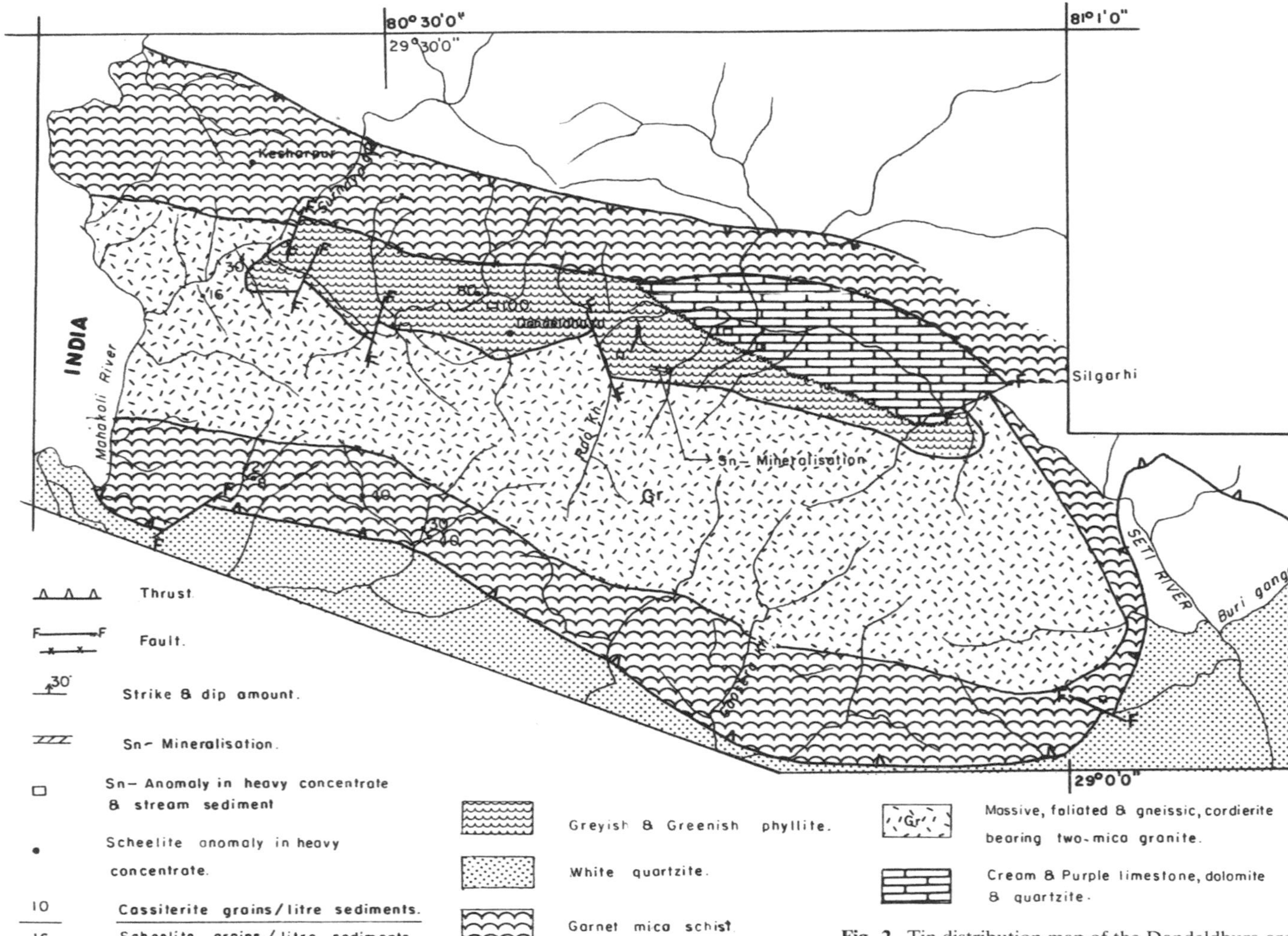


**Fig. 2.** Tin distribution map of the Dandeldhura area

hoots of the Ipa and Narayanthan granites. These pegmatites contain small amounts of minute idiomorphic and coloured cassiterite crystals occasionally with a columbite association. The small size and sporadic distribution of the pegmatite bodies suggest that the associated mineralization is economically insignificant.

## Status of Tin Exploration

Out of the extensive granitic terrain (8,500 km$^2$) favourable for Sn-mineralization, so far 700 km$^2$ have been explored, mainly around the Lesser Himalayan cordierite-bearing two-mica granites (Figure 2).

Around the Dandeldhura granite, 500 km$^2$ has been, to some extent, explored for tin. The cassiterite sulphide mineralization of the contact zone of the Dandeldhura granite at Meddi is considered to be the first tin prospect of Nepal, but has indicated a poor distribution of cassiterite-sulphide mineralized bodies, despite extensive (10 km wide) Sn-anomalies in the stream sediments and heavy concentrate surveys. In the stream sediments survey, there are anomalous tin-values of eightfold magnitude above background at a distance of eight km downstream from the source (Fig. 3). Significant tin anomalies in the heavy concentrates at a few other places did

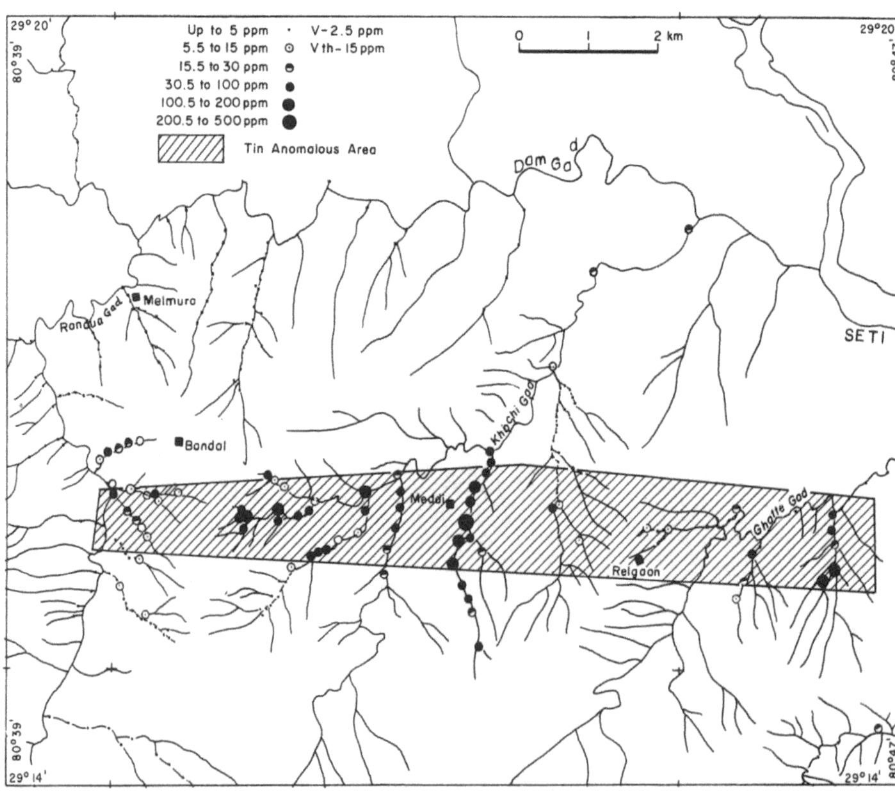


**Fig. 3.** Geochemical stream sediment survey of Dandeldhura area: tin discrete value map

not help to locate any mineralized zones (Joshi, 1980). The reconnaissance geochemical stream sediment survey carried out in 200 $km^2$ around the Palung and Udaipur granites did not reveal any Sn-anomaly (Adhikari, 1980).

## Conclusions

To-date, due to meagre geological investigation and erosion of productive mineralization, economically significant tin and other granite-related minerals have not been discovered in the granitic terrains of the Himalayas of Nepal. The presence of cassiterite-sulphide mineralization in the contact zones around the deeply eroded Dandeldhura granite is a positive indication for the possibility of other similar deposits in association with other Lesser Himalayan cordierite-bearing two-mica granites as well as with Higher Himalyan comparatively less eroded Tertiary granites. Among the known tin mineralizations, cassiterite-molybdenite-bearing quartz-tourmaline rocks, cassiterite-bearing pegmatites and cassiterite-rare earth-bearing pegmatites are not only insignificantly few in distribution but also relatively poor in tin content compared with the cassiterite-sulphide type. Although individual cassiterite-sulphide bodies are small in size, they are considerably rich. Hence more extensive attempts to explore for the top prized tin metal in the virgin areas may be rewarded by sizeable economic tin finds.

The possibility of encountering tin deposits in the Himalayan region has been considered by many workers. Sillitoe (1979) suggested that Bolivian type tin-tungsten deposits, known at the same elevations as the Himalayan granites, were generated at the down dip extremity of a subduction zone. The Lesser Himalayan cordierite-bearing two-mica granites, leucogranites of Higher Himalaya, and granitic rocks of the 'Central Crystallines' are of the anatectic type as the tin-tungsten granites of Cornwall, the Central Massif of France and North-West Spain, and also the Mesozoic granites of the Main Range of Malaysia, which were emplaced during a continental collision.

*Acknowledgements.* The author expresses sincere thanks to Mr. M.N. Rana, Director General, DMG for the kind permission to publish this paper. The author is also extremely grateful to the Ministry of Geology and Mineral Resources, People's Republic of China, for the invitation to attend the symposium and RMRDC for the financial support.

## References

Adhikari, P.P., 1980. Geochemical reconnaissance stream sediments survey with regional geological mapping, part of Dolkha and Ramechap. *Mineral Explor. Project. Unpubl. rept.*, Nepal.

Andrieux, J., Brunel, M. and Hamet, J., 1977. Metamorphism, granitization, and relations with the Main Central Thrust in Central Nepal, $^{87}Rb/^{87}Sr$ age determinations and discussion. *Himalaya Coloque Internat.* 268, C.N.R.S. ed. Paris, 31—40.

Frank, W., Thoni, M. and Purtscheller, F., 1977. Geology and Petrography of Kulu-South Lahaul area. *Himalaya Colloque Internat.*, 268, C.N.R.S. ed. Paris, 147—172.

Gansser, A., 1964. *Geology of the Himalayas.* Wiley-Interscience, New York, 289 pp.

Gansser, A., 1977. The great suture zone between Himalaya and Tibet, a preliminary account. *Himalaya Colloque Internat.*, 268, C.N.R.S. ed. Paris, 181—191.

Hagen, T., 1969. Report on the geological survey of Nepal 1, Preliminary reconnaissance. *Denkschr. Schweiz. Naturf. Ges.* 86, 185 pp.

Hamet, J. and Allègre, C.J., 1976. Rb-Sr systematics in granite from central Nepal (Manaslu): significance of the Oligocene age and high $^{87}Sr/^{86}Sr$ ratio in Himalayan orogeny. *Geology,* 4, 470–472.

Hashimoto, S., Ohta, Y., Akiba, C., 1973. *Geology of Nepal Himalaya.* Saikon Publ. Co. Japan, 286 p.

Heim, A. and Gansser, A., 1939. Central Himalaya. Geological observations of the Swiss expedition 1936. *Denkschr. Schweiz. Naturf. Ges.* 73, 245 p.

Joshi, P.R., 1977 and 1978. Report on geological and geochemical investigations of mineral resources of a part of Dandeldhura district, Mahakali Anchal. Vols. 1 and 2, *Dept. of Mines and Geol.* Nepal, unpubl. report.

Joshi, P.R., 1980. Report on tin-prospect, Meddi area Dandeldhura district. *Dept. of Mines and Geol.,* Nepal, unpubl. rept.

Kozminski, G., 1977. Report on geochemical mapping, Eastern Region. *Dept. of Mines and Geol.* Nepal, unpubl. rept.

Le Fort, P., 1975a. The anatectic Himalayan leuco-granites with emphasis on the Manaslu tourmaline granite. *Recent Researches in Geology,* 2, Hindustan Publ. Co., Dehli, 76–90.

Le Fort, P., 1975b. Himalayas: the collided range. Present knowledge of the continental arc. *Amer. Journ. Science,* 275, 1–44.

Le Fort, P., Deoban, F. and Sonet, J., 1980. The Lesser Himalayan cordierite granite belt, petrology and age of pluton of Mansehra, Pakistan. *Geol. Bull. Univ. Peshwar,* Spec. issue, 13, 51–62.

MEP, 1976. *Mineral Exploration Project,* unpubl. report GC/10/76.

Mitchell, A.H.G., 1979. Guides to metal provinces in the Central Himalaya Collision belt. *Memoir of the Geol. Soc. China,* 3, 167–194.

Mitchell, A.H.G., 1981. Himalayan and Trans-Himalayan granitic rocks in and adjacent to Nepal and their mineral potential. *Journ. Nepal Geol. Soc.,* 1, No. 1.

Pecher, A. and Le Fort, P., 1977. Origin and significance of the Lesser Himalaya Augen Gneiss. *Himalaya Colloque Internat.,* 268, C.N.R.S. ed. Paris, 319–329.

Sillitoe, R.H., 1979. Speculation on Himalayan metallogeny based on evidence from Pakistan. Abdul Farah and K.A. Dejong (eds.). *Geodynamics of Pakistan.* Geol. Surv. of Pakistan, Quetta.

Talalov, V.A., 1972. Geology and ores of Nepal. *U.N.D.P. unpubl. rept.,* 5 vols., Kathmandu.

Stöcklin, J., 1980. Geology of Nepal and its regional frame. *Journ. Geol. Soc. London,* 137, 1–34.

Stöcklin, J. and Bhattarai, K.D., 1977. Geology of Kathmandu area and central Mahabharat Range, Nepal Himalaya. *U.N.D.P. unpubl. rept., Kathmandu.*

## 6.11 Thailand

# 6.11.1 Geological Setting and Genesis of Primary Cassiterite and Scheelite Mineralization in the Nam Mae Lao Valley, Chiang Rai Province, Northern Thailand

H. GEBERT[1]

**Abstract**

Cassiterite and scheelite deposits of the Nam Mae Lao valley in northern Thailand are stratabound within Palaeozoic host rocks. Two types may be distinguished: metasediment- and amphibolite-hosted. There is a combination of the two types in one mining area.

The mineral deposits are members of the pre-Permian Palaeozoic metavolcano-sedimentary formation which developed in an island arc type of environment east of the Shan Thai Paraplatform. The deposits are interpreted as resulting from sea floor exhalative hydrothermal systems related to Palaeozoic volcanic activity. Post ore depositional folding and metamorphism took place in Carboniferous time followed by granite intrusion in the Triassic. Mineralization marginal to the granite developed where Triassic granites intruded into the metal-bearing Palaeozoic metavolcano-sedimentary host rocks.

## 1. Introduction

The occurrence in Thailand of stratigraphically-controlled tin and tungsten deposits was recognized for the first time in the course of detailed geological investigations by the authors during the German Geological Mission in Thailand from 1979 to 1982, selected results of which are given in this paper.

Stratabound cassiterite deposits became known only 20 years ago through the work of Baumann (1965) on the crystalline basement of the Erzgebirge in Saxony. The tin was discharged on the sea floor in a geosynclinal environment during a period of submarine volcanism. Based on the work of Lea and Rancourt (1958), Walker et al. (1975) and Hamilton et al. (1982), it is now recognized that cassiterite is a typical constituent in stratabound massive base metal sulphide ores. Ransom (1977) reports that the central part of the Sullivan orebody of Canada contains tin in excess of 500 ppm. The genesis of the stratabound cassiterite-pyrrhotite deposits of western Tasmania is subject to a major dispute that focusses on a granite-related replacement origin of the ores versus a sea floor discharge of metalliferous hydrothermal fluids (Hutchinson,

[1] ESCAP – RMRDC, Jalan Jenderal Sudirman 623, Bandung, Indonesia

1979, 1980, 1981a, 1981b, 1982; Plimer, 1980; Solomon, 1980; Patterson et al., 1981; Patterson, 1982; Newnham, 1981, 1985).

Stratabound scheelite ores rank amongst the world's giant deposits of tungsten. Recent significant discoveries of this type of deposit include the large orebodies at Felbertal near Mittersill in Austria and the important occurrences close to the coastal capital of Nunk in Greenland. At Felbertal the lens-shaped orebodies are 100–150 m wide and several 100 m long. A thickness of up to 20 m is attained in the centre of the lenses. The deposits are members of a metavolcano-sedimentary Palaeozoic formation and are richest within a horizon of alternating and interfingering layers of metalava and meta-tuff (Höll, 1971, 1975, 1977). There was some reworking and redepositing of tuff by turbidity or density currents on the sea floor. The metavolcanites are of ultramafic, tholeiitic, intermediate and acid composition. There are hydrothermal mineral formations. The scheelite near Nuk in Greenland is reported to be found in amphibolitic host rocks. The prospective area comprises 3,000 km$^2$ (Mining Journal, December 14, 1984).

## 2. Location

The area investigated by geochemical and geological methods totals 1,500 km$^2$ and is situated in Chiang Rai, Chiang Mai and Lampang provinces of northern Thailand (Fig. 1). The mineral deposits described in this paper are located in Wiang Pa Pao and Mae Suai districts of Chiang Rai province in close vicinity to national highway 1019 which connects Chiang Rai and Chiang Mai. Chiang Mai is linked with Bangkok by railroad. All locations are accessible at short distances by four-wheeldrive vehicles from this highway except for Mae Tho Mine which may be reached by footpath beginning near km 57.4 or from Ban Musoe, Muang Noi. The geographic locations of the mines are detailed in Table 1.

## 3. Regional Geological Overview

The results of the German Geological Mission to Thailand and the work of Bunopas (1981) led to the recognition of two north-south trending regional geologic units in northern Thailand developed west and east respectively of the line from Fang to Chiang Mai and Tak.

The western unit is the cratonic Shan Thai Paraplatform. It is tilted towards the west. It consists of a highly metamorphosed basement that is covered by formations ranging from Palaeozoic to Mesozoic. There was folding in late Mesozoic time which partially affected these covering sediments (Koch, 1973). On the paraplatform major faulting including horizontal movements resulted in a structural pattern of mosaic-type individual blocks.

The eastern unit is a segment of the Sukothai Fold Belt (Bunopas, 1981). Its weakly metamorphosed Variscian basement of geosynclinal sediments with intercalations of volcanites is overlain by late Palaeozoic to Mesozoic sediments. These sediments include a large proportion of molassetype strata. There are intercalations of basic volcanites. As in the western unit folding of strata overlying this basement rock

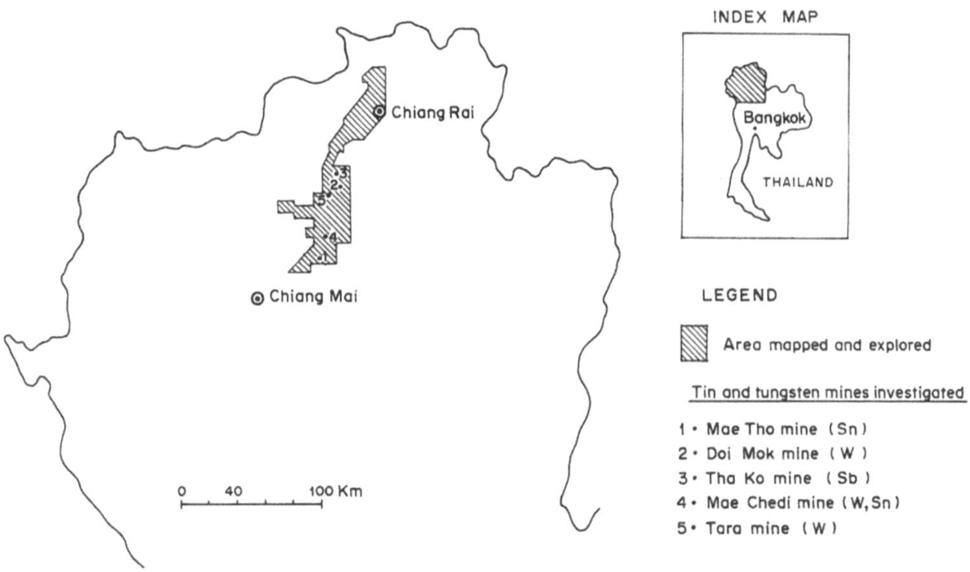


**Fig. 1.** Locality map of mines and area mapped in northern Thailand

took place in late Mesozoic time. This was followed by uplift and development of graben and basin structures.

There are high level intrusives of the anorogenic granite type in both regional units, not all emplaced at the same time.

The mineral deposits dealt with in this paper are members of the Variscian basement and occur at the western margin of the Sukothai Fold Belt. They are exposed in the Wiang Pa Pao Anticlinorium. Towards the east the Variscian basement is covered by late Palaeozoic and Mesozoic strata of the Lampang Synclinorium.

**Table 1.** Location of mineral deposits investigated

| | Mine | District | Co-ordinates | | Elevation above sea level |
|---|---|---|---|---|---|
| | | | N | N | |
| 1. | Mae Tho (cassiterite) | Wiang Pa Pao | 2103700 | 545800 | 1,300 m |
| 2. | Doi Mok (scheelite) | Wiang Pa Pao | 2145900 | 558600 | 1,000 m |
| 3. | Tha Ko (scheelite, stibnite) | Mae Suai | 2154100 | 554600 | 700 m |
| 4. | Mae Chedi (scheelite) | Wiang Pa Pao | 2113600 –2115500 | 549000 –550500 | 700 m |
| 5. | Tara (scheelite) | Wiang Pa Pao | 2145000 | 552200 | 550 m |

## 4. Stratigraphy

### 4.1 Pre-Permian Palaeozoic Strata

Geological mapping on the scale 1:50,000 led to the subdivision of Palaeozoic metamorphics into a metavolcano-sedimentary formation and a quartzitic formation. Stratabound tin and tungsten mineralization is restricted to the metavolcano-sedimentary formation. It appears that mineralization is confined to zones of predominance of basic volcanic members of this formation.

#### *4.1.1 The Metavolcano-Sedimentary Formation*

This is the oldest formation in the area investigated. It consists of amphibolites, metavolcano-sedimentary strata and interbedded carbonate and meta-clastic formations. The rock types identified are orthoamphibolite, diabase, metadacite, pillow lavas, meta-agglomerate, meta-tuffite, meta-silexite, phyllite, graphitic phyllite, quartz-mica schist, quartz schist, graphitic schist, quartzite, calc-silicate and marble. Biotite gneiss and hornblende gneiss have been noted in a narrow zone east of Doi Mok close to the faulted contact with Permian strata. MacDonald and Barr (1978) identified a volcanic arc assemblage of basalt, basalt porphyry, pyroxenite, tuff and agglomerate in the area of Chiang Rai and Wiang Pa Pao. A detailed account of the mineralogical and chemical composition of typical basic volcanic members of the area is given in Table 2. Four of the samples are of calc-alkaline basalt and two are low-K tholeiite, following the $SiO_2$-$K_2O$ relationships for island arc volcanic rocks (Basaltic Volcanism Study Project, 1981).

Attempts have been made to date the carbonate formations which occur as lenticular intercalations in the metavolcano-sedimentary sequence. However, no fossil record could be obtained. Therefore, the precise Palaeozoic age of the metavolcano-sedimentary formation remains unknown. A narrow pre-Permian Palaeozoic age range, exclusive of Cambrian is inferred by us.[2]

[2] On the Phyao and Chiang Rai sheets mapped by the German Geological Mission during 1965 to 1971, areas occupied by the metavolcano-sedimentary formation refered to several geological ages: Lower Carboniferous, hl, Carboniferous to Silurian (Ordovician), h-s (o), and Devonian to Silurian (Ordovician), d-s (o). The amphibolites are given an upper Carboniferous age. In the course of our investigation it was found that the amphibolites are a member of the metavolcano-sedimentary formation. They were in existence prior to folding and metamorphism which is proven by the same deformation patterns, e.g. microfolding, that prevail in the nonvolcanic members of the unit. For this reason the amphibolites and the associated members of this formation are older than Upper Carboniferous. This view is based on the assumption made by Baum et al. (1970) that folding and metamorphism in northern Thailand took place in Early Carboniferous time.

**Table 2.** Petrological and chemical record of basic volcanic members of the metavolcano sedimentary formation

| Sample No. | Co-ordinates X | Y | Rocktype | Minerals identified* | Chemical analyses** SiO₂ % | Al₂O₃ % | Fe₂O₃ % | MnO % | CaO % | MgO % | Na₂O % | K₂O % | P₂O₅ % | TiO₂ % | Ing. Loss % | Total % | Sample No. |
|---|---|---|---|---|---|---|---|---|---|---|---|---|---|---|---|---|---|
| R 150 | 629 | 487 | Diabase (dolerite basalt) | Sericitized plagioclase (55 %), augite, partly uralitized (27 %) chlorite (8.5 %), epidote (2 %), calcite (0.5 %), ore minerals (7 %), | 48.77 | 11.99 | 14.51 | 0.19 | 5.74 | 7.78 | 3.44 | 0.03 | 0.08 | 2.22 | 4.10 | 98.85 | R 150 |
| R 1 | 629 | 487 | | inclusions of tuffite | 42.78 | 12.60 | 14.66 | 0.12 | 0.07 | 6.26 | 3.17 | 1.14 | 0.22 | 2.78 | 2.50 | 98.12 | R 11 |
| R 107 | 662 | 488 | Orthoamphibolite | Sericitized plagioclase (57 %), hornblende (35.5 %), biotite (1 %), epidote, zoisite (1 %), chlorite (1 %), ore minerals (5 %) | 47.08 | 18.27 | 12.51 | 0.20 | 0.40 | 4.56 | 3.50 | 0.66 | 0.64 | 1.83 | 1.30 | 99.95 | R 107 |
| R 106 | 663 | 487 | | Hornblende (64 %), plagioclase (32 %), epidote, zoisite (1 %), chlorite (1 %), sericite (1 %), ore minerals (1 %) | 48.77 | 15.64 | 11.73 | 0.20 | 9.33 | 7.90 | 3.13 | 0.96 | 0.11 | 0.55 | 2.10 | 100.42 | R 106 |
| R 32 | 756 | 034 | | Amphibole epidote, zoisite, chlorite, albite, green spinel | 40.61 | 16.00 | 10.22 | 0.12 | 9.74 | 16.37 | 0.60 | 0.27 | 0.04 | 0.55 | 4.10 | 98.62 | R 32 |
| R 100 | 500 | 144 | | Hornblende (61 %), plagioclase (5 %), chlorite (3 %), diopside (6 %), titanite (6 %), epidote, zoisite (11 %), sericite (2 %), quartz (1 %), ore minerals (4 %) | 45.35 | 11.33 | 14.58 | 0.185 | 11.63 | 7.73 | 2.22 | 0.93 | 0.02 | 2.78 | 1.20 | 97.95 | R 100 |

* Analyst: M. Tarkian, Dept. of Mineralogy and Petrology, University of Hamburg
** Analyst: S. Vimonlohakarn, V. Sriroongnung, DMR

### 4.1.2 *The Quartzitic Formation*

This formation covers extensive terrain in the southern and in the western parts of the area where its thickness is inferred to exceed 500 m. It consists of slightly metamorphosed sandstones with minor intercalations of phyllite. The formation is devoid of volcanic intercalations. Due to the lack of a transitional facies, a disconformable relationship may be inferred between the metavolcano-sedimentary and the quartzitic formations. Erosion of the metamorphosed clastics commenced prior to Permian times. No fossil finds have yet been recorded within the limits of this area. Therefore, the age of the quartzitic formation is uncertain. According to Baum et al. (1970), "clastic sedimentation in basin areas continues from the Devonian to the Carboniferous without any conspicuous break".

## 4.2 Permian Strata

The metamorphosed Palaeozoic formations are unconformably overlain by Permian strata which have been mapped west of Wang Nua Chiang Rai. Three formations may be distinguished:

### 4.2.1 *The Conglomerate Formation*

The oldest formation is of conglomerate composed of large white quartz pebbles. In the area south of Doi Mok, large slabs of metasediments of the metavolcano-sedimentary formation are also part of this conglomerate. In places, there are intercalations of fine-grained clastic strata. The thickness of the formation varies from 0 to 200 m.

### 4.2.2 *The Transitional Formation*

Its facies is of red marl and siltstone and thinly bedded sandstone. Thinly bedded limestones occur at the top. In the sandstone, pelicipods *Bakevellia* sp[3] and *Merismopteria* sp[4] have been found. The former indicate an age range from Permian to Cretaceous but the latter proves a Permian age (Table 3). Both fossil groups have been found north of the road connecting Ban Mae Kachan and Wang Nua.

In the thin limestone beds, *Rugososchwagerina yabei* (STAFF) proving a Middle Permian age have been found close to the above-mentioned locality (Gebert, 1982). The thickness of this formation is estimated to be of the order of 300 m.

---

[3] Determined by E. Kemper, BGR, Hannover

[4] Determined by R. Ingavat and J. Chamnongthai, Geological Survey Division, DMR, Bangkok

**Table 3.** Palaeontological record, Nam Mae Lao valley area

| Fossil Sample No. | Co-ordinates X | Y | Rocktype | Permian stratigraphic unit | Fossil identified | Age |
|---|---|---|---|---|---|---|
| F 1 | 616 | 227 | Limestone | Transitional formation | *Fusulinids:* | |
| | | | | | * *Rugososchwagerina yabei* (STAFF) | Middle Permian |
| | | | | | ** *Robustoschwagerina tubina*** | Sakmarian |
| | | | | | ** *Pseudofusulina vulgaria* | |
| | | | | | ** *Nankinella* sp. | |
| | | | | | ** *Triticites* sp. | |
| | | | | | ** *Eoparafusulina* sp. | |
| | | | | | ** *Chusanella* sp. | |
| | | | | | ** *Boultonia* sp. | |
| | | | | | *Foraminifera:* | |
| | | | | | ** *Eotuberitina reitlingarae* | |
| | | | | | ** *Textularia* sp. | |
| | | | | | ** *Dimacammina* sp. | |
| | | | | | ** *Cribrogenerina* sp. | |
| | | | | | ** *Globivaloulina* sp. | |
| | | | | | ** *Pachyphloia* sp. | |
| | | | | | ** *Pachysphaerina* sp. | |
| | | | | | *Corals and Algae:* | |
| | | | | | ** *Vermiporella* sp. | |
| | | | | | ** *Macroporella* sp. | |
| | | | | | ** *Gyroporella* sp. | |
| F 2 | 632 | 183 | Grey shale | Transitional *formation* | *Pelicipods:* | |
| | | | | | * *Bakevillia* sp. | Permian to |
| | | | | | ** *Merismopteria* sp. | Cretaceous |
| F 3 | 587 | 198 | Red shale | Middle Series | | Permian |

Analysts: * E. Kemper, BGR. ** R. Ingawat, J. Chamnongthai, Geological Survey Division, DMR

### *4.2.3 The Limestone Formation*

The youngest Permian formation is of massive limestone (Ratburi Group). The massive limestone west of Chiang Rai has been dated as Middle Permian based on the identification of *Callowayinella meitienensis*, CHAN and *Palaeofusulina sinenses*, SHENG (Baum et al., 1970). The thickness of the limestone formation mapped west and south-west of Chiang Rai exceeds 100 m.

## 4.3 Cenozoic Strata

On the eastern slopes of the Nam Mae Lao valley, Neogene gravels and sands with intercalations of silt disconformably overlie the Permian and older formations. These deposits are likely to be of fluviatile origin. Extensive plains made of Quaternary clastic sediments have been mapped in the Nam Mae Lao valley and in the surroundings of Chiang Rai. In places, secondary cassiterite is mined.

The lower slopes of the steep granite mountains north of Chiang Rai are covered with thick colluvial deposits, including landslides.

## 5. Intrusives

Granites of Late Permian to Early Triassic age of northern Thailand are mainly S-type (Beckinsale et al., 1979). In the mapped area, biotite granite and leuco-granite may be distinguished. The term leuco-granite was applied by Boom et al. (1980) to granites which contain less than 0.6% MgO.

Granite contacts with the metavolcano-sedimentary formations are discordant and sharp. Contact metamorphic aureoles do not extend more than a few cm into the intruded metavolcano-sedimentary formation indicating a high level emplacement for the granite. A greisen facies has not been observed in the area investigated. Both the leucocratic and the biotite granites are cut by quartz veins. Their thickness varies from a few mm to several m. The preferred strike directions are NS and NE-SW.

### 5.1 Biotite Granites

The granitic country rock of the mapped area is part of the huge Fang-Mae Suai biotite granite pluton, outcrops of which extend over a distance of 250 km in a N-S direction. The granite continues below a cover of pre-Permian metavolcano-sedimentary strata, E of the Nam Mae Lao valley.

Boom et al. (1980) distinguish four principal types of biotite granite: Porphyritic biotite granite of the normal type, a biotite-rich variety, biotite granite with a low biotite content, and two-mica granite with a marked biotite content. There are also rare varieties of fine to medium grained biotite granite which contain hornblende. The typical biotite granite is porphyritic. The alkali feldspar porphyroblasts may reach 10 cm in size. There is a tendency towards the development of medium to fine grained textures near and at the contact zone. Xenoliths of metasediments occur throughout the granite pluton.

The Rb : Sr age of the Fang-Mae Suai granite pluton is 232 ± 31 Ma, according to Braun et al. (1976). The high initial $Sr^{87}/Sr^{86}$ ratio of 0.725 indicates its anatectic origin, from a continental basement.

### 5.2 Leucogranites

The minor acid plutonic rocks are medium to fine grained leucogranites. They intrude into the biotite granites along their eastern margins and form several stocks near Mae Chedi and Wiang Pa Pao. The age of the leucocratic granite stocks is unknown. It is inferred to be of the same age range as the biotite granites contrary to former interpretations (Gebert, 1980).

Boom et al. (1980) recognized a great variety of rocks such as leucocratic two-mica granites, muscovite granites and muscovite tourmaline granites. Within the leucocratic granites, a separate group of aplites has been identified.

The leucogranites exhibit the following features:

- extremely strong hydrothermal kaolinization of feldspar
- development of pyrite stockwork mineralization
- occurrence of black tourmaline
- presence of magnetite in variable amounts (Asnachinda, 1978)

The leucogranites occur at low morphologic levels, which is explained by more rapid weathering of the kaolinized granites compared with the unaltered biotite granites.

## 6. Structural Features

The major Variscian structure of the area investigated is the Wiang Pa Pao anticlinorium. The fold axes of the pre-Permian strata follow a N-S direction and gently plunge to the south. However deformation patterns differ greatly in the quartzitic and in the metavolcano-sedimentary formations. There are open folds with medium to steep dips of the limbs in the former. The latter is tightly folded and in general the limbs are isoclinal. The axial planes of the folds exhibit a "meiler" type of orientation. There is a westerly dip in the west and an easterly dip in the east.

Major differences in the degree of metamorphism may be observed with respect to both formations. Only phyllites are noted in the quartzitic formation. However schists, amphibolites and even gneisses are found in the metavolcano-sedimentary formation.

Both deformation and metamorphism may presumably be related to the collision in Carboniferous time of the volcanic arc and the cratonic Shan Thai Paraplatform. In the area investigated, strata of proven Middle Permian age are also folded. In general fold axes are parallel to those of the metamorphosed pre-Permian strata. The fold axial planes dip to the west and there is a gentle plunge of the axes towards the south. These structures originate from orogenic movements in Late Mesozoic time.

Major young faults have placed Permian sediments in contact with pre-Permian metamorphics. The faults developed in N-S and NNE-SSW directions and gave rise to the deposition of Neogene basin sediments, the thickness of which is known to exceed 1000 m elsewhere in Northern Thailand.

The youngest phase of tectonic movement detected in the area displaced Neogene sediments on the eastern side of the Nam Mae Lao valley. Therefore, tectonic movements continued into the Pleistocene.

Joints are strongly developed both in the Palaeozoic sedimentary strata and in the granites. Four major sets of joints have been observed. The predominant sets developed in the 50° to 70° and 300° to 330° directions, while the other sets are in the 340° to 360° and 80° to 90° directions. Another pattern developed in a N-S direction along the eastern margin of the plutonic belt.

## 7. Stratigraphically-Controlled Mineral Deposits of the Pre-Permian Metavolcano-Sedimentary Formation

There are two types of stratigraphically-controlled mineral deposits:

- Metasediment-hosted stratabound deposits
- Amphibolite-hosted stratabound deposits

They also occur in combination.

### 7.1 Metasediment-Hosted Stratabound Deposits

#### *7.1.1 Mae Tho Cassiterite Mine*

The cassiterite occurrence was discovered by Black Muser hilltribe people in 1980. Since then cassiterite ore has been extracted both from placer deposits and primary sources in a narrow valley of a tributary of the Nam Me Tho stream.

The primary cassiterite-bearing formation is exposed over a length of up to 300 m and a thickness of up to 25 m on the lower slopes on both sides of the valley. There is another zone of outcrops 400 metres to the W which appears to be the continuation of the former.

The mine area and its surroundings are made of deeply weathered and decomposed quartzites and schists of the metavolcano-sedimentary formation. A rhyolite was seen in a small exposure within quartzitic strata above the mine face on the western slope of the valley. Granodiorite porphyry samples were found in the placer deposits. No granite has been detected in the surroundings of the mine area. The nearest granite exposure, of biotite granite, is found in a deep valley at a distance of 1.5 km from the mine.

In the mine face a yellowish to brownish weathered fine grained quartzite and a grey to dark brown weathered, fine layered, argillaceous schist can be distinguished. These rock types occur in cyclic repetitions. The thickness of their beds ranges from few cm up to 4 m. The thinly laminated, grey quartzites contain accessory plagioclase, sericite, epidote, biotite and tourmaline according to thin section analyses of fresh rocks. Tourmaline quartzite has been noted at places. Thin section investigations of the fresh schistose rock type proved to be biotite schist.

In the mine area, the strata dip to the E at angles between 10° and 30°. There are two sets of joints striking E-W and N-S respectively.

Primary cassiterite and associated sulphides were detected in a 25 m thick lens-shaped zone within the series of alternating quartzite and biotite schist. Cassiterite is enriched in the quartzite layers. The intercalations of biotite schist contain traces of cassiterite. Hanging and footwall rocks of the mineralized zone are made of the same type of quartzite and schist as the stanniferous host-rocks.

Cassiterite is associated with sulphides both in the form of disseminated grains and as veinlet fillings in the quartzites. Microscopic studies have shown that disseminated cassiterite is xenomorphous and fine grained (10–50 $\mu$). Aggregates of cassiterite grains are enclosed by quartzite laminae parallel to the bedding planes. A few grains of scheelite have been noted. The veinlet type of mineralization is developed in an E-

W direction. The discordant crosscutting relationships are proof of the remobilisation and redeposition of metals during regional metamorphism that has affected the host rock formation (Hutchinson, 1984). Cassiterite crystals up to several mm in size, sometimes twinned, are found to occur together with sulphides and quartz. In general, fissures of the N-S direction lack fillings of quartz and ore minerals.

The Sn content of samples from the mineralized quartzitic layers ranges from 50 to 5000 ppm. The average grade of the mineralized quartzites on both sides of the valley approximates 1000 ppm Sn, based on the analyses of 100 pick and channel samples.

### *7.1.2 Doi Mok Scheelite Mine*

Since discovery of the deposit in 1970, some 5000 tonnes of scheelite concentrate have been extracted up to mid-1978 (Suensilpong and Jungyusuk, 1981). After that period production declined sharply and virtually ceased in 1980.

In the mine area, strata of the pre-Permian metavolcano-sedimentary formation and a biotite granite are exposed. The metavolcano-sedimentary formation at Doi Mok is of marble, calc-silicate, quartzite, quartzite schist, amphibolite and basic and acid metatuff and meta-tuffite. These strata strike N-S and dip to the E. There are also dacitic acid volcanic rocks in the Doi Mok mountain area (Tarkian, 1981). The exposure of the granite extends westerly from the mine and covers an area of 2 $km^2$. This mass is part of the huge N-S trending Mesozoic Fang-Mae Suai pluton.

There are 3 scheelite-bearing horizons concordantly interbedded with barren strata in the metavolcano-sedimentary formation (Fig. 2).

#### *Scheelite-Bearing Zone 1*

This zone is found on the Doi Mok mountain ridge close to the contact with the granite. Mine mapping has shown that the scheelite ore body had been fully extracted by the end of 1980.

The footwall of the former ore body is preserved only in few places. Thin section analyses of a fresh, dark grey schistose rock sample yielded quartz, hornblende and plagioclase as the main components indicating its volcanogenic source.

The hanging wall of the former ore body is a marble horizon which attains a thickness of up to 30 m and extends over a distance of 140 m in the mine area. The marble exhibits banded structures parallel to the general dip of 50° E. In the northern part of the mine area, pillow-shaped amphibolites are intercalated in the marble lens. There are also intercalations of calc-silicates in the marble. Thin greenish amphibolite layers are also developed. In the hanging wall of the marble acid metatuffites have been identified during microscopic studies. In some places joint surfaces are partly coated with scheelite. There is also a weak fine grained disseminated scheelite mineralization forming small lens-like and layered structures. The latter feature has been observed particularly in or near amphibolite and calc-silicate layers. Accessory minerals detected are sulphides and fluorite. No economic scheelite mineralization has been found in adits driven into this marble formation and associated rocks.

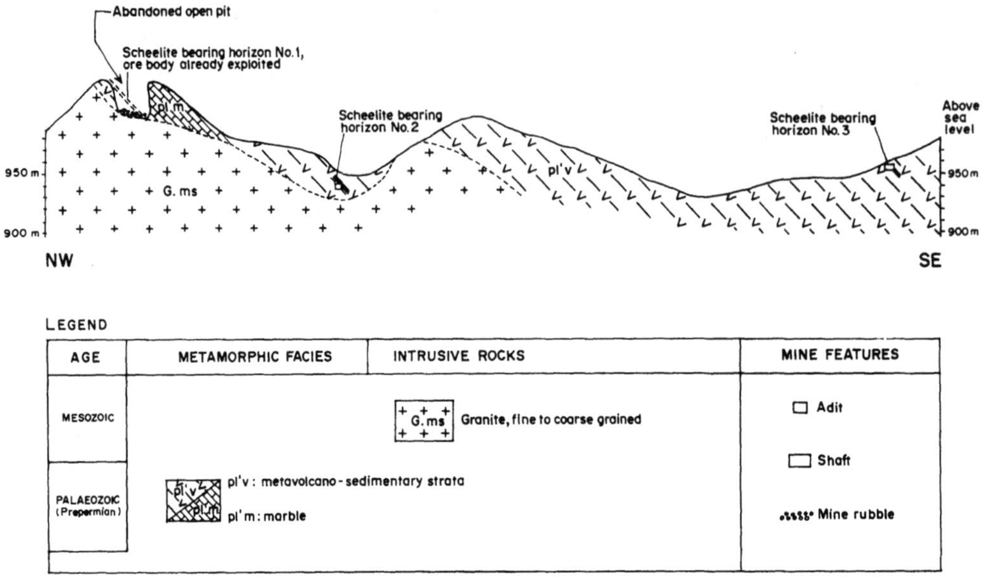


**Fig. 2.** Geological cross section through Doi Mok mine area, Wiang Pa Pao district, Chiang Rai province

From the old mining excavations it is seen that extraction of the ore body was conducted in the strike and dip direction of the strata which is N-S and 50°—60° E, respectively. Hence a concordant relationship between the former ore body and the enclosing rocks is being considered, indicating that the ore body was a mineralized member of the metavolcano-sedimentary formation. A thickness of the ore-bearing horizon of locally in excess of 2 m is inferred. Its length was about 100 m. Samples of the former ore body originate from a collection of the miners and from the dump. In general, scheelite is of the fine disseminated type. There are also coarse grained patches of scheelite enrichments. Many of the samples collected were high grade scheelite ores. The ores are greyish, sometimes with small light grey to brownish spots. A thinly laminated type of host-rock has also been observed. Sulphides are associated with scheelite. The bulk of the sulphides is of the fine to coarse grained disseminated type. Some of the samples show thinly layered structures. Massive sulphides have also been observed. Pyrrhotite, arsenopyrite, pyrite, marcasite and chalcopyrite have been recognized. Jivathanond (1981) and Suensilpong and Jungyusuk (1981) include sphalerite and galena as additional accessories.

Thin section analysis carried out at BGR headquarters by MÜLLER shows that the host-rocks of the scheelite/sulphide ore body were calc-silicates.

Previous descriptions of this scheelite deposit by Eberle (1982) refer to 3 different kinds of mineralization: the scheelite quartz vein, zones of scheelite mineralization and skarn mineralization. Jivathanond (1981) and Suensilpong and Jungyusuk (1981) adopt Eberle's (1972) views. The original ore body has been interpreted by them to be the scheelite quartz vein. However our investigations have shown that the original ore body was a scheelite and sulphide-mineralized calc-silicate member of the metavolcano-sedimentary formation. The scheelite quartz vein shown by Eberle in his Fig. 3

does not exist. The presence of the "zones of scheelite mineralization" in the granite is confirmed by us. But their occurrence is restricted to the near contact area of the original ore body with the granite. The "skarn mineralization" of Eberle (1972) is identical with the thin and weakly mineralized layers and joints developed in the marble of the hanging wall of the original ore body.

*Scheelite-Bearing Zone 2*

At a distance of 150 m SE of the abandoned pit, scheelite bearing zone No. 2 was identified in a small exploration adit made by the concession owner. This ore show is close to the granite. The scheelite and sulphide mineralization is restricted to a 0.5 m thick layer which is concordantly interbedded in the metavolcano-sedimentary formation. The host-rock is grey to brownish calc-silicates exhibiting thinly banded structures. In the footwall and the hanging wall of the ore horizon calc-silicates with thin intercalations of amphibolitic layers occur. The ore horizon is of several conformable bands up to 4 cm thickness with varying scheelite and sulphide contents. Scheelite and the associated sulphides are disseminated and fine to medium grained. The sulphide facies consists of pyrrhotite, pyrite, arsenopyrite and chalcopyrite.

*Scheelite-Bearing Zone 3*

At a distance of 500 m SE from the abandoned open pit mine site, scheelite was found by early explorers on the densely forested slopes of Doi Mok mountain. This occurrence has been recognized as an additional scheelite-bearing zone in the metavolcano-sedimentary formation at Doi Mok and has been named zone No. 3 by us. It is situated far from the granite. Corresponding to the previously described ore horizons, there is a concordant relationship between ore-bearing zone No. 3 and the enclosing strata. The scheelite-bearing horizon is low in grade and does not exceed a thickness of 1.5 metres in the abandoned pitting area. The host-rock and the enclosing foot- and hanging-walls of the ore bed are greyish, fine grained and thinly layered quartzites. The host-rock is characterized by tuffitic intercalations. Thin section analyses have proven high feldspar and chloritized biotite contents. Layers rich in tourmaline and sericite have also been identified. In places the quartzite is enriched with graphite.

The scheelite is of both the fine disseminated and the thinly banded type. The associated sulphides (pyrrhotite, arsenopyrite, pyrite, marcasite, chalcopyrite) exhibit the same disseminated and thinly layered structure. Minor scheelite and sulphide mineralization is also observed as thin veinlets, together with quartz, proving that the stratabound ore horizon was subjected to regional metamorphism (Hutchinson, 1984).

### *7.1.3 Tha Ko Scheelite-Stibnite Occurrence*

At Tha Ko mineral occurrence a small scale mining operation was conducted during 1979 and 1980. The site was abandoned in 1981 owing to depletion of the ore.

The country rock is made of metavolcano-sedimentary strata of pre-Permian age. Detailed geological mapping has delineated a 400 m long and up to 50 m thick marble lens intercalated in deeply weathered greenish metatuff and metatuffite. The marble is medium to coarse grained, whitish to grey and partly laminated. Intercalations of thin greyish calc-silicate layers have been observed in the marble close to its footwall. This marble lens may be correlated with the marble of Doi Mok. The strata strike N-S and dip to the E at an angle of 10 to 35° at and close to the open pit site.

Small amounts of scheelite and stibnite were extracted in 1979 and 1980 from soil and from cavities in the southern part of the marble lens. During mine mapping, no ore exposures were seen. It is inferred that the primary scheelite occurrence was identical to the type of minor scheelite mineralization found in the marble at Doi Mok in the hanging wall of the scheelite ore body already exploited. The sulphide mineral extracted at Tha Ko was stibnite. In contrast to Doi Mok mine, granite exposures are unknown at Tha Ko.

## 7.2 Amphibolite-Hosted Stratabound Deposits

### *7.2.1 Mae Chedi Scheelite-Cassiterite Mines*

In the general area SW of Ban Mae Kachan, mining was conducted in several places. One of the major open pits was the Kumplin mine which terminated operations in 1981 due to lack of ore reserves.

The general mine area of Mae Chedi is of a huge mass of amphibolite which is a member of the pre-Permian metavolcano-sedimentary formation. The amphibolite has been intruded by granite. Ore has been extracted in Mae Chedi mines from deeply weathered rocks of both amphibolite and the granite's marginal parts. South and SE of Kumplin mine, scheelite and cassiterite have also been mined from decomposed amphibolite, where granite had not been encountered. A basaltic ancestry is indicated for the amphibolite because it contains relics of pyroxene porphyroblasts. In the weathered amphibolite scheelite was seen in fracture fillings in the abandoned open pit 400 m E of Kumplin mine. The trend of these fractures is N-S, in general. A finely dispersed additional scheelite- and cassiterite-type of mineralization is inferred to be contained in parts of this decomposed rock type. This view is based on the locally high Sn and $WO_3$ values up to more than 1000 and 2000 ppm, respectively, of samples taken from amphibolitic terrain.

In the abandoned Mae Chedi open pits E of highway No. 1019, ore has also been extracted from amphibolite. Mining excavations terminated at or very close below the interface with the underlying granite after which the rocks were barren.

## 7.3 Combined Stratabound Deposit Types

### *7.3.1 Tara Scheelite Mine*

Scheelite and cassiterite had been recovered from nearby placer deposits during the past decade. In 1981 trial open pit mining operations were commenced to extract primary ore.

In the trial open pit, strata of the metavolcano-sedimentary formation are exposed up to a thickness of about 100 m over an area totalling 3000 $m^2$. The beds strike N-S and show a vertical dip. To the E of the pitting area a porphyritic biotite granite, which is a part of the huge Fang-Mae Suai granite pluton, is seen in contact with the metavolcano-sedimentary formation. In the surroundings of the zone of pits, both the granite and the metavolcano-sedimentary formation are covered by a thick blanket of alluvial sediments.

The strata of the metavolcano-sedimentary formation at Tara mine are of deeply weathered yellowish and brownish rocks. In rare cases small outcrops show dark grey and thinly layered fresh rocks of amphibolitic composition. The mineralization appears to be confined to individual horizons interbedded with barren strata. There is only low grade scheelite mineralization exposed in the current open pit. Thin section investigations indicate that the hostrocks are amphibolites with thin intercalations of calc-silicates. The minerals detected are plagioclase, green hornblende, diopside and epidote.

Scheelite occurs disseminated and in small veinlet fillings in the host-rock. In a few places small grains of cassiterite have been seen in veinlets, sometimes together with quartz. 40 samples, collected at random by means of channel and pick sampling from the mineralized metavolcano-sedimentary formation in the trial mining site, averaged 500 ppm $WO_3$ and 20 ppm Sn.

The biotite granite contains quartz-feldspar veinlets up to several mm thickness. There are traces of scheelite in veinlets near the granite contact with the mineralized metavolcano-sedimentary formation. 30 channel and pick samples from this part of the granite averaged 123 ppm $WO_3$ and 13 ppm Sn.

It is not known whether the mineralized zones in the metavolcano-sedimentary unit extend beyond the limits of the present pit and towards depth. Such extension may be inferred in view of the pattern of mineralization.

## 8. Mineralization Affiliated with Post-Palaeozoic Granites

In the Nam Mae Lao valley ore province tin and tungsten mineralization is also affiliated with granites that postdate regional metamorphism. This type of mineralization is restricted to sites where granites are in contact with metasediment- or amphibolite-hosted Palaeozoic stratabound mineralization. Such mineralization is restricted to granite margins, and there is no extension in the granite neither in depth nor laterally.

### 8.1 Doi Mok Scheelite Mine

The Doi Mok biotite granite is in contact with amphibolite, meta-tuffite, calc-silicates, the marble horizon and low grade remnants of ore-bearing zone No. 1. Xenoliths of calc-silicates and marble up to several metres in length occur in the granite. The contact between the granite and the metavolcano-sedimentary formation is discordant and sharp. The contact metamorphic alteration zone does not exceed 10 cm in all mine exposures.

**Table 4.** Whole rock analyses of samples from the Doi Mok granite

| Adit No. | % $SiO_2$ | % $Al_2O_3$ | % FeO | % MnO | % CaO | % MgO | % $Na_2O$ | % $K_2O$ | % $TiO_2$ | % LOI | % Total |
|---|---|---|---|---|---|---|---|---|---|---|---|
| 1 | 65.9 | 13.7 | 3.28 | 0.14 | 2.94 | 1.90 | 2.86 | 3.88 | 0.55 | 1.7 | 96.85 |
| 1 | 70.8 | 12.8 | 0.71 | 0.06 | 0.63 | 0.16 | 3.69 | 4.43 | 0.33 | 0.3 | 93.91 |
| 2 | 67.9 | 13.2 | 3.17 | 0.19 | 1.96 | 1.74 | 2.75 | 4.30 | 0.33 | 2.2 | 97.74 |
| 3 | 67.6 | 12.8 | 3.51 | 0.19 | 0.98 | 1.08 | 2.75 | 4.82 | 0.43 | 1.9 | 96.06 |
| 4 | 67.8 | 13.5 | 3.08 | 0.19 | 1.61 | 1.69 | 2.94 | 4.27 | 0.53 | 1.9 | 97.51 |

(LOI = loss on ignition)

Analyst: W. Wisch, BGR

The main constituent minerals of the granite are quartz, alkali feldspar, plagioclase and biotite. Near the contact zone different varieties of coarse to fine grained granites have been observed. There are also granites which lack biotite but contain chlorite. The chlorite is inferred to originate from biotite during the pneumatolytic and hydrothermal stages of alteration. By means of thin section investigation, microcline, tourmaline apatite, zircon, epidote, leucoxene, sericite, muscovite, carbonate, fluorite, and titanite have been recognized. Whole rock analyses of five granite samples collected from exploration adits are shown in Table 4.

Scheelite mineralization in the granite is of both veinlet and disseminated types. The disseminated type characterized by xenomorphic scheelite grains prevails. Associated sulphides are pyrrhotite, arsenopyrite and pyrite. Pyrrohotite occurs in globular aggregates; pyrite and arsenopyrite are found as idiomorphic crystals up to several mm in size. In addition a small amount of chalcopyrite has been detected in polished sections. The granite-hosted mineralization is known only where granite is in contact with the original ore body. Adits driven into the granite have proven that there is no extension of the mineralization into the granite beyond a few metres (max. 5 m) from its contact with the original ore body, neither laterally nor in depth.

## 8.2 Mae Chedi Scheelite-Cassiterite Mines

Granite varieties identified in thin sections from Kumplin mine are fine to medium grained two mica granite, porphyritic two mica granite and biotite-tourmaline granite. There are neither pegmatites nor greisen. Scheelite and cassiterite mineralization is contained in the quartz veins and veinlets in the granite. Accessory ore minerals are wolframite, pyrite, bismutite and sulphosalts. This type of mineralization is confined to the marginal zone of the granite and does not extend into the intrusive rock for more than 10 or 15 metres below the amphibolite cap as seen in the mine pits and shafts at all mining sites in the Mae Chedi area. Furthermore the mineralization within the granite is restricted to where the amphibolite is mineralized. In the Kumplin mine the veins strike ENE and dip vertically.

## 9. Interpretation of Ore Genesis

Previous work on tin and tungsten in northern Thailand (Bundesanstalt für Bodenforschung, Hannover, 1972; Eberle, 1972; Hosking, 1977; Mitchell, 1976, 1979; Asnachinda, 1978; Boom et al., 1980; Jivathanond, 1981; Suensilpong and Jungyusuk, 1981) has led to the conclusion that the primary mineral deposits formed as a result of granite intrusion. The deposits of northern Thailand are part of a mineralized belt which extends from eastern Burma through Thailand and Malaysia into Indonesia. Ore formation is restricted neither to a specific age of granite nor to a particular chemical composition. Braun et al. (1976) have shown that the tin- and tungsten-bearing granites of northern Thailand are of S-type. The ore deposits are located in the roof sections of granite cupolas and in adjacent host-rock formations. If granite had intruded into country rocks that are "impermeable to gas (volcanics), a possible Sn deposit would be expected within the granite, and in the presence of permeable (altered) country rock the possibility of deposit formation within the country rock must be taken into account" (Boom, 1980). The source of the metals is still in doubt.

Our own observations contradict the previously established model of a granitic origin of tin and tungsten ore deposits in northern Thailand, with respect to the Nam Mae Lao valley ore province:

1) The ore bodies and ore showings of the metavolcano-sedimentary formation are concordantly interbedded with barren members of this formation.
2) The amphibolite host-rocks are of the island arc type (MacDonald and Barr, 1978) and are intimately associated with the metasediments.
3) Scheelite, cassiterite and associated sulphides exhibit the same type of ore structures, i.e. fine to coarse grained disseminations and veinlet fillings, in addition to relic banding on both the micro and macro scales.
4) The scheelite, cassiterite and associated sulphidee ores of the metavolcano-sedimentary formation have undergone the same geological evolution as the enclosing rocks, i.e. diagenesis, folding and metamorphism. This is shown by folded ore layers, and by ore structures, resulting from regional metamorphism, such as veinlet fillings.
5) Folding and metamorphism pre-dates granite intrusion. This is shown by the pre-Permian age of the metavolcano-sedimentary formation and the radiometric Triassic age of the intrusive Mae Suai-Fang granite pluton.
6) There are sediment-hosted ore deposits where granites are absent.
7) Mineralization in granites is restricted to such places where granites are in contact with sediment or amphibolite-hosted mineral deposits. However such mineralization is confined to the granite margin. There are no extensions of mineralization in the granite neither in depth nor laterally.

These observations led to the conclusion that the sediment- and amphibolite-hosted tin and tungsten deposits of the Nam Mae Lao valley are not epigenetic replacement deposits affiliated with Triassic granite, but they originated from metal-depositing sea floor exhalative systems related to basic volcanic activity in an island-arc type of environment in pre-Permian Palaeozoic time. This type of environment is indicated by the geotectonic setting of the ore province and the calc-alkaline composition of the basic volcanites. Minor mobilisation and migration of the metals took place in pre-Permian Palaeozoic time. This led to the development of both the coarse grained disseminated and the veinlet types of sulphide, scheelite and cassiterite ores. In the latter case, discordant crosscutting relationships developed, proving the exposure of the ores to regional metamorphism (Hutchinson, 1984). The intrusion of Triassic granites into the metamorphic Palaeozoic ore beds gave rise to the assimilation of tin and tungsten by the granite and led to the development of the margin type granite mineralization.

Maucher (1965) recognized that the stratigraphically controlled type of scheelite deposit occurs mainly in Early Palaeozoic host-rocks and bears a genetic relationship to submarine basic volcanic activity. This type of volcanism "is observed to be associated with geotectonic lineaments characterized by Benioff zones of Cordilleran type or of island arcs" (Maucher, 1976). We interpret the stratigraphically-controlled ore bodies of the Nam Mae Lao valley area as deposits of the "Sb-W-Hg-Formation" in the sense of Maucher (1965). However, in this ore province Hg is substituted by Sn. Stibnite is found both within the area investigated (Tha Ko mine) and in adjoining areas of Wang Nua district in Lampang province. At Huai Hia, Huai Pong and Mae Taka, stibnite has been extracted from stratiform meta-tuffitic host-rocks that are associated with meta-silexites. At Sri Thoranee mine, stibnite and associated scheelite are volcanite-hosted.

According to Maucher (1976) "an augmented Au and Ag content of the scheelite-bearing hornblendites is typical, especially in horizons enriched in sulphides". In the Nam Mae Lao valley ore province, gold has been extracted by local prospectors both near Wang Nua and in proximity to Wiang Pa Pao.

*Acknowledgements.* The result presented were obtained from studies carried out by the advisory group of the Federal Institute of Geosciences and Natural Resources (BGR), Hannover within the framework of bilateral technical co-operation between the Federal Republic of Germany and the Kingdom of Thailand.

We wish to acknowledge support in the field by S. Chiemchindaratana, P. Sukvattanunt and B. Yokart, geologists of the Department of Mineral Resources, Bangkok. Sincere thanks are expressed to W. Wisch, BGR Hannover and V. Sriroongrueng, P. Suwimonprecha and S. Vimonlohakarn, DMR, Bangkok for the chemical analyses of rocks, ores and prospecting samples.

We are indepted to P. Müller and Th. Weiser of BGR, Hannover and to M. Tarkian of the University of Hamburg for the result of their microscopic and ore microscopic investigations. R. Ingawat, J. Chamnongthai, DMR, Bangkok and E. Kemper, BGR, Hannover identified the fossils collected by us. Special thanks are due to P. Wittekind of BGR, Hannover for his many helpful discussions on both the regional geology and the metallogeny of northern Thailand and for his critical review of the manuscript.

## References

Asnachinda, P., 1978. Tin mineralization in the Burmese-Malayan peninsula - a plate tectonic model. - *Proc. 3rd Reg. Conf. Geol. Mineral Res. SE Asia,* 293—299, Bangkok, Asian Instit. Techn.

Basaltic Volcanism Study Project, 1981. *Basaltic Volcanism on the Terrestrial Planets.* Pergamon Press, Inc., New York, 1286 pp.

Baum, F., Braun, E.V., Hahn, L., Hess, A., Koch, K.E., Kruse, G., Quarch, H. and Siebenhüner, M., 1970. On the geology of northern Thailand. - *Beih. Geol. Jb.* 102, Hannover, 23 pp, 1 table, 1 map.

Baum, F., Braun, E.V., Hahn, L., Hess, A., Koch, K.E., Kruse, G., Quarch, H. and Siebenhüner, M., 1970. On the geology of northern Thailand. - *Beih. Geol. Jb.* 102, Hannover, 23 pp, 1 table, 1 map.

Baumann, L., 1965. Zur Erzführung und regionalen Verbreitung des „Felsithorizontes" von Halsbrücke. - *Freib. Forsch.* Hft. C 186, 63–81.

Beckinsale, R.D., Suensilpong, S., Nakapadunrat, S. and Walsh, J.N., 1979. Geochronology and geochemistry of granite magmatism in Thailand in relation to a plate tectonic model. - *J. Geol. Soc. London,* 136, 529–540, 3 figs., 4 tables.

Boom, G.v.d., Rehder, S., Vetter, U. and Kottrup, G., 1980. Geochemical-petrographical evaluation of a hard rock survey in selected tin prospective granites in north Thailand. - *Federal Institute for Geosciences and Natural Resources,* Hannover, 32 pp. (unpublished report).

Braun, E.v., Besang, C., Eberle, W., Harre, W., Kreuzer, H., Lenz, H., Müller, P. and Wendt, I., 1976. Radiometric age determinations of granites in northern Thailand. - Geol. Jb. B. 21, 171–204, Hannover, 8 figs., 9 tables.

Bundesanstalt für Bodenforschung (Geological Survey of the Federal Republic of Germany), 1972. *Final report of the German Geological Mission to Thailand 1965–1971* - 94 pp., 17 figs., 2 maps, Hannover (unpubl.).

Bunopas, S., 1981. Palaeogeographic history of western Thailand and adjacent parts of South-East Asia. - A plate tectonics interpretation. - *Ph.D. thesis, Victoria University of Wellington, New Zealand,* 810 pp., 106 figs., 21 tables, 20 plates, 9 appendices.

Eberle, W.G., 1972. The scheelite deposit of Wiang Pa Pao, Chiang Rai province, northern Thailand (preliminary report). *Annex to newsletter of Geol. Soc. Malaysia,* 6–9, 3 figs., Kuala Lumpur.

Gebert, H., 1980. Tin and tungsten exploration in Chiang Mai and Chiang Rai provinces, northern Thailand. - Technical cooperation project No. 78.2208.3, *technical report* No. 7, 51 pp., 11 figs., Bangkok, July 1980 (unpublished report).

Gebert, H., 1982. Contributions to the geology of the Nam Mae Lao valley area and adjoining parts, Chiang Rai province, northern Thailand. - Technical cooperation project No. 78.2208.3, *technical report* No. 13, 11 pp., 5 figs., 2 tables, Hannover, August 1982 (unpublished report).

Hamilton, J.M., Bishop, D.T., Morris, D.G. and Owens, O.E., 1982. Geology of the Sullivan orebody. Kimberley, B.D., Canada. *In* Hutchinson, R.W., Spence, C.D. and Franklin, J.M. (eds.) Precambrian Sulphide Deposits. *Geol. Assoc. Canada Sp. Paper,* No. 25 (Robinson Mem. Vol.) 597–666.

Höll, R., 1971. Scheelitvorkommen in Österreich. - *Erzmetall* 24, 273–282.

Höll, R., 1975. Die Scheelitlagerstätte Felbertal und der Vergleich mit anderen Scheelitvorkommen in den Ostalpen. - Habilitationsschrift. *Bayer. Akad. Wiss., Math.-naturw.* Kl., Abh. N.F. 167 A, 114 pp.

Höll, R., 1977. Early Palaeozoic ore deposits of the Sb-W-Hg formation in the eastern Alps and their genetic interpretation. - *In: Time and Strata-Bound Ore Deposits.* Eds. Klemm, D.D. and Schneider, H.-J., Springer-Verlag, 169–198, 3 figs.

Hosking, K.F.G., 1977. The search for tungsten deposits. *Geol. Soc. Malaysia Bull.* 5, 70 pp., 18 figs., Kuala Lumpur.

Hutchinson, R.W., 1979. Evidence of exhalative origin for Tasmania tin deposits. *Canad. Inst. Min. Met. Bull.* Vol. 72, No. 808, 90–104.

Hutchinson, R.W., 1980. Evidence of exhalative origin for Tasmanian tin deposits. Reply to discussion. *Can. Inst. Min. Met. Bull.* Vol. 73, No. 815, 167–168.

Hutchinson, R.W., 1981a. Lode tin deposits of exhalative origin. *In* Hasbi A.b.Hj.H. and van Wees, H., (eds.): Complex tin ores and related problems. *SEATRAD Centre Tech. Pub.* No. 2, 107–136.

Hutchinson, R.W., 1981b. Lode tin deposits of exhalative origin. Reply to discussion. *In* Hasbi A.b.Hj.H. and van Wees, H. (eds.): Complex tin ores and related problems. *SEATRAD Centre Tech. Pub.* No. 2, 107–136.

Hutchinson, R.W., 1982. Geologic setting and genesis of cassiterite-sulfide mineralization at Renison Bell, western Tasmania. Discussion. *Econ. Geol.* Vol. 77, No. 1, 199–202.

Hutchinson, R.W., 1984. Massive sulfide deposits and their significance to other ores. Keynote paper, *GEOSEA V (Fifth Regional Congress on Geology, Mineral and Energy Resources of Southeast Asia),* Kuala Lumpur, April 9–13, 1984.

Jivathanond, S., 1981. Mineralography and geochemistry of the tungsten deposit at Doi Mok, Amphoe Wiang Pa Pao, Changwat Chiangrai. - *Master of Science Thesis, Chiang Mai University,* 78 pp., 17 figs., 7 tables, 17 plates (unpublished).

Koch, K.E., 1973. Geology of the Region Si Sawat-Thong Pha Phum - Sangkhlaburi (Kanchanaburi Province/Thailand). - *Bull. geol. Soc. Malaysia,* 6, 177—185, Kuala Lumpur.

Lea, E.R. and Rancourt, C., 1958. Geology of the Brunswick Mining and Smelting Orebodies. Gloucester County, *Can. Inst. Min. Met. Bull.,* Vol. 51, No. 551, 167—177.

MacDonald, A.S. and Barr, S.M., 1978. Tectonic significance of a late Carboniferous volcanic arc in northern Thailand. - *Proc. 3rd Reg. Conf. Geol. Mineral Res. SE Asia,* 151—156, Bangkok, Asian Instit. Technology.

Maucher, A., 1965. Die Antimon-Wolfram-Quecksilber-Formation und ihre Beziehungen zu Magmatismus und Geotektonik. - *Freiburger Forschungsh.* C 186, 173—188.

Maucher, A., 1976. The stratabound cinnabar-stibnite-scheelite deposits (discussed with examples from the Mediterranean region). *Handbook of Strata-Bound and Stratiform Ore Deposits,* Vol. VII, Elsevier, Amsterdam, 477—503.

Mining Journal, 1984. Greenland tungsten find. Vol. 303, No. 7791, 417.

Mitchell, A.H.G., 1976. Southeast Asian tin granites: magmatism and mineralization in subduction and collision related setting. *CCOP Newsl.* 3 (1—2), 10—14, Bangkok.

Mitchell, A.H.G., 1979. Rift, subduction and collision related tin belts. - *Geol. Soc. Malaysia Bull.* 11, 81—102.

Newnham, L.A., 1981. Lode tin deposits of exhalative origin. Discussion. *In:* Hasbi A.b.Hj.H. and van Wees, H., (eds.) Complex tin ores and related problems. *SEATRAD Centre Tech. Pub.* No. 2, 99—102.

Newnham, L.A., 1985. The western Tasmanian tin province with special reference to the Renison Mine (this publication).

Patterson, D.J., 1982. Geologic setting and genesis of cassiterite-sulfide mineralization at Renison Bell, western Tasmania. Reply to discussion. *Econ. Geol.* Vol. 77, No. 1, 203—206.

Patterson, D.J., Ohmoto, H., and Solomon, M.J., 1981. Geologic setting and genesis of cassiterite-sulfide mineralization at Renison Bell, western Tasmania. *Econ. Geol.* Vol. 76, No. 2, 393—438.

Plimer, I.R., 1980. Exhalative Sn and W Deposits Associated with Mafic Volcanism as Precursors to Sn and W Deposits Associated with Granites, *Mineral. Deposita* 15, No. 3, 275—289.

Ransom, P.W., 1977. Geology of the Sullivan ore body. Excursion guide, *Cominco Ltd., Kimberly,* 14 pp. (unpublished).

Solomon, M.J., 1980. Evidence of exhalative origin for Tasmanian tin deposits. Discussion. *Can. Inst. Min. Met. Bull.* Vol. 73, No. 815, 166—167.

Suensilpong, S. and Jungyusuk, N., 1981. Some aspects of scheelite-related granites. *Department of Mineral Resources,* Bangkok, 3 figs., 2 tables, 1 map (unpubl.).

Tarkian, M., 1981. Bericht über mikroskopische Untersuchung von Gesteinen aus Nord-Thailand. 6 pp. (unpublished).

Walker, R.R., Matulich, A., Amos, A.C., Watkins, J.J., and Mannard, G.W., 1975. The geology of the Kidd Creek Mine. *Econ. Geol.,* Vol. 70, No. 1, 80—89.

# 6.11.2 The Tin-Tungsten Granites of the Takua Pa Area, Southern Thailand

S. Nakapadungrat, N. Chulacharit, Y. Munthachit, T. Chotigkrai, and S. Sangsila[1]


## Abstract

The tin-tungsten granites of the Takua Pa area, southern Thailand, form a high level north-south elongated intrusion into Carbo-Permian pebbly mudstones. There are two groups of granites, the first group ($F_1$) consists of porphyritic biotite granite which has a coarse grained groundmass (G-1), fine to medium grained groundmass (G-2) and equigranular fine grained biotite granite (G-3); while the second group ($F_2$) consists of tourmaline-muscovite granite (G-4) and leucocratic granite (G-5).

$F_1$ granites have $SiO_2$ contents of 68–76%, 2.2–3.5% $Na_2O$ and 4.6–5.8% $K_2O$, whereas the $F_2$ granites have $SiO_2$ contents of 71–75%, 2.1–4.5% $Na_2O$ and 4.0–5.6% $K_2O$. $F_1$ granites are higher in CaO, FeO, Ba and Sr contents but lower in Rb, Li and Sn contents than the $F_2$ granites. Geochemical data reveal that $F_1$ granites were fractionated from G-1 to G-2 and to G-3 granites but $F_2$ granites have no fractionation trend.

S/W deposits, which are of pegmatite, quartz vein and disseminated types, are closely associated with tourmaline-muscovite (G-4) and leucocratic (G-5) granites. $F_1$ granites have no associated mineralization.


## Introduction

The Takua Pa area, southern Thailand, has been known as one of the richest tin-provinces for more than 50 years. From this province, about 150,000 t of tin concentrates have been produced during the last thirty years. However, the relation between the granites and tin-mineralization is poorly known. This paper presents the detailed geology of Khao Kata Khwam granites.

The study area (Fig. 1) is situated on the western side of the Thai Peninsula about 850 km south of Bangkok. It is located between latitude 8° 35′ N–8° 45′ N and longitude 98° 25′ E–98° 35′ E on the Royal Thai Survey topographic map sheets 4626II, Amphoe Thap Put, and 4626III, Amphoe Kapong, on a scale 1:50,000.

The granite batholith, which is elongated in a north-south direction, was emplaced in the Paleozoic strata of the Kang Krachan Formation. The sedimentary rocks consist of pebbly mudstone, laminated mudstone, shale, slate and sandstone. These rocks have been given an age of Permian-Ordovician (?) by Garson et al. (1975) or Carbo-

[1] Geological Survey Division, Department of Mineral Resources, Rama IV Road, Bangkok 10400, Thailand

Permian by Tantiwanit et al. (1983). These rocks have sharp contacts with the granites and are only slightly metamorphosed. This might indicate that the granites were emplaced at a high level.

## The Granites

The Khao Kata Khwam granites have been studied partly by Garson et al. (1975), Muenlek (1982) and Friedli and Jariyawat (1983). However, a detailed geological and geochemical study of the whole granite batholith was not made by them. The whole granite batholith has been mapped by Nakapadungrat et al. (1983) (as shown in Fig. 1). Five types of granites have been mapped, namely:

- Porphyritic biotite granite with coarse grained groundmass (G-1)
- Porphyritic biotite granite with fine to medium grained groundmass (G-2)
- Equigranular fine grained biotite (G-3)
- Tourmaline-muscovite granite (G-4)
- Leucocratic granite (G-5)

There are three types of primary tin deposit found in this area, i. e. dissemination, quartz-cassiterite-wolframite veins and cassiterite pegmatites. Disseminated cassiterite is found in the G-5 granite. Quartz-cassiterite-wolframite veins intrude both G-4 and G-5 granites. These veins, which trend in a NW-SE direction, range in width from 1 cm to 30 cm. The distribution of cassiterite pegmatites is similar to that of the quartz-cassiterite veins. They are found in the north and northwest of the granite batholith. The thickness of the pegmatites ranges from 20 cm to 2 m.

It is possible to classify the Khao Kata Khwam granites into two groups. The first group ($F_1$-granites) contains G-1, G-2 and G-3 granites. This group has no relation with tin-deposits. However, the second group ($F_2$-granites), that consist of the G-4 and G-5 granites, is closely associated with tin and tungsten deposits.

### Porphyritic Biotite Granite (G-1)

This granite is exposed at the middle and southern end of the batholith. It is porphyritic, with a medium to coarse grained groundmass (3—5 mm) containing subhedral to euhedral feldspar phenocrysts ranging in size from 1 x 3 cm to 3 x 5 cm. Quartz varies in size from 0.2 cm to 1 cm. Biotite, which is the major mafic mineral, ranges in size from 0.2 cm to 0.9 cm. Allanite, which occurs as blackish prismatic crystals, is occasionally seen. There are a number of mafic inclusions in this rock, ranging in size from 2 x 6 cm to 6 x 14 cm.

Petrographically, this granite has a hypidiomorphic-granular texture, and myrmekitic texture is also quite common. Most of the alkali feldspar is orthoclase and micro-perthite with small amounts of microcline. Plagioclase is oligoclase-andesine commonly with albite twinning. Zoning is normally observed. Quartz is mostly anhedral and commonly shows undulatory extinction. Biotite is redish brown to pale brown. Apatite and zircon occur commonly as inclusions in biotite. Allanite and sphene are important accessory minerals of this rock. Tourmaline, which is yellowish green, is occasionally seen. Chlorite, sericite and iron minerals occur as secondary minerals.

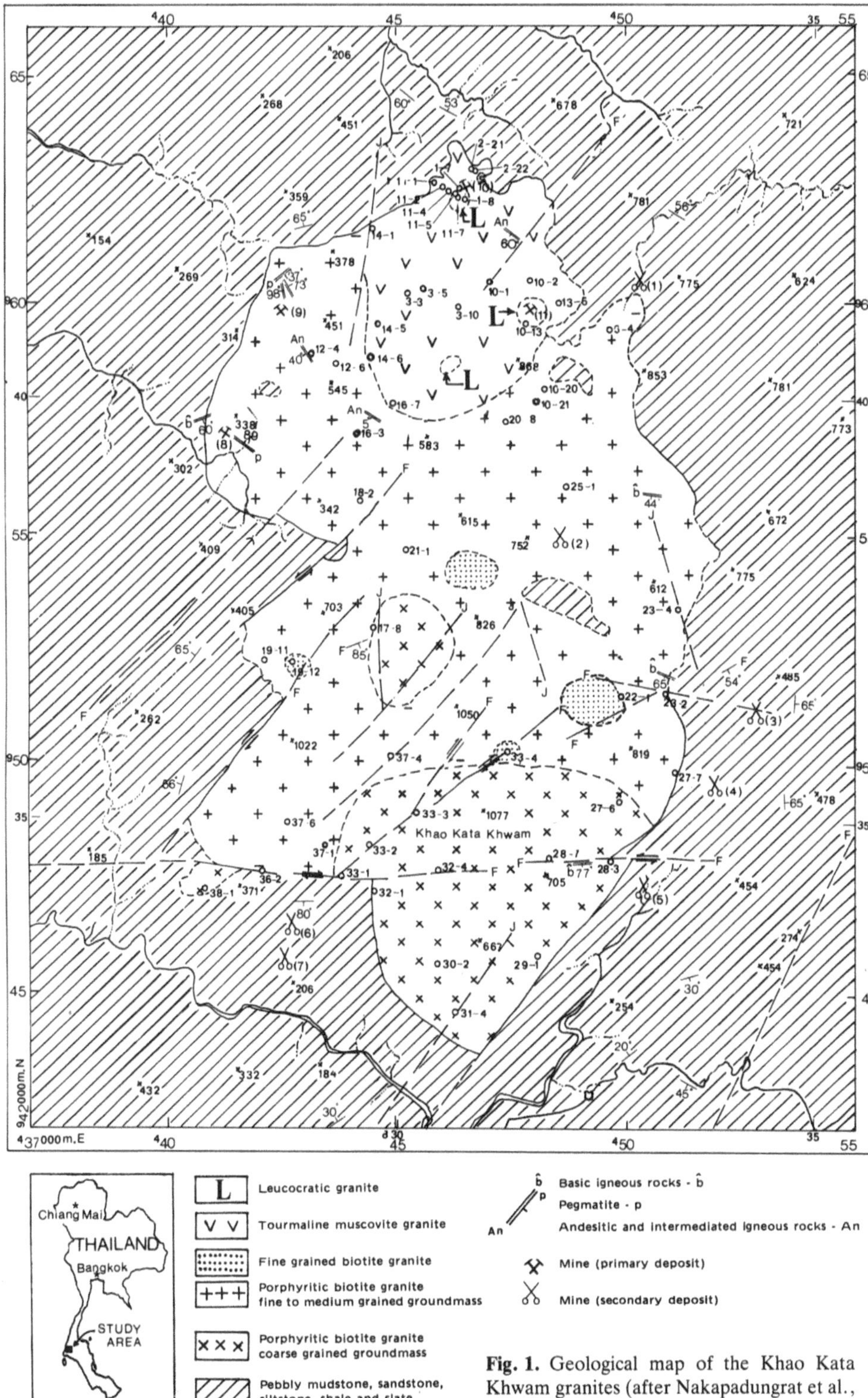


**Fig. 1.** Geological map of the Khao Kata Khwam granites (after Nakapadungrat et al., 1983)

### Porphyritic Biotite Granite (G-2)

This granite is the predominant type in the study area. This porphyritic rock has a fine to medium grained groundmass (1–3 mm) with feldspar crystals ranging in size from 0.4 x 1.2 cm to 1 x 2 cm. Quartz ranges in size from 0.1 cm to 0.6 cm with an average of 0.3 cm. Biotite occurs as clusters (0.4–0.6 cm) and as single crystals. It forms perfect hexagonal-shaped crystals (up to 0.4 cm) as well as small subhedral flakes.

In thin section, it has a hypidiomorphic-granular texture, and myrmekitic texture is also common. The alkali feldspar is microcline, microcline-perthite and microperthite. Plagioclase is zoned and strongly altered to sericite. Quartz is anhedral and shows undulatory extinction. Biotite, which is dark to pale brown, commonly has inclusions of zircon, monazite and apatite. The radioactivity of zircon and monazite make the pleochroic haloes completely black. A few biotite flakes are partially altered to chlorite and iron minerals. Muscovite commonly replaces biotite. Tourmaline is occasionally seen.

### Equigranular Fine Grained Biotite Granite (G-3)

This rock occurs as small stocks and dykes. It intrudes both the G-1 and G-2 granites. It is holocrystalline, and of a hypidiomorphic to allotriomorphic granular texture. The alkali feldspar is mostly microcline. Plagioclase is commonly altered to sericite and clay minerals. Quartz is anhedral. Biotite, which is subhedral to anhedral, is altered to chlorite, epidote and iron minerals. Apatite and zircon are accessory minerals and occur as inclusions in the biotite.

### Tourmaline-Muscovite Granite (G-4)

This rock is generally equigranular and medium to coarse grained. The feldspar ranges in size from 0.8 x 1.2 cm to 1.2 x 2.5 cm and is mostly subhedral to anhedral. Tourmaline crystals vary from 0.6 cm to 1.5 cm. The rock consists of quartz (27–342%), alkali feldspar (33–34%), plagioclase (22–24%), tourmaline (7–12%) and muscovite (3–5%). In some places, there is very little muscovite content so that the term tourmaline granite may be used.

It has a hypidiomorphic granular texture. The alkali feldspar is microcline and micro-perthite. Saussuritization after plagioclase is commonly observed. Quartz is anhedral and shows undulatory extinction. Tourmaline, which occurs as prismatic crystals, is yellowish to pale brown.

### Leucocratic Granite (G-5)

This rock is generally equigranular and fine to medium grained. All rock samples are white with a few spots of tourmaline. The alkali feldspar is microcline and orthoclase. The plagioclase is albite. Quartz, which shows undulatory extinction, and muscovite are essential minerals. Feldspar is commonly replaced by tourmaline. Cassiterite and opaque minerals are accessory minerals of this rock.

## Geochemistry

52 new analyses of major and trace elements from the Khao Kata Khwam granites are presented in Appendix 1 and 2, respectively.

The major elements were analysed by the wet or classical method but sodium and potassium were determined by the atomic absorption spectrophotometer at wavelengths 589 and 766.5 nm, respectively.

The trace elements were determined by the atomic absorption spectrophotometer after treating the rock samples with hydrofluoric and nitric acid. The U.S. Geological Survey Standard rock, G-2, $AGV_1$ and $GSP_1$ were used for calibration. The analysed trace element contents of these standard rocks are given in Table 1.

Co, Cr, Cu, Pb and Zn were determined at a wavelength of 240.7, 357.9, 324.7, 217 and 213.9 nm respectively. For Co and Pb, a background correction was necessary.

Li, Ni and Rb were determined at a wavelength of 670.8, 231.7 and 780.0 nm, respectively within an appropriate concentration range of potassium.

Ba and Sr were determined at a wavelength of 553.6 and 460.7 nm, respectively, by using boric acid, lanthanum and $(NH_4)_2EDTA$ as a releasing agent (Adams and Passmore, 1966).

Sn analysis was carried out separately by the ammonium iodide fusion method and extracted with TOPO-MIBK mixture and then determined at a wavelength of 286.3 nm.

W analysis was carried out by $KHSO_4$ fusion and forming a complex solution with Dithiol, extracted with $CCl_4$, and then determined by the spectrophotometer at a wavelength of 630 nm.

### Major Element Data

Variation in chemical composition is clearly demonstrated by a Harker variation diagram (Fig. 2). This plot shows that there is a distinct trend of the samples from the $F_1$-granites (G-1, G-2 and G-3), whereas the samples from the $F_2$-granites (G-4 and G-5) show no significant trend. Considering only the $F_1$-granites, the CaO, FeO, $Fe_2O_3$, and $Al_2O_3$ contents decrease gradually from G-1 to G-2 and to G-3 with increasing

**Table 1.** Analysis of selected trace elements in U.S. Geological Survey Standard rocks

| | G-2 | $AGV_1$ | $GSP_1$ |
|---|---|---|---|
| Ba | 1859*(1870) | 1189*(1208) | 1281*(1300) |
| Co | 8 ( 5.5) | 18 (14.1) | 6 ( 6.4) |
| Cr | 7 ( 7) | 13 (12.2) | 15 (12.5) |
| Cu | 11 (11.7) | 61 (59.7) | 31 (33.3) |
| Li | 31 (34.8) | 11 (12) | 31 (32.1) |
| Ni | 4 ( 5.1) | 25 (18.5) | 14 (12.5) |
| Pb | 32 (31.2) | 33 (35.1) | 54 (51.3) |
| Rb | 177 (168) | 68 (67) | 245 (254) |
| Sr | 473 (479) | 634 (657) | 250 (233) |
| Zn | 87 (85) | 94 (84) | 101 (98) |

* This study (---) from Flanagan, 1972

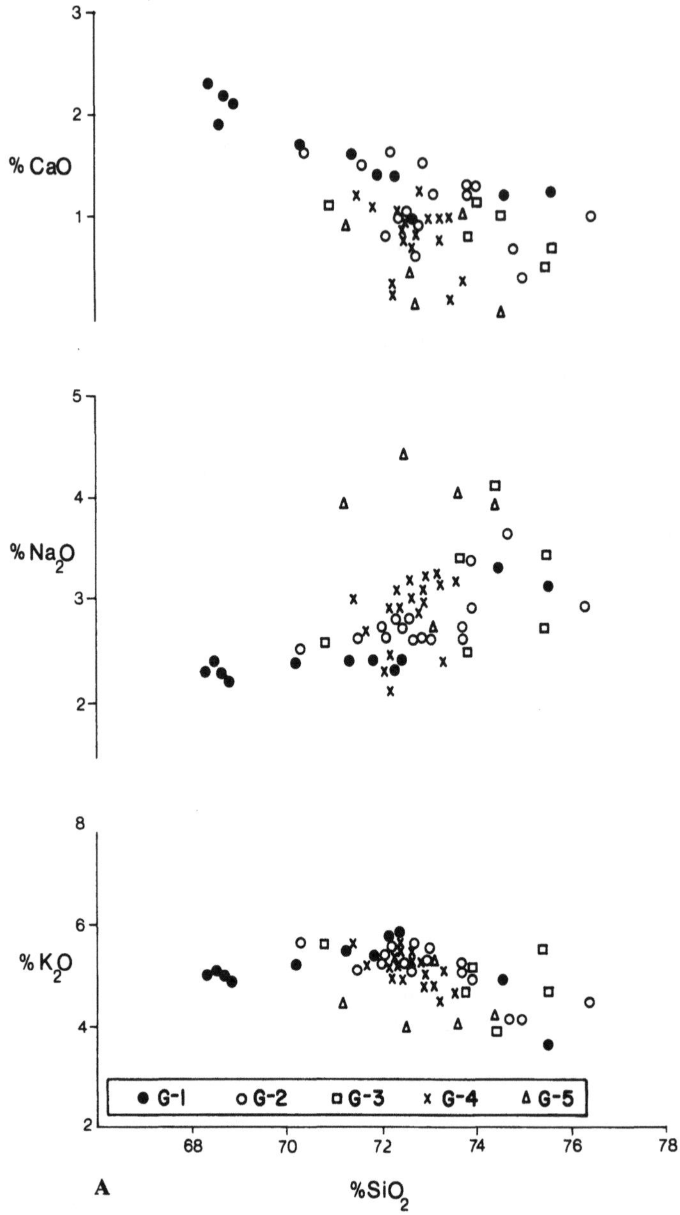


**Fig. 2A, B.** The Harker variation diagrams for the Khao Kata Khwam granites

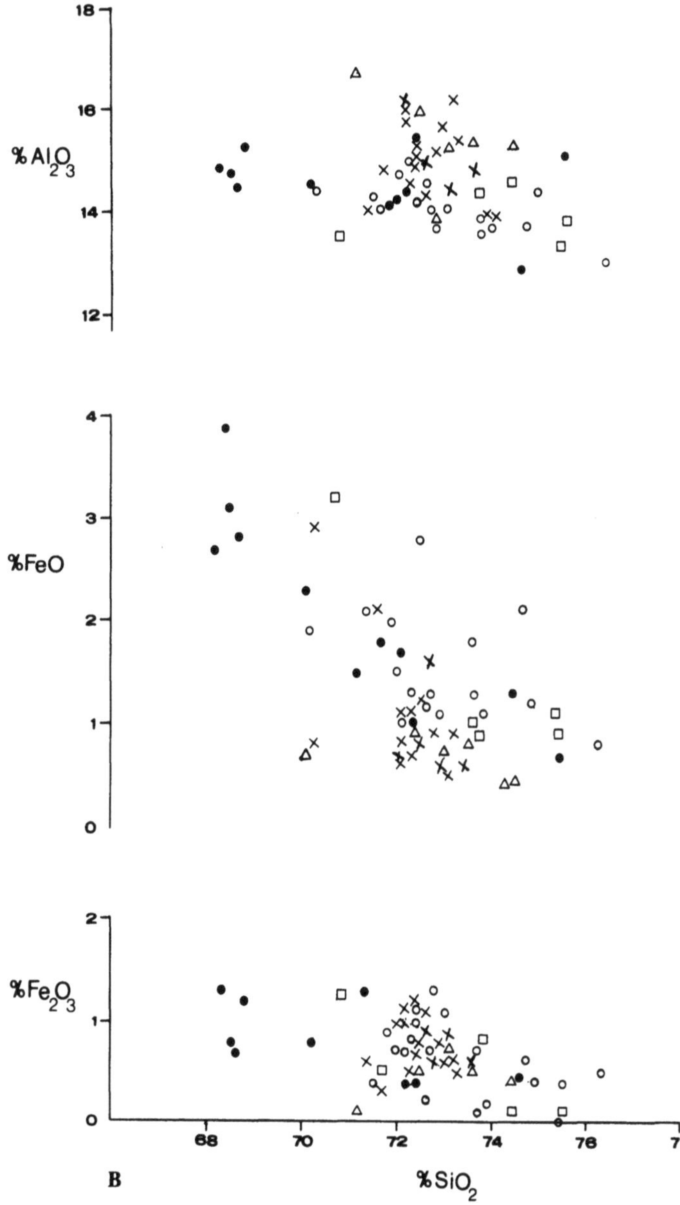


B

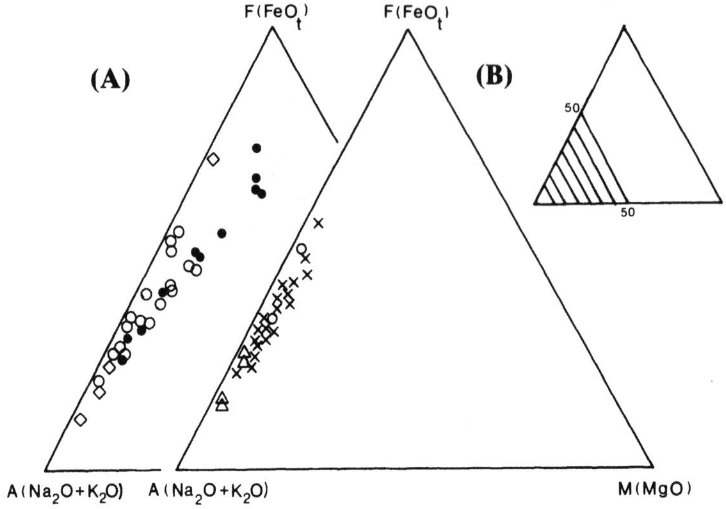


**Fig. 3A, B.** A($Na_2O+K_2O$) : $FeO_{(total)}$ : MgO diagram. Symbols as in Fig. 2

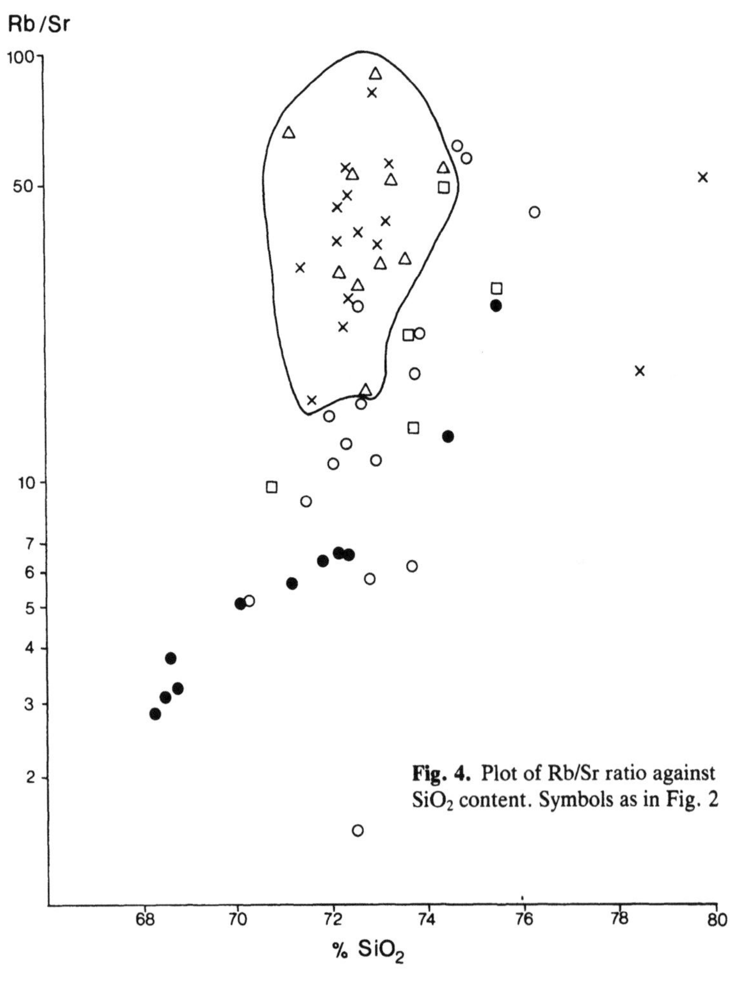


**Fig. 4.** Plot of Rb/Sr ratio against $SiO_2$ content. Symbols as in Fig. 2

$SiO_2$ content. However, $Na_2O$ increases with increasing $SiO_2$ content. It should be mentioned that samples 10-1 and 14-5, which belong to the $F_2$-granites, were not included in Fig. 2 since they were extremely altered and have very high $SiO_2$ contents (79.80 and 78.50% respectively).

The total alkalis ($Na_2O+K_2O$) : $FeO_{(total)}$ : MgO triangular diagram (Figs. 3A and B) shows well defined variations in both $F_1$ and $F_2$ granites. Both trends are towards alkali enrichment.

## Trace Element Data

Variation diagrams of Rb/Sr and K/Rb ratios versus $SiO_2$ are shown in Figures 4 and 5. The Rb/Sr ratio increases but the K/Rb ratio decreases from G-1 to G-2 and to G-3 with increasing $SiO_2$ content. The samples of G-4 and G-5 show an irregular distribution and are not of the same trend as those of G-1, G-2 and G-3. Thus, it is suggested that the parent magma of the $F_1$-granites was fractionated from G-1 to G-2 and to G-3 in accord with the increase of the Rb/Sr ratio and decrease of the K/Rb ratio. These trends are in general agreement with studies of Taylor (1965), Tauson and Kozlov (1973) and Nakapadungrat (1983). The relation between the Li content and the Rb/Sr ratio is shown in Figure 6. The Li content decreases slightly from G-1 to G-2 and to G-3 granites as the Rb/Sr ratio increases. It should be noted that the Li content of the $F_2$-granites (G-4 and G-5) is higher than the $F_1$-granites (G-1, G-2 and G-3). Figure 7 is a plot of Sn against the Rb/Sr ratio. For $F_1$-granites the Sn content increases gradually from G-1 to G-2 and to G-3 granites in accord with the increasing Rb/Sr ratio. Surprisingly, the Sn content of the $F_2$-granites increases sharply from G-4 to G-5 with a constant Rb/Sr ratio.

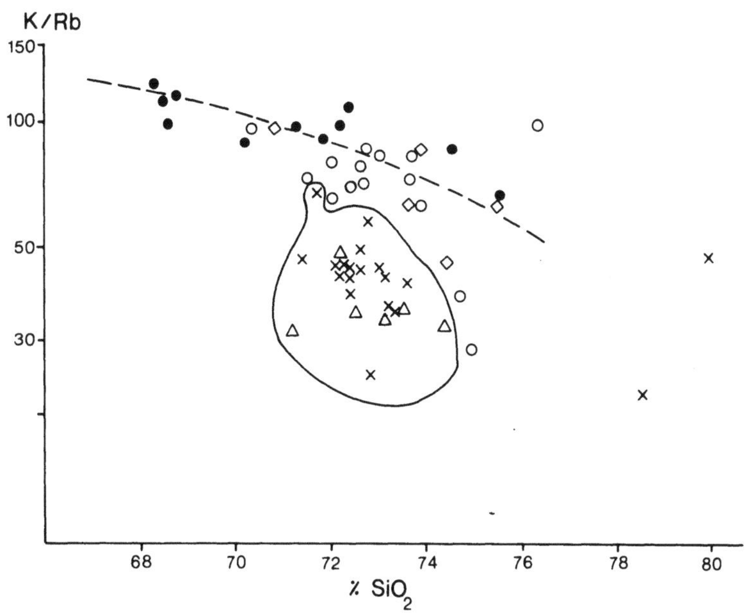


**Fig. 5.** Plot of K/Rb ratio against $SiO_2$ content. Symbols as in Fig. 2

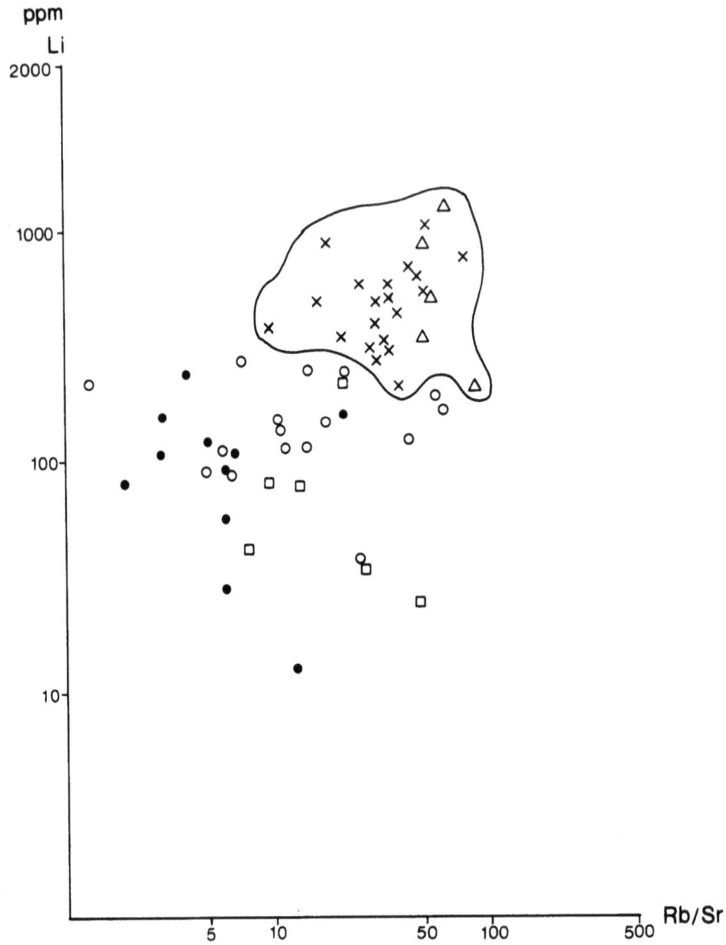


**Fig. 6.** Plot of Li content versus Rb/Sr ratio. Symbols as in Fig. 2

The plot of Rb versus Sr contents (Fig. 8) shows that the chemical variations of both $F_1$ and $F_2$ granites were probably dominated by alkali feldspar and plagioclase fractionation. However, the Ba versus Sr plot (Fig. 9) demonstrates clearly that chemical variation of the $F_1$-granites was dominated by alkali feldspar fractionation, while that of the $F_2$-granites did not follow the trend of feldspar fractionation. The Ba content of $F_2$-granites decreases from G-4 to G-5 with a constant content of Sr. This may imply that the $F_1$ and $F_2$ granites were not under the same conditions. The $F_1$-granites were in a closed system, but the $F_2$-granites were in an open system.

## Discussions and Conclusions

As described earlier, there are two groups of granites ($F_1$ and $F_2$). The composition of $F_1$-granites is $SiO_2$ (68–76%), $Na_2O$ (2.2–3.5%) and $K_2O$ (4.6–5.8%), whereas $F_2$-granites have $SiO_2$ (71–75%), $Na_2O$ (2.1–4.5%) and $K_2O$ (4.0–5.6%) (see Appendix 1). The $F_1$-granites are higher in CaO, FeO, Ba and Sr contents but lower

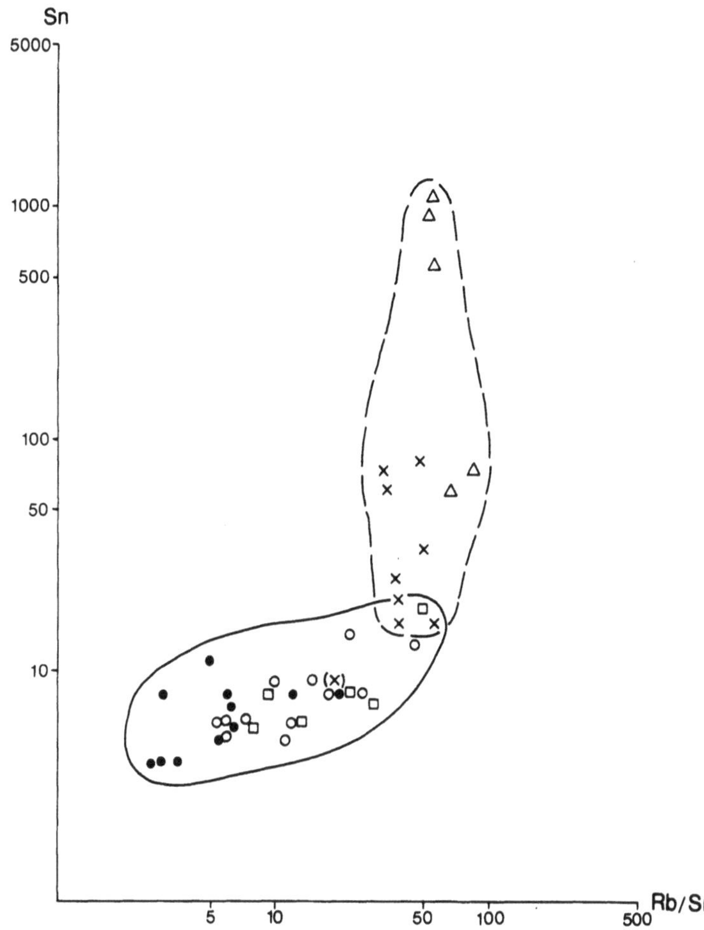


**Fig. 7.** Plot of Sn content against Rb/Sr ratio. Symbols as in Fig. 2

in Rb, Li and Sn than the $F_2$-granites. It is clearly seen that almost all granite types have high amounts of normative corundum. Very few samples contain normative corundum less than 1.1%. Thus, these granites could be classified as the S-type of Chappell and White (1974). This implies that the parent magmas could have formed by partial fusion of continental crustal material.

In addition, the CIPW normative quartz: albite: orthoclase plot (Fig. 10) shows that all Khao Kata Khwam granites fall close to the minimum isobar of 0.5 kilomar water pressure. This may indicate that the granites could have formed under a high geothermal gradient which induced partial melting in the crust. However, there are some differences between the Khao Kata Khwam and the granites of Chappell and White (1974). Whereas the Australian S-type granites all lie on the quartz-feldspar liquidus (White et al., 1977), almost all Khao Kata Khwam granites lie on the feldspar side (Fig. 10).

In can be concluded that only the $F_2$-granites are associated closely with tin and tungsten deposits. Hydrothermal alteration processes such as muscovitization, kaolinization and tourmalinization result in the tin-mineralization.

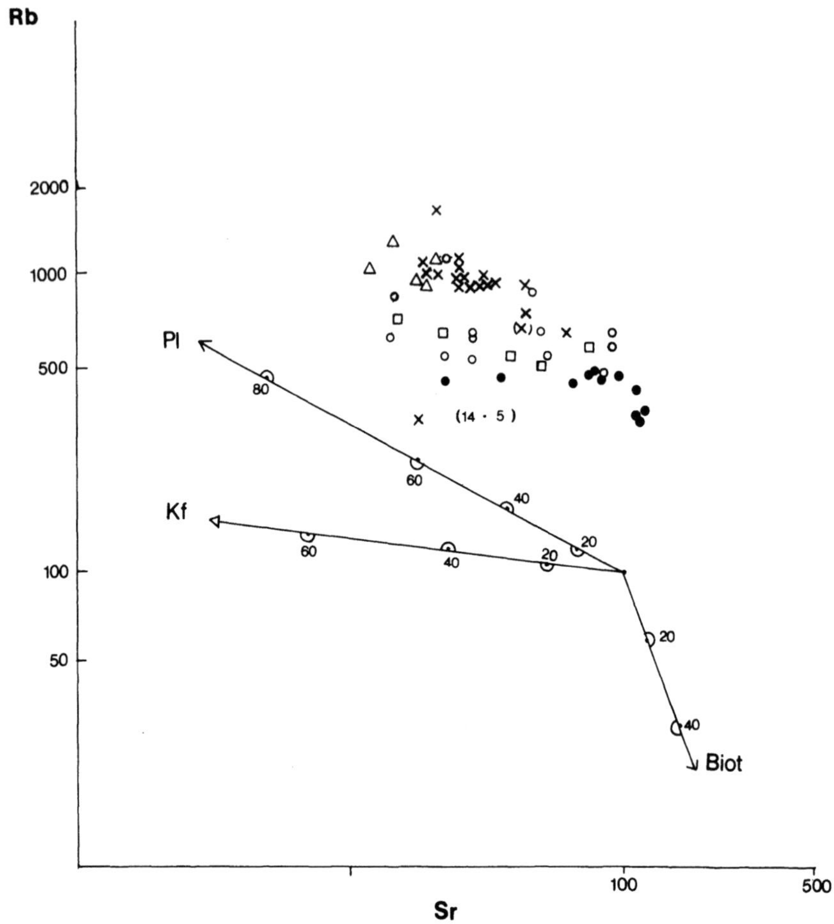


**Fig. 8.** Rb versus Sr variation in the Khao Kata Khwam granites

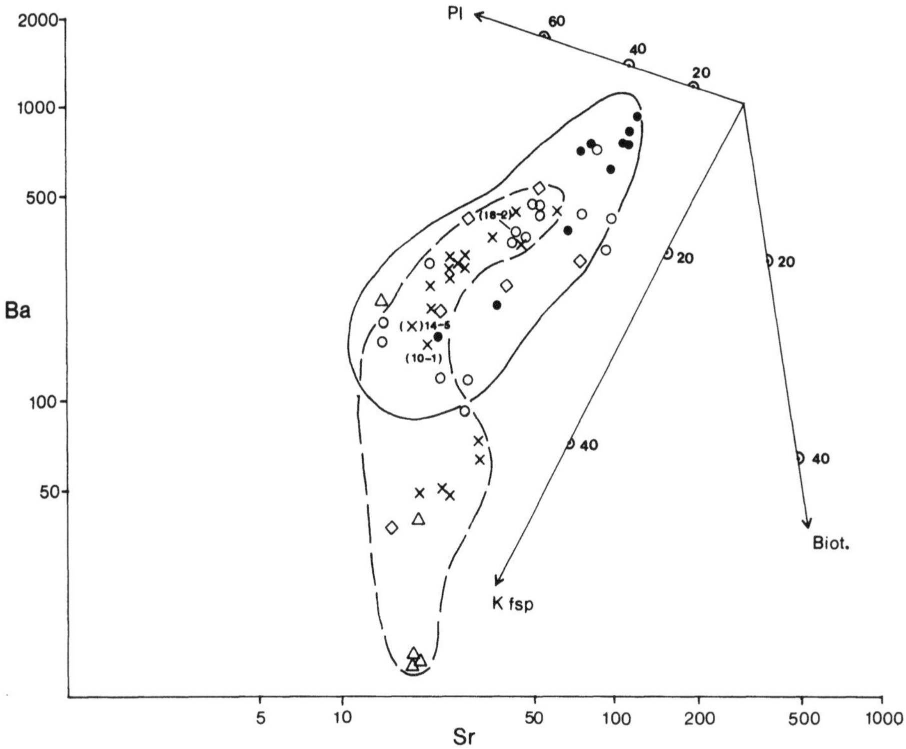


**Fig. 9.** Ba versus Sr variation in the Khao Kata Khwam granites. Symbols as in Fig. 2, and Fig. 8

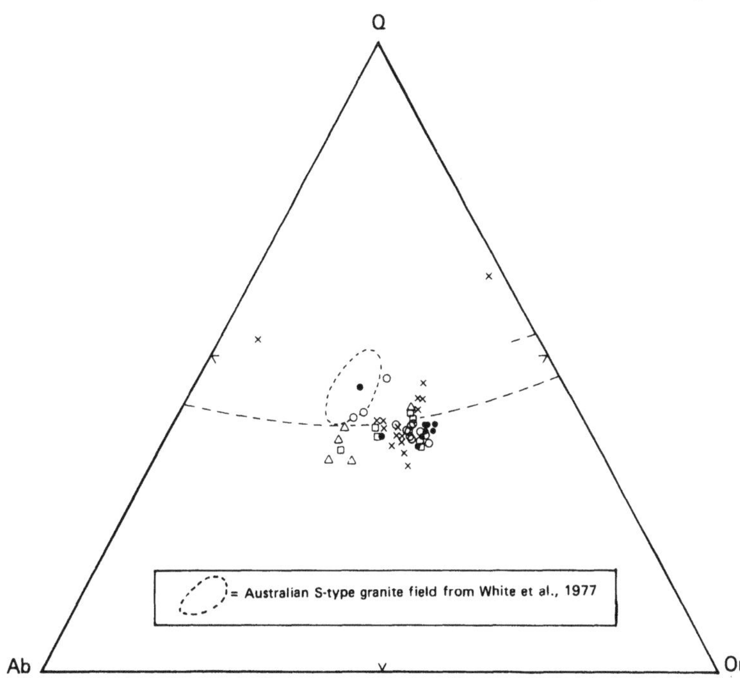


**Fig. 10.** CIPW normative Ab, Or and Q proportions of the Khao Kata Khwam granites plotted in the Q-Ab-Or-$H_2O$ system (after Luth et al., 1964)

**Appendix I.** Major elements and CIPW Norms of Khao Kata Khwam granites, Southern Thailand

| | G-1 | | | | | | | | | | |
|---|---|---|---|---|---|---|---|---|---|---|---|
| | 28-3 | 28-7 | 29-1 | 32-1 | 32-4 | 33-1 | 33-2 | 33-3 | 36-2 | 37-4 | 38-1 |
| *Major elements* (%) | | | | | | | | | | | |
| $SiO_2$ | 75.40 | 74.46 | 72.36 | 68.45 | 72.24 | 68.84 | 68.32 | 68.64 | 70.20 | 71.31 | 71.78 |
| $TiO_2$ | 0.11 | 0.22 | 0.12 | 0.57 | 0.23 | 0.58 | 0.62 | 0.57 | 0.43 | 0.36 | 0.37 |
| $Al_2O_3$ | 14.88 | 12.72 | 15.22 | 14.69 | 14.30 | 15.20 | 14.77 | 14.42 | 14.40 | 14.00 | 14.03 |
| $Fe_2O_3$ | 0.35 | 0.37 | 0.34 | 0.79 | 0.38 | 1.24 | 1.25 | 0.65 | 0.76 | 1.25 | 0.92 |
| FeO | 0.66 | 1.25 | 1.14 | 3.89 | 1.69 | 2.75 | 2.67 | 3.14 | 2.32 | 1.49 | 1.80 |
| MnO | 0.08 | 0.04 | 0.04 | 0.09 | 0.06 | 0.10 | 0.08 | 0.07 | 0.09 | 0.08 | 0.08 |
| MgO | 0.11 | 0.25 | 0.13 | 0.77 | 0.33 | 0.86 | 0.80 | 1.03 | 0.69 | 0.52 | 0.48 |
| CaO | 1.49 | 1.21 | 0.89 | 1.89 | 1.38 | 2.10 | 2.27 | 2.18 | 1.69 | 1.58 | 1.38 |
| $Na_2O$ | 3.07 | 3.28 | 2.36 | 2.42 | 2.32 | 2.20 | 2.29 | 2.31 | 2.37 | 2.40 | 2.41 |
| $K_2O$ | 3.63 | 4.88 | 5.79 | 5.06 | 5.72 | 4.88 | 4.96 | 5.00 | 5.19 | 5.45 | 5.32 |
| $P_2O_5$ | 0.02 | 0.02 | 0.05 | 0.15 | 0.06 | 0.07 | 0.21 | 0.18 | 0.13 | 0.15 | 0.15 |
| LOI | 0.72 | 0.72 | 0.98 | 0.38 | 0.69 | 0.62 | 0.84 | 0.59 | 0.80 | 0.45 | 0.49 |
| Total | 100.59 | 99.42 | 99.42 | 99.15 | 99.40 | 99.44 | 99.08 | 98.78 | 99.07 | 99.04 | 99.21 |
| *CIPW Norms* | | | | | | | | | | | |
| Q | 39.60 | 33.00 | 33.48 | 27.54 | 34.14 | 30.48 | 28.98 | 28.32 | 31.08 | 32.04 | 32.94 |
| C | 3.06 | - | 3.37 | 2.04 | 1.73 | 2.45 | 1.84 | 1.53 | 2.14 | 1.53 | 2.04 |
| Or | 21.68 | 28.91 | 34.47 | 30.02 | 33.92 | 28.91 | 29.47 | 29.47 | 30.58 | 32.25 | 31.69 |
| Ab | 26.20 | 27.77 | 19.91 | 20.44 | 19.39 | 18.34 | 19.39 | 19.39 | 19.91 | 20.44 | 20.44 |
| An | 7.51 | 5.56 | 4.45 | 8.62 | 6.95 | 10.56 | 10.29 | 10.01 | 7.51 | 13.90 | 6.12 |
| Hy | 1.22 | 2.08 | 2.02 | 7.58 | 3.31 | 4.74 | 4.90 | 7.09 | 4.70 | 2.49 | 3.18 |
| Hyen | 0.30 | 0.50 | 0.30 | 1.90 | 0.80 | 2.10 | 2.00 | 2.60 | 1.70 | 1.30 | 1.20 |
| Hyfs | 0.92 | 1.58 | 1.72 | 5.68 | 2.51 | 2.64 | 2.90 | 4.49 | 3.04 | 1.19 | 1.98 |
| Mt | 0.46 | 0.46 | 0.46 | 1.16 | 0.46 | 2.78 | 1.86 | 0.93 | 1.16 | 2.09 | 1.39 |
| Il | 0.15 | 0.46 | 0.30 | 1.06 | 0.46 | 1.06 | 1.22 | 1.06 | 0.76 | 0.61 | 0.76 |
| Hm | - | - | - | - | - | - | - | - | - | - | - |
| Ap | - | - | - | 0.34 | - | - | 0.34 | 0.34 | 0.34 | 0.34 | 0.34 |
| *Differentiation Index* (D.I.) | | | | | | | | | | | |
| D.I. | 87.48 | 89.68 | 87.86 | 78.00 | 87.45 | 77.73 | 77.84 | 77.18 | 81.57 | 84.73 | 85.07 |

| | G-2 | | | | | | | | | | |
|---|---|---|---|---|---|---|---|---|---|---|---|
| | 10-20 | 10-21 | 14-1 | 16-3 | 17-8 | 18-2 | 19-11 | 20-8 | 21-1 | 23-2 | 23-4 |
| *Major elements* (%) | | | | | | | | | | | |
| $SiO_2$ | 74.93 | 74.65 | 72.71 | 72.02 | 73.02 | 72.25 | 70.32 | 76.32 | 72.43 | 73.68 | 72.75 |
| $TiO_2$ | 0.07 | 0.07 | 0.22 | 0.22 | 0.25 | 0.23 | 0.44 | 0.14 | 0.25 | 0.23 | 0.26 |
| $Al_2O_3$ | 14.17 | 13.50 | 13.92 | 14.61 | 13.88 | 14.85 | 14.28 | 12.80 | 14.09 | 13.39 | 13.56 |
| $Fe_2O_3$ | 0.39 | 0.57 | 0.69 | 0.73 | 1.07 | 0.81 | 0.99 | 0.46 | 1.07 | 0.67 | 1.29 |
| FeO | 1.18 | 2.11 | 1.16 | 1.96 | 1.14 | 0.96 | 1.93 | 0.77 | 1.30 | 1.29 | 1.29 |
| MnO | 0.10 | 0.13 | 0.03 | 0.09 | 0.05 | 0.03 | 0.04 | 0.04 | 0.05 | 0.05 | 0.05 |
| MgO | 0.09 | 0.06 | 0.25 | 0.24 | 0.04 | 0.04 | 0.07 | 0.01 | 0.28 | 0.34 | 0.43 |
| CaO | 0.40 | 0.65 | 0.89 | 0.82 | 1.22 | 0.97 | 1.59 | 0.98 | 0.93 | 1.26 | 1.51 |
| $Na_2O$ | 3.43 | 3.56 | 2.58 | 2.74 | 2.60 | 2.81 | 2.45 | 2.71 | 2.66 | 2.61 | 2.64 |
| $K_2O$ | 4.14 | 4.08 | 5.63 | 5.20 | 5.52 | 5.45 | 5.60 | 4.24 | 5.30 | 5.23 | 5.18 |
| $P_2O_5$ | 0.01 | 0.01 | 0.08 | 0.08 | 0.04 | 0.12 | 0.11 | 0.01 | 0.11 | 0.08 | 0.08 |
| LOI | 0.67 | nil | 1.08 | 0.68 | 0.68 | 0.86 | 0.91 | 0.72 | 0.73 | 0.83 | 0.61 |
| Total | 99.58 | 99.39 | 99.24 | 99.39 | 99.51 | 99.38 | 98.73 | 99.20 | 99.20 | 99.66 | 99.65 |
| *CIPW Norms* | | | | | | | | | | | |
| Q | 37.50 | 35.52 | 33.54 | 33.36 | 33.54 | 33.12 | 30.30 | 42.12 | 34.14 | 34.56 | 33.06 |
| C | 3.37 | 2.04 | 2.14 | 3.26 | 1.33 | 2.96 | 1.53 | 2.24 | 2.55 | 1.12 | 0.82 |
| Or | 24.46 | 23.91 | 33.36 | 30.58 | 32.80 | 32.25 | 33.36 | 25.08 | 31.14 | 31.14 | 30.58 |

**Appendix I** (continued)

| | G-2 | | | | | | | | | | |
|---|---|---|---|---|---|---|---|---|---|---|---|
| | 10-20 | 10-21 | 14-1 | 16-3 | 17-8 | 18-2 | 19-11 | 20-8 | 21-1 | 23-2 | 23-4 |
| Ab | 28.82 | 29.87 | 22.01 | 23.06 | 22.01 | 23.58 | 20.96 | 23.06 | 22.53 | 22.01 | 22.53 |
| An | 1.95 | 3.34 | 3.61 | 3.34 | 6.12 | 3.89 | 6.95 | 4.97 | 3.89 | 6.12 | 7.51 |
| Hy | 2.05 | 3.63 | 1.79 | 3.24 | 1.02 | 0.76 | 2.18 | 0.92 | 1.89 | 2.38 | 2.16 |
| Hyen | 0.20 | 0.20 | 0.60 | 0.60 | 0.10 | 0.10 | 0.20 | – | 0.70 | 0.80 | 1.10 |
| Hyfs | 1.85 | 3.43 | 1.19 | 2.64 | 0.92 | 0.66 | 1.98 | 0.92 | 1.19 | 1.58 | 1.06 |
| Mt | 0.46 | 0.93 | 0.93 | 1.16 | 1.62 | 1.16 | 1.39 | 0.70 | 1.62 | 0.93 | 1.87 |
| Il | 0.15 | 0.15 | 0.46 | 0.46 | 0.46 | 0.46 | 0.91 | 0.30 | 0.46 | 0.46 | 0.46 |
| Hm | – | – | – | – | – | – | – | – | – | – | – |
| Ap | – | – | 0.34 | 0.34 | – | 0.34 | 0.34 | – | 0.34 | – | – |
| *Differentiation Index* (D.I.) | | | | | | | | | | | |
| D.I. | 90.78 | 89.30 | 88.91 | 87.00 | 88.35 | 88.95 | 84.62 | 90.26 | 87.81 | 87.71 | 86.17 |

| | G-2 | | | | | G-3 | | | | | |
|---|---|---|---|---|---|---|---|---|---|---|---|
| | 25-1 | 27-5 | 30-2 | 31-4 | 37-1 | 19-12 | 22-1 | 22-3 | 27-7 | 33-4 | 37-6 |
| *Major elements* (%) | | | | | | | | | | | |
| $SiO_2$ | 73.90 | 73.70 | 72.09 | 72.64 | 71.51 | 70.83 | 74.58 | 75.40 | 74.44 | 73.72 | 73.75 |
| $TiO_2$ | 0.16 | 0.12 | 0.36 | 0.14 | 0.35 | 0.35 | 0.05 | 0.11 | 0.02 | 0.07 | 0.22 |
| $Al_2O_3$ | 13.53 | 13.69 | 13.95 | 14.43 | 14.21 | 13.38 | 13.59 | 13.06 | 14.38 | 14.20 | 13.50 |
| $Fe_2O_3$ | 0.23 | 0.04 | 0.74 | 0.22 | 0.44 | 1.49 | 0.05 | – | 0.11 | 0.45 | 0.67 |
| FeO | 1.05 | 1.75 | 1.49 | 2.81 | 2.06 | 3.18 | 0.85 | 1.12 | 0.43 | 0.95 | 0.94 |
| MnO | 0.08 | 0.04 | 0.05 | 0.08 | 0.08 | 0.05 | 0.05 | 0.04 | 0.04 | 0.08 | 0.06 |
| MgO | 0.16 | 0.11 | 0.36 | 0.13 | 0.51 | 0.04 | 0.05 | 0.09 | 0.005 | 0.14 | 0.27 |
| CaO | 1.34 | 1.26 | 1.63 | 0.64 | 1.48 | 1.13 | 0.74 | 0.53 | 1.01 | 0.79 | 1.21 |
| $Na_2O$ | 2.92 | 2.73 | 2.60 | 2.74 | 2.61 | 2.61 | 3.42 | 2.72 | 4.13 | 3.38 | 2.52 |
| $K_2O$ | 4.89 | 5.16 | 5.41 | 5.34 | 5.10 | 5.55 | 4.69 | 5.46 | 4.10 | 4.81 | 5.14 |
| $P_2O_5$ | 0.03 | 0.03 | 0.07 | 0.08 | 0.16 | 0.07 | 0.02 | 0.03 | 0.05 | 0.03 | 0.11 |
| LOI | 0.64 | 0.48 | 0.57 | 0.17 | 0.56 | 0.14 | 0.48 | 0.54 | 0.72 | 0.41 | 0.65 |
| Total | 98.93 | 99.11 | 99.32 | 99.42 | 99.07 | 98.82 | 98.57 | 99.10 | 99.435 | 99.03 | 99.04 |
| *CIPW Norms* | | | | | | | | | | | |
| Q | 34.38 | 33.78 | 31.26 | 32.94 | 32.04 | 30.06 | 35.34 | 36.60 | 31.98 | 33.48 | 36.06 |
| C | 1.02 | 1.33 | 0.82 | 3.16 | 2.04 | 1.02 | 1.53 | 1.73 | 1.22 | 2.04 | 1.73 |
| Or | 28.91 | 30.58 | 32.25 | 31.69 | 30.02 | 32.80 | 27.80 | 32.25 | 24.46 | 28.36 | 30.58 |
| Ab | 24.63 | 23.06 | 22.01 | 23.06 | 22.01 | 22.01 | 28.82 | 23.06 | 35.11 | 28.30 | 21.48 |
| An | 6.67 | 6.12 | 8.06 | 2.22 | 6.39 | 5.56 | 3.61 | 2.50 | 5.00 | 3.89 | 5.28 |
| Hy | 2.12 | 3.47 | 2.62 | 5.18 | 4.34 | 5.32 | 1.68 | 2.18 | 0.66 | 1.62 | 1.62 |
| Hyen | 0.40 | 0.30 | 0.90 | 0.30 | 1.30 | 0.10 | 0.10 | 0.20 | – | 0.30 | 0.70 |
| Hyfs | 1.72 | 3.17 | 1.72 | 4.88 | 3.04 | 4.22 | 1.58 | 1.98 | 0.66 | 1.32 | 0.92 |
| Mt | 0.23 | – | 1.16 | 0.23 | 0.70 | 2.09 | – | – | 0.23 | 0.70 | 0.93 |
| Il | 0.30 | 0.15 | 0.61 | 0.30 | 0.61 | 0.61 | 0.15 | 0.15 | – | 0.15 | 0.46 |
| Hm | – | – | – | – | – | – | – | – | – | – | – |
| Ap | – | – | – | 0.34 | 0.34 | – | – | – | – | – | 0.34 |
| *Differentiation Index* (D.I.) | | | | | | | | | | | |
| D.I. | 87.92 | 87.42 | 85.52 | 87.69 | 84.07 | 84.87 | 91.96 | 91.91 | 91.55 | 90.14 | 88.12 |

| | G-4 | | | | | | | | | |
|---|---|---|---|---|---|---|---|---|---|---|
| | 11-1 | 11-2 | 11-4 | 12-4 | 12-6 | 13-5 | 14-1.1 | 14-5 | 14-10 | 16-7 |
| *Major elements* (%) | | | | | | | | | | |
| $SiO_2$ | 73.31 | 71.39 | 72.62 | 71.70 | 72.44 | 72.85 | 72.33 | 78.49 | 72.20 | 73.01 |
| $TiO_2$ | 0.17 | 0.16 | 0.12 | 0.34 | 0.25 | 0.16 | 0.17 | 0.13 | 0.19 | 0.11 |

**Appendix I** (continued)

| | G-4 | | | | | | | | | |
|---|---|---|---|---|---|---|---|---|---|---|
| | 11-1 | 11-2 | 11-4 | 12-4 | 12-6 | 13-5 | 14-1.1 | 14-5 | 14-10 | 16-7 |
| $Al_2O_3$ | 15.16 | 13.94 | 14.90 | 14.73 | 14.89 | 15.00 | 14.36 | 11.77 | 15.73 | 15.48 |
| $Fe_2O_3$ | 0.52 | 0.61 | 0.91 | 0.32 | 1.16 | 0.78 | 0.46 | 0.38 | 1.00 | 0.61 |
| FeO | 0.87 | 2.79 | 0.78 | 2.07 | 1.10 | 0.85 | 1.13 | 0.90 | 0.78 | 0.59 |
| MnO | 0.07 | 0.07 | 0.05 | 0.05 | 0.05 | 0.05 | 0.03 | 0.08 | 0.09 | 0.11 |
| MgO | 0.20 | 0.16 | 0.14 | 0.33 | 0.21 | 0.12 | 0.19 | 0.14 | 0.17 | 0.11 |
| CaO | 0.16 | 1.19 | 0.71 | 1.10 | 0.87 | 0.50 | 1.04 | 0.98 | 0.25 | 0.48 |
| $Na_2O$ | 2.41 | 3.01 | 3.04 | 2.73 | 2.94 | 3.07 | 2.88 | 4.34 | 2.08 | 3.15 |
| $K_2O$ | 5.07 | 5.56 | 5.40 | 5.16 | 5.00 | 5.13 | 5.24 | 0.89 | 4.97 | 4.83 |
| $P_2O_5$ | 0.07 | 0.10 | 0.11 | 0.10 | 0.07 | 0.08 | 0.06 | 0.10 | 0.09 | 0.08 |
| LOI | 1.15 | 0.47 | 0.92 | 0.62 | 0.70 | 0.82 | 1.02 | 0.85 | 1.62 | 0.90 |
| Total | 99.16 | 99.45 | 99.70 | 99.25 | 99.68 | 99.41 | 98.91 | 99.05 | 99.17 | 99.46 |
| *CIPW Norms* | | | | | | | | | | |
| Q | 38.58 | 28.08 | 32.46 | 32.10 | 33.84 | 33.90 | 32.64 | 47.52 | 40.26 | 35.10 |
| C | 5.41 | 1.12 | 3.06 | 2.86 | 3.06 | 3.77 | 2.04 | 2.14 | 6.73 | 4.49 |
| Or | 30.02 | 32.80 | 31.69 | 30.58 | 29.47 | 30.58 | 31.14 | 5.00 | 29.47 | 28.36 |
| Ab | 20.44 | 25.68 | 25.68 | 23.06 | 24.63 | 26.20 | 24.10 | 36.68 | 27.82 | 26.72 |
| An | 0.83 | 5.00 | 2.78 | 4.73 | 4.45 | 1.39 | 5.28 | 4.17 | 0.28 | 1.67 |
| Hy | 1.82 | 4.89 | 1.59 | 0.80 | 1.29 | 0.70 | 1.95 | 1.72 | 0.93 | 0.96 |
| Hyen | 0.50 | 0.40 | 0.40 | 3.12 | 0.50 | 0.30 | 0.50 | 0.40 | 0.40 | 0.30 |
| Hyfs | 1.32 | 49.49 | 1.19 | 3.92 | 0.79 | 0.40 | 1.45 | 1.32 | 0.53 | 0.66 |
| Mt | 0.70 | 0.93 | 0.23 | 0.46 | 1.62 | 1.16 | 0.70 | 0.46 | 1.39 | 1.16 |
| Il | 0.30 | 0.30 | 0.30 | 0.61 | 0.46 | 0.30 | 0.30 | 0.30 | 0.30 | 0.15 |
| Hm | – | – | – | – | – | – | – | – | – | – |
| Ap | – | 0.34 | 0.34 | 0.34 | – | 0.34 | – | 0.34 | 0.34 | 0.34 |
| *Differentiation Index* (D.I.) | | | | | | | | | | |
| D.I. | 89.04 | 86.56 | 89.83 | 85.74 | 87.94 | 90.68 | 87.88 | 89.20 | 87.55 | 90.18 |

| | G-5 | | | | |
|---|---|---|---|---|---|
| | 1-8 | 2-3 | 10-13 | 11-5 | 11-7 |
| *Major elements* (%) | | | | | |
| $SiO_2$ | 72.46 | 74.43 | 73.07 | 73.58 | 71.16 |
| $TiO_2$ | 0.01 | 0.01 | 0.14 | 0.03 | 0.03 |
| $Al_2O_3$ | 15.76 | 15.12 | 15.05 | 15.16 | 16.61 |
| $Fe_2O_3$ | 0.45 | 0.37 | 0.72 | 0.54 | 0.05 |
| FeO | 0.88 | 0.42 | 0.65 | 0.76 | 0.68 |
| MnO | 0.13 | 0.07 | 0.04 | 0.06 | 0.09 |
| MgO | 0.02 | 0.04 | 0.06 | 0.14 | 0.01 |
| CaO | 0.40 | 0.01 | 0.08 | 0.53 | 0.80 |
| $Na_2O$ | 4.42 | 3.88 | 2.65 | 4.04 | 3.94 |
| $K_2O$ | 3.91 | 4.05 | 5.21 | 3.96 | 4.38 |
| $P_2O_5$ | 0.06 | 0.01 | 0.10 | 0.06 | 0.09 |
| LOI | 0.76 | 1.17 | 1.12 | 0.81 | 1.32 |
| Total | 99.26 | 99.58 | 98.89 | 99.67 | 99.16 |
| *CIPW Norms* | | | | | |
| Q | 30.30 | 35.88 | 37.32 | 33.18 | 29.28 |
| C | 3.47 | 4.28 | 5.00 | 3.37 | 4.18 |
| Or | 23.35 | 23.91 | 30.58 | 23.35 | 26.13 |
| Ab | 37.20 | 33.01 | 22.53 | 34.06 | 33.54 |
| An | 1.95 | – | 0.28 | 2.50 | 3.06 |
| Hy | 1.45 | 0.76 | 0.73 | 0.93 | 1.32 |

**Appendix I** (continued)

| | G-5 | | | | |
|---|---|---|---|---|---|
| | 1-8 | 2-3 | 10-13 | 11-5 | 11-7 |
| Hyen | – | 0.10 | 0.20 | 0.40 | – |
| Hyfs | 1.45 | 0.66 | 0.53 | 0.53 | 1.32 |
| Mt | 0.70 | 0.46 | 0.93 | 0.70 | – |
| Il | – | – | 0.30 | – | – |
| Hm | – | – | – | – | – |
| Ap | – | – | – | – | 0.34 |
| *Differentiation Index* (D.I.) | | | | | |
| D.I. | 90.85 | 92.80 | 90.43 | 90.59 | 88.95 |

**Appendix II.** Trace elements and elemental ratios of Khao Kata Khwam granites, southern Thailand

| | G-1 | | | | | | | | | | |
|---|---|---|---|---|---|---|---|---|---|---|---|
| | 28-3 | 28-7 | 29-1 | 32-1 | 32-4 | 33-1 | 33-2 | 33-3 | 36-2 | 37-4 | 38-1 |
| *Trace elements* (ppm) | | | | | | | | | | | |
| Ba | 173 | 212 | 380 | 948 | 729 | 836 | 763 | 750 | 621 | 750 | 685 |
| Co | 102 | 114 | 108 | 84 | 175 | 97 | 91 | 131 | 156 | 138 | 79 |
| Cr | 60 | 66 | 64 | 110 | 42 | 71 | 73 | 52 | 50 | 56 | 43 |
| Cu | 6 | 8 | 4 | 20 | 6 | 7 | 5 | 7 | 4 | 4 | 4 |
| Li | 153 | 12 | 29 | 147 | 90 | 101 | 82 | 245 | 125 | 58 | 112 |
| Ni | 3 | 3 | 6 | 21 | 6 | 15 | 13 | 14 | 11 | 7 | 5 |
| Pb | 66 | 62 | 57 | 43 | 56 | 40 | 37 | 40 | 60 | 58 | 60 |
| Rb | 455 | 476 | 443 | 372 | 491 | 355 | 331 | 420 | 482 | 467 | 484 |
| Sr | 22 | 37 | 66 | 121 | 74 | 112 | 110 | 112 | 96 | 81 | 76 |
| Sn | 8 | 8 | 8 | 8 | 7 | 4 | 4 | 4 | 11 | 5 | 6 |
| Zn | 43 | 40 | 37 | 47 | 35 | 55 | 53 | 56 | 47 | 45 | 44 |
| W | 5 | 5 | 5 | 5 | 5 | 5 | 5 | 5 | 5 | 15 | 15 |
| *Elemental ratios* | | | | | | | | | | | |
| K/Rb | 66.23 | 85.11 | 108.50 | 112.92 | 69.71 | 114.21 | 124.40 | 98.83 | 89.39 | 96.88 | 91.25 |
| Ba/Rb | 0.38 | 0.45 | 0.86 | 2.55 | 1.48 | 2.35 | 2.31 | 1.76 | 1.29 | 1.61 | 1.00 |
| Rb/Sr | 20.68 | 12.86 | 6.71 | 3.07 | 6.64 | 3.17 | 2.83 | 3.75 | 5.02 | 5.77 | 6.37 |

| | G-2 | | | | | | | | | | |
|---|---|---|---|---|---|---|---|---|---|---|---|
| | 10-20 | 10-21 | 14-1 | 16-3 | 17-8 | 18-2 | 19-11 | 20-8 | 21-1 | 23-2 | 23-4 |
| *Trace elements* (ppm) | | | | | | | | | | | |
| Ba | 298 | 163 | 381 | 360 | 480 | 356 | 726 | 192 | 420 | 410 | 320 |
| Co | 128 | 131 | 119 | 147 | 116 | 87 | 81 | 159 | 122 | 78 | 84 |
| Cr | 64 | 156 | 30 | 86 | 41 | 36 | 44 | 26 | 66 | 49 | 71 |
| Cu | 12 | 19 | 4 | 16 | 2 | 3 | 12 | 3 | 12 | 3 | 9 |
| Li | 193 | 173 | 107 | 252 | 152 | 214 | 92 | 124 | 117 | 117 | 88 |
| Ni | 6 | 15 | 3 | 9 | 3 | 5 | 8 | 3 | 8 | 3 | 6 |
| Pb | 108 | 93 | 58 | 54 | 84 | 55 | 65 | 81 | 53 | 57 | 66 |
| Rb | 1,244 | 891 | 667 | 672 | 560 | – | 485 | 614 | 639 | 596 | 533 |
| Sr | 21 | 14 | 43 | 46 | 49 | 46 | 85 | 14 | 52 | 96 | 92 |
| Sn | – | – | – | 9 | 5 | 9 | 6 | 13 | 6 | 5 | 6 |
| Zn | 29 | 28 | 57 | 48 | 33 | 49 | 43 | 23 | 51 | 23 | 48 |
| W | 5 | 5 | 15 | 10 | 15 | 15 | 5 | 25 | 15 | 5 | 15 |

**Appendix I** (continued)

| | G-2 | | | | | | | | | | |
|---|---|---|---|---|---|---|---|---|---|---|---|
| | 10-20 | 10-21 | 14-1 | 16-3 | 17-8 | 18-2 | 19-11 | 20-8 | 21-1 | 23-2 | 23-4 |
| *Elemental ratios* | | | | | | | | | | | |
| K/Rb | 27.63 | 38.01 | 70.07 | 64.24 | 81.83 | – | 95.85 | 57.33 | 68.85 | 72.85 | 80.68 |
| Ba/Rb | 0.24 | 0.18 | 0.57 | 0.54 | 0.86 | – | 1.50 | 0.31 | 0.66 | 0.69 | 0.60 |
| Rb/Sr | 59.24 | 63.64 | 15.51 | 14.61 | 11.43 | 1.48 | 5.71 | 43.86 | 12.29 | 6.21 | 5.79 |

| | G-2 | | | | | G-3 | | | | | |
|---|---|---|---|---|---|---|---|---|---|---|---|
| | 25-1 | 27-5 | 30-2 | 31-4 | 37-1 | 19-22 | 22-1 | 22-3 | 27-7 | 33-4 | 27-6 |
| *Trace elements* (ppm) | | | | | | | | | | | |
| Ba | 96 | 144 | 450 | 125 | 446 | 520 | 202 | 300 | 38 | 413 | 250 |
| Co | 108 | 78 | 103 | 97 | 103 | 92 | 103 | 105 | 54 | 113 | 97 |
| Cr | 36 | 84 | 52 | 149 | 48 | 237 | 47 | 51 | 40 | 68 | 56 |
| Cu | 7 | 4 | 4 | 20 | 3 | 33 | 7 | 5 | 6 | 3 | 4 |
| Li | 255 | 164 | 139 | 37 | 280 | 82 | 33 | 43 | 24 | 231 | 79 |
| Ni | 3 | 8 | 7 | 15 | 9 | 23 | 3 | 5 | 3 | 8 | 4 |
| Pb | 64 | 69 | 52 | 57 | 69 | 45 | 62 | 62 | 106 | 54 | 63 |
| Rb | 660 | 521 | 568 | 575 | 580 | 508 | 635 | 608 | 750 | 642 | 531 |
| Sr | 29 | 29 | 52 | 22 | 74 | 53 | 22 | 74 | 15 | 29 | 39 |
| Sn | 15 | 8 | 9 | 8 | 6 | 8 | 7 | 6 | 19 | 8 | 6 |
| Zn | 31 | 36 | 49 | 45 | 44 | 32 | 12 | 21 | 4 | 38 | 31 |
| W | 5 | 5 | 5 | 5 | 15 | 5 | 10 | 15 | 636 | 5 | 5 |
| *Elemental ratios* | | | | | | | | | | | |
| K/Rb | 61.51 | 82.22 | 79.07 | 77.10 | 73.00 | 90.70 | 61.31 | 74.50 | 45.38 | 62.20 | 80.36 |
| Ba/Rb | 0.15 | 0.28 | 0.79 | 0.22 | 0.77 | 1.02 | 0.32 | 0.49 | 0.05 | 0.64 | 0.47 |
| Rb/Sr | 22.76 | 17.97 | 10.92 | 26.14 | 7.84 | 9.58 | 28.86 | 8.22 | 50.00 | 22.14 | 13.62 |

| | G-4 | | | | | | | | | |
|---|---|---|---|---|---|---|---|---|---|---|
| | 1-7 | 2-2.1 | 2-2.2 | 3-3 | 3-5 | 3-10 | 6-4 | 10-1 | 10-2 | 10-15 |
| *Trace elements* (ppm) | | | | | | | | | | |
| Ba | 50 | 50 | 50 | 310 | 287 | 300 | 345 | 163 | 295 | 310 |
| Co | 106 | 98 | 148 | 131 | 125 | 106 | 138 | 92 | 110 | 91 |
| Cr | 9 | 22 | 9 | 24 | 22 | 24 | 49 | 33 | 21 | 18 |
| Cu | 14 | 19 | 13 | 5 | 5 | 11 | 9 | 6 | 16 | 8 |
| Li | 670 | 550 | 510 | 324 | 276 | 388 | 468 | 603 | 418 | 202 |
| Ni | nil | nil | nil | 3 | 3 | 3 | 3 | 3 | 3 | 3 |
| Pb | 69 | 81 | 48 | 39 | 39 | 48 | 48 | 48 | 54 | 40 |
| Rb | 1,176 | 975 | 1,038 | 939 | 923 | 902 | 750 | 1,057 | 1,011 | 33 |
| Sr | 25 | 25 | 19 | 28 | 28 | 29 | 45 | 21 | 25 | 25 |
| Sn | 82 | 16 | 16 | – | – | – | – | 34 | – | – |
| Zn | 41 | 36 | 44 | 36 | 35 | 37 | 47 | 36 | 42 | 28 |
| W | – | – | – | 5 | 5 | 5 | 15 | 15 | 5 | 5 |
| *Elemental ratios* | | | | | | | | | | |
| K/Rb | 37.98 | 44.27 | 44.71 | 41.55 | 43.08 | 48.69 | 57.22 | 40.68 | 43.27 | 36.40 |
| Ba/Rb | 0.04 | 0.05 | 0.05 | 0.33 | 0.31 | 0.33 | 0.46 | 0.15 | 0.29 | 0.30 |
| Rb/Sr | 47.04 | 39.00 | 54.63 | 33.54 | 32.94 | 31.10 | 16.67 | 50.33 | 40.44 | 41.32 |

**Appendix II** (continued)

| | G-4 | | | | | | | | | |
|---|---|---|---|---|---|---|---|---|---|---|
| | 11-1 | 11-2 | 11-4 | 12-4 | 12-6 | 13-5 | 14-11 | 14-5 | 14-10 | 16-7 |
| *Trace elements* (ppm) | | | | | | | | | | |
| Ba | 213 | 75 | 63 | 452 | 369 | 250 | 452 | 283 | 260 | 267 |
| Co | 104 | 90 | 124 | 169 | 69 | 108 | 131 | 87 | 73 | 174 |
| Cr | 18 | 61 | 9 | 54 | 59 | 32 | 25 | 26 | 30 | 24 |
| Cu | 7 | 40 | 13 | 11 | 11 | 6 | 4 | 4 | 12 | 4 |
| Li | 1,004 | 494 | 304 | 376 | 558 | 745 | 342 | 856 | 502 | 307 |
| Ni | 3 | nil | nil | 9 | 9 | 5 | 6 | 5 | 4 | 5 |
| Pb | 56 | 88 | 68 | 52 | 44 | 44 | 55 | 11 | 45 | 46 |
| Rb | 1,192 | 992 | 907 | 644 | 975 | 1,756 | 986 | 338 | 927 | 904 |
| Sr | 21 | 31 | 31 | 61 | 36 | 21 | 43 | 18 | 25 | 25 |
| Sn | – | 63 | 73 | – | – | – | – | 9 | 20 | 25 |
| Zn | 41 | 75 | 32 | 61 | 38 | 45 | 38 | 16 | 41 | 33 |
| W | 5 | – | – | 5 | 10 | 5 | – | 10 | 10 | 5 |
| *Elemental ratios* | | | | | | | | | | |
| K/Rb | 35.31 | 46.53 | 49.42 | 66.52 | 42.57 | 24.25 | 44.12 | 21.86 | 44.51 | 44.35 |
| Ba/Rb | 0.18 | 0.08 | 0.07 | 0.70 | 0.38 | 0.14 | 0.46 | 0.54 | 0.28 | 0.30 |
| Rb/Sr | 56.76 | 32.00 | 29.26 | 10.56 | 27.08 | 83.62 | 22.93 | 18.78 | 37.08 | 36.16 |

| | G-5 | | | | |
|---|---|---|---|---|---|
| | 1-8 | 2-3 | 10-13 | 11-5 | 11-7 |
| *Trace elements* (ppm) | | | | | |
| Ba | 13 | 40 | 223 | 13 | 13 |
| Co | 126 | 132 | 156 | 132 | 47 |
| Cr | 13 | 2 | 22 | 14 | 2 |
| Cu | 68 | 139 | 18 | 17 | 32 |
| Li | 856 | 504 | 217 | 318 | 1,250 |
| Ni | nil | nil | 5 | nil | nil |
| Pb | 32 | 46 | 39 | 43 | 18 |
| Rb | 943 | 1,055 | 1,282 | 933 | 1,207 |
| Sr | 18 | 19 | 14 | 18 | 18 |
| Sn | 1,144 | 583 | – | 544 | 60 |
| Zn | 59 | 110 | 60 | 39 | 33 |
| W | – | – | 20 | – | – |
| *Elemental ratios* | | | | | |
| K/Rb | 34.42 | 31.87 | 33.74 | 35.23 | 30.12 |
| Ba/Rb | 0.01 | 0.04 | 0.18 | 0.01 | 0.01 |
| Rb/Sr | 52.39 | 55.53 | 91.57 | 51.83 | 67.06 |

*Acknowledgements.* We are deeply indebted to Mr. Sivavong Changkasiri, Director-General of the Department of Mineral Resources of Thailand who kindly gave permission to publish this paper. Grateful acknowledgements are also made to Mr. Thavat Japakasetr, the Director of the Geological Survey Division for his full support and encouragement.

We thank Mr. Pradit Boonsoong, Mr. Prapan Boonsoong and Mr. Somkiat Wongnuplod from Sombat Co. Ltd. for helping in manpower and accommodation.

We thank the following colleagues of the Geological Survey Division: Mr. Somchai Chaisen, Decha Maneenai, Amphorn Kamchu and Akachai Dilokmongkolkun for their help in preparing the manuscript.

## References

Adams, P.B. and Passmore, W.O., 1966. Critical factors in the determination of the alkaline earth elements in glass by Atomic Absorption Spectrometry. *Anal. Chem.*, v. 38, no. 4, 630–633.

Chappell, B.W. and White, A.J.R., 1974. Two contrasting granite types: *Pacific Geology*, v. 8, 173–174.

Flanagan, F.J., 1972. Descriptions and analyses of eight new USGS rock standards. *U.S. Geol. Surv. Prof. Paper*, 840, 171–172.

Friedli, R. and Jariyawat, P., 1983. Geochemical Prospecting Primary Study, Nok Hook and Ban She Beow Mines. *Draft report, SEATRAD*, Ipoh, Malaysia, 24 p. (unpubl.).

Garson, M.S., Young, B., Mitchell, A.H.G. and Tait, B.A.R., 1975. The geology of the tin belt in peninsular Thailand around Phuket, Phangnga and Takua Pa: *Overseas Mem. Inst. Geol. Sci. London 1*, 112 p.

Luth, W.C., Jahns, R.H., and Tuttle, O.F., 1964. The granite system at pressure of 4 to 10 kilobars. *Jour. Geophys. Res.*, v. 69, 759–773.

Muenlek, S., 1982. Personal communication.

Nakapadungrat, S., 1983. The Geochemistry and geology of the Thong Lang granite complex, central Thailand. *Proc. Conf. Geology and Mineral Resources, Thailand*, Nov. 1983, 11 pages.

Nakapadungrat, S., Manoonworawong, P. and Techawan, S., 1983. Geology of Tin-Tungsten granites around Nok Hook mining district. *Report of investigation, Geological Survey Division, Dept. Min. Resources, Thailand* (unpubl.).

Philpotts, J.A. and Schnetzler, C.C., 1970. Phenocryst-matrix partition coefficients for K, Rb, Sr and Ba, with applications to anorthite and basalt genesis: *Geochim. Cosmochim. Acta*, v. 34, 307–322.

Schnetzler, C.C. and Philpotts, J.A., 1970. Partition coefficients of some rare earth elements and barium between igneous matrix material and rock-forming mineral phenocrysts-I: in *'Origin and Distribution of Elements'* (Ahrens, L.H., ed.), Pergamon Press, London, 929–938.

Tantiwanit, W., Raksaskulwong, L. and Mantajit, N., 1983. The Upper Palaeozoic pebbly rocks in southern Thailand: *Proc. Workshop on Stratigraphic Correlation of Thailand and Malaysia*, Geol. Soc. Thailand, v. 1, 96–104.

Tauson, L.V. and Kozlov, V.D., 1973. Distribution functions and ratios of trace element concentrations as estimators of the ore-bearing potential of granites: *Proc. 4th Inter. Geochemical Exploration*, London, Inst. Mining Met. (Jones, M.J., ed.), 37–44.

Taylor, S.R., 1965. The application of trace element data to problems in petrology: *Phys. Chem. Earth*, v. 6, 133–213.

White, A.J.R., Williams, I.S., and Chappell, B.W., 1977. Geology of the Berridale, 1:100,000, sheet 8625; *Geol. Surv. N.S.W.*, 138 p.

# 6.11.3 Mineralogy of Tin and Niobium-Tantalum-Bearing Minerals in Thailand

J. Praditwan[1]


## Abstract

Mineralogical study of panned concentrates from the Omkoi area, Northern Thailand, and several products from the concentrator plant of Patana Mine, Central Thailand, and Chaiyayuth dredge, Tantikovit Mine and Loon Seng Mine in Southern Thailand was carried out by microscopic and electron microprobe examination of polished section. The results show that the common Nb-Ta-bearing minerals associated with cassiterite are columbite-tantalite and Nb-Ta rutile with some samarskite and tapiolite. The associated heavy mineral suite is ilmenite, garnet, monazite, rutile, zircon, tourmaline, maghemite, anatase, xenotime, topaz, scheelite, wolframite, martite, geothite, Mn-oxides, pyrite, arsenopyrite, spinel, lepidocrocite, marcasite and limonite, with some beryl, allanite, magnetite and hematite. In general cassiterite, columbite-tantalite, Nb-Ta rutile and samarskite from the three different areas have similar characteristics as follows: Cassiterite occurs mostly as free particles, anhedral crystals, brownish-grey in colour, weak bireflectance, commonly of twinned habit and showing reddish-white to brownish-white internal reflections. Some cassiterite is intergrown with columbite-tantalite, quartz, and to a lesser extent with Nb-Ta rutile in simple and mottled type intergrowths. Columbite-tantalite in general occurs as free particles, as small inclusions in cassiterite, and intergrown with cassiterite as well as quartz in simple and mottled type of intergrowth. Some columbite-tantalite shows replacement by microlite along the margin and/or along the fractures of the particles. From Omkoi area, columbite-tantalite replaced by tapiolite is also observed. Nb-Ta rutile is present as homogeneous and inhomogeneous particles. Homogeneous Nb-Ta rutile particles show a grey colour with weak bireflectance, strong anisotropism from reddish-grey to dark brown and a common twinned habit, while inhomogeneous Nb-Ta rutile particles consist of a grey Ta-rutile groundmass with dark grey patches of columbite-tantalite inclusions. These columbite-tantalite inclusions form irregular bodies in a wide size range and evenly distributed throughout the groundmass. Some Nb-Ta rutile is intergrown with cassiterite and/or quartz in simple intergrowth. Samarskite occurs as anhedral to subhedral, grey colour, with weak bireflectance and weak anistropism and having distinct light brown to dark brown internal reflections. A comparison of the general characteristics of Nb-Ta-bearing minerals in Thailand and Malaysia are as follows: Nb-Ta-bearing minerals are closely associated with tin-bearing pegmatites along


[1] Southeast Asia Tin Research and Development Centre, Tiger Lane, 31400 Ipoh, Malaysia

the Western Tin Belt of the Burma-Thai-Malaya orogen. In general columbite-tantalite in Thailand contains higher Ta than Nb, and tends to be high in Fe in the north, whereas in the south, Mn is predominant. In Malaysia, columbite as well as ferro-columbite is more common than tantalite. In the southern part (Bakri) Mn-tantalite with Ta-rich series such as wodginite and tapiolite are also observed. Strüverite (Ta-rutile) is more common than Nb-rutile in Thailand and is present as both homogeneous and inhomogeneous particles (with exsolution bodies of columbite-tantalite): whereas in Malaysia Nb-rutile, which occurs as inhomogeneous particles, is more common than Ta-rutile. Tungsten is present everywhere as a trace constituent in columbite-tantalite from Thailand, whereas in Malaysia tungstenian-columbite has been reported. No well-defined zoning of columbite-tantalite relative to cassiterite is apparent from the mineralogical data. From an operation point of view, the data reported here suggest that by carefully applying controlled magnetic and high tension separation to the magnetic products from the mine, is should be possible to obtain a columbite-tantalite concentrate with fairly high recoveries.

## Introduction

Tin, with minor amounts of niobium-tantalum, has been produced from several areas in Thailand. Niobium-tantalum in general is present as columbite-tantalite, Nb-Ta rutile and complex multiple oxides containing niobium-tantalum and titanium.

The association of cassiterite with niobium-tantalum-bearing minerals is most common in cassiterite-bearing pegmatites distributed throughout South-West Thailand in the so-called Western Tin Belt of the Burma-Thai-Malaya orogen, especially in Ranong and Phuket areas (Fig. 1). Most of the pegmatites are small with thickness of 0.5–1 m.

In the Omkoi area of Northern Thailand, cassiterite-bearing pegmatites cut all the Triassic to Lower Carboniferous granites as well as Precambrian (?) and Paleozoic country rocks. The pegmatitic lodes are relatively small, normally not greater than 3 m in width. Commonly tin-bearing pegmatites are characterized by the presence of garnet and muscovite. The most abundant constituents of pegmatites are alk.-feldspar megacrysts (mainly microcline perthite), quartz, albite and muscovite. Other minerals observed include columbite-tantalite, Nb-Ta rutile, tourmaline, biotite, zircon and beryl. The pegmatites in general are unzoned (Vichit, 1983).

In the central part of Thailand at Ban Rai, west of Uthai Thani, cassiterite is found in pegmatite dykes in association with columbite-tantalite (Nutalaya et al., 1979). These pegmatites consist mainly of feldspar, quartz and muscovite with accessory tourmaline, beryl, garnet and biotite. Columbite-tantalite and Nb-Ta rutile are also produced as by-products from the Patana tin mine, Kanchanaburi Province; and from Ratchaburi and Prachuap Khiri Khan provinces.

Southern Thailand (Ranong, Takua Pa, Phangnga, Trang and Phuket) is the most important and best known tin producing district of the country. The tin-bearing pegmatites and tin-bearing granites (biotite-muscovite granites) of this area are of Cretaceous-Tertiary age and have intruded slate, phyllite, quartzite and limestone of Palaeozoic age. Minerals associated with cassiterite are monazite, ilmenite, columbite-tantalite, Nb-Ta rutile, garnet, xenotime and wolframite.

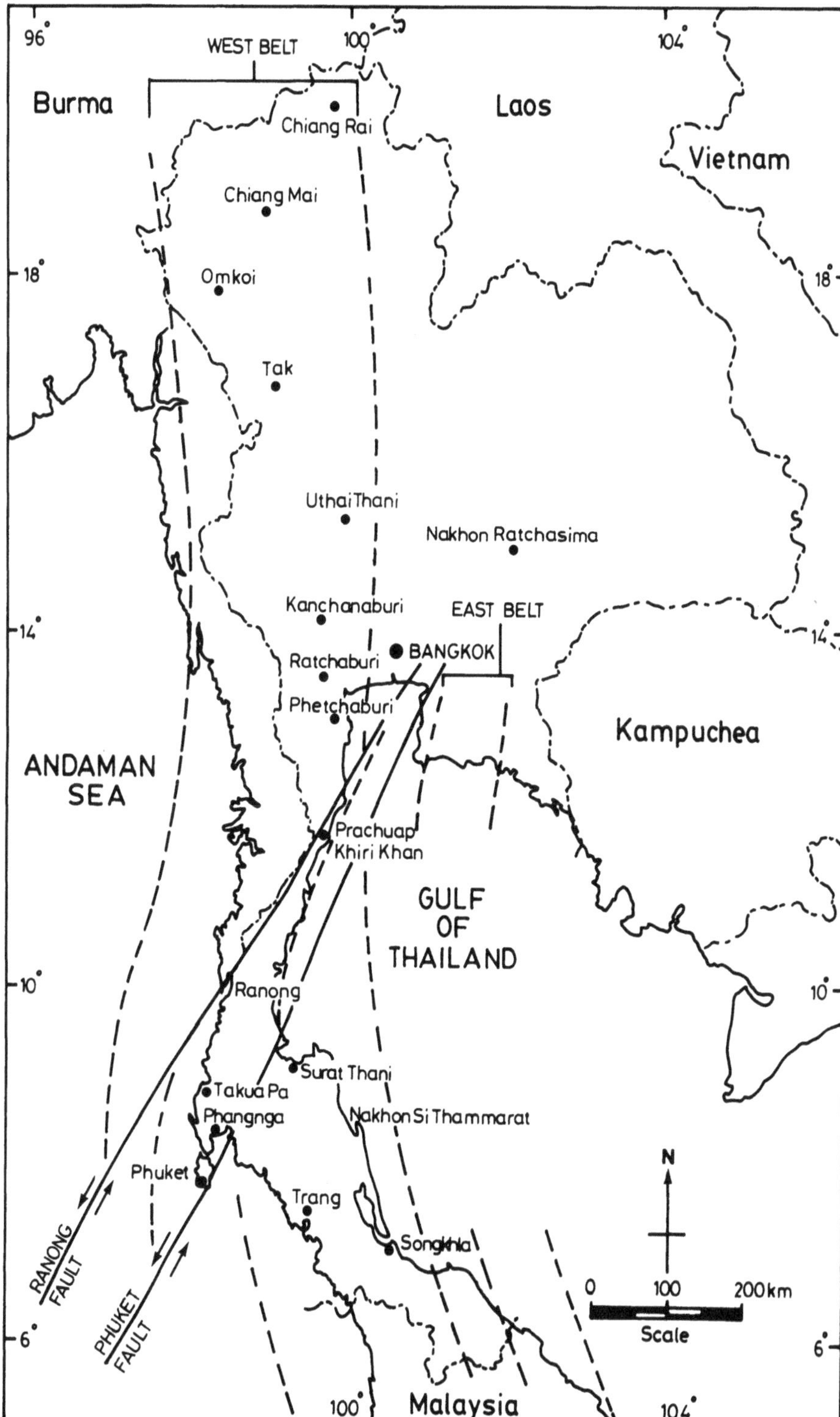


**Fig. 1.** Western Tin Belt of the Burma-Thai-Malaya orogen (after Hosking, 1977)

In the Phuket area the tin-bearing pegmatites form dykes, flat lenticular bodies and parallel vein swarms in metasedimentary rocks of the Phuket Group of Late Carboniferous to Permian age. In general, the pegmatites consist mainly of quartz, albite, microcline, kaolin and micas. Accessory minerals include cassiterite, wolframite, Nb-Ta and REE minerals, as well as zircon, rutile, ilmenite, and garnet. In the thick dykes, zoning of major minerals is less distinct, only the heavy mineral distribution indicates a zoning (Gocht and Pluhar, 1982).

The production of columbite-tantalite and strüverite (Ta-rutile) as by-products from tin mines in Thailand is shown in Tables 1 and 2.

The objective of this paper is to review the existing data on the mineralogy of tin and associated Nb-Ta minerals from several occurrences in Thailand.

Throughout this paper the term *particle* is used to include both grains as well as true particles.

**Table 1.** Production of Columbite-Tantalite as by-products from the tin mines of Thailand

| Production (tonnes) / Years | 1977 | 1978 | 1979 | 1980 | 1981 | 1982 | 1983 |
|---|---|---|---|---|---|---|---|
| *Northern Region* | | | | | | | |
| Uthai Thani | – | 25 | 47 | 36 | 29 | 37 | 47 |
| *Central Region* | | | | | | | |
| Kanchanaburi | 6 | 4 | 12 | 22 | 15 | 1 | 68 |
| Ratchaburi | – | – | 2 | 1 | – | – | – |
| *Southern Region* | | | | | | | |
| Phangnga | 82 | – | 45 | – | – | – | – |
| Phuket | 16 | – | 76 | 125 | – | – | – |
| Ranong | 12 | 35 | 216 | 172 | – | – | – |
| Trang | 40 | – | 9 | – | 4 | 1 | 4 |
| Total production | 156 | 64 | 407 | 356 | 49 | 39 | 549 |

**Table 2.** Production of Strüverite (Ta-rutile) as by-products from the tin mines of Thailand

| Production (tonnes) / Years | 1979 | 1980 | 1981 | 1982 | 1983 |
|---|---|---|---|---|---|
| *Southern Region* | | | | | |
| Nakhon Si Thammarat | – | 4 | – | – | – |
| Phangnga | 5 | – | – | – | – |
| Phuket | 177 | 80 | – | – | – |
| Ranong | 24 | 125 | 44 | 10 | 275 |
| Takua Pa | 25 | – | – | – | – |
| Trang | – | 92 | 0.1 | – | – |
| Total prodcution | 231 | 301 | 44 | 10 | 275 |

## The Occurrence of Cassiterite, Columbite-Tantalite, and Nb-Ta Rutile

### Omkoi Area

Omkoi district is approximately 185 km southwest of Chiangmai. Six samples were collected from selected eluvial and stream deposits in different locations. Panning was used to obtain the tin concentrates.

Examination of the polished sections under reflected light microscope and electron microprobe, indicates that the Nb-Ta-bearing minerals associated with cassiterite are columbite-tantalite, Nb-Ta rutile with some tapiolite and samarskite. Other associated heavy minerals are ilmenite, garnet (spessartite), monazite, rutile, zircon, maghemite, tourmaline, anatase, scheelite, wolframite with some lepidocrocite, xenotime, Mn-oxides, beryl and pyrite (Table 3).

Cassiterite in general occurs as free particles although some grains are intergrown with columbite-tantalite, quartz and to a lesser extent with Nb-Ta rutile.

Cassiterite as free particles is present mostly as anhedral crystals, brownish-grey in colour, weak bireflectance, very distinct anisotropism, a common twinned habit and showing reddish-white to brownish-white internal reflections. The size range of the cassiterite free particles varies from 2 mm down to 212 $\mu$ in diameter. Around 20% of these cassiterite particles contain small inclusions of columbite-tantalite.

The reddish-white to brownish-white internal reflections of cassiterite from this area may be caused by:

**Table 3.** Mineralogical composition of panned concentrate from Omkoi area, Chiangmai

| Sample No. (%) wt<br>Phase | 065 | 066 | 067 | 068 | 069 | 070 |
|---|---|---|---|---|---|---|
| Cassiterite | 76.2 | 33.8 | 46.0 | 70.2 | 81.2 | 65.7 |
| Columbite-tantalite | – | 4.8 | 1.0 | 6.4 | 6.2 | 5.2 |
| Nb-Ta rutile | – | 0.3 | – | 0.5 | 0.7 | 0.1 |
| Samarskite | 0.1 | – | – | tr | 0.2 | tr |
| Wolframite | 0.2 | 0.6 | – | – | 0.8 | 0.1 |
| Scheelite | 0.1 | 2.2 | – | 0.1 | 0.1 | – |
| Monazite | 0.9 | 16.8 | 4.6 | 11.2 | 3.7 | 13.8 |
| Ilmenite | 0.3 | 2.1 | 2.0 | 2.5 | 3.5 | 0.2 |
| Maghemite | 0.4 | 1.4 | 0.2 | 1.2 | 1.2 | 0.4 |
| Xenotime | – | – | 0.1 | tr | tr | tr |
| Anatase | 13.1 | tr | tr | – | tr | – |
| Tourmaline | 4.4 | 0.8 | 0.9 | 0.3 | 0.1 | 0.1 |
| Rutile | tr | 0.5 | 0.1 | 0.3 | 0.5 | 0.1 |
| Garnet | 0.5 | 14.9 | 0.7 | 5.0 | 0.8 | 4.2 |
| Zircon | 0.1 | 2.6 | 1.4 | 0.7 | 0.1 | 1.1 |
| Lepidocrocite | 0.7 | – | 0.1 | tr | tr | – |
| Quartz | 3.0 | 18.8 | 40.6 | 1.5 | 0.9 | 9.0 |
| Pyrite | – | 0.1 | 0.2 | – | – | – |
| Beryl | – | 0.3 | – | 0.1 | – | tr |
| Mn-oxides | – | – | 2.1 | – | – | – |
| | 100.0 | 100.0 | 100.0 | 100.0 | 100.0 | 100.0 |

1) the exsolution bodies of columbite-tantalite, and
2) the presence of iron, tantalum and tungsten in the crystal structure.

The latter possibility is indicated by electronprobe analysis which shows this cassiterite to contain W, Ta and Fe in that order of decreasing relative abundance (Praditwan, 1984).

Columbite-tantalite in general occurs in three different forms:

1) Free particles of columbite-tantalite are present mostly as euhedral crystals, grey coloured, distinctly anisotropic, with straight extinction, and with deep red internal reflections. The particles are mostly homogeneous, rarely containing exsolution bodies of cassiterite. The size range of this columbite-tantalite varies from 1.7 mm to 425 $\mu$. Some columbite-tantalite particles show replacement of microlite or tapiolite along the margin and/or along the fractures of the particles (Figs. 2 and 3). Results from electronprobe analysis (Fig. 4) show that some of columbite-tantalite particles are very high in Ta and Fe with small amounts of Nb and Mn; consequently the mineral should be called Tapiolite. This mineral is generally found in pegmaties poor in lithium occurring in late albite or quartz muscovite replacement units (Knorring and Fadipe, 1981).
2) Columbite-tantalite which occurs as small inclusions in cassiterite is present in two different forms:

   a) Blebs disseminated in the cassiterite (Fig. 5). The sizes of the blebs range from a few microns to ~ 50 microns in diameter. The largest blebs tend to develop crystal faces, some of which occupy the intergrain boundaries of the cassiterite. Some of these blebs are concentrated in specific growth zones in cassiterite (Fig. 6).

   b) Oriented lamellae, which are less common and much smaller than the disseminated blebs. The sizes of the lamellae range from a few microns to ~ 50 microns in length with preferred orientations along crystallographic directions of the cassiterite host (Fig. 7).

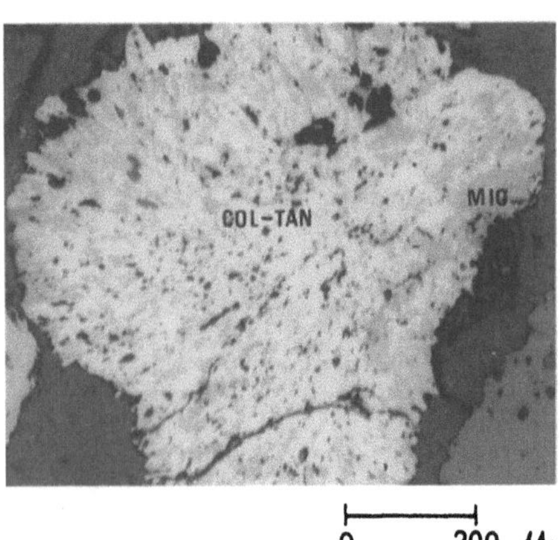


**Fig. 2.** Columbite-tantalite (col-tan) replaced by microlite (mic)

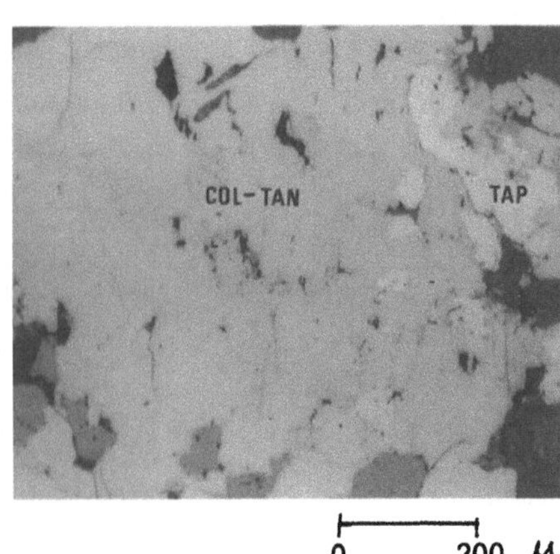


**Fig. 3.** Columbite-tantalite (col-tan) replaced by tapiolite (tap)

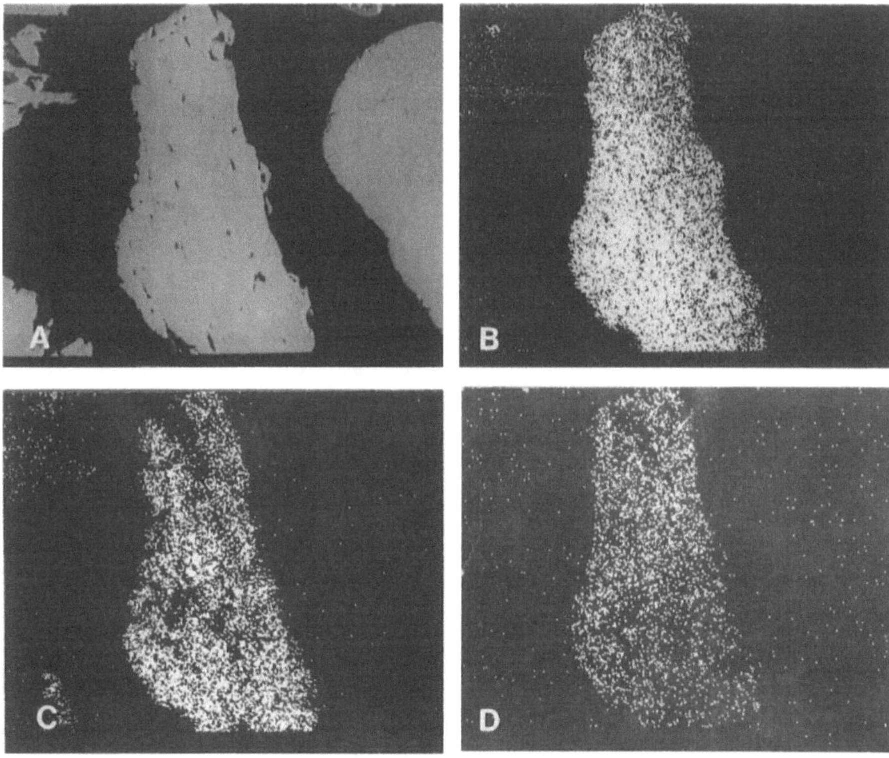


**Fig. 4A-D.** Tapiolite-free particles. **A** Back scattered electron image; **B** Ta-L$\alpha$ X-ray image; **C** Fe-K$\alpha$ X-ray image; **D** Nb-L$\alpha$ X-ray image

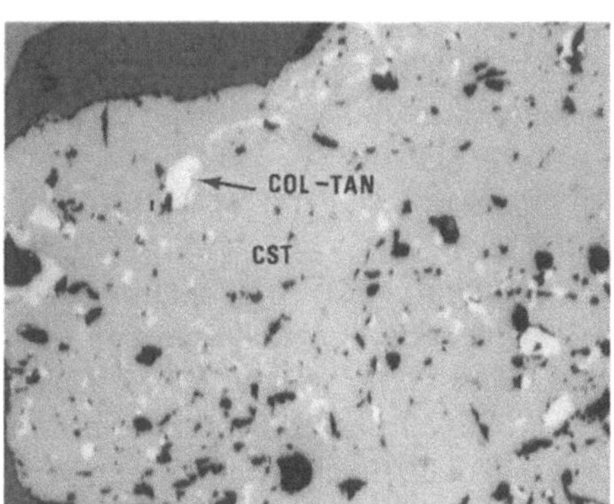


**Fig. 5.** Columbite-tantalite (col-tan) blebs disseminated in cassiterite (cst)

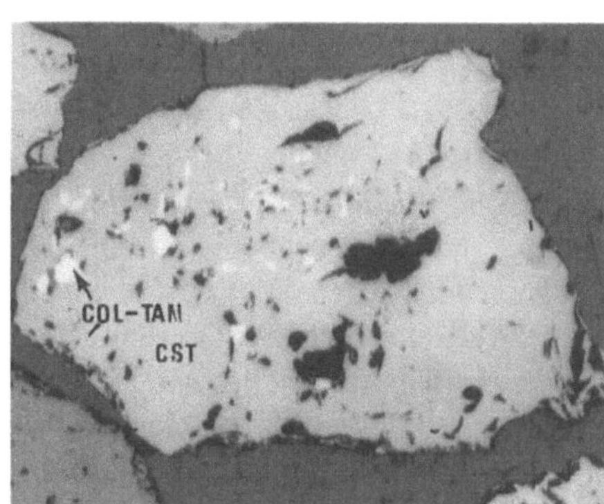


**Fig. 6.** Columbite-tantalite (col-tan) blebs are confined to growth zones in cassiterite (cst)

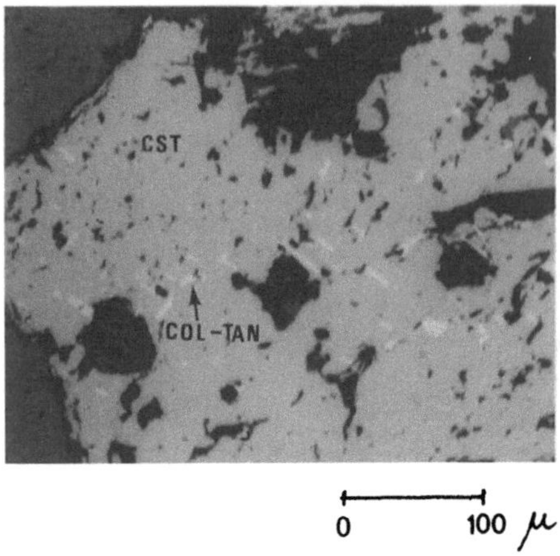


**Fig. 7.** Oriented lamellae bodies of columbite-tantalite (col-tan) in cassiterite (cst)

3) Columbite-tantalite intergrown with cassiterite as simple and mottled types of intergrowth (Amstutz, 1961). This columbite-tantalite is present as subhedral crystals, and is distributed in the size range 20–100 microns. Some of the columbite-tantalite is intergrown with quartz in simple intergrowth.

Nb-Ta rutile, less common than columbite-tantalite in this area, is present mostly as homogeneous particles which have grey colour, weak bireflectance and strong anisotropism from reddish-grey to dark brown; and commonly having a twinned habit. Less common inhomogeneous particles show grey to dark grey patches of the Nb-Ta rutile groundmass and columbite-tantalite inclusions. The columbite-tantalite inclusions in this Nb-Ta rutile are present as irregular bodies in a wide size range and are evenly distributed throughout the groundmass. According to Uytenbogaardt and Burke (1971), "to distinguish between Nb-rutile (ilmenorutile) and Ta-rutile (strüverite) in polished section: all samples containing columbite exsolution should be called Nb-rutile; homogeneous minerals are probably Ta-rutile, but chemical investigation is always necessary". Thus, the homogeneous and inhomogeneous Nb-Ta rutile as described above should be called strüverite and ilmenorutile respectively.

The other Nb-Ta-bearing mineral found is samarskite which everywhere in Southern Thailand is found in the final tin tailings.

According to Garson et al. (1975), the general formula for samarskite is $(Fe, Y, U)_2(Nb, Ti, Th)_2O_7$ with 32–48% of $Nb_2O_5$ and 5–12% of $TaO_2$ and for yttrotantalite is $(Y, Fe)_5[(Ta, Nb)_5O_7]_3$ with 15–21% of $Nb_2O_5$ and 37–47% of $TaO_2$. In Thailand, the mineral is probably intermediate in this series.

Samarskite in general occurs as anhedral to subhedral crystals, grey coloured, with weak bireflectance and weak anisotropism with distinct yellowish-brown to dark brown internal reflections. Alteration rims of samarskite to wiikite (?) (Winchell, 1951) are also observed. The electronprobe image (Fig. 8) shows that the samarskite from the Omkoi area is relatively higher in Ta than Nb. Other elements present are Ti, Fe, Y, U and Th, with trace constituents of W and Mn.

The results from electronprobe analysis of the ordinary rutile indicate the presence of trace constituents of Fe and Nb. This rutile shows a creamy-grey colour, weak bireflectance, very strong anisotropism, common twinned habit and yellowish-brown to reddish-brown internal reflections. Some rutile forms exsolution intergrowths with ilmenite.

## Patana Mine, Kanchanaburi

Patana Mine is located in Kanchanaburi province about 80 km west of Bangkok. Several products from the concentrator plant of this gravel pumping mine have been investigated with regard to mineralogy and liberation characteristics of the Nb-Ta minerals. The mineralogical study of the particulate samples was carried out by examination of polished sections with a polarizing microscope; selected polished sections were investigated semi-quantitatively using an electronprobe micro-analyzer.

The results indicate that the niobium-tantalum-bearing minerals, which are found associated with cassiterite in the area are columbite-tantalite and strüverite (Ta-rutile). The other heavy minerals include garnet, ilmenite, maghemite, lepidocrocite, tourmaline, zircon, monazite, rutile, Mn-oxides, spinel, anatase and allanite, with small amounts of Fe-hydroxides, scheelite, magnetite and hematite (Table 4).

In general, the columbite-tantalite is similar to that from the Omkoi deposit, and can be classified into three different types:

1) Columbite-tantalite which occurs as small inclusions in cassiterite is present in two different forms:

   a) As blebs disseminated in cassiterite. The size range of the blebs varies from a few microns to ~ 30 microns in diameter. The results from electronprobe analyses show that the relative amount of elements present in these columbite-tantalite blebs are Ta, Nb and W with trace contents of Fe, Mn and Y.

   b) As oriented lamellae which are less common than the disseminated blebs. The size of the lamellae ranges from a few microns to about 100 microns in length and are oriented following the crystallographic directions of the cassiterite.

2) Free particles of columbite-tantalite are mostly of euhedral crystals, have straight extinction and deep red internal reflections. Heart-shaped contact twins and radiating aggregates of platy crystals are also observed.

3) Columbite-tantalite intergrowths with cassiterite and quartz are present in two forms:

   a) As simple and mottled type intergrowths (Fig. 9) in which columbite-tantalite is present as irregular grains and distributed in the size range 30 to 100 microns.

   b) As radiating crystals intergrown with cassiterite, where the size of the individual columbite-tantalite crystals ranges from a few microns to about 100 microns (Fig. 10). From this figure it appears that cassiterite occurs as open space fillings between columbite-tantalite crystals. This would indicate a later stage of formation of cassiterite.

Strüverite occurs in small amounts as inhomogeneous particles where columbite-tantalite inclusions are set in a Nb-Ta rutile groundmass. The results from electronprobe analysis (Fig. 11) show that the groundmass is actually richer in Ta than Nb. Therefore, the mineral is strüverite (Ta-rutile) with inclusions of columbite-tantalite. In general these inclusions are present as irregular bodies, with a wide size range and are evenly distributed throughout the groundmass.

Strüverite from this area occurs mainly as free particles and is rarely intergrown with cassiterite.

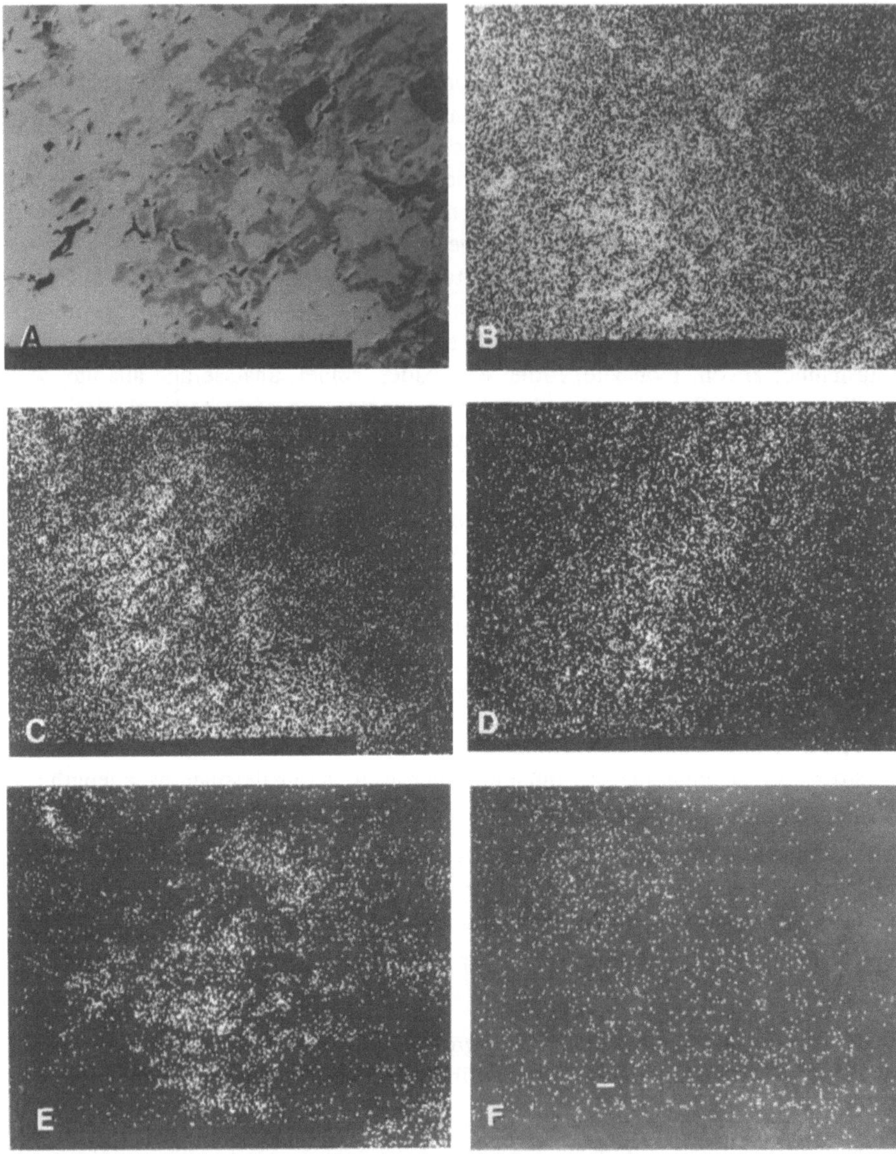


**Fig. 8A-F**

Phuket, Phangnga and Trang Areas

The mineralogical study of tin and Nb-Ta minerals from several products of the concentrator plants from different mines in Southern Thailand, namely Chaiyayuth dredge in Phangnga, Tantikovit Mine in Phuket and Loon Seng Mine in Trang, was carried out by examination of polished sections with a polarizing microscope. Selected polished sections were investigated semi-quantitatively using an electronprobe microanalyzer.

The results show that the niobium-tantalum-bearing minerals, associated with cassiterite, are columbite-tantalite, strüverite (Ta-rutile) with some samarskite. The

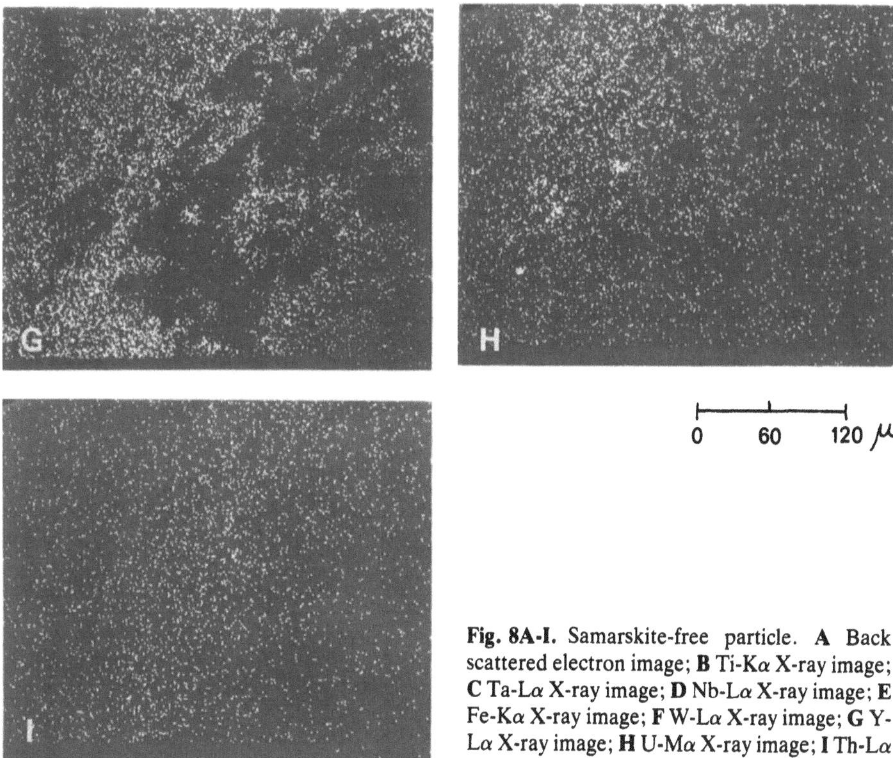


**Fig. 8A-I.** Samarskite-free particle. **A** Back scattered electron image; **B** Ti-Kα X-ray image; **C** Ta-Lα X-ray image; **D** Nb-Lα X-ray image; **E** Fe-Kα X-ray image; **F** W-Lα X-ray image; **G** Y-Lα X-ray image; **H** U-Mα X-ray image; **I** Th-Lα X-ray image

related heavy mineral assemblage includes: monazite, spinel, ilmenite, maghemite, hematite, Mn-oxides, rutile, anatase, xenotime, topaz, martite, wolframite, garnet, zircon, tourmaline, goethite, limonite, pyrite, marcasite and arsenopyrite as shown in Tables 5, 6 and 7 (Praditwan, 1983b).

Cassiterite, columbite-tantalite, Nb-Ta rutile and samarskite have similar characteristics to those from the northern and central part of Thailand.

Free particles of cassiterite are present mostly as anhedral crystals, having a brownish-grey colour, weak bireflectance, very distinct anisotropism, a common twinned habit and displaying brownish-white internal reflections. About 10% of the cassiterite free-particles contain small inclusions of columbite-tantalite.

Some cassiterite is intergrown with quartz in simple and mottled type intergrowths. From the Phuket area some cassiterite contains small inclusions of chalcopyrite, and some cassiterite is coated with a thin layer of Mn-oxides which causes it to become magnetic in character.

Columbite-tantalite in general occurs in three different forms:

1) Free particles of columbite-tantalite are present mostly as euhedral to subhedral crystals, grey coloured, distinctly anisotropic, with straight extinction and deep red internal reflections. Some columbite-tantalite shows replacement by microlite along the margin and/or along the fractures of the particles.

   Electronprobe analysis of a few columbite-tantalite particles from Trang area show that the mineral contains more Ta than Nb and more Mn than Fe (Fig. 12) so this

**Table 4.** Mineralogical composition of several products from the concentrator plant of Patana Mine, Kanchanaburi

| Product (%) | Conc. from Pa-long & Jig | Sluice box conc. | Lan-chute fine conc. | Coar-se tails woo-den sluice | Fine tails from Lanc-hute | Coar-sest size final conc. | Coar-se size mag | Coar-se size non-mag | Midd-le size mag | Midd-le size non-mag | Fine size mag | Fine size non-mag | Fine size non-mag con-ductor | Fine size non-mag non-cond | Fine size magnetite product from Frantz Isodynamic separator | | | | |
|---|---|---|---|---|---|---|---|---|---|---|---|---|---|---|---|---|---|---|---|
| | | | | | | | | | | | | | | | Mag 0.3A | Mag 0.5A | Mag 0.7A | Mag 0.95A | Non-Mag 0.95A |
| Phase | (1) | (2) | (3) | (4) | (5) | (6) | (7) | (8) | (9) | (10) | (11) | (12) | (13) | (14) | (15) | (16) | (17) | (18) | (19) |
| Cassiterite | 24.2 | 77.5 | 16.1 | 33.2 | 0.8 | 98.3 | 5.8 | 99.0 | 3.0 | 94.6 | 1.7 | 81.2 | 97.6 | 3.0 | 0.2 | 0.9 | 3.0 | 2.9 | 53.0 |
| Ilmenite | 23.3 | 7.4 | 24.8 | 9.0 | 18.4 | – | 39.8 | – | 30.8 | – | 43.7 | 0.2 | 0.5 | – | 44.0 | 8.1 | 0.7 | – | – |
| Garnet | 26.9 | 8.3 | 26.4 | 13.0 | 33.9 | – | 24.9 | – | 41.6 | – | 37.3 | 0.5 | 0.3 | 0.1 | 36.2 | 19.5 | 0.7 | – | 0.4 |
| Lepidocrocite | 6.0 | 0.1 | 0.2 | 18.7 | 3.9 | – | 4.5 | – | 1.4 | – | 0.3 | – | – | – | – | 3.4 | 1.8 | – | – |
| Tourmaline | 3.6 | – | 1.2 | 2.0 | 14.6 | – | 1.5 | – | 0.4 | – | 1.3 | 0.8 | – | 2.3 | – | 24.2 | 17.1 | – | 1.0 |
| Maghemite | 6.8 | 2.3 | 16.0 | 2.2 | 8.2 | – | 9.1 | – | 17.0 | 0.6 | 12.8 | – | – | – | 19.5 | 6.9 | 0.2 | 0.5 | – |
| Columbite-Tantalite | 0.3 | 0.2 | 0.2 | 0.3 | 0.3 | – | 8.6 | tr | 5.4 | 0.5 | 1.1 | – | 0.2 | 0.1 | 0.1 | 27.7 | 1.6 | – | 0.1 |
| Strüverite (Ta-rutile) | 0.1 | 0.1 | – | – | – | – | 1.6 | – | – | – | 0.1 | – | – | – | – | 0.2 | 2.1 | – | – |
| Allanite | 0.2 | – | – | 2.7 | 0.9 | – | 0.1 | – | – | – | – | – | – | – | – | – | – | – | – |
| Rutile | 0.7 | 0.4 | 1.8 | – | – | – | – | – | – | – | – | 1.1 | – | – | – | – | 0.2 | 0.3 | 4.3 |
| Anatase | 0.3 | – | – | – | – | – | – | – | – | – | – | 0.8 | 0.7 | 0.3 | – | – | – | – | 0.6 |
| Goethite | 0.2 | – | – | – | 0.4[*2] | – | – | – | – | 0.2 | – | – | – | – | – | – | – | – | – |
| Spinel | 0.4 | – | – | 0.3 | 1.0 | – | – | – | – | – | – | – | 1.4 | – | – | – | – | – | – |
| Mn-oxides | 0.5 | – | 0.3 | 12.2 | 0.3 | – | 0.3 | – | – | – | – | – | – | – | – | – | – | – | – |
| Fe-hydroxides | tr | – | 0.6 | 0.4 | 0.2 | – | – | – | – | – | – | 0.6 | – | – | – | – | – | 0.4 | 2.6 |
| Monazite | 0.8 | 1.0 | 1.1 | 0.5 | 2.5 | – | 1.4 | – | 0.3 | 2.6 | 1.7 | 3.8 | – | 5.7 | – | 8.6[*3] | 72.4 | 95.5 | 2.6 |
| Zircon | 2.2 | 0.9 | 10.8 | tr | 0.3 | – | – | – | – | 0.2 | – | 9.7 | 0.2 | 84.0 | – | 0.3 | 0.2 | 0.3 | 32.3 |
| Scheelite | tr | 0.2 | 0.1 | tr | – | – | – | 0.1 | – | 0.1 | – | 1.0 | 0.1 | 2.9 | – | – | – | – | 1.6 |
| Quartz | 3.6 | 1.5 | 0.1 | 3.5 | 14.0 | 1.7 | 0.8 | 0.9 | 0.1 | 1.2 | – | 0.3 | 0.4 | 0.2 | – | 0.1 | – | 0.1 | 0.2 |
| Magnetite | tr | 0.1 | – | 1.3 | – | – | 1.5 | – | – | – | – | – | – | – | – | – | – | – | – |
| Fluorite | tr | – | – | 0.7 | 0.3 | – | – | – | – | – | – | – | – | – | – | – | – | – | – |
| Arsenopyrite | tr | – | 0.3[*1] | tr | tr | – | 0.1 | – | – | – | – | – | – | – | – | – | – | – | 1.2[*1] |
| | 100.0 | 100.0 | 100.0 | 100.0 | 100.0 | 100.0 | 100.0 | 100.0 | 100.0 | 100.0 | 100.0 | 100.0 | 100.0 | 100.0 | 100.0 | 100.0 | 100.0 | 100.0 | 100.0 |

[*1] – Chalcopyrite and other sulphide minerals.
[*2] – Hematite.
[*3] – Including Xenotime 0.5%.

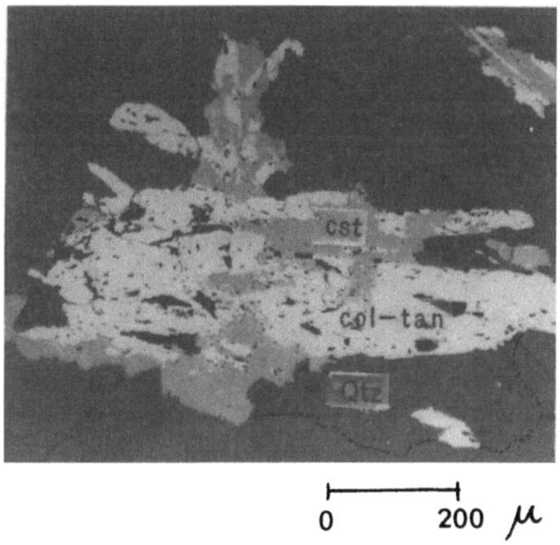


**Fig. 9.** Columbite-tantalite (col-tan) intergrown with cassiterite (cst) and quartz (qtz)

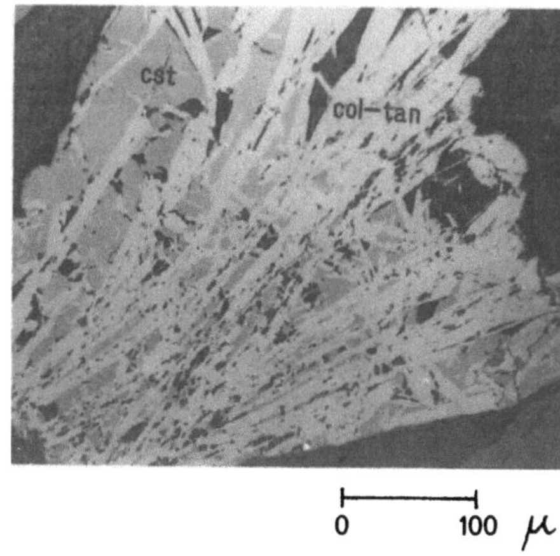


**Fig. 10.** Columbite-tantalite (col-tan) radiating crystals intergrown with cassiterite (cst)

columbite-tantalite series should be called "mangano-tantalite". According to Knorring and Fadipe (1981), this mineral may be regarded as characteristic of tantalum mineralization in lithium pegmatites. Mangano-tantalite also has been found associated with cassiterite in the Phuket tin field (Phetwaroon, 1984).

2) Columbite-tantalite as small inclusions in cassiterite is present in two forms:

a) As blebs disseminated in the cassiterite. The size of the blebs ranges from less than 10 microns to ~ 30 microns in diameter and is relatively small compared to the blebs from northern and central Thailand. In some cases the blebs are confined to growth zones in the cassiterite host.

b) Less commonly as oriented lamellae, the size of the lamellae ranges from <10 microns to ~ 50 microns in length (usually ~ 10 $\mu$), they are oriented in 1 or 2 crystallographic directions within the cassiterite.

3) Columbite-tantalite intergrown with cassiterite and quartz in simple type of intergrowth. A small proportion of columbite-tantalite grains from radiating aggregates with feldspar.

Nb-Ta rutile is present as homogeneous and inhomogeneous particles. The homogeneous Nb-Ta rutile particles show grey colour with weak bireflectance, strong anisotropism from reddish-grey to dark brown, and a common twinned habit. Electronprobe microanalysis indicates the following relative abundance of elements present in the homogeneous Nb-Ta rutile grains: Ti > Ta > Fe > Nb with some W (Fig. 13). This mineral contains relatively higher Ta than Nb so it should be called Ta-rutile or strüverite.

The inhomogeneous particles show grey colour for the Nb-Ta rutile groundmass with dark-grey patches of columbite-tantalite inclusions. The columbite-tantalite inclusions are present as irregular bodies in a wide size range that is evenly distributed throughout the groundmass. Textural relations of this columbite-tantalite suggests that it is an exsolved rather than an included phase (Wan Fuad Tuah, 1982). Electronprobe images were also taken for inhomogeneous Nb-Ta rutile. Results show that the groundmass is Ta-rutile whereas the exsolution phase is columbite-tantalite.

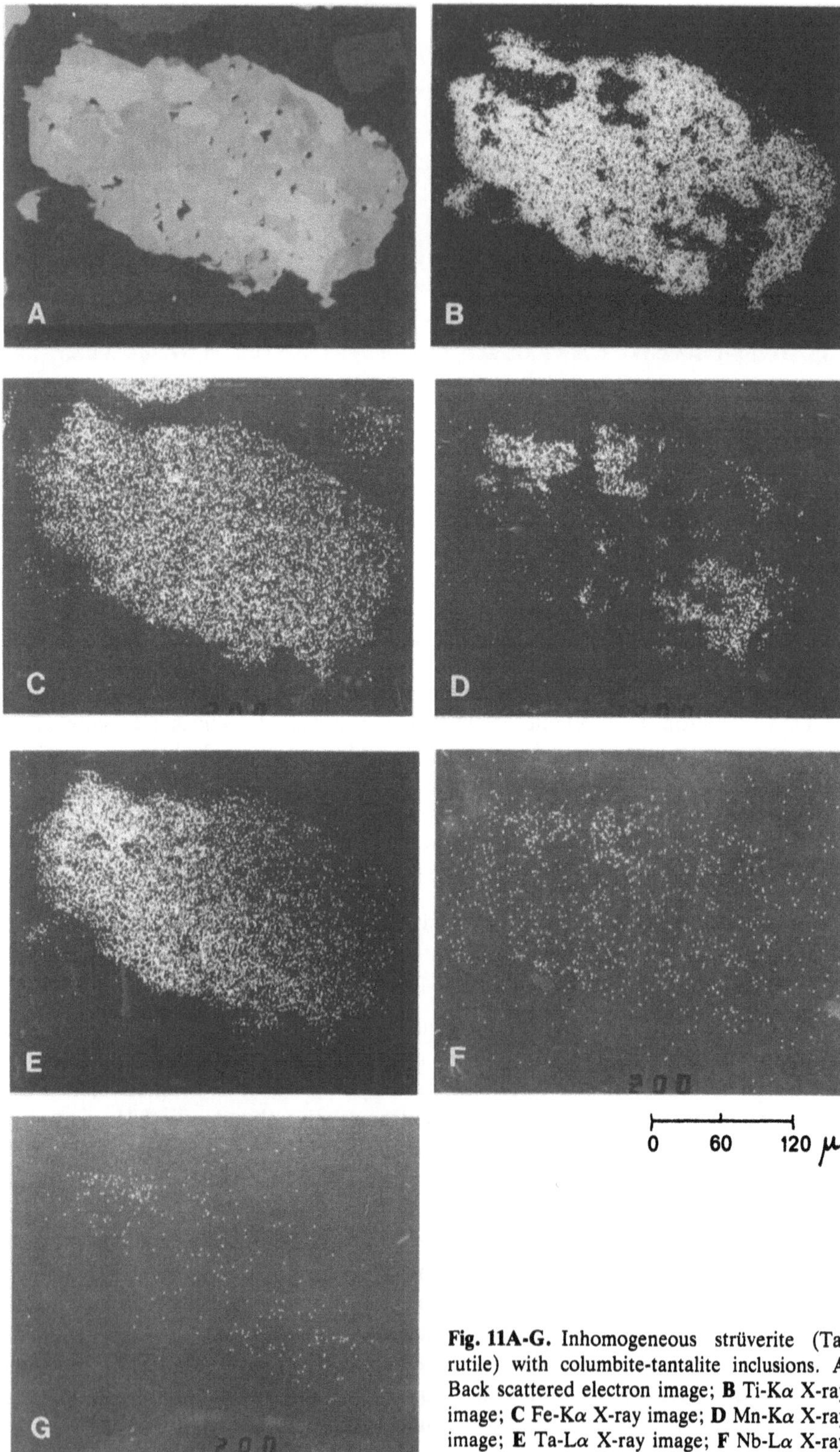


**Fig. 11A-G.** Inhomogeneous strüverite (Ta-rutile) with columbite-tantalite inclusions. **A** Back scattered electron image; **B** Ti-Kα X-ray image; **C** Fe-Kα X-ray image; **D** Mn-Kα X-ray image; **E** Ta-Lα X-ray image; **F** Nb-Lα X-ray image; **G** W-Lα X-ray image

**Table 5.** Mineralogical composition of several products from the concentrator plant of Chaiyayuth Dredge, Phangnga

| Product (%) wt<br>Phase | Feed to Plant | Coarse size from Willough-by | Fine size from Willough-by | Coarse tail from Lanchute | Coarse middling from Lanchute | Coarse final conc. | Coarse tailing from W & L | Fine tail from Lanchute | Fine tail from Willough-by | Fine tail from Lanchute | Fine middling from Lanchute | Fine final conc. |
|---|---|---|---|---|---|---|---|---|---|---|---|---|
| Cassiterite | 34.3 | 37.9 | 29.7 | 3.0 | 23.0 | 96.8 | 62.8 | 0.7 | 58.9 | 12.7 | 58.2 | 96.3 |
| Columbite-Tantalite | 1.4 | 0.4 | 0.6 | 0.8 | 0.4 | 1.5 | 1.7 | 0.5 | 1.7 | 1.3 | 3.7 | 1.6 |
| Strüverite (Ta-rutile) | 0.9 | 1.3 | 1.7 | 1.7 | 2.3 | – | 0.5 | 0.6 | 0.1 | 0.3 | 0.5 | – |
| Maghemite | 8.7 | 3.2 | 12.3 | 2.4 | 16.3 | – | 11.3 | 20.9 | 4.0 | 26.1 | 5.4 | 0.1 |
| Monazite | 7.7 | 12.9 | 2.5 | 7.0 | 1.3 | 0.8 | 2.5 | 1.5 | 3.0 | 3.2 | 3.6 | 0.8 |
| Tourmaline | 0.3 | 1.2 | 0.6 | 1.4 | tr | – | – | 0.5 | 0.2 | – | – | – |
| Ilmenite | 23.0 | 6.5 | 24.7 | 5.3 | 26.9 | 0.5 | 15.6 | 28.7 | 8.5 | 40.5 | 9.7 | 0.1 |
| Garnet | 13.7 | 22.9 | 18.6 | 43.0 | 25.9 | – | 3.5 | 24.0 | 10.6 | 10.3 | 17.3 | 1.1 |
| Topaz | 2.4 | 8.5 | 1.2 | 23.7 | 1.7 | – | – | – | – | – | – | – |
| Pyrite | 2.2 | 1.5 | 2.0 | 4.4 | 0.6 | – | 0.3 | 3.3 | 2.8 | 2.4 | 0.6 | – |
| Quartz | 3.6 | 2.5 | 4.2 | 4.1 | 0.3 | tr | 0.1 | 17.2 | 1.0 | tr | 0.1 | – |
| Mn-Oxides | 0.1 | 0.1 | – | | tr | – | – | – | – | – | – | – |
| Rutile | 0.4 | – | 0.6 | 0.9 | 0.3 | 0.2 | 0.7 | 0.6 | 0.5 | 1.5 | 0.3 | |
| Hematite | 0.1 | – | – | | – | – | – | – | 0.1 | 0.1 | 0.1 | – |
| Spinel | 0.1 | 0.1 | – | tr | tr | – | – | 0.4 | – | 0.1 | 0.1 | – |
| Zircon | 1.6 | tr | 0.7 | | – | – | 0.4 | – | 8.0 | 1.3 | 0.3 | – |
| Magnetite | 0.1 | 0.2 | – | 0.9 | 0.1 | – | – | 0.3 | – | – | – | – |
| Goethite | tr | 0.1 | – | 0.3 | – | – | – | – | – | – | – | – |
| Fe-hydroxides | tr | 0.6 | 0.2 | 0.9 | – | 0.2 | 0.4 | 0.2 | – | – | – | – |
| Anatase | tr | 0.1 | 0.4 | 0.1 | 0.6 | – | 0.2 | 0.6 | 0.6 | 0.2 | 0.1 | – |
| Scheelite | tr | – | – | 0.1 | 0.3 | – | – | – | – | – | – | – |
| | 100.6 | 100.0 | 100.0 | 100.0 | 100.0 | 100.0 | 100.0 | 100.0 | 100.0 | 100.0 | 100.0 | 100.0 |

**Table 6.** Mineralogical composition of several products from the concentrator plant of Tantikovit Mine, Phuket

| Phase / Product (%) wt | Feed to plant | Fine overflow from Willough-by | Oversize +1/8" | Coarse conc. from wooden sluice | Coarse tailing from wooden sluice | Coarse middling from wooden sluice | Fine conc. from Lanchute | Fine tailing from Lanchute | Dried conc. | Coarse size dried sample | Middle size dried sample | Fine size dried sample | Fine size magnetic product | Fine size non-mag-netic product | Middle size magnetic product | Middle size non-mag-netic product | Coarse size magnetic product | Coarse size non-mag-netic product |
|---|---|---|---|---|---|---|---|---|---|---|---|---|---|---|---|---|---|---|
| Cassiterite | 49.6 | 58.4 | 60.4 | 88.0 | 7.0 | 35.5 | 86.5 | 4.5 | 87.7 | 94.6 | 81.0 | 86.6 | 3.9 | 97.7 | 6.4 | 99.0 | 5.5 | 98.6 |
| Columbite-Tantalite | 0.3 | 0.6 | 0.2 | 0.7 | 0.1 | 0.1 | 0.9 | 0.3 | 1.0 | 0.6 | 0.6 | 1.8 | 7.9 | 0.6 | 5.5 | tr | 5.7 | – |
| Strüverite (Ta-rutile) | tr | – | – | – | – | – | – | – | – | – | – | – | 0.2 | – | 0.3 | – | – | – |
| Mn-Oxides | 13.1 | 4.3 | 14.8 | 0.7 | 35.0 | 40.9 | 0.4 | 9.3 | 0.7 | 1.6 | 0.6 | 0.4 | 1.1 | 0.1 | 2.7 | tr | 17.8 | 0.4 |
| Goethite | 0.8 | 0.3 | – | tr | 0.7 | 1.7 | – | tr | – | tr | 0.1 | 0.1 | – | – | – | – | 0.1 | – |
| Ilmenite | 0.4 | 0.6 | 0.3 | 0.1 | 0.5 | 2.7 | 0.2 | 0.4 | 0.3 | – | 0.4 | 0.3 | 2.1 | 0.1 | 1.8 | – | 5.9 | – |
| Garnet | 19.6 | 24.6 | – | 8.8 | 20.2 | 7.5 | 10.0 | 34.3 | 9.0 | 0.7 | 15.5 | 9.3 | 79.9 | 0.6 | 77.0 | – | 44.4 | – |
| Maghemite | 0.2 | 0.3 | – | 0.1 | 0.3 | 0.1 | 0.2 | 0.5 | 0.2 | 0.2 | 0.1 | – | 1.3 | – | 0.8 | – | 4.7 | – |
| Arsenopyrite | 0.9 | 0.4 | – | 0.4 | 0.4 | 0.1 | 0.3 | 0.2 | 0.2 | 0.5 | 0.4 | 0.3 | 2.2 | – | 4.9 | – | 16.3 | – |
| Tourmaline | 1.9 | 1.8 | – | tr | 6.5 | 1.2 | 0.1 | 6.4 | tr | tr | – | 0.1 | – | – | – | – | tr | – |
| Monazite | 0.1 | 0.5 | – | 0.1 | 0.1 | tr | 0.8 | 0.4 | 0.3 | tr | 0.4 | 0.6 | 1.2 | 0.5 | 0.4 | – | – | – |
| Pyrite | 0.1 | – | 0.3 | 0.1 | – | 0.3 | 0.2 | 0.2 | 0.1 | – | – | 0.1 | – | – | – | | tr | – |
| Quartz | 13.0 | 8.2 | 20.5 | 1.0 | 29.1 | 9.9 | 0.2 | 43.5 | 0.5 | 1.8 | 0.9 | 0.3 | – | 0.3 | 0.2 | 1.0 | 0.2 | 1.0 |
| Zircon | – | – | – | – | – | – | 0.1 | – | – | – | – | – | – | 0.1 | – | – | – | – |
| Rutile | tr | – | – | tr | 0.1 | tr | – | – | tr | – | – | 0.1 | – | – | – | – | tr | – |
| Hematite | tr | – | 3.4 | – | – | – | 0.1* | – | – | – | – | – | – | – | – | – | – | – |
| Chalcopyrite | – | – | – | – | – | tr | – | – | – | – | – | – | 0.2 | – | – | – | tr | – |
| | 100.0 | 100.0 | 99.9 | 100.0 | 100.0 | 100.0 | 100.0 | 100.0 | 100.0 | 100.0 | 100.0 | 100.0 | 100.0 | 100.0 | 100.0 | 100.0 | 100.6 | 100.0 |

* Included martitized magnetite.

**Table 7.** Mineralogical composition of several products from the concentrator plant of Loon Seng Mine, Trang

| Phase \ Product (%) wt | Feed to plant | Fine size from Willoughby | Middle size from Willoughby | Coarse size from Willoughby | Fine tail from Lanchute | Fine middling from Lanchute | Fine conc. from Lanchute | Middle size tailing from Lanchute | Middle size middling from Lanchute | Middle size conc. from Lanchute | Coarse tail from Lanchute | Coarse middling from Lanchute | Coarse conc. from Lanchute | Fine magnetic product | Fine non-magnetic product (Final conc.) | Middle size magnetic product | Middle size non-magnetic product (Final conc.) | Coarse magnetic product | Coarse non-m. product (Final conc.) |
|---|---|---|---|---|---|---|---|---|---|---|---|---|---|---|---|---|---|---|---|
| Cassiterite | 34.3 | 16.9 | 44.4 | 88.5 | 1.4 | 4.4 | 30.3 | 2.1 | 12.4 | 86.2 | 64.2 | 83.1 | 94.7 | – | 78.8 | 4.8 | 95.2 | 15.1 | 95.4 |
| Columbite-Tantalite | 1.8 | 6.1 | 2.3 | 3.4 | 0.1 | 0.9 | 7.6 | 0.1 | 2.7 | 5.9 | 0.3 | 3.9 | 2.7 | 0.6 | 0.4 | 43.1 | 0.8 | 23.2 | 2.3 |
| Strüverite (Ta-rutile) | 1.8 | 0.4 | 1.3 | 0.2 | 0.1 | 1.0 | 0.3 | 0.2 | 3.6 | 1.1 | 2.5 | 1.7 | 0.4 | – | 1.8 | 0.7 | 1.6 | 0.5 | 0.8 |
| Maghemite | 2.0 | 13.3 | 1.1 | 0.1 | – | 5.6 | 15.9 | 0.2 | 0.9 | 0.9 | tr | – | 0.1 | 37.1 | – | 7.4 | – | 3.5 | – |
| Monazite | 0.7 | 1.9 | 0.4 | 0.6 | – | 0.8 | 1.5 | 0.1 | 0.3 | 0.3 | 0.2 | – | tr | 0.2 | 1.6 | 0.6 | 1.0 | 0.8 | – |
| Tourmaline | 20.6 | 9.9 | 32.6 | 4.1 | 40.9 | 24.1 | 1.7 | 70.6 | 44.2 | 0.4 | 20.6 | – | 0.1 | – | 0.1 | 1.2 | – | 13.8 | 0.3 |
| Ilmenite | 9.2 | 23.3 | 2.0 | 0.8 | 3.3 | 11.4 | 23.5 | 2.3 | 5.4 | 1.5 | 0.5 | 1.2 | 0.4 | 43.3 | 0.1 | 14.0 | – | 10.7 | – |
| Garnet | 11.0 | 16.8 | 6.1 | 0.4 | 6.2 | 33.8 | 14.3 | 1.0 | 21.4 | 2.4 | 1.2 | 3.5 | 0.4 | 18.8 | 0.6 | 27.1 | – | 28.2 | – |
| Pyrite | tr | – | 0.1 | tr | – | – | – | – | 0.2 | 0.1 | tr | – | – | – | 0.4 | – | 0.1 | tr | 0.2 |
| Arsenopyrite | tr | 0.4 | 0.2 | tr | – | 0.2 | 0.7 | 2.5 | 0.9 | 0.1 | 0.2 | 0.7 | tr | – | 0.8 | – | 0.5 | 0.1 | – |
| Quartz | 14.7 | 6.7 | 6.0 | 1.4 | 46.4 | 14.3 | 0.4 | 16.1 | 5.3 | 0.2 | 5.6 | 4.0 | 0.7 | – | 0.6 | – | 0.4 | 1.6 | 0.4 |
| Rutile | 0.8 | 1.7 | 1.1 | – | 0.4 | 2.1 | 0.2 | 0.1 | 0.6 | 0.6 | tr | 0.4 | tr | – | 3.5 | 0.2 | 0.3 | 0.3 | 0.4 |
| Spinel | 1.6 | – | 0.1 | – | 0.2 | – | – | – | – | – | – | – | – | – | – | – | – | – | – |
| Zircon | 0.7 | 1.4 | – | tr | – | – | 2.4 | tr | tr | 0.3 | tr | – | tr | – | 11.1 | – | – | – | – |
| Goethite | tr | tr | 0.8 | tr | tr | tr | – | 1.7 | 0.2 | – | 0.1 | 0.9 | tr | – | – | – | 0.1 | 1.2 | – |
| Fe-hydroxides | 0.3 | 0.3 | 0.8 | 0.5 | 0.8 | 0.5 | 0.1 | 3.0 | 1.9 | – | 4.6 | 0.4 | 0.4 | – | 0.1 | 0.7 | – | 0.8 | – |
| Anatase | 0.5 | 0.9 | 0.1 | tr | 0.2 | 0.9 | 0.4 | tr | 0.1 | – | – | 0.2 | – | – | 0.1 | 0.2 | – | 0.2 | – |
| Samarskite | 0.1 | – | 0.6 | – | – | – | 0.7 | – | – | – | – | – | 0.1 | – | – | – | – | – | 0.2 |
| | 100.0 | 100.0 | 100.0 | 100.0 | 100.0 | 100.0 | 100.0 | 100.0 | 100.0 | 100.0 | 100.0 | 100.0 | 100.0 | 100.0 | 100.0 | 100.0 | 100.0 | 100.0 | 100.0 |

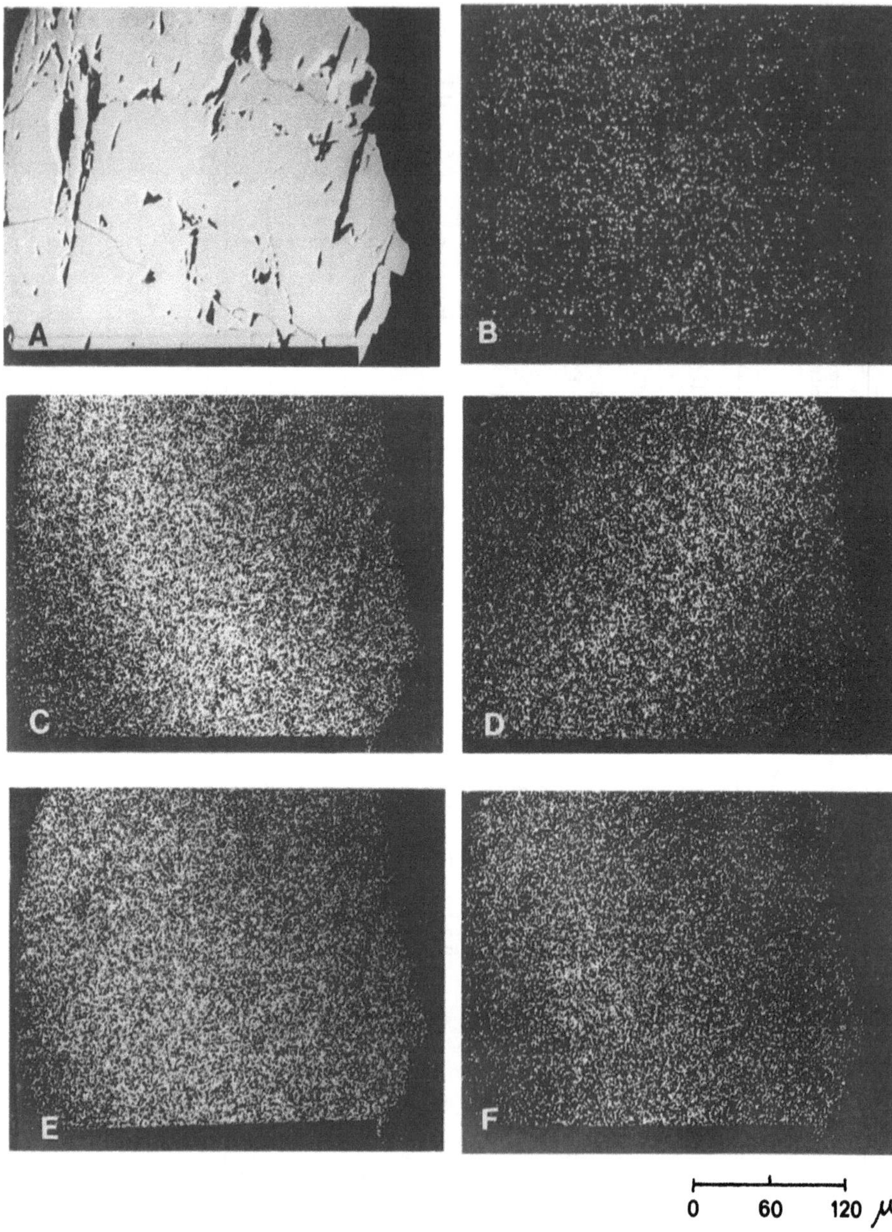


**Fig. 12A-F.** Manganotantalite X-ray images. **A** Back scattered electron image; **B** W-Lα X-ray image; **C** Ta-Lα X-ray image; **D** Nb-Lα X-ray image; **E** Mn-Kα X-ray image; **F** Fe-Kα X-ray image

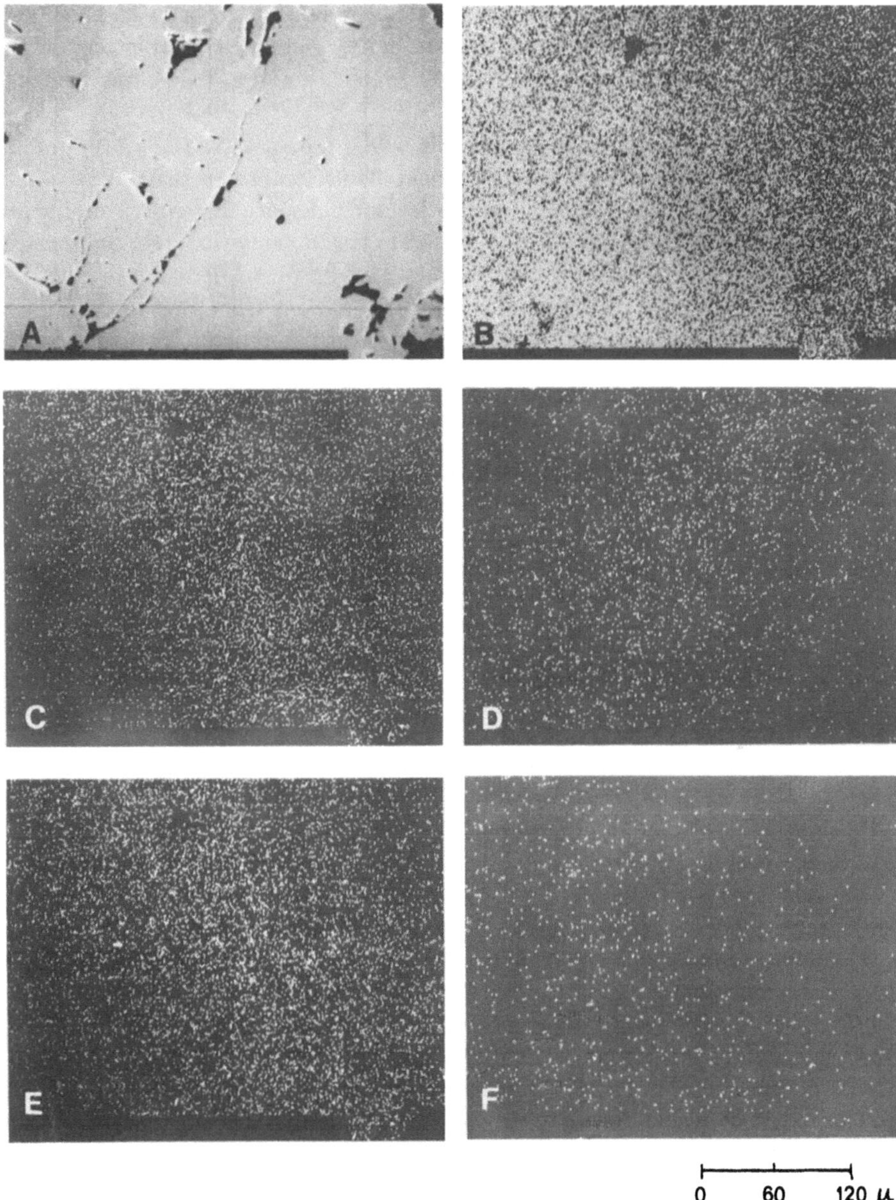


**Fig. 13A-F.** Homogeneous strüverite (Ta-rutile) particle. **A** Back scattered electron image; **B** Ti-Kα X-ray image; **C** Ta-Lα X-ray image; **D** Nb-Lα X-ray image; **E** Fe-Kα X-ray image; **F** W-Lα X-ray image

Intergrowths of Nb-Ta rutile with cassiterite as well as quartz are very common. Nb-Ta rutile and cassiterite are present as coarse masses (simple type of intergrowth); some show replacement texture with cassiterite replacing the Nb-Ta rutile forming uneven boundaries.

Rutile, present as blebs in the Nb-Ta rutile host, is also observed.

In some cases Nb-Ta rutile is more abundant than columbite-tantalite.

Another Nb-Ta-bearing mineral which is found associated with cassiterite is samarskite. It occurs as anhedral to subhedral crystals, grey-coloured, with weak bireflectance, weak anisotropism and distinct light brown to dark brown internal reflections. The particles all show patchy tonal variations from grey to dark grey, perhaps due to variations in the chemical composition. Alteration rims of samarskite to wiikite (?) are also observed.

## The Occurrence of Niobium-Tantalum-Bearing Minerals in Malaysia

The association of cassiterite and niobium-tantalum-bearing minerals in Malaysian tin deposits are also found mainly in pegmatitic tin deposits in the so-called "Western Tin Belt". According to Wan Fuad Tuah (1982), these pegmatitic tin deposits can be classified into two types:

1) columbite-tantalite-rich pegmatitic tin deposits, and
2) columbite-tantalite-poor pegmatitic tin deposits.

Columbite-tantalite-rich pegmatitic tin deposits are found only at Bakri and Semiling (Fig. 14). From Bakri, where columbite-tantalite was found in garnet-rich, quartz-feldspar-mica pegmatite, other niobium-tantalum minerals which have also been found are wodginite, fersmite, tapiolite, euxenite and fergusonite. The accessory minerals include: gahnite, ilmenite, Nb-Ta rutile, tourmaline, monazite, xenotime, garnet, pyrite, magnetite, anatase, allanite, siderite and rutile.

Columbite-tantalite-poor pegmatitic tin deposits; the pegmatites with small amounts of columbite-tantalite occur in Taiping, Chenderiang, southern Kinta Valley and Kuala Lumpur areas. Columbite-tantalite is present as exsolution bodies in the Nb-Ta rutile particles. The other minerals observed are topaz, rutile, tourmaline, beryl, fluorite, muscovite, alkali feldspar and lepidolite.

From their compositions, the Malaysian columbites and tantalites may be classified as follows:

| | |
|---|---|
| Bakri specimens | – Ferrocolumbite and Mn-tantalite with Ta-rich species such as wodginite and tapiolite |
| Semiling specimens | – Ferrocolumbite |
| Kinta Valley and Kuala Lumpur areas | – Ordinary columbite |

In addition, due to the high content of tungsten in the columbites from Kinta Valley and Kuala Lumpur, they may also be regarded as tungstenian-columbite.

Nb-Ta rutile is rather confined in distribution; it is mainly found in the Western Tin Belt of the peninsula, and is very rare in the Eastern Belt. It is particularly common in and around the Kinta Valley, especially at Tanjong Tualang, Kampong Gajah and

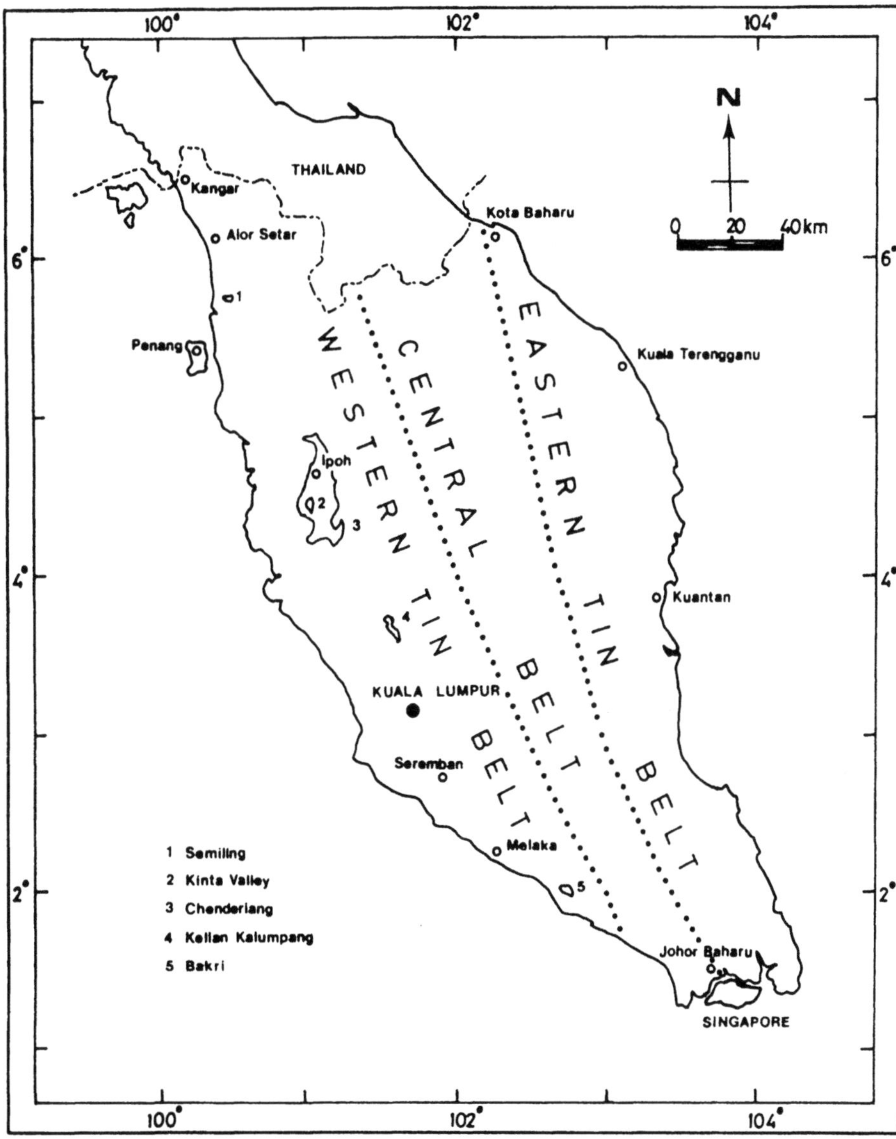


**Fig. 14.** Distribution of Nb-Ta bearing minerals associated with tin in Malaysia (after Wan Fuad Tuah, 1982)

Chenderiang. Examination of the polished sections shows that Nb-Ta rutile is mainly inhomogeneous and contains exsolved columbite-ixiolite, cassiterite, rutile and ilmenite inclusions.

The results from chemical analyses indicate a variation of the Nb-Ta ratio, which show that in most specimens Nb predominantes over Ta and confirms the bulk analyses that Nb-rutile is more common than Ta-rutile. The Ta-rich variety is observed in samples from Bakri, Kalumpang and Salak North.

## Practical Recovery Implication

According to Praditwan (1983a), the liberation study of columbite-tantalite from Patana Mine was performed in the feed material (concentrate from palong and jig) and magnetic products (Fig. 15). The columbite-tantalite content in the feed is 0.3% and becomes 8.6% in the coarse size magnetic product, 5.4% in the middle size magnetic product and 1.1% in the fine size magnetic product (Table 4).

Figure 16 shows the percentage columbite-tantalite distribution as free particles and intergrowth with other minerals in different types for different products. The columbite-tantalite liberation in the feed material is 32%. This amount is available for recovery without further grinding. Since the liberation in the magnetic products (coarse, middle and fine sizes) reaches around 70%, it can be predicted that a major amount of the columbite-tantalite should be recoverable. Consequently the fine size magnetic product (Fig. 15[11]) was passed through a Frantz Isodynamic separator at different intensities. The figures shown in Table 4 (Mag. 0.5A [16]) prove that it is possible to obtain a columbite-tantalite concentrate with fairly high recoveries. However, it should be noted that the field intensity during magnetic separation should be carefully adjusted, since columbite-tantalite reports in a fairly close range of intensities. Similar results could be obtained from coarse and middle size magnetic products, which show even higher contents of columbite-tantalite and about the same degree of liberation (Table 4, Fig. 16).

To further upgrade the columbite-tantalite concentrate, high tension separation should be tried.

Strüverite, which is present in small amounts and occurs mostly as free particles, could also be separated by carefully controlled magnetic separation (Mag. 0.7A, Table 4).

It should be noted that around 30% of cassiterite "free" particles from this area contain a small amount of inclusions of columbite-tantalite, the size ranges from a few microns to around 30 microns. It will be realized that there is little possibility of liberating the individual mineral through grinding, while retaining the particles in a size range suitable for mechanical separation. So some of the niobium and tantalum would report in the cassiterite concentrates for possible recovery during the tin smelting process.

## Conclusions

Mineralogical study of panning concentrates from Omkoi area and several products from concentrator plant of Patana Mine, Chaiyayuth dredge, Tantikovit Mine and Loon Seng Mine, was carried out by microscopic and electron microprobe examination of polished sections.

Results from the Omkoi area show that the common Nb-Ta-bearing minerals found associated with cassiterite are columbite-tantalite and Nb-Ta rutile with some tapiolite and samarskite. The associated heavy mineral suite is ilmenite, garnet (spessartite), monazite, rutile, zircon, maghemite, tourmaline, anatase, scheelite, wolframite, with some lepidocrocite, xenotime, Mn-oxides, beryl and pyrite.

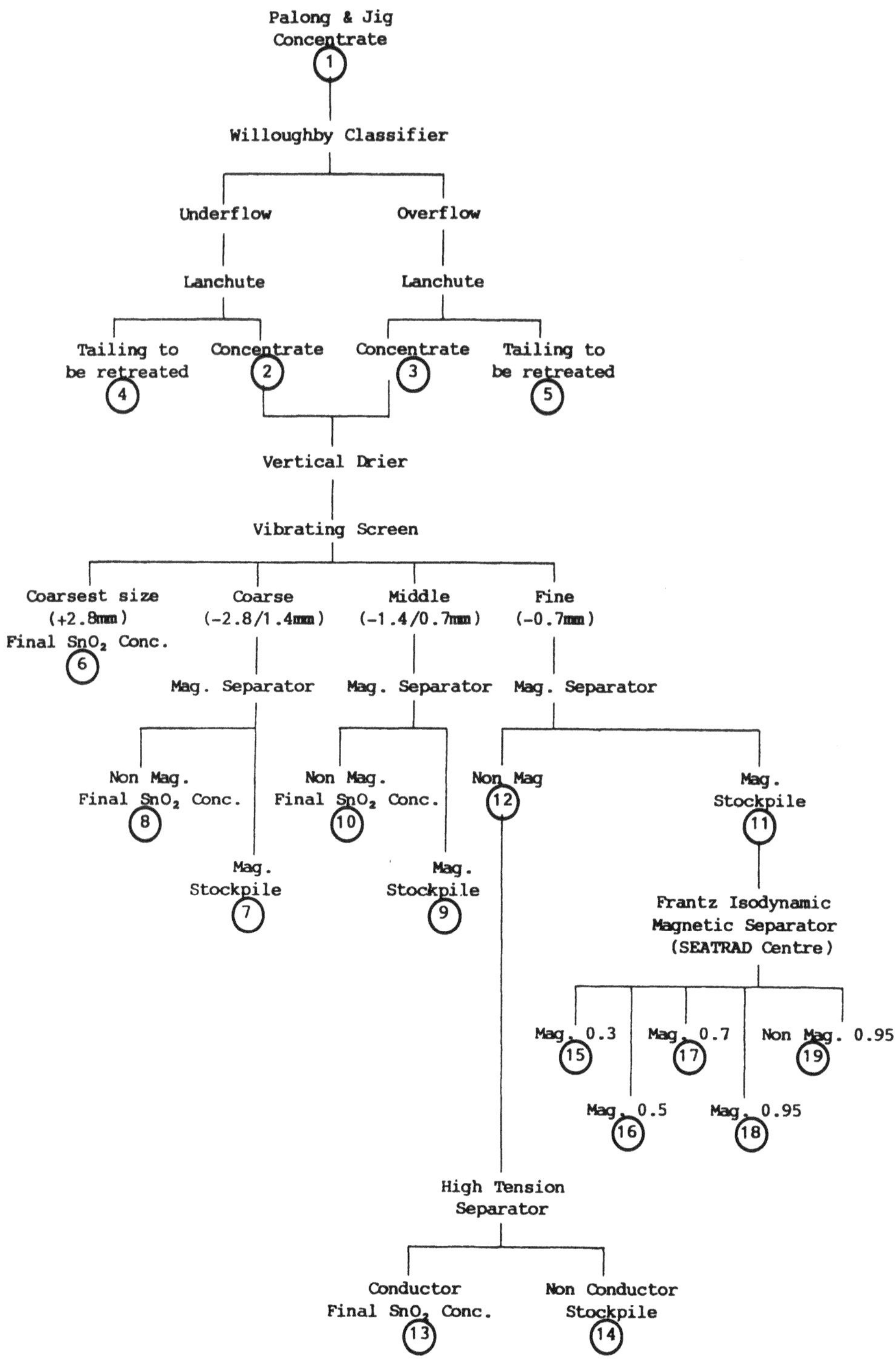


**Fig. 15.** Treatment plant flowsheet of Patana Mine

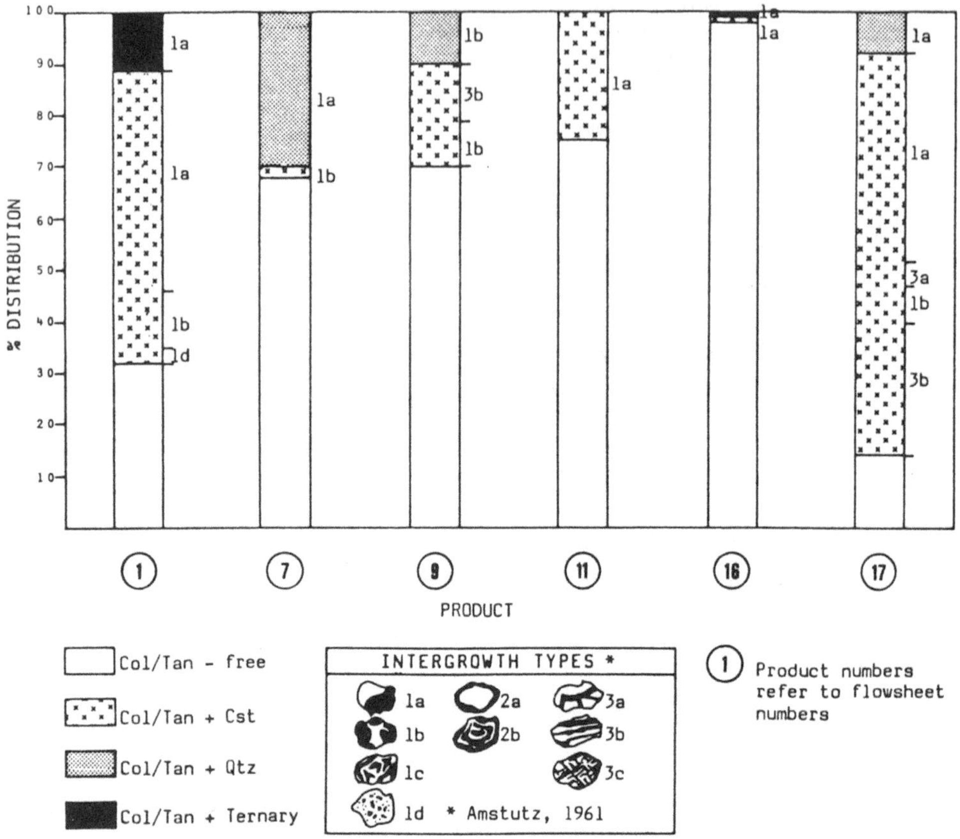


**Fig. 16.** % Columbite-tantalite distribution, as free particles and intergrown with other minerals in different intergrowth types, for different products

Niobium-tantalum-bearing minerals associated with cassiterite from Patana Mine are columbite-tantalite and Ta-rutile (strüverite). Other heavy minerals include garnet, ilmenite, maghemite, lepidocrocite, tourmaline, zircon, monazite, rutile, Mn-oxides, spinel, anatase and allanite, with small amounts of Fe-hydroxides, scheelite, magnetite and hematite.

In southern Thailand the cassiterite-related Nb-Ta minerals are columbite-tantalite, Ta-rutile (strüverite) with some samarskite. The heavy mineral assemblage also includes monazite, spinel, ilmenite, maghemite, hematite, Mn-oxides, rutile, anatase xenotime, topaz, martite, wolframite, garnet, zircon, tourmaline geothite, limonite, pyrite, marcasite and arsenopyrite.

In general cassiterite, columbite-tantalite, Nb-Ta rutile and samarskite from the three different areas have similar characteristics as follows.

## Cassiterite

Cassiterite in general occurs mostly as free particles. Some are intergrown with columbite-tantalite, quartz, and to a lesser extent with Nb-Ta rutile. From the Phuket area some cassiterite is coated with a thin layer of Mn-oxides which causes the cassiterite to become magnetic in character.

Free particles of cassiterite are present mostly as anhedral crystals, brownish-grey in colour, with weak bireflectance, very distinct anisotropism, commonly a twinned habit and showing reddish-white to brownish-white internal reflections. About 10–20% of the cassiterite particles contain small inclusions of columbite-tantalite.

The reddish-white to brownish-white internal reflections of cassiterite may be caused by:

1) exsolution bodies of columbite-tantalite, and
2) the presence of iron, tantalum and tungsten in the crystal structure.

## Columbite-Tantalite

Columbite-tantalite in general occurs in three different forms:

1) Free particles of columbite-tantalite are present mostly as euhedral crystals, grey coloured, with distinct anisotropism, straight extinction and deep red internal reflections. The particles are mostly homogeneous, rarely containing exsolution bodies of cassiterite.

Results from electronprobe microanalyses show that some of the columbite-tantalite series from the northern part is tapiolite (very high in Ta and Fe, with a small amount of Nb and Mn), which is generally derived from pegmatites poor in lithium, occurring in late albite or quartz muscovite replacement units. In contrast, the columbite-tantalite series from the southern part of the country is manganotantalite (higher Ta than Nb and higher Mn than Fe), which is characteristic of tantalum mineralization in lithium pegmatites.

Some columbite-tantalite shows replacement by microlite along the margin and/or along the fractures of the particles. From Omkoi area columbite-tantalite replaced by tapiolite is also observed.

2) Columbite-tantalite occurs as small inclusions in cassiterite, which is present in two different forms:

a) As blebs disseminated in the cassiterite. The size of the blebs is relatively coarse in Omkoi and Patana Mine (From a few microns to $\sim$ 50 microns) whereas elsewhere the size of the blebs is very small (from $<$ 10 microns to 30 microns). The coarse blebs tend to develop crystal faces, some of which occupy the intergrain boundaries of the cassiterite. Some of these blebs are concentrated in specific growth zones in cassiterite.
b) As oriented lamellae which are less common and much smaller than the disseminated blebs. Size of the lamellae ranges from a few microns to $\sim$ 50 microns in length and are oriented in 1 or 2 crystallographic directions of the cassiterite host.

3) Columbite-tantalite intergrown with cassiterite as well as quartz in simple and mottled type of intergrowths. In some of these cases columbite-tantalite was replaced by cassiterite.

## Nb-Ta Rutile

Nb-Ta rutile is present as homogeneous and inhomogeneous particles. Homogeneous Nb-Ta rutile particles show grey colour with weak bireflectance, strong anisotropism from reddish-grey to dark brown and a common twinned habit. Electronprobe microanalysis indicates that this homogeneous Nb-Ta rutile is strüverite (Ta-rutile).

The inhomogeneous Nb-Ta rutile particles consist of a grey Ta-rutile groundmass with dark grey patches of columbite-tantalite inclusions. The columbite-tantalite inclusions are present as irregular bodies in a wide size range and evenly distributed throughout the groundmass. Textural relations of this columbite-tantalite suggest that it is an exsolved rather than an included phase.

Intergrowths of Nb-Ta rutile with cassiterite and/or quartz are very common, except at Patana Mine where Nb-Ta rutile occurs mainly as free particles and is rarely intergrown with cassiterite.

## Samarskite

Samarskite occurs as anhedral to subhedral, grey coloured, with bireflectance and weak anisotropism and having distinct light brown to dark brown internal reflections. The particles all show patchy tonal variation from grey to dark grey possibly due to variations in the chemical composition. Alteration rims of samarskite to wiikite (?) are also observed.

A comparison of the general characteristics of Nb-Ta bearing minerals in Thailand and Malaysia is as follows:

- Nb-Ta-bearing minerals from these two countries are closely associated with tin-bearing pegmatites along the Western Tin Belt of the Burma-Thai-Malaya orogen.
- Columbite-tantalite in Thailand in general contains higher Ta than Nb, and tends to be high in Fe in the north, whereas in the south, Mn is predominant. In Malaysia columbite as well as ferrocolumbite is more common than tantalite. In the southern part (Bakri) ferrocolumbite and Mn-tantalite with Ta-rich species such as wodginite and tapiolite are also observed.
- In Thailand Ta-rutile is more common than Nb-rutile, and is present as both homogeneous and inhomogeneous particles (with exsolution bodies of columbite-tantalite); whereas in Malaysia Nb-rutile is more common than Ta-rutile.
- Tungsten is everywhere present as a trace constituent in columbite-tantalite from Thailand, whereas in Malaysia tungstenian columbite has been reported.
- No well-defined zoning of columbite-tantalite relative to cassiterite is apparent from the mineralogical data reported here.

From an operational point of view, data reported here suggest that by carefully applying controlled magnetic and high tension separation to the magnetic products from the mine, it should be possible to obtain a columbite-tantalite concentrate with fairly high recoveries.

*Acknowledgements.* The author wishes to express his appreciation to SEATRAD Centre for permission to present this paper. The author also gratefully acknowledges Dr. A.J. Sinclair for reading through the manuscript and giving invaluable advice in the preparation of this paper.

## References

Amstutz, G.C., 1961. Microscopy applied to mineral dressing. *Quarterly of the Colorado School of Mines,* 56(3), 443–481.

Garson, M.S., Young, B., Mitchell, A.H.G. and Tait, B.A.R., 1975. The geology of the tin belt in Peninsular Thailand around Phuket, Phangnga and Takua Pa. *Overseas Mem.* No. 1, Instit. Geol. Sci., London.

Gocht, W. and Pluhar, E., 1982. Types of tin-bearing pegmatites in Phuket, Thailand, with special reference to tantalum-rich ores: in *Metallization associated with acid magmatism* (Ed. by A.M. Evans), John Wiley, Chichester, 91–99.

Hosking, K.F.G., 1977. Known relationships between the 'hard-rock' tin deposits and the granites of Southeast Asia. *Bull. Geol. Soc. Malaysia* No. 9, 141–157.

Knorring, O. von and Fadipe, A., 1981. On the mineralogy and geochemistry of niobium and tantalum in some granite pegmatites and alkali granites of Africa. *Bull. Mineral,* 104, 496–507.

Nutalaya, P., Campbell, K.V., Macdonald, A.S., Aranyakanon, P. and Suthakorn, P., 1979. Review of the geology of Thai tin fields. *Bull. Geol. Soc. Malaysia* No. 11, 137–159.

Phetwaroon, W., 1984. The mineralogical study of heavy minerals from tin-mines in Kathu Valley, Changwat Phuket, *M.Sc. Thesis, Department of Geology, Graduate School, Chulalongkorn University.*

Praditwan, J., 1983a. Some characteristics of niobium-tantalum bearing minerals from Patana Mine, Kanchanaburi, Thailand. In: *Conf. on Geol. and Min.Res. of Thailand, Bangkok,* 12 p.

Praditwan, J., 1983b. The mineralogical study of tin and niobium-tantalum minerals from several products of the concentrator plants from Chaiyayuth dredge in Phangnga, Tantikovit Mine in Phuket and Loon Seng Mine in Trang, Thailand (unpublished).

Praditwan, J., 1984. Mineralogical study of tin and niobium-tantalum bearing minerals in Omkoi area, Chiangmai, Thailand (Report of Investigation No 33, SEATRAD centre, Ipoh).

Uytenbogaardt, W. and Burke, E.A.J., 1971. *Tables for microscopic identification of ore minerals.* Elsevier Scientific Publishing Company, Amsterdam.

Vichit, P., 1983. Tin deposits of Northern Thailand. In: *Conf. on Geol. and Min. Res. of Thailand, Bangkok,* 10 p.

Winchell, A.N., 1951. *Elements of optical mineralogy,* Part II, 4th Edition, John Wiley & Sons, New York.

Wan Fuad Tuah bin Wan Hassan, 1982. Aspects of geochemistry and mineralogy of tin and associated minerals in the Malaysian tin field. *Ph.D. Thesis, University of Leeds.*

# 6.11.4 The Geological Characteristics of the Pilok Sn-W-Mo Deposits, West Thailand

C. MAHAWAT[1]


## Abstract

The Pilok area, situated along the Thai-Burmese border, is one of the most productive tin-tungsten primary deposits in Thailand. The ore deposits occur predominantly as vein systems as well as localized carbonate replacement skarn. The environment of the deposit is post-tectonic within the plutonic arc-fold belt. The ore is closely and spatially associated with multiphased intrusions of 2-mica megacrystic granite, foliated granite, 2-mica equigranular non-foliated granite, 2-mica tourmaline-sulphide ore-bearing fine-grained granite as well as aplites and greisenized granites. The age of the oldest intrusive phase (2-mica megacrystic foliated granite) is 93 ± 4 Ma with an initial $^{87}Sr/^{86}Sr$ ratio of 0.738. The granites intruded discordantly into schist and quartzite of Middle Paleozoic age.

Ore deposits are mostly concentrated at the cusps and the contact zones between the granites and country rocks as well as along fault zones. Tin and tungsten commonly occur together in the vein system, whereas the stockworks and swarms of ore-bearing quartz veinlets contain mostly tin. Pyrite, pyrrhotite, molybdenite and fluorite sometimes are associated with tin-tungsten in the fault zone and contact skarn ores. Scheelite and molybdenite occasionally occur in the quartz-veins within the country rocks, whereas the tin-tungsten ore-bearing veins are limited to the granite along the contact zone.


## Introduction

The Pilok tin-tungsten mineralized province is located along the western part of central Thailand. This province is one of the most productive tin and tungsten regions in the nation. There are at least 50 active mines. More than half are mining primary ore deposits. This area is close to the famous Tavoy tin-tungsten province of Burma.

Since tin and tungsten are major economic commodities in Thailand, a number of research and exploration projects concerning ore deposits have been intensively carried out by the Department of Mineral Resources and the universities. The mineralized areas along the western part of Thailand are included in the number of projects (Fig. 1).

[1] Geological Survey, Department of Mineral Resources, Rama IV Road, Bangkok 10400, Thailand

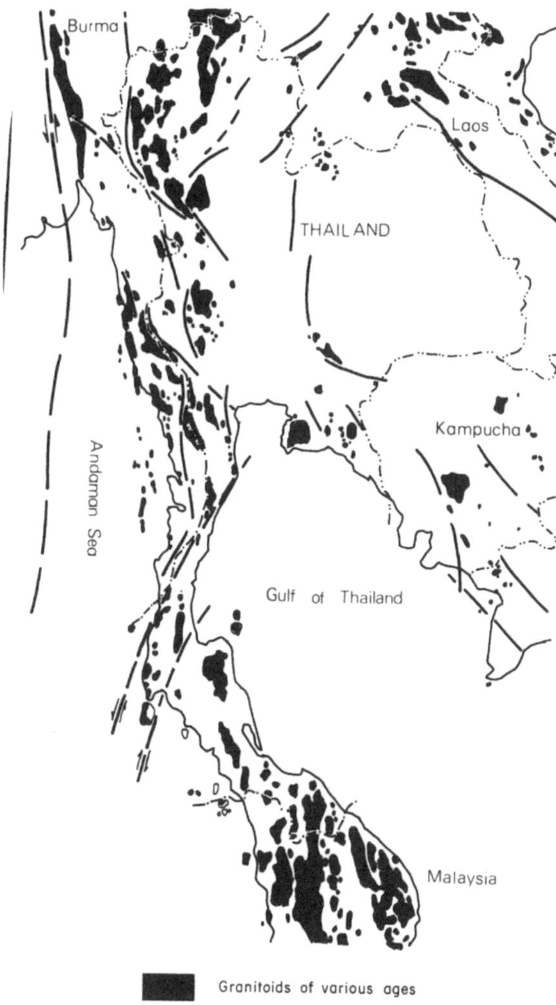


**Fig. 1.** Map show the distribution of granitic rocks (shaded area) in Thailand and adjacent area

## Regional Geological Setting

The general geology of Thailand, which is a part of Thai-Malay peninsula, is briefly outlined as follows:

The eastern zone of the country is mostly covered by Jurassic-Cretaceous continental red-beds, unconformably overlying Permian limestone and older metasediments.

The central zone and areas confined between the central and eastern zones are predominantly covered by flysch facies, volcanics, granites, Precambrian ? basement complexes and folded marine strata. The granites which are associated with volcanics, are hornblende-bearing; whereas the granites associated with folded metasediments and basement complexes are coarse-grained foliated alkali-feldspar megacrystic biotite-muscovite varieties.

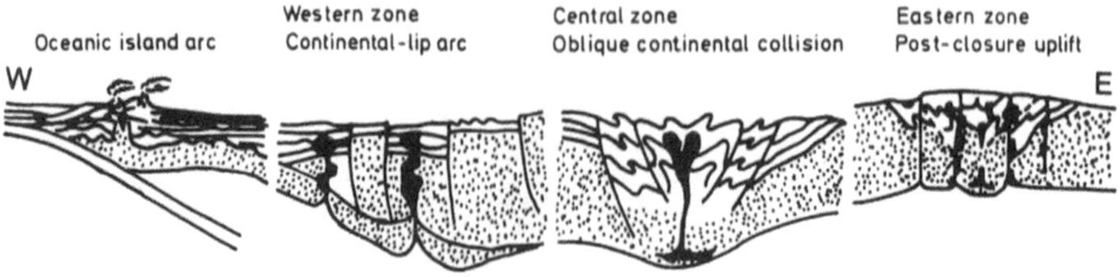


**Fig. 2.** Cartoon depicting the various geological environments of granitic rocks along E-W section across Thailand (modified from Pitcher, 1982)

The western zone which includes the mineralized field discussed in this paper, is partly covered by Silurian to Jurassic arenites, shales, limestone, turbidites and continental red-beds. The granites in this zone are predominantly of porphyritic biotite-muscovite varieties, intermingled locally with hornblende-biotite varieties. More details of the granite-related mineralization in this region are given below.

## The Granitoids of Thailand

The granitoids can be broadly subdivided, on the basis of their petrological and geochemical characteristics and geological environment, into central, eastern and western zones. The distribution of the granites and a schematic east-west cross-section depicting their location in the various geological environments of these zones are given in Figures 1 and 2.

### The Central Zone

The granites in the central zone are porphyritic biotite-muscovite varieties commonly characterized by predominent alkali-feldspar megacrysts. Most granites exhibit an orientation of the megacrysts and are commonly associated with foliated gneissic varieties, migmatitics or anatexites. The compositional range is restricted to high silica contents and most analyses have high normative corundum (peraluminous granite). The granites occur as a belt of several individual and extremely elongated plutons. In some places, they form complex multiple plutons (e.g. Rayong, Om Koi, and Thong Land plutons) which were emplaced at a high level in an intensively folded medium to high grade regional metamorphic terrain of Lower Paleozoic to Upper Triassic age. The granites are not associated with volcanics and xenoliths are almost entirely metasedimentary. The Rb/Sr ratio and U and Th contents are high. The rocks are enriched in LREE and show strongly developed negative Eu anomalies, and initial $^{87}Sr/^{86}Sr$ ratios are also high (0.72). The ages of the granites are mostly Upper Triassic to Lower Jurassic (Beckinsale et al., 1979). The granites are interpreted to occur along an encratonic ductile shear belt, similar to the S-types proposed by Chappell and White (1974), or Hercynotype of a continental collision zone as proposed by Pitcher (1983). The central zone granites are commonly related to Sn-W mineralization.

### The Eastern Zone

The granites in the eastern zone are mostly equigranular hornblende-biotite varieties and occur as isolated complexes of multiple plutons or small batholiths usually associated with diorite, andesite and basaltic dykes. Their enclaves are mixed appinitic and metasedimentary, and microgranular enclaves are abundant. The granitoids in this zone vary from tonalite to granodiorite through granite. Most are meta-aluminous with normative corundum less than 1.0%. The Rb/Sr ratios are mostly low and within the range of 0.1 to 10. U and Th contents in the hornblende predominant varieties are normal when compared with average silicic igneous rocks and granites in Scandinavia, Europe and America, but the biotite predominant varieties contain high U and Th, similar to those of the central zone.

The granites in the eastern zone commonly contain high LREE similar to those of the central zone, but have less developed negative Eu anomalies. These granites are not usually associated with migmatite, anatexite, foliated granite and high grade regional metasediments as is the case in the central zone; rather they are often associated with less folded, low grade regional metasediments. They commonly occur in the areas where molasse and vertical tectonics predominante. The granites in this zone are interpreted to have been emplaced in a post tectonic uplift regime similar to that of Calodonian I-type granites as proposed by Pitcher (1983). Cu and Mo mineralization is locally found in association with these rocks.

### The Western Zone

The granites in the western zone are characterized by multiple intrusions of equigranular hornblende granite and porphyritic biotite-muscovite granite with megacrystic alkali-feldspar, and have been found at Mae Sarieng and Mae Lama in the northwestern border region and at Phuket Island. The granites occur as small isolated plutons. Foliation is less prominent than in the central zone, and they are not associated with gneissic granite or migmatite. Enclaves are both metasedimentary and dioritic. The envelopes of these granites are of low grade regionally metamorphosed strata. The ages are mostly Cretaceous (Beckinsale et al., 1979). The initial $^{87}Sr/^{86}Sr$ ratios range from 0.707 to 0.73. Rb/Sr ratios are more or less similar to the biotite-predominant phases in the central and the eastern zones. U and Th contents in hornblende-bearing granite are equally as low as those of the granites in the eastern zone, whereas the porphyritic two mica granites have U and Th contents as high as the granites in the central zone. The tourmaline-muscovite and pegmatitic phases of granites in the western zone are associated with Sn-W mineralization. These rocks are interpreted to be due to mixing of I-Caledonian type and S-type granites, taking place along the marginal continental lip as defined by Pitcher (1983).

## Ore Deposits and Their Environment

The tin-tungsten mineralized province of the western zone is closely associated with plutonic complexes. The granitoids were emplaced as several individual elongated plutons which extend for several hundred km along the N-S trending western tectonic belt of Jurassic-Cretaceous age. The granitoids are predominantly of granite with minor leucogranite, aplites. Pegmatites and acidic dyke swarms are rather common.

The emplacement of the granitoids into the overlying country rock is passive and likely to be fault and fracture controlled. The evidence of sharp intrusive contacts, lack of migmatite or anatexite, and unfolded/sub-horizontal bedding of the country rocks indicate a passive, post tectonic, fault and fracture controlled high level emplacement.

Regarding the environment of ore deposition, this province is likely to be the plutonic-foldbelt type as described by Pollard and Taylor (1984), – tin deposits associated with granitoids which have a close spatial relationship with major folding, fracturing and uplifting. The structures and granitoid emplacement are predominantly of the late or post tectonic stage and are controlled by major fracture or suture zones.

Two types of cassiterite-wolframite deposits, vein and skarn, are found in this mineralized belt. They can be further subdivided into three vein and two skarn subtypes.

## Vein-Type Deposits

The primary tin-tungsten deposits in the Pilok mineralized field are mostly vein-type deposits. Most of the ore-bearing veins are confined to the cusps of granite stocks and to granite-country rock contacts.

Almost all of the ore-bearing veins are quartz veins. The pegmatite and quartz-feldspar veins are barren. From observations at some of the inactive mines, massive quartz veins predominate and wolframite is more abundant than cassiterite; whereas the swarms of quartz microveinlets are almost exclusively cassiterite-bearing. The quartz veins in the fault and shear zones commonly contain molybdenite, scheelite, pyrite, bismuthinite, arsenopyrite, galena ?, fluorite, cassiterite and wolframite. Most of the quartz veins apparently follow a sub-vertical NE-SW and NW-SE fracture system confined within the granitic host. The ore-bearing quartz veins are believed to be not deeper than 50 m from the roof contact.

The granite host is apparently complex and of multiple phases. The main granite masses are partly of coarse-grained biotite-muscovite porphyritic granite and partly of coarse-grained equigranular biotite-muscovite granite. These granites are locally greisenized and rich in quartz, feldspar, tourmaline, and muscovite. They are subsequently intruded by leucogranite, aplite, pegmatite and quartz veins.

Three vein-type tin-tungsten mineralized fields, which occur in the study area, namely Etong-Epu, Phapare, and Sombat, will be discussed. These three fields are likely to be related to small individual cupolas of the same pluton. They are aligned along the N-S trending granitic range of the western part of Thailand (Fig. 3).

### The Etong-Epu Field

This field covers an area approximately 12 km$^2$ and is located at latitude 14° 34' 00" N — 14° 43' 00" N and longitude 98° 21'29" E 98° 48' 19" E. There are more than 10

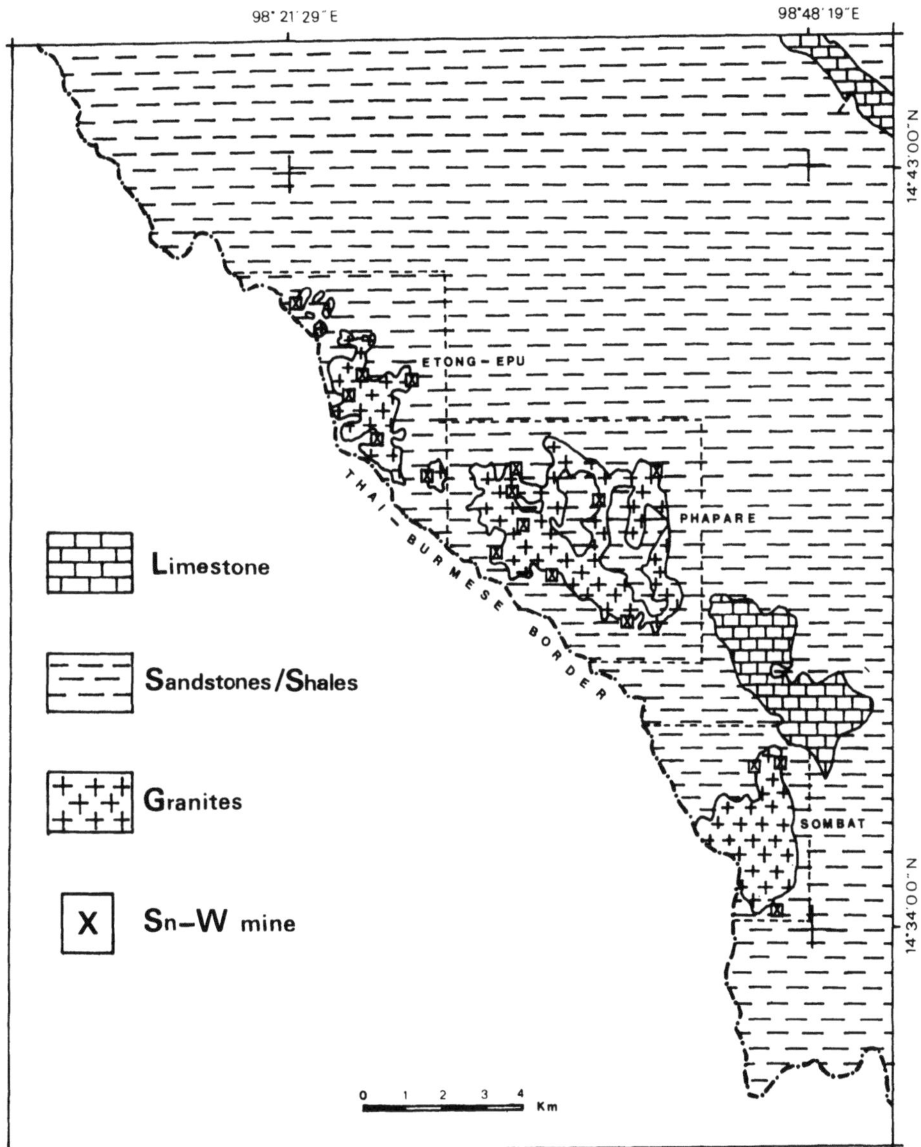


**Fig. 3.** Geological map of the tin-tungsten mineralized field in the Pilok area

mines which have operated in this area over that past 20 years. Mining has been carried out on two types of quartz veins: massive quartz veins and stockworks of swarms of ore-bearing quartz microveinlets. The diagrammatic cross-section of Figure 4a illustrates the relationship of the ore-bearing veins, granite host and the capping country rocks.

The swarms of sub-parallel sub-vertical ore-bearing quartz micro-veinlets are predominantly confined to the cusps of the granitic body which is of strongly albitized, chloritized and greisenized leucogranites. The ore-bearing massive quartz veins,

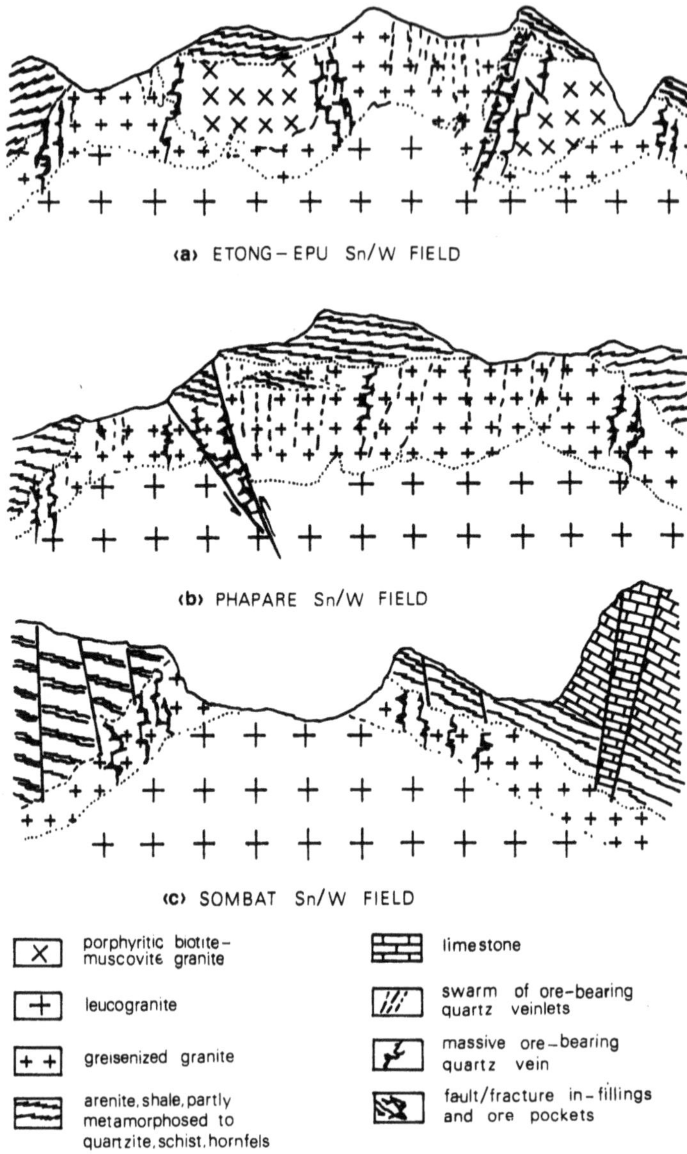


**Fig. 4a-c.** Ideal geological cross-section of the vein-type tin-tungsten mineralized fields in the Pilok area

which are commonly branched, are distributed elsewhere at the cupolas and the granite-country rock contact zones.

The sedimentary envelopes are of partly hornfelsed arenites and shales. The rocks are not folded and mostly show a sub-horizontal attitude. The main granitoid body is of coarse-grained biotite-muscovite granite subsequently cut by leucogranite. The greisenized granite is apparently confined along the granite – country rock contact.

The sub-vertical sub-parallel swarms of ore-bearing quartz microveinlets and massive ore-bearing quartz veins are likely to have resulted from post magmatic crystallization of the mineralized solutions which infilled sub-parallel fractures within the elongated granitic pluton.

## The Phapare Field

This field is located 3 km southeast of Eton-Epu and covers an area approximately 20 $km^2$. More than six mines have operated in this area and the mining has been carried out over 20 years.

The geological setting and the major ore-bearing veins of the Phapare field differ slightly from Etong-Epu. Almost all of the granite host of the ore-bearing veins is of greisenized granite and the veins are predominantly of sub-parallel subvertical swarms of microveinlets. Ore-bearing quartz veins locally in-fill fault and shear zones. Massive ore-bearing quartz veins are uncommon but quartz-feldspar-tourmaline pegmatites are commonly found in this area. The ore in the stockwork of the quartz microveinlets is almost wholly cassiterite. Massive pockets of mixed cassiterite-wolframite-scheelite-molybdenite-pyrite-chalcopyrite-bismuthinite-arsenopyrite-fluorite and other unidentified minerals are developed in the quartz veins which in-fill the fault and shear zones.

Wide zones of greisenized granite, composed of quartz, feldspars, muscovite, tourmaline and pyrite, are developed along the periphery of the granitic body, grading downward into the leucogranite in the mineralized regions.

The sedimentary envelope is of arenite and shale with sub-horizontal bedding. Roof pendants are common. Figure 4b is a diagrammatic cross-section across the mineralized field.

## The Sombat Field

This mineralized field is smaller than Phapare and Etong-Epu, and there were only three mines in operation during the past two years. This tin-tungsten field is located 3 km south of Phapare. It is likely that this field will be of considerable potential for vein and skarn-type deposits. The current mining has been of the ore-bearing quartz veins.

The environment of the mineralized field is related to the intrusion of a stock of biotite-muscovite granite and leucogranite into arenite, shale, and limestone country rocks. The granite-country rock contact zone is extensively greisenized. The greisenized granite contains many crusters of radiated tourmaline and muscovite. Numerous bands of sub-vertical sub-parallel ore-bearing quartz vein systems are embedded in the greisenized granite. The ore is composed mainly of cassiterite and wolframite; pyrite and chalcopyrite occasionally also occur in the quartz veins. Skarn-type deposits have not been explored, although the mineralized granite is not very far from the calcareous rocks. Figure 4c illustrates the diagrammatic cross-section of this mineralized field.

## Carbonate Replacement Sulphide Cassiterite Skarn

Two large scale carbonate replacement skarn type ore deposits are mined in this tin-tungsten field; namely Takopidtong Mine and Somporn Mine. Although these mines are both of pyrometasomatic-contact metasomatic ore deposits, they have different characteristics. The Takopidtong Mine is a large ore-body of multiple metallic oxides and sulphides, whereas the Somporn Mine is of simple ore skarn. The details of these two deposits are as follows:

### Takopidtong Mine

The mine is located at a latitude of 13° 40' 31" N and a longitude of 98° 10' 21" E, approximately 300 km west of Bangkok, and is situated adjacent to the Thai-Burmese border.

This cassiterite deposit occurs at the contact of granite with calcareous country rock (Fig. 5a). The ore is of massive polymetallic oxides and sulphides, forming a wide sub-vertical body within biotite-muscovite granite and greisenized granite. The diagrammatic cross-section (Fig. 5b) illustrates the relationship between the ore body, granite host, and the metasedimentary capping rocks.

The ore-body exhibits an apparent vertical zonation from a predominantly silicate-rich facies (garnet ± tremolite ± magnetite ± cassiterite ± stannite?) to an oxide-rich facies (magnetite ± hematite ± cassiterite ± pyrrhotite ± pyrite ± molybdenite and unidentified iron oxides) to a sulphide-rich facies (pyrite ± chalcopyrite ± bismuthinite ± molybdenite ± arsenopyrite ± galena? ± cassiterite/stannite ± pyrrhotite ± calcite and fluorite) successively from the lower to the upper levels.

The ore-body is confined within the granite host. Skarn devoid of cassiterite is locally formed along the contact between the greisenized granite and the calcareous capping rocks. Scheelite and wolframite are uncommon in this deposit.

The host exhibits an apparent outward zonation from two-mica granite in the central part of the pluton to greisenized granite at the contact with the country rocks. Both the biotite-muscovite granite and the greisenized granite are cut by leucogranite, aplite, feldspar pegmatite and quartz veins. It should be noted that less silicic granitoids such as diorite and granodiorite, which are usually expected to be related to skarn-type deposits (Stemprok, 1980), are not found here.

The calcareous capping rocks are locally transformed to marble and calc-silicate rocks.

### Somporn Mine

This mine is located at latitude 14° 01' 38" N and longitude 99° 02' 54" E, approximately 200 km west of Bangkok.

The mine is situated at the contact zone between limestone and granite (Fig. 6a). The ore body forms pockets of cassiterite ± scheelite ± pyrite ±chalcopyrite and other unidentified sulphide minerals, disseminated in skarn along the greisenized granite-marble contact zone.

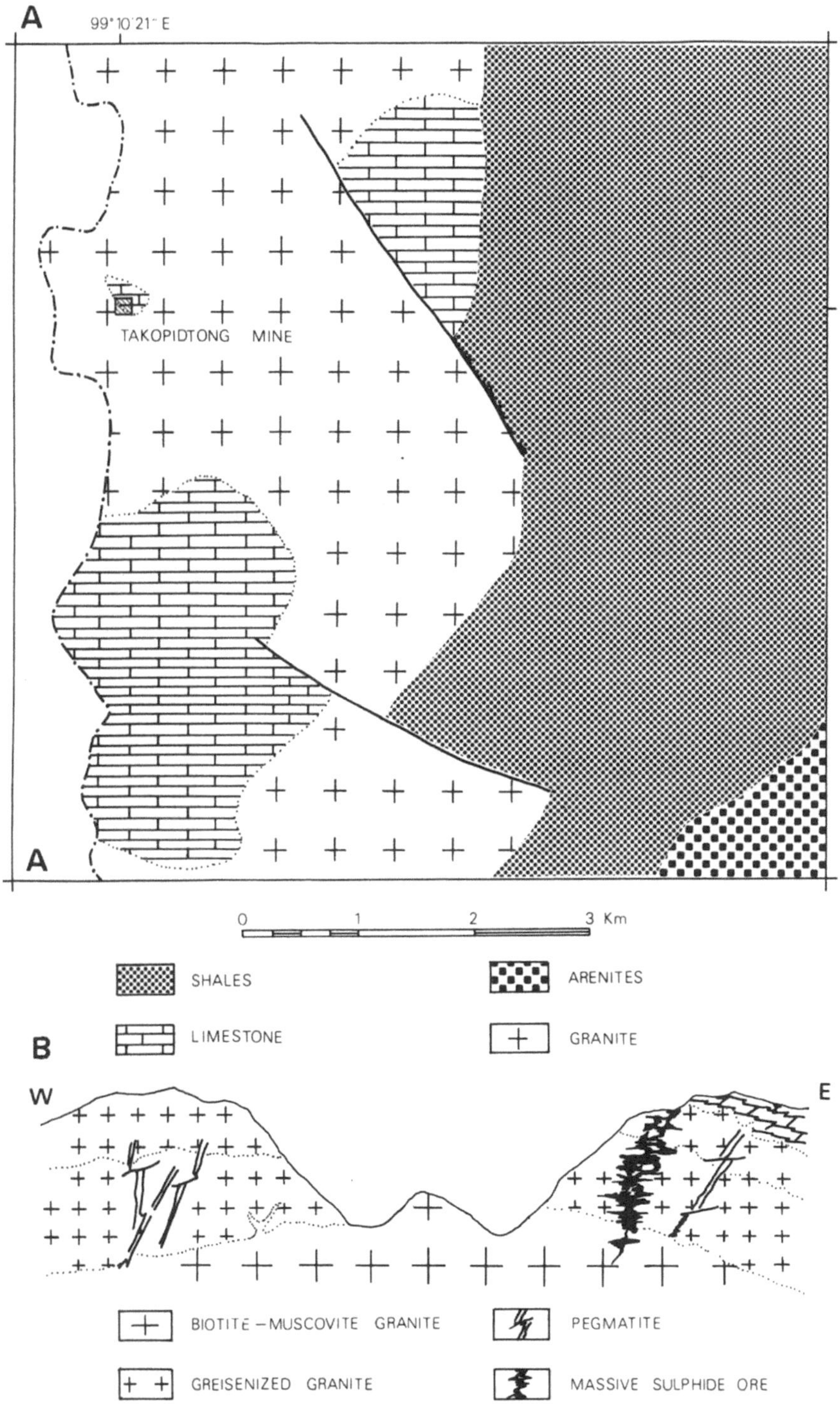


**Fig. 5A, B.** Geological map of the area of massive sulphide ore deposit at Takopidtong mine **A**, and the ideal cross-section of the mineralized area **B**

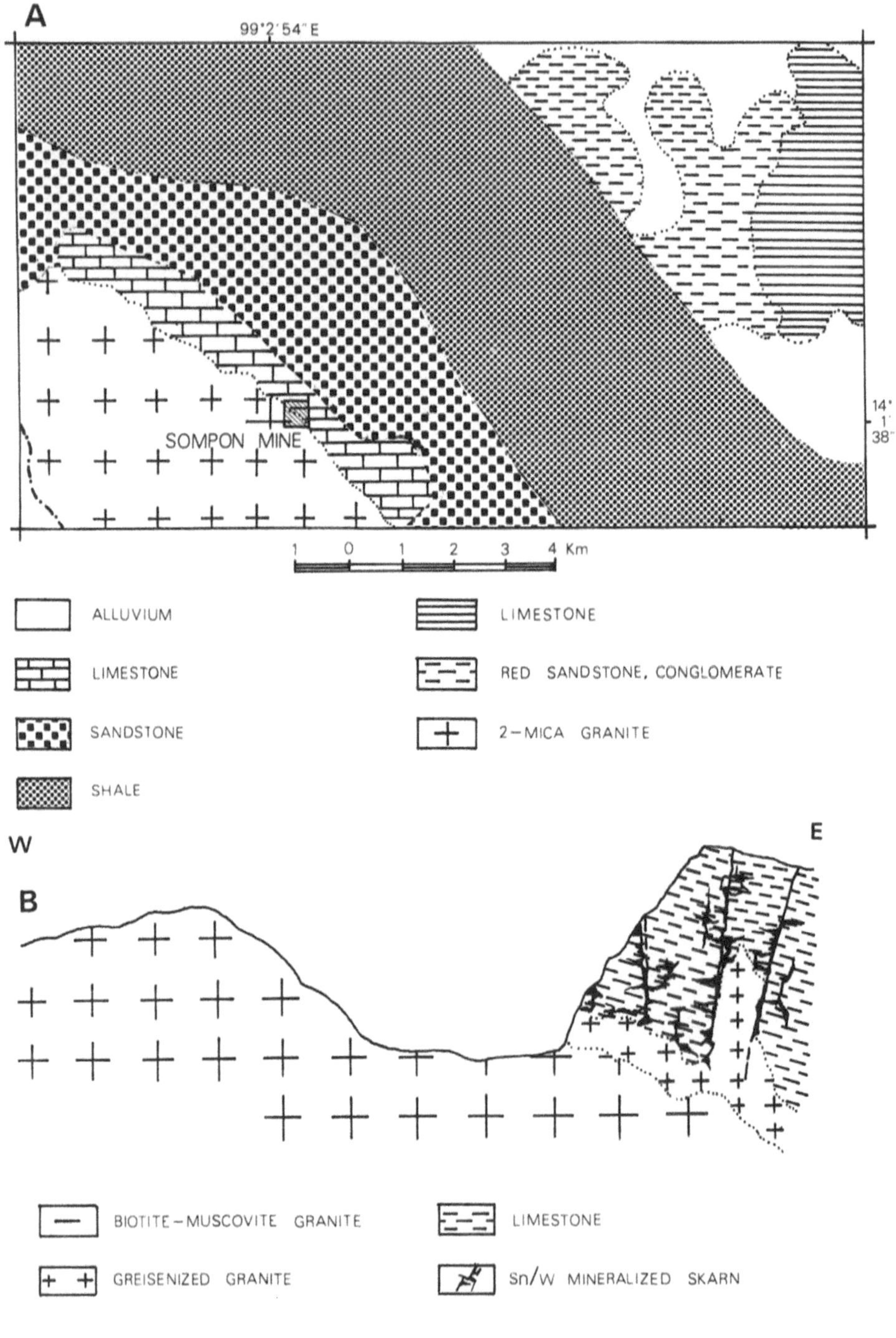


**Fig. 6A, B.** Geological map in the area of carbonate replacement skarn ore deposit at Somporn mine **A**, and the ideal cross-section of the mineralized area **B**

The diagrammatic cross-section (Fig. 6b) illustrates the relationships between the two-mica granite, the greisenized granite, the ore-body, and the capping rock. Mining is currently active both along the greisenized granite-limestone contact and within the limestone. Faults and fractures within the limestone are likely to be the channels of dykes and veins of granite to penetrate into the limestone and to cause skarn development. In the lower level, where the greisenized granite is in contact with the limestone, the skarn is predominantly of pyrite ± chalcopyrite ± garnet ± epidote ± muscovite ± cassiterite ± scheelite. In the higher level, the skarn is apparently less rich in sulphides and enriched in fluorite and calcite.

The pluton is predominantly of biotite-muscovite granite and exhibits extensive greisenization along the contact with the country rocks. The greisenized granite is distinctively rich in spheres of radiating tourmaline and muscovite. Both the greisenized granite and the biotite-muscovite granites are cut by leucogranite, pegmatite, and quartz veins. The pegmatite and quartz veins are cassiterite-bearing.

## Mineralization

The mineralization of the tin-tungsten deposits of this region can be summarized as follows.

Vein-type mineralization is for the most part restricted within the cusps of granite bodies. In the oxide stage, cassiterite ± wolframite ± iron oxides are concentrated in the lower parts of the veins, which are mostly confined within the zone of greisenized granite. In the sulphide stage, the crystallization of pyrite ± chalcopyrite ± molybdenite ± cassiterite ± wolframite gradually migrates upwards to the upper parts of the veins. Fluorite and calcite occur in the uppermost parts of the veins or in the veins which are confined in the granite-country rock contact. The lowermost extent of the veins is predominantly of feldspar ± quartz ± tourmaline ± muscovite. Figure 7a illustrates the variation and distribution of mineralization in the vein-type tin-tungsten deposit.

The mineralization of the skarn deposits in the silicate stage is of a metasomatic character with the crystallization of garnet ± tremolite ± epidote. It is restricted to the contact zone of the intrusive body or to the lowermost part of the ore body as the Takopidtong Mine. Passing into the oxide stage, mineralization of magnetite ± cassiterite takes place in the central part of the ore body or in the exomorphic skarn. In the sulphide stage, the crystallization of pyrrhotite ± pyrite ± arsenopyrite ± molybdenite ± chalcopyrite ± cassiterite ± scheelite is concentrated in the upper part of the massive sulphide ore body or from the contact inwards. Figure 7b illustrates the variation and distribution of mineralization in skarn-type deposits.

## Conclusions

1) Both vein-type and skarn-type tin-tungsten deposits in the Pilok mineralized fields of West Thailand, are closely spatially related with greisenized zones of the biotite-muscovite granite. Greisenized granites are extensive in the cupolas and cusps of several small isolated plutons.

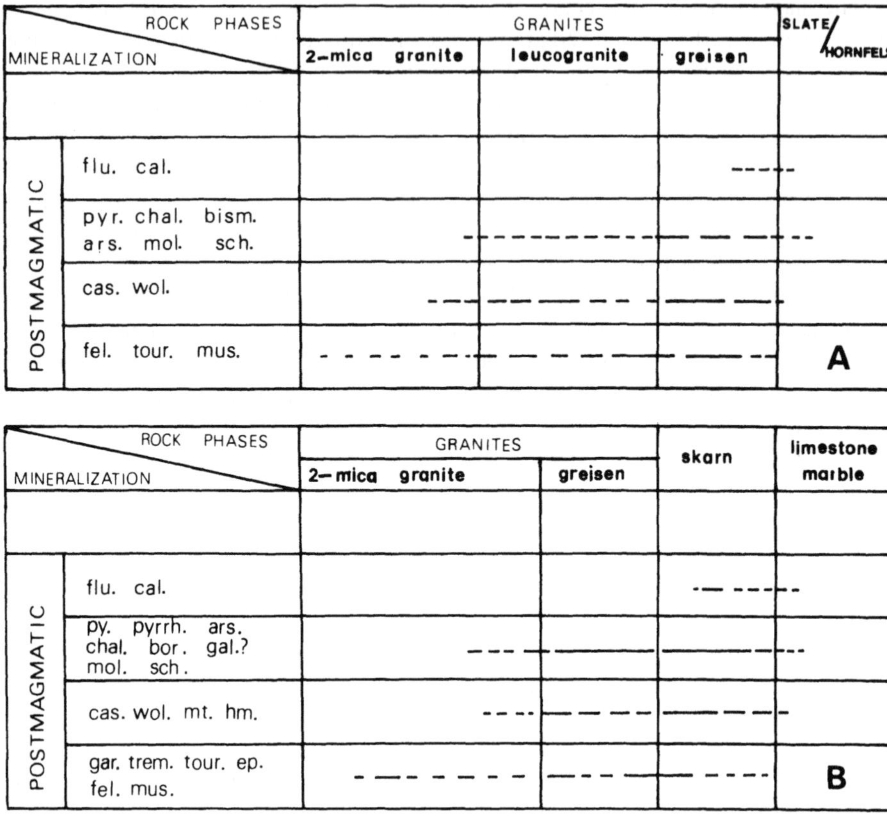


**Fig. 7A, B.** The variation of mineralization and distribution of ore formation in a tin-tungsten. **A** vein-type deposit; **B** skarn-type deposit

2) The cassiterite-predominant ore-bearing quartz microveinlets are commonly concentrated in the cusps, whereas tungsten-predominant ore-bearing massive quartz veins are randomly distributed.
3) The ore-bearing veins are confined within the granite hosts and ore-bearing pegmatites and ore disseminated in altered granites do not occur in this mineralized field.
4) The skarn-type deposits, which are developed at the contact between the two-mica/greisenized granite and the calcareous country rocks, are not related to subvolcanic or less silicic granitoids as have been observed in other parts of the world.

*Acknowledgements.* Acknowledgements are due to all my colleagues in the Geological Survey, for their ready assistance and constructive comments. Appreciation is extended to Dr. W. Pongsapich for his final review of the paper. The paper is presented with the permission of the Director General of the Mineral Resources Department, Thailand.

## References

Beckinsale, R.D., Suensilpong, S., Nakapadungrat, S., Walsh, J.N., 1979. Geochronology and geochemistry of granite magmatism in Thailand in relation to a plate tectonic model. *J. Geol. Soc. London,* 136, 529–540.

Chappell, B.W. and White, A.J.L., 1974. Two contrasting granite types. *Pacific Geology,* 8, 173–174.

Pitcher, W.S., 1983. Granite type and tectonic environment. Hsu, K.J. (ed.). *Mountain Building Processes.* Academic Press, London, 19–40.

Pollard, P.J. and Taylor, R.G., 1984. District analysis as an approach for target generation and exploration design - Sn/W deposits. *Fifth Regional Congress on the Geology, Mineral and Energy Resources of Southeast Asia, Kuala Lumpur, Malaysia.*

Stemprok, M., 1980. Tin and tungsten deposits of the West-Central Europe Variscides. *Proc. Fifth Quadrennial IAGOD Sym.*, Stuttgart, Germany, 495–512.

# Subject Index

MIX
Papier aus verantwortungsvollen Quellen
Paper from responsible sources
FSC® C105338

If you have any concerns about our products,
you can contact us on
**ProductSafety@springernature.com**

In case Publisher is established outside the EU,
the EU authorized representative is:
**Springer Nature Customer Service Center GmbH**
**Europaplatz 3, 69115 Heidelberg, Germany**

Printed by Libri Plureos GmbH
in Hamburg, Germany